2019 IEEE Asian Solid-State Circuits Conference (A-SSCC 2019)

Macao
4 – 6 November 2019

IEEE Catalog Number: CFP19SSC-POD
ISBN: 978-1-7281-5107-6

**Copyright © 2019 by the Institute of Electrical and Electronics Engineers, Inc.
All Rights Reserved**

Copyright and Reprint Permissions: Abstracting is permitted with credit to the source. Libraries are permitted to photocopy beyond the limit of U.S. copyright law for private use of patrons those articles in this volume that carry a code at the bottom of the first page, provided the per-copy fee indicated in the code is paid through Copyright Clearance Center, 222 Rosewood Drive, Danvers, MA 01923.

For other copying, reprint or republication permission, write to IEEE Copyrights Manager, IEEE Service Center, 445 Hoes Lane, Piscataway, NJ 08854. All rights reserved.

*** This is a print representation of what appears in the IEEE Digital Library. Some format issues inherent in the e-media version may also appear in this print version.

IEEE Catalog Number:	CFP19SSC-POD
ISBN (Print-On-Demand):	978-1-7281-5107-6
ISBN (Online):	978-1-7281-5106-9

Additional Copies of This Publication Are Available From:

Curran Associates, Inc
57 Morehouse Lane
Red Hook, NY 12571 USA
Phone: (845) 758-0400
Fax: (845) 758-2633
E-mail: curran@proceedings.com
Web: www.proceedings.com

2019 IEEE Asian Solid-State Circuits Conference (A-SSCC 2019)

Macao
4 – 6 November 2019

IEEE Catalog Number: CFP19SSC-POD
ISBN: 978-1-7281-5107-6

Table of Content

P 1

Plenary Session 1

1-1 What are the driving forces of DRAM?
Yiming Zhu
ChangXin Memory Technologies, China 1

1-2 Memory Centric Computing, The Foundation of Next Smart Society
KyoWon Jin
DRAM Development & Business, SK hynix, Korea 5

Industry 2

High Performance Technologies for Industrial Applications

2-1 (7025) A 45nm 76-81GHz CMOS Radar Receiver for Automotive Applications
Debapriya Sahu[4], Rittu Sachdev-Singh[4], Harikrishna Parthasarathy[4], Rohit Chatter-jee[4], Brian Ginsburg[3], Daniel Breen[3], Karan Bhatia[3], Sudhir Polarouthu[2], Vimal Edayath[4], Bhupendra Sharma[4], Meghna Agarwal[4], Karthik Subburaj[4], Anjan Prasad[1], Shankar Ram[4], Cathy Chi[3], Ross Kulak[3], Vijay Rentala[3], Neeraj Nayak[3]
[1]Analog Devices, India
[2]Intel Technology, India
[3]Texas Instruments, United States
[4]Texas Instruments India Pvt. Ltd., India 9

2-2 (7052) An Untrimmed PVT-Robust 12-bit 1-MS/s SAR ADC IP in 55nm Deeply Depleted Channel CMOS Process
Yingyun Zha[2], Loic Zahnd[1], Jian Deng[1], David Ruffieux[1], Komail Badami[1], Themis Mavrogordatos[1], Yoshihiko Matsuo[3], Stephane Emery[1]
[1]CSEM, Switzerland
[2]Formerly with CSEM, Switzerland
[3]Mie Fujitsu Semiconductor Limited, Japan 13

2-3 (7035) A Cost Effective Test Screening Circuit for embedded SRAM with Resume Standby on 110-nm SoC/MCU
Yoshisato Yokoyama, Kenji Goto, Tomohiro Miura, Yukari Ouchi, Daisuke Nakamura, Jiro Ishikawa, Shunya Nagata, Yoshiki Tsujihashi and Yuichiro Ishii
Renesas Electronics Corporation, Tokyo, Japan 17

2-4 (7094) A 2.666GT/s 128GB/s 14nm Memory I/O with Jitter and Crosstalk Cancellation
Harry Muljono, Kathy Peng, Linda Sun, Isaac Abraham, Charlie Lin, Yanjie Zhu, Chunrong Song
Intel Corporation, Santa Clara, United States 21

ETA+MEM 3

Secure and Smart Computing Systems

3-1 (7037) A Si-Backside Protection Circuits Against Physical Security Attacks on Flip-Chip Devices
Takuji Miki[2], Makoto Nagata[2], Hiroki Sonoda[2], Noriyuki Miura[2], Takaaki Okidono[1], Yuuki Araga[3], Naoya Watanabe[3], Haruo Shimamoto[3], Katsuya Kikuchi[3]
[1]ECSEC, Japan
[2]Kobe University, Japan
[3]National Institute of Advanced Industrial Science and Technology, Japan 25

3-2 (7135) A 28nm 512Kb adjacent 2T2R RRAM PUF with interleaved cell mirroring and self-adaptive splitting for extremely low bit error rate of cryptographic key

Xiaoyong Xue[1], Jianguo Yang[2], Yuejun Zhang[1], [3], Mingyu Wang[1], Hangbing Lv[2], Xiaoyang Zeng[1], Ming Liu[2]
[1]*Fudan University, China*
[2]*Chinese Academy of Sciences, China*
[3]*Ningbo University, Ningbo, China.* 29

3-3 (7172) A 0.5V Real-Time Computational CMOS Image Sensor with Programmable Kernel for Always-On Feature Extraction

Tzu-Hsiang Hsu, Yen-Kai Chen, Tai-Hsing Wen, Wei-Chen Wei, Yi-Ren Chen, Fu-Chun Chang, Ren-Shuo Liu, Chung-Chuan Lo, Kea-Tiong Tang, Meng-Fan Chang, and Chih-Cheng Hsieh
National Tsing Hua University, Taiwan 33

3-4 (7090) A 16K SRAM-Based Mixed-Signal In-Memory Computing Macro Featuring Voltage-Mode Accumulator and Row-by-Row ADC

Hyunjoon Kim, Qian Chen, and Bongjin Kim
Nanyang Technological University, Singapore 35

ACS 4

Power Management Techniques

4-1 (7176) A 69.3% Efficiency, 6.78-MHz Wireless Power Delivery System with 0X/1X Regulating Rectifier and Reconfigurable Power Amplifier

Jonathan Fuh, Fu-Bin Yang, and Po-Hung Chen.
National Chiao Tung University, Taiwan 37

4-2 (7189) An 80mA Capacitor-Less LDO with 6.5µA Quiescent Current and No Frequency Compensation Using Adaptive-Deadzone Ring Amplifier

Bohui Xiao[2], Praveen Kumar Venkatachala[2], Yang Xu[4], Ahmed Elshater[2], Calvin Lee[2], Spencer Leuenberger[3], Qadeer Ahmad Khan[1], Un-Ku Moon[2]
[1]*Indian Institute of Technology Madras, India*
[2]*Oregon State University, United States*
[3]*Skyworks Solutions Inc., United States*
[4]*Texas Instruments, United States* 39

4-3 (7082) A DCM ZVS Class-D Power Amplifier for Wireless Power Transfer Applications

Xinyuan Ge[1], Lin Cheng[2] and Wing-Hung Ki[1]
[1]*Hong Kong University of Science and Technology, Hong Kong*
[2]*University of Science and Technology of China, China* 43

4-4 (7012) Time-Based Digital LDO Regualtor with Fractionally Controlled Power Transistor Strength and Fast Transient Response

Jin-Gyu Kang, Min-Gyu Jeong, and Jeongpyo Park, and Changsik Yoo
Hanyang University, Korea 45

4-5 (7087) A Single-Inductor Triple-Output Converter with an Automatic Detection of DC or AC Energy Harvesting Source for Supplying 93% Efficiency and 0.05mV/mA Cross Regulation to Wearable Electronics

T. Nagateja[1], Shao-Qi Chen[1], Li-Cheng Chu[1], Ke-Horng Chen[1], Ying-Hsi Lin[2], Shian-Ru Lin[2], Tsung-Yen Tsai[2]
[1]*National Chiao Tung University, Taiwan*
[2]*Realtek Semiconductor Corp., Taiwan* 49

SOC 5
Low Power Deep Learning Processor

5-1 (7062) A 0.7mm2 8.54mW FocusNet Display LSI for Power Reduction on OLED Smart-phones
Tsu-Ming Liu, Chang-Hung Tsai, Shawn Shih, Chih-Kai Chang, Jia-Ying Lin, Wayne Hsieh, Yung-Chang Chang, and Chi-Cheng Ju
MediaTek Inc. Taiwan 53

5-2 (7033) A 47.4µJ/epoch Trainable Deep Convolutional Neural Network Accelerator for In-Situ Personalization on Smart Devices
Seungkyu Choi, Jaehyeong Sim, Myeonggu Kang, Yeongjae Choi, Hyeonuk Kim, and Lee-Sup Kim
KAIST, Korea 57

5-3 (7074) A Sparse-Adaptive CNN Processor with Area/Performance balanced N-Way Set-Associate PE Arrays Assisted by a Collision-Aware Scheduler
Zhe Yuan[1], Jingyu Wang[1], Yixiong Yang[1], Jinshan Yue[1], Zhibo Wang[1], Xiaoyu Feng[1], Yanzhi Wang[2], Xueqing Li[1], Huazhong Yang[1], Yongpan Liu[1]
[1]Tsinghua University, China
[2]Northeastern University, United States 61

5-4 (7179) A 2.25 TOPS/W Fully-Integrated Deep CNN Learning Processor with On-Chip Training
Cheng-Hsun Lu, Yi-Chung Wu, and Chia-Hsiang Yang
National Taiwan University, Taiwan 65

RF 6
RF & mm-Wave Front-End Circuits and Clock Generator

6-1 (7178) A Ka-Band CMOS Phase-Inverting Amplifier with 0.6 dB Gain Error and 2.5° Phase Error
Chenyu Xu, Dixian Zhao
Southeast University, China 69

6-2 (7054) A High-Performance Low Complexity All-Digital Fractional Clock Multiplier
Nahla T. Abou-El-Kheir[1], Ralph D. Mason[2], Mingze Li[3], M.C.E. Yagoub[4]
[1]University of Ottawa, Canada
[2]Carleton University, Ottawa, Canada 73

6-3 (7131) A DC-43.5 GHz CMOS Switched-Type Attenuator with Capacitive Compensation Technique
Peng Gu and Dixian Zhao
Southeast University, China 77

6-4 (7111) A 28-GHz Compact SPDT Switch Using LC-Based Spiral Transmission Lines in 65-nm CMOS
Xiangyu Meng[1], Zhenpeng Zheng[1], Jiaqi Zhang[1] and Patrick Yue[2]
[1]Sun Yat-sen University, China
[2]Hong Kong University of Science and Technology, China 79

6-5 (7132) A 20-GHz Ultra-Low-Power LNA Using gm-Boosted and Current-Reuse Techniques in 65-nm CMOS for Satellite Communication Terminals

Jiajun Zhang and Dixian Zhao

Southeast University, China 81

WLN 7
Advanced Clock Generators

7-1 (7092) A 4-GHz Sub-harmonically Injection-Locked Phase-Locked Loop with Self-Calibrated Injection Timing and Pulsewidth

Xuefan Jin, Dong-Seok Kang, Youngjun Ko, Kee-Won Kwon, and Jung-Hoon Chun

Sungkyunkwan University, South Korea 83

7-2 (7104) A 9.4MHz-to-2.4GHz Jitter-Power Reconfigurable Fractional-N Ring PLL for Multi-Standard Applications in 7nm FinFET CMOS Technology

Sangdon Jung, Jaehong Jung, Byungki Han, Seunghyun Oh, and Jongwoo Lee

Samsung Electronics, Korea 87

7-3 (7153) A 2.4-GHz 500-µW 370-fsrms Integrated Jitter Sub- Sampling Sub-Harmonically Injection-Locked PLL in 90-nm CMOS

Chun-Yu Lin, Yu-Ting Hung, and Tsung-Hsien Lin

National Taiwan University, Taiwan 91

7-4 (7002) A Sub-Sampling PLL with Robust Operation under Supply Interference and Short Re-Locking Time

Yuan Cheng Qian, Yen-Yu Chao, and Shen-Iuan Liu

National Taiwan University, Taiwan 95

PD 8
Panel Discussion

Are Analog and Mixed Mode Circuits the future solution of AI SoCs?

P 9
Panel Session 2

9-1 AI and IoT for Social Value Creation

Yasunori Mochizuki

NEC Corporation, Japan 99

9-2 Millimeter-Wave System-on-Chip Applications from Space Explorations to Contactless Connectivity

Mau-Chung Frank Chang

National Chiao Tung University, Taiwan 103

ACD+DC 10

Sensor Interfaces and Data Converters

10-1 (7064) An Energy-Efficient BJT-Based Temperature-to- Digital Converter with ±0.13°C (3σ) Inaccuracy from -40 to 125°C
Rushil K. Kumar, Hui Jiang, and Kofi A. A. Makinwa
Delft University of Technology, The Netherlands 107

10-2 (7085) A 16.1-b ENOB 0.064mm2 Compact Highly-Digital Closed-Loop Single-VCO-based 1-1 SMASH Resistance-to-Digital Converter in 180nm CMOS
Elisa Sacco[1], Johan Vergauwen[2] and Georges Gielen[1]
[1]*Katholieke Universiteit Leuven, Belgium*
[2]*Melexis Technologies, Belgium* 109

10-3 (7089) A Low-Noise Sub-Bandgap Reference with a ±0.64% Untrimmed Precision in 16nm FinFET
Matthias Eberlein[1], Harald Pretl[2]
[1]*Intel Germany, Germany*
[2]*Intel Austria, Austria* 113

10-4 (7005) A 1.2V 86dB SNDR 500kHz BW Linear-Exponential Multi-Bit Incremental ADC Using Positive Feedback in 65nm CMOS
Biao Wang[1], Sai-Weng Sin[1], Seng-Pan U[1], Franco Maloberti[3], R. P. Martins[2]
[1]*University of Macau, Macau*
[2]*University of Macau / Universidade de Lisboa, Macau*
[3]*University of Pavia, Italy* 117

10-5 (7134) A Multi-Slice VCO-based Quantizer for On-Chip Power Supply Noise Analysis Achieving 0.11 (mV)2/sqrt(MHz) Noise Floor
Pengfei Zhai, Xiong Zhou, Yan Cai, Zheng Zhu, Fan Zhang, Zixiao Lin, Qiang Li
University of Electronic Science and Technology of China, China 121

10-6 (7086) A 265μW Continuous-Time 1-2 MASH ADC Achieving 100.6 dB SNDR in a 24 kHz Bandwidth
Sujith Billa, Suhas Dixit and Shanthi Pavan
Indian Institute of Technology Madras, India 123

DCS 11

Advanced Energy-efficient Digital Circuits

11-1 (7116) Drop-In Energy-Performance Range Extension in Microcontrollers Beyond VDD Scaling
Saurabh Jain, Longyang Lin, Massimo Alioto
National University of Singapore Singapore 125

11-2 (7041) A 28nm fully digital voltage monitor with 16.5uV/°C accuracy and 0.8mV quantized error from -40 to 160°C for ISO26262 ASIL-D capable MCU
Toshifumi Uemura, Yuko Kitaji, Kazuki Fukuoka
Renesas Electronics Corpora, Japan 129

11-3 (7163) HyCUBE: A 0.9V 26.4 MOPS/mW, 290 pJ/op, Power Efficient Accelerator for IoT Applications

Bo Wang, Manupa Karunarathne, Aditi Kulkarni, Tulika Mitra, Li-Shiuan Peh

National University of Singapore, Singapore 133

11-4 (7023) A 54% Power-Saving Static Fully-Interruptible Single- Phase-Clocked Shared-Keeper Flip-Flop in 14nm CMOS

Amit Agarwal, Steven Hsu, Monodeep Kar, Mark Anders, Himanshu Kaul, Raghavan Kumar, Vikram Suresh, Sanu Mathew, Ram Krishnamurthy, Vivek De

Intel Corporation, United States 137

11-5 (7113) 0.54 pJ/bit, 15Mb/s True Random Number Generator Using Probabilistic Delay Cell for Edge Computing Applications

Fei Li, Ming Ming Wong, Aarthy Mani, Vishnu Paramasivam, and Anh Tuan Do

Agency for Science, Technology and Research, Singapore 141

11-6 (7205) A Smart Hardware Security Engine Combining Entropy Sources of ECG, HRV and SRAM PUF for Authentication and Secret Key Generation

Sai Kiran Cherupally[1], Shihui Yin[1], Deepak Kadetotad[1], Chisung Bae[2], Sang Joon Kim[2], and Jaesun Seo[1]

[1]Arizona State University, , United States
[2]Samsung Advanced Institute of Technology, Korea 145

RF 12

RF, mm-Wave & THz Transmitters and Receivers

12-1 (7119) A Packaged Fully Digital 390GHz Harmonic Outphasing Transmitter in 28nm CMOS

Alexander Standaert, Patrick Reynaert

Katholieke Universiteit Leuven, Belgium 149

12-2 (7020) A Fully Integrated 27.5-30.5 GHz 8-Element Phased-Array Transmit Front-end Module in 65nm CMOS

An'an Li[1], Yingtao Ding[1], Zipeng Chen[2], Wei Wang[2], Sijia Jiang[2], Shiyan Sun[1], Zhiming Chen[1], and Baoyong Chi[2]

[1]Beijing Institute of Technology, China
[2]Tsinghua University, China 153

12-3 (7199) A 2Mbps sub-100μW Crystal-less RF Transmitter with Energy Harvesting for MultiChannel Neural Signal Acquisition

Heng Huang, Milin Zhang, Guolin Li, Zhihua Wang

Tsinghua University, China 157

12-4 (7031) Direct-Conversion Receiver Front-End for 180 GHz with 80 GHz Bandwidth in 130 nm SiGe

Paul Stärke, Andres Seidel, Corrado Carta, Frank Ellinger

Technische Universität Dresden, Germany 161

12-5 (7070) A Blocker-Tolerant Direct Sampling Receiver for Wireless Multi- Channel Communication in 14nm FinFET CMOS

Barosaim Sung, Chilun Lo, Jaehoon Lee, Sangdon Jung, Seungjin Kim, Jaehong Jung, Seungyong Bae, Youngsea Cho, Yong Lim, Dooseok Choi, Myeongcheol Shin, Soonwoo Choi, Byungki Han, Seunghyun Oh and Jongwoo Lee

Samsung Electronics, Korea 165

12-6 (7109) A Sub-0.6V, 330 μW, 0.15 mm2 Receiver Front-End for Bluetooth Low Energy (BLE) in 22 nm FD-SOI with Zero External Components

Ehsan Kargaran[1], Carl Bryant[2], Danilo Manstretta[1], Jon Strange[1], Rinaldo Castello[1]
[1]*University of Pavia, Italy*
[2]*MediaTek, Kent, United Kingdom* 169

ETA 13
Energy-Aware Circuits & Systems

13-1 (7166) A 4-Mbps 41-pJ/bit On-off Keying Transceiver for Body-channel Communication with Enhanced Auto Loss Compensation Technique

Jian Zhao[1], Jingna Mao[3], Wenyu Sun[2], Yuxuan Huang[2], Yixiong Yang[2], Huazhong Yang[2] and Yongpan Liu[2]
[1]*Shanghai Jiao Tong University, China*
[2]*Tsinghua University, China*
[3]*Chinese Academy of Science, China* 173

13-2 (7193) A battery-less 31 mW HBC receiver with RF energy harvester for implantable devices

Jihee Lee, Jaeeun Jang, Jaehyuk Lee, and Hoi-Jun Yoo
KAIST, Korea 177

13-3 (7154) A Piezoelectric Energy Harvesting Interface for Irregular High Voltage Input with Partial Electric Charge Extraction with 3.9× Extraction Improvement

Muhammad Bilawal Khan, Hassan Saif, and Yoonmyung Lee
Sungkyunkwan Unversity, Korea 181

13-4 (7061) A 100-pA Adaptive-FOCV MPPT Circuit with >99.6% Tracking Efficiency for Indoor Light Energy Harvesting

Peng-Chang Huang and Tai-Haur Kuo
National Cheng Kung University, Taiwan 185

DC 14
ADC Techniques

14-1 (7133) A Single-Supply Buffer-Embedding SAR ADC with Skip-Reset having Inherent Chopping Capability

Min-Jae Seo[1], Dong-Hwan Jin[1], Ye-Dam Kim[1], Jong-Pal Kim[2], Dong-Jin Chang[1], Won-Mook Lim[1], Jae-Hyun Chung[1], Chang-Un Park[1], Eun-Ji An[1] and Seung-Tak Ryu[1]
[1]*KAIST, Korea*
[2]*Samsung Advanced Institute of Technology, Korea* 189

14-2 (7055) A 68 dB SNDR Compiled Noise-Shaping SAR ADC With On-Chip CDAC Calibration

Harald Garvik[2], Carsten Wulff[1] and Trond Ytterdal[2]
[1]*Nordic Semiconductor, Trondheim, Norway*
[2]*Norwegian University of Science and Technology, Norway* 193

14-3 (7128) A Digitally-Calibrated 70.98dB-SNDR 625kHz-Bandwidth Temperature-Tolerant 2nd-order Noise-Shaping SAR ADC in 65nm CMOS

Jae Sik Yoon, Jiyoon Hong, and Jintae Kim
Konkuk University, Korea 195

14-4 (7069) 8.6fJ/step VCO-Based CT 2nd-Order ΔΣ ADC

Akshay Jayaraj[2], Abhijit Das[1], Srinivas Arcot[2] and Arindam Sanyal[2]
[1]*Texas Instruments, Dallas, United States*
[2]*University at Buffalo, United States* 197

FPGA 15

Intelligent System on FPGA

15-1 (7206) 110.3-bits/min 8-Ch SSVEP-based Brain-Computer Interface SoC with 87.9% Accuracy
Wooseok Byun[1], Dokyun Kim[3], Sung Yeon Kim[3], Ji-Hoon Kim[2]
[1]Chungnam National University, Korea
[2]Ewha Womans University, Korea
[3]Seoul National University of Science and Technology, Korea 201

15-2 (7076) FPGA-Based Sparsity-Aware CNN Accelerator for Noise-Resilient Edge-Level Image Recognition
Seungsik Moon, Hyunhoon Lee, Younghoon Byun, Jongmin Park, Junseo Joe, Seokha Hwang, Sunggu Lee, and Youngjoo Lee
Pohang University of Science and Technology, Korea 205

15-3 (7141) Flexible Low Power CNN Accelerator for Edge Computing with Weight Tuning
Miaorong Wang, Anantha P. Chandrakasan
Massachusetts Institute of Technology, United States 209

15-4 (7122) An Asynchronous Reconfigurable SNN Accelerator With Event-Driven Time Step Update
Jilin Zhang[1], Hui Wu[1], Jinsong Wei[2], Shaojun Wei[1], Hong Chen[1]
[1]Tsinghua University, China
[2]University of Science and Technology of China, China 213

MEM 16

Intelligent Memory

16-1 (7160) A 55nm 1-to-8 bit Configurable 6T SRAM based Computing-in-Memory Unit-Macro for CNN-based AI Edge Processors
Zhixiao Zhang[2], Jia-Jing Chen[3], Xin Si[3], Yung-Ning Tu[3], Jian-Wei Su[2], Wei-Hsing Huang[3], Jing-Hong Wang[3], Wei-Chen Wei[3], Yen-Cheng Chiu[3], Je-Min Hong[3], Shyh-Shyuan Sheu[1], Sih-Han Li[1], Ren-Shuo Liu[3], Chih-Cheng Hsieh[3], Kea-Tiong Tang[3], Meng-Fan Chang[3]
[1]Industrial Technology Research Institute, Taiwan
[2]National Tsing Hua Univeristy / Industrial Technology Research Institute, China
[3]National Tsing Hua University, Taiwan 217

16-2 (7182) A 24 kb Single-Well Mixed 3T Gain-Cell eDRAM with Body-Bias in 28 nm FD-SOI for Refresh-Free DSP Applications
Jonathan Narinx, Robert Giterman, Andrea Bonetti, Nicolas Frigerio, Cosimo Aprile, Andreas Burg, and Yusuf Leblebici
École Polytechnique Fédérale de Lausanne, Switzerland 219

16-3 (7122) Configurable BCAM/TCAM Based on 6T SRAM Bit Cell and Enhanced Match Line Clamping
Jongeun Koo, Eunhwan Kim, Seunghyun Yoo, Taesu Kim, Sungju Ryu, and Jae-Joon Kim
Pohang University of Science and Technology, Korea 223

16-4 (7142) Sub-ns Access Sub-mW/GHz 32 Kb SRAM with 0.45 V Cross-Point-5T Cell and Builtin Y_ Line

Chen-Yu He[2], Kuei-Hua Tang[2], Tsung-Shen Chen[1], Kang-Yu Chang[2], Chien-Hung Lin[2], Katsuyuki Sato[2], Shyh-Jye Jou[2], Po-Hung Chen[2], Hung-Ming Chen[2], Bor-Doou Rong[1], Kiyoo Itoh[2]

[1]*Etron Technology Inc., Taiwan*
[2]*National Chiao Tung University, Taiwan* 227

16-5 (7114) Privacy-Aware Data-Lifetime Control NAND Flash System for Right to be Forgotten with In-3D Vertical Cell Processing

Shun Suzuki, Kyoji Mizoguchi, Hikaru Watanabe, Toshiki Nakamura.
Yoshiaki Deguchi, Keita Mizushina and Ken Takeuchi

Chuo University, Japan 231

WLN17

High-speed Wireline Receiver Techniques

17-1 (7125) A 1.64mW Differential Super Source-Follower Buffer with 9.7GHz BW and 43dB PSRR for Time- Interleaved ADC Applications in 10nm

Yizhak Shifman, Yoel Krupnik, Udi Virobnik, Ahmad Khairi, Yosi Sanhedrai and Ariel Cohen

Intel Corporation, Israel 235

17-2 (7138) A 4.8pJ/b 56Gb/s ADC-Based PAM-4 Wireline Receiver Data-Path with Cyclic Prefix in 14nm FinFET

Gain Kim[5], Lukas Kull1, Danny Luu[1], Matthias Braendli[3], Christian Menolfi1, Pier-An- drea Francese[3], Hazar Yueksel[4], Cosimo Aprile[2], Thomas Morf[3], Marcel Kossel[3], Alessandro Cevrero[3], Ilter Ozkaya[2], Hyeon-Min Bae[5], Andreas Burg[2], Thomas Toifl[1], Yusuf Leblebici[2]

[1]*Cisco Systems (Switzerland) GmbH, Switzerland*
[2]*École Polytechnique Fédérale de Lausanne, Switzerland*
[3]*IBM Research Zurich Laboratory, Switzerland*
[4]*IBM T. J. Watson Research Center, United States*
[5]*KAIST, Korea* 239

17-3 (7045) A 32-Gb/s 0.46-pJ/bit PAM4 CDR Using a Quarter-Rate Linear Phase Detector and a Low-Power Multiphase Clock Generator

Zhao Zhang. Guang Zhu, Can Wang, Li Wang and C. Patrick Yue

HKUST-Qualcomm Lab, Department of Electronic and Computer Engineering
Hong Kong University of Science and Technology, China. 241

17-4 (7024) A Maximum-Eye-Tracking CDR with Biased Data-Level and Eye Slope Detector for Optimal Timing Adaptation

Hye-Yoon Joo[1,2] and Deog-Kyoon Jeong[1]

[1]*Seoul National University, Korea*
[2]*Samsung Electronics, Korea* 243

ACS 18

Capacitive Power Converters

18-1 (7158) A 200-MHz Wide Input Range CMOS Passive Rectifier with Active Bias Tunning

Xiaofei Li[1], Fangyu Mao[1], Pyungwoo Yeon[2], Yan Lu[1], Maysam Ghovanloo[2], Rui P. Martins[3]

[1]*University of Macau, Macau*
[2]*Georgia Institute of Technology, United States*
[3]*University of Macau / Universidade de Lisboa, Macau* 245

18-2 (7052) A 7.5 - 42V Input High-VCR Monolithic DC-DC Converter Using Stacked Isolated SC Cores

Elly De Pelecijn and Michiel Steyaert

Katholieke Universiteit Leuven, Belgium 247

18-3 (7112) A 918MHz Wide-Range CMOS Rectifier with Diode-Feeding and Switch-Capacitor-Based Load Modulation Technique

Chen-Yi Kuo, Chun-An Lu, Yu-Te Liao

National Chiao Tung University, Taiwan 251

18-4 (7140) A CMOS Switched-Capacitor Boost Mode Envelope Tracking Regulator with 4% Efficiency Improvement at 7.7dB PAPR for 20MHz LTE Envelope Tracking RF Power Amplifiers

Neha Kumari[1], Shang-Hsien Yang[1], Ke-Horng Chen[1], Ying-Hsi Lin[2], Shian-Ru Lin[2], Tsung-Yen Tsai[2]

[1]*National Chiao Tung University, Taiwan*

[2]*Realtek Semiconductor Corp, Taiwan* 255

18-5 (7101) A Conversion-Ratio-Insensitive High Efficiency Soft-Charging-Based SC DC-DC Boost Converter for Energy Harvesting in Miniature Sensor Systems

Junyoung Park[2], Hyungmin Gi[2], Seungchul Jung[1], Sang Joon Kim[1], Yoonmyung Lee[2]

[1]*Samsung Advanced Institute of Technology, Korea*

[2]*Sungkyunkwan University, Suwon, Korea* 259

SOC 19

Low Power SoC for IoT

19-1 (7181) 33us, 94uJ Optimal Ate Pairing Engine on BN Curve over 254b Prime Field in 65nm CMOS FDSOI

Makoto Ikeda, Tadayuki Ichihashi, Hiromitsu Awano

University of Tokyo, Japan 263

19-2 (7080) An IoT Sensor Node SoC with Dynamic Power Scheduling for Sustainable Operation in Energy Harvesting Environment

Yuji Yano[3], Seiya Yoshida[3], Shintaro Izumi[3], Hiroshi Kawaguchi[3], Tetsuya Hirose[4], Masaya Miyahara[2], Teruki Someya[5], Kenichi Okada[5], Ippei Akita[1], Yoshihiko Kurui[6], Hideyuki Tomizawa[6], and Masahiko Yoshimoto[3]

[1]*Advanced Industrial Science and Technology, Japan*

[2]*High Energy Accelerator Research Organization, Japan*

[3]*Kobe University, Japan*

[4]*Osaka University, Japan*

[5]*Tokyo Institute of Technology, Japan*

[6]*Toshiba Corporation, Japan* 267

19-3 (7066) A 3.01 mm2 65.38Gb/s Stochastic LDPC Decoder for IEEE 802.3an in 65 nm

Qichen Zhang[1], Yun Chen[1], Xiaoyang Zeng[1], Keshab K. Parhi[3], Borivoje Nikolic[2]

[1]*Fudan University Shanghai, China*

[2]*University of California, Berkeley Berkeley, United States*

[3]*University of Minnesota Minneapolis, United States* 271

19-4 (7159) A Millimeter Wave Digital CMOS Baseband Transceiver for Wireless LAN Applications

Kang-Lun Chiu[3], Hsun-Wei Chan[3], Wei-Che Lee[3], Chang-Ting Wu[3], Henry Lopez[3], Hung-Chih Liu[3], Meng-Yuan Huang[2], Chun-Yi Liu[3], Tsai-Hua Lee[3], Hsin-Ting Chang[3], Chih-Wei Jen[3], Nien-Hsiang Chang[1], Pei-Yun Tsai[2], Yen-Cheng Kuan[3], Shyh-Jye Jou[3]

[1]*National Applied Research Laboratories, Taiwan*
[2]*National Central University, Taiwan*
[3]*National Chiao Tung University, Taiwan* 275

RF 20

RF Signal Synthesizers

20-1 (7171) A 23-mW 60-GHz Differential Sub-Sampling PLL with an NMOS- Only Differential-Inductively-Tuned VCO

Bingwei Jiang, Howard C. Luong
Hong Kong University of Science and Technology, Hong Kong 279

20-2 (7093) A 0.003-mm2 440fsRMS-Jitter and -64dBc-Reference-Spur Ring-VCO-Based Type-I PLL Using a Current-Reuse Sampling Phase Detector in 28-nm CMOS

Zunsong Yang[1], Yong Chen[1], Pui-In Mak[1], Rui P. Martins[2]
[1]*University of Macau, China*
[2]*University of Macau / Universidade de Lisboa, China* 283

20-3 (7126) A 360-456 MHz PLL frequency synthesizer with digitally controlled charge pump leakage calibration

Peilin Yang, Yanshu Guo, Hanjun Jiang and Zhihua Wang
Tsinghua University, China 285

20-4 (7100) A 12-GHz All-Digital Calibration-Free FMCW Signal Generator Based on a Retiming Fractional Frequency Divider

Zhengkun Shen, Heyi Li, Haoyun Jiang, Zherui Zhang, Junhua Liu and Huailin Liao
Key Laboratory of Microelectronic Devices and Circuits (MOE)
Peking University, China 287

20-5 (7177) A 100Mb/s 3.5GHz Fully-Balanced BFOOK Modulator Based on Integer-N Hyrbid PLL

Cong Ding, Haixin Song, Woogeun Rhee, and Zhihua Wang
Tsinghua University, China 291

ETA 21

Biomedical Sensor Interfaces

21-1 (7071) A 15-Ch. 0.019 mm2/Ch. 0.43% Gain Mismatch Orthogonal Code Chopping Instrumentation Amplifier SoC for Bio-Signal Acquisition

Jeong Hoan Park[2], Tao Tang[2], Lian Zhang[2], Kian Ann Ng[1], Jerald Yoo[3]
[1]*N.1 Institute for Health, Singapore*
[2]*National University of Singapore, Singapore*
[3]*National University of Singapore / N.1 Institute for Health, Singapore* 295

21-2 (7053)

A 10μW -74.6 dB THD Arterial Pulse Waveform Sensing System with Automatic Bridge-Offset Calibration and Super Class-AB Output Stage

Yu-Pin Hsu, Zemin Liu, Mona Hella
Rensselaer Polytechnic Institute, United States 297

21-3 (7063) A 0.012 mm2, 1.5 GΩ ZIN Intrinsic Feedback Capacitor Instrumentation Amplifier for Bio-Potential Recording and Respiratory Monitoring

Lian Zhang[1], Tao Tang[1], Jeong Hoan Park[1], Jerald Yoo[2]

[1]*National University of Singapore, Singapore*

[2]*National University of Singapore / N.1 Institude for Health, Singapore* 301

21-4 (7010) T/R-Switch Composed of 3 High-Voltage MOSFETs with 12.1 μW Consumption that can Perform Per-channel TX to RX Self-Loopback AC Tests for 3D Ultrasound Imaging with 3072-channel Transceiver

Shinya Kajiyama, Yutaka Igarashi, Toru Yazaki, Yusaku Katsube, Takuma Nishimoto, Tatsuo Nakagawa, Yohei Nakamura, Yoshihiro Hayashi and Taizo Yamawaki

Hitachi, Ltd., Kokubunji, Japan 305

21-5 (7118) A High DR High-Input-Impedance Programmable-Gain ECG Acquisition Interface with Non-inverting Continuous Time Sigma-Delta Modulator

Junhao Liang[2], Sai-Weng Sin[2], Seng-Pan U[2], Franco Maloberti[4], Rui P. Martins[3], Hanjun Jiang[1]

[1]*Tsinghua University, China*

[2]*University of Macau, Macau*

[3]*University of Macau / Universidade de Lisboa, Macau*

[4]*University of Pavia, Italy* 309

Author Index 313

Panel Discussion 320

Committees 325

Welcome Message

On behalf of the organizing committee, it is my great pleasure to extend to you all a very warm welcome to the IEEE Asian Solid-State Circuits Conference (A-SSCC) on November 4 to 6, 2019. A-SSCC is an international electronics forum that takes place in Asia with the support of the IEEE Solid-State Circuits Society. Since its inception in 2005, A-SSCC 2019 will be the 15th edition of the conference where the most updated and advanced chips and circuit designs in solid-state and semiconductor fields will be presented.

The theme of A-SSCC 2019 is "Silicon System for Next Smart Society". Solid-state circuits have improved our daily life for more than 50 years. There is no doubt that we have benefited from a wide variety of electronic equipment such as mobile computing devices, car electronics and digitalized social infrastructures. The rapid progress of artificial intelligence accelerated by deep learning, in conjunction with big data collected by the Internet of Things, will continue changing our lives significantly in the near future. As a consequence, we will use smarter mobile devices, drive or be driven by, smarter cars and live on a smarter infrastructure, leading to a totally different quality of life. There, silicon systems will face new challenges, which our professional community should overcome, but human kind shall take the advantage of new opportunities that will arise.

To embrace the conference theme of this year, a full and rich three-day program consisting of 4 outstanding plenary talks, 4 key technology tutorials, 1 panel discussion, 1 industry session and various regular paper sessions covering more than 80 regular papers in the areas of analog circuits, data converters, digital circuits and systems, emerging technology and applications, memory, radio-frequency circuits, system-on-chip and signal processing, wireline and mixed signal circuits have been organized. The Student Design Contest and FPGA exhibition will also include demonstrations from the best student papers as well as the presentation of advanced technologies with real-time discussion. In addition, the warm welcome reception, the social hour, the banquet with performances and the farewell event plus the post-world-heritage site visit will enable the interactive communication among experts and students from all over the world.

We are pleased to have the presence of a number of world renowned experts joining this conference and sharing their precious experience and knowledge with us. A-SSCC 2019 will provide a valuable opportunity for academics, students, research scientists, industry specialists and decision-makers to engage and interact with each other, and hopefully to come up with new innovations to collaborate on smart mobile devices and more, towards building the next smart society.

I would like to take this opportunity to express my sincere gratitude and appreciation to the members of the organizing committee chaired by Prof. Seng-Pan U, the steering committee chaired by Prof. Tadahiro Kuroda, and the technical program committee chaired by Prof. Mototsugu Hamada and assisted by Prof. Chen-Hao Chang (Co-Chair), Jun Deguchi (Vice-Chair), Po-Hung Chen (Vice-Co-Chair), as well as all speakers, authors, and sponsors.

Once again, thank you very much for joining us in A-SSCC 2019. I am sure that you will have fruitful and rewarding exchanges in this conference. Please enjoy it and share your positive experience with colleagues and friends. Your continuous support is of paramount importance for the continuing success of A-SSCC.

Prof. Rui Martins, IEEE Fellow

A-SSCC 2019, Conference Chair
State-Key Laboratory of Analog and Mixed-Signal VLSI, Director
University of Macau, Vice-Rector (Global Affairs)

Foreword

I would like to welcome you to the IEEE Asian Solid State Circuits Conference (A-SSCC) 2019. The conference is held in Macau, China from Nov. 4th to Nov. 6th, 2019. Being one of the five conferences fully sponsored by the IEEE Solid State Circuits Society, A-SSCC has grown to be a leading conference in the field of integrated circuits and systems design.

The conference theme for this year is "Silicon Systems for Next Smart Society." Solid-state circuits have improved our lives for more than 50 years. There is no doubt that we have benefited from a wide variety of electronic equipment such as mobile computing devices, car electronics and digitalized social infrastructures. Rapid progress of artificial intelligence accelerated by deep learning, in conjunction with big data, collected by IoT, may change our lives non-linearly. As a consequence, we will use smarter mobile devices, drive or be driven by, smarter cars and live on a smarter infrastructure, leading to a totally different quality of life. There, we will face new challenges and opportunities for silicon systems, which our community should overcome and take advantage of.

This year we received 199 submissions from 25 countries and regions around the world. Among all submissions, 92% of papers presented measurement results with silicon chips. After rigorous review process, including on site meeting on July 26, the Technical Program Committee (TPC) selected 84 high quality papers from 17 countries and regions. The acceptance rate is 42%.

The conference starts with 4 tutorials on Nov. 4, 2019. Prof. Kenneth K. O, University of Texas at Dallas presents "On-chip millimeter wave voltage measurements for debugging, built-in self-test and self-healing", Prof. Masato Motomura, Tokyo Institute of Technology presents "AI Computing: What it is about & How hardware can help it out", Prof. Seung-Tak Ryu, Korea Advanced Institute of Technology presents "Bringing back pipelined ADCs in the era of SAR ADCs", Prof. Meng-Fan Chang, National Tsing Hua University presents "Nonvolatile logic and computing-in-memory for AI edge chips."

Four plenary speeches are presented by distinguished scholars and industry leaders. Mr. Yiming Zhu, Chairman & CEO, ChangXin Memory Tech., China, presents "China IC industry development: past, now and future of memory industry" and Mr. KyoWon Jin, Exec. VP & General Manager, DRAM Dev & Biz, SK Hynix, Korea, addresses "Memory centric, the foundation of next smart society" in the morning of Nov. 5, 2019. On Nov. 6, Dr. Yasunori Mochizuki, NEC Fellow, NEC, Japan, shares with us his view on "AI and IoT for Social Value Creation", and Dr. Mau-Chung Frank Chang, President, National Chiao Tung University, Hsinchu, Taiwan, talks about "CMOS Terahertz circuits for radar, radio, imager and spectrometer system applications."

A panel discussion is held on Tuesday evening with the topic "Are Analog and Mixed Mode Circuits the future solution of AI SoCs?", moderated by Prof. Noriyuki Miura, Kobe University, Japan. The panel invites experts from UNIST, Carnegie Mellon University, KAIST, Tsinghua University, National Tsinghua University, and Osaka University to discuss issues related to the future of AI circuits and systems. The industry session, held also on Tuesday, highlights advances in high performance techniques for industrial applications. Four outstanding industry papers are presented by speakers from Texas Instruments, CSEM, Renesas, and Intel. The regular conference papers are grouped in 18 sessions in 4 parallel tracks. The Student Design Contest provides live demos from the top 15 student-authored papers. Three winners are selected and recognized at the conference banquet. FPGA papers also showcase their system demonstrations.

A-SSCC 2019 TPC consists of 105 members divided into 10 technical subcommittees. The members come from both industry and academia around the world. This year, TPC members gathered in Qingdao on late July to select excellent papers. Their contributions to maintain a high-quality A-SSCC are highly appreciated. Furthermore, I would like to acknowledge the leadership of the technical subcommittee chairs: Prof. Po-Chiun Huang of National Tsing Hua University (Analog Circuits and Systems), Dr. Kazuko Nishimura of Panasonic Corp. (Data Converters), Prof. Jun Zhou of University of Electronic Science and Technology of China (Digital Circuits and Systems), Dr. Chi-Cheng Ju of Mediatek (SoC and Signal Processing), Prof. Minoru Fujishima of Hiroshima University (RF), Prof. Chulwoo Kim of Korea University (Wireline and Mixed-Signal Circuits), Prof. Woogeun Rhee of Tsinghua University (Emerging Technology and Applications), Dr. Junghwan Choi (Memory) of Samsung Electronics, Dr. Shigeki Tomishima of Intel (FPGA) and Dr. Stefan Rusu of TSMC (Industry Program).

I would also like to acknowledge Prof. Baoyong Chi of Tsinghua University and Prof Jung-Hoon Chun of Sungkyunkwan University for organizing the Student Design Contest, Prof. Hoi-Jun Yoo of KAIST for preparing the plenary program and panel, and Prof. Hoi-Jun Yoo of KAIST and Woogeun Rhee of Tsinghua University for the tutorial planning.

I would like to extend my sincere appreciation to all authors and speakers, conference organizers, committee members, moderators, panelists, and, last but not least, all the participants. I hope you will enjoy the technical program of A-SSCC 2019, take this opportunity to network with experts around the world, and bring back good memories with you!

Mototsugu Hamada

Technical Program Committee Chair of A-SSCC 2019
Keio University, Japan

PROCEEDINGS

2019 IEEE Asian Solid-State Circuits Conference (A-SSCC)
PROCEEDINGS OF TECHNICAL PAPERS

November 4 - 6, 2019
The Parisian Macao, Macao SAR, China

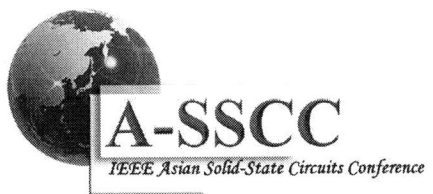

What are the driving forces of DRAM?

YiMing Zhu
CXMT, CEO
Hefei, Anhui, China
YiMing@cxmt.com

Abstract— this paper describes some challenges of DRAM technology and some prospects driven by internet environments and 5G network. New business based on high speed and high quality networking environment, such as contents services, e-sports and services business based on big data will be a strong driving force of DRAM industry.

Keywords—DRAM, Memory, Mobile, Internet, Cloud, System, IoT, Automotive

I. INTRODUCTION

There have been many concerns on the future of DRAM industry. The limit of technology scaling have difficulties of packing more memory cell into small area and thereby DRAM business well known as the features of low cost/high performance is losing active motivation in further driving technology development. However, DRAM is still one of the most important part of state-of-the art computing environments in that alternative solution is not ready or far behind in technology maturity[1]. Accordingly, industry does its best to continuously seeking for technical solutions that give new active power to DRAM society. As shown in Fig.1, current DRAM technology development speed is being slow down drastically since 2y technology node. As technology advances, DRAM industry faces a lot of challenges to overcome the hurdle to achieve Moore's Law. The time of moving to next technology node DRAM is 2x times longer than before(18 months → 32 months).

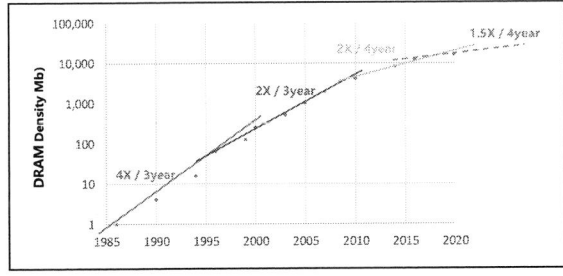

Fig. 1 DRAM technology development speed

In addition to the demand on high density, the demand on speed and power is also increasing. In early 2000, the appearance of mobile & smart phone, the connected things including IoTs and new computing environments such as cloud service, AI and ML, demand the importance of power as well as performance. Fig 2 shows the trend of DRAM bandwidth required. The applications based on big data and AI

are newly rising and asking for much higher performance and energy efficiency than the traditional DRAM technology has supported [2].

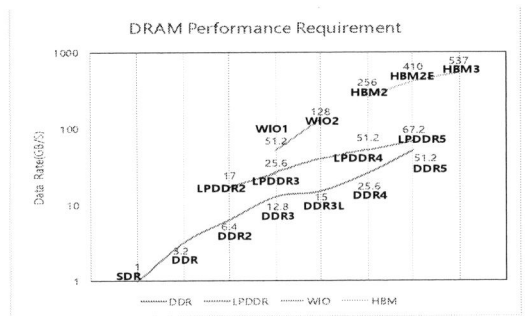

Firg.2 DRAM performance trend

In this paper, we will review the challenges DRAM industry face, the driving force of DRAM industry and the memory system trend following the demand of memory performance.

II. MEMORY CHALLEENGE & SOLUTION

In this section, we discuss about the challenges and prospects facing memory technology. The challenge of memory technology is that scaling down is approaching its limit. As already mentioned in the introduction, many technical efforts have been made on the process technology, circuit design and system design side as we reach the more advanced technology node.

A. Process scaling limit

Stable DRAM operation depends on amount of charges stored inside cell capacitor. The charge retention time is decreasing with process scaling due to capacitor size reduction and transistor leakage current increase. Gate-induced drain leakage GIDL) and sub-threshold leakage current increased in access transistors. Building higher aspect ratio capacitor gets limited with process scaling. Fail bit counts caused by data retention increases as technology advances. To help further process scaling, on-die ECC will be used in DDR5 as a JEDEC standard and new cell design is being considered and experimented[1][2][3].

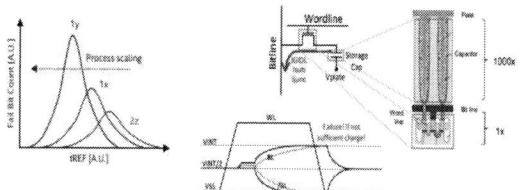

Fig.3 Process scaling challenge

B. Sensing Margin

Besides DRAM cell capacity scaling, BL sense amplifier sensing margin is degraded (Fig.4). DRAM cell capacity had been maintained around 20-30ff until 2y node and now starts to reach at 10fF in 1x node. Small DRAM cell capacity and large BL sense amp offset leads to almost zero sensing margin[4]. New circuit technique and new material are used to enhance or improve sensing margin.

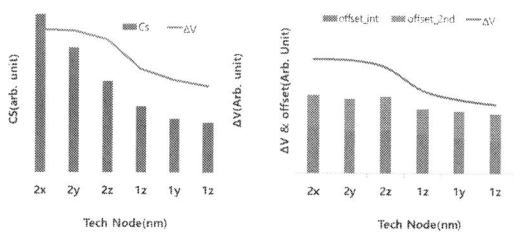

Fig.4 Cell capacitor and sensing signal

C. Power scaling limit

DRAM supply voltage has scaled down over the years from 3.3V down to 1.2V in DDR4 and 1.1V in DDR5. For further supply voltage scaling, low threshold voltage and low leakage device are necessary. New gate material, Hi-K metal gate and new device structure, GAA are considered as an option to reduce supply voltage[5][6].

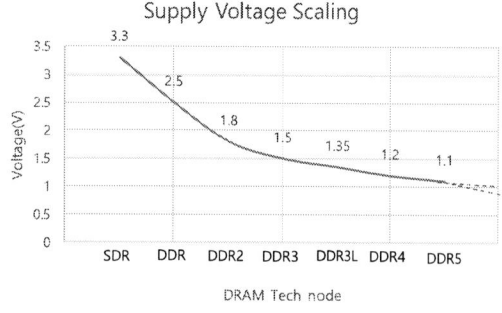

Fig. 5 Supply voltage scaling

D. Manufacturing cost

DRAM process technology to handle ultra small feature size is getting complex and requires more CaPEX and longer turn-around time, making it more difficult to achieve ROI(return of investment) in each new generation technology. Fig.6 shows returning advantages when the process technology has a transition to next generation technology. Since 2z, new process technology advantage reduces. The process cost overhead increases due to an increase of process step and the net die increase of per wafer is limited. New technology to reduce the process step such as EUV will improve ROI.

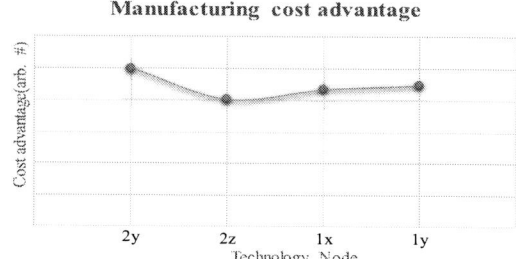

Fig.6 Manufacturing cost trend

III. DRIVING POWER OF DRAM

A. Convergence to Smartphone

Since 2010, smart phone starts to be widely spread and the functions of digital camera, MP3 and cell phone have been integrated into smartphone. Most of the users use smart phone to send and receive text messages, images and moving pictures. Therefore the internet traffic by smart phone increase very fast. The cloud service that stores/processes user data and creates new service based on user-generated data, is getting more popular and prosperous. As the number of smart phone unit increases and the performance of smart phone improves, the memory consumption of smart phone reaches more than 40% of total memory consumption (Fig.7). At the same time, cloud service provider starts to increase their CaPEX and to bring in more server memory consumptions.

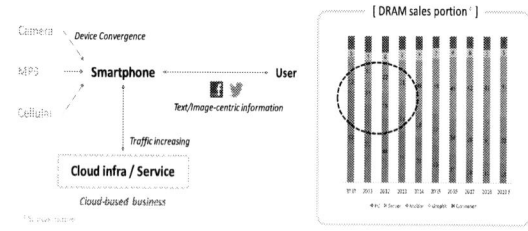

Fig.7 Convergence to smart phone and memory consumption

Until the mid-2010, the increase rate of smart phone and its market cap increase were 40% and 25% respectively. After then, the smart phone market increase rate dropped to a sing digit and the unit increase rate records minus values. As the convergence to smart phone proceed, the differentiation of each generation from the previous one is not so easy. The replacement period of smart phone is getting prolonged[7].

978-1-7281-5107-6/19 $31.00 © 2019 IEEE

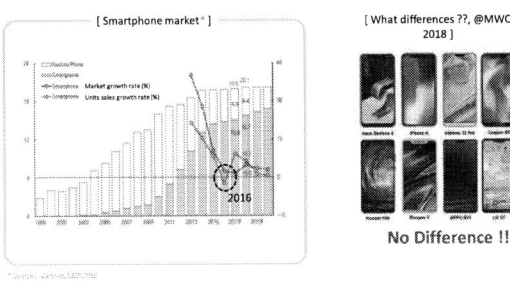

Fig.8 Saturation of the smartphone market

B. Video contents & new business model by network technology

When the converged smartphone led Internet traffic by mid-2010, new video-centric business models, combined with advanced streaming technology and content business from mid-term onwards, were emerged, resulting in more network traffic. Thus, in 2020, video becomes a large part of the increased web traffic, and hence a large investment in the public cloud to deal with these services.

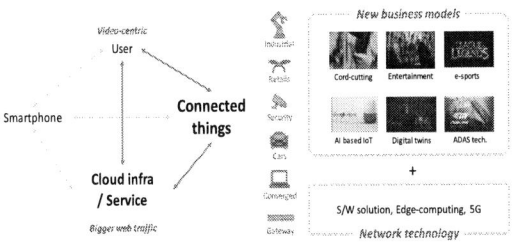

Fig.9 Video contents and new business model

Netflix services as a content provider needs to provide a very reliable service to users even in high traffic situation by new adaptive bandwidth technology and cloud-based synchronization with global cloud system, expanding content locally and offering low-latency services.

5G will bring enhanced mobile broadband, massive machine type communication and utra-reliable low latency communication. The emergence of 5G network will help to enrich/transform the entire internet based business by enabling diverse forms of product or business models. As a result, 5G technology is expected to be one of the key tools to boost the needs of the cloud infrastructure /service and the large capex.

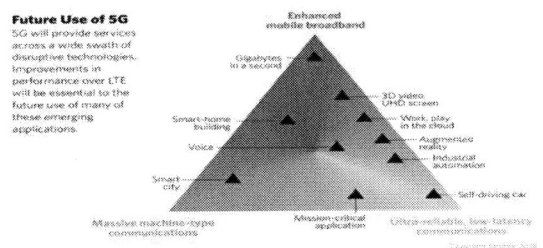

Fig.10 5G network effects on internet business

C. Automotive & IoT

The entry of connected things (automotive, IoT, wearable & sensor) is rapidly increasing. About 30 percent of the market share is expected in 2021. Especially automotive and IoT market growth rate is most noticeable. IoT devices sense, or collect data and store in cloud data center for next service. Vision IoT device with vision sensor even performs deep learning or inference processing and requires high performance & relatively larger memory size.

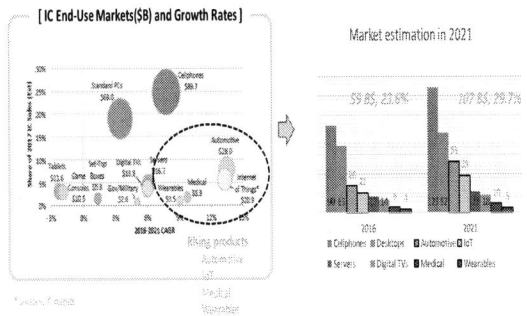

Fig.11 Automotive & IoT growth rate

By 2025, most of new cars are expected to be connected to the Internet. Connected cars are rapidly becoming a part of our daily lives and providing a self-driving capability soon. The autonomous vehicle's driving system relies on AI driving system helped by 5G network. The carmakers are also developing platforms that provide automotive-related services similar to smart phone case. Thus, autonomous driving brings about strong demand on high performance computing in the cloud and edge computing.

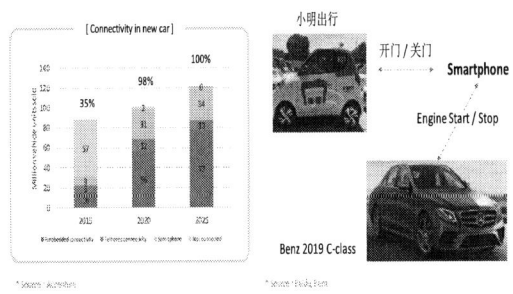

Fig.12 Connected car service

IV. COMPUTER MEMORY SYSTEM

A. Memoory capacity & Bandwidth requirement

Mobile services processes the large data generated by the connected devices and asks very fast response, efficient and reliable operation. For those, in-memory computing, server virtualization and machine learning are increasingly adapted in cloud and edge computing system. Apparently, the system volume in the data center is strongly boosted and the memory capacity per system is also increasing as memory intensive workload such as HPC, graphics and finance & gaming has to be processed. The data center processor growth speed is so fast. Simultaneously, the demand for memory bandwidth is growing rapidly to meet high quality display of smart phone, high quality of pc gaming and big data analytics.

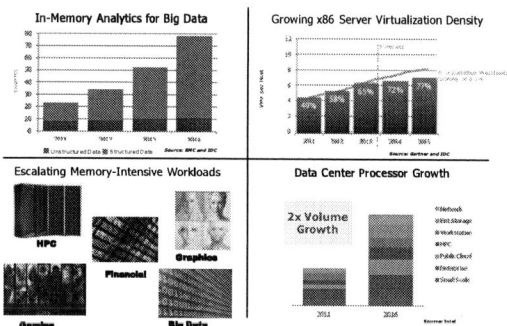

Fig. 13 Cloud data center computing environment

B. Memory subsystem

Changes are taking place in the traditional memory hierarchy to meet the aforementioned memory bandwidth needs. WIO based memory system was proposed instead of LPDDR in mobile system, and near memory computing system was introduced in high end server(Fig.14 a). HBM is kept very close to GPU/SOC and is used to provide high memory bandwidth and high data traffic of 10 times of DDR4 and 4 times higher than gddr5, proving to be a key element for new computing system such as large data analytics, high end graphics and AI/ML(Fig14.b)[9]. It is expected that HBM will be used more widely in low power high performance graphics such as AR/VR besides high-end applications in the future.

Another example is the big data analytics & in-memory computing system using SCM(Fig.14 c). Big data application is proposed as a solution to the large memory solution using SCM since it has to process a large number of unstructured data generated by nature, and the relatively limited memory size makes it difficult to obtain the desired computer performance because it causes the frequent data traffic between memory and storage. The solution using Intel's Optane memory is being tried by many in-memory database or virtualization as a solution for memory capacity expansion[10].

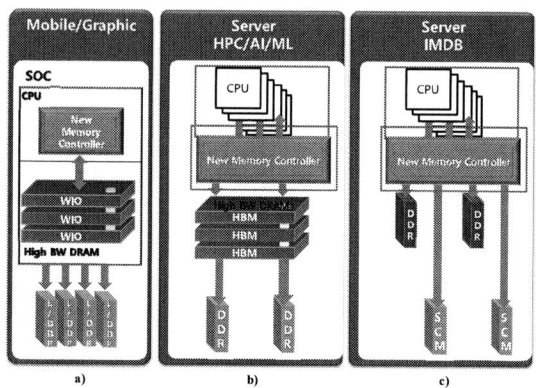

Fig. 14 Memory subsystem change

V. CONCLUSION

We reviewed the challenges of the memory industry, as well as the impact on memory industry of the internet environments, as a driving force of the development of the memory industry. Memory subsystems are also evolving in line with various computing environments and are following system needs. Now & future, the internet-based contents service and connected things along with PCs and mobile smart phones are the driving force of the memory industry. The 5G technology is expected to further boost these businesses by providing the high performance and reliability of network for new internet based service.

ACKNOWLEDGMENT

Appreciate for the support and preparation of data and material by CXMT team.

References

[1] SungKye Park, "Technology scaling challenge and future prospecs of DRAM and NAND flash memory," Proceedings of SPIE - The International Society for Optical Engineering

[2] Meng-Qi Wen et, "J. Clerk Maxwell, A high performance DRAM design using U-FinFET as access transistor," 2018 14th IEEE International Conference on Solid-State and Integrated Circuit Technology (ICSICT)

[3] https://www.jedec.org/ddr5details

[4] S. Lee, "Technology Scaling Challenges and Opportunities of Memory Devices," IEDM'16

[5] A. Spessot et, "Optimized process simulation of USJ for HKMG DRAM periphery transistors

[6] Hyunwoo Chung, et, "Novel 4F2 DRAM celll with vertical pillar transistor(VPT) in Proceeding of the European Solid State Device, 2011

[7] https://www.cisco.com/

[8] https://www.mwc.com/

[9] https://www.nvidia.com/en-us/deep-learning-ai/solutions

[10] https://www.intel.com/content/www/us/en/products/memory-storage/optane-dc-persistent-memory.html

Memory Centric Computing,
The Foundation of the Next Smart Society

KyoWon Jin
DRAM Development & Business
SK hynix
Icheon, South Korea
kyowon1.jin@sk.com

Abstract—ICT industry is taking a new turn driven by the development of artificial intelligence and 5G network. A smart society is already being here in everyday life, and the world we live in is becoming ever smarter. However, many tasks remain to be tackled. In the future, the artificial intelligence computing system will play a pivotal role in the "next smart society", which explains why a lot of research is underway in related industries, aiming for dramatic performance improvement.

While various approaches are being developed to utilize systems for a smart society, data transfer between processors and memories is still hindering system performance. In addition, increased power consumption due to increased bandwidth remains a problem. The memory industry is currently developing high-bandwidth memory that can meet system requirements to contribute to improving system performance, but it is time to develop more innovative memory technologies to meet the requirements for a smart society.

In order to achieve this goal, we need to not only improve the memory itself but also combine computing with memory to minimize unnecessary bandwidth usage. In other words, we need to place the memory close to the processor using diverse System-in-Package assembly technology. Furthermore, it is necessary to change the architecture of computing system, which makes cross-industry collaboration essential.

Keywords—ICT industry; data processing; memory; logic processor; System in Package; power consumption; bandwidth

I. INTRODUCTION

Artificial intelligence is already being used broadly in everyday life, and the world we live in is becoming ever smarter. A smart society consists of smart factories, smart cars, smart homes, smart buildings, smart healthcare, and much more. A smart society can be represented by three characteristics: autonomy, connectivity, and cognition. For example, the data that are used in a smart society are connected to everything everywhere, and such data makes up the cognitive IoT. Therefore, data are an essential element of a smart society.

The data used in a smart society can be characterized by four features: volume, variety, velocity, and connectivity as shown in Figure 1.

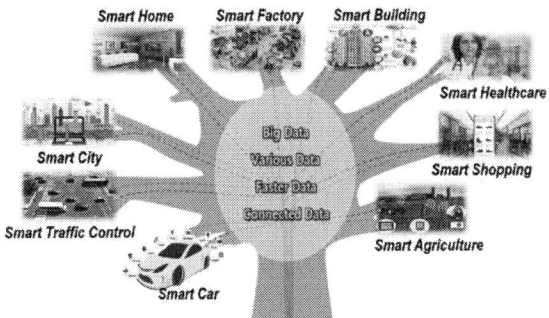

Figure 1. Characteristics of New Data Paradigm

To put that into perspective, 188 million emails are being sent around the world every single minute, and the amount of data generated in the factories of SK hynix has increased 112 times since 2012 as shown in Figure 2.

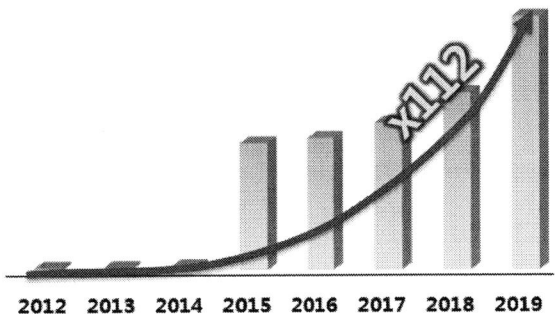

Figure 2. Factory Data from SK hynix

This is why we call today "the era of Big Data." Big data are being treated as meaningful information such as structured data, unstructured data, and cognitive data. In addition, the amount of such data has skyrocketed since the '90s. Given the variety of information contained in big data, producing meaningful data from it can take a long time, but markets wish to use all information in real time. However, it is a big challenge to process and analyze the big amount of data in real time, although processing and analyzing time improved.

II. TECHNOLOGY FOR SMART SOCIETY

There are many technologies that facilitate data management in a smart society. In order to deal with large and diverse data into meaningful data, a technology that handles the data must be based on big data processing. As with all other industries, the semiconductor industry is also using big data processing, especially in the developing and manufacturing memory devices as shown in Figure 3.

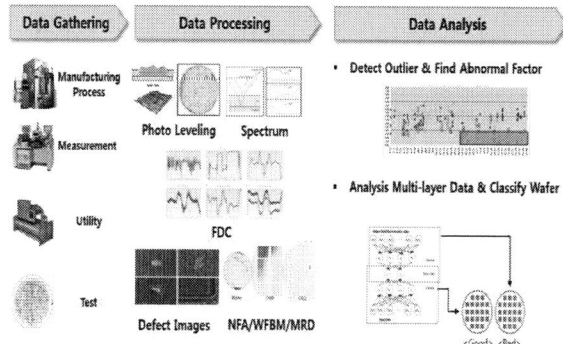

Figure 3. Steps for Big Data Processing in SK hynix

Big data processing technology can handle large and diverse data, but it cannot address the needs for fast and accurate data or provide the autonomy to judge and act on things. In order to solve this problem, machine learning, which is an infinite iterative learning method that repeatedly compares and analyzes data, has been developed, follow by deep learning algorithm for better accuracy. Today, AI is used as an umbrella term for various algorithms and frameworks that have since been developed [1].

Data traffic can be congested in the current infrastructure and facilities, and it could be getting worse to follow the scale and development speed of the cloud server itself. Therefore, large and diverse data do not act as meaningful data in all fields and areas and serve the results in real time. Edge computing technology has been introduced to avoid data traffic congested problem by processing big amount of necessary data locally.

Despite the introduction of edge computing, the amount of remote data that need to be processed is steadily increasing to a point where it cannot be handled by the current network system. This is why 5G technology has emerged. Indeed, 5G technology is optimized for maximizing real time responses.

III. CHALLENGES STEMMING FROM CHANGES

In stated previously, the existing problems could be solved by technologies that enable efficient data movement. However, these technologies demand improved performance through high bandwidth implementation, power reduction, higher capacity, and higher reliability.

Although AI, which is based on very fast and simple iterative computation, has been solved by distributed computing, accelerators using GPUs that perform parallel processing are in the spotlight due to limitations caused by the inefficiency of data usage between existing systems. Parallel process still has a memory bandwidth limit due to amount and speed of data supported by the memory.

To increase the performance of repetitive learning or inference, parallel computing using accelerators and distributed computing using multiple GPUs simultaneously would raise power and thermal problems [3].

To handle large and diverse data, more memory capacity to process may be needed. Big data processing could affect the performance due to movement of huge data.

Edge computing and devices to minimize data movement could require greater reliability than that of conventional memory devices depending on their environment and purpose. For example, in the case of car, the reliability of memory is the biggest problem that is also related to fatality as shown in Figure 4 [4].

Allowed Failures at the Edge: ZERO

High reliability is the most important requirement for some edge devices

	Automotive	Consumer	Industry
Temperature	-40~150°C	-0~70°C	-40~85°C
Failure Rate	0ppm	<100~300ppm	<100ppm
Longevity	Over 10 Year	1~3 Year	5~10 Year
Humidity	~100%	Low	High

Figure 4. Reliability of Automotive Device

There is another challenge from an economic perspective. Capital investment in the semiconductor memory industry has skyrocketed over the last few years, but its bit growth is not proportionately reflecting the efforts anymore. Also, the rate of cost reduction is gradually coming to a point of saturation, and many industry experts have started to believe that the scaling limit of semiconductor memory is not a physical limit, but a cost limit as shown in Figure 5 [5].

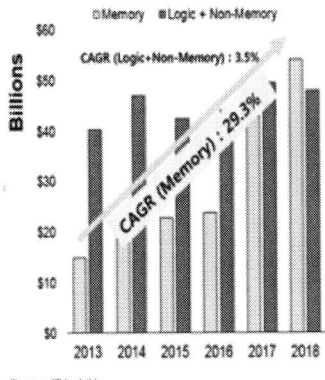

Figure 5. Increasing Capital Investment in Memory Industry

IV. MEMORY CENTRIC COMPUTING, THE FOUNDATION OF NEXT SMART SOCIETY

It can be stated that the key to the next smart society is data, and the key to data movement is memory. Limitations of minimizing data movement and improving its efficiency are from the inefficiency of data movement between memory tiers. Moreover, there is a limitation of expanding DRAM's density and capacity that stores data. To create a smart society, data movement should be efficient, and the implementation of such efficient data movement can be achieved with memory technologies as mentioned above.

The memory industry is preparing solutions to meet the needs of the market and customers. In an attempt to live up to their expectations, the industry has come up with many new ideas such as ultra-high bandwidth, ultra-low power, ultra-high capacity memory, processing in memory and cost effective memory.

SK hynix introduced High Bandwidth Memory (HBM) for bandwidth-hungry customers in 2013. The most recent version of HBM has 4.8 times higher bandwidth than that of GDDR6 as shown in Figure 6. For customers seeking power efficiency, Ultra-low Power (ULP) memory will be ready for the market from 2020.

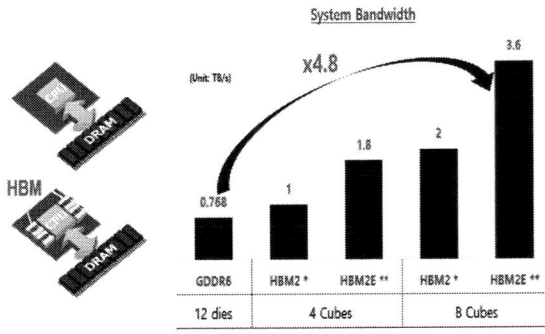

Figure 6. HBM Bandwidth

SK hynix is also preparing for high-capacity memory devices. Although three-dimensional stacking memory is already available in the market, SK hynix is developing a Managed DRAM Solution (MDS) and a non-volatile memory solution. An MDS is designed to have a configuration of 512GB per DIMM, offering eight times more capacity than the existing one at a lower cost, made possible by increasing the net die on 300mm wafer.

Many accelerators are researched for solutions for the AI era to improve data processing performance. Unfortunately, in the current computing architecture, there is a bottleneck due to the limited data transfer speed between accelerators and memory devices. This bottleneck issue can be solved by adding Multiplication & Accumulation (MAC) operation to memory and supporting simple operations from outside the logic chip. SK hynix is doing research on products in which the Processing in Memory (PiM) concept can be utilized as shown in Figure 7.

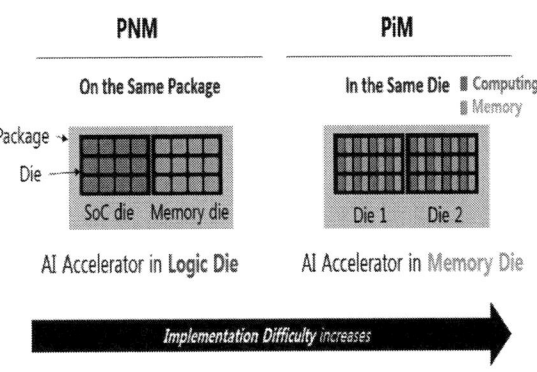

Figure 7. PiM Concept

This new memory will likely be HBM-based; so if HBM satisfies the bandwidth requirements, which have rapidly increased in the AI era, PiM will be able to provide greater value through the memory-based architecture in the AI accelerator area.

V. CONCLUSION

Traditionally, the memory industry has produced commodity products, and they still are playing a critical role in terms of generating revenues and presenting technological competitiveness in transistor geometry. Especially, ever smaller transistor geometry has been the core of memory firms' business model based on production of standardized products in massive scale.

However, in the era of industry revolution 4.0 the paradigm of technological requirements is being shifted such that who or what can best process data in most efficient manner, and these requirements may be emphasized differently according to the ICT customers' demands, such as bandwidth, power, capacity, and footprint. How to carefully balance among ingredients would eventually manifest into a product with customized characters.

978-1-7281-5107-6/19 $31.00 © 2019 IEEE

These trends have been already influencing the historical hierarchy of the memory subsystem as shown in Figure 8 [2]. Depending on what characters need stressed, customized segments in the hierarchy may be created. This presents the memory industry new opportunities, the opportunities that not only increase one's business but actually change the ICT industry by enhancing the overall systems' performance, breaking through the memory wall problems. SK hynix has always been at the forefront of technological exploration, and HBMx, ULP, MDS, and PiM are believed to be able to meet various needs of the changing industry.

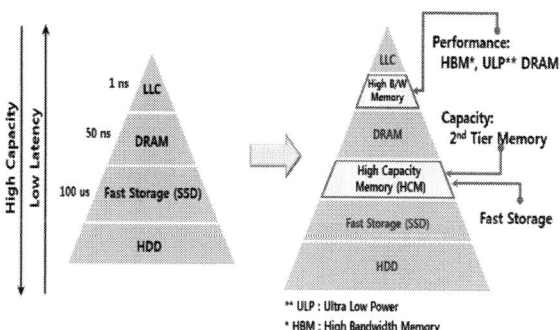

Figure 8. Memory Hierarchy

....

REFERENCES

[1] EuiCheol Lim, "Near Data Processing for AI & Big Data," ISCA 2019 [Online]. Available: https://iscaconf.org/isca2019/workshops.html

[2] EuiCheol Lim, "Rethinking on the Memory Hierarchy," ITC-CSCC 2019 [Online]. Available: http://itc-cscc2019.org/2019/pages/program.VM

[3] Y. H. Chen, J. Emer, and V. Sze, "Eyeriss: A Spatial Architecture for Energy-Efficient Dataflow for Convolutional Neural Networks," in ISCA 2016

[4] YongJae Park, "Delivering Autonomous Vehicle Performance with DRAM" in AUTOMOTIVE SYMPOSIUM 2017

[5] K. Rupp, "Computing Beyond Moore's Law: Architecture and Device Innovations" in Fujitsu Forum 2016

A 45nm 76-81GHz CMOS Radar Receiver for Automotive Applications

Debapriya Sahu[#], Rittu Sachdev-Singh[#], Harikrishna Parthasarathy[#], Rohit Chatterjee[#], Brian Ginsburg[$], Daniel Breen[$], Karan Bhatia[$], Sudhir Polarouthu[*], Vimal Edayath[#], Bhupendra Sharma[#], Meghna Agarwal[#], Karthik Subburaj[#], Anjan Prasad[^], Shankar Ram[#], Cathy Chi[$], Ross Kulak[$], Vijay Rentala[$], Neeraj Nayak[$]

[#]Texas Instruments, Bengaluru 560093, India.

[$]Texas Instruments, Dallas 12500, USA.

[*]Intel Technology, Bengaluru 560103, India.

[^]Analog Devices, Bengaluru 560093, India.

Email: d-sahu@ti.com, rittu@ti.com, harip@ti.com, rohit@ti.com, bginzz@ti.com, dbreen@ti.com, kbhatia@ti.com, sudhir.polarouthu@intel.com, vimal-e@ti.com, bhups@ti.com, meghna@ti.com, skarthik@ti.com, anjanprasad.eswaran@analog.com, nshankar@ti.com, cathychi@ti.com, rkulak@ti.com, rvijay@ti.com, neeraj.nayak@ti.com

Abstract—This paper describes the design and performance of a 45nm CMOS (f_T=160GHz) 76-81GHz radar receiver used in Advanced Driver-Assistance Systems which utilizes Frequency-Modulated Continuous Wave (FMCW) synthesis. The receiver achieves 18dB noise figure and -7dBm 1dB compression point consuming 225mW power. The design includes programmable gain for linearity-noise trade-off, 15MHz wide bandwidth baseband filters, 1.8G samples per second 3rd order Sigma-Delta ADC with 70dB Spurious free dynamic range. In-built safety architecture enables meeting stringent automotive safety requirements.

Keywords—ADAS, millimeter wave communication, receiver, automotive electronics, FMCW, system-on-chip.

I. INTRODUCTION

Millimeter-wave (mmW) radar sensor technology for Advanced Driver-Assistance Systems (ADAS) offers many benefits over other technologies like Video, LIDAR and Ultrasound. Some of them are - (1) Improved precision and reliability, (2) Robustness against weather conditions (rain, fog, storm), and brightness or darkness, (3) Flexible integration capability (sensor is mounted behind the vehicle bumper), (4) Better detection range than Ultrasound (10cm to 10m for Ultrasound, 5cm to >300m for Radar), and (5) Excellent accuracy at both narrow-field view for city driving and wide-field view for country driving. A fully integrated CMOS Radar Transceiver gives added benefits - (1) Improved functionality by integrating mmW front-end and digital on a single integrated chip (IC), enabling autonomous fault monitoring, failure diagnosis and prevention [1]-[2], (2) Low cost, and (3) Compact solution size (compared to multi-module 77GHz or integrated 24GHz solution with antenna). The small form factor also enables easier integration for corner sensing radars in automobiles.

II. SYSTEM ANALYSIS AND ARCHITECTURE

A. FMCW System

Main advantages of a Frequency-Modulated Continuous Wave (FMCW) system are good close-in target detection and high resolution between closely spaced objects. The FMCW transmitter (Tx) generates and transmits a frequency ramp $f_T(t)$ shown in Fig. 1. This signal gets reflected from objects in the path and reaches the receiver (Rx) after a round trip time delay t_d. Received signal $f_R(t)$ is down-converted with the same FMCW source frequency, to a low frequency offset signal $f_{IF}(t)$. This offset frequency is constant for an object, and proportional to its distance from antenna. Transmitted signal $f_T(t)$ ramps from 76GHz to 80GHz (B = 4GHz) in few microseconds, rendering even the fastest cars static for one ramp cycle. Faster Tx ramp or larger RF bandwidth B gives better frequency resolution for close-by objects.

The Rx FFT engines then map $f_{IF}(t)$ to the object distance. The FFT output shows multiple tones corresponding to the objects in the signal path. Reflected signal from nearer objects is stronger in power and lower in frequency than that from farther ones (Received Power α 1/distance[4]). The components with largest power and lowest frequency are from the Tx-Rx antenna coupling and reflection from the car's own bumper. These tones (jammers) can be very strong (-5dBm to -10dBm); they are unavoidable in FMCW radar due to the full duplex operation. They make the linearity requirement of the Rx very stringent, and performance gets heavily limited by sensitivity-linearity tradeoff. High dynamic range, large baseband bandwidth and High Pass Filtering (HPF) enable detection of the desired low power signals in presence of these jammers. This helps achieve the target range, resolution and maximum object velocity performance.

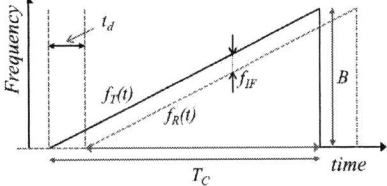

Fig. 1. A FMCW ramp over time.

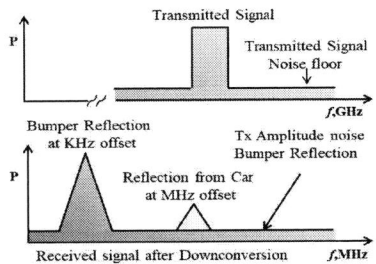

Fig. 2. Amplitude Noise due to Bumper Reflection.

978-1-7281-5107-6/19 $31.00 © 2019 IEEE

Another issue caused by the jammers is that the Tx amplitude noise, transmitted along with the FMCW signal, can mask low power desired signals. Fig. 2 shows this phenomenon affecting the Rx Signal to Noise Ratio (SNR). Overall Rx sensitivity, and hence the maximum range of the radar, is limited by the Noise Figure (NF) of the Rx and the Tx amplitude noise. The total system noise is budgeted assuming equal contributions from Rx and Tx. Rx NF also degrades in the presence of the jammer, which up-converts the Rx's own in-band noise by the same process as shown in Fig. 2.

B. Receiver Block Diagram

Our radar transceiver consists of 4 receivers, 3 transmitters, and a Local Oscillator (LO) synthesizer on a single chip. Fig. 3 shows the block diagram of one radar Rx. The mmW signal from the antenna is converted by an on-chip balun into a differential signal to drive the first differential Low Noise Amplifier (LNA). The balun offers ESD protection for both HBM and CDM and is co-designed with the package to offer 50Ω matching to the antenna. The last LNA stage splits the signal into In-phase (I) and Quadrature (Q) paths. The LNA stages are followed by passive mixers which down-convert the mmW signal to baseband (BB). A 77-81GHz FMCW Phase Locked Loop (PLL) feeds the LO into a Quadrature generator (IQGEN), which generates I and Q LO phases to drive the mixers. Mixers are followed by baseband amplifier-filters, which provide the required HPF, and variable gain to accommodate the high dynamic range of the input signal. The BB output is digitized by a 3rd order $\Sigma\Delta$ ADC.

III. RECEIVER DESIGN

A. LNA

The Rx front-end (FE) consists of three gain stages. While the architecture of individual stages is similar, the performance requirement is slightly different. The strong bumper reflection and Tx-Rx coupling gets filtered only in the BB stages, and all preceding stages need to accommodate these high power tones. The front-end should be linear and not get compressed by this tone. This tone also causes up-conversion of the low frequency (1/f) noise, which would limit the overall NF. To reduce this up-conversion, the 1dB compression point (P1dB) of the LNA has to be ~10dB higher than the reflected tone power. This constraint puts an upper bound on the LNA gain. On the other hand, we need to maximize the gain for better NF and range. A trade-off is achieved by choosing degeneration inductors for the individual stages, with the 1st stage having the lowest and the 3rd stage the highest.

Fig. 4 shows the circuit diagram of the 1st & 2nd LNA stage. Each stage is conjugate matched at the input and output to avoid loss due to internal signal reflection. Power gain in 45nm CMOS at 80GHz is extremely low and all possible ways to maximize gain are employed. These include use of Transmission Lines (TL) to match the impedance at each interface and NMOS C_{gd} neutralization using cross coupled capacitors to reduce power loss. Transformers are used to give passive voltage gain at the NMOS gates and increase the overall power gain of the stage.

Another challenge for the Rx front end is the PVT (Process, Voltage supply and Temperature) variation of the gain and hence NF. At low temperature the gain is higher, causing NF degradation due to the up-conversion phenomenon explained above. To overcome this, gain programmability options are provided in the 2nd and 3rd stages. This is implemented through a MOS resistance, connected through coupling capacitors, in parallel to the output load of the LNA as shown in Fig. 5. The resistance can be varied by programming V_{CTRL}.

Automotive applications require safety diagnostics [2] to ensure the system is always functional. Monitoring functions in Rx ensure that gain and NF are within limit. A Tx-Rx mmW loop back checks the Rx parameters on field. Power Detectors (PD_1 in Fig. 4) monitor signal power at multiple places in the Rx chain. This information is used in safety diagnostic and for programming gain across PVT variations.

B. Mixer and IQGEN

The receiver uses I/Q architecture to reject the image band and reduce Tx amplitude noise by 3dB. It provides additional advantages like crossing interferer detection and digital correction of delay errors between the 4 Rx channels. Output of the 2nd LNA is split into I and Q paths. A 3rd LNA drives each of these components into a passive mixer as shown in Fig. 6. The mixers down-convert the input signal to a 0-15MHz band and drives the baseband stages. The mixer I/Q LO ports are driven by the IQGEN, which converts the FMCW LO signal into I and Q phases. The RF and the LO ports of the mixer are conjugate matched to the previous stages using TL and transformers.

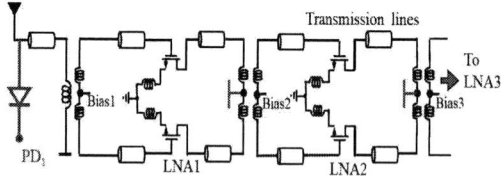

Fig. 4. Low Noise Amplifier.

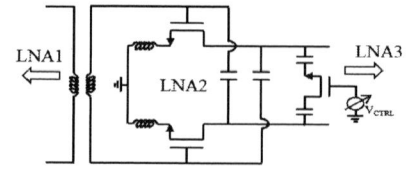

Fig. 5. LNA2 Gain Programming.

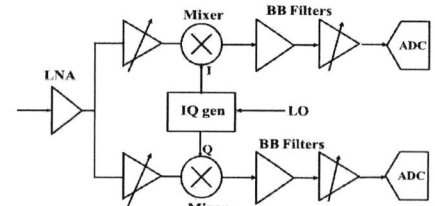

Fig. 3. Receiver Block Diagram.

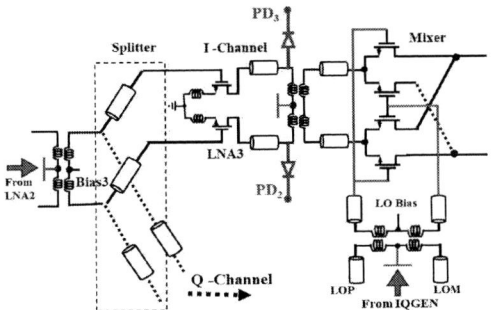

Fig. 6. LNA3 and passive Mixer Stage.

The conventional divide by 2 based IQGEN scheme would mandate the LO generator and buffers work at ~160GHz, where 45nm CMOS offers very little power gain. A λ/4 delay line is used to delay the LO by 90° to generate the Q phase. At 80GHz, λ/4 delay-line is very long (470µm) hence a slow wave branch line coupler is employed to reduce area [3]. Floating strips of metal are placed periodically below the TLs, perpendicular to the direction of current flow, increasing the effective permittivity and reducing the physical length. Further area reduction is achieved by meandering the λ/4 delay line.

C. Baseband Filter

The Base Band stages provide 40dB/decade high pass filtering for the jammers compensating the 40dB/decade drop in signal power with distance. They also provide variable gain to accommodate the high dynamic range of the signal. Fig. 7 shows two band-pass filter stages (BB1 and BB2). Each baseband stage is an opamp-based active RC filter, with programmable HPF and LPF corners. All opamps are class AB amplifiers, biased using Monticelli method, with output common mode maintained using Common Mode Feedback (CMFB) loops. The two BB stages achieve a maximum 1dB BW of 15MHz. Limited RF gain and low output impedance of the passive mixers makes the noise requirement from the BB filters extremely stringent. The RF frontend gains up the jammer demanding high linearity from the baseband stages.

BB1 provides a virtual ground to the signal current from the mixer, which is critical for mixer linearity and conversion gain. The HPF corner is realised with an active RC filter placed in the feedback of the Main Amp (M_{amp}). The Feedback Amp (FB_{amp}) provides a low impedance path to low frequency current (freq < HPF1), thereby reducing the swing at BB1 output. BB1 supports HPF 3dB corners of 0.15-0.7MHz, and LPF 3dB corners of 18-45MHz. Both the M_{amp} and the FB_{amp} are stabilized using Miller cap compensation. With large signal swing, the mixer output impedance becomes asymmetric and common mode noise shows up as differential noise. The CMFB amp bandwidths are kept high to suppress this noise. A clip detector measures the BB1 output and adjusts the RF front-end gain over PVT variation to maintain optimal signal strength at BB1.

BB2 serves as the variable gain amplifier and ADC driver. It has a gain range of 24dB. HPF 3dB corners of 0.35-15MHz and LPF 3dB corners of 12-45MHz are supported. In order to maintain amp bandwidth and stability across gain settings, the frequency compensation of the Opamp is done through a feed-forward stage. The input resistor for BB2 Opamp (R_{IN}) is a major contributor to BB2 noise, but a lower R_{IN} value results in a higher capacitance and hence higher area for the same HPF. The passive values have been chosen as a compromise between NF and area.

D. ADC

Fig. 8 shows the ADC block diagram, a 3rd order modulator using CIFB (cascaded integrator with distributed feedback) is used to meet 10.5 bit ENOB with a sampling clock of 1.8GHz. An innovative chopping technique helps achieve 70dB spurious free dynamic range (SFDR), to avoid false detection due to ghost tones (harmonic distortion). The ADC uses a single-bit quantizer due to its several attractive features such as linear 1-bit feedback DAC and simple 1-bit quantizer with relaxed offset requirement. A passive integrator (R_1C_1) at the input is followed by two active Gm-C integrators. The resistors R_{FF1} and R_{FF2} implement a feed-forward path by feeding the input signal to the cascode nodes (virtual grounds) of Gm_1 and Gm_2. This reduces the signal swing at Gm_1 input which helps in achieving good linearity. The feedback RZ (return-to-zero) DACs (I_{DAC1}-I_{DAC4}) are implemented as 1-bit PMOS current-steering DACs. RZ DAC is inherently linear because every bit interval has a rising and a falling edge. A 3-tap FIR filter is used to achieve good jitter immunity. The innermost loop (I_{DAC4}) adds a direct path around the quantizer to compensate for excess loop delay caused by comparator and feedback circuit.

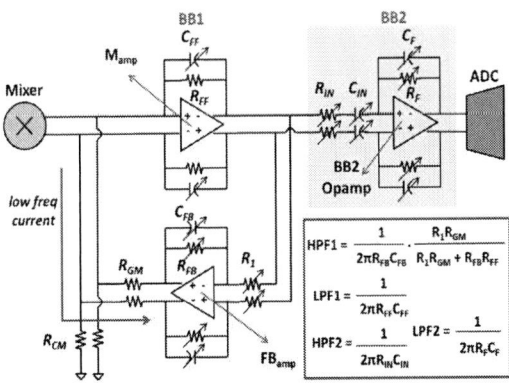

Fig. 7. IF Amplifier Stages.

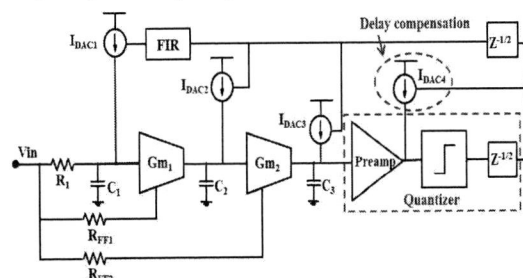

Fig. 8. ADC Architecture.

The quantizer performs the comparison and provides the RZ waveform to 1st, 2nd and 3rd I_{DAC} feedbacks in one clock period. When the comparator samples the input, I_{DAC4} feedback starts just after half clock. All these delays along with FIR filter delay are compensated to maintain the noise transfer function similar to that of an ideal 3rd order modulator. In feedback DACs, mismatch errors such as non-fifty percent clock duty-cycle and DAC switch mismatch directly contribute to even order harmonic distortion. DAC switches (SW_1 and SW_2) in Fig. 9, are biased in saturation to avoid any inadvertent shorting of virtual ground. Due to threshold voltage mismatch of SW_1 and SW_2, some part of DAC current (I_{DAC1}) flows into current source parasitic capacitor (C_P) and shows up as even harmonics in output FFT. To reduce this distortion, a new chopping technique is employed, which moves the DAC errors to a higher frequency and helps achieve 70dB SFDR. Fig. 9 shows the scheme to up-convert the DAC errors. SW_1 gets connected to V_P in phase A, while SW_2 gets connected to V_P in phase B. In addition to correcting even harmonics, this scheme also up-converts DAC switch flicker noise. Since correction happens in real time, errors across temperature can be corrected without external intervention.

IV. MEASUREMENT RESULTS & CONCLUSION

The radar transceiver is implemented in a 9 metal layer 45nm CMOS process. Fig. 10 shows the chip micrograph highlighting the single receiver, occupying 2.5mm². Total area for the four receivers including the IQGEN, baseband filters & ADC is 10mm². The die is flip-chip assembled into a standard BGA package (FCBGA). The performance numbers quoted here are measured on packaged devices. Fig. 11. shows the measured NF across frequency and temperature. The receiver achieves typical NF of 16dB with worst case of 18dB and P1dB of -7dBm in the 76-81GHz. Table I. summarizes the performance numbers of our radar receiver and gives a comparison with other published and commercial automotive radar data.

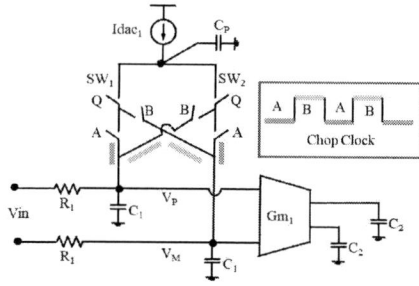

Fig. 9. Chopping technique to remove even order non-linearity.

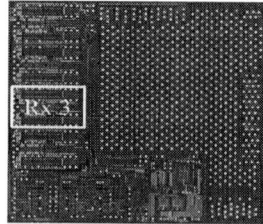

Fig. 10. Chip Micrograph highlighting a single radar Receiver.

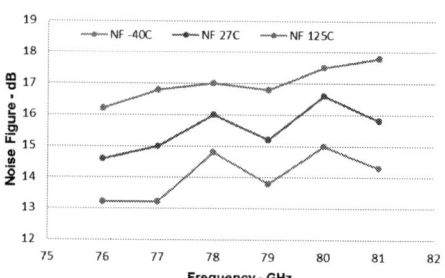

Fig. 11. Noise Figure Data across Frequency and Temperature.

TABLE I.   PERFORMANCE COMPARISON

	This Work	[4]	[5]	[6]
No. of Rx	4	4	3	6
Package	FCBGA	eWLB	RCPBGA	FCBGA
Technology	45nm CMOS	130nm SiGe	180nm SiGe	130nm SiGe
Freq (GHz)	76-81	76-81	76-77	76-81
IF BW (Hz)	15M (1dB)	10M (3dB)	5M (3dB)	-
HPF order	2	1	2	-
ADC ENOB	10.5	10.5	No ADC	No ADC
NF Typ/Max	16/18 dB	13/- dB	14/16 dB	12/- dB
P1dB (dBm)	-7	-10	-5	3
Power (mW)	900ᵃ	-	800ᵇ	618ᶜ
Temp (ºC)	-40 to 125	-40 to 125	-40 to 125	25 to 125

ᵃ· 4 Rx with ADC and IQGEN, ᵇ· 3 Rx, no ADC with freq doubler and LO buffers, ᶜ· 6 Rx, no ADC

The receiver, including IQGEN but excluding LO and distribution buffers, consumes 225mW from 1/1.4V supply. All four receivers combined consume 900mW for continuous operation. CMOS gives the advantage of a fully integrated digital system along with fault monitoring and failure diagnosis. Our CMOS solution in a standard package compares well with other published and commercial SiGe solutions, proving mmW capability in a low f_T (160GHz) technology.

REFERENCES

[1] B. Ginsburg, et al., "A Multimode 76-to-81GHz Automotive Radar Transceiver with Autonomous Monitoring," *ISSCC*, 2017.

[2] K. Subburaj, et al., "Monitoring Architecture for a 76-81GHz Radar Front End," *IEEE RFIC Symposium*, 2018.

[3] C.-Y. Kuo, et al., "Miniature 60GHz slow-wave CPW branch-line coupler using 90nm digital CMOS process," *IEEE Electronics Lett.*, vol. 47, Issue 16, pp. 924–925, Aug. 2011.

[4] (2018) 76 - 81GHz MMIC transceiver (4 RX / 3 TX) for automotive radar applications, Rev. 4.0, 10/2018 [Online] Available: https://www.st.com/resource/en/data_brief/strada770.pdf Accessed on 5 Apr 2019.

[5] (2016) 76-77 GHz RF receiver front-end for W-band radar applications Rev. 5.0, 9/2016 [Online]. Available: https://www.nxp.com/docs/en/data-sheet/MC33MR2001R.pdf Accessed on 5 Apr 2019.

[6] T. Fujibayashi, et al., "A 76- to 81-GHz Packaged Single-Chip Transceiver for Automotive Radar," *IEEE BCTM*, pp. 166-169, 2016.

An Untrimmed PVT-Robust 12-bit 1-MS/s SAR ADC IP in 55nm Deeply Depleted Channel CMOS Process

Y. Zha*, L. Zahnd*, J. Deng*, D. Ruffieux*, K. Badami*, T. Mavrogordatos*, Y. Matsuo**, S. Emery*

**Mie Fujitsu Semiconductor Limited, Yokohama, Japan

*CSEM, Switzerland

Abstract—This paper presents an industry-ready PVT-robust 12-bit 1 MS/s untrimmed SAR ADC IP operating from 0.5/0.9V supply voltage for a sub-threshold sensor interface. The ADC exploits Fujitsu's 55 nm Deeply Depleted Channel (DDC) technology to dynamically regulate the bulk voltage of the NMOS and PMOS transistors to compensate for the PVT variations. This dynamic regulation of the the bulk voltage is enabled by a technology-assisted replica-biasing based design strategy. This enables a PVT-robust comparator operation up to 14MHz frequency from a 0.5V supply voltage to allow the ADC to achieve 68 dB $\pm$ 1.1 dB SNDR and 88 dB $\pm$ 3.5 dB THD over P(SS,TT,FF) - V(0.45V to 0.55V) - T(-40°C to 90°C) variations at an average 12fJ/CS efficiency at $1/10^{th}$ of the sampling frequency.

I. INTRODUCTION

Over the last decade, capacitive DAC based successive approximation (SAR) ADCs have demonstrated a tremendous improvement (>10X) in the power efficiency [1]–[3]. This improvement has been enabled by a combination of the following 3 factors: (a) innovative and efficient DAC switching schemes which reduce the energy loss during charge re-cycling (b) reduction of supply voltage and current to thermal noise limited values which results in a highly efficient deep sub-threshold operation of transistors and (c) aggressive reduction in the unit capacitor (C_u) value which results in a proportional reduction in the DAC capacitance. However, barring the first factor, the other two factors often result in very large spread of SNDR and THD metrics over PVT corners. This is because sub-threshold operation results in a performance which highly sensitive to PVT variations due to exponential variation of the drain-current w.r.t. gate-source voltage. Further, aggressive reduction of the (C_u) limits the DAC matching yield and hence has a detrimental effect on the SNDR and THD performance. Typically, such performance characterization over a wide range of PVT corners is often not pursued in the academic works [1]–[3].

This work reports a 1MS/s PVT-robust, 68 dB $\pm$ 1.1 dB SNDR / 88 dB $\pm$ 3.5 dB THD and a highly power efficient industrial grade SAR ADC IP in Fujitsu's 55 nm Deeply Depleted Channel (DDC) technology. SNDR and THD robustness over a wide range of P(SS,TT,FF)-V(0.45V to 0.55V)-T(-40°C to 90°C) corners is guaranteed through a DDC technology-assisted replica-biasing based bulk voltage regulation loop and is highlighted next in the Section II. This bulk voltage regulation is then utilized to mitigate the

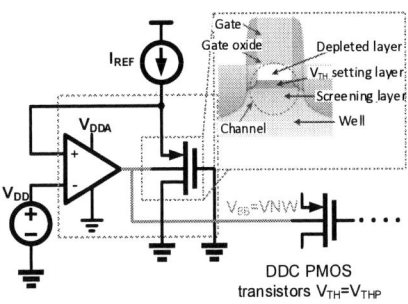

Fig. 1. Dynamic threshold voltage regulation of the PMOS transistors by controlling the bulk voltage to compensate for the PVT variations in 55nm DDC process. Similar regulation circuitry exists for NMOS transistor

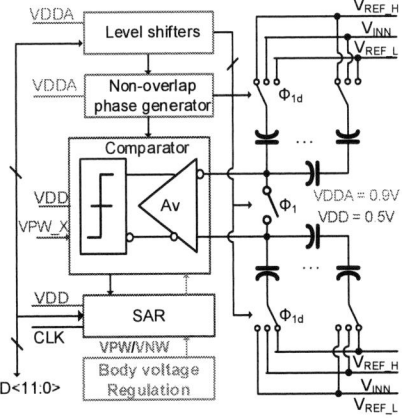

Fig. 2. Architecture of the 1MS/s 12b SAR ADC highlighting the different building blocks and power domains. VPW and VNW control the bulk voltages of the transistors in the ADC to enable replica-biasing based bulk voltage regulation for PVT compensation

impact of PVT variations and hence limit the performance spread of the proposed SAR ADC. This is discussed in Section III. Finally, Section IV discusses the PVT robustness of the measurement results of the proposed SAR ADC IP and is compared to both academic designs and to industrial SAR ADC IPs. The paper conclusions are provided in Section V.

978-1-7281-5107-6/19 $31.00 © 2019 IEEE

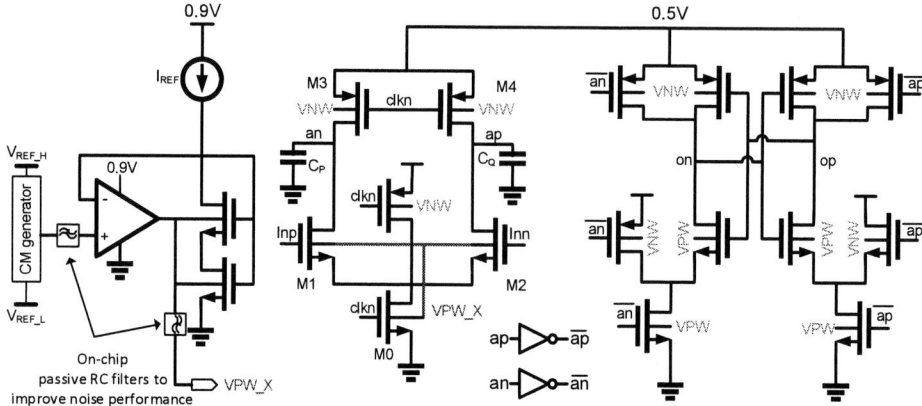

Fig. 3. The dedicated bulk-biasing regulator with on-chip filtering (left) and the Architecture of the 0.5V supply 14MS/s double tail comparator (right)

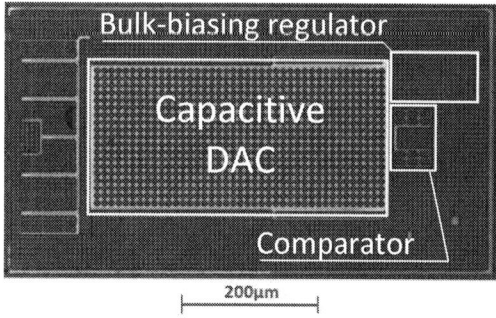

Fig. 4. Die micro-graph of the SAR ADC IP highlighting the important building blocks and the ADC dimensions

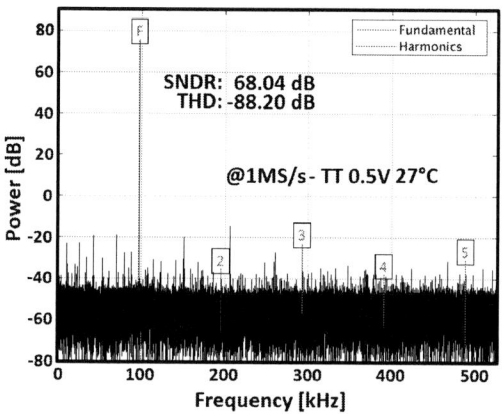

Fig. 5. Measured SNDR and the THD at $1/10^{th}$ of the sampling frequency

II. DDC TECHNOLOGY-ASSISTED REPLICA-BIASING LOOP

DDC transistors (illustrated in the Fig. 1 have two major advantages over bulk CMOS: (a) A large body-factor (sensitivity of the threshold voltage to the bulk voltage defined as $\Delta V_{TH}/\Delta V_{BB}$) up to 375mV/V in the current 55nm technology where V_{TH} and V_{BB} are the threshold voltage and the bulk voltage of the MOSFET (b) reduced local mismatch due to lower random dopant fluctuations. The large body-factor enables to design a V_{TH} regulating loop as shown in Fig. 1. While the Fig. 1 shows the bulk voltage regulation loop for PMOS transistor only, a similar regulation loop has been designed for the NMOS transistor as well. This loop dynamically adjusts the V_{BB} to regulate V_{TH} so that, for a given gate voltage and constant I_{REF}, the drain current I_D and g_m/I_D of the replica-biasing transistor remain constant with respect to PVT variations. This PVT robustness is replicated over the complete ADC design by biasing all PMOS transistors with the regulated V_{BB} while taking advantage of the reduced local mismatch. A similar bulk-voltage regulating loop using the NMOS transistor controls the V_{TH} of the NMOS transistors in the design to mitigate the impact of PVT variations.

III. BULK-BIASED SAR ADC ARCHITECTURE AND IMPLEMENTATION

The architecture of the SAR ADC along with the replica-biasing is shown in Fig. 2. The 12b 1MS/s ADC requires 14 cycles for each sample conversion including the sampling cycle and a dedicated reset cycle. The ADC operation at 1MS/s necessities the comparator to meet the speed and the noise specification across PVT corners at 14MHz. Further, this comparator operates from a reduced supply voltage of 0.5V to improve the ADC's power-efficiency. As it can be seen from Fig. 3 (right), the speed of the utilized double-tail dynamic comparator, is a function of V_{THP} (the PMOS threshold) of the input pair, $C_{P,Q}$ and I_{CM} (the common-mode current through M1,2), while the input-referred noise depends on V_{THP}, the ratio $g_{m1,2}/I_{CM}$ and $C_{P,Q}$. Generally, in the deep sub-threshold operation regime, PVT variations result in a wide spread on V_{THP}, $g_{m1,2}$ and I_{CM}, thus deteriorating

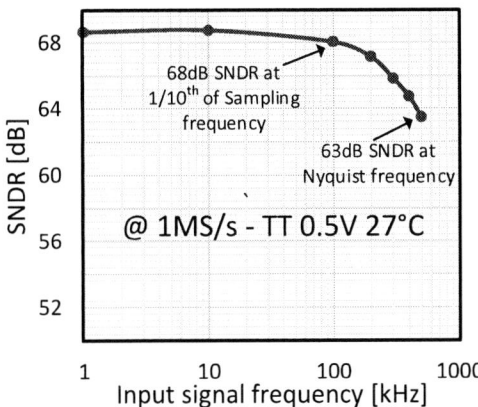

Fig. 6. Measured variation of the SNDR w.r.t. input signal frequency

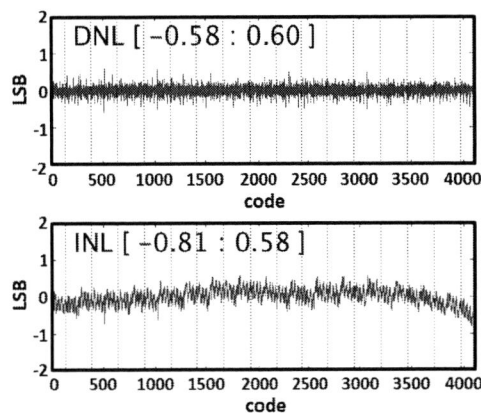

Fig. 7. Measured DNL (top) and INL (bottom) characteristics of the SAR ADC IP

the SNDR, the THD and the speed of the ADC and hence limiting the ADC yield.

To mitigate the impact of these PVT variations induced design spread on the speed and matching performance of the comparator, a dedicated bulk-biasing regulator is implemented for the comparator as shown in Fig. 3 (left). The replica-biasing loop generates the bulk voltage of M0-M2, VPW_X to compensate the V_{TH} so that the ratios I_{CM}/I_{REF} and $g_{m1,2}/I_{CM}$ remain constant across PVT corners and only depend on I_{REF}. This implies that, for a fixed I_{REF}, $g_{m1,2}$ and the discharge current through M0-M2 I_{CM} are PVT-insensitive, thus allowing the ADC to reliably maintain the desired speed and accuracy across the PVT corners.

Further, as shown in Fig. 2, the SAR ADC uses a split capacitor DAC array comprising of 8b main and 4b sub DAC and a unit bridge capacitor. The DAC is implemented based on a unit MIM capacitor C_u of 33 fF. This relatively large C_u avoids linearity degradation due to process variations. The peripheral blocks, including the phase generator and the DAC run at 0.9V to allow rail-to-rail input swing swing up to 1.8Vpp-d.

IV. MEASUREMENTS

This section details the characterization results for the SAR ADC IP over a wide range of PVT corners. The die micrograph of the industry-ready SAR ADC IP is shown in Fig. 4 and it occupies an area of 0.27 mm^2. As illustrated in the Fig. 5, at $1/10^{th}$ of the sampling frequency, the ADC achieves a measured SNDR and THD performance of 68 dB and -88 dB respectively in TT corner at 27°C. The variation of the SNDR w.r.t the input signal frequency is shown in Fig. 6. The static linearity of the ADC is characterized by DNL and the INL measurements and are shown in Fig. 7 for TT corner at 27°C. It can be seen from the Fig. 7 that DNL is limited to -0.58 to 0.6 LSBs while the INL is limited to -0.81 to 0.58 LSBs.

To benchmark the SAR ADC IP for industry readiness, the ADC performance is characterized over all combinations of process (SS,TT,FF), voltage (0.45V, 0.5V, 0.55V) and temperature (-40°C, 0°C, 27°C, 40°C, 90°C). The ADC achieves 68 dB $\pm$ 1.1 dB SNDR and -88 dB $\pm$ 3.5 dB THD over the complete set PVT variations. A comprehensive box plot highlighting the SNDR and THD robustness to PVT variations is illustrated in Fig. 8. The box plot is made for sample size of 10 chips in each corner. The total measured power dissipation from 0.5V and 0.9V supply is 30 μW including the power consumption of all the replica-biasing loops. This results in a FoM of 24.5 fJ/CS when computed at Nyquist input frequency and 12 fJ/CS when computed at $1/50^{th}$ of the sampling frequency. Comparison of the proposed ADC IP to the state-of-the-art academic [1]–[3] ADC designs and industrial [4], [5] SAR ADC IPs is illustrated in Table I. It can be seen that It can be seen the DDC technology assisted replica-biasing strategy enables the ADC to achieve a very high degree of PVT robustness with the best power efficiency amongst the PVT robust ADC IPs.

V. CONCLUSIONS

This work introduced a PVT robust 12b 1MS/s untrimmed SAR ADC IP in Fujitsu's 55 nm Deeply Depleted Channel CMOS process. The ADC's robustness to PVT variations is guaranteed through a bulk voltage biasing based threshold voltage regulation loop. This generated bulk voltage is then replica biased across the entire ADC to mitigate the impact of PVT variations. The efficacy of the bulk biasing based threshold regulation is validated with the ADC SNDR and THD measurements over a wide range of PVT corners. The designed SAR ADC achieves the best FoM amongst the reported SAR ADC IPs with PVT robustness.

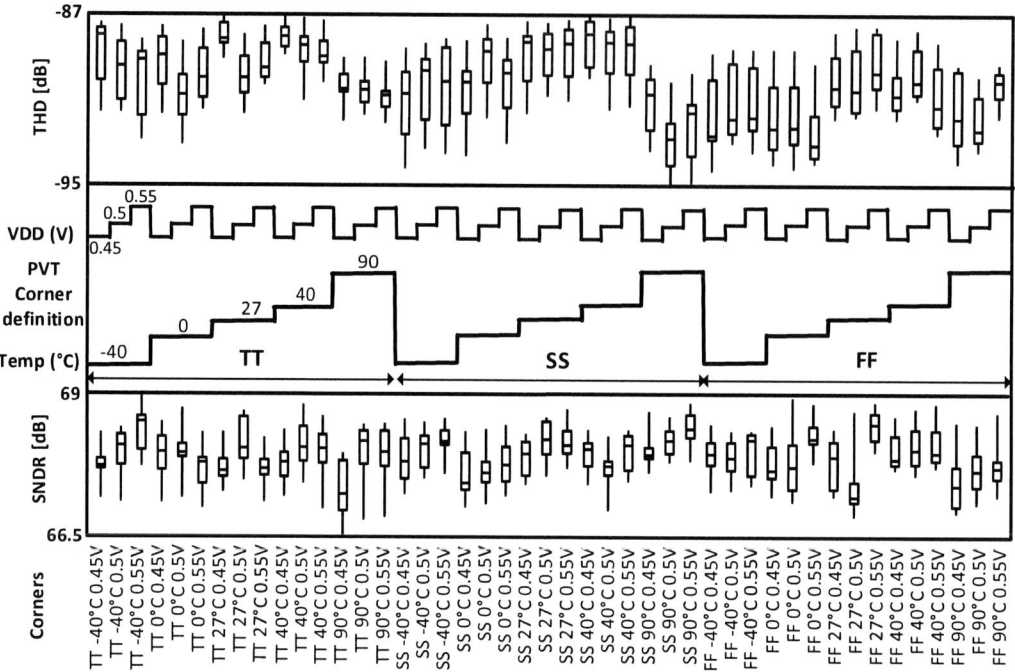

Fig. 8. Measured SNDR and THD performance across PVT corners illustrating the robustness of the design. Box plot was made for a sample size of 10 chips in each corner

	This work	VLSI 16 [2]	ISSCC 15 [3]	ISSCC 14 [1]	Industrial IP [4]	Industrial IP [5]
Tech.	Fujitsu 55nm	90 nm	65 nm	40 nm	180 nm	40 nm
Area (mm^2)	0.27	0.035	0.06	0.065	0.15	0.11
Supply voltage (V)	0.5/0.9	0.3	0.6	0.45	1.8	1.1
Sampling rate (kS/s)	1000	600	100	200	100k	15000
SNDR (dB)	68	57	57	54	65	63
THD (dB)	-86	-73	-75	-73	-80	-75
PVT robustness	Yes	NA	NA	NA	Yes	Yes
FoM (fJ/conv-step)	24.5* / 12$^\#$	0.44*	1.5*	0.85*	50$^\#$	55$^{\#\#}$
* f$_{in}$ @ Nyquist,		$^\#$ f$_{in}$ @ 1/50th of Nyquist,		$^{\#\#}$ includes reference power	

TABLE I
COMPARISON OF THE PROPOSED SAR ADC IP TO THE STATE-OF-THE-ART ACADEMIC DESIGNS AND INDUSTRIAL IPS

REFERENCES

[1] H. Tai, Y. Hu, H. Chen and H. Chen, "11.2 A 0.85fJ/conversion-step 10b 200kS/s subranging SAR ADC in 40nm CMOS," 2014 *IEEE International Solid-State Circuits Conference Digest of Technical Papers (ISSCC)*, San Francisco, CA, 2014, pp. 196-197.

[2] S. Hsieh and C. Hsieh, "A 0.44fJ/conversion-step 11b 600KS/s SAR ADC with semi-resting DAC," 2016 *IEEE Symposium on VLSI Circuits (VLSI-Circuits)*, Honolulu, HI, 2016, pp. 1-2.

[3] P. Harpe, H. Gao, R. van Dommele, E. Cantatore and A. van Roermund, "21.2 A 3nW signal-acquisition IC integrating an amplifier with 2.1 NEF and a 1.5fJ/conv-step ADC," 2015 *IEEE International Solid-State Circuits Conference - (ISSCC) Digest of Technical Papers*, San Francisco, CA, 2015, pp. 1-3.

[4] SAADC 10MS/s12b https://www.teledynedalsa.com/en/home/

[5] S3ADS122MD12BT40LPB: https://www.s3semi.com/ip/

A Cost Effective Test Screening Circuit for embedded SRAM with Resume Standby on 110-nm SoC/MCU

Yoshisato Yokoyama, Kenji Goto, Tomohiro Miura, Yukari Ouchi, Daisuke Nakamura,
Jiro Ishikawa, Shunya Nagata, Yoshiki Tsujihashi and Yuichiro Ishii

Renesas Electronics Corporation, Tokyo, Japan
yoshisato.yokoyama.jx@renesas.com

Abstract—We demonstrate an ultra-low standby power embedded SRAM with cost effective test screening circuits on 110-nm SoC/MCU. A source bias design technique has been applied to reduce SRAM standby power in the resume standby mode where the stored data is retained. Meanwhile the testing time is increased due to the long pause time to screening the retention failures, because the source bias level will be shifted slightly by small SRAM leakage current. Therefore, a 2-stage test screening method for resume standby mode is newly proposed. It is applied for only low-temperature test loop to minimize the total testing time without any test coverage loss. Test chips including 130kbit SRAMs on 110-nm show full reading and writing operations and 0.28 pW/cell resume standby power at 1.5 V typical supply voltage and 25°C. The testing time is reduced to 1/50 by proposed 2-stage screening testing circuits with only 0.03% area overhead.

Keywords—SRAM; Low standby power; Data retention; Testing cost; Resume standby; IoT

I. INTRODUCTION

Internet-of-Things (IoT) devices are demanded both low active power and low standby power in the sleep mode with cost effective solutions. Especially, reducing a total BOM cost is much important in low-end SoC/MCUs for consumer/industry IoT applications. Those devices are typically line-upped on mature technology nodes around 90 nm to 180 nm for reducing wafer cost. It is because that analog blocks are dominant, not needed huge core logics and embedded SRAMs. In addition, it tends to be required at 150 °C high-temperature operation and standby even non-automotive products because of the cost reduction by heat sink/fan less solutions.

Meanwhile, the testing cost per transistor increases year by year [1]. Reducing test cost is also an important factor in such low-end SoC/MCUs. Fig. 1 depicts a diagram of an SoC/MCU chip with built-in self-tests (BISTs) for logic and memory. On memory-BIST for SRAMs, techniques to screen out failures exactly and to shorten test time have already been reported [2] - [4]. However, there is no discussion on reducing the embedded SRAM retention test time for low leakage standby modes.

In this work, we implement an ultra-low standby power embedded SRAM macro with resume standby circuits on a 110-nm SoC/MCU. Besides, we newly propose a 2-stage test screening method for the resume standby SRAM. It is applied for only low-temperature test loop in the memory BIST flow to minimize the total testing time without any test coverage loss.

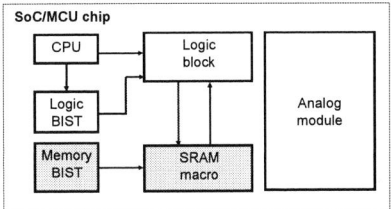

Fig. 1 Embedded SRAM with memory BIST on an SoC/MCU chip.

II. RESUME STANDBY SRAM ON 110-NM

SRAM source bias control techniques in the low-standby mode are effective for reducing standby power [5] - [12]. Fig. 2(a) shows a 6T SRAM bitcell circuit with source bias control on 110-nm technology. To apply independent bias in the resume standby mode, each power source line (ARVDD) and ground source line (ARVSS) in bitcells is separated with VDD and VSS, respectively. Fig. 2(b) shows effects of leakage reductions in each bias technique at 25°C and 150°C. We apply all bias techniques, which are 1)Bitline lowering, 2)ARVSS rising and 3)ARVDD lowering, to reduce much leakage power. In this case, enough cell bias at resume standby mode should be kept. Fig. 3(a) plots the temperature dependencies of the read static noise margin (SNM) [13] and write margin (WM) defined by write-trip-point [14] by considering with local variations. The worst process condition for SNM is FS (NMOS: fast, PMOS: slow) corner. Whereas the worst process condition for WM is SF (NMOS: slow, PMOS: fast). The Monte Carlo simulation results show that the bitcell has less WM than SNM, but each μ/σ exceeds 6, which corresponds no failure bits up to around 1Gbit. It is found that both are enough margins at typical 1.5 V -10% operation. A mature process usually has no special process steps for SRAM bitcell to reduce the wafer cost, so the pull-up/pull-down/pass-gate MOSs are same characteristics of core MOSs. The large on-current of pull-up PMOS is the cause of the decrease of WM and the increase of standby leakage current. That is the reason why we apply ARVDD source bias as well as bitline and ARVSS biases. Fig. 3(b) plots the estimated minimum operating voltage (V-min) for resume standby mode with source biases at each temperature of -40/25/150°C. It is found that the retention margin at low temperature (LT) become worse than high temperature (HT), so the data retention test should be done at LT as described later section. Fig. 4 shows each test time of embedded SRAM with and without resume standby circuit. Memory BIST proceeds the IDDQ test, read/write function test, DC stress screening test and data retention test. The function test takes a short time for small SRAM capacity with typical test clock frequency. Neither DC screening nor IDDQ take significant testing times. Meanwhile, the testing time of data retention with resume standby circuit is increased due to the long pause time to screening the retention failures. Because, the source bias level will be shifted slightly with long time constant by small SRAM leakage current in 110nm mature process technology.

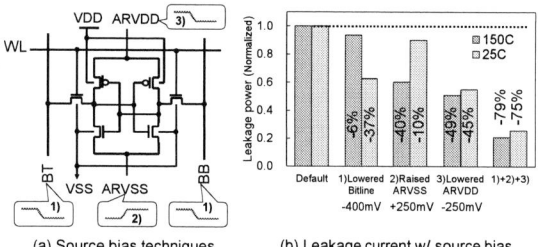

(a) Source bias techniques   (b) Leakage current w/ source bias

Fig. 2 110-nm 6T SRAM bitcell circuit with source biases and estimated leakage power reduction at 25°C and 150°C by SPICE simulation.

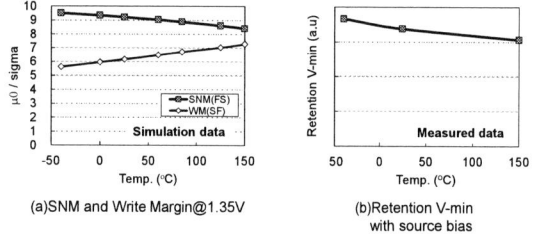

(a)SNM and Write Margin@1.35V   (b)Retention V-min with source bias

Fig. 3 a) Simulated temperature dependencies of the static-noise margin (SNM) and write margin (WM) at 1.35 V, and b) Estimated retention V-min based on measured data.

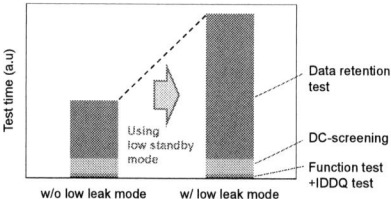

Fig. 4 Each test time of embedded SRAM w/ and w/o resume standby.

The source bias technique with diode connected MOS shifts the source line voltage by its Vth. In precisely, the voltage of source line is determined by the balances between the diode connected MOS current and leakage current of the bitcell array. The source bias techniques are applied to both ARVDD and ARVSS with keeping the voltage difference between ARVDD and ARVSS (cell bias), which is proportional to the data retention margin. Fig. 5(a) illustrates the proposed source bias control circuit with 2-stage data retention test mode. SRAM macro transits to standby mode with holding stored data if the signal RS becomes high. Both of ARVDD and ARVSS are

biased by the circuit to reduce standby power without any on-die regulators. As descried in the previous paragraph, PMOS has large leakage current, therefore it is also important to bias ARVDD to reduce the pull-up PMOS leakage current. However, the bitcell has smaller data retention margin at LT than HT due to unbalance of leakages as shown in Fig. 3 (b). Then, we apply the always-on NMOS (N_{AO} in Fig. 5(a)) connected to ARVSS to the circuit in order to keep the cell bias [10]. N_{AO} has a weak drivability which is smaller than bitcell leakage at HT, and is larger than bitcell leakage at LT. Fig. 5(b) shows the temperature dependency of voltages of ARVDD, ARVSS and cell bias carried out by SPICE simulation. N_{AO} discharges ARVSS to 0 V at LT, then the cell bias is increased with small standby power overhead at HT. Therefore, the proposed circuit can reduce the standby leakage current not only at HT but also at room temperature (RT) and LT with enough cell bias.

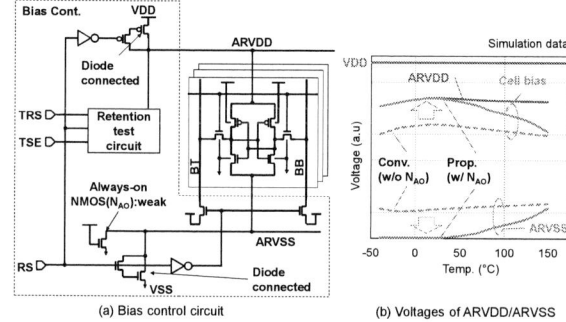

(a) Bias control circuit   (b) Voltages of ARVDD/ARVSS

Fig. 5 Circuit diagrams of bias control circuit for resume standby and SPICE simulation results of the cell bias with ARVDD and ARVSS bias levels.

III. 2-STAGE RETENTION TEST CIRCUIT

Target SRAM leakage current is less than 1 pA/cell for low-power SoC/MCU applications in 110-nm technology nodes. Besides, due to the large parasitic capacitance of the mature process, discharging ARVDD takes long time at LT. It increases testing cost significantly. Fig. 6 shows the voltage dependency of failure bitcell count in the retention mode with source bias in cases of 1 sec, 10 sec and 30 sec of pause time. The fail count of 10 sec and 30 sec are larger than that of 1 sec. It means 1 sec pause time is not enough to stabilize the ARVDD bias level. If 10 sec is needed for retention test, the testing cost becomes significant high.

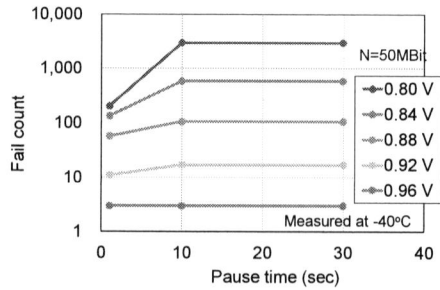

Fig. 6 Pause time dependency of retention test with source bias.

978-1-7281-5107-6/19 $31.00 © 2019 IEEE

Next, we discuss the 2-stage retention test method to reduce pause time. Fig. 7 shows the proposed retention test circuit. The test circuit has a pull-down weak NMOS (N_{PD}). N_{PD} is connected to ARVDD, and it is controlled by additional control pins TRS and TSE. The retention test circuit operates in two modes. One is "strict test", the other is "medium test". Both test modes are controlled by memory-BIST circuit. When RS input signal is asserted, ARVDD is discharged by the leakage current of the bitcell array, and it takes long time to saturates ARVDD to a certain voltage V_{RET} shown in Fig. 7. In the strict test mode, N_{PD} immediately discharges ARVDD to the voltage V_{STR} which is lower than V_{RET} and pulls down ARVDD continuously. The strict test might have some over-screened chips, those are not failure samples, but it can certainly screen out failure chips with very short pause time. In the medium test mode, when the internal node RPLAVD is charged to Vth of the inverter, N_{PD} is turned off and stop discharging ARVDD, and then ARVDD bias level saturates to V_{RET}. It is expected that the results of medium test can well reproduce the true V-min compared to the strict test and the medium test can avoid over-screening. Proposed test flow is shown in Fig. 8. The combination of strict test and medium test are applied for LT test condition, and can reduce the average pause time with preventing over screening. The area overhead of proposed retention test circuit is less 0.03% of the SRAM macro.

Fig. 9 shows SPICE simulation waveforms of the proposed SRAM macro at process SS and -40°C, confirming expected operation and bias levels. In this condition, ARVDD in medium test reaches to V_{RET} earlier than the conventional test.

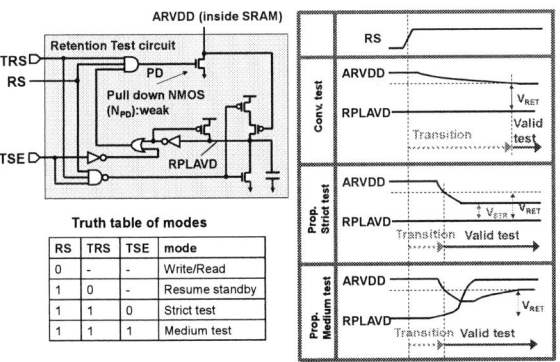

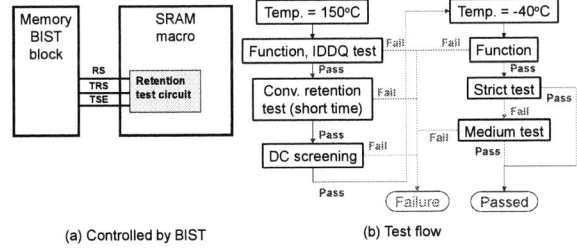

Fig. 8 Proposed 2-stage retention test flow with memory-BIST.

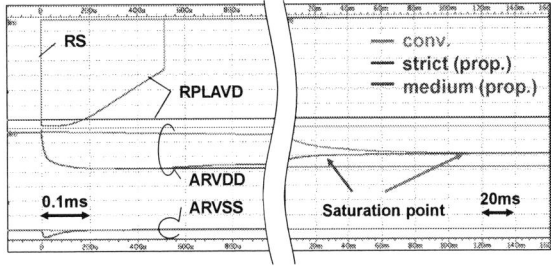

Fig. 9 SPICE simulation waveforms of retention test circuit of 4k-word x 40-bit SRAM at worst condition (1.35 V, process SS and -40°C).

IV. DESIGN AND FABRICATION OF TEST CHIP

Fig. 10 portrays a die photograph of a test chip and layout plot of a SRAM macro on 110-nm platform. 160-kbits SRAM macro with proposed retention test circuit are implemented in the test chip with 16 instances. Table 1 summarizes the test chip features. We observed full read/write operation, and stable retention at -40°C to 150°C. The dynamic powers in read (write) operation are 90 (105) μW/MHz at 1.5 V typical supply voltage and 25°C. Fig. 11 shows a typical SHMOO plot of the read access time vs V-min. The read access time is 4.74 ns at the typical condition of 1.5 V. Fig. 12(a) shows cumulative distribution functions (CDFs) of V-min at -40°C and 150°C. The median values of V-min at process TT (NMOS: typical, PMOS: typical) and SF (NMOS: slow PMOS: fast) corners are 0.95 V and 1.14 V respectively, those values are enough margin for typical operating voltage of 1.5 V. Fig. 12(b) shows CDFs of standby power without source bias and with source bias conditions at 25°C. The median values are 2.3 μW without source bias and 0.73 μW with source bias respectively. The standby power per bitcell is 0.28pW at 25°C, reduced by 70% by applying source bias technique.

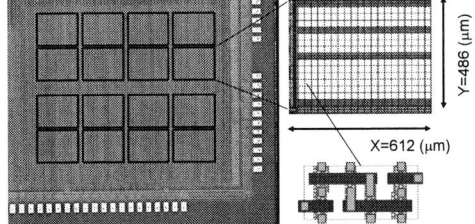

Fig. 10 Photograph of the test chip and layout plot of SRAM macro.

Table 1 Features of the test chip.

	Features
Technology	110-nm process technology
Macro configuration	2.5-Mbit (4096 word x 40 bit x 16)
Macro size	612 μm x 486 μm
	0.302 mm²@160-kbit
Bit density	0.518 Mbit/mm²
Access time @typ	4.7 ns
Cycle time @typ	6.8 ns(147 MHz)
Dynamic power	Read: 90 μW/MHz
@typ	Write:105 μW/MHz
Standby power	114.3 μW @150°C
@2.5-Mbit	0.73 μW @25°C

Fig. 7 Proposed retention test circuit, truth table and waveforms.

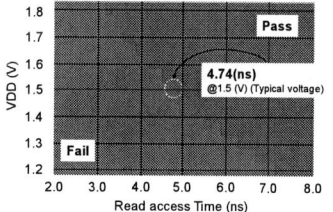

Fig. 11 Shmoo plot V-min vs. access time at 150°C.

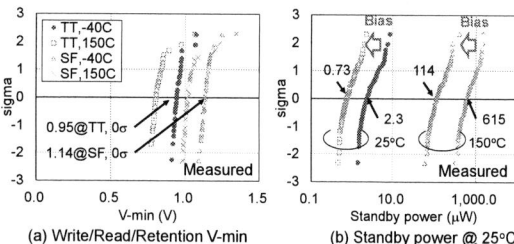

(a) Write/Read/Retention V-min (b) Standby power @ 25°C

Fig. 12 Distribution of measured V-min at -40/150°C and leakage power at 25°C.

Fig. 13(a) and (b) show CDFs of the retention V-min of 0.2 sec conventional test (conv. 0.2s), 10 sec conventional one (conv. 10s), proposed strict test and medium test at SS/-40°C and TT/-40°C respectively. Both strict test and medium test are executed for 0.2 sec pause time. The conv. 10s can be assumed as the true V-min of each chip due to its enough long pause time. The conv. 0.2s distributes smaller than conv. 10s. The median values of conv. 0.2s is shifted -330 mV with respect to conv. 10s at SS/-40°C. It goes out failure chips. The V-min of the strict test and medium test are distributed larger than the conv. 10s. Therefore, it can be screen-out failure chips correctly. The strict test is distributed larger than the medium test, then the strict test might excessively screen out pass chips. Probabilistically, if there are 1/100 of over-screened chips at the 1st strict test and followed by 2nd medium test, the 2-stage retention test time at LT is almost same as 1st strict test time in average. As a result, the proposed test flow can reduce the data retention test time to 1/50 as shown in Fig. 14.

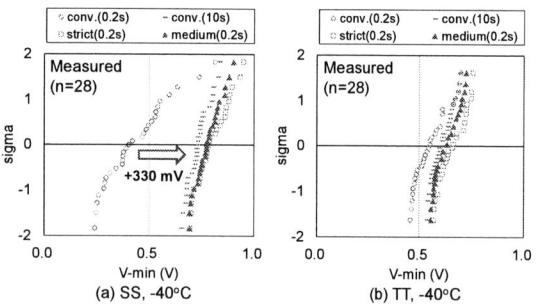

(a) SS, -40°C (b) TT, -40°C

Fig. 13 Measured retention V-min at -40°C and effects of proposed retention test modes.

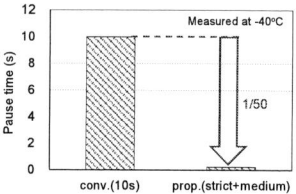

Fig. 14 Data retention time comparison with the conventional method.

V. SUMMARY

We implemented an ultra-low standby power embedded SRAM macro with resume standby circuits on a 110-nm SoC/MCU. It can reduce standby power at HT and RT with enough cell bias. a 2-stage test screening method for resume standby mode were proposed. Test chips including 130kbit SRAMs on 110-nm confirmed full reading and writing operations. Measured resume standby power at 1.5 V typical supply voltage and 25°C was reduced by 70% compared to conventional one, achieving 0.28 pW/cell. The test time was reduced to 1/50 with only 0.03% area overhead.

REFERENCES

[1] ITRS2001, "https://www.semiconductors.org/".

[2] Q. Xu et. al, "Test scheduling for built-in self-tested embedded SRAMs with data retention faults", IET Journals & Magazines, vol. 1, no. 3, pp.256-264, 2007.

[3] Y. Yokoyama et. al, "A cost effective test screening method on 40-nm 4-Mb embedded SRAM for low-power MCU", A-SSCC, pp1-4, 2015

[4] Y. Ishii et.al, "A 28 nm Dual-Port SRAM Macro With Screening Circuitry Against Write-Read Disturb Failure Issues", IEEE Journal of Solid-State Circuits, vol. 46, no. 11, pp.2535-2544

[5] Toshikazu Fukuda et. al, "13.4 A 7ns-access-time 25µW/MHz 128kb SRAM for low-power fast wake-up MCU in 65nm CMOS with 27fA/b retention current", ISSCC, pp.236 - 237, 2014.

[6] T. Kamei et. al, "A resume-standby application processor for 3G cellular phones", ISSCC, pp.336-531, Vol.1, 2004

[7] N. Maeda et. al "A 0.41 µA Standby Leakage 32 kb Embedded SRAM with Low-Voltage Resume-Standby Utilizing All Digital Current Comparator in 28 nm HKMG CMOS", IEEE Journal of Solid-State Circuits, pp.917-923, Vol.4, 2013

[8] Y. Yokoyama et. al, "40nm Ultra-low leakage SRAM at 170 deg.C operation for embedded flash MCU", ISQED, pp.24-31, 2014.

[9] Y. Yokoyama et. al, "40-nm 64-kbit Buffer/Backup SRAM with 330 nW Standby Power at 65°C Using 3.3 V IO MOSs for PMIC less MCU in IoT Applications", A-SSCC, pp.9-12, 2018.

[10] M. Yamaoka et. al, "A 300-MHz 25-/spl mu/A/Mb-leakage on-chip SRAM module featuring process-variation immunity and low-leakage-active mode for mobile-phone application processor", IEEE Journal of Solid-State Circuits, pp.186-194, Vol.40, 2005.

[11] K. Osada et. al, "16.7-fA/cell tunnel-leakage-suppressed 16-Mb SRAM for handling cosmic-ray-induced multierrors", IEEE Journal of Solid-State Circuits, pp.1952-1957, Vol. 38, no.11, 2003.

[12] R. Islam et. al, "Low power SRAM techniques for handheld products", ISLPED, pp.198-202, 2005.

[13] E. Seevinck, F. J. List, and J. Lohstroh, "Static-Noise Margin analysis of MOS SRAM cells", IEEE Journal of Solid-State Circuits, vol. SC-22, no. 5, pp. 748-754, Oct. 1987.

[14] R. Heald and P. Wang, "Variability in sub-100 nm SRAM designs", ICCAD Digest, pp. 347-352, 2004.

978-1-7281-5107-6/19 $31.00 © 2019 IEEE

A 2.666GT/s 128GB/s 14nm Memory I/O with Jitter and Crosstalk Cancellation

Harry Muljono, Kathy Peng, Linda Sun, Isaac Abraham, Charlie Lin, Yanjie Zhu, Chunrong Song

Scalable Performance CPU Development Group, Intel Corporation, Santa Clara, USA

Email: harry.muljono@intel.com

Abstract— A DDR interface that serves DDR4 and Optane DC Persistent Memory DIMMs achieves 2.666GT/s transfer rate on a state of the art 14nm Server CPU with a total of 6 channels, yielding 128GB/s data bandwidth required to fulfill high bandwidth demand of server applications. To support 2 DIMM per Channel (DPC) configuration, it utilizes three novel techniques, namely Phase Based Crosstalk Cancellation (PXC), Data Dependent Jitter Cancellation (DDJC) and un-clocked DFE architecture.

I. INTRODUCTION

The new Intel® Xeon® Processor Scalable family codenamed Cascadelake (CLX) is a 28-core processor built on 14nm 2nd Generation Tri-gate Transistor technology with 11 metal layers [1][2]. It has 6-channel DDR capable of supporting 2-DIMM per channel configuration as shown in Figure 1. Each channel operates at 2666MTs speed providing a total of 128GB/s memory bandwidth. Daisy-chain interconnect topology connecting CPU and the two DIMMs introduces inter-symbol interference (ISI) and signal reflections. In addition, the large number of high speed DDR pins clustered in close proximity generates crosstalk noise. These noise components pose significant challenges in meeting platform margin. This paper discusses an innovative crosstalk and data-dependent jitter cancellation schemes to reclaim the margin loss.

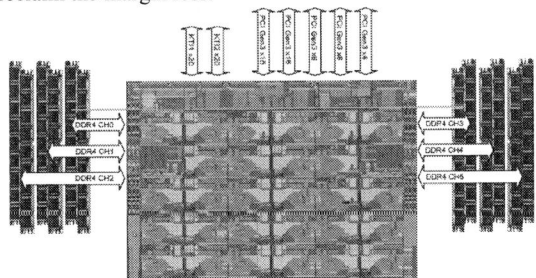

Figure 1 – Cascade Lake CPU I/Os

II. FLOORPLAN

The 6-channel DDR interface is split in two independent and identical layout sections containing 3 channels each. They are mirrored and laid out on opposite sides of the CPU die.

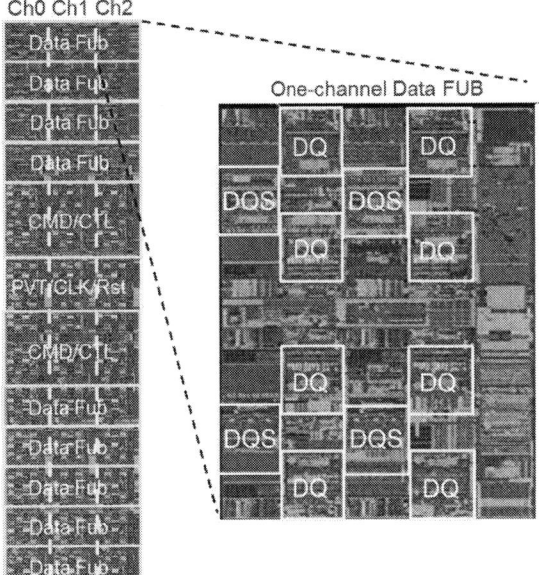

Figure 2 – DDR I/O Section Floorplan

The section laid out on the west side of the SKX die, hosting channels 0, 1 and 2, is shown on Figure 2. Its center portion hosts the PVT compensation and reset Functional Unit Blocks (FUBs), common to all 3 channels. Also in the center, there is one clock FUB per channel, each driving up to 2 differential clock pairs. These FUBs are located in the center so as to minimize routing distance to the PLL in the Memory Controller (MC), which helps with jitter. The large FUBs driving command and control signals are located close to the center, so as to minimize routing distance to the MC, which helps with latency. The data FUBs are laid out on the outside, North and South of the command and control FUBs. Each data FUB manages a byte of data and two differential strobes. FUB location enables pinout ordering match between CPU and DIMM card, thus minimizing package and board layer transition to eventually improve signal integrity.

III. TRANSMITTER DESIGN

The DDR I/O transmitter (TX) design is a voltage mode driver that uses a shared resistor topology with cascoded pullup (PMOS) and pulldown (NMOS) devices. Shown in Figure 3, the transmitter is divided into 3 segments to provide slew rate control with all segments having equal weight that can be configured to either be ODT (for READ) or WRITE mode. Slew rate compensation (SCOMP) is achieved by having the 3 segments turning on in a staggered fashion separated by a Process, Voltage and Temperature (PVT) compensated target delay. To combat asymmetric pullup vs. pulldown paths which may lead to substantial rise/fall TCO mismatches, TCO compensation is implemented using preset TCO codes which can support a correction range of up to 80ps. The transmitter also utilizes a 2-tap, de-emphasis style equalization scheme using independently controlled static legs. A constant impedance scheme is maintained where every leg subtracted from one side (ex: pullup) is added to the other side (ex: pulldown), reducing the swing while still maintaining a constant Reff. In addition, there is a current mode swing boost (IMODE) feature which is a programmable current DAC used to inject additional up to 8mA of current from the pad node to VSS, therefore providing a configurable increase in the voltage swing during data transitions from 1->0.

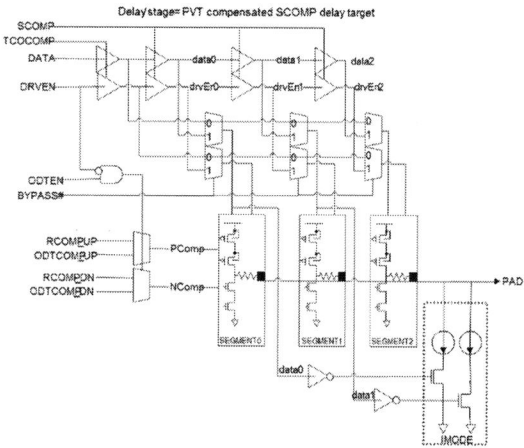

Figure 3 – Cascade Lake DDR I/O Transmitter Block Diagram

A. Phase-Based Crosstalk Cancellation

Phase-based Crosstalk Cancellation (PXC) is an innovative technique utilized to reduce crosstalk noise generated by data bit lane coupling, particularly along vertical through-hole via regions in the CPU socket, thus improving write margins. An in-depth analysis of crosstalk induced jitter can be found in [3]. PXC modulates the timing of a victim signal at the TX according to the predicted crosstalk jitter due to an aggressor's bit pattern, i.e. the additional timing modulation at the TX nulls out cross talk jitter as the victim signal propagates through the channel.

As shown on Figure 4, PXC victim-attacker selection logic establishes a relationship between a victim lane and its within-byte attacker lanes. The pairing is based on the proximity of their locations as indicated on the package pin map, as supported by pre-silicon simulation. Each victim lane

can have up to 2 attackers programmed. The subsequent logic processes the victim and attacker(s) transition information for each cycle, calculating the expected phase shift for the victim lane PICLK at that cycle, then sends the modified PI code to the DLL/PI AIP block. Outbound data is pipelined to align with the shifted clock edge. The PICLK phase shifts translates into the victim lane data phase shift, resulting in a dynamic phase correction to counter the noise due to the attackers' transition pattern.

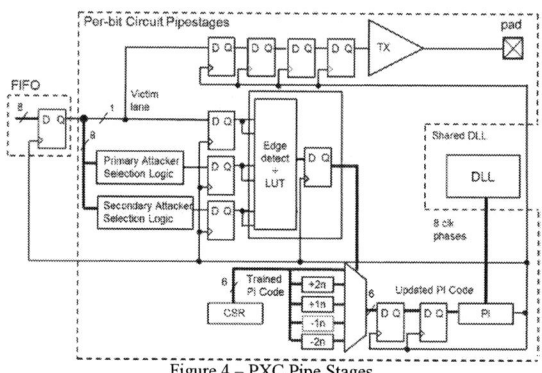

Figure 4 – PXC Pipe Stages

B. TX DDJC

The transmitter also incorporates a new feature called Data Dependent Jitter Cancellation (DDJC). Its objective is to cancel the Inter symbol interference (ISI) created by previous bits since the Vref crossing time of the current bit is dependent on the previous bit values [4]. The TX delay is adjusted by varying slew rate to compensate for the expected jitter. As illustrated in Figure 5, if the previous 2UIs (t-2, t-1) are the same, then the current data transition is pulled in by speeding up the data slew rate; and if the previous 2UIs (t-2, t-1) are switching then the current data transition is pushout by slowing down the data slew rate.

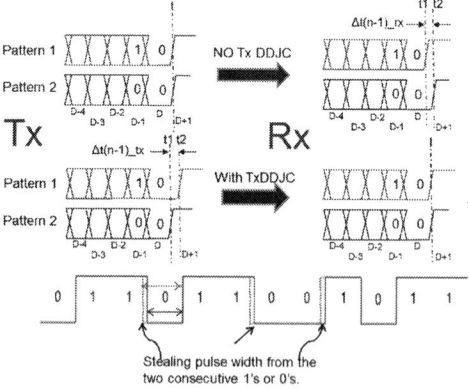

Figure 5 – DDJC Operation Principle

DDJC logic captures previous data transactions and enacts slope adjustments by utilizing a control bit ('BYPASS#') which can bypass the SCOMP delay chain when set to '0', as shown on Figure 3. This causes all driver segments to turn on simultaneously, thereby enabling a faster data slew rate.

IV. RECEIVER DESIGN

The receiver (RX) architecture satisfies the contemporary need to support increasing data rates (bandwidth) and yet provide sufficient signal amplification on a multi-voltage domain design such as DDR. It contains three analog stages, as shown on Figure 6.

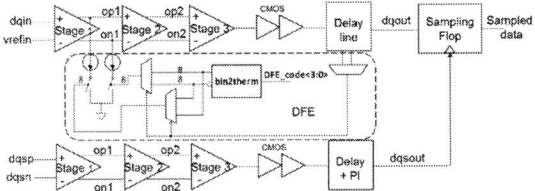

Figure 6 – Cascade Lake DDR I/O Receiver Block Diagram

Stage 1 is primarily a level shifting stage with a modest gain of about 2dB and it downshifts the signal from the 1.2V IO domain to the 0.95V domain. Stage 1 also contains the mechanisms to implement Voltage Offset Cancellation (VOC) and Continuous Time Linear Equalization (CTLE)[5-6]. Figure 7 shows stage 1 receiver with M1, M2 and M3 forming current sources. M4 and M5 implement the primary input arm of the cascode. The VOC, CTLE and finally the output load network (outn/outp) are implemented with parallel, differential segments. Stage 2 is a fully differential amplifier and stage 3 is a differential to single ended amplifier. The CTLE could support up to 7dB AC gain with upto 1.1GHz bandwidth. Post silicon the VOC was tested to be effective in canceling upto 20mVpp in offsets.

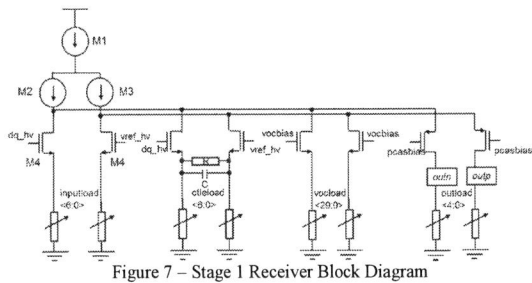

Figure 7 – Stage 1 Receiver Block Diagram

The receiver adopts a delay-matching structure typically seen in source synchronous transmission (particularly with single-ended signaling) in order to cancel out jitter between data and clock due to their common supply noise. Illustrated in Figure 7, the single-ended data signal "DQ" is amplified and delayed to match the latency of strobe, DQSP/DQSN, up to the sampling flop. In addition to jitter cancellation, the match-delay structure guarantees there is enough range during training and testing for the sampling strobe to margin the data eye for +/- 0.5UI.

A. Decision Feedback Equalization

The SKX DDR receiver implements DFE by decoupling the sampling and decision making slicer since the delay matching circuitry is a limiting amplifier chain[5]. As shown in Figure 6, the delay matching circuitry functions as a slicer which quantifies incoming signal to its binary symbol value. ISI cancellation is achieved by an 8-bit thermometer coded,

current Digital to Analog Converter (iDAC) steering offset current at the CTLE amplifier output. The iDAC polarity is controlled by DFE feedback, while the offset magnitude is controlled by DFE code. As long as the total loop delay is well-controlled such that the summing device subtracts ISI at desired timing within the incoming data UI, DFE can be achieved. In addition, fractional delay can be implemented to further optimize the DFE performance.

V. POWER DOMAINS

To optimally balance circuit performance and power consumption, there are three power supplies in DDRIO as shown in Figure 8. One is the nominal 1.2V I/O supply used for the DDR I/O TX and RX AFE. Another is a quiet analog power supply (Vccddra) for noise and jitter sensitive circuits such as the DLL's and PI's. The last is the digital supply, called Vccddrd, which is used by the MC and most of the digital logic and circuits. Each of these last two supplies are provided by an on-die Fully Integrated Voltage Integrator (FIVR).

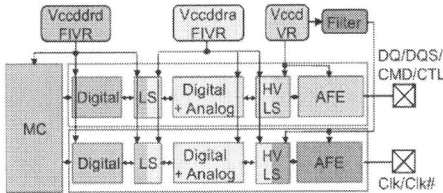

Figure 8. The Three Power Domains of MC and DDRIO

Vccddra is targeted at 0.95V fixed while Vccddrd will be adjusted between 0.8 & 1.1V based on binning and frequency of operation. This offers a good compromise between power-savings and analog performance. A filtered Vccd supply (Vccdq), generated through on-package inductor, supplies the Clk and Clk# AFEs to improve jitter at their pad outputs.

VI. CLOCKING

SKX implements two separate balanced clock distributions for MC and DDRIO from the same PLL output. Signals can pass between the two clock domains seamlessly since these two distributions are on the same vccddrd power domain and have matched delays.

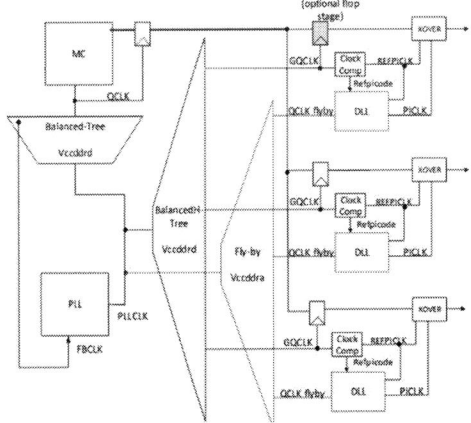

Figure 9. SKX Clocking Scheme for MC and DDRIO

As illustrated in Figure 9, the balanced distribution is implemented with a CMOS H-tree topology to save power, unlike the one in MC which has the mesh topology for better matching. This fits better to the tall and skinny DDRIO floorplan. At the end of this distribution, a full rate system clock named gqclk, is distributed to each IO functional block to run major logic functions, including those for transmitter, receiver and IO PVT Compensation. DDRIO has another full rate clock called piclk. Piclk is generated from a phase interpolator (PI) which mixes clocks 1/8 of cycle apart from a DLL near the IO circuit. The DLL gets its clock from the PLL output through a CMOS fly-by distribution. Piclk phase can be adjusted with a PI code, which is trained by firmware, to the accuracy of $1/64^{th}$ of cycle. A carefully designed cross-over circuit (Xover) provides safe signal passing mechanism between gqclk and piclk clock domains with minimum latency. A version of piclk called refpiclk is matched with gqclk at each IO block through firmware training. The input signal crosses clock domains from gqclk to refpiclk in the first stage of the Xover. This is validated as if gqclk and refpiclk are aligned but with special skew to account for training error and temperature and voltage drift. The signal crosses domains from refpiclk to piclk in the second Xover stage. The piclk code is an offset code from the refpiclk code determined during IO functional training. These source synchronized clocks in DDRIO and MC guarantee minimum latency delay between MC and DDRIO.

VII. LAYOUT

Figure 10 shows the layout of TX and RX circuits. The outbound signal generated from the final flop arrives at the top edge of the block and propagates down through TX DDJC, TCOCOMP, SCOMP, predriver, level shifter and finally TX AFE. The signal bump pad is located right on top of the D1D2 diodes. TX AFE, IMODE and the RX AFE first stage amplifier are placed next to the diodes to minimize pad capacitance. The output of the stage 1 receiver travels through the second and third stage amplifier and 1-tap DFE structure before leaving the left edge of the block.

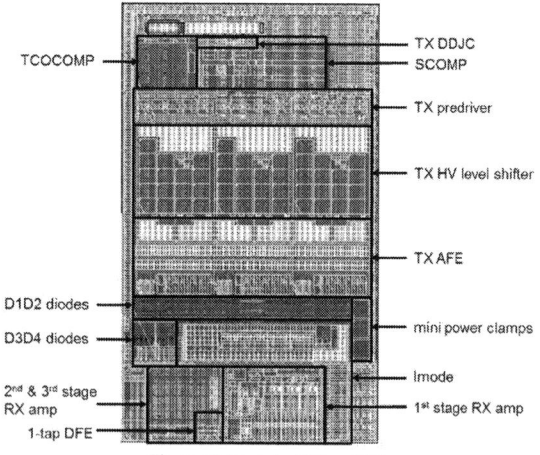

Figure 10. Layout of TX and RX

VIII. SILICON RESULTS

Figure 11 shows typical data write eye diagram at 2666 MT/s, probed at a DRAM device in a 2 DPC configuration. RX

margin obtained via BSSA 2-D Margining tool is shown in Figure 12. Thanks to the circuit enablers, the eye opening is relatively large with sufficient margins at 2666MTs. Improvement was measured to be approximately 5-10ps from PXC, 5ps from DDJC, 10-20ps from RxDFE and 10-20mV from RxCTLE. Figure 11 and 12 show eye diagrams with enablers.

Figure 11. Data Write Eye Diagram at DIMM Connector, 2667MT/s.

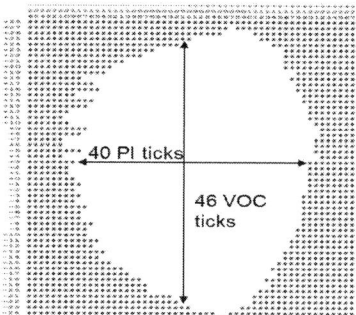

Figure 12. BSSA 2-D Read Eye Margin

IX. CONCLUSION

A 14nm DDR I/O interface achieves 2.666GT/s transfer rate, yielding 128GB/s data bandwidth required to fulfill the high bandwidth demand of server applications. The design utilizes three novel techniques, namely Phase Based Crosstalk Cancellation (PXC), Data Dependent Jitter Cancellation (DDJC) and un-clocked DFE architecture. Future higher DDR rates are expected to shrink eye margins and those new techniques will help open up the eye and continue providing sufficient margins.

X. ACKNOWLEDGMENT

The authors acknowledge the contribution of the entire Broadwell Server & Xeon Phi DDR design team.

XI. REFERENCES

[1] S. Rusu, *et al.*, "A 22nm 15-Core Enterprise Xeon® Processor Family", ISSCC Digest of Technical Papers, Page 102-103, 02/2014.

[2] S. Natarajan, *et al.*, "A 14nm Logic Technology Featuring 2nd – Generation FinFET Transistors, Air-Gapped Interconnects, Self-Alighned Double Patterning and a 0.0588um^2 SRAM cell size", IEDM Tech. Digest, December 20014.

[3] J. Buckwalter, A. Hajimiri, "Cancellation of Crosstalk-Induced Jitter", IEEE JSSCC, Vol.41, No.3, March 2006.

[4] J. Buckwalter, A. Hajimiri, "Analysis and Equalization of Data-Dependent Jitter", IEEE JSSCC, Vol41, No.3, March 2006.

[5] F. Spagna, et al., "A 78mW 11.8Gb/s Serial Link Transceiver with Adaptive RX Equalization and Baud-Rate CDR in 32nm CMOS", IEEE ISSCC, DOI 10.1109/ISSCC.2010.5433823, February 2010.

[6] T. Sümesaglam, "An 11-Gb/s Receiver With a Dynamic Linear Equalizer in a 22-nm CMOS", IEEE Trans. Circuits Syst., II, Exp. Briefs, Vol. 61, No.4, April 2014.

A Si-Backside Protection Circuits Against Physical Security Attacks on Flip-Chip Devices

Takuji Miki[*], Makoto Nagata[*], Hiroki Sonoda[*], Noriyuki Miura[*], Takaaki Okidono[†],

Yuuki Araga[‡], Naoya Watanabe[‡], Haruo Shimamoto[‡], Katsuya Kikuchi[‡]

[*]Kobe University, Kobe, Japan

[†]ECSEC, Tokyo, Japan

[‡]National Institute of Advanced Industrial Science and Technology, Tsukuba, Japan

Email: miki@cs26.scitec.kobe-u.ac.jp, nagata@cs.kobe-u.ac.jp

Abstract—Flip-chip semiconductor packaging reduces the footprint of IC chips while leads to serious vulnerabilities against emerging backside physical security attacks. The proposed backside buried metal (BBM) structure forming a meander wire pattern on the Si backside detects unexpected disconnection of the meander wire and warns malicious attempts. Besides prevention is also achieved against both passive probing attacks and active laser fault injection as well. Unlike other conventional laminate-based protection, this backside monolithic approach does not require frontside wiring resources or additional packaging layers, resulting in only 0.0025 % area-overhead. The BBM meander was formed on the backside of a 0.13 μm CMOS cryptographic chip by wafer-level via-last backside buried metal processing.

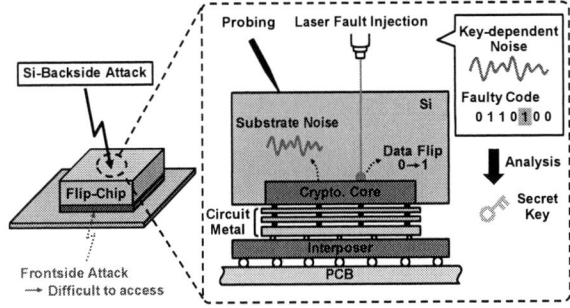

Fig. 1. Physical attacks on flip-chip.

I. INTRODUCTION

Internet of things (IoT) applications prefer the smaller form factor of devices even with more functionality and lower power, and thus push the use of flip-chip assembly technologies for thinner and almost chip scale or even multiple chip packaging. On the other hand, cryptography is a key for data security among distributed IoT edge devices [1], based on digital encryption algorithms for data protection and digital signature algorithms for device authentication. Physical security attacks on such devices explore side-channel information leakages, through passively by measuring power noise or electromagnetic (EM) radiation [2], or actively by injecting logical faults with laser or high-power EM irradiation [3]. To prevent from these attacks, various countermeasures have been reported. An active secure shield proposed in [4] detects physical intrusion inside the chip by monitoring encrypted signal-carrying wires on the top metal layers. A probing detection scheme reported in [5] also detects the physical approach by sensing capacitance variation. However, these techniques only counter the attacks from the frontside (surface) of an IC chip. Fig. 1 shows physical attacks on flip-chip devices. Flip-chip implementation exposes the Si-backside of IC chip, which allows malicious attackers to directly access to the target through Si substrate, and easily estimate a secret key of cryptography circuit by analyzing the acquired side-channel leakage [6], [7]. The active top-metal shield technique can be also used for backside protection by bonding two dies back to back [4], and also, a breakable die with an exotic film lamination structure impedes backside

physical attacks [8]. However, they require additional die or layer, which causes an increase in the size of an IC chip.

In this paper, a chip-size-efficient countermeasure against backside physical attacks is presented. A backside buried metal (BBM) structure forming a meander pattern is newly proposed to detect physical backside attacks such as laser-cutting, polishing and milling without chip size overhead. The BBM also protects a secret key of cryptography from direct probing attacks by reducing substrate noise leakage. Since a laser cannot penetrate the Cu meander wiring, this structure can protect the cryptographic core from laser fault injection (LFI) attack. A prototype cryptographic chip with the meander BBM demonstrates secure characteristics against above attacks. The rest of this paper is organized as follows. Section II presents the detail of the proposed protection structure and each mechanism of countermeasure against physical backside attacks including disconnection, probing and LFI. Section III shows the chip measurement results. Finally, Section IV gives the conclusion.

II. BACKSIDE PROTECTION AGAINST PHYSICAL ATTACKS

A. Si-Backside Protection Structure

Fig. 2 shows the protection structures from the backside physical attacks. The conventional structure shields noise or laser from/to outside of the chip with a backside laminating film as shown in Fig.2(a). This approach induces chip-size overhead, which makes a non-negligible impact on IoT devices

978-1-7281-5107-6/19 $31.00 © 2019 IEEE

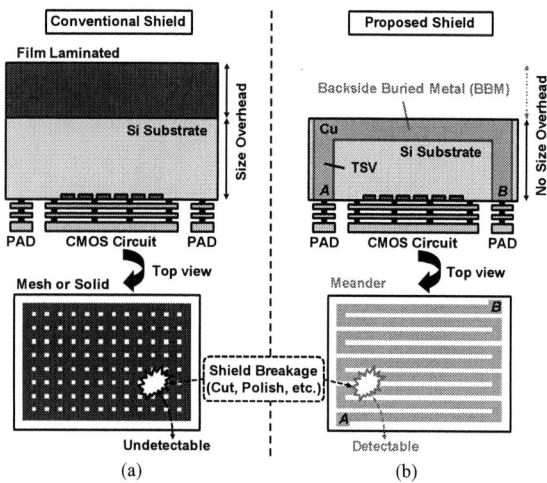

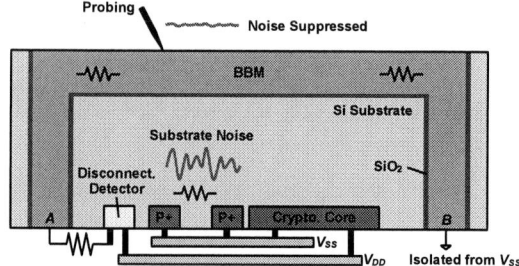

Fig. 4. Side-channel leakage suppression.

Fig. 2. Countermeasures against Si-backside physical attacks (a) conventional and (b) proposed shield structure.

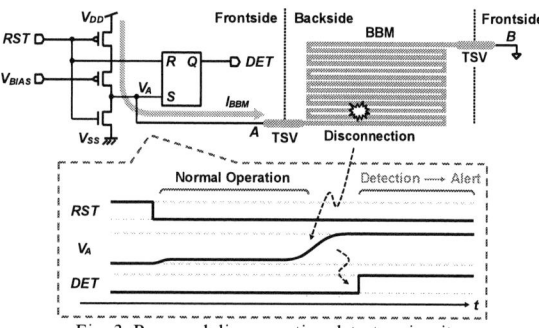

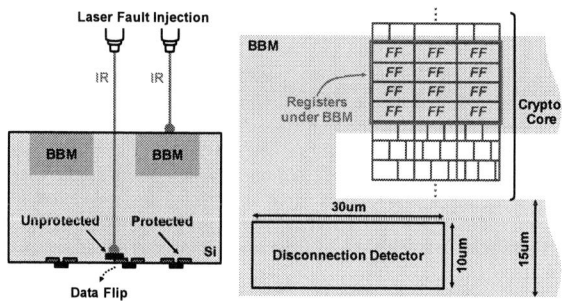

Fig. 5. Protection from laser fault injection attack.

Fig. 3. Proposed disconnection detector circuit.

required to be as small as possible. Besides, a mesh or solid pattern is often applied to the shield, however, it cannot detect a breakage caused by attacker's attempt to remove the shield using strong laser or focused ion beam (FBI) irradiation. Fig. 2(b) shows the proposed shield structure. Cu metal is buried in the backside of the existing Si substrate, and connected to the surface CMOS circuit through Si vias (TSVs) under IO PAD which results in no chip-size overhead. The BBM forms a meander pattern with current flow in the wire to detect physical attacks since the current stops at once as the wire is disconnected wherever by the laser or mechanical polishing. Moreover, the BBM meander pattern can be drawn with fine pitch, thus, the monolithic shield structure remains the blocking effect of noise and laser as described in the following subsections.

B. Disonnection Detector

Fig. 3 depicts the circuit schematic and operation waveforms of the proposed disconnection detector circuit. Only three transistors and one standard-cell are used to realize the detection function. A driver starts to flow current I_{BBM} over a meander to the biased node B after the reset RST is released, and an RS-latch continuously monitors its voltage level at the set node V_A which is normally low enough for the threshold of

the latch. When the BBM meander is disconnected, the V_A rises up to V_{DD} because the current I_{BBM} stops. Then, the RS-latch changes its internal logical state DET and warns the system for the advent of potential attacks. The static current I_{BBM} during normal operation is as small as 150 nA, since it can be controlled by V_{BIAS} and the BBM is well isolated from the Si substrate. Attackers may try to avoid the detection scheme by bypassing the meander. However, the proposed BBM can hide the current path by using frontside metal connection and form multiple dummy patterns. This "obfuscated meander" structure can prevent the bypass attack by detecting the undesired connection of true current path and dummy patterns.

C. Probing Attack Resistance

Probing on backside Si substrate discloses a secret key by analyzing the key-dependent substrate noise [6]. This substrate attack can capture side-channel information even with a leakage suppression technique such as an on-chip power line isolation [9], since it can directly probe the substrate closest to the cryptography circuit. Substrate current flows through the P+ contact at the logic operation, and it produces voltage variation due to the resistive property of silicon substrate as shown in Fig. 4. This noise can be acquired by probing the Si-backside, however, the BBM makes the noise not appear on the surface since it is isolated from substrate ground. Hence, the side-channel leakage through the BBM can be suppressed. Though the meander pattern exposes Si space, this gap is narrow enough compared to the tip area of the probe needle.

D. Laser Fault Injection Resistance

The laser of 1064 nm (Infrared) wavelength passes through the Si substrate and generates electron-hole pairs when it hits the diffusion area of a transistor. It potentially flips data held in

IEEE Asian Solid-State Circuits Conference
November 4 – 6, 2019
The Parisian Macao, Macao SAR, China

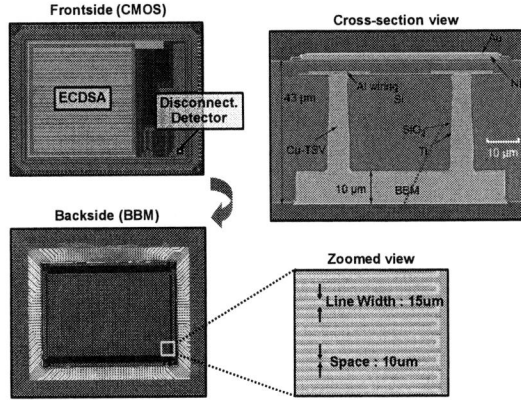

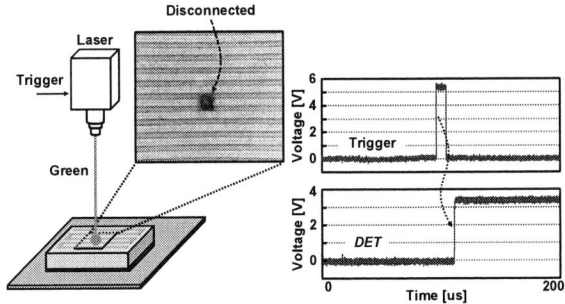

Fig. 6. Die photo of frontside and backside of prototype chip and SEM image of BBM.

Fig. 7. Evaluation of laser disconnection attack on BBM.

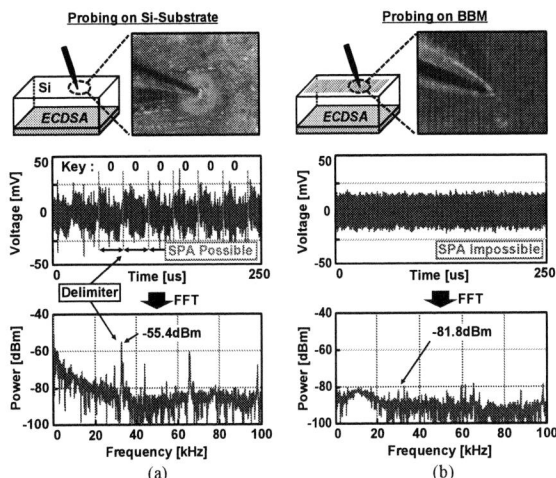

Fig. 8. Backside probing attack on ECDSA chip
(a) w/o BBM and (b) w/ BBM.

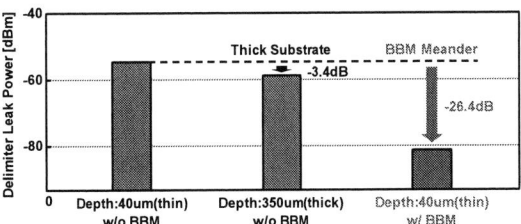

Fig. 9. Comparison of delimiter leak power among
Si-backside structures.

registers [10], which is used for the subsequent analysis to estimate the secret key. The Cu metal blocks IR laser injection, thus, the target registers of accumulation circuits in the cryptographic core can be protected from LFI attack by hiding them under the Cu BBM as shown in Fig. 5. Although IR laser can penetrate into the inside of a chip through the gap of BBM inherent to a meander pattern, the BBM line-width of 15 μm is wide enough to cover the limited number of registers of cryptographic importance. Fig. 5 also shows the protection of the disconnection detector circuit itself from the breaking attack. The detector can be also placed under BBM thanks to its small size of 10 μm x 30 μm. Thus, it can prevent malicious attack that destroys the detector before BBM breakage.

III. SI EXPERIMENTS

A Si test vehicle was developed with a 0.13 μm CMOS technology. Fig. 6 shows a die micrograph of both front and backside of the prototype chip. The chip embeds an elliptic curve digital signature algorithm (ECDSA) accelerator which occupies 2 mm x 2 mm area with 200 kgate and consumes 21.6 mA at the operation clock of 33 MHz. The wafer completed by the Si fab was then processed with the via-last, Cu BBM technology [11]. The BBM meander has the line widths, space and thickness of 15 μm, 10 μm and 10 μm, respectively, and is connected to TSVs with the radius and height of 5 μm and 40 μm, respectively. The cross-sectional view of the prototype

chip in Fig. 6 confirms that the meander BBM and the TSVs to frontend circuits are seamlessly formed.

The protection against a laser cutting attack is testified as shown in Fig. 7. Once the Cu BBM meander is disconnected by the focused laser of 532 nm (Green) wavelength, the detector immediately asserts the warning of physical attacks. This unification of the disconnection detector in a Si IC chip eliminates the need of additional protective structures in subsequent packaging and assembly stages. Note that "cut then restored" attack is invalid since the probing or LFI must be executed during circuit operation, thus the detector works at the moment when the BBM is disconnected.

Fig. 8 demonstrates the simple power analysis (SPA) attack during the operation of ECDSA at 33 MHz by probing voltage bounce directly on the backside Si substrate using a probing station. The number of clock cycles in processing scalar multiplications is visibly exposed in the waveform at the probing on Si-substrate as shown in Fig. 8(a). To measure a side-channel leakage quantitatively, we solved the leakage power in the frequency of "delimiter" that indicates the processing time of scalar multiplications. In this evaluation, the key of ECDSA is set to all 0 to make it easy to see the delimiter frequency as with a single tone. The direct probing on the Si-substrate induces a peak power of -55.4 dBm at 32 kHz delimiter in FFT spectrum, which suggests the key estimation

978-1-7281-5107-6/19 $31.00 © 2019 IEEE   27

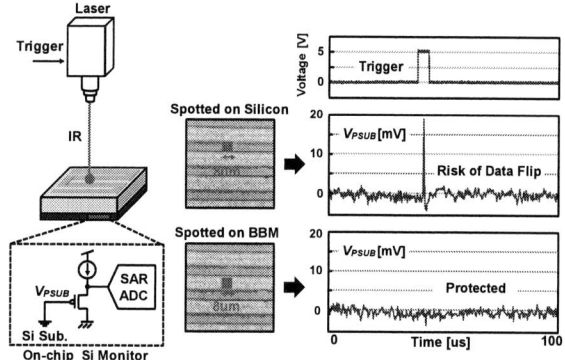

Fig. 10. Measured Si-substrate voltage at IR laser irradiation.

Table 1. Comparison with the other protection techniques against backside physical attacks.

		[4]	[8]	This work
Structure		Top-metal shield + Back to back bond.	Breakable die + Film laminating	Backside Buried Metal (BBM)
Backside protection against	Probing	YES	YES	YES
	LFI	YES	YES	YES
	Disconnect (Cut, Polish)	YES	YES	YES
	Bypass	No	YES	YES
Overhead		+10% Si area Top metal occupied Two dies needed	+120μm thickness (Film layer)	+0.0025% Si area (Detector)

with SPA is possible. On the other hand, when the direct probing is executed to the chip with BBM structure, the delimiter disappears in the waveforms and its power is sufficiently reduced to -81.8 dBm as shown in Fig. 8(b). The comparison of the delimiter power among Si-backside structures is given in Fig. 9. The delimiter leakage of -55.4 dBm is measured at the probing on Si-backside of ECDSA chip with the thickness of 40 μm. Although the thicker substrate of 350 μm attenuates -3.4 dB of the delimiter power, it still causes the key leakage. However, the proposed BBM structure drastically reduces the delimiter leakage by -26.4 dB even while keeping thin substrate thickness of 40 μm. This is beneficial to low height packaging.

Fig. 10 shows the measured Si substrate voltage fluctuation at the moment of IR laser irradiation. The Si substrate voltage is acquired and digitized on-chip by a source follower and a SAR ADC embedded in the prototype chip. IR laser focused with the aperture of 8 μm is completely blocked by the Cu BBM when it is spotted on its metal. On the other hand, the Si substrate voltage bounces due to electron-hole pair induction [12], once the beam is spotted on the space between BBM stripes. This experiment indicates the core part of cryptographic accelerator can be effectively protected from laser irradiation by placing them under the BBM since the size of logic cells are much smaller than the widths of BBM stripe.

Table 1 shows the comparison with the state-of-the-art countermeasures against backside physical attacks. Although manufacturing costs of secure structure increase, cryptographic circuit can be protected from various attacks. Since the BBM and TSV are formed in the existing backside Si-substrate and under IO pads, respectively, our technique only requires the disconnection detector with extremely small size of 300 μm², which is only 0.0025 % of chip area on the frontside. Compared to other works which require additional frontside physical layers, the proposed structure achieves less overhead while keeping backside physical attack resistances.

IV. CONCLUSION

The BBM Cu meander was monolithically unified with CMOS detector circuits and successfully demonstrated the prevention of physical security attacks including probing and

LFI, and also the detection of shield breakage, with little area overhead. The scheme is generally applicable to secure devices in diversified frontside technology nodes.

ACKNOWLEDGMENT

This work was also partly supported by the Cabinet Office (CAO), Cross-ministerial Strategic Innovation Promotion Program (SIP), "Cyber-Security for Critical Infrastructure" (funding agency: NEDO).

REFERENCES

[1] I. Verbauwhede, *et al.*, "Circuit Challenges from Cryptography," *IEEE ISSCC Dig. Tech. Papers*, pp. 428-429, Feb. 2015.

[2] P. Kocher, *et al.*, "Differential power analysis," *CRYPTO, Lecture Notes in Computer Science*, vol. 1666, pp. 388–397, Aug. 1999.

[3] K. Sakiyama, *et al.*, "Information-Theoretic Approach to Optimal Differential Fault Analysis," *IEEE Trans. on Information Forensics and Security*, vol. 7, no. 1, pp. 109-120, Feb. 2012.

[4] J.-M. Cioranesco, *et al.*, "Cryptographically Secure Shields," *IEEE International Symposium on Hardware-Oriented Security and Trust (HOST)*, pp. 25-31, May 2014.

[5] S. Manich, *et al.*, "Detection of Probing Attempts in Secure ICs," *IEEE International Symposium on Hardware-Oriented Security and Trust (HOST)*, pp. 134-139, 2012.

[6] D. Fujimoto, *et al.*, "Side-Channel Leakage on Silicon Substrate of CMOS Cryptographic Chip," *IEEE International Symposium on Hardware-Oriented Security and Trust (HOST)*, pp. 32-37, May 2014.

[7] C. Helfmeier, *et al.*, "Breaking and entering through the silicon," *Proc. of the 20th ACM Conference on Computer and Communications Security*, pp. 733-744, Nov. 2013.

[8] S. Borel, *et al.*, "A Novel Structure for Backside Protection Against Physical Attacks on Secure Chips or SiP," *IEEE 68th Electronic Components and Technology Conference (ECTC 2018)*, pp. 515-520, May 2018.

[9] C. Tokunaga and D. Blaauw, "Secure AES Engine with A Local Switched-Capacitor Current Equalizer," *IEEE ISSCC Dig. Tech. Papers*, pp. 64-65, Feb. 2009.

[10] J. G. J. van Woudenberg, *et al.*, "Practical optical fault injection on secure microcontrollers," in *Proc. Workshop Fault Diagnosis Tolerance Cryptogr. (FDTC)*, pp. 91-99, Sep. 2011.

[11] Y. Araga, *et al.*, "A Thick Cu Layer Buried in Si Interposer Backside for Global Power Routing," *IEEE Trans. on Components, Packaging and Manufacturing Technology*, vol. 9, no. 3, pp. 502-510, 2019.

[12] K. Matsuda, *et al.*, "A 286F²/Cell Distributed Bulk-Current Sensor and Secure Flush Code Eraser Against Laser Fault Injection Attack," *IEEE ISSCC Dig. Tech. Papers*, pp. 352-353, Feb. 2018.

978-1-7281-5107-6/19 $31.00 © 2019 IEEE

A 28nm 512Kb adjacent 2T2R RRAM PUF with interleaved cell mirroring and self-adaptive splitting for extremely low bit error rate of cryptographic key

Xiaoyong Xue[1]*[+], Jianguo Yang[2+], Yuejun Zhang[1, 3+], Mingyu Wang[1+], Hangbing Lv[2]*, Xiaoyang Zeng[1], Ming Liu[2]

[1]ASIC and System State Key Laboratory, School of Microelectronics, Fudan University, China; [2]Key Laboratory of Microelectronics Devices and Integrated Technology, Institute of Microelectronics of the Chinese Academy of Sciences, China; [3]School of Electrical Engineering and Computer Science, Ningbo University, Ningbo, China.
*Email: xuexiaoyong@fudan.edu.cn, lvhangbing@ime.ac.cn

Abstract—Reliability and cost efficiency are of utmost importance for cryptographic key generation and storage in IoT applications. This paper presents a 2T2R RRAM PUF scheme with assisting circuit techniques for dense and reliable cryptographic key generation in 28nm logic process. The set time mismatch in two adjacent RRAM devices is exploited for PUF output with better tolerance against PVTA variations. Array architecture featuring interleaved cell mirroring improves the randomness by eliminating the systematic deviation. Self-adaptive splitting with current limiter eliminates the negative effect of mismatch in the peripheral circuits and suppresses the undue duration of large operation current. Silicon data demonstrates our RRAM PUF passes NIST test with intra-chip Hamming distance (HD) of 0, inter-chip HD of 0.496 and bit error rate of $<1\times10^{-5}$@0-120 °C. The cell size of 2T2R PUF is 0.54 μm^2 and can be reduced when the selector is shared or the RRAM set/reset voltages are optimized below the core supply voltage.

Keywords—*PUF, RRAM, interleaved cell mirroring, self-adaptive splitting, low bit error rate, IoT*

I. INTRODUCTION

Ubiquitous Internet of Things (IoT) endpoint devices such as wearables, home appliances and various sensors bring the concern of information security, which makes cryptography an absolute necessity in practical applications. High security level calls for increasing key length of up to 15 Kb and sophisticated encryption algorithms like Elliptic Curve Cryptography (ECC) and Rivet-Shamir-Adleman (RSA) algorithm (Fig. 1) [1]. The requirements for secure key generation and storage are: (i) a source of randomness to ensure the keys unpredictable and unique; (ii) a protected memory that reliably stores the key [2].

Physically unclonable function (PUF) is able to tackle both the above-mentioned requirements for cryptographic key [3]. However, it is challenging to obtain an area-efficient and reliable PUF. The state-of-the-art PUFs based on propagation

This work was supported by National Natural Science Foundation of China (61704029, 61874028, 61834009 and 17ZR1446800), Huawei Innovation Research Program (HIRP), the MOST of China under Grant (2016YFA0203800) and the Strategic Priority Research Program of the Chinese Academy of Sciences under Grant No. XDPB12. [+]equal contribution

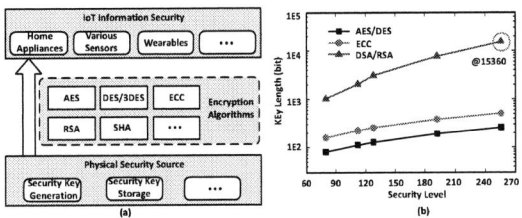

Fig. 1. (a) Typical IoT chip architecture; (b) Key length for different cryptography algorithms. High security level calls for increasing key length of up to 15 Kb as well as sophisticated encryption algorithms like ECC and RSA.

delay, ring oscillator, and threshold voltage of SRAM, etc. are sensitive to PVTA variations, resulting in poor bit error rate (BER). For instance, the SRAM-PUF provides the best area efficiency for BER less than 10^{-5} [2]. However, since IoT endpoints are area-constrained, increasing key length realized by SRAM is unsatisfying for large area cost. A PUF based on gate-oxide rupture permits zero BER [4], but it is one-time programmable (OTP) and needs high voltage (~5.5V), making it not impractical in advanced process calling for re-configurable keys. PUFs based on emerging resistive memory devices may have higher density. However, they are mostly based on variations in high resistance state (HRS) or low resistance state (LRS), and the cell resistance drift (R-drift) owing to read/write (R/W) disturbance and PVTA variations easily affects the stability of PUF output [5]. A 2T2R PUF was proposed for low BER based on comparing the high resistance state (HRS) of two 1T1R cells from different arrays [6]. However, a concern exists that the mismatch of sense amplifier and the systematic cell variation between different arrays may deviate the resistance splitting to generate more "1"s or "0"s. Stabilization and error-correction techniques such as reliability screening, masking algorithms and error correcting code can improve the output reliability but cause considerable costs in area and power consumption.

This paper presents a 2-transistor-2-resistor (2T2R) RRAM PUF scheme with assisting circuit techniques to deal with the above-mentioned challenges. The set time mismatch in two adjacent RRAM devices is exploited for PUF output, achieving better tolerance against PVTA variations. Array architecture with interleaved cell mirroring (ICM) of 2T2R PUF cell improves the randomness by eliminating the effect of systematic deviation. Self-adaptive splitting with current limiter (SAS-CL) is employed to avoid the negative effect of mismatch in peripheral circuits and suppress the undue duration of large operation current. Silicon data from 512Kb PUF test chip verified the design.

II. PROPOSED PUF SCHEME

A. Adjacent 2T2R PUF cell for stable output and high density

Fig. 2 shows the proposed 2R-based PUF cell where the selector can be a transistor, a diode, a bidirectional diode, or inexistent, i.e. x-bar. The selector can also be shared by the two RRAM devices (e.g. 1T2R) to save area. In this work, the 2T2R PUF cell is implemented for verification with consideration of compatibility with the 1T1R memory function test to save cost. Fig. 3 shows the schematic, cross section and layout of the 2T2R PUF cell as well as the I-V curve of the used TaOx RRAM. The two 1T1R RRAM cells are adjacently placed within the same array to alleviate the systematic deviation between different arrays, which may cause the PUF to generate more "1"s or "0"s. The IO device is used as the selector to reliably withstand the set/reset voltages (>1.05V) but results in poor density for PUF. The area of 2T2R PUF cell is 0.54 μm², still achieving >8X improvement in density compared to the state-of-the-art SRAM (4.6μm² @ 22 nm) already in practical use [2, 10]. Because the RRAM cell size is mainly determined by the selector, the PUF cell size will not vary too much since the size of IO device is relatively constant for different process nodes. However, the RRAM set/reset

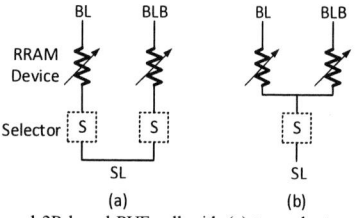

Fig. 2. Proposed 2R-based PUF cell with (a) two selectors and (b) one selector to save area. The selector can be a transistor, a diode, a bidirectional diode, or inexistent, i.e. x-bar.

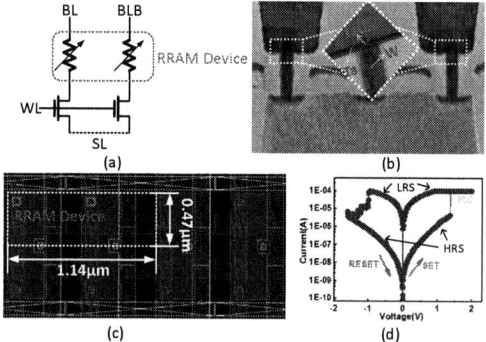

Fig. 3. (a) Schematic, (b) cross section and (c) layout of 2T2R PUF cell, and (d) IV curve of TaOx RRAM (ICC is the compliance current). The inset in (b) shows the integration of Ta/TaOx/W RRAM device. The area of 2T2R PUF cell is 0.54 μm²@28nm, i.e., >8X in density compared to the state-of-the-art SRAM (4.6μm²@ 22nm) [2, 10].

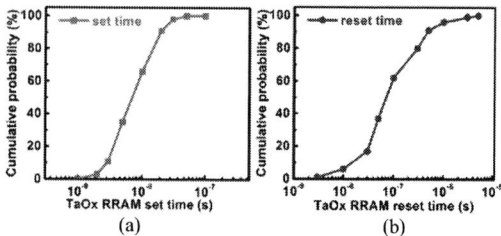

Fig. 4. Distribution of TaOx RRAM (a) set time and (b) reset time. Because set is abrupt, it is extremely rare for two RRAM devices to have the CFs formed simultaneously. Reset is gradual, and the CFs in two RRAM devices may both experience different extents of rupture at the end. Therefore, set is used in our PUF for large resistance window.

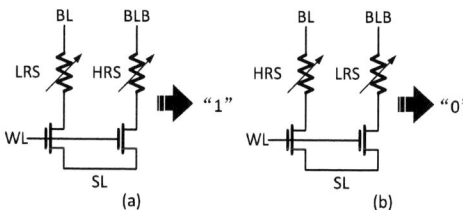

Fig. 5. Random output of 2T2R PUF: (a) "1", (b) "0". Due to differential storage, large resistance window between LRS and HRS in a PUF cell can overcome R-drift caused by R/W disturbance and PVTA variations to ensure stable output, meeting the low BER requirement of ID and key generation. However, 1R-based PUF cell (e.g. 1T1R) suffers greatly from R-drift issue.

voltages can be optimized below the core supply voltage by device and process techniques in future work, then the PUF cell size will be reduced remarkably.

As the set/reset time for conductive filament (CF) formation/ rupture is subject to large distribution, as shown in Fig. 4, the competition of CF formation/ rupture during set/reset can be employed to generate stable PUF output. The set process is abrupt and it is extremely rare for two RRAM devices to have their CFs formed simultaneously. In contrast, the reset process is gradual, which may make the CFs in two RRAM devices both experience different extents of rupture at the end, causing poor resistance window. Therefore, the set time variation is chosen in 2T2R PUF for large resistance window. The initial state for the two RRAM devices is HRS. When applying a same set voltage on both RRAM devices, one will switch to LRS first. If the set voltage is taken off at this point, the PUF cell will have one RRAM device in LRS and the other in HRS. Since it is unpredictable which RRAM device switches first, a random bit can be obtained by binarizing the resistive switching behaviors of the two cells, as shown in Fig. 5. The differential resistance storage for cryptographic key provides sufficient resistance window (>10X) to tolerate R-drift and good immunity against side-channel attack when using symmetrical sense amplifier for readout. The selection of set voltage (Vset) has to meet the majority (99%) of RRAM devices, about 1.5V in this work.

For key regeneration, a reset operation is needed to bring the two RRAM devices in a PUF cell to HRS, and then a new random bit is expected to be generated thanks to the cycle-to-cycle variation of RRAM set time.

B. Array architecture with interleaved cell mirroring of adjacent 2T2R PUF cell to eliminate systematic deviation

Fig. 6 shows the PUF array architecture with peripheral decoder and read/write (R/W) modules. The complementary bitlines (e.g. BLm & BLBm) are connect to the corresponding data ports of PUF cells. For 3-terminal selector like transistor, the wordline (e.g. WLn) is connected to the gate terminal of the selectors and the source line (e.g. SLn) is connected all the source terminal of the selectors along the same row. For 2-terminal selector like diode, the wordlines (e.g. WLn) are connected to the bottom of the PUF cell along the same row. Owing to the local comparison of set time between two adjacent RRAM devices in a PUF cell, the global spatial variation of set time hardly affects the randomness of the PUF output. However, there may still exist a systematic deviation between the left RRAM cell and the right one, which causes the probabilities of getting "1" and "0" to deviate from 0.5. In order to alleviate this effect, interleaved cell mirroring (ICM) of the adjacent 2T2R PUF cell is proposed for array arrangement where the PUF cells in the neighboring columns are mirrored by exchanging the connection of the complementary bitlines to the column selector. Thus, the

probabilities of getting "1" and "0" are averaged. Fig. 7 compares the simulation results of Hamming weight (HW) of 1024 keys of 512b realized by 2T2R PUFs which are far-placed 2T2R, i.e. two 1T1R cells with each from a separate array, and adjacent 2T2R without & with ICM. The array-to-array systematic deviation is assumed to be 5% and the column-to-column systematic deviation is 0.05%. The proposed adjacent 2T2R PUF cell with ICM achieves the closest probability to 0.5 for getting "1"/ "0".

C. Self-adaptive splitting with current limiter to avoid mismatch effect and undue duration of large current

For accurate key generation, it is critical to take off the set voltage from the PUF cell rightly after the CFs are formed in one RRAM device. This can be realized by self-adaptive write techniques to capture the switch point [7, 8]. However, the large parasitic capacitance (tens to hundreds of fF) of bitlines may still induce a long duration of compliance current through the LRS device, which is detrimental to the RRAM distribution and increases power consumption. To suppress this effect, self-adaptive splitting with current limiter (SAS-CL) is proposed as shown in Fig. 8(a). A current limiter is placed at the top of previous self-termination write (STW) circuit [7]. The gate terminal of M2 is controlled by a current mirror which is formed by a current source Icomp and a diode-

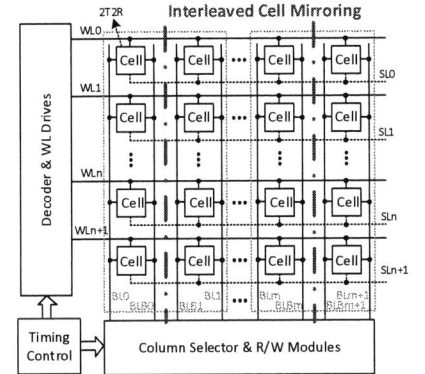

Fig. 6. PUF cell array architecture featuring interleaved cell mirroring with peripheral decoder, controller and read/write modules.

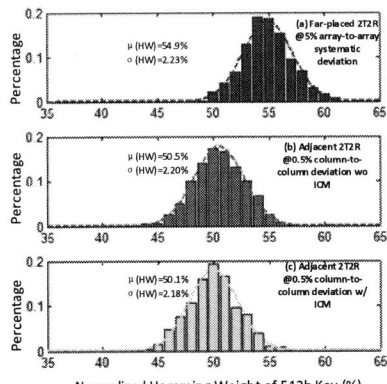

Fig. 7. Comparison of Hamming weight of 1024 keys of 512 bits realized by 2T2R PUFs. (a) far-placed 2T2R, i.e. two 1T1R cells with each from a separate array; (b) adjacent 2T2R wo ICM; (c) adjacent 2T2R w/ ICM.

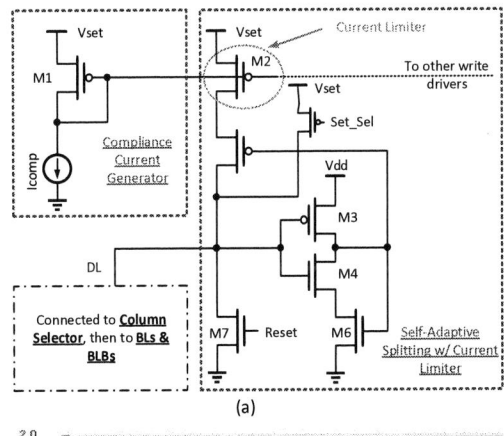

(a)

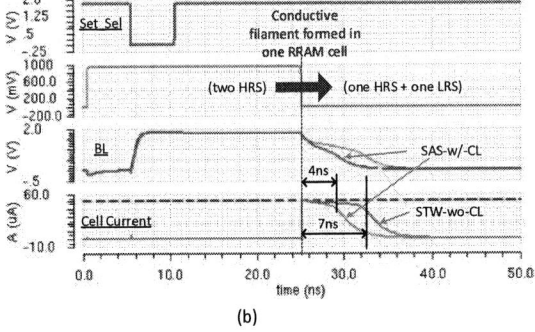

(b)

Fig. 8. (a) Schematic and (b) simulation results of proposed SAS-CL. SAS-CL reduces the duration of compliance current by >40% (7ns to 4ns). The write energy is calculated by integration of I*V with the write cycle time and is reduced by ~40% (175fJ/bit to 105fJ/bit).

connected PMOS transistor M1. Simulation results in Fig. 8(b) show the duration of compliance current is reduced by >40% (7ns to 4ns). The write energy is also reduced by ~40% (175fJ/bit to 105fJ/bit). Moreover, because the BL&BLB of 2T2R are connected to the same driving signal DL during splitting, the mismatch in peripheral circuits is no longer an issue for inducing systematic deviation to PUF output.

III. TEST CHIP

The proposed adjacent 2R-based (2T2R) PUF was verified on a 28nm 512Kb test chip using TaOx RRAM (Fig. 9). The HW distribution of 1024 keys of 512b indicates nearly ideal randomness of the adjacent 2T2R PUF cell with ICM (Fig. 10). The average BER is $<10^{-5}$@0-120°C thanks to the large resistance window of the 2R-based PUF cell and SAS-CL (Fig. 11). The intra-chip HD is 0 with σ=0 while the inter-chip HD is 0.496 with σ=0.042 (Fig. 12). The count of measured chips is 20. Table I shows the 2T2R PUF passed the NIST test. Table II compares this work with the state-of-the-art [4, 5, 6, 9, 10].

REFERENCES

[1] Standards for Efficient Cryptography Group [Online]. Available: http://www.secg.org/

[2] R. Maes, *Physically Unclonable Functions*: Springer, 2013.

[3] R. Pappu, B. Recht, J. Taylor, and N. Gershenfeld, "Physical one-way functions," *Science*, vol. 297, no. 5589, pp. 2026-2230, 2002.

[4] M. Wu, et al., "A PUF scheme using competing oxide rupture with bit error rate approaching zero," *ISSCC*, 2018, pp. 130-131.

[5] Y. Pang, et al. "A novel PUF against machine learning attack: Implementation on a 16 Mb RRAM chip," *IEDM*, 2017, pp.12.2.1-4.

[6] P. Yachuan, et al. "A Reconfigurable RRAM Physically Unclonable Function Utilizing Post-Process Randomness Source With< $6×10^{-6}$ Native Bit Error Rate," *ISSCC*, 2019, pp. 402-403.

[7] M. Chang, et al. "Embedded 1Mb ReRAM in 28nm CMOS with 0.27-to-1V read using swing-sample-and-couple sense amplifier and self-boost-write-termination scheme," *ISSCC*, 2014, pp. 332-333.

[8] X. Xue, et al. "A 0.13 μm 8 Mb Logic-Based CuxSiyO ReRAM With Self-Adaptive Operation for Yield Enhancement and Power Reduction." *JSSC*, vol. 48, no. 5, 1315-1322, 2013.

[9] Y. Su, et al. "A 1.6pJ/bit 96% stable chip-ID generating circuit using process variations", *ISSCC*, 2007, pp. 406-407.

[10] S. K. Mathew, et al. "A 0.19pJ/b PVT-variation-tolerant hybrid physically unclonable function circuit for 100% stable secure key generation in 22nm CMOS," *ISSCC*, 2014, pp. 278-279.

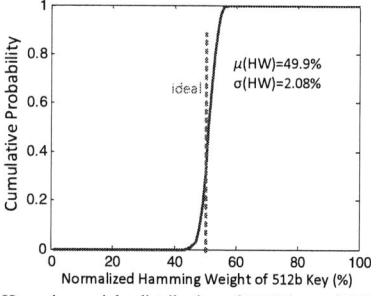

Technology	28 nm logic +TaOx RRAM
Supply voltage	1.8V IO, 1.05V Core
Temperature	0~120°C
Die size	1650 μm * 1630 μm
Array size	1024*1024
Capacity	512Kb PUF (1Mb RRAM)
PUF cell structure	2T2R
PUF cell size	0.54(0.14)μm² for IO(Core) device selector
BER (for PUF)	<1e-5
Intra-chip HD	μ=0, σ=0
Inter-chip HD	μ=0.496, σ=0.042

Fig. 9. Diagram of 512Kb PUF test chip and key features.

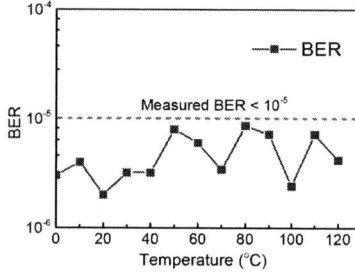

Fig. 10. Hamming weight distribution of 1024 keys of 512b indicates nearly ideal randomness.

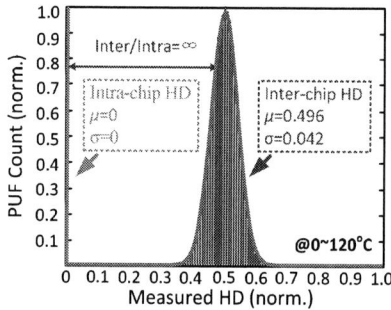

Fig. 12. Measured intra-chip HD with μ=0 and σ=0 and inter-chip HD with μ=0.496 and σ=0.042 indicate good randomness.

Table I. PROPOSED 2T2R PUF PASSED NIST TEST.

Test Name	Stream Length	No. of Run	P-value	Pass
Frequency	10240	40	0.883171	Yes
Block Frequency	10240	40	0.779188	Yes
Longest Run of Ones	10240	40	0.096578	Yes
Cumulative Sums	10240	40	0.657933	Yes
Linear Complexit	10240	40	0.108791	Yes
Serial	10240	40	0.616305	Yes
Rumns	10240	40	0.153763	Yes
Non-Overlapping Template	10240	40	0.699313	Yes

Table II. COMPARISON WITH THE STATE-OF-THE-ART.

	ISSCC07 [9]	ISSCC14 [10]	IEDM17 [5]	ISSCC18 [4]	ISSCC19 [6]	This work
PUF Type	SRAM-based	SRAM-based	Arbiter-based	Gate oxide	RRAM (HRS)	RRAM (Set Time)
Technology	130nm	22nm	55nm	55nm	130nm	28nm
Area/bit(μm²)	73.83	4.66	NA	0.66	2.86	0.54 for IO selector 0.14 for core selector
Density	128	100K	16Mb	64K	8K	512K
Voltage (V)	1	0.9	NA	5.5	1.2/3.3	1.05/1.8
BER	0.0378	0.0097	NA	0	<6.1e-6	<1e-5
Inter HD	0.50125	0.49	0.5	0.499999	0.499	0.496
Intra HD	NA	0.0258	0	0	0	0
Op. Temp. (°C)	NA	25~50	25~125	-40~150	25~150	0~120

Fig. 11. Measured BER is less than 10^{-5} from 0 to120°C, satisfying the low BER requirement (10^{-5}) of cryptographic key.

A 0.5V Real-Time Computational CMOS Image Sensor with Programmable Kernel for Always-On Feature Extraction

Tzu-Hsiang Hsu, Yen-Kai Chen, Tai-Hsing Wen, Wei-Chen Wei, Yi-Ren Chen, Fu-Chun Chang,
Ren-Shuo Liu, Chung-Chuan Lo, Kea-Tiong Tang, Meng-Fan Chang, and Chih-Cheng Hsieh
National Tsing Hua University, Hsinchu, Taiwan
Email: cchsieh@ee.nthu.edu.tw

Abstract—This paper presents a 0.5V computational CMOS image sensor (C^2IS) with array-parallel computing capability for always-on feature extraction. By applying the developed pulsed-width modulation (PWM) pixel and switch-current integration (SCI), the in-sensor 8-directional matrix-parallel multiply-accumulate (MAC) operation is realized. Moreover, the analog-domain convolution-on-readout (COR) operation, the programmable 3x3 kernel with ±3-bit weights, and the tunable-resolution column-parallel ADC (1b to 8b) are implemented to achieve the real-time feature extraction without use of additional memory. The C^2IS prototype has been fabricated and verified to demonstrate the raw and feature images at 480fps with a power consumption of 77/91 (uW) and the resultant FoM of 9.8/11.6 (pJ/pix/frame), respectively.

Keywords—*Always-on, convolution, computational CMOS image sensor, feature extraction, processing-in-sensor*

I. INTRODUCTION

To bring the intelligent vision into the IoT edge device, power constraint, data arrangement with latency and hardware cost must be considered. Recently, application-driven CMOS image sensor (CIS) with feature extraction capability using processing-in-sensor (PIS) operation shows a great potential for the low power always-on applications [1-4]. Since the feature extraction of images relies on the operation of matrix convolution of sub-array, the multiply-accumulate (MAC) capability with fully-connected (8-directional) network and programmable weighting are essential for PIS realization. Considering the required area and power penalty of PIS implementation, the reported works [1-3] demonstrated some specific features with either fixed weight [1-2] or not fully-connected operation (1-directional) [3], sacrificing the programmability and accuracy. The other powerful solution is using 3D-stacking architecture to solve the area limitation issue with considerable power and cost [4]. Moreover, all these works need extra memory for intermediate value storage during calculation, and the achievable frame rates of feature output are low with limited resolution.

In this work, we propose a low-voltage and energy-efficient computational CIS (C^2IS) featuring 0.5V operated pulse-width-modulation (PWM) pixel, 8-directional analog MAC operation, programmable 3x3 kernel, tunable feature output resolution (1b~8b), and convolution-on-readout (COR) operation to achieve the real-time, accurate, and arbitrary feature imaging output. As shown in Fig. 1, the proposed C^2IS provides a real-time always-on solution for instant-event detection and awakes the afterward CNN processor only when necessary. Consequently, the required data transfer and buffer for image feature extraction are highly reduced to achieve a low-latency energy-efficient inference solution in AI application.

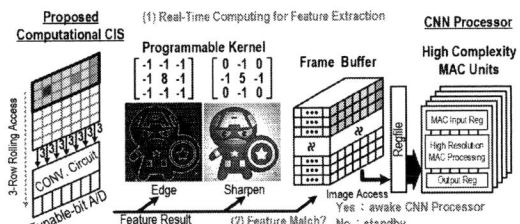

Fig. 1. The concept using the proposed C^2IS technique

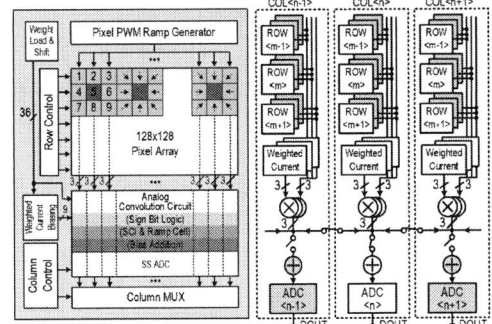

Fig. 2. The chip architecture

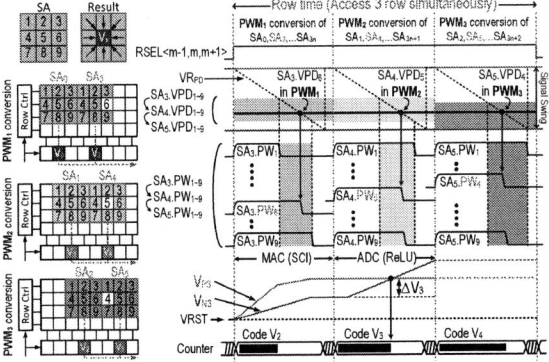

Fig. 3. Circuit implementation of matrix-parallel analog convolution

Fig. 4. Timing diagram of the proposed convolution-on-readout operation

978-1-7281-5107-6/19 $31.00 © 2019 IEEE

II. ARCHITECTURE AND CONCEPT

Fig. 2 shows the chip architecture of the C²IS prototype. It consists of the 128x128-pixel array, matrix-parallel analog-domain convolution circuit, current-domain weight generator with kernel access control, column-parallel single-slope (SS) ADCs, and peripheral supporting circuity. A 0.5V-operated linear-response PWM pixel [5] is adopted to provide a power-efficient imaging solution. The in-pixel voltage-to-pulse conversion circuit provides a robust and accurate signal for analog-domain convolution. To implement the 3x3 convolution, the 3x3-pixel subarrays are selected for matrix-parallel analog MAC operation by turning on 3 rows (<m-1>, <m>, <m+1>) simultaneously. For analog convolution of pixel <n, m>, the 9-pixel outputs from 3 adjacent columns (<n-1>, <n>, <n+1>) are multiplied by the 9 weighted currents and accumulated. The convolution result is then quantized by the following column-parallel SS ADC<n> with tunable resolution. In normal imaging mode, the PWM output pulses are quantized by the same counter in SS ADCs without area overhead.

III. CIRCUIT IMPLEMENTATION

Fig. 3 shows the circuit implementation of the matrix-parallel analog convolution with 3x3 kernel and ±3-bit weights (-8 to +8). The 4T PWM pixel consists of in-pixel comparator (MCP/MCN) with ramp voltage (VR). To access the pixels of 3 rows simultaneously, 3 output buses (PW<m-1>, <m>, <m+1>) are implemented in each column for the corresponding rows. The MAC function is implemented by applying the switching-current-and-integration (SCI) operation [6] controlled by the signal-dependent pulse width and weight-dependent current level. The weights of 3x3 kernel are applied to control the array-shared 9 current DACs for precise current level. In each column, 3 pulses (PW<m>) from 3 selected rows are gated to be ADD<m>/SUB<m> signal according to the positive/negative weight. The weight-dependent currents I_P<m>/I_N<m> are then switched on and summed up as a total current I_P<n>/I_N<n>. By turning on "SUM" of the neighboring 2 columns (<n+1>&<n-1>), the positive/negative weight-dependent currents of 3x3 pixel subarray and 3x3 kernel are summed up as I_P/I_N and integrated on the corresponding capacitor (C_P/C_N) respectively to achieve the MAC operation of 3x3 pixel subarray and 3x3 kernel. The judging logic (JG) is implemented to check the signal polarity of convolution result (V_P>V_N?) with a sign bit output (SIGN). When SIGN=1 (V_P>V_N), V_N is ramped up using the ramp cell controlled by VB_R and $RAMP_N$ to cross V_P and trigger counter to digitize V_P-V_N to Dout as SS ADC. On the contrary, when SIGN=0 (V_P<V_N), V_P is ramped up to cross V_N to accomplish AD conversion. With the differential input capability, the effective convertible swing of V_P-V_N is efficiently doubled under 0.5V supply. The ReLU function is also implemented by quantizing the positive convolution result only and ignoring the negative one depends on SIGN level with power saving.

Fig. 4 shows the timing diagram of the proposed convolution-on-readout (COR) operation. To accomplish all the MAC operations of a specific row with 3x3 kernel in a row time, 3 PWM conversions (PWM_{1-3}) are activated sequentially with subarray shifting as noted in SA_{3n}, SA_{3n+1} and SA_{3n+2}. With the matrix-parallel MAC operation, all the convolution results of subarrays in the same color (SA_{3n}, SA_{3n+1}, or SA_{3n+2}) are generated simultaneously. Take the MAC operation for the convolution result of V_3 from subarray SA_3 as an example, the PWM conversions of VPD (SA_3.VPD_{1-9}) are activated in PWM_1 period to generate 9 pulses (SA_3.PW_{1-9}) with VPD-node-referred threshold VR_{PD} (from VR). The integrated

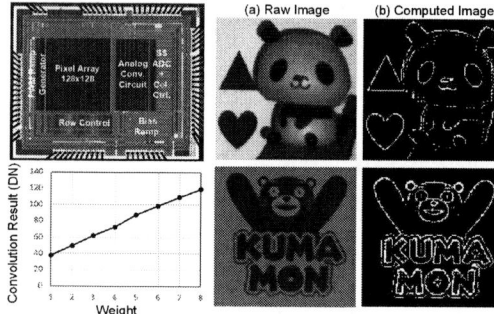

Fig. 5. Measurement results of imaging mode and computation mode

TABLE I. Performance Comparison

	[1] 2017 ISSCC	[2] 2018 VLSI	[3] 2017 VLSI	[4] 2017 ISSCC	This Work
Process	65nm	180nm	180nm	90nm/60nm	180nm
Supply	2.5/0.5~0.8V	1.2/0.8V	3.3/1.8V	3.3/2.9/1.8/1V	0.5V
Pixel Size	7um x 7um	7.9um x 7.9um	31um x 31um	3.5um x 3.5um	7.6um x 7.6um
Pixel Type	N.A.	1.75T-pinned	3T+SC	BSI pixel	PWM
Fill Factor	N.A.	55%	19%	N.A.	36%
Array Size	320x240	256x216 128x108 (LBP/ED)	256x256 (Spat.) 64x84 (Tempo)	1296x976	128x128
Frame Rate	Feature: 1fps	Image: 15fps Feature: 30fps	Feature: 30fps	Image: 60fps Feature: 1000fps	Image: 480fps Feature: 480fps
Feature	Haar-like filtering	LBP²/ED³	Spatial-Temporal processing	Spatial-Temporal processing	1st-layer CNN, ED, Blur, Sharpen..
Weight	Fixed	Fixed	Fixed	Programmable	Programmable
Operation	Image subarray Summation/decision	Column-parallel 8-directional 4 Pixel comparison	Column-parallel 1-directional Pixel difference	Column-parallel 8-directional bit-scan ALU⁴	Matrix-parallel MAC 8-directional 3x3 kernel convolution
Processing	Mixed-mode	Analog	Mixed-mode	4b/1b Digital	Analog
Memory	Yes (analog)	Yes (analog)	Yes (analog)	Yes (digital)	No
Proc. result resolution	1b (pass/fail)	1b (binary)	2b (Spat) 1b (Tempo)	4b (Spat) 1b (Tempo)	1~8b
Power	23.8uW	Image 32uW LBP/ED 6.5/12.7uW	Spat: 2.19mW Tempo: 29.94uW	Image: 230mW 4b feature: 363mW	Image: 77uW Conv.: 91uW
FoM⁵	309	Image: 38.5 LBP/ED: 16/31	134	Image 3030 4b feature: 286	Image: 9.8 Conv: 11.6

SCI: Switch Capacitor, LBP²: Local Binary Pattern, ED³: Edge Detection
ALU⁴: Arithmetic Logic Unit, FoM⁵: pJ/pixel/frame

voltage (V_{P3}/V_{N3}) after SCI operation is then quantized to code V_3 by activating the corresponding ramp cell. Similarly, the convolution results V_4 and V_5 are generated in period PWM_2 and PWM_3 with the 9 pulses from SA_4 and SA_5 respectively. The quantized convolution results are then readout sequentially like rolling-shutter operation. As a result, the proposed COR operation achieves the full-rate (480fps) real-time analog convolution without latency penalty.

IV. MEASUREMENT RESULT

Fig. 5 shows the measurement results of C²IS prototype. In imaging mode, the 128x128 raw image is successfully captured at 0.5V supply with a power consumption of 77uW at 480fps. In computation mode, the convolution result of a constant input versus 3-bit weight (1~8) shows a linear response without distortion. The convolution function is successfully verified by applying the "edge extraction" kernel. Table I shows the performance comparison. With the proposed real-time analog-domain COR operation, the prototype 0.5V C²IS achieves an 8-directional matrix-parallel MAC operation with a FoM of 11.6pJ/pix/frame at 480fps.

REFERENCES

[1] K. Bong, et al., "A 0.62mW ultra-low-power convolutional-neural-network face-recognition processor and a CIS integrated with always-on haar-like face detector," IEEE ISSCC, pp. 248-249, 2017.

[2] X. Zhong, et al., "A 2pJ/Pixel/Direction MIMO Processing Based CMOS Image Sensor for Omnidirectional Local Binary Pattern Extraction and Edge Detection," IEEE Symposium on VLSI Circuits, pp. 247-248, 2018.

[3] K. Lee, et al., "A 272.49 pJ/pixel CMOS Image Sensor with Embedded Object Detection and Bio-Inspired 2D Optic Flow Generation for Nano-Air-Vehicle Navigation," IEEE Symposium on VLSI Circuits, pp. 294-295, 2017.

[4] T. Yamazaki et al., "4.9 A 1ms High-Speed Vision Chip with 3D-Stacked 140GOPS Column-Parallel PEs for Spatio-Temporal Image Processing," IEEE ISSCC, pp. 82-83, 2017.

[5] A. Y. Chiou et al., "An ULV PWM CMOS Imager with Adaptive Multiple-Sampling Linear Response, HDR Imaging, and Energy Harvesting," IEEE JSSC, vol. 54, pp. 298-306, 2019.

[6] K. Korekado, et al., "An Image Filtering Processor for Face/Object Recognition Using Merged/Mixed Analog-Digital Architecture," IEEE Symposium on VLSI Circuits, pp. 220-223, 2005.

A 16K SRAM-Based Mixed-Signal In-Memory Computing Macro Featuring Voltage-Mode Accumulator and Row-by-Row ADC

Hyunjoon Kim, Qian Chen, and Bongjin Kim

School of Electrical and Electronic Engineering, Nanyang Technological University, 50 Nanyang Avenue, Singapore, 639798
Email: {kimh0003, e170029}@e.ntu.edu.sg, bjkim@ntu.edu.sg

Abstract- **This work proposes an SRAM-based mixed-signal in-memory computing macro using a pseudo-differential voltage-mode accumulator and a row-by-row ADC. The macro consists of 128 parallel mixed-signal dot-product computing units, each with 128 bitcells (64 for synapses, 32 for ADC reference, and 32 for offset calibration). A bitcell comprises of a 6T SRAM cell, an XNOR-based binary multiplier, and a pseudo-differential voltage-mode driver. A 65nm test-chip is fabricated, and the measured energy efficiency is 87TOPS/W with 5bit ADC at 0.5V supply.**

Keywords- *in-memory computing, dot-product, SRAM, multiply and accumulate, ADC, voltage-mode accumulator*

I. INTRODUCTION

Artificial neural networks (ANNs) have been adopted in a variety of machine learning tasks such as image classification and speech recognition. Recently, significant research efforts have been devoted to embedding small ANNs in edge devices with growing concerns of latency, bandwidth, and privacy in cloud-based computing. One of the recent efforts is binarized neural networks (BNNs) reducing the hardware overhead with binary constraints in weight and/or activation bit-precisions [1]. However, integrating BNNs in battery-operated mobile devices is challenging due to a significant energy overhead for off-chip DRAM access. Prior works [2-5] embedded BNNs in or near memory and achieved moderate accuracy for low-complexity tasks while achieving relatively high energy/area efficiency by using mixed-signal circuits. Despite the recent achievements, performance degrading factors of mixed-signal computing such as nonlinearity and PVT variation as well as the overhead of ADCs remain as major concerns. To address these concerns, we propose two key features for the efficient mixed-signal computation in-memory: 1) pseudo-differential voltage-mode accumulator; 2) row-by-row ADC using replica bitcells.

II. PROPOSED IN-MEMORY COMPUTING MACRO

Fig. 1 shows the block diagram of the proposed in-memory

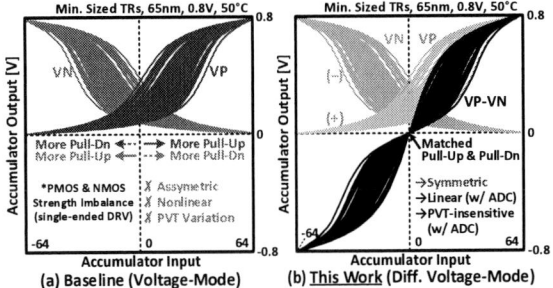

Fig. 2. Transfer characteristics of voltage-mode accumulators.

computing unit with 128 bitcells implementing a mixed-signal dot-product between 64 inputs and synapses (Fig. 1, left). The synapse weights stored in SRAM cells of the first 64 bitcells are multiplied by the inputs using XNOR gates embedded in each bitcell. Note that dedicated vertical routings for inputs are not required since SRAM bitlines are reused for external inputs (e.g., X_0 and $/X_0$) while the weights stored in SRAM cells (e.g., W_0 and $/W_0$) are directly connected to the XNOR gates from the internal SRAM nodes. The multiplied results from the 64 bitcells on the left of Fig. 1, right are accumulated using a pair of inverters (P/N) in each bitcell which drive a positive (VP) and a negative (VN) output voltage node. While the 64 bitcells are used for the in-memory dot-product operation, the other 64 replica bitcells are used for ADC reference (32×) and offset-calibration (32×). Finally, the pseudo-differential accumulator output voltages based on pull-up/down ratio is converted to a digital output at the proposed row-by-row ADC by sweeping reference levels using replica bitcells. For instance, the replica bitcells are swept from -30 to +30 (=ΣR_i, where i=0 to 31) at 5bit mode. Note that ADC resolution is reconfigurable (1-5bit) and it takes 2^N-1 cycles per conversion at N-bit mode. To save power, the voltage-mode drivers are power-gated (EN=0) after each cycle using a comparator output decision detector (DET).

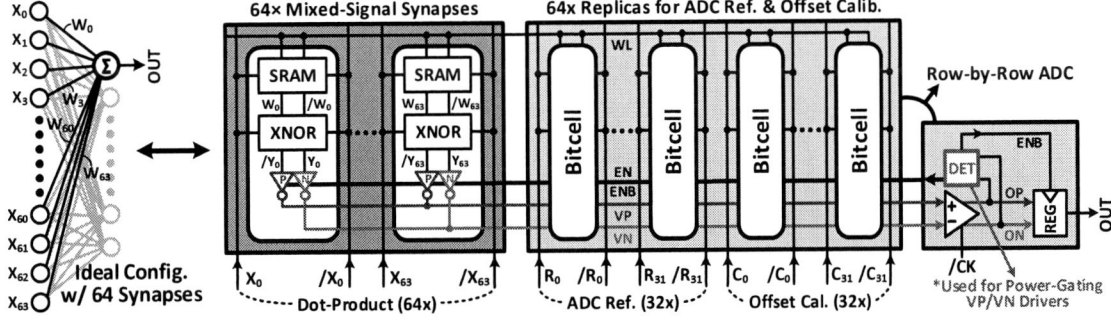

Fig. 1. Block diagram of the proposed mixed-signal in-memory computing unit with 128 bitcells for 64-input dot-product operation.

IEEE Asian Solid-State Circuits Conference
November 4 – 6, 2019
The Parisian Macao, Macao SAR, China

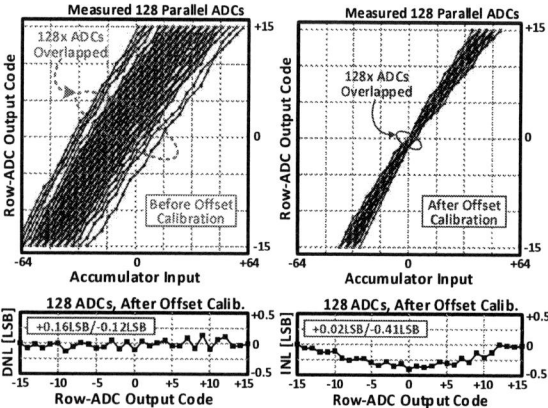

Fig. 3. Measured linearity of the proposed ADC at 5bit mode.

Fig. 2 compares the simulated transfer characteristics of the baseline single-ended [2] and the proposed pseudo-differential voltage-mode accumulator. The baseline accumulator with 64× parallel single-ended drivers (i.e., an inverter as a unit) suffers from nonlinear and asymmetric transfer curves with wide PVT variation. This work proposes a pseudo-differential driver (i.e., a pair of inverters as a unit) which compensates a strength imbalance between PMOS and NMOS and provides symmetric transfer characteristics with the better linearity and the reduced variation. Besides, the proposed voltage-mode accumulator has benefits including 2× output range and the reduced sensitivity to supply noise.

Despite the advantages mentioned above, the residual PVT variation and nonlinearity of the proposed accumulator are still significant issues, especially for achieving multi-bit output resolution. Another critical challenge of the mixed-signal in-memory computing is the overhead of ADC for converting the analog dot-product outputs to the digital code to interface with the external digital domain. To deal with the large footprint of the parallel ADCs, the conventional mixed-signal in-memory computing macros with parallel dot-product units rely on time-multiplexing. As an example, 64-to-1 [2] or 16-to-1 [3] time-multiplexing has been used for prior mixed-signal in-memory computing works, and hence results in the lower (64× or 16×) throughput compared to the ideal case (i.e., an ADC per each dot-product unit). In this work, we present a compact row-by-row ADC with replica bitcells for compensating both residual nonlinearity and process variation. The compact row-by-row

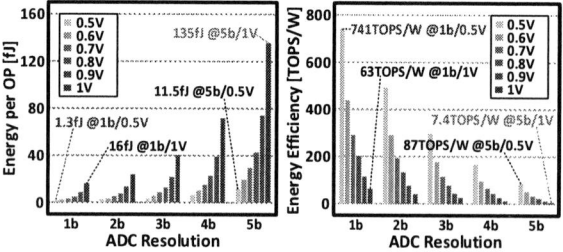

Fig. 4. Measured energy per OP and energy efficiency.

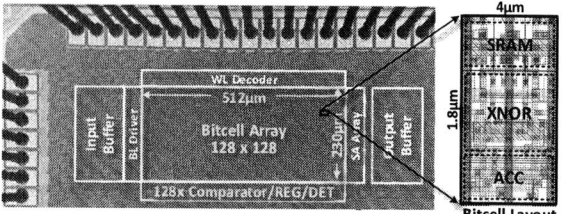

Fig. 5. Die micrograph and bitcell layout.

	[2] VLSI'18	[3] ISSCC'18	[4] ISSCC'18	[5] JSSC'18	This Work
Technology	65nm	65nm LP	65nm	65nm GP	65nm GP
Supply Voltage	0.6-1.0V	1.2V	1.0V	0.55-1.0V	0.5-1.0V
ML Algorithms	MLP/CNN	CNN	SVM	DNNs	MLP/CNN
Accumulator Type	Single-Ended Voltage	Voltage Average	BL Discharging	Digital Logic	Pseudo-Diff. Voltage
Bitcell Array	256 x 64	64 x 256	512 x 256	N/A	128 x 128
Weight/Output Bit	1b/3.46b	1b/7b	4b/1b	1b/1.59b	1b/1-5b
ADC Type	Flash ADC	Charge-Sharing ADC	N/A (No ADC)	N/A (Digital)	Row-by-Row ADC
# ADCs/Neurons	1/64	16/256	N/A	N/A	128/128
Energy Efficiency	139 TOPS/W	28.1 TOPS/W	3.13 TOPS/W	2.3-6 TOPS/W	87 TOPS/W*
Area Efficiency	0.8 TOPS/mm²	N/A	N/A	0.37 TOPS/mm²	3.97 TOPS/mm²

Table I. Performance comparison table.

ADC comprising of a single comparator and 32 replica bitcells is embedded in each row of bitcell array, and the conversion of the analog accumulator output into digital code is sequentially done by sweeping ADC references driving the replica bitcells as shown in Fig. 1. The proposed ADC sequentially generates a thermometer code bit-by-bit, and they are finally converted to a binary code using a simple thermometer-to-binary decoder.

III. EXPERIMENTAL RESULTS AND SUMMARY

A test-chip is fabricated using a 65nm CMOS technology. The 65nm test-chip comprises 128×128 (16Kb) bitcells, and each row works as a 64-input mixed-signal dot-product unit by using 64 synapses and an embedded row-by-row ADC with 64 replica bitcells. For N-bit ADC, $128/(2^N-1)$ parallel outputs are generated from 128 parallel dot-product units since it takes 2^N-1 cycles per conversion. Fig. 3 shows the measured 5bit ADC linearity from 128 dot-product outputs before and after offset calibration. The measured DNL and INL are +0.16/-0.12LSB and +0.02/-0.41LSB after offset calibration. Fig. 4 shows the measured performance of the test-chip. The measured energy efficiencies are 7.4-87TOPS/W and 63-741TOPS/W at 0.5-1V supply with 5 and 1bit ADC, respectively. The 65nm test-chip die micrograph with bitcell layout is shown in Fig. 5 and the proposed 16Kb bitcell array occupies 0.118mm². Comparison with state-of-the-arts is summarized in Table I.

REFERENCES

[1] I. Hubara et al., NIPS, pp. 4107–4115, 2016.
[2] Z. Jiang et al., VLSI Tech, pp. 173-174, Jun. 2018.
[3] A. Biswas et al., ISSCC, pp. 488-489, Feb. 2018.
[4] S. Gonugondla et al., ISSCC, pp. 490-491, Feb. 2018.
[5] K. Ando et al., JSSC, vol. 53, no. 4, Apr. 2018.

978-1-7281-5107-6/19 $31.00 © 2019 IEEE

A 69.3% Efficiency, 6.78-MHz Wireless Power Delivery System with 0X/1X Regulating Rectifier and Reconfigurable Power Amplifier

Jonathan Fuh, Fu-Bin Yang, and Po-Hung Chen, *Member, IEEE*
Institute of Electronics, National Chiao Tung University, Hsinchu, Taiwan
Email: hakko@nctu.edu.tw

Abstract—In this paper, we present a 6.78-MHz high-efficiency wireless power transfer system with a reconfigurable class-D power amplifier and a 0X/1X regulating rectifier. The proposed reconfigurable power amplifier delivers 0.25X/1X output power by adjusting the power-stage configuration to improve system efficiency. To enhance the receiver efficiency, the proposed 0X/1X regulating rectifier with adaptive offset compensation achieves both voltage rectification and voltage regulation in a single conversion stage. The measurement results demonstrate a 90.1% peak receiver efficiency and a 69.3% peak system efficiency. The system provides the load circuit with 5-V regulated output and 250-mW maximum output power.

Keywords—wireless power transfer, regulating active rectifier, resonant coupling, reconfigurable power amplifier

I. INTRODUCTION

In recent years, wireless power transfer (WPT) systems have been developed for use in implantable medical devices (IMDs). WPT systems can be applied in pacemakers, retinal prostheses, and in deep brain stimulators to reduce the frequency of battery replacement. However, the poor efficiency of WPT systems increases the surface temperature of IMDs and damages tissue. Previous works [1-3] have reported the development of rectifier techniques to promote receiver efficiency. However, system performance is significantly degraded when the coupling condition varies. The authors in [4] demonstrated the use of global-loop control, which adjusts the supply voltage of the power amplifier to overcome positioning errors, but an additional dc–dc converter is required. In this work, we propose a wireless power delivery system operating at 6.78 MHz with a reconfigurable power amplifier and an efficient regulating rectifier to obtain a higher power conversion efficiency (PCE). Figure 1 shows the proposed system architecture.

II. PROPOSED WIRELESS POWER TRANSFER SYSTEM

A. 0X/1X Regulating Rectifier

The proposed regulating rectifier consists of a comparator-based active rectifier and a hysteresis comparator, as shown in Fig. 2. The operation mode automatically selects either the 1X mode or 0X mode depending on the output voltage (V_{OUT}) conditions. During the 1X mode, comparators CMP_1 and CMP_2 with adaptive offset compensation are activated to convert the sinusoidal input voltage V_{AC} to a dc level with high efficiency. In the 0X mode, the rectifier adopts a free-wheeling state and

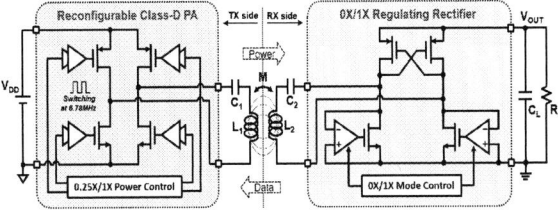

Fig. 1. System architecture of the proposed wireless power delivery system.

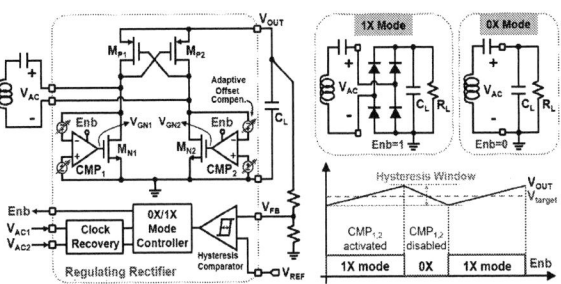

Fig. 2. Proposed 0X/1X regulating rectifier.

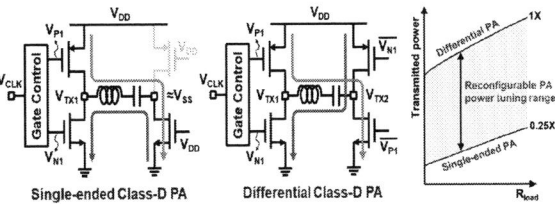

Fig. 3. Proposed reconfigurable class-D power amplifier.

no real power is delivered to the output capacitor. Therefore, the switching-based 0X/1X regulating rectifier with local-loop control regulates the output voltage within the target hysteresis window without using an additional dc–dc converter.

B. Reconfigurable Class-D Power Amplifier

Figure 3 shows the proposed reconfigurable class-D power amplifier, which combines differential and single-ended operations. The available output power of a differential class-D power amplifier is four times greater than that of a single-ended amplifier. This effectively increases the output power range but reduces the PCE in light load conditions, since the receiver frequently operates in the 0X mode. Energy is transferred back and forth between the receiver coil and the capacitor, which degrades the system's efficiency. Therefore, we developed a global-loop power control that uses a reconfigurable power amplifier with 0.25X/1X power selection. With this power control, the transmitter power is reduced in light load conditions to reduce the duration of the 0X mode without using a dc–dc converter.

III. MEASUREMENT RESULTS

The proposed power transmitter and power receiver chips are fabricated using a TSMC 0.25-μm CMOS process. Figure 4 shows the chip micrograph, as well as the developed transmitter coil, detection coil, and receiver coil. Figure 5 shows the measured waveforms of the proposed regulating rectifier. The output voltage is regulated at 5V by automatically selecting the 0X/1X operating mode of the rectifier.

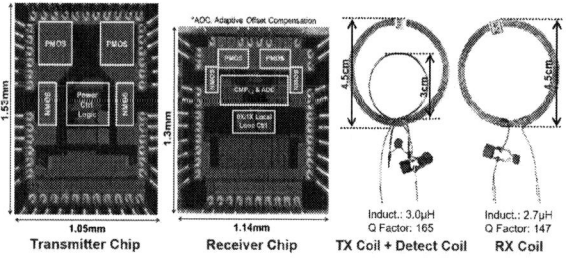

Fig. 4. Chip micrograph and hand-made coils for power and data transmission.

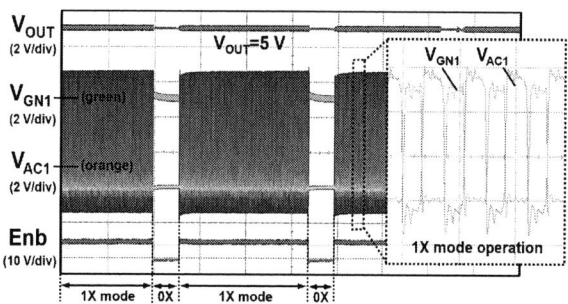

Fig. 5. Measured waveform of local-loop regulation by the proposed 0X/1X regulating rectifier.

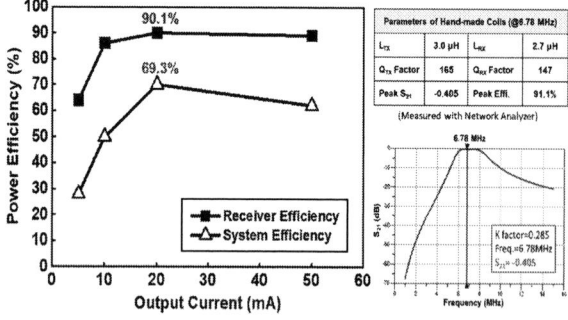

Fig. 6. Measured efficiency under different output current and the performance measurement of hand-made coils.

TABLE I. PERFORMANCE COMPARISON

	JSSC'13 [1]	JSSC'15 [2]	TBCAS'15 [4]	ISSCC'17 [5]	JSSC'17 [3]	This work
Technology	BCD 0.35 μm	CMOS 0.35 μm	CMOS 0.35 μm	CMOS 0.35 μm	CMOS 0.35 μm	CMOS 0.25 μm
Receiver Structure	3R Rectifier	R^3 Rectifier	R^3 Rectifier	Rectifier Charger	3-Mode Rectifier	Regulating Rectifier
Resonant Frequency	6.78 MHz	13.56 MHz	13.56 MHz	6.78 MHz	6.78 MHz	6.78 MHz
Output Voltage	5 V	3.6 V	3.7 V	4.2 V	5 V	5 V
Maximum Output Power	6 W	102 mW	234 mW	1.65 W	6 W	250 mW
Off-chip Components	5 diodes, 3 capacitors	2 capacitors	2 capacitors	1 capacitor	1 capacitor	1 capacitor
Peak Receiver Efficiency	86%	92.6%	92.5%	91.5%	92.2%	90.1%
Peak System Efficiency	55%	50%	62.4%	58.6%	N/A	69.3%

Figure 6 shows the measured efficiencies in different loading conditions and the parameters of hand-made coils. The maximum efficiencies of the power receiver and WPT system are 90.1% and 69.3%, respectively. Table I compares the performance of our proposed system with those of state-of-the-art WPT systems, from which we can see that the proposed architecture achieves the highest peak system efficiency.

IV. CONCLUSION

In this paper, we presented a WPT system with a reconfigurable class-D power amplifier and a 0X/1X regulating rectifier. The proposed regulating rectifier achieves voltage rectification and voltage regulation in one stage, while the proposed reconfigurable power amplifier controls the transmission power level to enhance efficiency in light load conditions. As a result, the proposed WPT system obtains a 69.3% peak system efficiency, 90.1% peak receiver efficiency, and 250-mW maximum output power.

ACKNOWLEDGMENT

This work was supported by the Ministry of Science and Technology, Taiwan, under Grant 108-2636-E-009-009 (Young Scholar Fellowship Program). We thank TSRI for chip fabrication.

REFERENCES

[1] J. H. Choi, et al, "Resonant Regulating Rectifiers (3R) Operating for 6.78 MHz Resonant Wireless Power Transfer (RWPT)," *IEEE J. Solid-State Circuits*, vol. 48, no. 12, pp. 2989–3001, Dec. 2013.

[2] X. Li, et al, "A 13.56 MHz wireless power transfer system with reconfigurable resonant regulating rectifier and wireless power control for implantable medical devices," *IEEE. J. Solid-State Circuits*, vol. 50, no. 4, pp. 978–989, Apr. 2015.

[3] L. Cheng, et al, "A 6.78-MHz single-stage wireless power receiver using a 3-mode reconfigurable resonant regulating rectifier," *IEEE J. Solid-State Circuits*, vol. 52, no. 5, pp. 1412–1423, May 2017.

[4] X. Li, et al, "Reconfigurable resonant regulating rectifier with primary equalization for extended coupling- and loading-range in bio-implant wireless power transfer," *IEEE Trans. Biomed. Circuits Syst.*, vol. 9, no. 6, pp. 875-884, Dec. 2015.

[5] M. Huang, et al, "A Resonant Bidirectional Wireless Power Transceiver with Maximum-Current Charging Mode and 58.6% Battery-to-Battery Efficiency," *ISSCC*, pp. 376-377, Feb. 2017.

4-2 (7189)

An 80mA Capacitor-Less LDO with 6.5µA Quiescent Current and No Frequency Compensation Using Adaptive-Deadzone Ring Amplifier

Bohui Xiao[1], Praveen Kumar Venkatachala, Yang Xu, Ahmed ElShater, Calvin Yoji Lee,
Spencer Leuenberger, Qadeer Ahmad Khan[*], and Un-Ku Moon

School of Electrical Engineering and Computer Science, Oregon State University, Corvallis, OR USA
[*]Department of Electrical Engineering, Indian Institute of Technology Madras, Chennai, TN India
[1]Now with Skyworks Solutions Inc., Hillsboro, OR USA

Abstract—**This paper presents a capacitor-less low dropout (LDO) regulator that requires no frequency compensation, with the use of an adaptive-deadzone ring amplifier. Due to the dynamic behavior of the ring amplifier depending on the input voltage, the proposed LDO features a hybrid (digital or analog) operation and achieves both fast transient response and high output accuracy with a low quiescent current. Moreover, an adaptive deadzone biasing scheme is employed to ensure high stability for a wide range of load current operating conditions. A 1V LDO prototype with 80mA maximum load current is implemented in a 0.18µm CMOS process, consuming only a quiescent current of 6.5µA with a dropout voltage of 90mV. It achieves 172–432ns settling time for load current transitions between 1mA and 81mA. The active area is 0.1mm².**

Keywords—low dropout (LDO) regulator; capacitor-less; frequency compensation; ring amplifier; quiescent current

I. INTRODUCTION

Fully-integrated capacitor-less analog low dropout (LDO) regulators are required to avoid the bulky external capacitor at the output node for cost-efficient on-chip power management. However, the absence of a large off-chip output capacitor results in several design challenges in terms of both loop stability and load transient response. To guarantee the loop stability in the capacitor-less LDO regulator under various load current operating conditions, the dominant pole is moved from the LDO output node to the output of the error amplifier or inside it, and numerous frequency compensation techniques have been proposed [1–3]. Nevertheless, those compensation techniques still need large on-chip capacitors and consume quite some static current. On the other hand, to optimize the overall power efficiency, the quiescent current and the dropout voltage need to be reduced; however, a low quiescent current undesirably slows the transient response of a LDO regulator.

Ring amplification has been demonstrated as a scalable and power-efficient amplifier technique to drive large capacitive loads in a switched-capacitor system, such as pipelined analog-to-digital converters (ADCs) [5]. The slew-based charging and dynamic behaviors of the ring amplifier (RA) lead to a fast transient response with little power consumption compared to conventional operational transconductance amplifiers (OTAs). Those features raise great interest in incorporating a ring amplifier in a continuous-time feedback system that demands fast loop response and low power consumption. So far such a promising application has not been explored yet.

This work was supported by Semiconductor Research Corporation (SRC) TxACE and in part by Center for Design of Analog-Digital Integrated Circuits and Analog Devices.

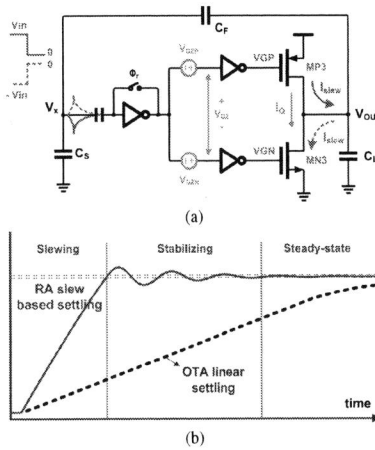

(a)

(b)

Fig. 1. Ring amplifier in a switched-capacitor feedback circuit (a) basic structure, and (b) transient response.

A capacitor-less LDO regulator that exploits an adaptive-deadzone ring amplifier is presented in this paper. The dynamic behavior of the ring amplifier depending on the input voltage (i.e. large-signal or small-signal) enables a hybrid (digital or analog) LDO that achieves both a fast transient response and high output accuracy with a low quiescent current and with no need of frequency compensation for stability. Moreover, an adaptive deadzone biasing scheme is proposed to ensure a stable LDO while operating across a wide range of load currents. A LDO prototype regulating a 1V output voltage with a 90mV dropout voltage is designed. It achieves 172–432ns settling time for load current transitions between 1mA and 81mA with only a 6.5µA quiescent current.

II. OVERVIEW OF RING AMPLIFIER OPERATIONS

The basic structure of a three-stage ring amplifier applied in a switched-capacitor feedback circuit is shown in Fig. 1(a) [4]. The two inverters in the second stage are embedded with offset voltages (i.e. V_{DZP} and V_{DZN}) at their inputs, and the difference between them is defined as the deadzone voltage V_{DZ}. Without the deadzone voltage, the circuit tends to oscillate due to the absence of frequency compensation in the negative feedback loop. However, with an appropriate deadzone voltage, the output stage is biased with a very low quiescent current I_Q. This

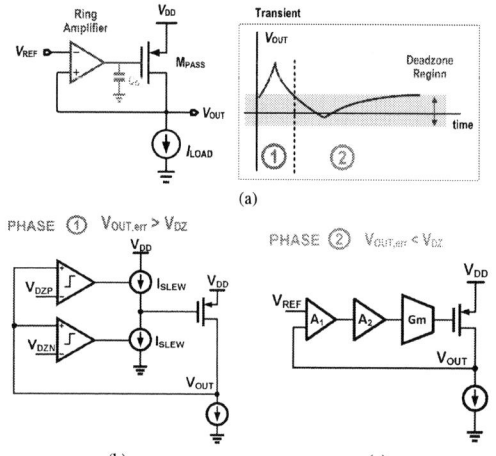

Fig. 3. Implementation of the ring amplifier based LDO: (a) complete schematic, and (b) replica biasing circuit.

Fig. 2. Incorporating a ring amplifier in a capacitor-less LDO: (a) concept diagram, (b) and (c) equivalent diagrams in two operation phases.

results in a dominant pole at the output of the ring amplifier that ensures the small-signal stability.

To illustrate the dynamic behavior of the ring amplifier, Fig. 1(b) shows the transient response with a step input. The major advantage of the ring amplifier over conventional OTAs will be evident from understanding its large-signal and small-signal operations. The entire transient response is divided into three phases: the initial slewing phase, the intermediate stabilizing phase and the final steady-state phase. During the slewing phase when the input voltage of the ring amplifier is quite large, the first two stages of the ring amplifier act as digital inverters, resulting in rail-to-rail gate voltages at the inputs of the third stage. It turns on the third stage which charges/discharges the load capacitor with a large slew current I_{slew}. As the charge across the load capacitor builds up and the output voltage approaches close to the steady-state value, the amplifier enters the stabilizing phase. Then the negative feedback reduces the voltage swing at the virtual ground (i.e. node V_X), and the ring amplifier transitions from the fast large-signal slewing to a low bandwidth small-signal steady-state

operation. Finally, since the major part of the transient settling is complete, the steady-state bandwidth requirement of the amplifier is significantly relaxed, and the output stage is biased with a low quiescent current in contrast to a conventional OTA. In this way, the dominant pole is placed at the output of the ring amplifier, ensuring the small-signal stability. Moreover, due to the multi-stage structure of the ring amplifier, high gain is achieved resulting in high settling accuracy.

In summary, the dynamic current consumption, the initial fast slew-based settling capability and the high gain during the steady state make the ring amplifier highly power efficient in achieving a fast transient response and high accuracy with no need of frequency compensation.

III. PROPOSED CAPACITOR-LESS LDO REGULATOR USING AN ADAPTIVE-DEADZONE RING AMPLIFIER

Fig. 2 illustrates an alternative way to analyze the analog LDO regulator when incorporating a ring amplifier as the error amplifier. The deadzone voltage of the ring amplifier creates a range of input voltage (called as deadzone region) for which the amplifier operates in its steady state. As shown in Fig. 2(b), for input voltages beyond the deadzone region (i.e. phase 1), the ring amplifier operates in an equivalent digital mode, therefore charging the large capacitance at the gate of the pass transistor with high dynamic current. Once the input voltage is within the deadzone region (i.e. phase 2), as shown in Fig. 2(c), the ring amplifier behaves as a high gain and low bandwidth amplifier. Meanwhile, the dominant pole of the entire LDO loop is located at the output of the ring amplifier. This dynamic behavior of the ring amplifier depending on the input voltage with respect to the deadzone region enables a hybrid LDO with both a fast transient response and high output accuracy, unlike the conventional analog LDOs in which two separate loops for high gain and fast transient are used [1–3].

A. Implementation of Proposed LDO

Fig. 3(a) shows the schematic of the proposed capacitor-less LDO regulator, in which a ring amplifier with an adaptive deadzone biasing scheme is employed. The first stage of the ring amplifier (i.e. RA-STG1) is implemented with a simple single-stage differential common-source amplifier, and local common-mode feedback (CMFB) is realized to set the output

978-1-7281-5107-6/19 $31.00 © 2019 IEEE

CM voltage of RA-STG1. Two current-starved inverters are utilized in the second stage of the ring amplifier (i.e. RA-STG2), and the deadzone control voltages V_{DZP} and V_{DZN} are applied through the two current sources biased by a replica biasing circuit. As a result, it makes the operation of this ring amplifier based LDO insensitive to process, voltage and temperature (PVT) variations.

To allow for low dropout voltage, the pass transistor (i.e. MP$_{PASS}$) is designed with PMOS device and sized to ensure a driving capability of 150mA maximum load current under a 1.1V supply voltage. The huge pass transistor results in large parasitic capacitance at the output of the third stage of the ring amplifier (i.e. RA-STG3). In the case of load transition from high to low current, the high gain and large delay through the pass transistor causes overshoot and ringing at the output. Expanding the deadzone region by simply increasing the difference of deadzone control voltages in RA-STG2 could mitigate the output ringing, but at the expense of a slower transient response. In this work, an adaptive deadzone biasing scheme is proposed that dynamically extends the deadzone region based on the LDO load current. In this way, it enables a gradual transition between the two phases of LDO operations while retaining a fast transient response.

B. Replica Biasing for RA-STG2

The replica biasing circuit, as shown in Fig. 3(b), sets the deadzone control voltages in the second stage of the ring amplifier and makes it insensitive to PVT variations [5]. The input voltage of RA-STG2 is set to the output CM voltage of RA-STG1. A diode-connected PMOS (or NMOS) transistor with a size proportional to that in RA-STG3 is biased by a fixed current source. A negative feedback loop formed through a differential amplifier ensures that the output voltage of RA-STG2 tracks this gate voltage of the PMOS transistor by appropriately adjusting V_{DZP} and V_{DZN} in RA-STG2. In this way, during the steady-state operation, a well-defined quiescent current in the third stage of the ring amplifier is maintained. Since this feedback loop is to set the steady-state bias voltages in the second and third stages of the ring amplifier, a low bandwidth single-stage differential-to-single-ended amplifier topology is employed here.

C. Adaptive Deadzone Biasing for RA-STG3

An NMOS transistor MN$_{ADZ}$ acting as a voltage controlled resistor is inserted in series with the bottom NMOS transistor MN$_3$ in the third stage of the ring amplifier. The small-signal parameters of the ring amplifier remain unchanged by ensuring that the turn-on resistance of MN$_{ADZ}$ is relatively small. A load current sensing transistor MP$_{SENSE}$ is mirrored with the pass transistor MP$_{PASS}$, and therefore its current is proportional to the LDO load current. Then the control voltage of MN$_{ADZ}$ (i.e. V_{ADZ}) is generated through a diode-connected NMOS transistor MN$_{SENSE}$ that is biased by the sensing current.

In the case of the load transition from high to low current, the sensing current flowing through MP$_{SENSE}$ is reduced, which makes V_{ADZ} also become smaller accordingly. In this way, MN$_{ADZ}$ dominates the resistance of MN$_3$, therefore it weakens MN$_3$ from pulling down the output of RA-STG3. As a result, it creates an effective damping to eliminate any large-signal ringing. While in the case of the load transition from low to high current, V_{ADZ} is then increased due to a larger sensing current through MP$_{SENSE}$ and MN$_{SENSE}$. The turn-on resistance

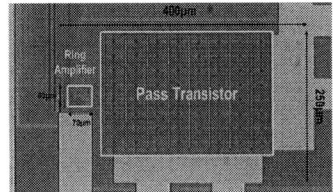

Fig. 4. Die micrograph.

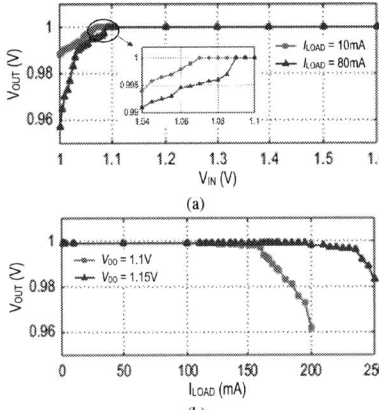

Fig. 5. Measured dc characteristic: (a) line regulation, and (b) load regulation.

of MN$_{ADZ}$ is minimized, so that it does not affect the fast transient response defined by the ring amplifier. Equivalently, this biasing scheme creates an adaptive deadzone range for the ring amplifier while retaining a fast transient response based on the LDO load current operating condition.

An additional PMOS transistor MP$_{ADZ}$ controlled by the LDO output voltage is placed in parallel with MP$_{SENSE}$. It defines the width of the extended deadzone range during the high to low load transition. Trimming of the source voltage V_{TR} or configuration in width of MP$_{ADZ}$ is enabled in this work to allow the option of defining a proper deadzone region for a wide range of regulated output voltages.

The small-signal loop gain of the LDO with load currents of 100μA and 80mA is simulated to illustrate the loop ac stability during the stead state. A dc gain of higher than 90dB is achieved across the light and heavy load current conditions. The unity-gain bandwidth is around 100kHz, and the phase margin is above 80°.

IV. MEASUREMENT RESULTS

A LDO regulator prototype with a 1V output voltage is implemented in a 0.18μm CMOS process. The die micrograph is shown in Fig. 4, and the active area is 0.1mm². As can be seen, the ring amplifier occupies a quite small area of 70μm×60μm, and the total area is dominant by the pass transistor along with wide metal routing.

The measured dc line and load regulation of the capacitor-less LDO is depicted in Fig. 5. The LDO achieves a dropout voltage of 90mV while regulating a 1V output voltage for the supply line voltage up to 1.6V. Thanks to the replica biasing scheme in the ring amplifier, the LDO output voltage is not vulnerable to supply voltage variation. On the other hand,

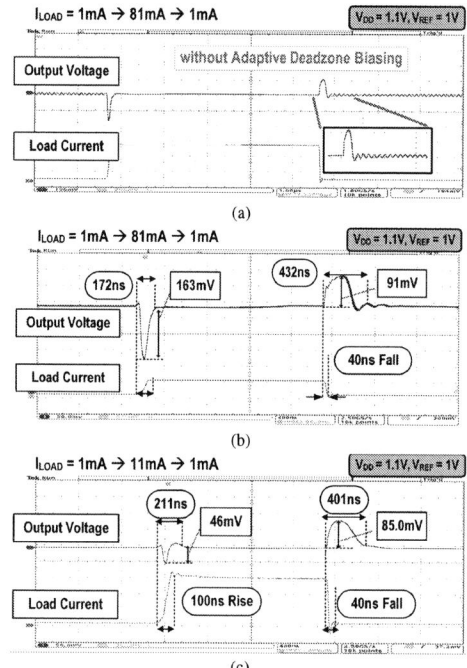

Fig. 6. Measured load transient response: (a) without the adaptive deadzone biasing scheme, (b) and (c) with the proposed biasing enabled for different load current transition sizes.

TABLE I. PERFORMANCE SUMMARY AND COMPARISON WITH STATE-OF-THE-ART

	This work	[1] TCAS'07	[3] TCAS'13	[2] ESSCIRC11	[6] ISSCC'17	[7] JSSC18
Technology	0.18μm	0.35μm	0.11μm	0.35μm	65nm	65nm
Control method	Hybrid, ring amp.	Analog, OTA	Analog, OTA	Analog, OTA	Analog, inv. based	Digital, SR + AA
V_{OUT} (V)	1	2.8	2	1.5	0.55	0.95
Dropout volt. (mV)	90	200	200	142	50	100
Max I_{LOAD} (mA)	80	50	200	100	50	10
Quies. curr. I_Q (μA)	6.5	65	41.5	27	32	3.2
C_{OUT} (pF)	50	100	40	100	40	100
Edge time Δt (ns)	100	1000	500	1000	150	1
ΔI_{OUT} (mA)	10, 80	50	200	100	10	10
ΔV_{OUT} (mV)	46, 163	90	385	25	133.9	105
Settle time (ns)	172–432	~ 15000	650	~ 1000	~ 3000	~ 5000
PSRR@1kHz (dB)	-40	-57	N/A	-60.6	~ -20	N/A
FoM (ps)	0.15, 0.008	0.23	0.016	0.007	1.71	0.34
Active area (mm²)	0.1	0.29	0.21	0.2	0.016	0.034

designs [1–3], [6–7]. This work favorably compares with prior art. It achieves a faster load transient response while dissipating a more than 5× lower quiescent current than [1], [3] and [6]. A figure-of-merit, defined as FoM = $(C_{OUT} \cdot \Delta V_{OUT} \cdot I_Q)/\Delta I_{OUT}^2$, is used to evaluate the LDO performance. This work achieves a competitive FoM of 0.008ps compared to other designs.

V. CONCLUSION

This work presents a capacitor-less LDO regulator using an adaptive-deadzone ring amplifier. Due to the dynamic behavior of the ring amplifier, the LDO prototype achieves a fast load transient response with a low quiescent current and with no frequency compensation. In addition, an adaptive deadzone biasing scheme is employed to enable this ring amplifier based LDO to operate across a wide range of load current operating conditions. It regulates a 1V output voltage with a quiescent current of 6.5μA and a 90mV dropout voltage. The inverter-based structure of the ring amplifier allows compatibility with process scaling for this proposed LDO implementation.

ACKNOWLEDGMENT

The authors would like to thank Asahi Kasei Microdevices for providing chip fabrication and packaging.

REFERENCES

[1] R. Milliken, J. Silva-Martínez, and E. Sánchez-Sinencio, "Full on-chip CMOS low-dropout voltage regulator," *IEEE Trans. Circuits Syst. I: Reg. Papers*, vol. 54, no. 9, pp. 1879–1890, Sep. 2007.

[2] C. Chen, and C. Hung, "A fast self-reacting capacitor-less low-dropout regulator," *IEEE European Solid-State Circuits Conf. (ESSCIRC)*, pp. 375–378, Nov. 2011.

[3] Y. Kim and S. Lee, "A capacitorless LDO regulator with fast feedback technique and low-quiescent current error amplifier," *IEEE Trans. Circuits Syst. II: Exp. Briefs*, vol. 60, no. 6, pp. 326–330, Jun. 2013.

[4] B. Hershberg, S. Weaver, K. Sobue, S. Takeuchi, K. Hamashita and U. Moon, "Ring amplifiers for switched capacitor circuits," *IEEE J. Solid-State Circuits*, vol. 47, no. 12, pp. 2928–2942, Dec. 2012.

[5] P. Venkatachala, S. Leuenberger, A. ElShater, *et al.*, "Process invariant biasing of ring amplifiers using deadzone regulation circuit," *IEEE Int. Symp. Circuits Syst. (ISCAS)*, May 2018.

[6] F. Yang and P. Mok, "A 65nm inverter-based low-dropout regulator with rail-to-rail regulation and over 20dB PSR at 0.2V lowest supply voltage," *IEEE ISSCC Dig. Tech. Papers*, pp. 106–107, Feb. 2017.

[7] M. Huang, Y. Lu, S. U and R. Martins, "An analog-assisted tri-loop digital low-dropout regulator," *IEEE J. Solid-State Circuits*, vol. 53, no. 1, pp. 20–34, Jan. 2018.

operating under a 1.1V supply voltage, the LDO is capable of driving a 150mA maximum load current while the output voltage drops by 2mV off the nominal voltage. When the supply voltage is increased to 1.15V, the maximum load current can reach 200mA with a 2mV output voltage drop.

Fig. 6 shows the measured load transient response for load current steps with 100ns rise and 40ns fall time with and without the proposed adaptive deadzone biasing technique in the ring amplifier. It is noted that a total capacitance of 50pF is estimated at the LDO output during the measurements. The LDO is operating under a 1.1V supply voltage, and dissipates a quiescent current of 6.5μA under no external load condition. As illustrated in Fig. 6(a), without the adaptive deadzone biasing scheme, visible ringing is observed in the LDO output voltage. When this proposed biasing scheme is enabled, as shown in Fig. 6(b), the settling time is 172ns and 432ns for load current transitions from 1mA to 81mA and from 81mA to 1mA, respectively. Meanwhile, a maximum undershoot voltage of 163mV and a maximum overshoot voltage of 91mV are observed, respectively. In addition, under the condition that the load current transition step size is smaller, for instance, from 1mA to 11mA and from 11mA to 1mA, the settling time is 211ns and 401ns, respectively. The maximum undershoot and overshoot voltages become smaller, as shown in Fig. 6(c).

The power supply rejection ratio (PSRR) of the LDO is measured at a load current of 80mA. The PSRR for frequencies below 1kHz is better than -40dB, and degrades to about -19dB at 300kHz.

Table I summarizes the performance of the capacitor-less LDO in this work, and compares it with other state-of-the-art

A DCM ZVS Class-D Power Amplifier for Wireless Power Transfer Applications

Xinyuan Ge[1], Lin Cheng[2] and Wing-Hung Ki[1]

[1]The Hong Kong Univ. of Sci. & Tech., Hong Kong, China; [2]Univ. of Sci. and Tech. of China, Hefei, China
Email: xgeab@connect.ust.hk, eecheng@ustc.edu.cn, eeki@ust.hk

Abstract—A discontinuous conduction mode (DCM) zero-voltage switching (ZVS) scheme for a class-D power amplifier (PA) for wireless power transfer is presented. The sizes of the ZVS inductor and capacitor are reduced by 7.5 and 5 times and ZVS is achieved even subject to PA supply voltage and output current variations. The class-D PA was fabricated with a 0.35 μm high-voltage CMOS process. The ZVS LC resonant tank consists of a 39 nH inductor and a 200 nF capacitor. The switching frequency was 6.78 MHz and the supply voltage was 20 V. Measurement results show that the PA delivered an output power of 7.18 W with a peak efficiency of 87.0%.

Keywords—Class-D, power amplifier, zero-voltage switching (ZVS), wireless power transfer

I. INTRODUCTION

Class-D power amplifiers (PAs) are widely used in wireless power transfer (WPT) systems. However, due to large parasitic capacitance at the switching nodes of the PA, a zero-voltage switching (ZVS) scheme is needed to improve power conversion efficiency (PCE). In [1], additional capacitors are added in parallel across the LC series-resonant tank of the WPT receiver (RX) such that the equivalent loading at its transmitter (TX) side appears to be inductive. Consequently, the PA output current lags behind that of the switching node voltage V_{SW1} and ZVS is achieved. However, once the PA output current changes, the ZVS condition will be violated. In [2], [3], an auxiliary ZVS tank formed by an inductor (L_Z) in series with a capacitor (C_Z) is connected in parallel to the PA switching node to generate a 90° phase lagged ZVS current (I_Z), as shown in Fig. 1(a). This ZVS scheme is robust w.r.t. variations of the supply voltage (V_{DD}) and the output current (I_1) of the PA. For I_Z to work in the continuous conduction mode (CCM) and to lower the peak of I_Z, L_Z and C_Z have to be large and bulky. In this work, a discontinuous-conduction-mode (DCM) ZVS scheme is proposed. Compared to the conventional CCM ZVS scheme, the proposed scheme reduces the sizes of the ZVS inductor and capacitor by better than five times while still maintaining a high PCE.

II. PROPOSED DCM ZVS SCHEME

A. Review of Conventional CCM ZVS Scheme

Fig. 1(b) shows the working principle of the conventional CCM ZVS scheme [2], [3]. When at resonance, I_1 is in phase with V_{SW1} and does not change the charge on C_{OSS} after a deadtime transition. C_Z is chosen to be sufficiently large such that V_Z is constant in the steady state. The waveform of I_Z is triangular with the slopes given by

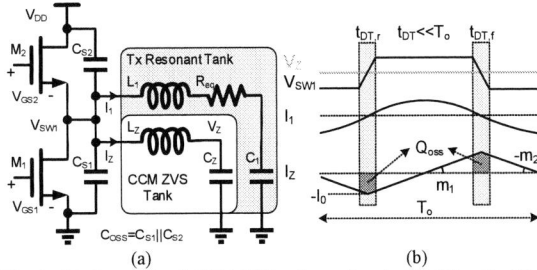

Fig. 1. (a) Conventional CCM ZVS scheme for class-D PA with (b) its working principle.

$$m_1 = (V_{DD} - V_Z)/L_Z, \ -m_2 = -V_Z/L_Z. \quad (1)$$

Assume steady state operation with $I_Z(0) = -I_0$. By volt-second balance we have

$$V_Z = V_{DD}/2, \ m_1 = m_2 = V_{DD}/(2L_Z), \ I_0 = V_{DD}T_o/(8L_Z). \quad (2)$$

During the deadtime duration t_{DT} ($\ll T_o$), I_Z can be approximated as a DC current I_0. For ZVS, the shoot-through current should be 0, and the total charge transferred between C_{OSS} and C_Z within t_{DT} is equal to $C_{OSS}V_{DD}$, which yields

$$L_Z = t_{DT}/(8f_oC_{OSS}). \quad (3)$$

Eq. (2) and Eq. (3) show that L_Z is proportional to t_{DT} and inversely proportional to I_0. In practice, I_0 has to be limited such that t_{DT} cannot be too short, because the power transistors $M_{1,2}$ have parasitic capacitors $C_{GD1,2}$, and their drivers have output resistances. A large slewing of dV_{SW1}/dt may turn on M_1/M_2 wrongly during the rising and falling transitions of V_{SW1} [4]. Besides, I_0 should not exceed the current rating of $M_{1,2}$ to protect them from overcurrent damage. Moreover, a large CCM ZVS current will increase conduction loss in $M_{1,2}$. To lower the peak of I_Z, L_Z is chosen to be large.

B. Proposed DCM ZVS Scheme

In the above CCM ZVS scheme, although I_Z is only needed during t_{DT} for charge transfer, it has to ramp up and down for the entire T_o, and a large L_Z has to be used to lower I_Z. Ideally, we only need to generate a short current pulse within t_{DT} to charge and discharge C_{OSS}, and for the rest of T_o, the current should be zero to reduce conduction loss and switching noise injected to the power rails. Our remedy is the DCM ZVS scheme shown in Fig. 2. Two small bidirectional switches $M_{Z1(2)}$ are added. During t_{DT}, M_{Z1} is ON and M_{Z2} is OFF. If V_{SW1} swings from H to L, C_Z absorbs the charge through L_Z; if V_{SW1} swings from L to H, C_Z returns the charge through L_Z. For the rest of T_o, M_{Z2} is

IEEE Asian Solid-State Circuits Conference
November 4 – 6, 2019
The Parisian Macao, Macao SAR, China

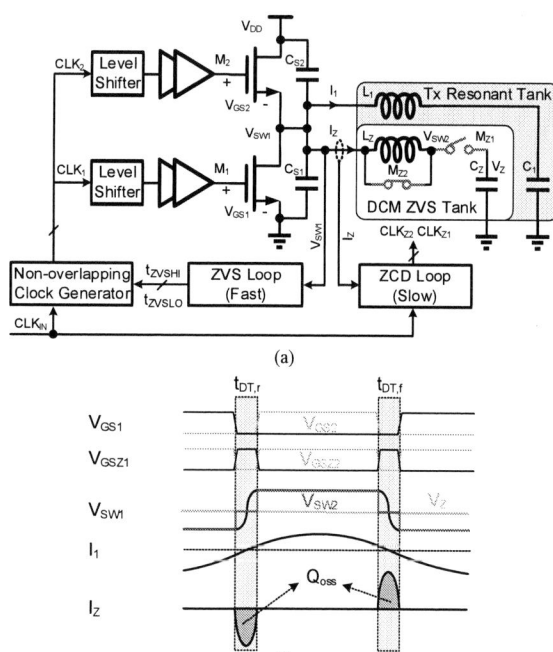

(a)

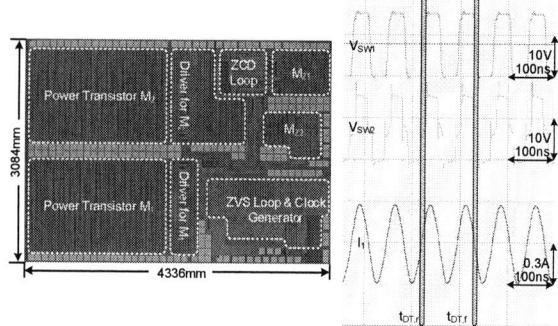

(a)   (b)

Fig. 3. (a) Die photo. (b) Measurement of steady-state operation waveforms.

TABLE II PERFORMANCE COMPARISON AMONG DIFFERENT SCHEMES

	Proposed DCM ZVS[#]	Conventional CCM ZVS[#]	Hard Switching Mode[*]
L_Z (nH)	**39**	300	Not use
C_Z (nF)	**200**	1000	Not use
Peak η_{pwr}	**94.0%**	94.7%	84.3%
Peak η_{tot}	**87.0%**	88.3%	78.9%
$t_{DT,r}/t_{DT,f}$ (ns)	**15/14**	16/17	9/9

[#] Measurement results.
[*] Simulation results.
η_{pwr}: Power conversion efficiency for the power stage.

(b)

Fig. 2. (a) System architecture of the proposed DCM ZVS class-D PA; (b) working principle.

Table I: SUMMARY OF DESIGN PARAMETERS IN DCM/CCM ZVS SCHEMES.

	Proposed DCM ZVS	Conventional CCM ZVS	Scaling factor (DCM/CCM)
L_Z	$\dfrac{t_{DT}^2}{\pi^2 C_{OSS}}$	$\dfrac{t_{DT}}{8 f_o C_{OSS}}$	$\dfrac{8}{\pi^2}\dfrac{t_{DT}}{T_o}$
C_Z	$\dfrac{C_{OSS} V_{DD}}{\Delta V}$	$\dfrac{T_o}{4 t_{DT}}\dfrac{C_{OSS} V_{DD}}{\Delta V}$	$\dfrac{4 t_{DT}}{T_o}$

ΔV is acceptable voltage ripple on C_Z.

ON and the residual charge freewheels through L_Z, while M_{Z1} is OFF and I_Z is then 0. The class-D PA uses dual-loop control to enhance system robustness. In the fast ZVS loop, the voltage difference between V_{SW1} and V_{DD} (V_{SS}) is sampled and held at the end of the V_{SW1} rising (falling) transition and is further used to adjust the turn-on time of M_2 (M_1) to ensure ZVS. The slow zero current detection (ZCD) loop senses I_Z to regulate the turn-off (turn-on) time of M_{Z1} (M_{Z2}) such that L_Z is shorted when $I_Z=0$. Suppose the deadtime transition starts at t=0, I_Z and V_{SW1} are determined by

$$\begin{cases} L_Z \cdot dI_Z(t)/dt = V_{SW1}(t) - V_Z \\ V_{SW1}(t) = V_{SW1}(0) - 1/C_{OSS} \cdot \int_0^t I_Z(\tau) d\tau \end{cases} . \quad (4)$$

In accounting for boundary conditions, it can be shown that

$$I_Z(t) = \begin{cases} -V_Z/(L_Z \omega_z) \cdot \sin(\omega_z t) & \text{rising} \\ (V_{DD} - V_Z)/(L_Z \omega_z) \cdot \sin(\omega_z t) & \text{falling} \end{cases}, \quad \omega_z t_{DT} = \pi. \quad (5)$$

where $\omega_z = 1/\sqrt{L_Z C_{OSS}}$ ($\geq \omega_o$). In the steady state, $V_Z = V_{DD}/2$ and $V_{SW1}(t_{DT}) = V_{DD}$ (0) at its rising (falling) transition, which indicate that ZVS is completed. The DCM ZVS scheme needs a much smaller L_Z as given by

$$L_{Z,DCM} = t_{DT}^2/(\pi^2 C_{OSS}) = (8 t_{DT})/(\pi^2 T_o) \cdot L_{Z,CCM}. \quad (6)$$

Moreover, since I_Z is in DCM, the amount of charge transferred between C_{OSS} and C_Z during T_o is much smaller than I_Z is in CCM, and a much smaller C_Z can be used. Comparisons of design parameters between the proposed DCM ZVS scheme and the conventional CCM ZVS scheme are listed in Table I.

III. MEASUREMENT RESULTS

The proposed class-D PA with DCM ZVS was fabricated with a 0.35 µm high-voltage CMOS process and the die photo is shown in Fig. 3(a). The switching frequency is 6.78 MHz. A 39 nH inductor (L_Z) together with a 200 nF capacitor (C_Z) forms the DCM ZVS resonant tank. Fig. 3(b) shows measured steady-state waveforms of V_{SW1}, V_{SW2} and I_1 when V_{DD} was 15 V. V_{SW1} shows smooth transitions both at its rising and falling edges, which demonstrates the effectiveness of the scheme. The DCM ZVS class-D PA achieves a peak efficiency of 87.0% when delivering 7.18 W output power at 20 V supply voltage. Table II summarizes and compares the performance of this work with the conventional CCM ZVS scheme and also the hard-switching scheme. The proposed DCM ZVS scheme for class-D PA reduces sizes of ZVS inductor and capacitor by 7.5 and 5 times while still maintains a high PCE.

REFERENCES

[1] F. Mao, et al., "A reconfigurable cross-connected wireless-power transceiver for bidirectional device-to-device charging with 78.1% total efficiency," *IEEE Int. Solid-State Circ. Conf.*, pp. 140–142, 2018.

[2] M. A. de Rooij, "The ZVS voltage-mode class-D amplifier, an eGaN® FET-enabled topology for highly resonant wireless energy transfer," *APEC*, pp. 1608–1613, 2015.

[3] H. Tebianian, et al., "A 13.56 MHz full-bridge class-D ZVS inverter with dynamic dead time control for wireless power transfer systems," *IEEE Trans. Ind. Electron.*, pp. 1–1, 2019.

[4] M. Berkhout, "A class D output stage with zero dead time," *IEEE Int. Solid-State Circ. Conf.*, pp. 134-135, 2003.

978-1-7281-5107-6/19 $31.00 © 2019 IEEE

4-4 (7012)

Time-Based Digital LDO Regualtor with Fractionally Controlled Power Transistor Strength and Fast Transient Response

Jin-Gyu Kang, Min-Gyu Jeong, and Jeongpyo Park, and Changsik Yoo
Department of Electronic Engineering
Hanyang University, Seoul, Korea
Email: csyoo@hanyang.ac.kr

Abstract—The number of enabled power transistors and their ON-time of a digital low-dropout (LDO) regulator are controlled by a time-based controller. By the time-based controller consisting of two voltage-controlled oscillators (VCO) and a phase-detecting switch driver, the effective number of enabled power transistors and therefore their strength can be fractionally controlled. The transient response is greatly improved by a transient detector which selectively disables one of the two VCOs during a transient condition. At steady state, the number of switching power transistor is either zero or one, minimizing the ripple of the output voltage. Implemented in a 65 nm CMOS technology, the time-based digital LDO regulator can provide 0.5~1.1 V output from 0.9~1.2 V input. For the maximum load current of 19 mA, the peak current efficiency is 99.3 %. For both the step-up and -down changes of the load current by 3 mA, the voltage level of the output is stabilized in less than 90 ns.

Keywords—Low-dropout regulator (LDO), digital LDO, time-based control, voltage-controlled oscillator (VCO), CMOS.

I. INTRODUCTION

Digital LDO regulator is gaining popularity over analog one for its excellent process scalability and ease of frequency compensation [1]-[5]. The transient response of a conventional digital LDO regulator, however, is limited by its clock frequency because the output voltage level is compared with a reference level and corrected only once per clock cycle. Higher clock frequency and/or multi-bit quantization of the output by an analog-to-digital converter (ADC) can improve the transient response but at the cost of poor power efficiency. To break the trade-off between the transient response and the power efficiency, a complex event-driven control scheme can be used for an ADC-based digital LDO regulator [2]-[3]. In [4], fast transient response and good power efficiency are simultaneously achieved by controlling the clock frequency adaptively with a beat frequency quantizer. Although promising, a complex digital controller is required.

This paper describes a time-based digital LDO regulator which offers (1) fractionally controlled strength of power transistor, (2) fast transient response, and (3) small output voltage ripple. Only two VCOs and a phase detecting switch driver (PDSD) are required to perform the time-based control of the digital LDO regulator and there is no complex digital control logic.

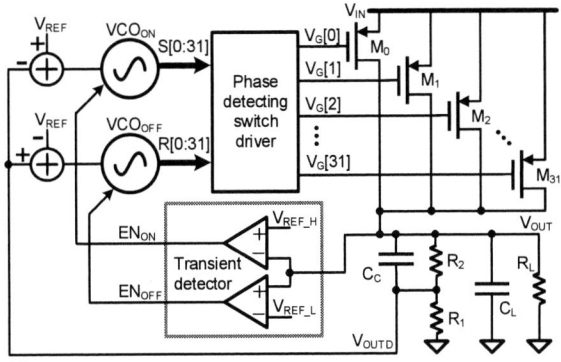

Fig. 1. Time-based digital LDO regulator.

II. TIME-BASED DIGITAL LDO REGULATOR

A. Architecture

The proposed time-based digital LDO regulator shown in Fig. 1(a) is controlled by VCO_{ON} and VCO_{OFF} without any complex digital control logic. During the normal operation, the enable signals EN_{ON} and EN_{OFF} of the VCOs are all "1", enabling the oscillation of the two VCOs. The oscillation frequencies of VCO_{ON} and VCO_{OFF} are proportional to $V_{REF}-V_{OUTD}$ and $V_{OUTD}-V_{REF}$, respectively where V_{OUTD} is the scaled output voltage level and V_{REF} is the reference level.

The PDSD performs the phase detection between the multiphase outputs S[0:31] of VCO_{ON} and R[0:31] of VCO_{OFF} and generates the gate driving signals $V_G[0:31]$ for the pMOS power transistors $M_0\sim M_{31}$ whose pulse widths are proportional to the detected phase difference $\Phi_{ERR}=\Phi_{VCO\_ON}-\Phi_{VCO\_OFF}$ between VCO_{ON} and VCO_{OFF}. The PDSD pulls down the gate driving signal $V_G[k]$ at the rising edge of S[k] to turn on the power transistor M_k and pulls it up at the rising edge of R[k] to turn off M_k. Therefore, the ON-time of the power transistors is proportional to the phase difference Φ_{ERR} which is a function of the output V_{OUT} and the load current I_{OUT}.

978-1-7281-5107-6/19 $31.00 © 2019 IEEE   45

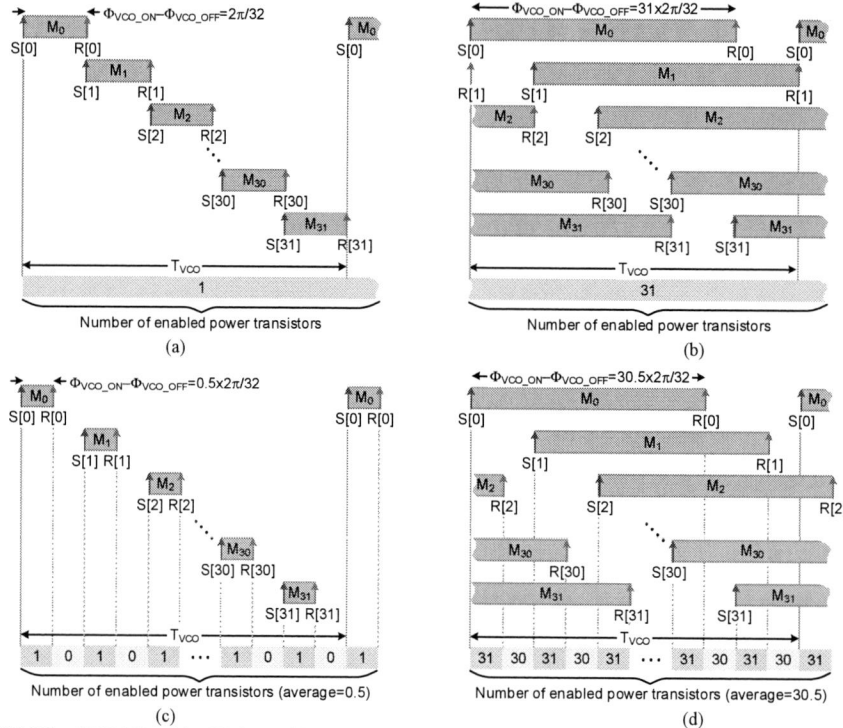

Fig. 2. Rising edges of S[0:31] and R[0:31] and the ON-times of the power transistors shown as shaded rectangles with transistor names when the phase error Φ_{ERR} is (a) $2\pi/32$, (b) $31\times2\pi/32$, (c) $0.5\times2\pi/32$, and (d) $30.5\times2\pi$.

When the voltage level of the output moves outside the window defined by V_{REF_H} and V_{REF_L}, the transient detector enables the transient mode as will be explained later.

The capacitor C_C (=0.5 pF) generates a zero to ensure the stability of the feedback loop which has two poles, one generated by the load capacitor C_L and the other one by the VCOs.

B. Number of enabled power transistors and their ON-times

Fig. 2 illustrates how the rising edges of S[0:31] and R[0:31] determine the number of enabled power transistors and their ON-times (shown as shaded rectangles with transistor names) for different values of the phase difference Φ_{ERR}. At steady state, the two VCOs VCO$_{ON}$ and VCO$_{OFF}$ have constant phase difference Φ_{ERR} and the same period (T_{VCO}) of course. The multi-phase outputs S[0:31] and R[0:31] have constant phase spacing of $2\pi/32$.

When $\Phi_{ERR}=N\times2\pi/32$ (N is an integer), N power transistors among $M_0\sim M_{31}$ are enabled sequentially with the ON-time equivalent to the phase of $\Phi_{ERR}=N\times2\pi/32$. When $\Phi_{ERR}=2\pi/32$, for example, one power transistor among $M_0\sim M_{31}$ is enabled one by one for $T_{VCO}/32$ (equivalent to the phase of $2\pi/32$) as shown in Fig. 2(a), which is equivalent to having one power transistor enabled at all times. When $\Phi_{ERR}=31\times2\pi/32$, 31 power transistors are enabled at all times while each power transistor is enabled with the ON-time equivalent to the phase of

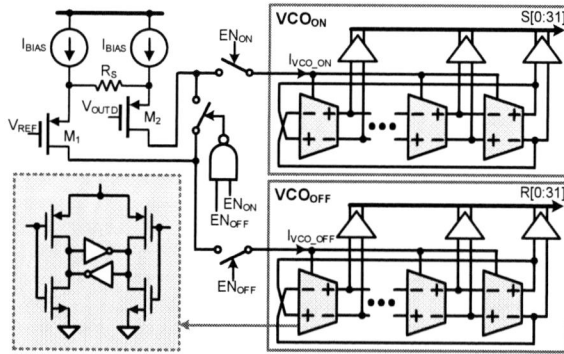

Fig. 3. Voltage controlled oscillators; VCO$_{ON}$ and VCO$_{OFF}$.

$31\times2\pi/32$ as shown in Fig. 2(b). When $\Phi_{ERR}=(N+\alpha)\times2\pi/32$ where $0<\alpha<1$, the number of enabled power transistors toggles between N and ($N+1$), making the effective number of enabled power transistors equal to ($N+\alpha$). When $\Phi_{ERR}=0.5\times2\pi/32$, for example, one or zero power transistor is enabled with the ON-time equivalent to the phase of $0.5\times2\pi/32$ as shown in Fig. 2(c), which is equivalent to having 0.5 power transistor enabled at all times. When $\Phi_{ERR}=30.5\times2\pi/32$, 30 or 31 power transistors are enabled with the ON-time equivalent to the phase of $30.5\times2\pi/32$ as shown in Fig. 2(d). This is equivalent to having 30.5 power transistor enabled at all times.

978-1-7281-5107-6/19 $31.00 © 2019 IEEE

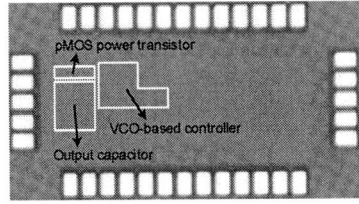

Fig. 4. Chip microphotograph.

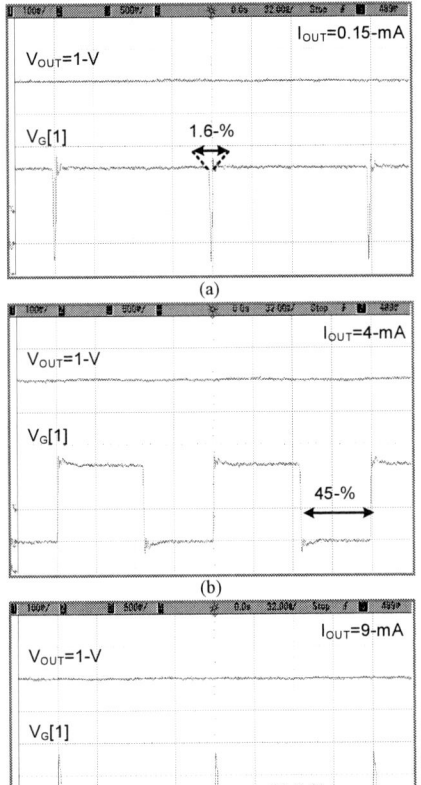

Fig. 5. Duty cycles of the gate driving signal $V_G[1]$ when I_{OUT} is (a) 0.15 mA, (b) 4 mA, and (c) 9 mA.

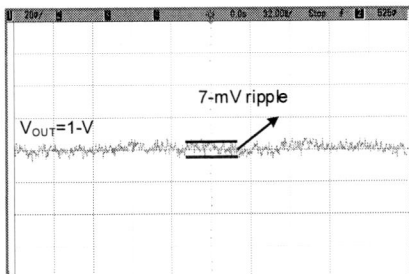

Fig. 6. Output voltage ripple.

In summary, both the number of enabled power transistors and their ON-time are controlled by the VCOs. This allows the time-based digital LDO regulator to have fractionally controlled strength of power transistor. The ripple of the output voltage is small because the number of switching power transistors is either zero or one at steady state.

C. Voltage controlled oscillator

As shown in Fig. 3, the VCOs VCO_{ON} and VCO_{OFF} are differential ring oscillators with 16 stages to generate 32 phase outputs $S[0:31]$ and $R[0:31]$, respectively. The bias currents I_{VCO_ON} and I_{VCO_OFF} of VCO_{ON} and VCO_{OFF} and therefore their oscillation frequencies are proportional to $V_{REF}-V_{OUTD}$ and $V_{OUTD}-V_{REF}$, respectively. During the transient mode, one of the two VCOs is disabled by the enable signals EN_{ON} and EN_{OFF} to improve the transient speed as explained below.

D. Transient detection

When V_{OUT} moves outside the window defined by V_{REF_H} and V_{REF_L} of the transient detector, either EN_{ON} or EN_{OFF} becomes "0" and the digital LDO regulator enters the transient mode.

If $V_{OUT}>V_{REF\_H}$, VCO_{ON} is disabled by EN_{ON}="0" and only the rising edges of $R[0:31]$ are generated to turn off power transistors one by one at every $T_{VCO\_OFF}/32$ where T_{VCO_OFF} is the period of VCO_{OFF}. Because power transistors are turned off at every rising edge of $R[0:31]$, the voltage level of the output V_{OUT} decreases quickly to a desired level.

If $V_{OUT} < V_{REF\_L}$, VCO_{OFF} is disabled by EN_{OFF}="0" and only the rising edges of $S[0:31]$ are generated to turn on power transistors one by one at every $T_{VCO\_ON}/32$ where T_{VCO_ON} is the period of VCO_{ON}. Because power transistors are turned on at every rising edge of $S[0:31]$, the voltage level of the output V_{OUT} increases quickly to a desired level.

During the transient mode, the oscillation frequency of an enabled VCO is much higher than that during the normal operation because all the bias current is steered to it as shown in Fig. 3, which further improves the transient response.

III. Experimental Results

The time-based digital LDO regulator has been implemented in a 65 nm CMOS technology and occupies 0.059 mm² active silicon area including the 200 pF output capacitor as shown in Fig. 4. With the minimum dropout voltage of 0.1 V, the output V_{OUT} can be regulated from 0.5 V to 1.1 V with the input V_{IN} ranging from 0.9 V to 1.2 V. The line and load regulations are measured to be better than 2.63 mV/V and 0.15 mV/mA, respectively. The peak current efficiency is 99.3 % when the load current I_{OUT} is 19 mA.

Fig. 5 is the measured waveforms of V_{OUT} and the gate driving signal $V_G[1]$ of the power transistor M_1 when I_{OUT} is 0.15 mA, 4 mA, and 9 mA for V_{IN}=1.2 V and V_{OUT}=1.0 V. It can be seen the ON-time (duty cycle) of power transistor becomes larger to supply more current for larger load current. The output voltage ripple is smaller than 7 mV at steady state as shown in Fig. 6.

978-1-7281-5107-6/19 $31.00 © 2019 IEEE   47

IEEE Asian Solid-State Circuits Conference
November 4 – 6, 2019
The Parisian Macao, Macao SAR, China

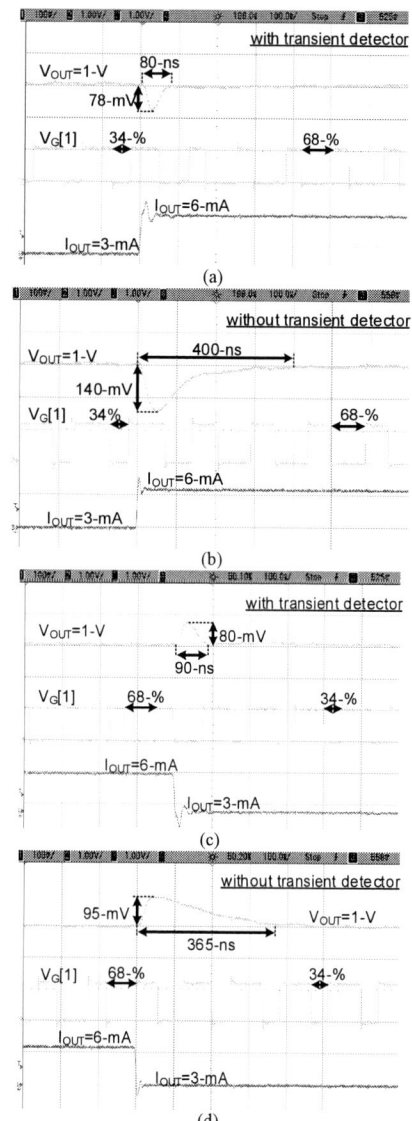

(a)

(b)

(c)

(d)

Fig. 7. Load transient responses; (a) step-up with the transient detector, (b) step-up without the transient detector, (c) step-down with the transient detector, and (b) step-down without the transient detector.

Fig. 7 shows the load transient response with and without the transient detector. When the transient detector is enabled, V_{OUT} recovers its desired level in less than 90 ns with the peak ripple smaller than 80 mV. Without the transient detector, it takes more than 365 ns for V_{OUT} to recover its desired level and the peak ripple is 140 mV. The performance of the digital LDO regulator is summarized and compared with other works in Table I. For the fair performance comparison, the figure of merit (FoM) is defined as;

$$FoM = C_{OUT} \times \frac{\Delta V_{OUT}}{I_{OUT}} \times \frac{I_Q}{I_{OUT}} \qquad (1)$$

TABLE I
PERFORMANCE SUMMARY AND COMPARISON

	[2]	[3]	[4]	[5]	This work
CMOS tech. [nm]	65	65	65	130	**65**
Active area [mm²]	0.03	0.029	0.0374	0.114	**0.059**
V_{IN} [V]	0.45~1	0.15~1	0.6~1.2	0.5~1.2	**0.9~1.2**
V_{OUT} [V]	0.4~0.95	0.45~0.95	0.4~1.1	0.45~1.14	**0.5~1.1**
Load current I_{OUT} [mA]	0.014 ~3.36	0.0072 ~3.51	<100	0.1~4.6	**0.15~19**
C_{OUT} [pF]	100	400	40	1,000	**200**
Quiescent current (I_Q) [µA]	8.1~258	216	100~ 1,070	221	**131**
Peak current efficiency (E_{PEAK}) [%]	99.2	96.3	99.5	98.3	**99.3**
$V_{OUT,RIPPLE}$ [mV]	<1.5	N.A.	<25*	<20*	**<7**
Line regulation [mV/V]	N.A.	N.A.	N.A.	N.A.	**2.63**
Load regulation [mV/mA]	N.A.	N.A.	N.A.	<10	**0.15**
Load transient response — ΔI_{OUT} [mA]	1.4.	0.4	50.	1.4.	**3**
ΔV_{OUT} [mV]	24.8	40	108	90	**80**
T_{SETTLE} [ns]	11,200	80	1,240	1,100	**90**
FoM [ps]	20	1,250	36	1,960	**5.8**

(* estimated value)

where C_{OUT} is the capacitance at the output node, ΔV_{OUT} is the output voltage ripple during the load transient, I_{OUT} is the load current, and I_Q is the quiescent current consumption. This work shows the smallest (best) value of the FoM.

IV. CONCLUSION

The controller of a digital LDO regulator is realized by only time-based circuits such as VCOs and PDSD which enables the strength of power transistor to be fractionally controlled. The transient response is greatly improved by the transient detector. Implemented in a 65 nm CMOS technology, the time-based digital LDO regulator shows the best FoM.

ACKNOWLEDGMENT

This work was supported by the Samsung Research Funding & Incubation Center of Samsung Electronics under Project Number SRFC-IT1501-52. The CAD tools were provided by IDEC.

REFERENCES

[1] Y. Okuma, et al., "0.5-V input digital LDO with 98.7% current efficiency and 2.7-µA quiescent current in 65nm CMOS," *Proc. IEEE Custom Integrated Circuits Conf.* 2010.

[2] D. Kim, et al., "A 0.5V-V_{IN} 1.44mA class event-driven digital LDO with a fully integrated 100pF output capacitor," *Dig. Tech. Papers, IEEE Int. Solid-State Circuits Conf.* pp. 346-348, 2017.

[3] D. Kim, et al., "Fully integrated low-drop-out regulator based on event-driven PI control," *Dig. Tech. Papers, IEEE Int. Solid-State Circuits Conf.* pp. 146-147, 2016.

[4] S. Kundu, et al., "A fully integrated 40pF output capacitor beat-frequency-quantizer-based digital LDO with built-in adaptive sampling and active voltage positioning," *Dig. Tech. Papers, IEEE Int. Solid-State Circuits Conf.* pp. 308-309, 2018.

[5] S. B. Nasir, et al., "A 0.13µm fully digital low-dropout regulator with adaptive control and reduced dynamic stability for ultra-wide dynamic range," *Dig. Tech. Papers, IEEE Int. Solid-State Circuits Conf.* pp. 98-99, 2015.

A Single-Inductor Triple-Output Converter with an Automatic Detection of DC or AC Energy Harvesting Source for Supplying 93% Efficiency and 0.05mV/mA Cross Regulation to Wearable Electronics

T. Nagateja[1], Shao-Qi Chen[1], Li-Cheng Chu[1], Ke-Horng Chen[1], Ying-Hsi Lin[2], Shian-Ru Lin[2], Tsung-Yen Tsai[2]

[1]National Chiao Tung University, [2]Realtek Semiconductor Corp., Hsinchu, Taiwan

Abstract—the proposed single-inductor triple-output (SITO) converter can accept both DC battery voltage and AC energy harvesting voltage to provide a triple step up and down outputs. Storing the energy to the H-bridge capacitor first and pumping it later can ensure a continuous inductor current for reducing the output voltage ripple by 67%. Due to the H-bridge, the inductor current value is reduced to 1/5X to achieve high efficiency of 93% under the current loading of 120mA. Moreover, the proposed PWM and the adaptive on-time hybrid controller uses different modulation methods in a single controller to meet the requirements of different multiple outputs.

Keyword: single-inductor triple-output (SITO) converter, pulse width modulation (PWM), adaptive on-time (AOT).

I. INTRODUCTION

For wearable electronics, the input energy source can be either battery or energy harvesting AC or DC sources. Conventional designs are not efficient because AC and DC sources need two-stage structure including one full bridge and one DC-DC converter as shown in Fig. 1. In this paper, to accept AC or DC input energy source, a capacitor C_H replaces the load of the full bridge to store energy in Fig. 2. The proposed AC/DC detector can decide whether the input energy source is AC or DC. Fig. 3 lists the operation modes for AC or DC source. Since the voltage across the C_H is charged by an AC or DC source, the SITO in Fig. 2 can have sufficiently high input voltage to generate three step down or up outputs. The inductor current value can be kept lower by 5X to achieve high efficiency compared to conventional single-inductor multiple-output (SIMO) converters [3]. Moreover, to meet the requirements of multiple outputs, the SITO controller contains the pulse width modulation (PWM) and adaptive on-time (AOT) hybrid (PAH) modulator.

The paper is organized as follows. Section II describes the ARC technique. Circuit implementation is illustrated in Section III. Experimental results are presented in Section IV. Finally, conclusions are made in Section V.

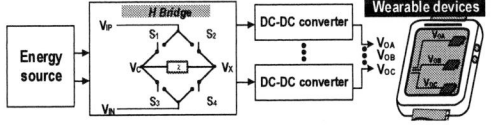

Fig. 1. Conventional two-stage power management for wearable devices.

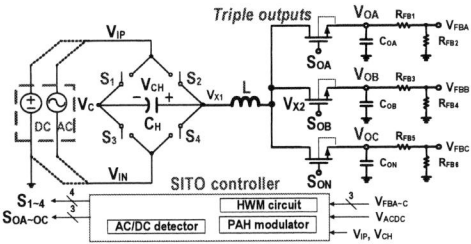

Fig. 2. The proposed SITO converter accepts AC and DC voltage sources.

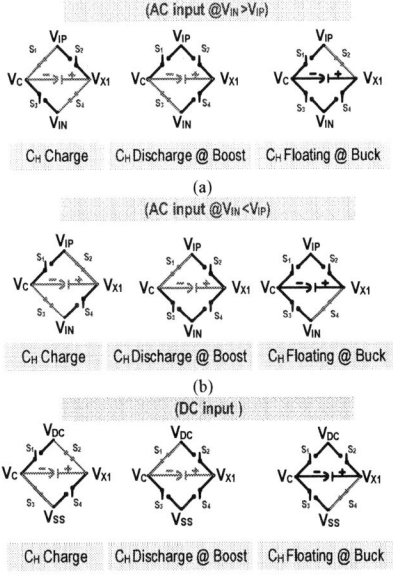

Fig. 3. On/off status in different operation modes. (a) and (b) under AC input. (c) Under DC input.

II. PROPOSED PAH MODULATOR

The PAH operation waveform of the inductor current is shown in Fig. 4. Since V_{OC} has a continuous loading request in Fig. 4(a), the PAH modulator uses the PWM to decide the duty of V_{OC} by comparing the V_{REF} and V_{FBC} for getting small load regulation of 10mV/100mA. On the other hand, the other two outputs, V_{OB} and V_{OA}, are modulated by the AOT due to light load conditions in normal mode. Once one or both outputs require energy, the AOT operation adds one or two adaptive on-times on the error

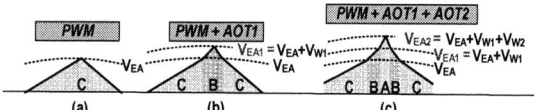

Fig. 4. Waveform of the inductor current in the PAH operation. (a) PWM only. (b) PWM and one AOT. (c) PWM and two AOT modes.

amplifier output V_{EA} to form the inductor current waveforms in Fig. 4(b) and (c), respectively. The AOT can achieve a fast transient response of 3μs in case of 100mA load step and improved light load efficiency of 93%. In Fig. 5, the SITO controller uses the modulation reference generator (MRG) to stack two AOT inductor currents (for V_{OB} and V_{OA}) on the PWM inductor current (for V_{OC}) by the addition of V_{W1} and V_{W2} to the V_{EA} to form V_{EA1} and V_{EA2}, respectively. V_{EA1} and V_{EA2} can decide the duties of V_{OB} and V_{OA}, separately. Compared to the single control technique in [1]-[3], the PAH modulator provides PWM and AOT techniques to meet the different output requirements in wearable electronics.

III. CIRCUIT IMPLEMENTATION

A. HWM Circuit

In Fig. 5, V_A and V_B indicate V_{OA} and V_{OB}, respectively, by comparing the feedback voltages V_{FBA} and V_{FBB} with the reference V_{REF}. In the HWM of Fig. 6(a), V_A and V_B will be sampled by the CLK signal and transform into the 2-bit M[1:0]. Each bit of M[1:0] will be separately sent to an individual mode counter. If the digital signal remains high for four consecutive periods, the energy distributed to the output is insufficient. Thus, the on-time will be increased by the hysteresis controller to enhance light-load efficiency and reduce switching loss. Simultaneously, the hysteresis controller increases the transient speed by modulating the on-time rapidly through the transient detector. In the case of light-to-heavy load change at V_{OA}, the V_{W2} will be set to a higher level to provide more power. On the contrary, if I_{OA} steps from heavy to light, the V_{W2} will be decreased to reduce energy transmission. The timing diagram in Fig. 6(b) shows the sampling signal V_{sample} detects the energy mode signals V_{AM} and V_{BM} every 4 switching periods. When the V_{AM} or V_{BM} changes from low to high, V_{W1} or V_{W2} will increase from V_{R1} to V_{R2}, or V_{R3} to get a larger on-time. In the case of a heavy-to-light load change, the V_{W1} will rapidly decrease through the reset of a 2-bit counter by the signal V_{HL} to reduce the on-time immediately. The MRG circuit in Fig. 7(a) decides the V_{mod} by the signal M[1:0] to control the on-time of the H bridge. The comparison of the current sensing signal V_{CS} with V_{EA1}, V_{EA2}, and V_{mod} can generate the control signals for three output switches V_{SWA}~ V_{SWC}, and the control signals V_{S1}~V_{S4} for the H bridge structure. The timing diagram in Fig. 7(b) shows the on-time T_{ON} in case of increasing inductor current under four different conditions.

B. AC/DC Detector

To determine the input voltage is AC or DC signal, the detection of the input voltage is necessary. Fig. 8(a) shows the structure of the AC/DC detection circuit. The frequency detector

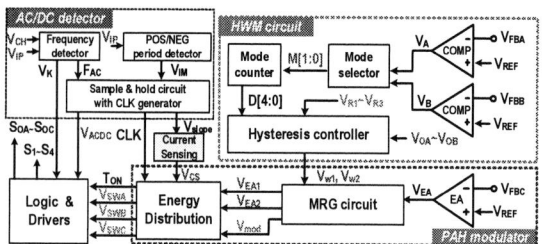

Fig. 5. Proposed SITO controller.

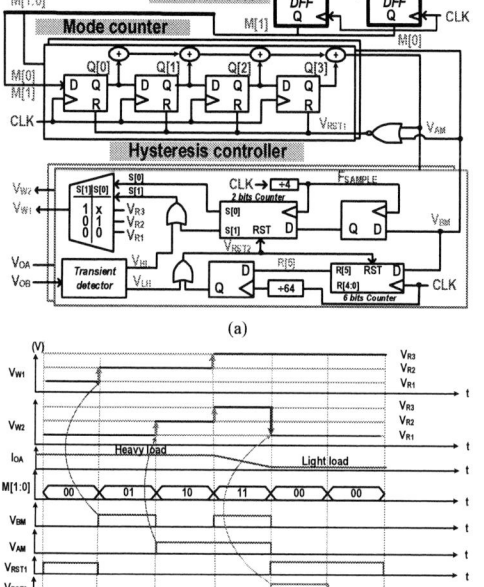

Fig. 6. (a) HWM circuit. (b) Timing diagram.

detects the peak voltage of the V_{IP} and generates the clock signal F_{AC} for AC input conversion. Moreover, the valley point of V_{CH} will also be detected to generate signal V_K. The two signals F_{AC} and V_K determine the on-time for AC input conversion. Moreover, since the operation for AC input signal has two conditions: positive period ($V_{IP} > V_{IN}$) and negative period ($V_{IP} < V_{IN}$). The positive and negative period detector determines the condition by comparing the V_{IP} signal with zero voltage to generate the digital control signal V_{IM}. If the V_{IM} is high, the condition is in a positive period. Contrarily, the condition is in a negative period when V_{IM} is low. The two aforementioned blocks decide the frequency of the input voltage and differentiate the two conditions for AC input conversion. By the combination of the signals F_{AC} and V_{IM}, the signal V_{ACDC} determines the type of the input signal.

The timing diagram in Fig. 8(b) shows if the input source switches from AC to DC, the F_{AC} retains low and fails to reset

978-1-7281-5107-6/19 $31.00 © 2019 IEEE   50

IEEE Asian Solid-State Circuits Conference
November 4 – 6, 2019
The Parisian Macao, Macao SAR, China

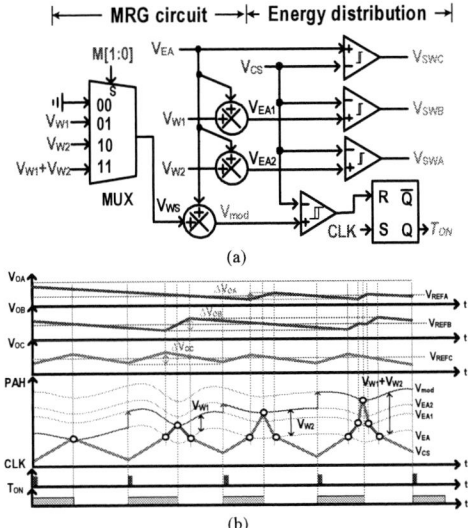

(a)

(b)

Fig. 7. (a) On-time generator. (b) Timing diagram.

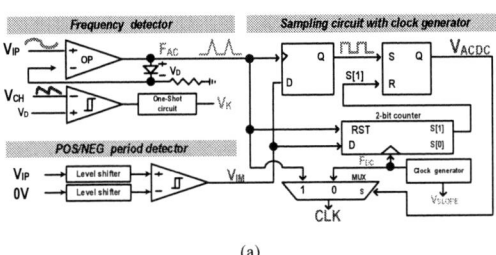

(a)

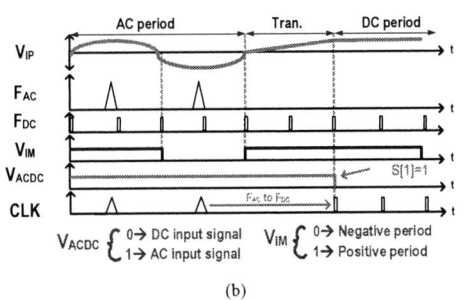

(b)

Fig. 8. AC/DC detector and timing diagram.

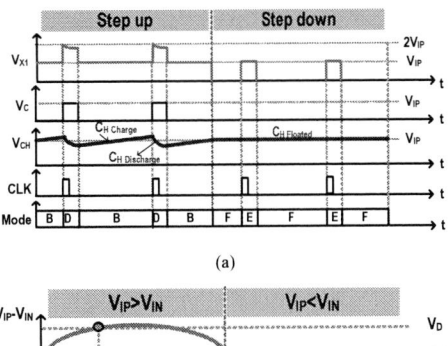

(a)

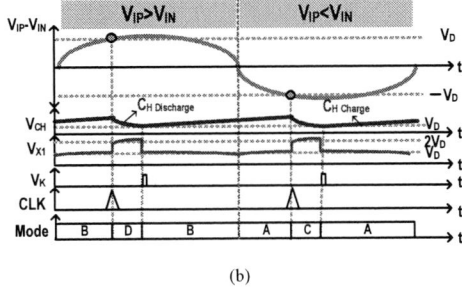

(b)

Fig. 9. Timing diagram. (a) DC input source. (b) AC input source.

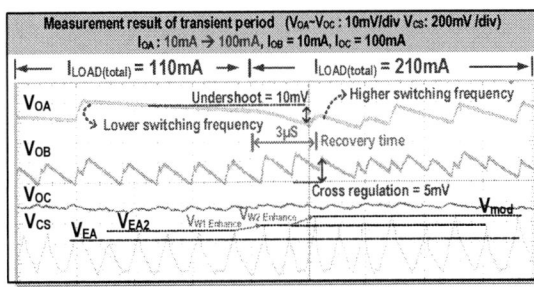

Fig. 10. Measured waveforms under AC energy harvesting input source in case of I_{OA} changing from 10mA to 100mA.

IV. EXPERIMENTAL RESULTS

Fig. 10 shows the transient response when I_{OA} steps from 10mA to 100mA under the AC energy harvesting source. V_{OA} operates at a lower switching frequency at light loads but the switching frequency increases at heavy loads. The recovery time of 3μs demonstrates fast transient response while the overshoot voltage is 10mV with a 5mV cross regulation effect. Fig. 11(a) shows the steady state performance. The output voltage ripple remains as low as 15mV at the AOT output and 5mV at the PWM output. With the aid of PAH controller, the comparison shows efficiency enhancement of 20% at light loads in Fig. 11(b). Fig. 12(a) shows the comparison of I_L/I_{load} versus the conversion ratio. The design shows 5X smaller low inductor current compared to the prior designs [2], [4] – [5].

The comparison of efficiency versus the conversion ratio in different input voltages is shown in Fig. 12(b). The peak efficiency is 93% when the input voltage is 3.3V with an 86% conversion ratio. Table I shows the comparison table with state-

the 2-bit counter. The counter starts the accumulation by the DC switching frequency F_{DC}, which is a fixed frequency clock signal for DC conversion produced by the clock generator. The V_{ACDC} drops eventually and switches the CLK from F_{AC} to F_{DC} by the multiplexer. Fig. 9(a) shows the step up/down conversion of the DC input voltage. The V_{X1} swings from V_{IP} to 2 times V_{IP} in the step up conversion and swings from 0 to V_{IP} in the step-down conversion. In the AC input signal of Fig. 9(b), the V_{X1} is pumped to high voltage for the step-up voltage conversion during C_H discharging period.

978-1-7281-5107-6/19 $31.00 © 2019 IEEE

of-the-art [1]-[3]. The off-chip components are a 2.2µH inductor, 4.7µF output capacitors, and 500nF C$_H$. The chip photo is shown in Fig. 13 and the area of 1.8 * 2.1mm².

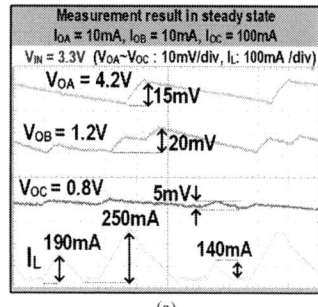

(a)

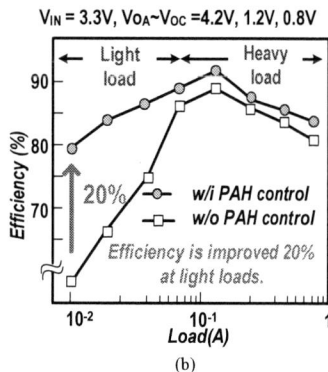

(b)

Fig.11. (a)Measurement results in steady state. (b) Efficiency versus I$_{load}$.

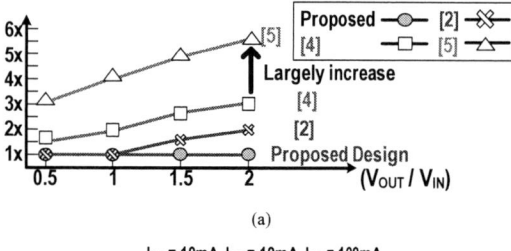

(a)

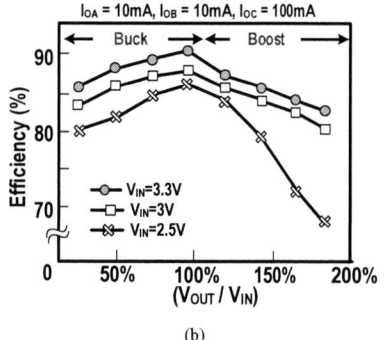

(b)

Fig. 12. (a) Comparison of I$_L$/I$_{load}$ versus conversion ratio. (b) Comparison of efficiency versus DC input source V$_{IN}$.

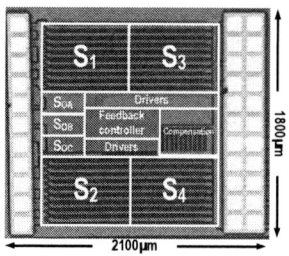

Fig. 13. Chip micrograph.

Table I: Specifications and comparison table with state-of-the-arts.

Comparison Table				
Performance \ Designs	[1]	[2]	[3]	This work
Process	0.25µm	65nm	0.35µm	40nm
Input voltage(V)	2.5~5	3.4~4.3	2.1~3.3	2.5~3.3
Output voltage(V)	1~5	1.2~2.8	1.8~3.3	4.2V, 1.2V, 0.8V
Number of channels	2	5	4	3
Control method	EPWM	AERC	Tail-current	AOT & HWM
Switching frequency(MHz)	2	1.2	0.5	1
Inductor(µH)	2.2	2.2	4.7	2.2
Capacitor(µH)	20	4.7	10	4.7
Peak efficiency(%)	90	83.1	91	93
Cross regulation(mV/mA)	0.02	0.067	0.05	0.05
Load regulation(mV/mA)	0.92	N/A	0.5	0.1

V. CONCLUSION

The proposed single-inductor triple-output (SITO) converter can accept both DC battery voltage and AC energy harvesting voltage to provide triple step up and down outputs. Storing the energy to the H-bridge capacitor first and pumping it later can ensure a continuous inductor current for reducing the output voltage ripple by 67%. Due to the H-bridge, the inductor current value is reduced to 1/5X to achieve high efficiency of 93% under the current loading of 120mA. Moreover, the proposed PWM and adaptive on-time hybrid controller uses different modulation methods in a single controller to meet the requirements of different multiple outputs.

REFERENCES

[1] Weiwei Xu, Ye Li, Zhiliang Hong, "A 90% Peak Efficiency Single-Inductor Dual-Output Buck-Boost Converter with Extended-PWM Control," *IEEE ISSCC*, pp. 394 - 396, Feb., 2011.

[2] Chien-Wei Kuan, Hung-Chih Lin, "Near-Independently Regulated 5-Output Single-Inductor DC-DC Buck Converter Delivering 1.2W/mm2 in 65nm CMOS" *IEEE ISSCC*, pp. 274 - 276, Feb., 2012.

[3] Yanqi Zheng, Marco Ho, "A Single-Inductor Multiple-Output Auto-Buck–Boost DC–DC Converter with Tail-Current Control" *IEEE Trans. on Power Electronics*, pp.7857–7875, Nov. 2016.

[4] Xiang-En Hong, Jian-Fu Wu, "98.1%-Efficiency Hysteretic-Current-Mode Noninverting Buck–Boost DC-DC Converter with Smooth Mode Transition" *IEEE Trans. on Power Electronics*, pp. 2008 - 2017, Mar., 2017.

[5] Piero Malcovati, Massimiliano Belloni, "A 0.18µm CMOS 91%-Efficiency 0.1-to-2A Scalable Buck-Boost DC-DC Converter for LED Drivers" *IEEE ISSCC*, pp. 280-282, Feb., 2012.

A 0.7mm² 8.54mW FocusNet Display LSI for Power Reduction on OLED Smart-phones

Tsu-Ming Liu, Chang-Hung Tsai, Shawn Shih, Chih-Kai Chang, Jia-Ying Lin,
Wayne Hsieh, Yung-Chang Chang, and Chi-Cheng Ju
Multimedia Development
HsinChu 300 Taiwan
mingle.liu@mediatek.com; cc.ju@mediatek.com;

Abstract—the first-reported FocusNet display LSI integrating an 11-layer fully convolutional neural network and supporting color, contrast, and sharpness visual quality enhancement engine is designed. A FocusNet deep learning (DL) model is proposed to predict the human's focus region applying a tone mapping curve so as to decrease the pixel luminance, leading to 19.5mW of system power saving on OLED smart-phone. Implemented in 10nm CMOS technology, the proposed DL-assisted FocusNet display LSI costs 3.21M gates and 52kB SRAM with 0.7mm² area. Operated at 0.9V and 205MHz, this LSI consumes 8.54mW of power dissipation.

Keywords— tone mapping; deep learning; display; OLED;

I. INTRODUCTION

Power consumption is regarded as a critical issue for battery-operated mobile devices. Among all components of mobile devices, display subsystems consume more than 50% of total energy [1]. Therefore, saving power for display which consumes a lot of power without decreasing the user experience becomes necessary.

Compared to existing LCD display using back-light, OLED display has different operating structure and cannot reduce power consumption by existing dimming technology. Several low-power techniques for OLED displays are explored [2]-[4]. OLED display power consumption is highly dependent on the image content (more precisely, the pixel values). An intuitive way is to darken the contents that are not of interest to the user since one never looks at the whole screen at one point in time. However, existing design dims a fixed region (top or bottom) of interest on the screen by a pre-define UI applications [4] and therefore cannot be applied to difference video scenario. In this work, a region of interest (ROI) is detected by a FocusNet deep learning neural network model. The main idea is to keep quality within the ROI while decreasing the pixel value in the background. Usually, this implies a dimmed result at the picture level. However, we enhance the contrast so as to compensate for the reduced pixel brightness. The experimental results reveal that 19.5mW power saving can be achieved.

Fig. 1 depicts the proposed DL-assisted FocusNet display processing pipeline for video playback system. It is evident that deep learning approaches provide superior results for recognition problems [5]. Recently, standalone deep learning processors [5]-[7] have been reported, but an integrated deep learning solution with display processor is rarely discussed. In this work, the proposed FocusNet deep learning model is trained to predict and output human's focus region information by a large scale of training dataset. The model takes the video reconstructed image as an input and outputs the focus information, which indicates the probability distribution of the visual attention area. Take the heatmap in Fig. 1 for example, the FocusNet DL accelerator predicts the region of lighthouse as the visual attention area. The processed map adaptively decreases the pixel luminance through the tone-curve table to achieve better power efficiency while preserving the subject image visual quality.

In particular, the FocusNet is implemented by a fully convolutional neural network with 11-layer which consists of 2 basic feature layers, 1 shallow focus feature layer, 1 deep focus feature layer, and 2 focus information extraction layers. To reduce the computational complexity, a channel reduction layer is designed in the first layer to transfer the input color image to single channel image. In low-power mode, contrast enhancement adjusts the image tone mapping curve so as to reduce the complexity on non-visual contact region, leading to 19.5mW system power saving in OLED FHD resolution and 60Hz while maintaining the same mean opinion score (MOS) in terms of subjective visual quality [8].

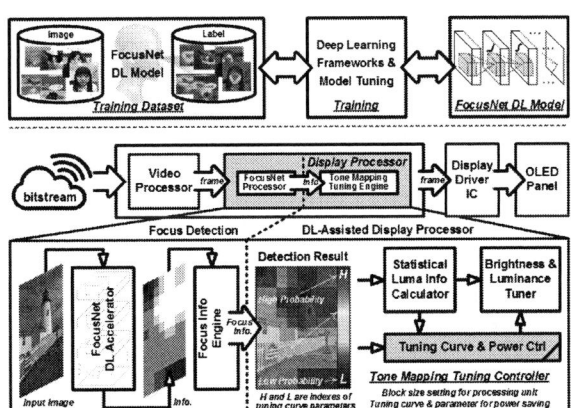

Fig. 1. Proposed DL-assisted FocusNet display processor for video playback system with contrast enhancement.

The rest of this paper is organized as follows. Section II presents the associated architecture of the proposed FocusNet display processor. Section III describes the implementation and measurement results. Finally, concluding remarks are made in Section IV.

II. PROPOSED DL-ASSISTED FOCUSNET DISPLAY PROCESSOR

A. System Block Diagram

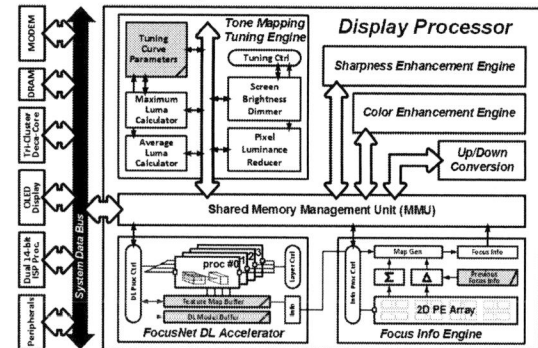

Fig. 2. FocusNet display processor in OLED smart-phone SoC.

Fig. 2 depicts the system architecture of this smart-phone SoC. The host CPU controls all modules and the memory usage for different video-based scenario. There are several blocks (e.g. modulation/demodulation, image processor, video processor, display processor, graphic engine) interfaced to DRAM and CPU via system and CPU buses. Taking a video playback as an example, software OS moves the video streams to the corresponding position in DRAM. The video processor reads the data, performs pixel reconstruction, and therefore output video frame buffer to external DRAM. The display processor fetches the previously reconstructed video frame and executes the image enhancements, and performs up/down converter so as to meet display panel resolution.

In a display processor core, it incorporates the FocusNet result to adjust the contrast enhancement locally. It analyzes the focus region to indicate the human interest in a frame. Based on the focus region results, the color, contrast and sharpness enhancement can decrease the pixel luminance while keeping the image quality in the same subjective level. Experimental results reveals that we can save 19.5mW of system power without significantly affecting the perceived video quality.

B. FocusNet network

In this work, the proposed FocusNet deep learning model is trained to predict human's focus region information and adaptively adjust the proposed DL-assisted display processor. To train the FocusNet model, pixels in the displayed image are labeled to either focus pixels or non-focus pixels. The PSNR metric is used as the objective function for error backpropagation during the model learning process. Fig. 3 shows the network architecture of the proposed FocusNet which is implemented by a fully convolutional neural network with 11 layers.

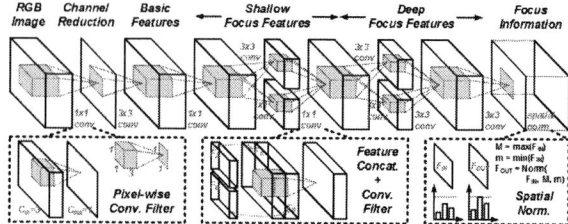

Fig. 3. FocusNet neural network architecture.

To predict human focus information, the image pixel data is directly inputted to the network for analysis, and basic features are detected in first two layers with 1×1 and 3×3 convolutional kernel size. In shallow and deep focus feature layers, 2 convolutional filters are applied to extract features in different kernel size, and two 64-feature maps are concatenated in channel wise to output a 128-feature map for further computation. Last two layers with 3×3 convolutional filter and spacial normalization, Eq. (1)-(3), are designed to output FocusNet information for visual quality enhancement.

$$M = MaxNeuronValue(\,F_{input}\,) \qquad (1)$$

$$m = MinNeuronValue(\,F_{input}\,) \qquad (2)$$

$$F_{output} = Normalization(\,F_{input},\,M,\,m\,) \qquad (3)$$

C. FocusNet DL Accelerator

To efficiently perform deep learning computation and achieve FHD@60Hz real-time requirements with the proposed 11-layer FocusNet neural network model, a FocusNet DL accelerator is designed as shown in Fig. 4. It comprises of four deep learning processors and is connected to model and feature map buffers. We adopt convolution layer with filter size of 1×1 and 3×3, and two kinds of common activation functions include Sigmoid and ReLU are implemented by the time-sharing multiplexing. Moreover, this network aggregates the features by channel wise feature concatenation to improve the detection capability for different feature sizes.

As a typical task scheduling strategy, we divide time into equal time slots, and schedule newly arrived tasks in a time slot as a batch. The duration of a time slot should be significantly shorter than the average job execution time. In our system, maximal 4 convolutional filters are processed in a time slot. During the inference stage, the data of input layer is loaded from the feature map buffer and broadcasted to deep learning processors (cores #0-3) for parallel computation with maximal 4 convolution kernels. Moreover, the neuron value is activated by the activation function before output to feature map buffers to further save data accessing time, as shown in Fig. 5.

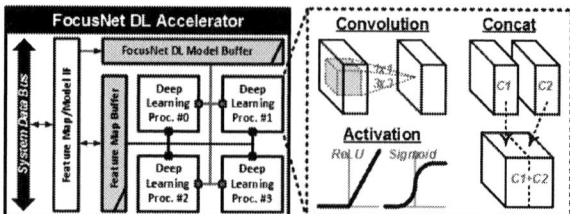

Fig. 4. FocusNet neural network accelerator.

IEEE Asian Solid-State Circuits Conference
November 4 – 6, 2019
The Parisian Macao, Macao SAR, China

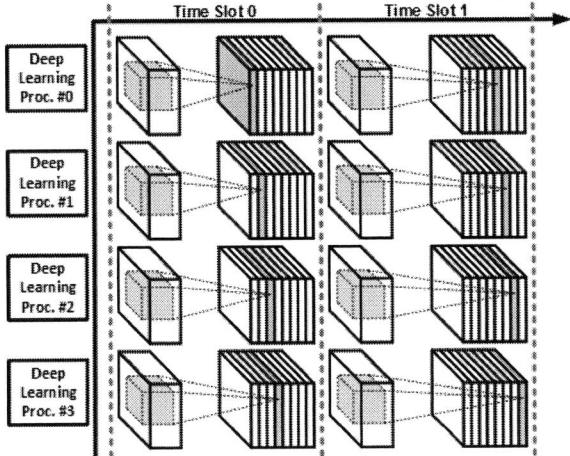

Fig. 5.   Task scheduling strategy to perform deep learning computation.

D. *Focus Information Engine*

To further improve the energy efficiency of FocusNet DL accelerator, a focus information engine is designed in Fig. 6(a). During the post-processing, a 2D processing element (PE) array calculates the statistics of focus information including active area and difference between current ($FocusInfo_t$) and previous ($FocusInfo_{t-1}$) detection result. Fig. 6(b) shows the experimental result with a live video playback scenario. In this case, the ROI object/region is moving with slow motion in the beginning. By adopting the proposed focus information post-processing, the *FocusInfo* can be reused to save both computational time and power consumption. When the object is moving with fast motion, the *FocusInfo* is frequently re-calculated to update focus information to keep detection accuracy. According to the measurement result, the average focus information update rate achieves 34%. That is, we can reduce the FocusNet DL accelerator power by 66%.

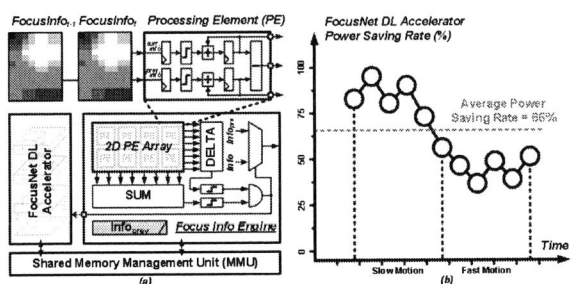

Fig. 6.   (a) Focus information engine and (b) measurement result of dynamic map update mechanism.

E. *Tone-curve Mapping Tuning Engine*

We use tone mapping to change the brightness levels of an image while preserving the image contrast. To achieve visually lossless during video playback, a scalable tone mapping architecture is required. Fig. 7 addresses the quality adjustment in tone mapping engine. In particular, the proposed DL-assisted

display processor adaptively adjusts quality through grid table and power table. Both tables decide the block size that tone curve applied and the tone curve mapping points. According to focus information detected by the proposed FocusNet deep learning model, higher level in focus information map indicates the visual contact and therefore allocates smaller block size and keep more linear of tone curve than lower level in focus information map.

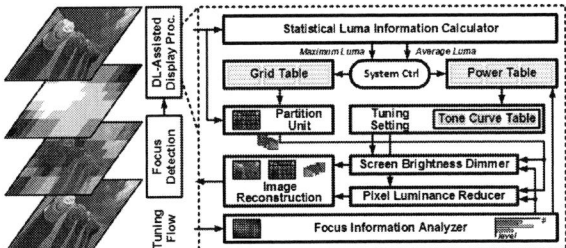

Fig. 7.   Tone mapping curve adjustment during video playback.

III.   IMPLEMENTATION AND MEASUREMENT RESULTS

Fig. 8(a) shows the chip features and specification. This chip is fabricated in a 10nm FinFET process and costs 3.21M gates and 52kB internal SRAM with 0.7mm² area. This chip integrates quad-core FocusNet DL accelerator to efficiently perform the proposed FocusNet fully convolutional neural network with 11 layers, focus information post-processing engine to dynamically update/re-calculate focus information map based on the analysis result, tone mapping tuning engine, sharpness/color enhancement engine, and up/down conversion module. Operated at 0.9V and 205MHz, the proposed LSI consumes 8.54mW to achieve FHD@60Hz for video playback and display. The evaluation board and die photo are shown in Fig. 8(b).

Features		FocusNet Display LSI
Process		10nm FinFET
Supply Voltage		core 0.9V
Spec.	FocusNet	Fully Convolutional Network (11 Layers)
	Display Engine	Color / Sharpness Tone Mapping
Logic Gate Count		3.21M gates
Core Size		0.7mm²
Max. Working Freq.		205MHz
Memory	Internal	52kB SRAM
	External	32-b LPDDR3-1866
Power	1080p@60fps	8.54mW

Fig. 8.   (a) Chip features and specification and (b) evaluation board and chip die photo of image tuning modules and FocusNet computation modules.

To clarify the system performance and the visual quality, an evaluation board is designed with a 4.7" FHD (1920×1080) OLED display panel. Since power consumption is proportional to the sum of brightness of each pixel. The proposed display processor decreases the pixel value according to the region of non-focus and keep the quality on the region of human focus. A subjective assessment based on human opinion score

978-1-7281-5107-6/19 $31.00 © 2019 IEEE

scheme, a.k.a. MOS, is used to measure the visual quality level. The quality level of a video sequence based on MOS model is rated on a scale from 1 to 5, where 5 is the best possible score. Fig. 9 shows the results of images w/ and w/o FocusNet contrast enhancement, a subjective visual quality achieves near lossless of 4 in MOS [8] measurement. To further compare the objective visual quality in the focus region, we use CA-310 display analyzer [9] for OLED panel brightness measurement. In ROI region, our solution achieves 0.0 nits (candelas per square meter) and 1.5 nits of brightness level difference between w/ and w/o FocusNet display in portrait and non-portrait case, respectively. And 5.6 nits and 20.4 nits of brightness are dimmed in non-ROI region for power saving.

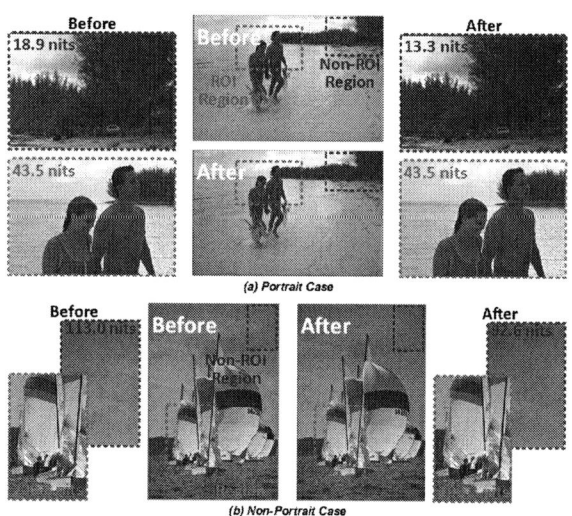

Fig. 9. Performance evaluation and comparison.

Table I shows the OLED quality and system power evaluation in terms of delta nits in ROI/non-ROI region and power measurements where the power is measured from the battery side of OLED smart-phone. Four images are selected from publicly available data set (Kodak [10]). Compared to the conventional design without exploring FocusNet display enhancement, our proposal achieves 19.5mW of power saving in average while keeping the brightness level the same (1.8 nits difference in average) as the conventional one in ROI region. And the detailed experimental results are shown in Table II.

TABLE I. OLED QUALITY AND SYSTEM POWER EVALUATIONS

Image	Δnits ROI Region	Δnits nonROI Region	Power Saving
Test Image #1	2.8 cd/m²	30.0 cd/m²	12.2 mW
Test Image #2	3.0 cd/m²	16.2 cd/m²	11.2 mW
Test Image #3	1.5 cd/m²	20.4 cd/m²	28.7 mW
Test Image #4	0.0 cd/m²	5.6 cd/m²	26.0 mW
Average	**1.8 cd/m²**	**18.1 cd/m²**	**19.5 mW**

*Δnits means the measured nits difference between with and without applying the proposed FocusNet Display

TABLE II. DETAILED EXPERIMENTAL RESULTS

Test Image	#1	#2	#3	#4
Before Processing (Conventional)				
After Processing (w/ FocusNet)				
Focus Information				

IV. CONCLUSION

In this paper, a 10nm DL-assisted FHD display processor with FocusNet DL accelerator LSI is first-reported. The 11-layer FocusNet deep learning model, tone mapping tuning engine, and sharpness/color enhancement engine, up/down conversion are incorporated into display processor to achieve 19.5mW power reduction measured from the battery side of OLED smart-phone. Moreover, the subjective and objective quality in terms of MOS and brightness level average difference is 4 and 1.8 nits in ROI region which achieve nearly lossless as compared to conventional one (i.e. without exploring FocusNet display). It is very power-efficient and enables significant power saving for OLED with viewing experience guarantee, allowing users to enjoy extended battery life without sacrificing viewing experience.

REFERENCES

[1] A. Carroll and G. Heiser, "An analysis of power consumption in a smartphone," in Proc. USENIX Annual Technical Conference, pp. 21-21, Jun. 2010.

[2] M. Dong, Y.S. K. Choi, and L. Zhong, "Power-Saving Color Transformation of Mobile Graphical User Interfaces on OLED-based Displays," in Proc. IEEE/ACM ISLPED, pages 339-342, Aug. 2009.

[3] D. Shin, Y. Kim, N. Chang, and M. Pedram, "Dynamic Voltage Scaling of OLED Displays," in Proc. IEEE/ACM DAC, pages 53-58, Jun. 2011.

[4] K. Tan, T. Okoshi, A. Misra, and R. Balan, "Focus: A usable & effective approach to OLED display power management," in Proc. ACM International Joint Conference Pervasive Ubiquitous Computing, pp. 573-582, Sep. 2013.

[5] B. Fleischer et al., "A scalable multi-teraOPS deep learning processor core for AI training and inference," in Proc. IEEE Symposium on VLSI Circuits, pp. C35-C36, Jun. 2018.

[6] S. Yin et al., "A 1.06-to-5.09 TOPS/W reconfigurable hybrid-neural-network processor for deep learning applications," in Proc. IEEE Symposium on VLSI Circuits, pp. C26-C27, Jun. 2017.

[7] C.H. Tsai et al., "A 41.3/26.7pJ per neuron weight RBM processor supporting on-chip learning/inference for IoT applications," IEEE J. Solid-State Circuits, vol. 52, no. 10, pp. 2601–2612, Oct. 2017.

[8] "Subjective video quality assessment methods for multimedia applications," ITU-T Recommendation P.910, Apr. 2008.

[9] CA-310 Display Color Analyzer, website available: https://www.konicaminolta.com/instruments/download/catalog/display/pdf/ca310_catalog_eng.pdf

[10] Koak Image Dataset, website available: http://www.cs.albany.edu/~xypan/research/snr/Kodak.htm

A 47.4μJ/epoch Trainable Deep Convolutional Neural Network Accelerator for In-Situ Personalization on Smart Devices

Seungkyu Choi, Jaehyeong Sim, Myeonggu Kang, Yeongjae Choi, Hyeonuk Kim, and Lee-Sup Kim

School of Electrical Engineering
Korea Advanced Institute of Science and Technology (KAIST)
Daejeon, Republic of Korea
Email: skchoi@mvlsi.kaist.ac.kr

Abstract—**A scalable deep learning accelerator supporting both inference and training is implemented for device personalization of deep convolutional neural networks. It consists of three processor cores operating with distinct energy-efficient dataflow for different types of computation in CNN training. Two cores conduct forward and backward propagation in convolutional layers and utilize a masking scheme to reduce 88.3% of intermediate data to store for training. The third core executes weight update process in convolutional layers and inner product computation in fully connected layers with a novel large window dataflow. The system enables 8-bit fixed point datapath with lossless training and consumes 47.4μJ/epoch for a customized deep CNN model.**

I. INTRODUCTION

Deep learning can be accessed from a wide range of applications in various smart devices. Among the deep neural networks, convolutional neural networks (CNNs) have shown the highest performance in decision making for diverse vision tasks, hence many deep learning accelerators to run inference on embedded systems have been developed [1], [2].

Meanwhile, the necessity of additional training of deep neural networks in smart device applications has been increased due to the performance degradation by different characteristics of user's environment [3]. Previous methodologies prefer cloud-centric training by the cause of high energy consumption and only provide inference mode in dedicated accelerators. However, sending data to train on servers is vulnerable to the user privacy, and moreover its long update latency can be severe weakness in future smart devices [5]. Thus on-device training to allow device users for more personalized experience is desired as shown in Fig. 1. The device itself can further tune the deep learning model with user's personal data to fit into his/her behavior.

There are some previous designs of deep learning accelerators supporting training [4], [5]. However, they do not consider the detailed behaviors of training but only supports higher precision data types compared to the inference mode and follows the same design methodology of the inference processors. Therefore, we analyze the dataflow in all types of multiply-and-accumulate (MAC) operations in CNN training depicted in Fig. 2 and also optimize the massive additional memory accesses occurred by the training process.

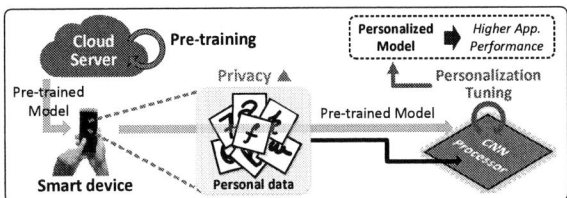

Fig. 1. Training on devices with personal data.

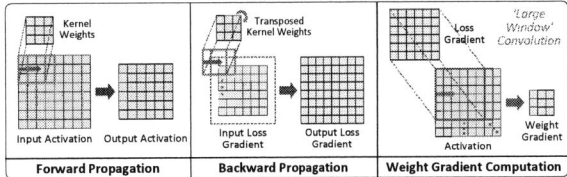

Fig. 2. MAC operations in inference and training of a convolutional layer.

In this paper, we propose a CNN processor facilitating both inference and training for low power embedded systems consuming 47.4μJ/epoch. The processor contains re-configurable cores for inference/training and the cores having different dataflow separately run the MAC operations of CNNs. The reconfigurable processor is designed with three key features: 1) a masking method reducing 88.3% of intermediate data storage at the training phase, 2) a separate weight gradient computing engine with large window dataflow, 3) a modified 8-bit fixed point datapath with lossless training.

II. PERSONALIZATION APPLICATION ANALYSIS

To verify the effectiveness of on-device training in Fig. 1, we construct a device personalization application by customizing a CNN model 'HachNet' to classify handwritten characters of 62 classes composed of 10 digits, 26 upper alphabets, and 26 lower alphabets. We first train the HachNet model depicted in Fig. 3 (a) with tons of typical dataset to obtain a well-trained model. From the pre-trained model, we further train with personal data which are user's own handwritings.

IEEE Asian Solid-State Circuits Conference
November 4 – 6, 2019
The Parisian Macao, Macao SAR, China

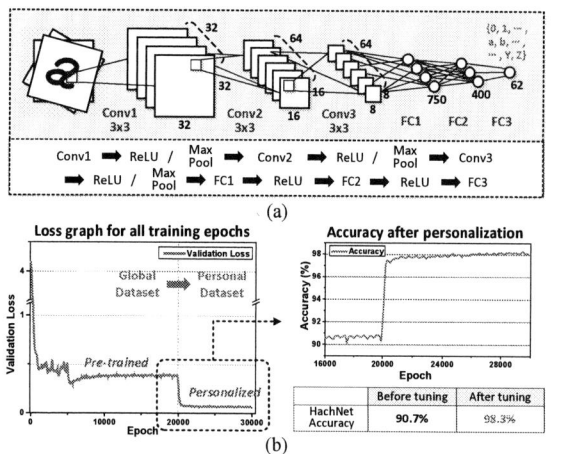

(a)

(b)

Fig. 3. (a) The customized CNN model 'HachNet' classifying 62 classes of handwritten charaters. (b) Training and accuracy analysis for the handwriting personalization.

As shown in the graph of Fig. 3 (b), the loss of validation data decreases near to 0 to fit into the user's data, which clarifies the effectiveness of the personalization. Consequently, the classification accuracy of the handwritings achieved by the HachNet increases from 90.7% to 98.3%. This evaluation result certifies the strength of on-device training for future smart devices to employ enhanced deep learning applications. Thus, the development of an accelerator supporting both inference and training with low energy consumption is needed.

III. HARDWARE ARCHITECTURE

A. CNN Inference/Training Reconfigurable Architecture

The processor is designed with three reconfigurable cores for inference and training as illustrated in Fig. 4. Two propagation (P) cores operate forward propagation of a convolutional layer in inference/forward-training phase and operate back-propagation of a convolutional layer in backward-training phase. A fully connected/weight update (FW) core runs all the inner products of fully connected layers and the weight update process of a convolutional layer in backward-training phase. The computing area is designed with near threshold voltage (NTV) enduring cells to operate in extremely low voltage.

The P core is composed of 128 MACs and a forward/backward pooling/ReLU (FB-PRE) engine. Reconfiguring the datapath by streaming either activation (AT) or loss gradient (LG) bus from the feature memory (FMEM) enables both forward and backward convolution in the same MAC array. For forward convolution, the AT bus is streamed directly to the MAC array and compute with the corresponding kernel weight values in output stationary manner [6]. The final accumulated sum is sent to the FB-PRE engine to operate ReLU and pooling layer. In the case of backward convolution, the LG bus is first streamed to FB-PRE engine to prepare the input of the backward convolution. The prepared input is streamed to the MAC array to produce the final accumulated sum.

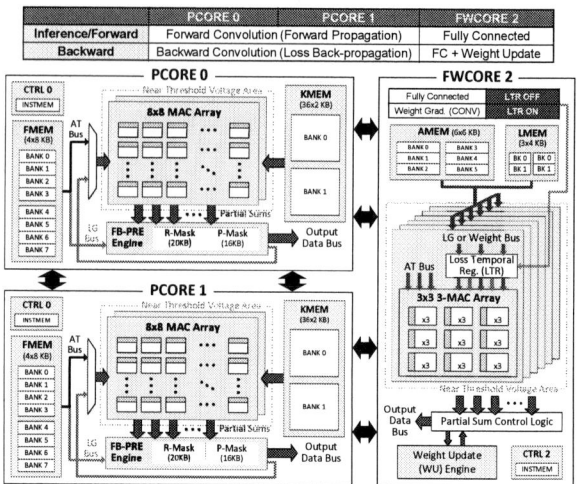

Fig. 4. Overall architecture.

The FW core contains 162 (27×6) MACs, 4 loss temporal register files (LTRs), a partial sum control logic, and a weight update (WU) engine. To compute the weight gradients of convolutional layers, a separate dataflow architecture for efficient MAC operation is implemented. Also, the core can change the computing mode for MAC operations of fully-connected layers by turning off read/write of the LTRs.

B. Propagation Core

1) Data loading with double memory frequency: The forward and backward convolution follow the same dataflow as shown in Fig. 2. The feature data memory is filled with activation data for forward convolution and loss gradient data for backward convolution.

Different feature data is streamed to each MAC and a single weight value is globally streamed to a MAC array as depicted in Fig. 5. By setting memory read cycle twice faster than the computing cycle, we can stream two groups of feature data bus which allows all the data to be sent to every MAC with no stalls in the computing cycle. In spite of extra power consumption due to the high memory frequency, we can hide the latency occurred by the limited feature memory bandwidth. Eliminating idle MACs due to the latency shortens the computing time which leads to drastic energy reduction in the total process.

2) Masking logic for intermediate data: Fig. 6 describes the masking scheme to reduce index data generated from the

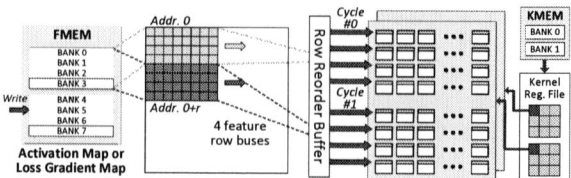

Fig. 5. Data loading sequence in the P core.

978-1-7281-5107-6/19 $31.00 © 2019 IEEE   58

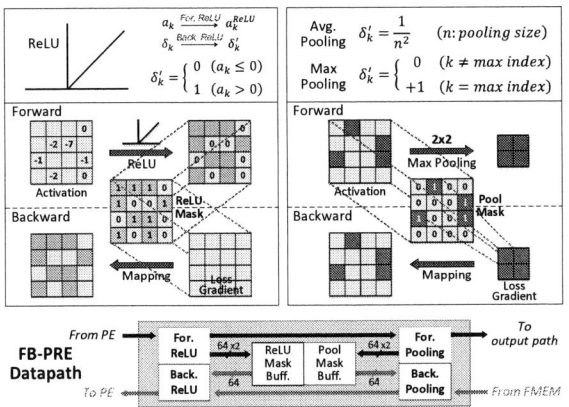

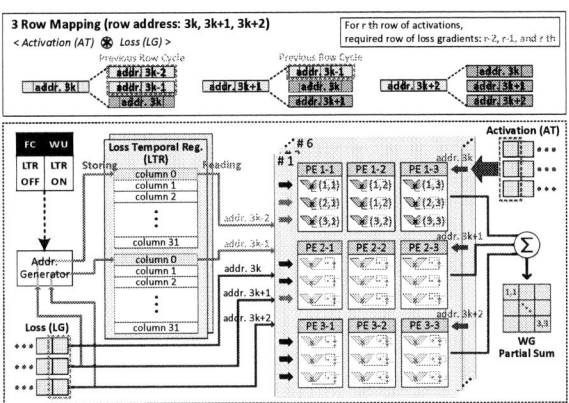

Fig. 6. Masking scheme for ReLU and pooling layers and the datapath of FB-PRE engine.

Fig. 7. The large window dataflow architecture implementation for 3x3 kernel weight gradient computation.

forward phase of pooling/ReLU (PRE) layers and required in the backward phase of PRE layers. From the equation in Fig. 6, index data of the outputs must be stored to process backward PRE. The capacity to store all the index data is very large resulting in a considerable increase in off-chip memory access since the data must be held during the whole training phase.

Accordingly, we develop a FB-PRE engine to encode the index from the forward PRE into a 1-bit masking data then store them to an additional buffer. Then the backward PRE reads the masking data and the masks are mapped into the inputs to proceed to the backward convolution. By compressing the index data small enough to store in an on-chip buffer, we can eliminate frequent off-chip memory accesses for storing the produced intermediate data in the forward training phase and reading in the backward training phase.

C. Fully Connected/Weight Update Core

There are another massive MAC operations for calculating weight gradients to update the weights. As shown in Fig. 2, the weight gradient computation can be also regarded as 'large window convolution' where the computing kernel is loss gradients and their size is as large as the activation data size [7]. The computation lacks the number of reusing data compared to the forward/backward convolution, hence following the same structure of P core leads unnecessary SRAM and register accesses. We implement the large window dataflow architecture and share it with the computation of fully connected layers since it also shows similar dataflow style.

The objective of the architecture is to allow full reuse of both activations and loss gradients in weight gradient computation to make only a single access per data from the SRAM. For the case of computing three consecutive rows of activation, corresponding loss gradient rows must be prepared to compute as described in Fig. 7. If the two previous rows of loss gradients are temporarily stored in processing elements (PEs), then all data are prepared with no additional access from the memory. Therefore, we implement loss temporal register files (LTRs) to store the present data and simultaneously read

the previously stored data. This allows to stream previous and present loss gradient rows to the PE at the same time.

Since a single activation row is computed with three rows of loss gradients in the 3x3 kernel weight case, MAC grouping to share the same activation data is also progressed. Implementing LTRs and grouping MACs enable full reuse of both inputs and save massive amount of energy consumption from the on-chip memory accesses. Furthermore, by just switching off the LTR address counter, the architecture can compute MAC operations of fully connected layers.

D. Lossless 8-bit Fixed Point Datapath

Fig. 8 presents the datapath where entire weights, activations, and loss gradients are computed in 8-bit dynamic fixed point. Simply configuring all the datapath in 8-bit fixed point results in divergence of training loss as shown in Fig. 9 (a) which is not properly trained. Even 12-bit width of fixed point training shows some accuracy degradation. The reason is that

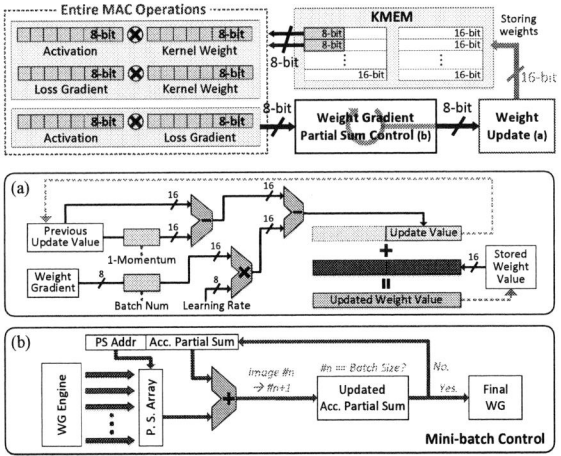

Fig. 8. The proposed 8-bit fixed point computing datapath. (a) Weight update process. (b) Image-wise partial sum control for weight gradients.

IEEE Asian Solid-State Circuits Conference
November 4 – 6, 2019
The Parisian Macao, Macao SAR, China

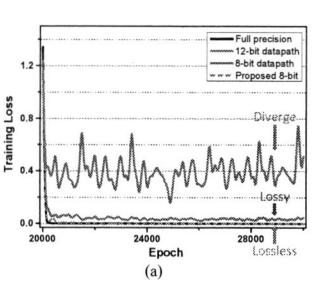
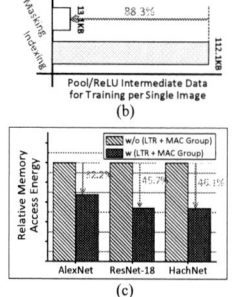

Fig. 9. (a) Training loss with different bit-width datapath. (b) Memory size of the on-chip storage for intermediate data between convolutions. (c) Energy consumed by on-chip memory accesses in weight gradient computation.

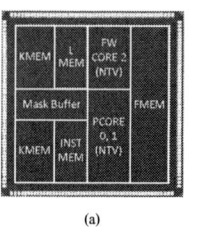
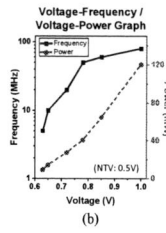

Fig. 10. (a) Chip micrograph. (b) Voltage vs. frequency/power. (c) Summary.

TABLE I
COMPARISON WITH TRAINABLE DEEP LEARNING PROCESSORS

		This Work	**Jetson TX2**	**VLSI18 [4]**	**ISSCC19 [5]**
Technology		65nm	16nm	14nm	65nm
Die Area [mm²]		10.24	43.6	9	16
Bit Precision		8 fixed	16 float	16 float	8/16 float
Max Freq. [MHz]		160	1377	1500	200
GOPS/mm² ¹		13.1	2.1	7.7	18.8
Power [mW]		9.5-120.5	7500 ²	N/A	43.1-367
Energy ³	H	47.4μ	554.4μ	N/A	N/A
[J/epoch]	A	13.7m	123.0m		

¹ Peak performance per area normalized for 65nm tech. ² Board power. ³ One epoch means a whole training iteration of an image, H: HachNet, A: AlexNet w/o 1st layer.

47.4μJ/epoch for a single image in HachNet about 11.7x less than the state-of-the-art embedded GPU Jetson TX2.

V. CONCLUSION

In conclusion, a 40.7mW, 47.4μJ/epoch energy-efficient CNN training accelerator is proposed. A masking scheme in the propagation core and a distinct dataflow architecture for weight gradient computation achieves drastic energy reduction from the memory accesses occurred in the training phase. By only extending the fraction bit-width in the weight updating process, 8-bit fixed point datapath in whole MAC operations is available without any accuracy degradation in the HachNet training process. The proposed accelerator shows 11.7x better energy efficiency compared to the previous embedded GPU.

the updating weight value is very small which needs longer bits than expected to represent the weights. To prevent lossy training, we extend 8-bit of fraction part only in the weight updating process until storing in the weight buffer. In Fig. 8 (a), the 8-bit weight gradient is computed to produce the 16-bit weight updating value. The final weight values are stored back in 16-bit but only reads the upper 8-bits so the whole computing datapath of MACs maintains 8-bit fixed point.

Moreover, a weight gradient partial sum controller in Fig. 8 (b) arranges the accumulation of weight gradients from different images inside a mini-batch. The accumulation supports image-wise stacking to produce final weight gradients and it minimizes the off-chip memory accesses occurred along multiple images in a single mini-batch.

IV. IMPLEMENTATION & MEASUREMENT RESULTS

The chip micrograph and summary are shown in Fig. 10. The accelerator is fabricated in a 65nm CMOS technology and occupies 10.24mm² with total 364KB of on-chip SRAM. It runs with the supply voltage of 0.63V to 1.0V and NTV of 0.5V for the computing cells. The measured power consumption is 9.55mW at 5MHz and 120.5mW at 80MHz displaying the lowest range compared to previous training processors which mostly fits in low power smart devices.

Fig. 9 (b) presents the amount of reduced intermediate data to hold for whole training process by the masking logic in the P core. For the HachNet of a single image, the produced index data from forward pooling/ReLU is 112.1KB which becomes a considerable overhead for the on-chip storage. By appling the masking logic, the intermediate data is squeezed by 88.3% which can hold all the data in only 13.1KB without any off-chip memory accesses. From Fig. 9 (c), the large window dataflow architecture saves the energy consumed from the on-chip memory accesses of SRAM and registers by up to 46.1% in the weight gradient computation. Both schemes to reduce the additional memory accesses from the training process save up quite a large amount of the total energy consumption.

Table I describes the comparison with the previous trainable processors. Since previous works do not report the power or only report the power efficiency, comparison of training energy is only done with the embedded GPU. This work consumes

REFERENCES

[1] J. Lee, C. Kim, S. Kang, D. Shin, S. Kim and H. Yoo, "UNPU: An energy-efficient deep neural network accelerator with fully variable weight bit precision," in *IEEE Journal of Solid-State Circuits*, vol. 54, no. 1, pp. 173-185, Jan. 2019.

[2] Z. Yuan *et al.*, "Sticker: A 0.41-62.1 TOPS/W 8bit neural network processor with multi-sparsity compatible convolution arrays and online tuning acceleration for fully connected layers," *2018 IEEE Symposium on VLSI Circuits*, Honolulu, 2018, pp. 33-34.

[3] M. Song *et al.*, "In-situ AI: Towards autonomous and incremental deep learning for IoT systems," *IEEE International Symposium on High Performance Computer Architecture (HPCA)*, Vienna, 2018, pp. 92-103.

[4] B. Fleischer *et al.*, "A scalable multi-teraOPS deep learning processor core for AI training and inference," *2018 IEEE Symposium on VLSI Circuits*, Honolulu, 2018, pp. 35-36.

[5] J. Lee, J. Lee, D. Han, J. Lee, G. Park and H. Yoo, "LNPU: A 25.3TFLOPS/W sparse deep-neural-network learning processor with fine-grained mixed precision of FP8-FP16," *IEEE International Solid-State Circuits Conference (ISSCC)*, San Francisco, 2019, pp. 142-144.

[6] Y. Chen, J. Emer and V. Sze, "Eyeriss: A Spatial architecture for energy-efficient dataflow for convolutional neural networks," *ACM/IEEE 43rd Annual International Symposium on Computer Architecture (ISCA)*, Seoul, 2016, pp. 367-379.

[7] S. Choi, J. Sim, M. Kang and L. Kim, "TrainWare: A memory optimized weight update architecture for on-device convolutional neural network training," *ACM International Symposium on Low Power Electronics and Design (ISLPED)*, Seattle, 2018, pp. 19:1-19:6.

978-1-7281-5107-6/19 $31.00 © 2019 IEEE

A Sparse-Adaptive CNN Processor with Area/Performance balanced N-Way Set-Associate PE Arrays Assisted by a Collision-Aware Scheduler

Zhe Yuan[1], Jingyu Wang[1], Yixiong Yang[1], Jinshan Yue[1], Zhibo Wang[1], Xiaoyu Feng[1],
Yanzhi Wang[2], Xueqing Li[1], Huazhong Yang[1], Yongpan Liu[1]
[1]Tsinghua University, Beijing, China
[1]Tsinghua National Laboratory for Information Science and Technology
[1]Beijing National Research Center for Information Science and Technology
[2]Northeastern University, Boston, United States
ypliu@tsinghua.edu.cn*

Abstract—**Convolutional Neural Networks give heavy storage and computation burden to accelerators, whose energy efficiency can be improved by leveraging their sparsity. However, using sparsity in networks will introduce large overhead especially when networks have various sparse situations and multiple quantization. This work named STICKER-II firstly proposes an area/performance balanced chip with simultaneously sparsity/quantization adaptive capability, enabled by multi-sparsity and multi-quantization compatible storage and computation circuits. Further more, N-way set-associate PE architecture is explored to trade off its performance and area with the collision-aware hardware scheduler. This chip achieves 4.64 TOPS/W energy efficiency on Alexnet, 1.65x better than the state-of-the-art sparse processors, along with 19.4% PE area reduction.**

I. INTRODUCTION

Convolutional Neural Networks (CNNs) have millions of weights and activations leading to critical challenges for the chip computing power and the battery life. Many previous works [1]–[7] tried to alleviate those challenges by using sparsity. Although sparsity can potentially reduce computation and memory overhead, directly implementation of sparse accelerators will introduce large overhead especially when networks have various sparse situations and multiple quantization bit. Previous sparse accelerators suffered from either area-expensive or high-collision PE arrays meanwhile cannot support multi-quantization bit. An area and performance balanced sparse accelerator is highly needed.

To design an area/performance balanced accelerator with simultaneously sparsity/quantization adaptive capability (shown in Fig. 1), two challenges remain: 1) Directly support of reconfigurable quantization modes in sparse networks will introduce large index overhead. Two types of PEs are adopted in previous works. Sparse PEs [1, 6] cannot support configurable quantization. Multi-bit PEs [2-5] cannot support sparse index; 2) Lack of an area/performance balanced architecture to support irregular accumulation of sparsity networks. In previous

This work was supported in part by National Key R&D Program 2018Y-FA0701500,NSFC Grant 61934005,61674094 and Beijing Innovation Center for Future Chip.

works, SCNN [1] and STICKER [6] can directly process both sparse activations and weights. However, SCNN [1] uses a hash solution with too much area overhead while STICKER [6] blocks PEs when collisions occur, which degrades the performance.

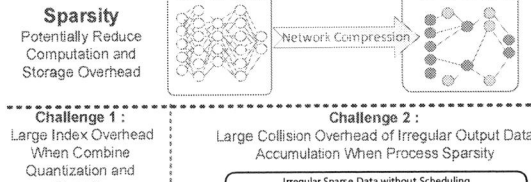

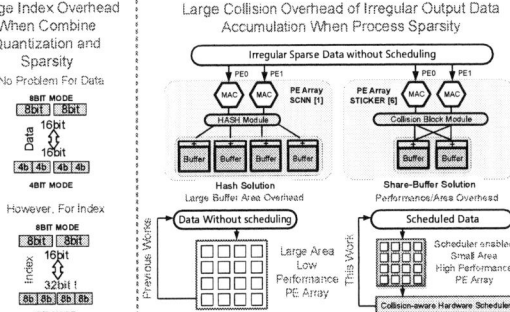

Fig. 1. Introduction and Challenges for Efficiently Processing Sparse CNNs.

This work, STICKER-II, proposes an area/performance balanced CNN accelerator achieving 3.96x/1.65x peak/average energy efficiency by three key techniques list as follows:

- A processor architecture supporting automatic adaptation among different sparsity and quantization modes. Up to 20.7x energy efficiency is achieved thanks to the combination of sparsity and quantization adaptation.
- Multi-Sparsity Multi-Quantization Bit (MSMB) compatible storage format and MSMB memory, achieving 82% memory access reduction for CNNs with 10% sparsity.
- Area/performance balanced N-way set-associate PE arrays enabled by a collision-aware hardware scheduler, which can process irregular outputs with 19.4% smaller PE area and 37.1% performance improvements (N=4).

IEEE Asian Solid-State Circuits Conference
November 4 – 6, 2019
The Parisian Macao, Macao SAR, China

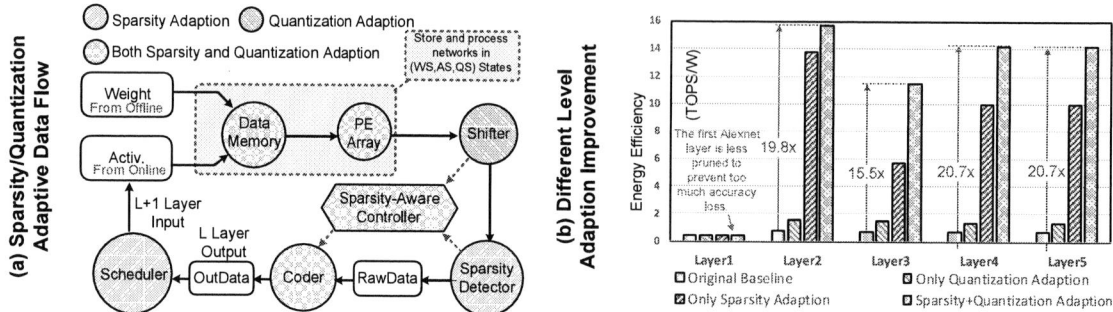

Fig. 2. Sparsity/Quantization Adaptive Work Flow and Improvement. (a)Sparsity/Quantization Adaptive Data Flow. (b) Different Level Adaption Improvement.

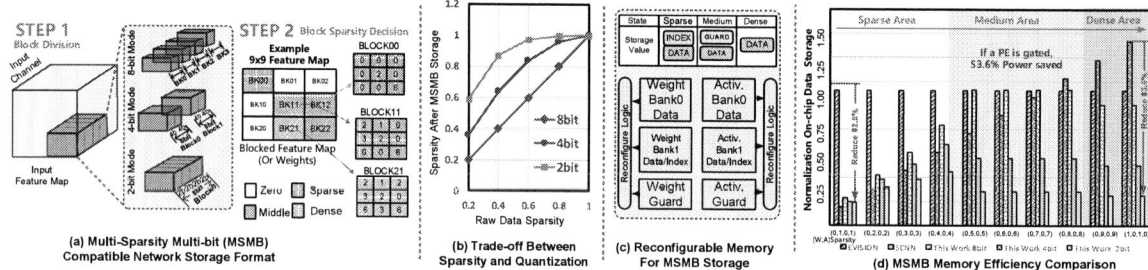

Fig. 3. Multi-Sparsity Multi-Quantization Bit (MSMB) Compatible Solution. (a) Multi-Sparsity Multi-bit Compatible Network Storage Format. (b) Trade-off Between Sparsity and Quantization. (c) Reconfigurable Memory for MSMB Storage. (d) MSMB Memory Efficiency Comparision.

II. SPARSITY-AWARE MULTI-QUANTIZATION PROCESSOR

In this section, we introduce hardware design of this chip. The top-level hardware architecture is shown in Fig. 4. In this chip, we use a 16x16 PE array. The details of each module will be introduced part by part.

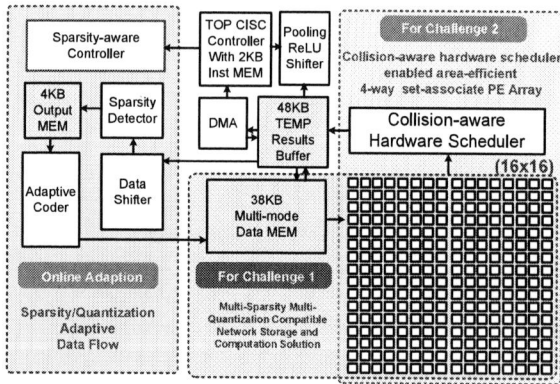

Fig. 4. Top-Level Chip Architecture Overview.

A. Sparsity and Quantization Adaptation Data Flow

This chip can adaptively cover different sparsity and quantization bits. The data flow of this chip is shown in Fig. 2. Weight sparsity and quantization are trained and analyzed offline. An online detector is designed to calculate the activation sparsity in real time. After that, activations go through a data-length shifter, an adaptive encoder and the collision-aware

hardware scheduler to be automatically encoded and scheduled for the next layer.

As shown in Fig. 2b, on pruned Alexnet, the first layer cannot have much energy efficiency gain due to less pruned and 8-bit quantization to prevent too much accuracy loss. However, in other layers, 15.5x-20.7x energy efficiency gain is obtained via simultaneously sparsity and quantization support.

B. Multi-Sparsity Multi-Quantization bit compatible solution

Multi-sparsity and multi-quantization storage format is designed to store and process both activations and weights in different states. As shown in Fig 3a, there are two steps to store data into the MSMB format. In step 1, the data is cut into different blocks based on the specified quantization mode. In 2/4/8-bit mode, each block has 4/2/1 channels. The data in the same location of different channels share an index, which makes different modes have the same index overhead. In step 2, the sparsity of each weight block will be analyzed offline, and the activation blocks are classified into 3 states (Sparse/Medium/Dense) online. As shown in 3b, there exists a trade-off between sparsity and quantization. Due to index share scheme, the networks are less sparse in 2/4 bit-mode compared with raw networks in 8 bit-mode. The trade-off is reasonable because according to our observation, some layers are dense with short bit data and other layers are on the contrary. It is hard to find a layer has not only very sparse data but also low bit precision without accuracy loss.

An MSMB reconfigurable memory is designed for this format (Fig. 3c). Dense blocks only store data without indexes.

978-1-7281-5107-6/19 $31.00 © 2019 IEEE   62

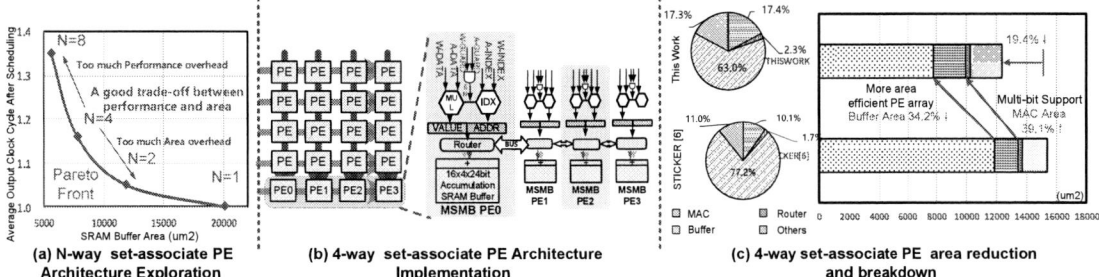

(a) N-way set-associate PE Architecture Exploration

(b) 4-way set-associate PE Architecture Implementation

(c) 4-way set-associate PE area reduction and breakdown

Fig. 5. N-way Set-associate PE Array Design. (a) N-way Set-associate PE Architecture Exploration. (b) 4-way set-associate PE Architecture Implementation. (c)4-way Set-associate PE Area Reduction and Breakdown.

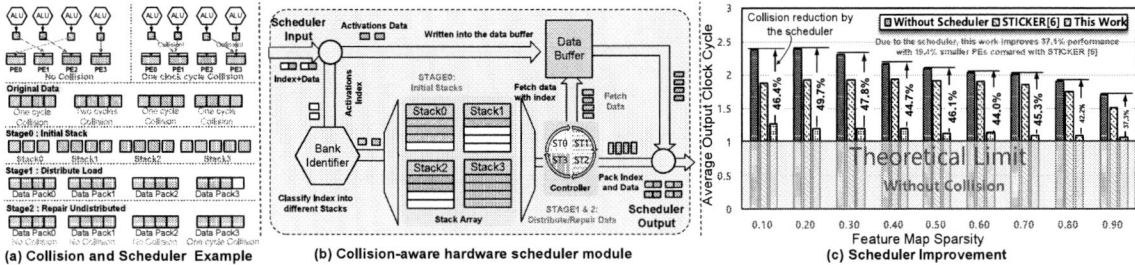

(a) Collision and Scheduler Example

(b) Collision-aware hardware scheduler module

(c) Scheduler Improvement

Fig. 6. Collision-aware hardware schedule. (a) Collision and Scueduler Example. (b) Collision-aware hardware shceduler module. (c) Scheduler Improvement.

Zero-guard is stored in middle blocks to turn off 53.6% power consumption of a PE. Sparse blocks only store non-zero data and indexes, which reduces 82.8% memory storage under 10% sparsity at 2-bit (Fig. 3d). In 2/4/8-bit mode, the MAC in PE will be configured to finish 4/2/1 MAC operations of adjacent channels in one clock cycle.

C. N-way set-associate PE array exploration

In the PE array, PEs in a column share an activation, and PEs in a row share a weight. After data scheduling, collisions are reduced so that every adjacent N PEs in a row can share N buffers and construct N-way set-associate PEs. There exists a trade-off between performance and area (Fig. 5a). Larger N means that more buffers are shared and each buffer is smaller. However, a larger N gives more burden to the scheduler and leads to more collisions. Different N forms the Pareto front. N=4 is implemented in this work based on the chip area constraint (Fig. 5b). Irregular output data are transmitted among PEs by routers. The routers work with the scheduler to detect collisions and block PE sets. Compared with the state-of-the-art sparse NN processors [6], the 4-way set-associate PE array can reduce 34.2% buffer areas. Because this work adopts more powerful multi-bit compatible MACs with 39.1% area overhead, 19.4% average PE area is reduced (Fig. 5c).

D. Collision-aware hardware scheduler

The scheduler improves the performance of the 4-way set-associate PE array by removing the potential collisions caused by buffer reusing. An example is shown in Fig. 6a and the hardware design is shown in Fig. 6b. The scheduler changes data sequences in three stages to make sure that the input data

of the next layer have as few collisions as possible. In stage 0, the indexes for different buffers are classified into 4 stacks. In stage 1, the indexes are distributed one by one from stack 0 to stack 3, to make sure that every 4 continuous data belong to different accumulator buffers. If some stacks are already empty and some undistributed indexes still remain, these indexes will be distributed with the fewest collisions in stage 2. Although this work uses more aggressive buffer design, the adoption of the proposed collision-aware scheduler reduces as much as 49.7% average output clock cycle(Fig. 6c). In total, the scheduler enabled PE arrays improve 37.1% performance with 19.4% smaller area compared with STICKER [6].

III. FABRICATION AND MEASUREMENT RESULTS

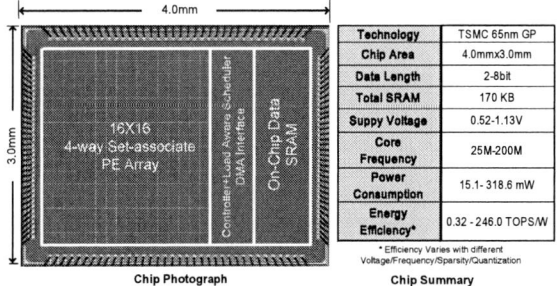

Technology	TSMC 65nm GP
Chip Area	4.0mmx3.0mm
Data Length	2-8bit
Total SRAM	170 KB
Suppy Voltage	0.52-1.13V
Core Frequency	25M-200M
Power Consumption	15.1- 318.6 mW
Energy Efficiency*	0.32 - 246.0 TOPS/W

*Efficiency Varies with different Voltage/Frequency/Sparsity/Quantization

Chip Photograph   **Chip Summary**

Fig. 8. Die Photo and Chip Summary.

This chip (Fig. 8) was implemented in a 65nm CMOS technology and runs at 200M@1.13V. Fig. 9 highlights its wide-range model compression supporting capacity. Peak energy efficiency reaches 246 TOPS/W when sparsity ratios of both activations and weights are 5% at 2bit data length.

	Conference	General: Tape-out	Tech. (nm)	Core Area (mm2)	Supply voltage (V)	Function – Sparsity Supporting: Memory Save	Clock Skip	Onchip Scheduling	Multi-Bit Support	Efficiency and Performance: Power (mw)	Max Clock Freq. (MHz)	Data Length (bits)	MACs Num	Efficiency (TOPS/W) Alexnet	Peak
SCNN[1]	ISCA2017	NO	16	7.90	-	A+W	A+W	x	x	-	1000	16	1024	-	-
EYERISS[2]	ISSCC2016	YES	65	12.3	0.82-1.17	x	x	A	x	235-332	250	16	168	0.24	0.3
EVISION[3]	ISSCC2017	YES	28	1.9	0.65-1.1	x	x	A+W	o	8-300	200	1-16	Nx256	0.53	10.0
QUEST[4]	ISSCC2018	YES	40	121.6	0.77-1.1	A+W	x	x	o	3300	300	1-4	-	0.59	2.3
UNPU[5]	ISSCC2018	YES	65	11.6	0.63-1.1	A	x	x	o	3-297	200	1-16	-	-	50.6
STICKER[6]	VLSI2016	YES	65	7.8	0.67-1.0	A+W	A+W	A+W	x	21-248	200	8	256	2.82	62.1
Samsung[7]	ISSCC2019	YES	8	5.5	0.5-0.8	W	W	x	x	39-1553	933	8,16	1024	-	11.5
This Work	-	YES	65	7.8	0.52-1.13	A+W	A+W	A+W	o	15-318	200	2-8	Nx256*	4.64	246.0

- No Report   **2bit : N=4 4bit: N=2 8bit: N=1
* [2][3][4] are reported data on their different fine-tuned models. For fair comparison, this work and [6] are tested on the same pruned model.

Fig. 7. Energy Efficiency Comparision between Different Sparsity NN Processor.

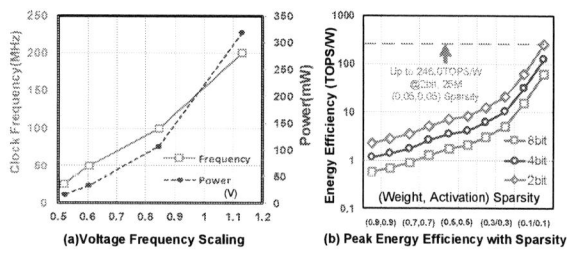

Fig. 9. Voltage Frequency Scaling and Peak Efficiency. (a) Voltage Frequency Scaling. (b) Peak Energy Efficiency with Sparsity.

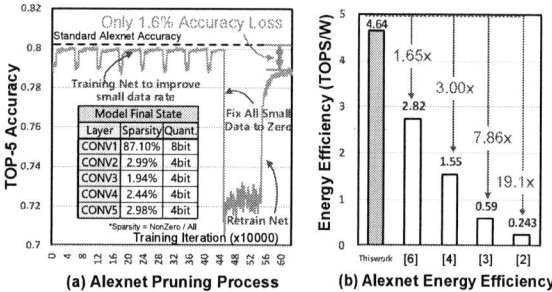

Fig. 10. Alexnet pruning and energy efficiency. (a) Alexnet Pruning Process. (b) Alexnet Energy Efficiency.

Real-world CNNs can be pruned and quantized to very sparse and short length with less than 2% top-5 accuracy loss (Fig. 10a). The 2-5 convolutional layers of Alexnet can be quantized into a 4-bit mode with less than 3% sparsity and reduce more than 97% of the model. This work can achieve 4.64 TOPS/W energy efficiency on the Alexnet model, 1.65x better than the state-of-the-art work [6] due to its complete sparsity and quantization support with a hardware scheduler 10b).

Among recent NN processors optimized for sparsity and quantization (Fig. 7), this work is the first to realize simultaneous sparsity and quantization adaption and achieves the highest peak energy efficiency with area/performance balanced N-way associate PE arrays.

IV. CONCLUSION

In conclusion, this work fabricates an area/performance balanced convolutional neural network accelerator STICKER-II with simultaneously sparsity/quantization adaptive capability. The chip is enabled by three key features: 1) A processor architecture supporting automatic sparsity and quantization adaption; 2)Multi-sparsity multi-quantization compatible storage and computation solution; 3)Area/performance balanced N-way set-associate PE architecture with a on-chip collision-aware scheduler. With all these features, this chip achieves 4.64 TOPS/W energy efficiency on Alexnet, 1.65x better than the state-of-the-art sparsity processors, along with 19.4% PE area reduction.

REFERENCES

[1] A. Parashar, M. Rhu, A. Mukkara, A. Puglielli, R. Venkatesan, B. Khailany, J. Emer, S. W. Keckler, and W. J. Dally. Scnn: An accelerator for compressed-sparse convolutional neural networks. In *Proc. ACM/IEEE 44th Annual Int. Symp. Computer Architecture (ISCA)*, pages 27–40, June 2017.

[2] Y. Chen, T. Krishna, J. Emer, and V. Sze. Eyeriss: An energy-efficient reconfigurable accelerator for deep convolutional neural networks. In *Proc. IEEE Int. Solid-State Circuits Conf. (ISSCC)*, pages 262–263, January 2016.

[3] Bert Moons, Roel Uytterhoeven, Wim Dehaene, and Marian Verhelst. envision: A 0.26-to-10tops/w subword-parallel dynamic-voltage-accuracy-frequency-scalable convolutional neural network processor in 28nm fdsoi. In *Solid-State Circuits Conference (ISSCC), 2017 IEEE International*, pages 246–247. IEEE, 2017.

[4] K. Ueyoshi, K. Ando, K. Hirose, S. Takamaeda-Yamazaki, J. Kadomoto, T. Miyata, M. Hamada, T. Kuroda, and M. Motomura. Quest: A 7.49Tops multi-purpose log-quantized dnn inference engine stacked on 96MB 3D SRAM using inductive-coupling technology in 40nm CMOS. In *Proc. IEEE Int. Solid - State Circuits Conf. - (ISSCC)*, pages 216–218, February 2018.

[5] J. Lee, C. Kim, S. Kang, D. Shin, S. Kim, and H. Yoo. Unpu: A 50.6Tops/w unified deep neural network accelerator with 1b-to-16b fully-variable weight bit-precision. In *Proc. IEEE Int. Solid - State Circuits Conf. - (ISSCC)*, pages 218–220, February 2018.

[6] Zhe Yuan, Jinshan Yue, Huanrui Yang, Zhibo Wang, Jinyang Li, Yixiong Yang, Qingwei Guo, Xueqing Li, Meng-Fan Chang, Huazhong Yang, et al. Sticker: A 0.41-62.1 tops/w 8bit neural network processor with multi-sparsity compatible convolution arrays and online tuning acceleration for fully connected layers. In *2018 IEEE Symposium on VLSI Circuits*, pages 33–34. IEEE, 2018.

[7] J. Song, Y. Cho, J. Park, J. Jang, S. Lee, J. Song, J. Lee, and I. Kang. An 11.5Tops/w 1024-MAc butterfly structure dual-core sparsity-aware neural processing unit in 8nm flagship mobile SoC. In *Proc. IEEE Int. Solid-State Circuits Conf. - (ISSCC)*, pages 130–132, February 2019.

A 2.25 TOPS/W Fully-Integrated Deep CNN Learning Processor with On-Chip Training

Cheng-Hsun Lu, Yi-Chung Wu, and Chia-Hsiang Yang

Graduate Institute of Electronics Engineering, National Taiwan University, Taipei, Taiwan

Abstract—This paper presents a deep learning processor that supports both inference and training for the *entire* convolutional neural network (CNN) with any size. The proposed design enables on-chip training for applications that ask for high security and privacy. Techniques across design abstraction are applied to improve the energy efficiency. Re-arrangement of the weights in filters is leveraged to reduce the processing latency by 88%. Integration of fixed-point and floating-point arithmetics reduces the area of the multiplier by 56.8%, resulting in an unified processing element (PE) with 33% less area. In the low-precision mode, clock gating and data gating are employed to reduce the power of the PE cluster by 62%. Maxpooling and ReLU modules are co-designed to reduce the memory usage by 75%. A modified softmax function is utilized to reduce the area by 78%. Fabricated in 40nm CMOS, the chip consumes 18.7 mW and 64.5 mW for inference and training, respectively, at 82 MHz from a 0.6V supply. It achieves an energy efficiency of 2.25 TOPS/W, which is 2.67 times higher than the state-of-the-art learning processors. The chip also achieves a 2×10^5 times higher energy efficiency in training than a high-end CPU.

Index Terms—Deep learning, convolutional neural network, specialized processor, CMOS digital integrated circuits.

I. INTRODUCTION

Deep learning has been successfully applied to many applications and even demonstrates beyond-human capability in some cases, such as image recognition. Convolutional neural network (CNN), which mainly consists of convolutional layers and fully-connected layers, is one of the most commonly used models for deep learning. Generally, training for CNNs is conducted in the cloud and inference can be performed on edge devices. Specialized processors for CNN inference have been well developed [1]–[4].

For applications associated with security and privacy, training on the edge device is preferred to prevent data loss and privacy leakage. Additionally, on-chip adaptation, which relies on incremental training, can enhance the classification performance according to the user-specific data. Fig. 1 shows such an application scenario. Given a pre-trained model, a custom model can be created on the user end through on-chip adaptation. The potential applications include ID verification and healthcare monitoring.

Learning processors with a training capability for fully-connected layers have been presented in [5]–[7]. Training for essential convolutional layers, however, has not been investigated. Considering the computational complexity of convolutional layers is 16.5× higher than that of fully-connected layers for the VGG-16 model, acceleration in training for convolutiaonal layers is necessary.

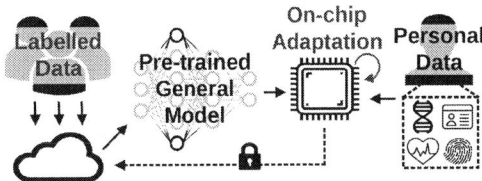

Fig. 1. Potential applications through on-chip training.

This paper presents a deep learning processor that supports both fully-connected and convolutional layers. Power and area are minimized at the algorithm, architecture, and circuit levels. The chip was fabricated in a 40nm CMOS technology with core area of 5.0 mm². It delivers an energy efficiency of 2.25 TOPS/W and 649 GOPS/W for inference and training, respectively. It demonstrates a 2.67× higher energy efficiency in inference than state-of-the-art design, despite the capability in training for the entire CNN. It also achieves a 210,032× higher energy efficiency than a high-end CPU.

II. CONVOLUTIONAL NEURAL NETWORK

A CNN is mainly composed of convolutional layers and fully-connected layers. In addition, maxpooling function, rectifier linear unit (ReLU), and softmax function are also necessary. For example, the VGG-16 model for image recognition [8] includes 13 convolutional layers with 3×3 filters cascaded by 3 fully-connected layers. Each convolutional layer is followed by a ReLU. Maxpooling functions are placed after the 2nd, 4th, 7th and 10th convolutional layers. The first two fully-connected layers are followed by ReLU. The softmax function is needed for the last fully-connected layer.

Fig. 2(a) illustrates the computation flow for CNN inference and training. In the inference phase, each convolutional layer receives an input feature map with N_c channels. The input feature map is convolved with N_r filters. Convolution of the input feature map and a filter generates one channel in the output feature map. The matrix-vector multiplications in fully-connected layers can be viewed as convolutions of an input feature map and filters with height and width of one.

Training is performed in a back-propagation manner through stochastic gradient descent. For each layer of the neural network in each iteration, a weight is updated with the product of a learning rate and its weight gradient. To obtain all the weight gradients, computations of feature map gradients (FGs) and weight gradients (WGs) are performed in each layer. At

978-1-7281-5107-6/19 $31.00 © 2019 IEEE

IEEE Asian Solid-State Circuits Conference
November 4 – 6, 2019
The Parisian Macao, Macao SAR, China

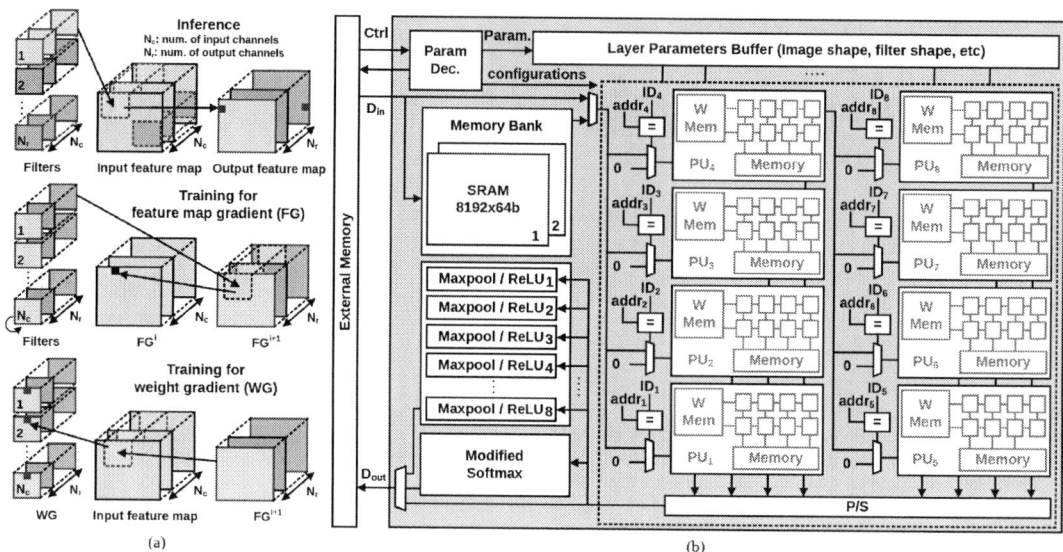

Fig. 2. (a) Computation flows for inference and training and (b) system architecture of the proposed deep learning processor.

the i-th layer, the FGs are first computed by performing convolutions between the FGs of the $(i+1)$-th layer and the filters. After FGs are computed, WGs are calculated by operating convolutions between the FGs of the $(i+1)$-th layer and the input feature map. Then, each weight is updated with the product of its WG and a learning rate.

To perform training efficiently, a flexible architecture needs to be devised to realize convolutions that are shared in both the inference and training phases. Fixed-point arithmetic is sufficient for inference, but it is not applicable to gradients with a high dynamic range for back-propagation. Additionally, a large memory for storing interim maxpooling and ReLU outputs introduces high hardware complexity. Direct mapping of the essential softmax function is not area efficient, which is also challenging for hardware realization.

III. SYSTEM ARCHITECTURE

Fig. 2(b) shows the architecture of the proposed deep learning processor. The processor includes eight processing units (PUs), eight maxpooling-ReLU modules, and a modified softmax module. The proposed design supports both inference and training for the entire CNN. Convolutional layers are processed in parallel across the channel domain. Fully-connected layers are processed by treating the matrix-vector multiplications as convolutions of 1×1 filters. Each PU computes convolutions for eight filters. The interim outputs of PUs are sent either to the maxpooling-ReLU modules or to the modified softmax module, according to the type of the layer.

A. Processing Unit

A PU consists of two memory banks and eight processing element clusters (PECs). The eight PECs in a PU perform convolutions over eight filters in parallel. Although filters are

required for inference and training, weights in the filters are arranged in different orders. If weights in the filters for training are arranged as for inference when computing FGs, outputs from all PUs need to be accumulated to produce one channel of FGs. This leads to a long processing latency. In this work, data re-arrangement for storing weights is proposed by exchanging the contents of channels and filters in the memory. Fig. 3(a) shows that the computation of FGs becomes identical and the cycle count for computing FGs can be reduced.

B. Processing Element Cluster

Fig. 3(b) shows the details of a PEC, which includes four processing elements (PEs) and four accumulators. A PE can be reconfigured to support two modes. In the first mode, it supports convolution across four channels for inference and computes FGs for training. In the second mode, it performs convolution of input feature map and WGs and updates weights. To accommodate the high dynamic range of gradients, 16b floating-point (FP16) arithmetic is adopted for training, and 10b/5b fixed-point (INT10/INT5) arithmetic is used for inference. Integrating floating-point and fixed-point arithmetics reduces the hardware area. Combining the floating-point and fixed-point integrated multiplier and adder into an unified PE unit, a large shifter can be eliminated. For low-precision modes, INT5 data format is applied to save power through clock gating and data gating.

C. Maxpooling-ReLU Module

Originally, all the interim outputs of each layer need to be stored for back-propagation (BP). However, storing only the index and sign digit of the max element of the pooling function is sufficient, which can be leveraged to reduce the

978-1-7281-5107-6/19 $31.00 © 2019 IEEE

IEEE Asian Solid-State Circuits Conference
November 4 – 6, 2019
The Parisian Macao, Macao SAR, China

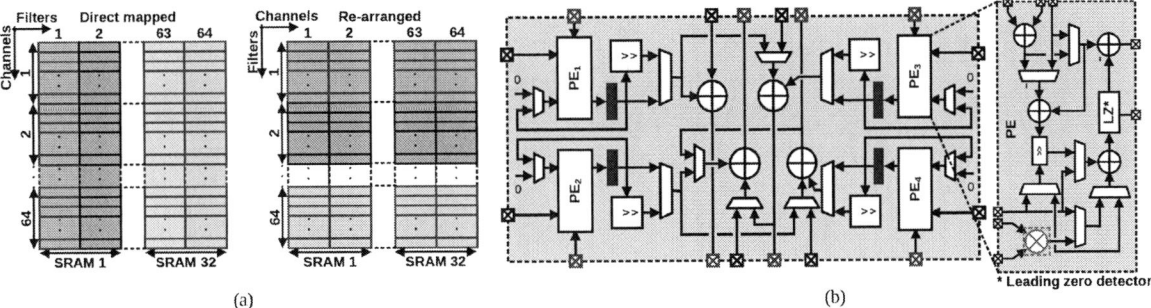

(a)   (b)

Fig. 3. (a) Data re-arrangement for storing weights and (b) architecture of a PEC that includes four PEs.

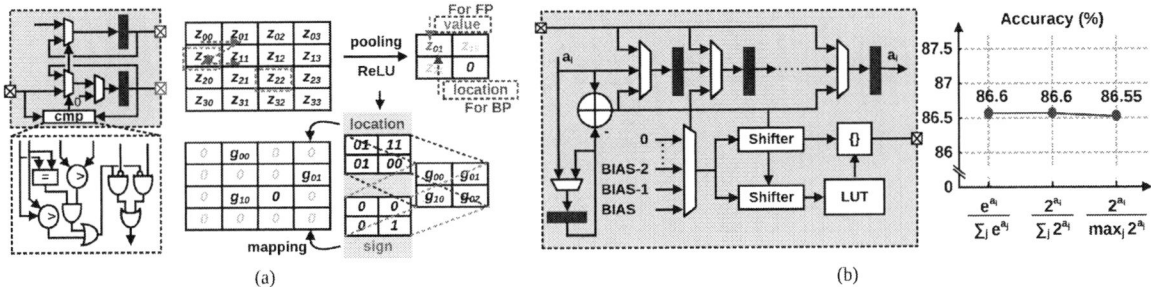

(a)   (b)

Fig. 4. (a) Maxpooling-ReLU module and (b) modified softmax module with testing accuracy on the CIFAR-10 dataset.

memory usage. Fig. 4(a) shows the proposed maxpooling-ReLU module. As an example, a 2×2 pooling window is applied to a 4×4 feature map. For forward propagation (FP) in the inference phase, the max elements, z_{01}, is chosen and its attributes (index '01' and sign digit '0') are stored. In the training phase, the feature map gradient, g_{00}, is mapped to the location '01' accordingly since the sign digit indicates a positive value. In contrast, the max element, z_{22} (out of z_{22}, z_{23}, z_{32} and z_{33}) is blocked in the inference phase because it is negative (with a sign digit of '1') and the associated gradient, g_{02}, is therefore masked.

D. Modified Softmax

The original formulation of the softmax function has high computational complexity since exponential operations are involved. In this work, a modified softmax is proposed for efficient hardware mapping. The base e is replaced by 2 and the output is normalized by the max term instead of the summation. Fig. 4(b) shows the architecture of the modified softmax module. The numerical performance of the modified softmax function is tested on CIFAR-10, an image classification dataset, using the VGG-16 model. As can be seen, the accuracy loss is merely 0.05%, indicating the effectiveness of the modified design.

IV. CHIP IMPLEMENTATION

The proposed deep learning processor was designed and implemented in a 40nm technology. Fig. 5 shows performance

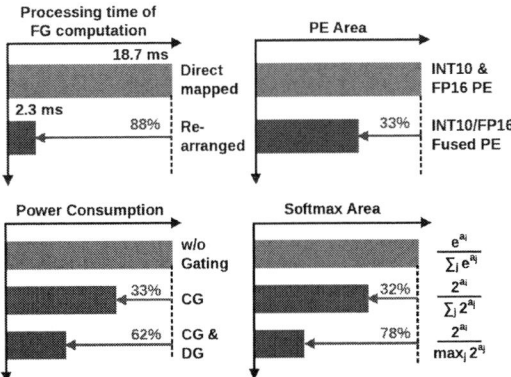

Fig. 5. Performance enhancement through the proposed techniques.

enhancement by employing the proposed design techniques. Re-arranging weights in filters for computing FGs, by exchanging the contents of channels and filters, reduces the processing latency by 88%. Integration of floating-point and fixed-point also reduces the areas of multiplier and adder by 56.8% and 17.3%, respectively. With such multipliers and adders, the area of the PE is reduced by 33%. Extensive clock gating and data gating are deployed. If low-precision arithmetic is allowed, the power can be saved by 62%. By storing the data attributes (location and polarity) of the maxpooling-ReLU

978-1-7281-5107-6/19 $31.00 © 2019 IEEE   67

IEEE Asian Solid-State Circuits Conference
November 4 – 6, 2019
The Parisian Macao, Macao SAR, China

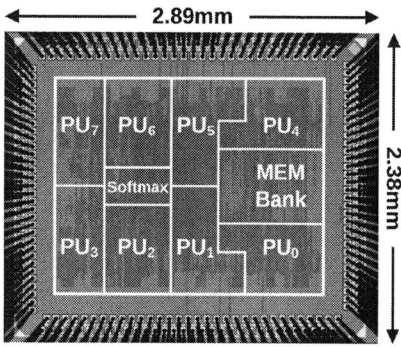

Fig. 6. Chip micrograph.

TABLE I
PERFORMANCE OF DEEP LEARNING PROCESSORS

	TETC'17 [5]	TCAS-I'18 [6]	VLSI'18 [7]	This Work
Inference	FC	FC+CONV	FC+CONV	FC+CONV
Training	FC	FC	FC	FC+CONV
Softmax Support	-	-	-	YES
Techonolgy	130nm	65nm	65nm	40nm
Core Area [mm²]	13.96	3.52	12	5.00
Max Freq. [MHz]	200 (1.2V)	200 (1.2V)	200 (1V)	232 (0.9V)
Data Format	INT8, INT16	INT13, INT16	INT8	INT5, INT10, FP16
SRAM [KB]	96	119	170	336
Power [mW]	2210	126	20.5-248.4	18.7
Area Eff. [GOPS/mm²]	2.75	14.55	-	*12.48
**Energy Eff. [GOPS/W]	56.49	660.3	†843.4	*2250

** Normalized to INT16 and 40nm   * Operated at MEP (82MHz, 0.6V)
†For a practical network

TABLE II
PERFORMANCE COMPARISON FOR CNN TRAINING

	Intel i7-2600	This Work
Techonolgy	32nm	40nm
Core Area [mm²]	351.6	5.0
Max Freq. [MHz]	3,400	232
Data Format	FP32	FP16
Power [mW]	95,000	64.5
Area Eff. [GOPS/mm²]	0.021	*12.48
**Energy Eff. [GOPS/W]	3.09x10⁻³	*649

** Normalized to FP16 and 40nm   * Operated at MEP (82MHz, 0.6V)

V. CONCLUSION

Deep learning processors that can support training find applications in which the security and privacy are critical, but previous designs only support training for a fraction of the entire CNN. This work presents an energy-efficient deep learning processor that can support training for the entire CNN. A processing element cluster is designed to support both inference and training with lower area overhead through hardware sharing. A processing element is designed to perform fixed-point and floating-point operations for inference and training, respectively, for efficient computations. The memory usage for maxpooling and ReLU is reduced by only storing the location and sign digit of the interim outputs. The essential softmax function is modified for efficient hardware mapping with negligible performance loss. The chip achieves a higher energy efficiency than the state-of-the-art learning processors, despite the training capability for the entire CNN. It provides an energy-efficient solution for intelligent edge devices that require both inference and training.

ACKNOWLEDGMENT

This work is supported by MOST 107-2218-E-002-035. The authors would like to thank TSRI for chip fabrication and technical support.

REFERENCES

[1] D. Shin, *et. al.*, "DNPU: An 8.1TOPS/W Reconfigurable CNN-RNN Processor for General-Purpose Deep Neural Networks," *ISSCC*, pp. 240–242, Feb. 2017.
[2] B. Moons, *et. al.*, "Envision: A 0.26-to-10 TOPS/W subword-parallel dynamic-voltage-accuracy-frequency-scalable convolutional neural network processor in 28nm FDSOI," *ISSCC*, pp. 246–257, Feb. 2017.
[3] S. Bang, *et. al.*, "A 288μW Programmable Deep-Learning Processor with 270KB On-Chip Weight Storage Using Non-Uniform Memory Hierarchy for Mobile Intelligence," *ISSCC*, pp. 250–252, Feb. 2017.
[4] J. Lee, *et. al.*, "UNPU: A 50.6TOPS/W unified deep neural network accelerator with 1b-to-16b fully-variable weight bit-precision," *ISSCC*, pp. 218–219, Feb. 2018.
[5] D. Kim, *et. al.*, "A Power-Aware Digital Multilayer Perceptron Accelerator with On-Chip Training based on Approximate Computing, " *IEEE TETC*, pp. 164–178, 2017.
[6] D. Han, *et. al.*, "A Low-Power DNN Online Learning Processor for Real-Time Object Tracking Application, " *IEEE TCAS-I*, pp. 1794–1804, 2018.
[7] Z. Yuan, *et. al.*, "STICKER: A 0.41-62.1 TOPS/W 8bit Neural Network Processor with Multi-Sparsity Compatible Convolution Arrays and On-line Tuning Acceleration for Fully Connected Layers, " *Symp. VLSI Circuits*, pp. 139–140, 2018.
[8] K. Simonyan, *et. al.*, "Very Deep Convolutional Networks for Large-Scale Image Recognition," *ICLR*, 2015.

module, a 75% area saving is achieved. Compared with the direct-mapped implementation, the modified softmax module reduces the area by 78%.

Fig. 6 shows the micrograph of the chip. The core area of the proposed deep learning processor is 2.5×2.0 mm². The processor can operate at the maximum frequency of 232 MHz from a 0.9V supply voltage. It dissipates 18.7 mW and 64.5 mW for inference and training, respectively, at the minimum energy point (MEP: 82MHz, 0.6V). Table I shows the performance of deep learning processors with on-chip training. This work supports inference and training for both fully-connected and convolutional layers, which is not achievable in the previous designs. It also incorporates the essential softmax function. The proposed design achieves a competitive area efficiency and an energy efficiency of 2.25 TOPS/W for inference, yielding a $2.67\times$ performance improvement in energy dissipation. Note that the energy efficiency for a practical neural network (AlexNet) of [7] is used for comparison. Table II shows the training performance compared with a high-end CPU. This work achieves a $594\times$ higher area efficiency and a $210,032\times$ higher energy efficiency.

A Ka-Band CMOS Phase-Inverting Amplifier with 0.6 dB Gain Error and 2.5° Phase Error

Chenyu Xu[1,2], Dixian Zhao[1,2]

[1]National Mobile Communication Research Laboratory, Southeast University, China
[2]Purple Mountain Laboratories, Nanjing, China
dixian.zhao@seu.edu.cn

Abstract—This paper demonstrates the design of a Ka-band 0°/180° phase-inverting amplifier (PIA) based on a 3-port transformer in 65-nm CMOS technology. Cascode topology with common-gate (CG) switch is employed to increase the reverse isolation and gain. Layout optimization techniques including double parallel inductors and self-resonant capacitor are used to improve the balance performance. The proposed PIA achieves 10.4 dB gain at 26.5 GHz with 3-dB bandwidth of over 6 GHz (i.e., 24-30 GHz). The maximum measured gain and phase errors are 0.6 dB and 2.5°, respectively. The DC power consumption is only 2.4 mW with 1-V supply.

Keywords—phase-inverting amplifier (PIA), millimeter-wave, double parallel inductors, self-resonant capacitor, CMOS

I. INTRODUCTION

With the growing demand of high data-rate transmission and medium-range communication, millimeter-wave phased-array transceivers (TRX) become important [1], [2]. Passive phase shifters (PS) play an essential role in the large-scale phased-array systems as they provide phase shift for each channel with zero DC power consumption [3]. However, their passive nature leads to high insertion loss (8-9 dB) and large silicon area when 360° phase-shift range is required [4], [5]. In contrast, the structure shown in Fig. 1(a) can achieve full 360° phase-shift range and provide power gain simultaneously. The phase-inverting amplifier (PIA) takes the function of discrete 0°/180° phase shift while providing additional gain. Besides, the phase-shift demand for the former PS module is now eased to 180°, which potentially leads to lower insertion loss and smaller chip size. In addition to being applied in PS, PIA can be utilized in the transmitter (TX) of multiple input multiple output (MIMO) radar as well. Binary phase modulation (BPM) MIMO technique uses the complete transmission capabilities of devices compared with the conventional time division modulation (TDM) MIMO [6]. As shown in Fig. 1 (b), the PIA achieves BPM while relaxing the requirement of local oscillator (LO) drive capability.

The work in [4] and [7] report the design of two PIAs at 60 GHz and 94 GHz, both based on balun and transmission-line (T-L). However, similar architecture is not suitable for the design at 28 GHz due to the large area of the λ/4 T-L. In this work, the design of a Ka-band PIA based on the common-gate (CG) switch

This work was supported in part by the National Key Research and Development Program of China under Grant 2018YFB180088, in part by the National Nature Science Foundation of China under Grant 61674035, in part by the Nature Science Foundation of Jiangsu Province under Grant BK20160690, and in part by the Fundamental Research Funds for the Central Universities. (*Corresponding author: Dixian Zhao*)

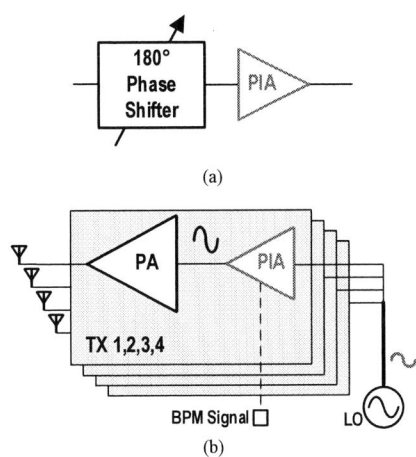

Fig. 1. PIA applications in (a) phased-array TRX and (b) BPM-MIMO radar.

topology and the 3-port transformer is elaborated. Layout optimization techniques including double parallel inductors and self-resonant capacitor are proposed to offset the undesired coupling and reduce the adverse effect of supply on balance, respectively. Fabricated in 65-nm CMOS technology, the proposed PIA achieves the measured peak gain of 10.4 dB at 26.5 GHz with 3-dB bandwidth of over 6 GHz. The measured gain/phase error between the 0° and 180° phase shift states is less than 0.6 dB/2.5° across the 6 GHz band (i.e., 24 – 30 GHz).

II. DESIGN CONSIDERATIONS

In Fig. 2(a) and 2(b), the two PIA topologies are shown, including the common-source (CS) PIA and the cascode PIA. For both topologies, the discrete 0°/180° phase shift is achieved by using two parallel switches (M_1 and M_2) that commutate the current in the primary coil of the 3-port transformer. For the 0° phase-shift state, M_1 is tuned on and M_2 is tuned off (i.e., $V_{0°} = 1$ V and $V_{180°} = 0$ V). As a result, Port 1 is selected and the current of RF signal flows into the dotted terminal of the primary coil ending to the ground. For the 180° phase-shift state, M_1 is tuned off and M_2 is tuned on. In this state, Port 2 is selected and the current flows in the opposite direction compared with the 0° state. Therefore, for the two states, the coupled currents flowing in the secondary coil will be opposite in direction and be

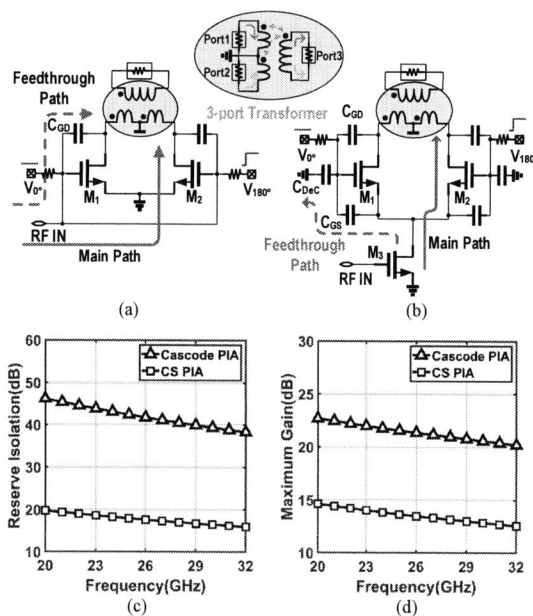

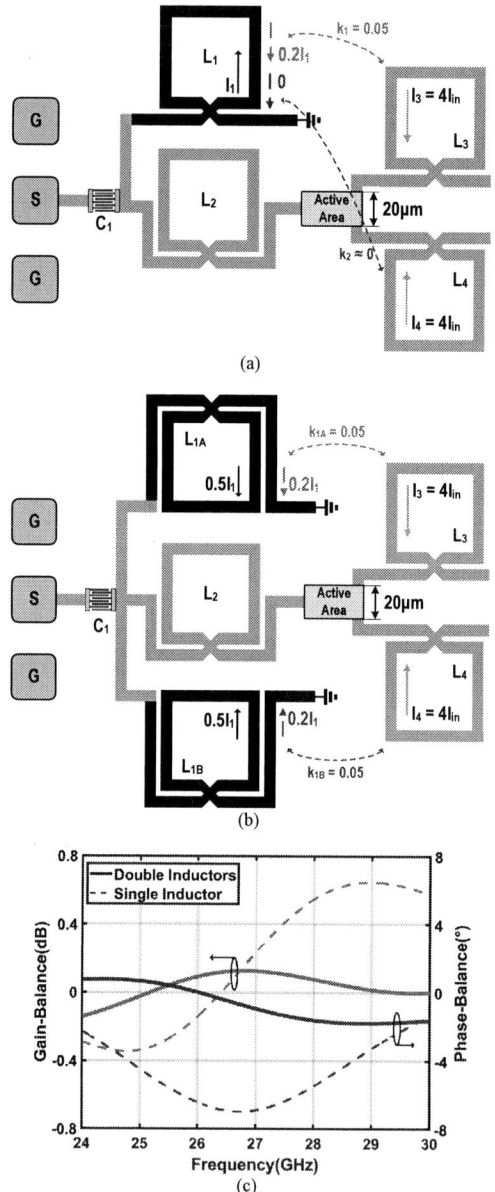

Fig. 2. The topology of PIA: (a) CS PIA, (b) cascode PIA. (c) The reverse isolation and (d) the maximum gain of the 2 topologies versus frequency.

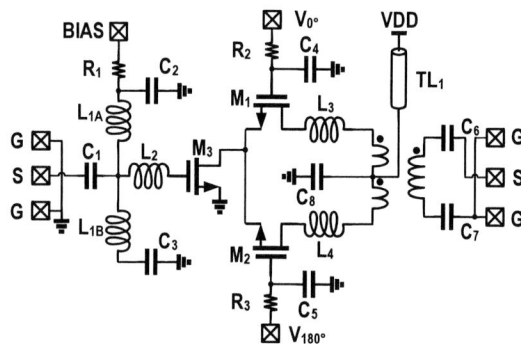

Fig. 3. The schematic of the 0°/180° PIA.

identical in magnitude, which means discrete 0°/180° phase shift is achieved.

The CS topology which adds both RF and switching signals at the gate usually features enough voltage margin, medium power gain and high output power, whereas the design of a PIA is also concerned on reserve isolation at the off state. As shown in Fig. 2 (a), when M_1 is off and M_2 is on, the RF signal will leak to the primary coil of the 3-port transformer directly through the gate-to-drain capacitor (C_{GD}). In contrast, the CG switch of the cascode shown in Fig. 2 (b) can short the parasitic capacitors connected to the gate by the decoupling capacitor (C_{DeC}). Neglecting the coupling through bulk, the feedthrough path will end to the ground and no RF signal will leak to the transformer. Fig. 2(c) and 2(d) depict the simulated reverse isolation and maximum gain of the 2 topologies, which illustrates that the

Fig. 4. The layout of L-L-C match network: (a) the conventional single parallel inductor, (b) the proposed double parallel inductors. (c) The simulated pahse (blue) and gain balance (red) results of these 2 layouts versus frequency.

cascode PIA can enhance the reserve isolation while providing a higher maximum gain.

III. CIRCUIT IMPLEMENTATION

In Fig. 3, the schematic of the proposed PIA is depicted. The 3-port transformer along with the L-C matching network is employed to ensure a compact design at Ka band. The L-L-C

978-1-7281-5107-6/19 $31.00 © 2019 IEEE

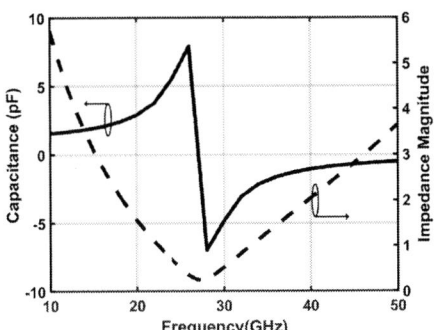

Fig. 5. The capacitance (solid) and the impedance magnitude (dashed) of the self-resonant capacitor.

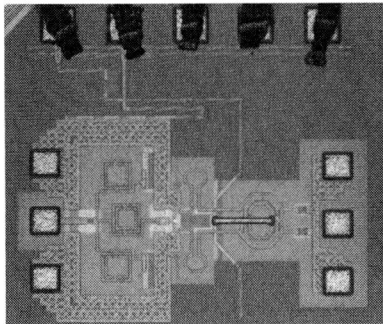

Fig. 6. Micrograph of the 0°/180° PIA.

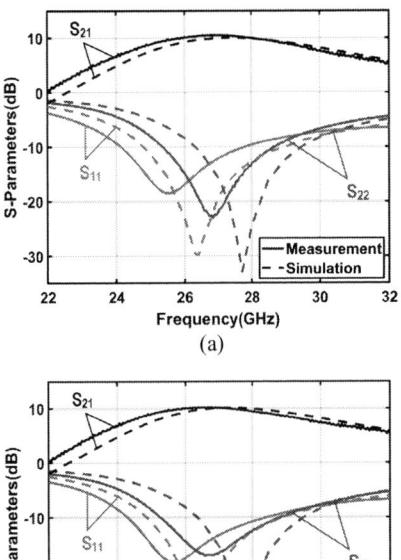

Fig. 7. Simulated (dashed) and measured (solid) S-parameters: (a) 0° state, (b) 180° state.

(L_2, L_1, C_1) matching network is used for broadband input impedance matching. Two inductors (L_3, L_4) in series with the primary coil simplifies the implementation of the 3-port transformer. The passive devices are co-optimized to broaden the bandwidth. For the balance performance, layout optimization techniques including double parallel inductors (L_{1A}, L_{1B}) and self-resonant capacitor (C_8) are discussed as follows.

A. Double parallel inductors technique

Due to the unbalanced structure of L-L-C matching network, its layout is asymmetric as shown in Fig. 4 (a). According to EM simulations, the coupling coefficient k_1 of L_1 and L_3 is 0.05, k_2 is approximately 0 and the output current flowing in L_3 or L_4 is four times the current in L_1. When the transistor connected to L_3 is switched on, I_3 will couple a current equal to $0.02I_1$ in L_1. In contrast, the coupling current is approximately 0 when the other path including L_4 is switched on. As shown by the dotted line in Fig. 4 (c), the unbalanced coupling will deteriorate the balance performance greatly.

For the proposed double parallel inductors technique, L_1 is replaced by 2 identical inductors (L_{1A}, L_{1B}) to ensure a symmetrical layout, as depicted in Fig. 4 (b). Although the unexpected coupling still exists, it has been offset by the proposed technique (i.e., $k_{1A} = k_{1B}$). Note that the value of L_{1A} and L_{1B} is twice than L_1. Fig. 4 (c) shows that the phase/gain balance of the proposed double parallel inductors technique are improved a lot compared with the single inductor.

B. Self-resonant capacitor technique

In single-ended circuits, decoupling capacitors are used to provide a low impedance path for RF signal [8]. The value of the capacitor is generally several pF. However, limited by the dimension of the active area (i.e., 20 μm in Fig. 4), the decoupling capacitor of VDD is hard to exceed 1 pF. Therefore, the self-resonant capacitor is implemented in this work for the decoupling of VDD. Metal-oxide-metal (MOM) capacitors have parasitic inductance in addition to capacitance, thus adjusting the capacitor size can achieve self-resonance across the desired bandwidth. According to the EM simulations depicted in Fig. 5, the self-resonant capacitor presents a considerably small in-band impedance magnitude (i.e., sub-1 from 24-30 GHz). The self-

resonant capacitor provides a low impedance path for RF signal which reduces the leak of RF signal through the line of VDD supply. Therefore, the unbalanced coupling is reduced, and the balance performance is improved.

IV. MEASUREMENT RESULTS

The circuit is implemented in 65-nm CMOS technology and tested on a high-frequency probe station. Fig. 6 is the micrograph of the 0°/180° PIA with a core chip of 450×360 μm². From a 1-V supply, the design consumes 2.4 mW power in both 0° and 180° phase-shift states.

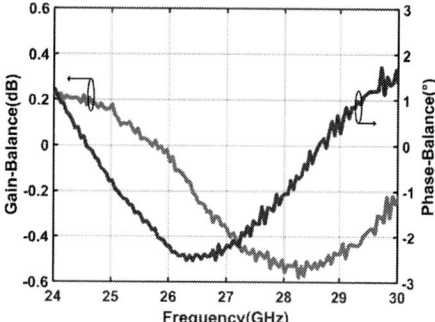

Fig. 8. Measured gain (red) and phase (blue) balance results across the bandwidth.

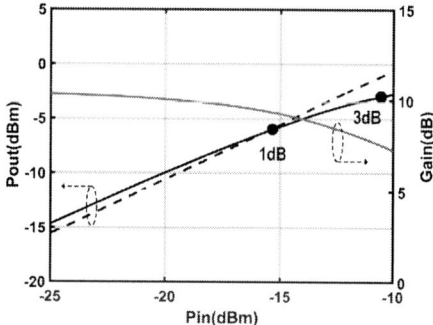

Fig. 9. Measured power gain, output power at 26.5 GHz

Fig. 7 shows the measured and simulated S-parameters of the 2 states. The maximum gain is 10.4 dB at 26.5 GHz with 3-dB bandwidth of over 6 GHz (i.e., 24-30 GHz). The input and output return losses (RL) are higher than 5 dB across the 6 GHz bandwidth. Fig. 8 shows the measured gain/phase balance performance. The gain and phase errors are less than 0.6 dB and 2.5° across the bandwidth, respectively.

The measured power performance is plotted in Fig. 9. The OP_{1dB} is -5.9 dBm and the OP_{3dB} is -2.7 dBm with 2.4 mW DC power consumption.

In Table I, the performance of the proposed PIA is summarized and compared with the prior-art designs. Implemented in the backward technology, this work achieves the highest output power and gain with the least gain error and comparable phase error.

V. CONCLUSION

In this paper, a Ka-band 0°/180° PIA based on a 3-port transformer in 65-nm CMOS technology has been presented.

TABLE I

PERFORMANCE SUMMARY AND COMPARISON

	This work	T-MTT 2015 [4]	JSSC 2016 [7]
Passive Devices	L-C + Transformer	T-L + Balun	T-L + Balun
Technology	65-nm CMOS	120-nm SiGe BiCMOS	45-nm SOI CMOS
Frequency (GHz)	24-30	82-98	55-65
VDD (V)	1	2.7	2.1
Gain (dB)	10.4	4	8
OP_{1dB} (dBm)	-5.9	-10	N/A
Phase Error (°)	2.5	8	2
Gain Error (dB)	0.6	2	1
P_{DC} (mW)	2.4	N/A	N/A

The double parallel inductors technique and the self-resonant capacitor technique are introduced to optimizing the balance performance. The proposed PIA achieves 10.4 dB gain at 26.5 GHz with 3-dB bandwidth of over 6 GHz (i.e., 24-30 GHz). The maximum gain and phase errors are less than 0.6 dB and 2.5° across the bandwidth, respectively. The DC power consumption is only 2.4 mW with 1-V supply.

REFERENCES

[1] U. Kodak, et al, "Bi-directional flip-chip 28 GHz phased-array core-chip in 45nm CMOS SOI for high-efficiency high-linearity 5G systems," IEEE Radio Frequency Integrated Circuits Symposium (RFIC), pp. 61-64, June 2017.

[2] Y. Tousi and A. Valdes-Garcia, "A Ka-band digitallycontrolled hase shifter with sub-degree phase recision," IEEE Radio Frequency Integrated Circuits Symposium (RFIC), pp. 356-359, May 2016.

[3] P. Gu, et al, "Ka-Band CMOS 360° Reflective-Type Phase Shifter with ±0.2 dB Insertion Loss Variation Using Triple-Resonating Load and Dual-Voltage Control Techniques," IEEE Radio Frequency Integrated Circuits Symposium (RFIC), pp. 152-155, June 2018.

[4] A. Natarajan, et al, "W-Band Dual-Polarization Phased-Array Transceiver Front-End in SiGe BiCMOS," IEEE Transactions on Microwave Theory and Techniques, pp. 1989-2002, June 2015.

[5] M. Tsai and A. Natarajan, "60GHz passive and active RF-path phase shifters in silicon," IEEE Radio Frequency Integrated Circuits Symposium (RFIC), pp. 223-226, June 2009

[6] B. Ginsburg, et al, "A multimode 76-to-81GHz automotive radar transceiver with autonomous monitoring," IEEE International Solid - State Circuits Conference - (ISSCC), pp. 158-160, Feb. 2018.

[7] T. Dinc, et al, "A 60 GHz CMOS Full-Duplex Transceiver and Link with Polarization-Based Antenna and RF Cancellation," IEEE Journal of Solid-State Circuits, pp. 1125-1140, May 2016.

[8] B. Razavi, "A Millimeter-Wave CMOS Heterodyne Receiver With On-Chip LO and Divider," IEEE Journal of Solid-State Circuits, pp. 477-485, Feb. 2008.

978-1-7281-5107-6/19 $31.00 © 2019 IEEE

A High-Performance Low Complexity All-Digital Fractional Clock Multiplier

Nahla T. Abou-El-Kheir [*1], Ralph D. Mason [#2], Mingze Li [#3], M.C.E. Yagoub [*4]

School of Electrical Engineering and Computer Science, University of Ottawa, Canada
#*Electronics Department, Carleton University, Ottawa, Canada*

[1]nabou081@ uottawa.ca, [2]rmason@ doe.carleton.ca, [3]mingzeli4@ cmail.carleton.ca, [4]myagoub@uottawa.ca

Abstract—**A fractional Injection Locked Clock Multiplier (ILCM) based on an injection locked Digitally Controlled Ring Oscillator (DCRO) is proposed in this work. Process Voltage Temperature (PVT) calibration was performed using a replica oscillator and a high speed digital counter. The proposed design was fully synthesized and designed in a TSMC 65 nm GP CMOS process with no analog or RF enhancements. The measured phase noise (1 MHz offset) and RMS jitter (1 kHz to 30 MHz) for $3.125 \times f_{ref}$ injection locking were found to be -126 dBc/Hz and 251 fs, respectively. The ILCM consumes 13.25 mW and has a fraction resolution of $f_{ref}/32$.**

I. INTRODUCTION

IN communication and digital systems, clock multipliers play an essential role in the overall system performance by providing precise low jitter clock signals, which can be used in clock generation, clock data recovery, and frequency synthesis [1][2]. Clock multipliers based on Ring Oscillators (ROs) have drawn a lot of interest recently for various reasons, including small silicon area, wide tuning range, and low power. Injection Locked Clock Multipliers (ILCMs) have been investigated intensively to enhance RO spectral purity [1]-[4]. To date, several methods have been used to realize fractional ILCMs as resolution/phase noise trade-off found in integer ILCMs limits their applications. Sequential injection method, which makes use of the multiphase output of the RO [4][5], is generally less complex; however, the frequency resolution is limited compared to other approaches. Digital-to-time converter based methods, wherein the reference signal is shifted in time to be aligned with the oscillator output transition [6], can provide improved frequency resolution but at the expense of higher complexity. A soft injection method, in which the injection amplitude controls the RO phase adjustment [7], can also provide improved resolution and low spurious levels using two PLLs stages to overcome degraded injection efficiency.

In this design, we propose an all-digital ILCM based on a pseudo differential Digitally Controlled Ring Oscillator (DCRO) structure [8]. A novel fractional injection operation utilizes a low complexity sequential injection technique and achieves a resolution of $f_{ref}/32$. In traditional sequential ILCMs [4][5], the fractional resolution is limited to the inverse of the number of RO stages (N_{osc}) or the inverse of the number of $2 \times N_{osc}$. Hence, for better resolution, N_{osc} has to be increased, consequently, this reduces the oscillating frequency of the RO, increases power consumption at a given frequency, and increases jitter due to delay cell mismatch [4].

However, this is not the case in the proposed fractional ILCM as the achieved resolution using only 4 stages RO is $f_{ref}/(8 \times N_{osc})$. To the best of the authors' knowledge, this is the first implementation of a fully synthesizable fractional ILCM using a differential DCRO that achieves phase noise performance comparable to fully customized ILCMs. The following sections describe the proposed ILCM structure and operation, then the measurement results are presented.

II. PROPOSED ILCM ARCHITECTURE

The proposed ILCM architecture is shown in Fig. 1. It consists of a main and a replica DCRO (DCRO1 and DCRO2, respectively). Each DCRO is controlled by a Frequency Code Word (FCW), which is provided to it by the microprocessor (μP). Injector and select signals are only provided to DCRO1, as DCRO2 operates in free running mode when calibration is required. Select signals include SEL_{sign} (used to switch REF_n and REF_p polarity), SEL_{cell} (selects which column of delay cells is used for injection in fractional mode at each reference clock) and $SEL[0:3]$ (decoded version of the binary encoded SEL_{cell} signal). The injector is implemented using four main blocks. First, a pulse generator is used to generate the injection pulse signal for DCRO1 when in injection mode. The reference signal (REF) is fed into an AND gate and a chain of inverters to provide a narrow injection pulse (Pulse) signal. Second, a Single-to-differential signal converter is used to provide differential reference signals (REF_n and REF_p) using two XOR gates and cross coupled inverters to generate symmetric signals. Third, an edge selection block swaps the differential reference signals when SEL_{sign} is high. Fourth, a clock tree to distribute Pulse, REF_n and REF_p signals among the delay cells with minimum clock skew so that injection occurs at the same time for all delay cells in the selected column (i.e. whichever column of cells has their $SEL[i]$ signal high).

The main and replica DCROs are each composed of parallel delay cells as shown in Fig. 2, where the first row is always enabled unless the oscillator is turned off. The rest of delay cells are enabled one after the other to increase the frequency using enable signals $EN[0:255]$. Each DCRO delay cell consists of pseudo differential delay logic and injection logic for forcing the internal delay cell signals to align with the differential reference clock signals. For DCRO2, the injection logic is included so that both DCROs will have similar free running frequencies; however, the logic is disabled by connecting the injection logic inputs to logic 0.

978-1-7281-5107-6/19 $31.00 © 2019 IEEE

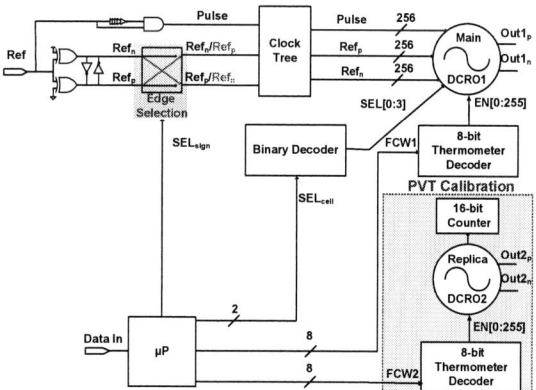

Fig. 1. ILCM diagram.

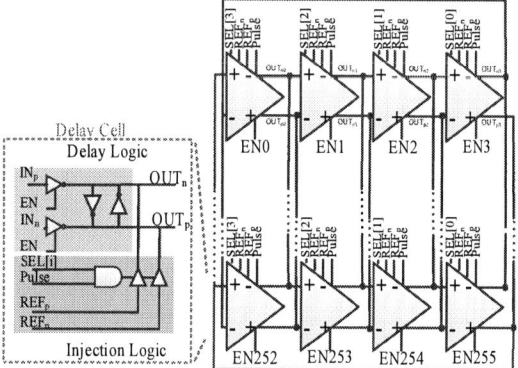

Fig. 2. DCRO structure.

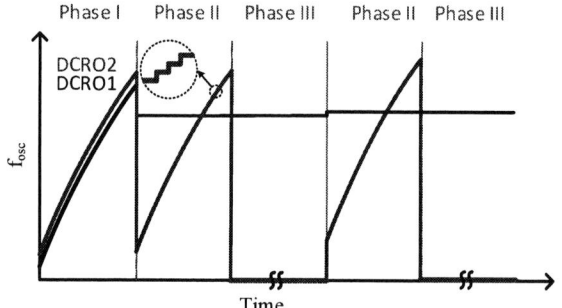

Fig. 3. Calibration algorithm timing diagram.

In the proposed calibration method, the DCRO2 and a high-speed counter are used to track real time Process Voltage Temperature (PVT) variations. The timing diagram of the proposed calibration method is shown in Fig. 3. In phase I, DCRO1 and DCRO2 are operated sequentially to obtain the tuning curves from which the frequency differences between DCRO1 and DCRO2 are calculated. The frequency tuning curves are discrete and DCROs' frequency readings are not exactly the same due to variations in the delay cells and different Place and Route (P&R) parasitics. In phase II, the DCRO2 tuning curve is obtained and FCW1 is updated based

on the frequency differences obtained at phase I and the DCRO2 frequency reading in phase II. This phase takes less than 200 μs. In phase III, the calibration is paused and DCRO2 is turned off for a programmable period of time (typically longer than the phase II duration). Phase II and phase III are repeated at some regular update intervals. The proposed technique fully compensates for PVT variations between DCRO1 and DCRO2.

III. FRACTIONAL INJECTION LOCKING

In the proposed injection technique, the transitions of delay cell outputs (OUT_p and OUT_n) are forced to occur at the same time relative to REF_n and REF_p at each reference clock cycle. As shown in Fig. 2, when Pulse and SEL[i] are high, REF_n and REF_p are injected into OUT_p and OUT_n, respectively. Only one column can be selected for injection on each reference clock cycle based on the values of the SEL[0:3] signals. This injection technique does not require a precise pulse signal generator since injection occurs within the pulse window and is not controlled by the pulse directly. The injection point depends on the differential reference clocks transition time not the injection pulse timing. However, injection methods which shorts the outputs when injection locked [2][3] require generating a precise narrow pulse signal at high frequencies, which is difficult to achieve over PVT. Hence, the proposed injection method overcomes this challenge and relaxes the requirements on the injection pulse at the expense of additional injection circuitry.

To achieve fractional resolution, the rising edge of the reference signal is injected into different delay cells in a rotational sequence to achieve a fractional injection ratio. As can be seen in Fig. 1, the SEL[i] signals are responsible for selecting different columns of delay cells that are to be injected at the next reference edge. Whereas, the SEL_{sign} determines whether the positive or negative edge of OUT_n and OUT_p will be injected by swapping the REF_p and REF_n signals. This feature further enhances the frequency resolution. FCW1 controls the integer multiplication (N), while SEL[i] and SEL_{sign} control the fractional multiplication (M) of DCRO1. Hence, the output frequency is a fractional multiple (K) of the reference signal. Fig. 4 shows the timing diagram for the REF, SEL[i] and Pulse signals with N=3 and M=1/8. It can be noted that the SEL[i] and SEL_{sign} signals change on the negative edge of REF to allow the signals to settle before the next injection pulse and minimize switching noise. Moreover, they are stored in a 3×128 shift register that provides a programmable fractional frequency multiplication factor (K) with a frequency resolution of $1/32 \times f_{ref}$. The fractional frequency can be expressed as:

$$f_{osc} = Kf_{ref} = (N + M)f_{ref} \qquad (2)$$

and M is given by:

$$M = \frac{PH_{av}}{2 \times N_{osc} \times i} \qquad (3)$$

978-1-7281-5107-6/19 $31.00 © 2019 IEEE

where i is the number of consecutive injections per stage and PH_{av} the average phase shift per reference cycle, which depends on SEL_{sign} and the number of delay cells that are skipped per reference cycle. As an example, if 9/16 fraction is chosen, i is set to 1 (because the fraction is greater than 1/8) and PH_{av} must be set to 4.5 to achieve the required fraction. It can be set to 4.5 by choosing the injection points to rotate each reference cycle as $0{\rightarrow}5{\rightarrow}1{\rightarrow}6{\rightarrow}2{\rightarrow}7{\rightarrow}3$ and so on, as shown in Fig. 5. Moreover, i is set to 1 by injecting each of these points once.

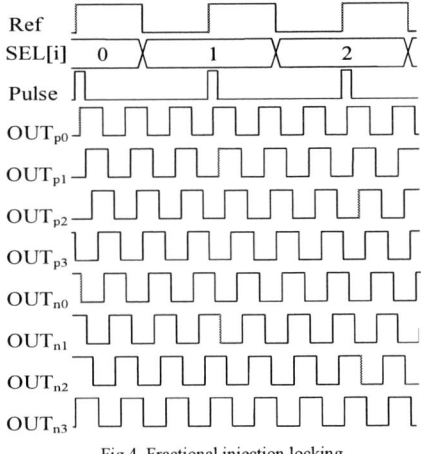

Fig.4. Fractional injection locking.

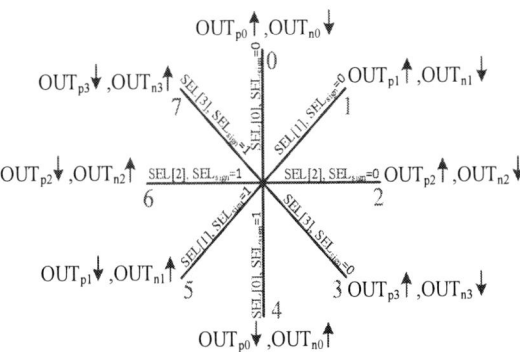

Fig. 5. Fractional phase diagram

IV. MEASUREMENT RESULTS

The proposed design was implemented with the aid of commercially available CAD tools using a standard cell digital design flow. It was fabricated in TSMC 65 nm, the die micrograph is shown in Fig. 6. The ILCM, including replica DCRO, µP, injector, decoders and counter, has a total die area of 0.021 mm². The DCRO lowest output frequency is 1 GHz and the highest output frequency is 1.8 GHz. The use of a differential structure and decoupling capacitors placed close to the DCROs, as well as physically separating the DCROs on the die, helped reduced frequency pulling between the DCROs. A square wave reference signal was generated using a R&S SMA100B source generator and

R&S FSWP was used for measuring the phase noise and jitter. The phase noise and jitter performance at K=3.125 and K=3.0625, as well as the source generator, are shown in Fig. 7. It can be seen that at 1.1 V and 1.7 GHz, the phase noise at 1 MHz offset and RMS jitter from 1 kHz to 30 MHz for the above respective K values were found to be -126 and -125 dBc/Hz, and 251 and 253 fs, respectively. The frequency spectrum for K = 3.125 can be seen in Fig. 8 where the largest and closest fractional spur (68 MHz offset) is at -48 dBc. The fractional spurs would contribute 517 fs of jitter if found in band. Furthermore, the power consumption at 1.7 GHz is 13.25 mW. Fig. 9 shows the RMS integrated jitter from 1kHz-30MHz (including spurs if in band) and fractional spurs for fractions (M) from 1/32 up to 16/32.

Fig. 10 shows the maximum absolute frequency error between estimated free running values of DCRO1 and its free running frequency values, across all frequencies, using the PVT calibration algorithm discussed in section II. The initial calibration was done at 1 V and 30 °C then the voltage was swept by 20% and the temperature was increased to 50 ☐ and 70 ☐. The worst-case error across all frequencies was 4.3 MHz (0.25 %), which is less than the smallest step size of DCRO1 (5 MHz). This indicates that the calibration technique used in this work is capable of estimating DCRO1 free running frequency effectively under PVT variations.

The ILCM performance is listed in Table 1 and compared to prior fractional injection designs. This work shows good performance in terms of jitter, phase noise, area, FOM[a] and FOM[b].

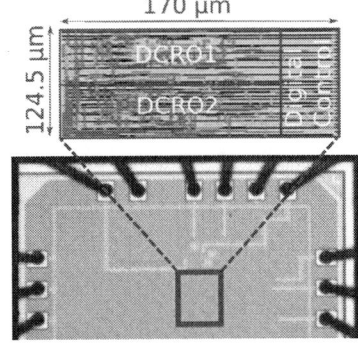

Fig. 6. Die Microphotograph of ILCM.

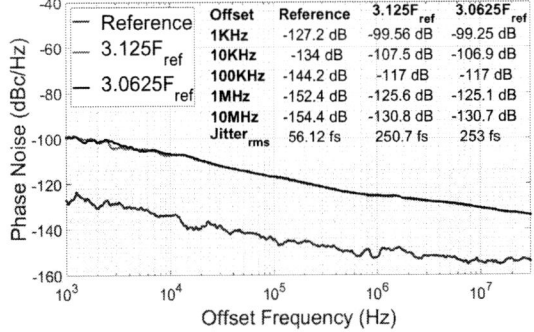

Fig. 7. Measured phase noise for K=3.125 and K=3.0625 at 1.7 GHz.

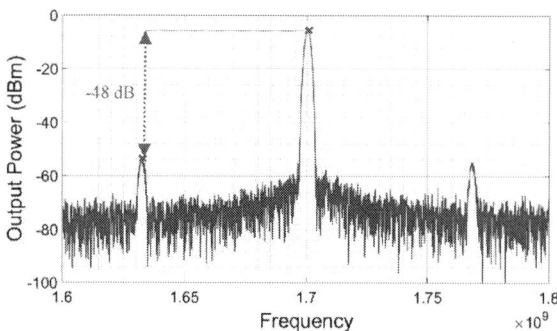

Fig. 8. Frequency spectrum at 1.7 GHz, 1.1 V and K=3.125

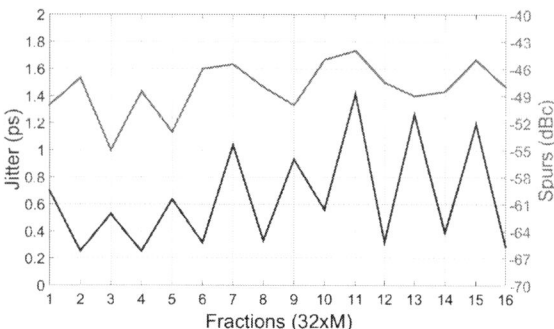

Fig. 9. Jitter and fractional spurs for M from 1/32 to 16/32

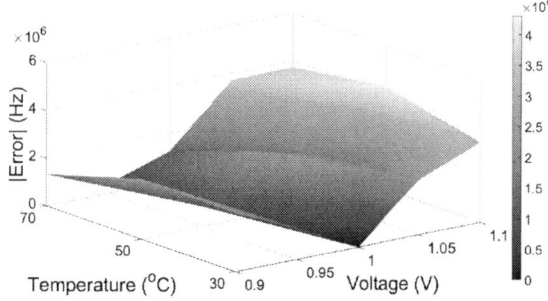

Fig. 10. Maximum absolute frequency error between estimated and free running DCRO1 frequencies.

V. CONCLUSION

This paper proposes a low complexity high-performance ILCM. All design cells were synthesized employing vendor supplied digital cells and automatically placed and routed to improve scalability and reduce design time. The proposed design has the advantage of reducing the fractional resolution by at least four times the traditional sequential fractional designs without reducing the oscillating frequency or sacrificing the jitter. Finally, the proposed design occupies only 0.021 mm² and shows excellent jitter, phase noise, and FOM when compared to other recent designs.

TABLE I
PERFORMANCE SUMMARY

	[4] ISSCC'12	*[5]* JSSC'16	*[6]* ISSCC'14	*[7]* ISSCC'15	*This work*
Freq.(GHz)	0.58	1.2-2	1.6-1.9	0.8-1.7	1-1.8
Reference Frequency (MHz)	32	400	50	380	544
Power (mW)	10.5	3.6	3	3	13.25
RMS jitter min/max (ps)	2.42*/8.05* 100- 40M	0.461/NA 1k-30MHz	1.15*/1.4* 30k-30M	3.6*/NA 1k-100M	0.251/1.4* 1k-30M
PN at 1 MHz (dBc/Hz)	-113 @ 0.58 GHz	-123 @ 1.6 GHz	-115 @ 1.6 GHz	-114 @ 1.5 GHz	-126 @ 1.7 GHz
Fractional Spur (dBc)	NA	-43	-47	NA	-48
Frac. Resolution	$f_{ref}/32$	$f_{ref}/10$	$f_{ref}/2^{18}$	$f_{ref}/172$	$f_{ref}/32$
Area(mm^2)	0.158	0.032	0.4	0.048	0.021
Topology	IL-PLL	IL-DLL	MDLL	Soft IL-PLL	ILCM
FOM[a] (dB)	-222	-241	-234	-224	-241
FOM[b] (dB)@1MHz	-158	-181	-174	-173	-179
Fully Synthesizable?	No	No	No	Yes	Yes

*Jitter including fractional spurs in band

$$FOM^a = 20\log\left(\frac{T_t}{1s}\right) + 10\log\left(\frac{P}{1mW}\right)$$

$$FOM^b = L\{f_{offset}\} - 20\log\left(\frac{f_o}{f_{offset}}\right) + 10\log\left(\frac{P}{1mW}\right)$$

REFERENCES

[1] K. H. Teng and C. H. Heng, "A 370-pj/b multichannel BFSK/QPSK transmitter using injection-locked fractional-n synthesizer for wireless biotelemetry devices," IEEE J. Solid-State Circuits, vol. 52, no. 3, pp. 867–880, Jan. 2017.

[2] D. Cooms, A. Kholy, R. K. Nandwana, A. Elmallah, and P. K. Hanumolu, "A 2.5-to-5.75GHz 5mW 0.3psrms-jitter cascaded ring-based digital injection-locked clock multiplier in 65nm CMOS," International Solid-State Circuits Conference (ISSCC) , Feb. 2017.

[3] R. Wang and F. f. Dai, "A 0.8~1.3 GHz multi-phase injection-locked PLL using capacitive coupled multi-ring oscillator with reference spur suppression," Custom Integrated Circuit Conference (CICC), TX, May 2017.

[4] P. Park, J. Park, H. Park, and S. H. Cho, "An all-digital clock generator using a fractionally injection-locked oscillator in 65nm CMOS," International Solid-State Circuits Conference (ISSCC), Feb. 2012.

[5] M. Kim, S. Choi, and J. Choi, "A low-jitter and fractional-resolution injection-locked clock multiplier using a DLL-based real-time pvt calibrator with replica-delay cells," IEEE J. of Solid-State Circuits, vol. 51, no. 2, pp. 401-411, Feb. 2016.

[6] G. Marucci, A. Fenaroli, G. Marzin, S. Levantino, C. Samori, and A. Lacaita, "A 1.7GHz MDLL-based fractional-N frequency synthesizer with 1.4ps RMS integrated jitter and 3mW power using a 1b TDC," International Solid-State Circuits Conference (ISSCC), Feb. 2014.

[7] W. Deng, D.Yang, A. T. Narayanan, K. Nakata, T. Siriburanon, K. Okada and A. Matsuzawa "A 0.048mm2 3mW synthesizable fractional-N PLL with a soft injection-locking technique," International Solid-State Circuits Conference (ISSCC), Feb. 2015.

[8] N. T. Abou-El-Kheir, R. D. Mason, M. Li and M. C. E. Yagoub, "A 65 nm Compact High Performance Fully Synthesizable Clock Multiplier Based on an Injection Locked Ring Oscillator," IEEE International Symposium on Circuits and Systems (ISCAS), Florence, 2018, pp. 1-5.

A DC-43.5 GHz CMOS Switched-Type Attenuator with Capacitive Compensation Technique

Peng Gu[1,2] and Dixian Zhao[1,2,*]

[1] National Mobile Communication Research Laboratory, Southeast University, Nanjing 210096, China
[2] Purple Mountain Laboratories, Nanjing 211100, China
* Email: dixian.zhao@seu.edu.cn

Abstract—This paper presents an ultra-broadband 5-bit switched-type attenuator (STA). The resistor values of the five attenuation cells are optimized to ensure accurate amplitude tuning. Capacitive compensation technique is adopted to reduce phase error and enhance the bandwidth. Fabricated in 65-nm CMOS technology, the 5-bit STA occupies a core chip area of 0.036 mm^2. It exhibits ultra-broadband (i.e., DC to 43.5 GHz) operation with 15.5-dB amplitude tuning range and 0.5-dB tuning step. The insertion loss of the reference state is 1.5 – 5.3 dB from DC to 43.5 GHz. The RMS amplitude error and phase error are < 0.19 dB and < 3.1° over the whole measured band.

Keywords—*phased-array system, amplitude tuning, switched-type attenuator, capacitive compensation, CMOS*

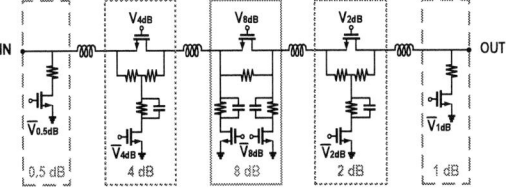

Fig. 1. Schematic of the 5-bit STA with capacitive compensation technique. Compensation capacitors are empoyled by the 2-, 4- and 8-dB cells.

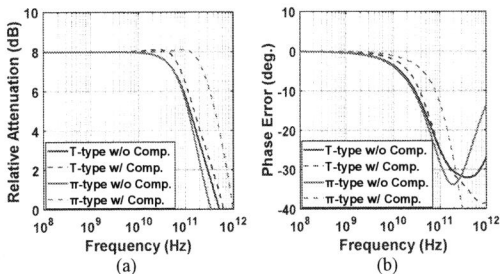

Fig. 2. Calculated (a) relative attenuation and (b) phase error of an 8-dB cell using the T-type or the Π-type topology, with and without capacitive compensation.

I. INTRODUCTION

Gain control blocks are employed in each element of a phased-array system to compensate for gain variations between elements and reduce side lobes [1]. For a linear array using Taylor synthesis [2], large amplitude tuning range is needed to better suppress the side lobes. For example, more than 12 dB amplitude tuning is required for a -30 dB side-lobe level (SLL).

Variable gain amplifiers (VGAs) and attenuators have been widely used to accomplish amplitude control [3]-[7]. VGAs are capable of tuning amplitude and providing power gain at the same time, but with power consumption. In contrast, the attenuators consume zero DC power and can operate over a wide band. Among the passive attenuators [4], the switched-type attenuator (STA) has the advantage of compact chip size, direct digital control and relatively low insertion loss. However, challenges remain for the STA design. (1) Attenuation cells are sensitive to load impedance, which potentially degrades amplitude tuning accuracy. (2) The operation frequency is limited by the parasitic capacitance of the switches. (3) Phase variation among all the states should be carefully compensated to reduce the phase error of the phased-array system.

This paper presents an ultra-broadband 5-bit STA with 15.5-dB amplitude tuning range and 0.5-dB tuning step. The resistor values of each cell are optimized to ensure broadband impedance matching, which is essential for accurate amplitude tuning. Capacitive compensation is employed to reduce phase error and enhance the bandwidth. Measured results show that the proposed 5-bit STA achieves ultra-broadband operation

from DC to 43.5 GHz and accurate amplitude tuning with < 0.19 dB RMS amplitude error and < 3.1° RMS phase error.

II. CIRCUIT DESIGN

STA is composed of cascaded attenuation cells to satisfy the demand of large amplitude tuning range. Fig. 1 shows the schematic of the proposed 5-bit STA. The five cells are designed to achieve 0.5-, 1-, 2-, 4- and 8-dB attenuation so as to cover 15.5-dB amplitude tuning range with 0.5-dB step.

The 0.5- and 1-dB cells are accomplished by the simplified T-type topology. Without series switch or series resistor in the signal path, the two attenuation cells introduce ultra-low insertion loss. When the attenuation gets high, the impedance mismatch of the simplified T-type topology becomes intolerable. Thus, it is not suitable for high attenuation cells. In contrast, the T-type topology (see the 2-dB cell in Fig. 1) or the Π-type topology (see the 8-dB cell in Fig. 1) can provide high attenuation and ensure impedance matching simultaneously, since they employ both series and shunt resistors. To achieve broadband impedance matching and thus ensure accurate

---

This work was supported in part by the National Key Research and Development Program of China under Grant 2018YFB180088, in part by the National Nature Science Foundation of China under Grant 61674035, in part by the Nature Science Foundation of Jiangsu Province under Grant BK20160690, and in part by the Fundamental Research Funds for the Central Universities. (*Corresponding author: Dixian Zhao.*)

IEEE Asian Solid-State Circuits Conference
November 4 – 6, 2019
The Parisian Macao, Macao SAR, China

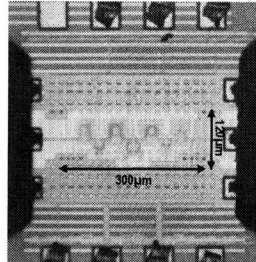

Fig. 3. Micrograph of the 5-bit STA.

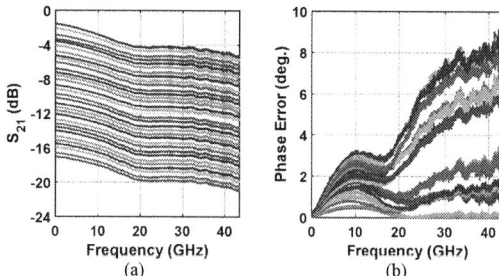

Fig. 4. Measured (a) insertion loss and (b) phase error versus frequency for all the 32 states.

TABLE I.   COMPARISON OF THE PRIOR-ART ATTENUATORS

Reference	This work	[5]	[6]	[7]
Technology	65-nm CMOS	0.13-μm BiCMOS	0.25-μm BiCMOS	55-nm BiCMOS
Bandwidth (GHz)	DC-43.5ᵃ	DC-20	6-12.5	71-76
Att. range (dB)	15.5	31.5	16.5	7.1
Att. step (dB)	0.5	0.5	0.26	0.25
Insertion loss (dB)	1.5-5.3	1.7-7.2	< 12.7	< 2.2
Return loss (dB)	> 14	> 12	> 13	> 18
RMS amplitude error (dB)	0.19	0.37	0.26	NA
RMS phase error (deg.)	3.1	4.0	3.5	0.9
Core area (mm²)	0.036	0.14	0.29	0.053

ᵃ Limited by the measurement setup.

III. MEASUREMENT RESULTS

The STA prototype is implemented in 65-nm CMOS technology. The micrograph is shown in Fig. 3, with a core chip area of 0.036 mm². The chip has been measured on a high-frequency probe station. The S-parameters are measured by a vector network analyzer operating up to 43.5 GHz.

In Fig. 4(a), the measured S_{21} of the 5-bit STA is shown. As indicated, the proposed STA achieves 15.5-dB amplitude tuning range and 0.5-dB tuning step over DC–43.5 GHz (limited by measurement setup). In the reference state, the insertion loss over the measured band is between 1.5 and 5.3 dB. Fig. 4(b) depicts the phase error among all the states. As expected, the phase error becomes large when the frequency increases. Though, it is small with a maximum value of 9°. The calculated RMS amplitude error and RMS phase error is < 0.19 dB and < 3.1° across the whole band, indicating that accurate amplitude tuning is achieved.

In Table II, the performance of the proposed STA is summarized and compared with the prior-art designs. The proposed STA exhibits ultra-broadband operation (DC–43.5 GHz) with the most compact core chip area. It achieves accurate amplitude tuning with low insertion loss (1.5–5.3 dB), low RMS amplitude error (< 0.19 dB), low RMS phase error (< 3.1°) and broadband impedance matching simultaneously.

amplitude tuning, the series and shunt resistors are optimized by solving $S_{11} = 0$ for the T-type and Π-type topologies.

The off-capacitance of the switches degrades the amplitude tuning accuracy at high frequencies and introduces phase error between the reference and attenuation states. To enhance the operation bandwidth and reduce the phase error, the capacitive compensation technique is employed by the T-type topology for the design of the 2- and 4-dB cells (see Fig. 1). Though, for the design of the 8-dB cell, the Π-type topology with capacitive compensation is preferred. In Fig. 2, a comparison is made between the T-type and Π-type topologies with and without compensation. The calculated relative attenuation is shown in Fig. 2(a). As the operation frequency increases, the relative attenuation becomes less than 8 dB due to the impact of the off-capacitance. While the operation bandwidth can be improved by using capacitive compensation for both topologies, the Π-type topology provides higher operation frequency. Shown in Fig. 2(b) is the phase error versus frequency. As depicted, the phase error becomes large when the operation frequency increases. With capacitive compensation, the phase error can be reduced by both topologies. Note that the Π-type topology with compensation can better reduce the phase error than the T-type topology. Thus, the 8-dB attenuation cell is implemented by the compensated Π-type topology.

Since the attenuation cells exhibit capacitive impedance due to the parasitic capacitance as well as the compensation capacitance, series inductors are inserted between the cells for impedance matching (see Fig. 1). Note that the impedance matching is carried out at 30 GHz to enhance the high-frequency performance.

REFERENCES

[1] B. Sadhu, J. F. Bulzacchelli and A. Valdes-Garcia, "A 28GHz SiGe BiCMOS phase invariant VGA," in *IEEE RFIC*, 2016, pp. 150-153.

[2] J. -Y. Li, Y. -X. Qi and S. -G. Zhou, "Shaped Beam Synthesis Based on Superposition Principle and Taylor Method," *IEEE Trans. Antennas Propag.*, vol. 65, no. 11, pp. 6157-6160, Nov. 2017.

[3] Y. Yi, D. Zhao and X. You, "A Ka-band CMOS Digital-Controlled Phase-Invariant Variable Gain Amplifier with 4-bit Tuning Range and 0.5-dB Resolution," in *IEEE RFIC*, 2018, pp. 152-155.

[4] B. Ku and S. Hong, "6-bit CMOS Digital Attenuators With Low Phase Variations for$X$-Band Phased-Array Systems," *IEEE Trans. Microw. Theory Tech.*, vol. 58, no. 7, pp. 1651-1663, July 2010.

[5] I. Song, *et al.*, "Design and Analysis of a Low Loss, Wideband Digital Step Attenuator With Minimized Amplitude and Phase Variations," *IEEE J. Solid-State Circuits*, vol. 53, no. 8, pp. 2202-2213, Aug. 2018.

[6] M. Davulcu, *et al.*, "7-Bit SiGe-BiCMOS Step Attenuator for X-Band Phased-Array RADAR Applications," *IEEE Microw. Wireless Compon. Lett.*, vol. 26, no. 8, pp. 598-600, Aug. 2016.

[7] T. N. Ross, K. T. Ansari, S. Tiller and M. Repeta, "A 5-bit, 0.25 dB Step Variable Attenuator at E-Band," in *IEEE RFIC*, 2018, pp. 156-159.

978-1-7281-5107-6/19 $31.00 © 2019 IEEE

6-4 (7111)

A 28-GHz Compact SPDT Switch Using LC-Based Spiral Transmission Lines in 65-nm CMOS

Xiangyu Meng[1], Zhenpeng Zheng[1], Jiaqi Zhang[1] and Patrick Yue[2]
1 SEIT, SUN YAT-SEN University, Guang Zhou, China
2 HKUST-Qualcomm Optical Wireless Lab, Hong Kong University of Science and Technology, Hong Kong, China
Email: mengxy26@mail.sysu.edu.cn

Abstract—A novel LC-based spiral transmission line consisting of only one three-turn inductor and two capacitors is proposed for the first time. Based on this structure, the function of the quarter-wavelength transmission line can be realized with smaller area and less insertion loss compared with conventional micro-strip transmission lines. By using the LC-based spiral transmission line, a compact Single Pole Double Throw (SPDT) switch is implemented in 65-nm CMOS. The SPDT switch occupies only 0.028 mm² with measured 1.3-dB insertion loss and 24-dB isolation at 28 GHz.

Keywords—SPDT, transmission line, millimeter-wave, CMOS

I. INTRODUCTION

On-chip integrated RF switches enable transceiver systems to reduce the number of antennas by half, thereby reducing package size and system cost. In recent years, high-performance millimeter-wave CMOS SPDT switches have been presented with the continuous development of CMOS technology. Among the millimeter-wave SPDT designs, quarter-wave shunt switch has been the dominant topology [1]. However, the on-chip quarter-wavelength becomes too long in K and Ka bands, which results in a bulky layout. Although the layout area could be reduced by folding the transmission lines, considering the spacing leaved for crosstalk minimization, the area of the SPDT switch using meander transmission lines is still large [2]. The SPDT switch using inductors to replace quarter-wavelength transmission lines has smaller area but suffers from reduced bandwidth and more insertion loss [3][4]. In this paper, a novel structure of quarter-wavelength transmission line is created based on lumped inductors and capacitors. The proposed transmission line structure helps the SPDT switch reduce the area while maintaining great performance.

II. DESIGN OF BUILDING BLOCK

A. LC-based spiral transmission line

The schematic of the proposed LC-based spiral transmission line is shown in Fig. 1 (a). The transmission line consists of two parallel capacitors C_{C1}, C_{C2} and three inductors L_{C1}, L_{C2}, L_{C3}, in which the three inductors are coupled by

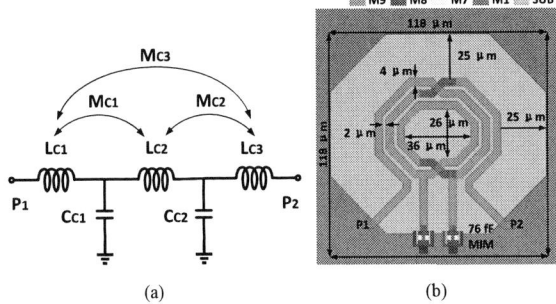

Fig. 1. (a) Schematic and (b) Layout of the proposed LC-based spiral transmission line.

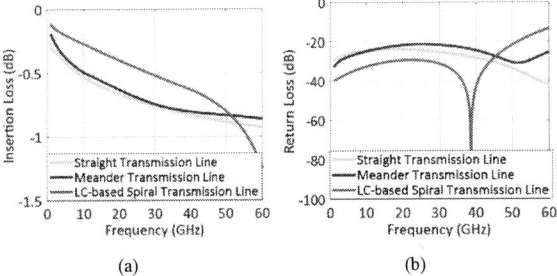

Fig. 2. Simulated (a) Insertion loss and (b) Return loss of the straight transmission line, the meander transmission line and the LC-based transmission line.

mutual inductance M_{C1}, M_{C2} and M_{C3}, respectively. At first glance, the proposed LC-based structure looks like a Chebyshev LC low pass filter. However, the mutual coupling among the inductors helps the structure maintain the transmission line characteristics in a wide frequency range. Meanwhile, mutual coupling is introduced by overlapping the three inductors, which reduces the total area. Considering the inductance of the three inductors is set to L, the capacitance of the two capacitors is set to C, the mutual inductance is set to M and the load impedance connected to port P_2 is Z_0, the input impedance of port P_1 is given by

$$Z = \frac{C^2 s^5 (L^3 - 3LM^2 + 2M^3) + Z_0 C^2 s^4 (L^2 - M^2) + 4Cs^3 (L^2 + LM - 2M^2) + Z_0 Cs^2 (3L + 2M) + 3s(L + 2M) + Z_0}{C^2 s^4 (L^2 - M^2) + Z_0 C^2 s^3 L + Cs^2 (3L + 2M) + 2Z_0 Cs + 1} \quad (1)$$

978-1-7281-5107-6/19 $31.00 © 2019 IEEE

79

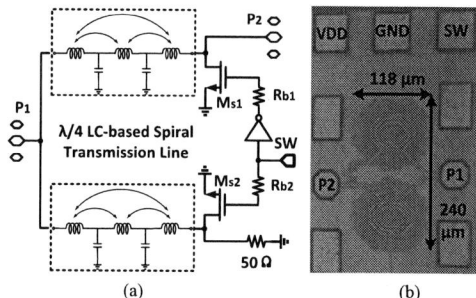

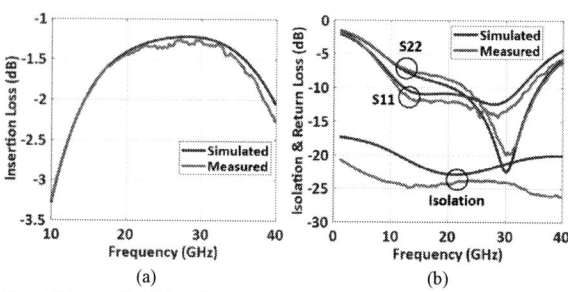

Fig. 3. (a) Schematic and (b) Micrograph of the proposed SPDT switch.

Fig. 4. Measured and Simulated results of (a) insertion loss, (b) isolation and return loss.

It could be derived that $Z=Z_0$ holds for transmission lines within half wavelength when the coupling coefficient k is set to 0.625 and the LC value is defined by $Z_0^2 = 3.38L/C$. Meanwhile, the phase shift of the proposed transmission line is 90 degrees and the impedance transformation function is realized simultaneously at frequency $f_0 = \sqrt{0.2438/LC}/2\pi$. Based on the proposed transmission line model, a 30-GHz quarter-wavelength transmission line is realized in 65-nm CMOS process, as shown in Fig. 1 (b). The three inductors in the model are routed as a three-turn inductor and the inner two taps are connected to two MIM capacitors, respectively. The three-turn inductor is mainly implemented by the top thick metal with 4-μm width and 670-μm length. The area of the proposed quarter-wavelength transmission line is 118 μm × 118 μm. For reference, a straight micro-strip transmission line and a meander micro-strip transmission line are realized by the top thick metal with the same 4-μm width in 65-nm CMOS. The transmission line length is set to 1230 μm and 1286 μm respectively to maintain 90 degrees phase shift at 30 GHz. The simulated small signal parameters are plotted in Fig. 2. It can be denoted from the figures that the return loss of the proposed transmission line is superior to that of the other two below 44 GHz and the insertion loss performs better within 50 GHz.

B. SPDT switch

Fig. 3. (a) shows the schematic of the proposed SPDT switch. Two quarter-wavelength transmission lines are implemented based on the LC-based spiral transmission line structure shown in Fig. 1. (b). One side of both transmission lines is connected to port P_1 and the other side of the two transmission lines is connected to two shunt NMOS transistors M_{S1} and M_{S2} (W/L=60μm/60nm), respectively. The on-off state of the two transistors is controlled by reverse voltage 0 V and 1.2 V. The control voltage is generated by a on-chip inverter and fed via two resistors R_{b1} and R_{b2} (R=10 kΩ). Due to the symmetry of the SPDT switch, the drain of transistor M_{S2} is connected to a 50 ohm on-chip resistor, while the drain of transistor M_{S1} is connected to port P_2. The two-port network is supposed to present the insertion loss, isolation, and return loss performance of the SPDT switch.

III. MEASUREMENT RESULTS

The proposed compact SPDT switch using LC-based spiral transmission lines was fabricated in a 65-nm CMOS process (Fig. 3(b)). The core size of the SPDT switch is 118 μm × 240 μm. The measured and simulated results of insertion loss,

isolation and return losses are shown in Fig. 4. The measured minimum insertion loss is 1.3 dB at 28 GHz. The measured isolation is above 23.7 dB from 20 GHz to 35 GHz. The input and output return losses are measured to be < -14 dB and < -15 dB, respectively. Table I summarized the measured results and compared with the state-of-the-art performances. The proposed SPDT switch has competitive performance of bandwidth, insertion loss, isolation and linearity. Meanwhile, the SPDT switch implemented by the proposed LC-based spiral transmission lines consumes one tenth of chip area of the SPDT switch using conventional λ/4 transmission lines.

TABLE I. PERFORMANCE COMPARISON

	This work	[1]	[3]	[4]
Technology	65-nm CMOS	90-nm CMOS	65-nm CMOS	65-nm CMOS
Topology	λ/4-shunt	λ/4-shunt	Lumped combiner	Lumped combiner
Freq. (GHz)	20-35	50-70	57-66	65-77
BW (%)	55%	33%	15%	17%
IL (dB)	1.3-1.5	1.5-1.9	1.7-2	1.8-2.6
ISO (dB)	23.7-25.7	25-27	21.1-22	25-30
IP$_{1dB}$ (dBm)	10.8*	13.5*	13.4*	10
Area (mm²)	0.028	0.28	0.02	0.015

*simulated results

ACKNOWLEDGMENT

The authors would like to thank the Department of Science and Technology of Guangdong Province for funding support under the Key Areas Research & Development Program (HKUST contract no. GDST19EG05).

REFERENCES

[1] M. Uzunkol and G. Rebeiz, "A Low-Loss 50-70 GHz SPDT Switch in 90 nm CMOS," in IEEE Journal of Solid-State Circuits, vol. 45, no. 10, pp. 2003-2007, Oct. 2010.

[2] J. G. Yang, M. Kim and K. Yang, "An InGaAs PIN-diode based broadband traveling-wave switch with high-isolation characteristics," 2009 IEEE International Conference on Indium Phosphide & Related Materials, Newport Beach, CA, 2009, pp. 207-209.

[3] J. He, Y. Xiong and Y. P. Zhang, "Analysis and Design of 60-GHz SPDT Switch in 130-nm CMOS," in IEEE Transactions on Microwave Theory and Techniques, vol. 60, no. 10, pp. 3113-3119, Oct. 2012.

[4] R. Shu and Q. J. Gu, "A Transformer-Based V-Band SPDT Switch," in *IEEE Microwave and Wireless Components Letters*, vol. 27, no. 3, pp. 278-280, March 2017.

A 20-GHz Ultra-Low-Power LNA Using g_m-Boosted and Current-Reuse Techniques in 65-nm CMOS for Satellite Communication Terminals

Jiajun Zhang[1,2] and Dixian Zhao[1,2,*]

[1] National Mobile Communication Research Laboratory, Southeast University, China
[2] Purple Mountain Laboratories, Nanjing, China
[*] Email: dixian.zhao@seu.edu.cn

Abstract—A 20-GHz low-power low-noise amplifier (LNA) in 65-nm CMOS is presented. The LNA is cascaded with a single-ended g_m-boosted common-gate (CG) stage and a differential neutralized common-source (CS) stage. The current-reuse technique is employed to save the power consumption. The LNA achieves a measured power gain of 14.9 dB at 21 GHz with a –3-dB bandwidth of 4.8 GHz. The lowest noise figure (NF) is 3.3 dB at 19.5 GHz. The LNA consumes 1.9 mW from a 1-V supply, with a chip area of 600 μm × 700 μm.

Keywords—*CMOS, current-reuse, K-band, low-noise amplifier (LNA), low-power, satellite communication.*

I. INTRODUCTION

The next generation high throughout satellite (HTS) communication will use *K*-band (17.7–20.2 GHz) and *Ka*-band (27.5–30 GHz) as downlink and uplink spectrum for high data-rates [1] and demand the application of large scale phased-array technique. The power consumption of phased-array RF front-end is an essential issue. In addition, low noise figure (NF) is required to guarantee the communication quality. Hence, low-power low-noise amplifiers (LNA) play an important role in phased-array RF front-end for HTS. Traditionally, the RF front-ends for satellite communication are mainly based on III-V technologies [2]. However, advanced CMOS technology is now widely applied to millimeter-wave communication because of its high speed, low cost, low power and high integration. In this paper, an ultra-low-power LNA in 65-nm CMOS is presented. The power consumption is minimized while gain and NF performances are maintained thanks to the proposed g_m-boosted and current-reuse techniques.

II. LNA DESIGN CONSIDERATION AND IMPLEMENTATION

Common-source (CS) stage with source degeneration is the most widely used topology for simultaneous noise and input matching, but it is normally intended for narrowband applications. To broaden the bandwidth, high-order input matching network should be utilized with the penalty of insertion loss and chip area. Moreover, the relatively high input impedance of CS stage is a problem in low-power design. The common-gate (CG) LNA has an input impedance of about $1/g_m$, which makes it attractive despite relatively low gain and high NF. Fig. 1(a) shows the schematic of the proposed CG-CS LNA. A single-ended g_m-boosted CG stage is designed to interface with the antenna, cascaded with a differential neutralized CS pair. Transformer XF_2 is placed between the two stages for inter-

This work was supported in part by the National Key Research and Development Program of China under Grant 2018YFB180088, in part by the National Nature Science Foundation of China under Grant 61674035, in part by the Nature Science Foundation of Jiangsu Province under Grant BK20160690, and in part by the Fundamental Research Funds for the Central Universities. (*Corresponding author: Dixian Zhao*).

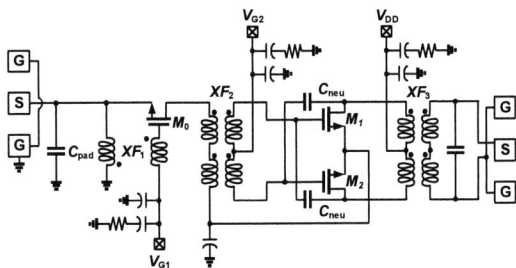

Fig. 1. Schematic of the proposed CG-CS LNA.

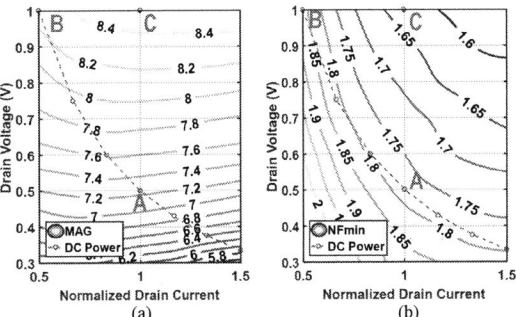

Fig. 2. (a) MAG and (b) NF_{min} of a standalone CG transistor for different bias conditions.

stage matching and current-reuse function. The 1-V DC supply V_{DD} is fed through the center tap of the output matching transformer XF_3 and shared by the two stages. Transformer XF_1 is adopt for g_m-boosted function. A pair of cross-coupled capacitors C_{neu} at CS stage are utilized for neutralization.

Fig. 2(a) and (b) shows the simulation results of maximum available gain (MAG) and minimum noise figure (NF_{min}) of a standalone CG transistor under different bias conditions (i.e., drain voltage and drain current). Conventionally, the transistor is biased with normal drain voltage and current (marker C). In this design, transistor with half drain voltage and normal current is utilized and it is marked with A. In addition, normal voltage and half current is marked with B. Among these bias conditions, marker C is of the best performance (1.2 dB higher MAG and 0.15 dB lower NF_{min} than marker A), but it consumes twice the DC power. This design (marker A) is with half power consumption and little deterioration in overall RF performances.

TABLE II. COMPARESION OF STATE-OF-THE-ART K-BAND CMOS LNAS

Reference	This work	MWCL'09 [3]	TMTT'13 [4]	TMTT'11 [5]	MWCL'12 [6]	MWCL'17 [7]	MWCL'19 [8]	MWCL'10 [9]
CMOS	65-nm	45-nm	65-nm	65-nm	65-nm	65-nm	65-nm	0.18-μm
Frequency (GHz)	17.2–22.0	21–25*	19.4–26.7	21.5–26.0*	17.5–26.0	15.8–30.3	22–25	19.0–22.0*
Bandwidth (GHz)	4.8	4	7.3	4.5	8.5	14.5	3	3
Gain (dB)	14.9	7.1	18.9	14.3	17.9	10.2	24	13.2
NF (dB)	3.3	4	4.7	2.8	3.3	3.3	3.3	4.1
Power (mW)	1.9	3.6	12	7	5.6	12.4	12	7.1
FoM#	13.6	3.2	2.6	4.1	9.5	5	2.1	1.6

* Graphically estimated $^{\#} \mathrm{FoM} = \dfrac{|\mathrm{Gain}| \times \mathrm{BW[GHz]}}{(F-1) \times \mathrm{P_{DC}[mW]}}$.

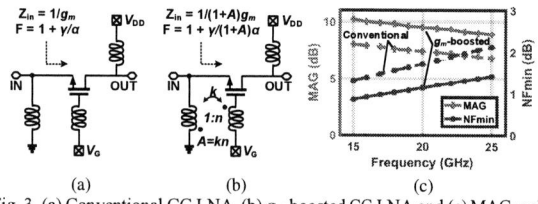

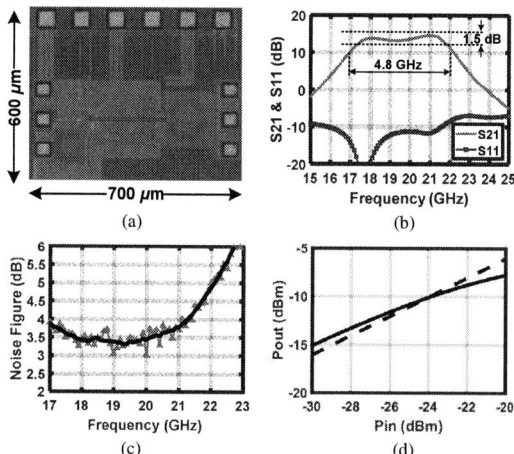

Fig. 3. (a) Conventional CG LNA, (b) g_m-boosted CG LNA and (c) MAG and NF$_{min}$ of conventional CG LNA and g_m-boosted CG LNA ($A = kn = 1$).

Fig. 3 shows the comparison of a conventional CG LNA and a g_m-boosted CG LNA. With g_m-boosted technique, magnetic coupling between the coils greatly boosts the effective transconductance G_m of the transistor by an inverting gain factor $A = k \cdot n$, where n is the turn ratio and k is the coupling coefficient of the transformer. This technique raises the voltage gain and thus reduces the impact of the channel noise. It also lowers the input impedance, which helps reduce the transistor size and save the power consumption. However, a large value of n is not feasible because the transformer "amplifiers" the gate noise current to the input by a factor of n and because of the non-idealities of the on-chip transformer. Moreover, a large n will cause a reduction on the CG transistor size and thus drain DC current, which may be insufficient for the second stage (as the DC current is reused). Hence, a transformer with a turn ratio $n = 1$ and $k = 0.7$ is implemented in this design.

III. MEASUREMENT RESUTLS

The proposed LNA has been fabricated in 65-nm CMOS technology. As shown in Fig. 4(a), the LNA occupies a chip area of 600 μm × 700 μm, including G-S-G pads, DC pads and decoupling capacitors. The overall circuit consumes 1.9 mW from a 1-V supply. S-parameters and NF are measured from 10 to 26.5 GHz with Keysight N5242B PNA-X Network Analyzer. Measured S_{21} and S_{11} are shown in Fig. 4(b). The LNA achieve a peak gain of 14.9 dB at 21GHz, with a −3-dB bandwidth of 4.8 GHz. Measured NF is shown in Fig. 4(c). A lowest NF of 3.3 dB is achieved at 19.5 GHz and the NF is below 4 dB from 17 to 21 GHz. Fig. 4(d) shows the P$_{1dB}$ is -24 dBm at 21 GHz. Table II summarizes the measured performance of the proposed LNA and compares it with other state-of-the-art K-band CMOS LNAs. This LNA achieves the lowest power consumption with competitive gain, NF and bandwidth performances, leading to a figure of merit (FoM) as high as 13.6.

Fig. 4. (a) Chip photo, (b) measured S_{21} and S_{11}, (c) measured NF and (d) measured P$_{1dB}$.

REFERENCES

[1] G. de Jong, et, al, "A Fully Integrated Ka-Band VSAT Down-Converter," *Solid-State Circuits. IEEE Journal of*, vol. 48, no. 7, pp. 1651-1658, July, 2013.

[2] T. LaRocca, et al, "Secure Satellite Communication Digital IF CMOS Q-Band Transmitter and K-band Receiver" *IEEE J. Solid-State Circuits*, vol. 54, no. 5, pp. 1329–1338, May 2019.

[3] W.-C. Wang, et al, "A 1 V 23 GHz Low-Noise Amplifier in 45 nm Planar Bulk-CMOS Technology with High-Q Above-IC Inductors," *IEEE Microw. Wireless Compon. Lett.*, vol. 19, no. 5, pp. 326–328, May 2009.

[4] H.-Y. Chang, et al, "65-nm CMOS Dual-Gate Device for Ka-Band Broadband Low-Noise Amplifier and High-Accuracy Quadrature Voltage-Controlled Oscillator," *IEEE Trans. Microw. Theory Techn.*, vol. 61, no. 6, pp. 2402–2413, June. 2013.

[5] M.-H. Tsai, et al, "ESD-Protected K-Band Low-Noise Amplifiers Using RF Junction Varactors in 65-nm CMOS," *IEEE Trans. Microw. Theory Techn.*, vol. 59, no. 12, pp. 3455–3462, Dec. 2011.

[6] M.-H. Tsai, et al, "A 17.5–26 GHz Low-Noise Amplifier with Over 8 kV ESD Protection in 65 nm CMOS," *IEEE Microw. Wireless Compon. Lett.*, vol. 22, no. 9, pp. 483–485, Sep. 2012.

[7] P. Qin, et al, "Compact Wideband LNA With Gain and Input Matching Bandwidth Extensions by Transformer," *IEEE Microw. Wireless Compon. Lett.*, vol. 22, no. 9, pp. 483–485, Sep. 2012.

[8] Y. Ding, et al, "Design and Implementation of an Ultracompact LNA With 23.5-dB gain and 3.3-dB Noise Figure," *IEEE Microw. Wireless Compon. Lett.*, vol. 29, no. 6, pp. 406–408, June. 2019.

[9] T.-P. Wang, "A Low-Voltage Low-Power K-Band CMOS LNA Using DC-Current-Path Split Technology," *IEEE Microw. Wireless Compon. Lett.*, vol. 20, no. 9, pp. 519–522, Sep. 2010.

978-1-7281-5107-6/19 $31.00 © 2019 IEEE

A 4-GHz Sub-harmonically Injection-Locked Phase-Locked Loop with Self-Calibrated Injection Timing and Pulsewidth

Xuefan Jin, Dong-Seok Kang, Youngjun Ko, Kee-Won Kwon, and Jung-Hoon Chun
College of Information & Communication Engineering, Sungkyunkwan University, Suwon, South Korea
Email: {cicd819, jhchun}@skku.edu

Abstract—A 4-GHz sub-harmonically injection-locked phase locked loop (ILPLL) with on-chip calibration is presented. The injection timing and pulsewidth of the injected pulse are self-calibrated to achieve a low phase noise. The phase noise of the proposed ILPLL was -110.1 dBc/Hz at 1-MHz offset frequency, while that of the conventional PLL was -104.8 dBc/Hz. The measured reference spur was also reduced from -44.36 dBc to -52.87 dBc with the proposed calibration technique. Fabricated in a 28-nm CMOS process, the proposed ILPLL occupies 0.09 mm². Operating at 4 GHz, it consumes 11.4 mW from a 1.0-V power supply.

Keywords—Ring oscillator, phase-locked loop (PLL), injection-locked PLL (ILPLL), injection-locked ring oscillator (ILRO), self-calibrated

I. INTRODUCTION

Compared with LC-oscillator phase-locked loops (PLLs), PLLs based on a ring oscillator offer a much wider tuning range, smaller area, and better power efficiency in producing multiple output phases. However, a wider loop bandwidth is needed to suppress the noise of the ring voltage-controlled oscillator (VCO) and improve jitter performance. The loop bandwidth of the PLL is typically less than one-tenth of the reference frequency due to the loop stability requirements, which results in a poor out-band phase noise. Recently, many different approaches have been used to improve jitter performance, such as a multiplying delay-locked loop (MDLL), sub-sampling PLL, and sub-harmonically injection-locked PLL (ILPLL).

In an MDLL [1], the clean reference clock is used to periodically clean up the accumulated jitter of the ring oscillator. However, it employs a multiplexer to switch back and forth between the clean reference clock and the noisy VCO clock. This multiplexer's switching timing control is a critical issue in determining the jitter performance of an MDLL because it may cause large reference spurs or deterministic jitters.

In a sub-sampling PLL [2], the charge pump noise is suppressed by the sub-sampling phase detector (PD). However, because it lacks a divider, an additional frequency-locked loop (FLL) is needed to avoid harmonic locking. In addition, the accuracy of phase detection may be degraded due to the

sub-sampling PD's non-linear characteristic as the operating frequency increases.

The operation principle of a sub-harmonically ILPLL [3]-[4] is similar to that of an MDLL. A clean low-frequency clock pulse is injected to reduce the VCO's accumulation jitter. However, if the injection timing is not controlled properly, large reference spurs may be pronounced at the PLL output or the ILPLL may fail to lock. Prior works [5]-[6] used an injection timing calibration circuit with a controllable delay line. However, the additional calibration loop operating in the background complicates the structure of ILPLL. On the other hand, the phase noise can be improved by optimizing the pulsewidth of the injected pulse [6] because the pulsewidth of the injected pulse affects the $2N$th harmonic power of injection. Although the pulsewidth of the injected pulse can be controlled in the prior work [6], it is done so manually in the background. Also, there has been no research on controlling the pulsewidth and injection timing of the injected pulse automatically in the PLL-based injection-locked clock multiplier (ILCM) [7]-[9].

In this work, a PLL-based ILCM with self-calibrated injection timing and pulsewidth is presented. This paper is organized as follows. In Section II, the technique of self-aligned injection timing with maximum injection strength is introduced. In Section III, the proposed ILPLL architecture and detail of this works are presented. The measurement results are shown in Section IV, followed by the conclusions in Section V.

II. PROPOSED TECHNIQUE OF SELF-ALIGNED INJECTION TIMING WITH MAXIMUM INJECTION STRENGTH

Fig. 1(a) shows a simplified block diagram of the proposed ILPLL. For the pulse generator (PG) used in the proposed ILPLL, the rising edge of the injected pulse (INJ) is controlled by $\triangle$T1, and the falling edge is controlled by $\triangle$T2. Therefore, by regulating $\triangle$T1 and $\triangle$T2, we can simultaneously control the injection timing and pulsewidth. Fig. 1(b) shows the timing diagram of the proposed ILPLL, where the cross point of the VCO clocks, CK45 and CK225, is aligned to the center of the INJ, showing that the injection timing is self-aligned correctly.

When the injected pulsewidth is equal to $0.25T_{VCO}$, the maximum $2N$th harmonic power is achieved to improve the

978-1-7281-5107-6/19 $31.00 © 2019 IEEE

IEEE Asian Solid-State Circuits Conference
November 4 – 6, 2019
The Parisian Macao, Macao SAR, China

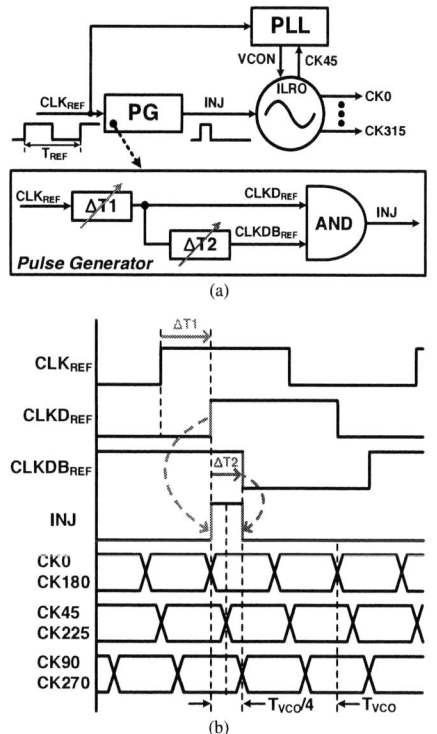

Fig. 1. (a) Simplified block diagram and (b) timing diagram of the proposed ILPLL.

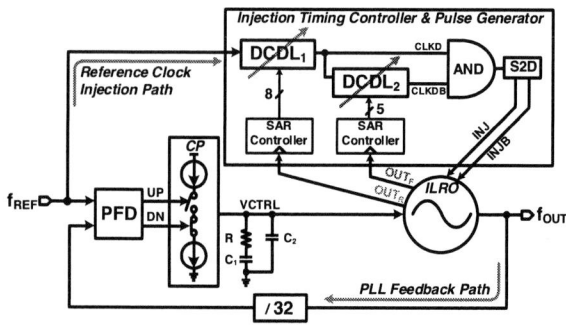

Fig. 2. Block diagram of the proposed ILPLL.

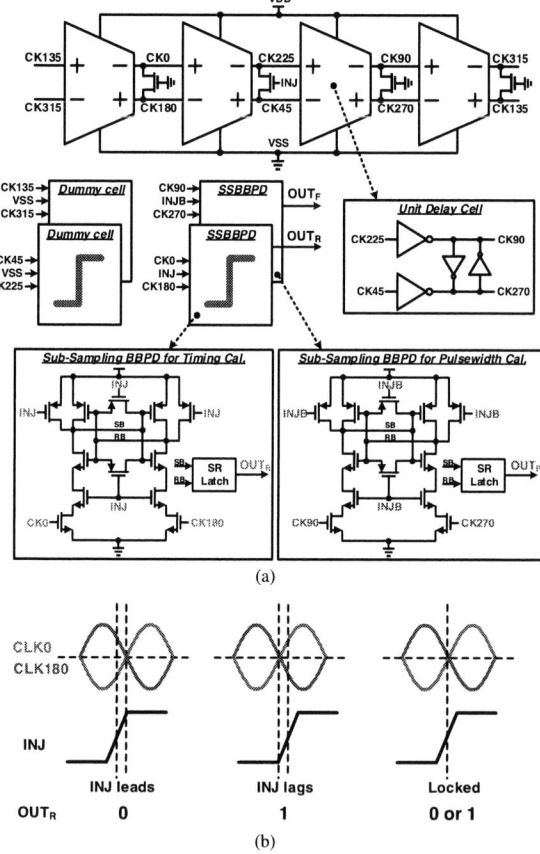

Fig. 3. (a) Block diagram of the ILRO with sub-sampling BBPD and (b) Operation waveform of the SSBBPD.

phase noise [6]. As shown in Fig. 1(b), if the delay of $\triangle T1$ is regulated until the rising edge of INJ is aligned to the cross point of CK0 and CK180 and the delay of $\triangle T2$ is regulated until the falling edge of INJ is aligned to the cross point of CK90 and CK270, the pulsewidth of INJ will be self-calibrated to $0.25T_{VCO}$, and the cross point of CK45 and CK225 will be self-aligned to the center of INJ.

III. PROPOSED ARCHITECTURE AND IMPLEMENTATION

A. Proposed ILPLL and ILO architecture

Fig. 2 shows the proposed ILPLL architecture based on a ring oscillator. It consists of the PLL feedback path and the reference clock injection path. The PLL feedback path is composed of the conventional PLL building blocks such as a divide-by-32 divider, a phase and frequency detector (PFD), a charge pump (CP) and a ring oscillator with injection. The pulse generator in the reference clock injection path generates a pulse signal (INJ) to be injected into the injection-locked ring oscillator (ILRO) using a clean reference clock (f_{REF}). The timing delay and pulsewidth of the INJ are controlled by $DCDL_1$ and $DCDL_2$, respectively.

As shown in Fig. 3(a), the ILRO employs sub-sampling bang-bang phase detector (SSBBPD) to detect the polarity of the injection timing error between CK0 and INJ. Then, the $DCDL_1$ is updated by the SAR controller to minimize the

injection timing error. The detailed operation waveforms of the SSBBPD are shown in Fig. 3(b). Once the calibration of the injection timing is completed, the polarity of the timing error between CK90 and INJB is detected, and the delay of $DCDL_2$ is controlled to calibrate the pulsewidth of INJ.

978-1-7281-5107-6/19 $31.00 © 2019 IEEE

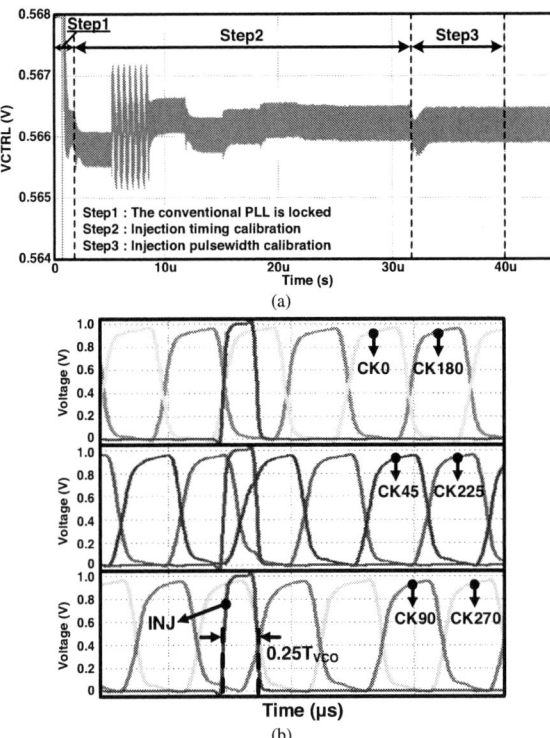

Fig. 4. Simulation results of (a) control voltage (VCTRL) and (b) clock waveform when the calibration is finished in the propose ILPLL.

B. Operation flowchart

The operation procedure is as follows:

Step 1 : The reference clock injection path is disconnected, and the conventional PLL is activated. The output frequency (f_{OUT}) is set to $N \times f_{REF}$ by the conventional PLL.

Step 2 : After the PLL feedback path is locked, the reference clock injection path is enabled to start the calibration of injection timing. In this step, the $DCDL_1$ code is updated until the INJ is aligned to CK0. In the injection timing controller, the reference clock is divided by 512 to generate a low frequency clock, $f_{REF}/512$, for the SAR controller. Therefore, the update rate of the $DCDL_1$ is much slower than the lock time of the PLL.

Step 3 : Once the injection timing calibration is finished, the injection pulsewidth calibration is enabled. The $DCDL_2$ code is updated until the INJB is aligned to CK90.

C. Simulation Results

Fig. 4(a) shows how the VCO's control voltage (VCTRL) changes during the injection timing calibration. As the injection timing calibration is completed in step 2, the injection timing mismatch is reduced, and the VCTRL is also stabilized. As shown in Fig. 4(b), the rising edge of injected pulse (INJ) is aligned to the cross point of CK0 and CK180, and the falling edge is aligned to the cross point of CK90 and CK270.

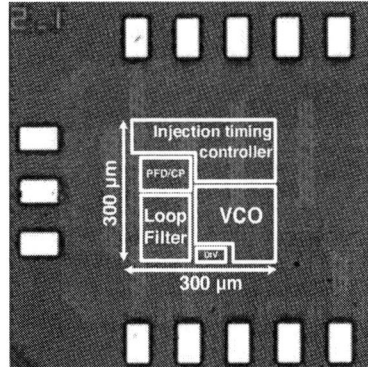

Fig. 5. Die photograph.

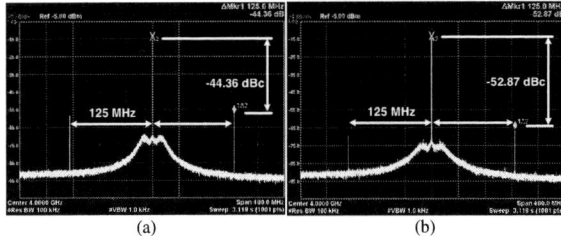

Fig. 6. Measured spectrum of the proposed ILPLL output (a) without and (b) with timing calibration.

Therefore, the cross point of CK45 and CK225 is self-aligned to the center of INJ when all calibration is finished.

IV. MEASUREMENT RESULTS

The proposed ILPLL was implemented in a 28-nm CMOS process, and the die photograph is shown in Fig. 5. It occupies an active area of 300 μm × 300 μm, and consumes 11.4 mW from a 1.0-V supply while operating at 4 GHz. Fig. 6 shows the measured output spectrum with and without timing calibration. When injection timing calibration was used, the reference spur was reduced from -44.36 dBc to -52.87 dBc. As shown in Fig. 7, without the injection, the conventional PLL achieves the phase noise of -104.8 dBc/Hz at 1-MHz offset. With the injection, the measured phase noise was reduced to -110.1 dBc/Hz at a 1-MHz offset. Moreover, the phase noise of the proposed ILPLL was reduced near the loop bandwidth of the conventional PLL.

Table I summarizes the performance of the proposed ILPLL and compares it with that of other ILCMs based on an analog PLL. The figure of merit (FOM) in this table is defined as [10]

$$FOM = 20 \log \left(\frac{\sigma_t}{1s} \right) + 10 \log \left(\frac{P}{1mW} \right)$$

Despite having a large divider ratio N, this work showed good RMS jitter and FOM performance compared to the other ring oscillator-based ILPLLs [6], [9]. In addition, the proposed

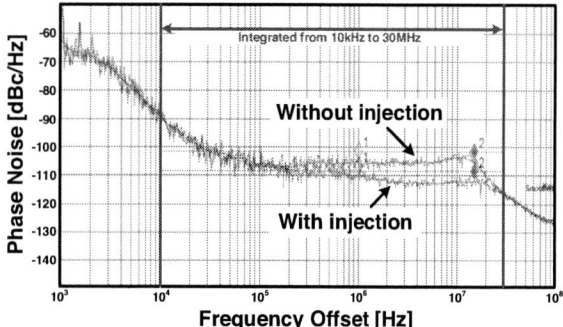

Fig. 7. Measured phase noise of the proposed ILPLL output with and without injection at 4 GHz.

TABLE I
PERFORMANCE SUMMARY AND COMPARISON

	[9]	[8]	[6]	This Work
Technology (nm)	65	65	40	28
VCO Type	Ring	LC	Ring	Ring
Topology	Analog ILPLL+PNF	Analog ILPLL	Analog ILPLL	Analog ILPLL
Output freq. (GHz)	1.2	3	1.6	4
Divider ratio	24	15	32	32
PN @1MHz (dBc/Hz)	-109.4	-127	-106.7	-110.1
RMS jitter (ps)	1.48 (100k-100M)	0.15 (1k-30M)	2.29 (1k-30M)	0.73 (10k-30M)
Ref. spur (dBc)	-57	-47.6	-44	-53.9
Power (mW)	19.8	9.1	1.49	11.4
Area (mm^2)	0.6	0.6	0.14	0.09
FOM (dB)	-223.6	-247	-231	-232.2
Injection timing	Self calibrated	Self calibrated	manual	Self calibrated
Injection pulsewidth	-	-	manual	Self calibrated

ILPLL achieved the smallest active area while simultaneously performing two types of self-calibration.

V. CONCLUSION

In this paper, an ILPLL with self-calibrated injection timing and pulsewidth was proposed, and its implementation was described. In the proposed architecture, the cross point of the VCO clock can be self-aligned to the center of the injected pulse, and the $2N$th harmonic power of the injected pulse can also be maximized by the automatic pulsewidth calibration. The proposed ILPLL can be used either as a low-jitter clock generator in digital systems or as a frequency synthesizer in wireline communication system.

ACKNOWLEDGMENT

This work was supported by Samsung Electronics Co. Ltd.

REFERENCES

[1] Ramin Farjad-rad et al., "A 0.2- 2 GHz 12 mW multiplying DLL for low-jitter clock synthesis in highly-integrated data-communication chips," *IEEE Integrated Solid-State Circuits Conf. (ISSCC) Dig. Tech. Papers*, pp. 55-57, Feb. 2002.

[2] Xiang Gao et al., "A Low Noise Sub-Sampling PLL in Which Divider Noise is Eliminated and PD/CP Noise is Not Multiplied by N2," *IEEE Journal of Solid-State Circuits*, vol. 44, no. 12, pp. 3253-3263, Dec. 2009.

[3] Jri Lee, Huaide Wang, "Study of Subharmonically Injection-Locked PLLs," *IEEE Journal of Solid-State Circuits*, vol. 44, no. 5, pp. 1539-1553, May 2009.

[4] Sheng Ye, Lars Jansson, and Ian Galton, "A Multiple-Crystal Interface PLL With VCO Realignment to Reduce Phase Noise," *IEEE Journal of Solid-State Circuits*, vol. 37, no. 12, pp. 1795-1830, Dec. 2002.

[5] Yi-Chieh Huang, Shen-Iuan Liu, "A 2.4GHz sub-harmonically injection-locked PLL with self-calibrated injection timing," in *IEEE Integrated Solid-State Circuits Conf. (ISSCC) Dig. Tech. Papers*, pp. 338-340, Feb. 2012.

[6] Chih-Li Wei, Ting-Kuei Kuan, and Shen-Iuan Liu, "A sub-harmonically Injection-Locked PLL With Calibrated Injection Pulsewidth," in *IEEE Transactions on Circuits and Systems II: Express Briefs*, vol. 62, no. 6, pp. 548-552, June 2015.

[7] Che-Fu Liang, Keng-Jan Hsiao, "An Injection-Locked Ring PLL with Self-Aligned Injection Window," in *IEEE Integrated Solid-State Circuits Conf. (ISSCC) Dig. Tech. Papers*, pp. 90-92, Feb. 2011.

[8] Zhao Zhang, Liyuan Liu, and Nanjian Wu, "A Novel 2.4 to 3.6 GHz Wideband Subharmonically Injection-Locked PLL with Adaptively-Aligned Injection Timing," in *IEEE Asian Solid-State Circuit Conf. (ASSCC)*, pp. 369-372, Nov. 2014.

[9] Alvin Li, Yue Chao, Xuan Chen, Liang Wu, and Howard C. Luong, "A Spur-and-Phase-Noise-Filtering Technique for Inductor-Less Fractional-N Injection-Locked PLLs," in *IEEE Journal of Solid-State Circuits*, vol. 52, no. 8, pp. 2128-2140, Aug. 2017.

[10] Xiang Gao et al., "Jitter analysis and a bechmarking figure-of-merit for phase-locked-loops," in *IEEE Transactions on Circuits and Systems II: Express Briefs*, vol. 56, pp. 117-121, Feb. 2009.

A 9.4MHz-to-2.4GHz Jitter-Power Reconfigurable Fractional-N Ring PLL for Multi-Standard Applications in 7nm FinFET CMOS Technology

Sangdon Jung[1], Jaehong Jung[2], Byungki Han, Seunghyun Oh, and Jongwoo Lee

Samsung Electronics

[1]sangdon.jung@samsung.com, [2]j0214.jung@samsung.com

Abstract— **This paper proposes a ring-VCO-based fractional-N PLL with a noise-power reconfigurable ring-VCO and a self-chopped reference frequency doubler for multi-standard applications. To cover the various specifications of SoC chips, the jitter and power of the proposed PLL can be reconfigurable through R and C adjustments of the ring-VCO. Moreover, the reference frequency doubler provides 6 dB improvements on DSM quantization noise without reference spurs. In addition, the PLL also guarantees a wide frequency range from 9.4 MHz to 2.4 GHz while maintaining optimum loop-bandwidth against process, voltage, and temperature variations through an adaptive loop-bandwidth technique. The measured integrated RMS-jitter and power consumption of this PLL are 6.2 ps and 6 mW, respectively for the low-jitter specification, which translated to -216.4 dB FoM, and 13 ps and 2.1 mW, respectively for the low-power specification, which achieves 0.85 mW/GHz efficiency.**

Keywords— **Fractional-N PLL, phase noise, reconfiguration, ring-VCO, doubler, quantization noise.**

I. INTRODUCTION

As the demand of highly integrated multi-standard applications has increased, many phase-locked loops (PLLs) are used in a single SoC chipset, and each PLL has different jitter and power specification [1]. Fig. 1 shows the specifications of the jitter requirement versus the power consumption about PLLs for various applications. The high-speed data-converters such as WiFi and LTE ADCs need the low-jitter performance for the sampling operation. On the other hand, the power consumption is more important than the jitter specification for the low-speed digital blocks.

To reduce the design resource and period, there have been attempts to substitute the multiple PLLs with a single PLL [2, 3] for multiple applications. However, these attempts have a disadvantage that the jitter and power of the PLL must be designed for the most tight jitter specification; thus, the power sacrificing is inevitable in scenarios where only digital blocks operate. Moreover, as time division technique in the wireless communication dynamically adjusts bandwidth of the data-converters, it is more efficient if the PLLs can be optimized for each specification of data-converter. Considering the digital blocks, the wide-frequency range is an essential specification of the PLL for dynamic frequency scaling (DFS), which adjusts the operating frequency to minimize the power consumption [4-7]. As timing requirement relaxed in the low speed, the power reduction with relaxed jitter at the same PLL can yield an additional power efficiency in the DFS systems.

To address these challenges, this paper proposes a wide-frequency range fractional-N ring PLL in 7nm FinFET CMOS. The jitter and power of this PLL is adjusted by reconfiguring R and C of the ring-VCO according to noise-power control bit, N_{CB}, for multi-standard applications.

II. ARCHITECTURE

The ring-VCO is generally used over a LC-VCO as multi-standard PLLs due to compact size, technology independence, and wide frequency range for the DFS system. Fig. 2 shows the architecture of the proposed ring PLL, which is composed of a reference frequency doubler (RF-D), an adaptive loop bandwidth (L-BW) controller, and a noise-power reconfigurable ring-VCO. The L-BW of the PLL is determined by considering an overall noise characteristic. Especially, to operate the wide-frequency range, the L-BW must be compensated across process, voltage, and temperature (PVT) variations for optimized jitter performance. The proposed PLL uses the adaptive L-BW technique [2], which can maintain the L-BW by tracking control voltage, V_{CTRL}. A resistor-based V-to-I converter in Fig. 2 converts the sensed V_{CTRL} to the current, I_V. After that, I_V determines the current of charge pump (CP) and the unit current of 6-bit binary-weighted DAC through mirroring factors (α and β, respectively); therefore, the L-BW of the PLL is only proportional to the reference clock, F_{REF}, as shown in Fig. 2. Furthermore, the 6-bit automatic frequency calibration (AFC) technique makes that V_{CTRL} is locked at nearly a half of the supply voltage, V_{DD}, to minimize the mismatch of CP current and accomplish wide-frequency cover range across the PVT variations.

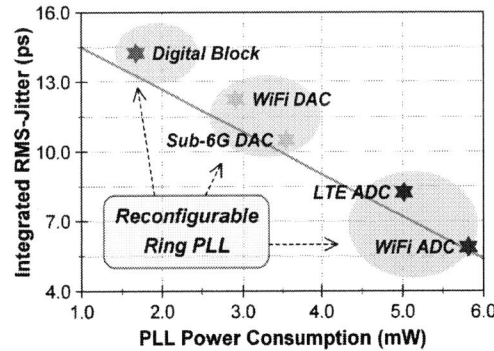

Fig. 1. PLL specifications about the jitter and power requirements.

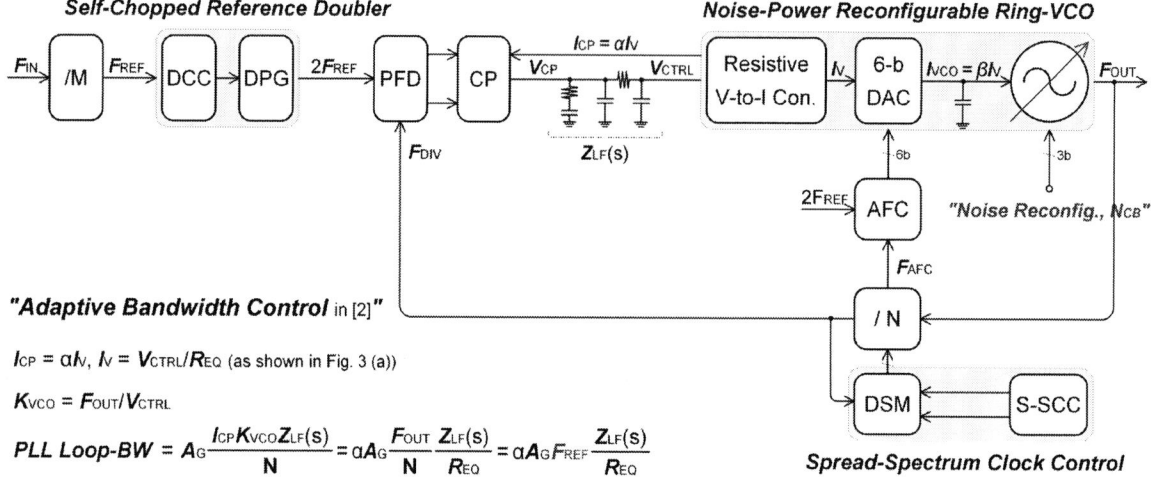

Fig. 2. Architecture of the proposed ring PLL using a self-chopped reference doubler and a noise-power reconfigurable ring-VCO.

The RF-D, which consists of a duty-cycle corrector (DCC) and double pulse generator (DPG), generates two times faster frequency reference clock, $2F_{REF}$. The DCC using the self-chopping technique reduces the duty-cycle variations of the input clock, F_{IN}, to minimize the reference spurs of the PLL. Additionally, the higher reference frequency suppresses the quantization noise of the DSM by 6.0 dB in the fractional operation. As the PLL inherently alleviates the phase noise of the ring-VCO through a high-pass filtering, the thermal noise of the ring-VCO determines the PLL noise performance; thus, the proposed PLL utilize the noise-power reconfigurable ring-VCO through the trade-off relationship between noise and power. The thermal noise characteristics of the ring-VCO can be adjusted by controlling the VCO bias current, βI_V, through the resistor-based V-to-I converter, and the load condition of a delay cell in the ring-VCO. Consequently, the proposed ring-VCO can configure the thermal noise according to the noise-power control bit, N_{CB}, maintaining the PLL output frequency, F_{OUT}.

III. CIRCUIT IMPLEMENTATION

A. Noise-Power Reconfigurable Ring-VCO

Generally, the thermal phase noise of the ring-VCO is determined by the noise of the transistors in delay cells. Fig. 3 (a) shows the proposed ring-VCO. Here, the bias current of the ring-VCO, βI_V, is generated by the resistor-based V-to-I converter and 6-bit binary weighted DAC.

$$\mathcal{L}(f) = \frac{2kT}{\beta I_v}\left(\frac{1}{(V_{DD}-V_{th})}\eta + \frac{1}{V_{DD}}\right)\left(\frac{f_0}{f}\right)^2 \tag{1}$$

Equation (1) shows the relationship between a single-side band (SSB) thermal phase noise [8] and the VCO bias current, βI_V. The thermal phase noise of the ring-VCO is significantly improved according to βI_V; thus, the out-band characteristic of the ring-VCO can be controlled by adjusting the βI_V. As V_{CTRL} and the equivalent resistor, R_{EQ}, in the regulation loop of V-to-I converter generates I_V, the resistor trimming that utilizes R_C

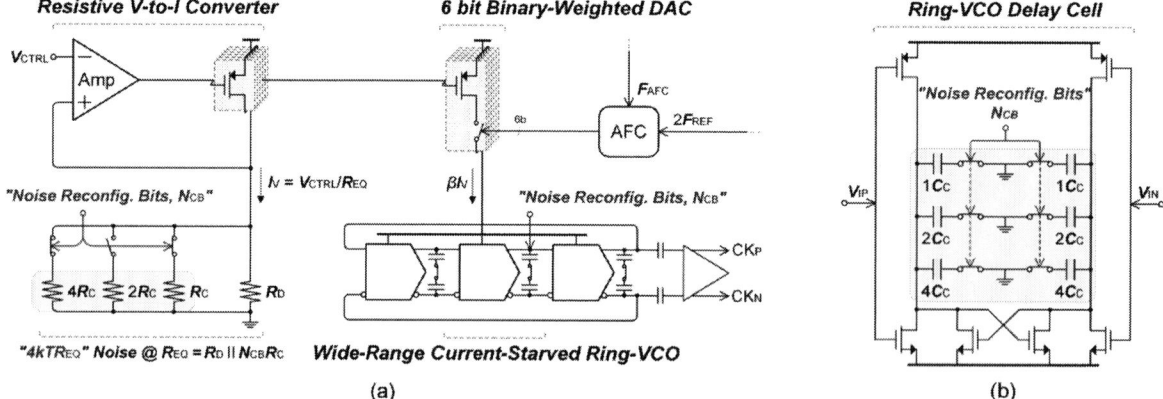

Fig. 3. Circuit diagrams of (a) wide-range current-starved ring-VCO including noise-power reconfiguration and (b) its delay cell unit.

Self-Chopped Continuous Duty-Cycle Corrector (DCC) **Double Pulse Generator** **Current Controlled Delay Line**

(a) (b)

Fig. 4. Circuit diagrams of (a) self-chopped reference frequency doubler and (b) its current-controlled delay unit cell.

according to N_{CB} results in the phase noise reconfiguration of the ring-VCO.

However, since the increased βI_V results in the unintentional frequency drift due to increased transconductance of the delay cells at the same mirroring coefficient (β), the load conditions of the delay cells should also be adjusted. Thus, the proposed delay cell in Fig. 3 (b) uses the controllable capacitor banks (C_C). As the 3-bit N_{CB} signal of the ring-VCO simultaneously controls the resistor, R_C, and capacitor, C_C, the frequency can be maintained constant without the additional locking process and the AFC operation. Moreover, since the noise configuration of the ring-VCO is controlled only by the passive devices, its output frequency is relatively tolerant to the PVT variations. Considering that the L-BW of the proposed PLL is out of 1 MHz, the out-band phase noise of the ring-VCO is reconfigurable from -107.2 to -114.8 dBc/Hz at 10 MHz frequency offset to satisfy the target specifications.

B. Self-Chopped Reference Frequency Doubler

The proposed self-chopped RF-D generates two times reference clock, 52 MHz, by using the 26 MHz input clock. As the operating frequency of the DSM is 2x faster than before, the quantization noise is also improved by 6 dB. The proposed ring PLL selectively uses the RF-D to cover the tight jitter (i.e. RMS-jitter < 7 ps) of WiFi ADC.

Fig. 4 (a) describes the detailed circuit diagram of the self-chopped RF-D. The self-chopped DCC compensates the duty-cycle error of the input reference clock, F_{REF}. After that, XOR gate produces two times faster reference frequency output, $2F_{REF}$. This DCC is composed of six-chain current-controlled delay lines (CCDLs), - RC filters, an V-to-I converter, and an amplifier (Amp). Two differential clocks, CK_{XP} and CK_{XN}, are generated from the single input clock having the duty-cycle error information. The duty-cycle information is converted to nearly DC voltages, V_{PI} and V_{NI}, by the RC filters, and the difference between two filtered voltages is directly proportional to the duty-cycle error. The amplifier compares these filtered voltages, and its output, V_{CN}, controls the V-to-I converter. After that, the bias voltages, V_{CN} and V_{CP}, generated

from the V-to-I converter control the rising and falling slopes using the 6-chain CCDLs in Fig. 4 (b), until the V_{PI} is equivalent to V_{NI}.

However, the DCC's flicker and thermal noise is added in the reference clock path. To mitigate noise effect and compensate the systematic and random mismatches, the self-chopping

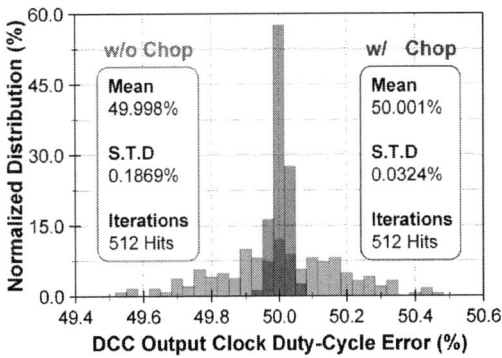

Fig. 5. Monte-Carlo simulation results of DCC duty cycle error.

Fig. 6. Die micrograph.

technique is conducted to the DCC loop as shown in Fig. 4 (a). The Monte-Carlo simulation results in Fig. 5 (total number of iterations = 512 Hits) show that when the chopping technique of DCC is enabled and disabled, the standard deviations of the duty-cycle error are significantly reduced from 0.19 to 0.03 %, which is 5.8x improvement.

IV. CONCLUSION

The proposed ring PLL is implemented in the 7nm FinFET technology. Fig. 6 shows the die micrograph with the active area of 0.032 mm^2. Fig. 7 shows the measured phase noise of the best jitter performance (N_{CB} = 7) at 2.4 GHz PLL's output frequency. Thanks to the RF-D, the quantization noise is simultaneously improved about 6 dB; therefore, the integrated RMS-jitter of the proposed PLL from 10^2 Hz to 10^2 MHz is achieved from 6.2 ps (N_{CB} = 7) to 13.2 ps (N_{CB} = 0), while consuming the power from 2 mW (N_{CB} = 0) and 6 mW (N_{CB} = 7). At N_{CB} = 7, the proposed PLL yields a jitter figure-of-merit (FoM) of -216.4 dB with the best jitter performance. In addition, at N_{CB} = 0, it exhibits the excellent power efficiency of 0.85 mW/GHz with the lowest power consumption.

Fig. 8 shows the trade-off relationship between the power consumption and integrated RMS-jitter of the proposed PLL

according to the 3-bit noise-power control bit, N_{CB}. This proves; the proposed jitter-power reconfigurable PLL can provide optimized performances for the multi-standard applications.

Compared with the state-of-the-art PLLs, as shown in TABLE I, this work achieves an outstanding jitter FoM at N_{CB} = 7 and an excellent power efficiency at N_{CB} = 0 with the wide frequency range from 9 MHz to 2.4 GHz with relatively low reference clock frequency of 26 MHz.

TABLE I. PERFORMANCE COMPARISON OF FRACTIONAL-N RING PLLs

	This Work		[3] '13 ISSCC	[4] '14 ISSCC	[5] '15 ISSCC
Technology	7nm FinFET		28nm	20nm	14nm
Synthesizer Type	Frac.-N		Frac.-N	Frac.-N	Frac.-N
Oscillator Type	Ring		Ring	Ring	Ring
L-BW Compensation	Yes		No	No	No
Freq. Range (GHz)	**0.009 – 2.4**		0.032 – 2.0	0.025 – 1.6	0.032 – 2.0
Reference (MHz)	26		10 – 30	-	32
Power (mW)	2.0 @2.4G	6.0 @2.4G	5.3 @2.0G	3.1 @1.6G	2.1 @2.0G
Integrated Jitter (ps)	12.2	**6.2**	19.3	28	18.8
Efficiency (mW/GHz)	**0.85**	2.50	2.05	1.94	1.03
Architecture	CP-PLL (Jitter-Power Reconfigurable)		BB-DPLL	BB-DPLL	BB-DPLL
FoM* (dB)	**-215.3**	**-216.4**	-207.0	-206.1	-211.4
Area (mm^2)	0.032		0.026	0.012	0.009

* FoM = $20log(Jitter/1s) + 10log(P/1mW)$

REFERENCES

[1] N. Kurd *et al.*, "A Family of IA 22nm Processors," *ISSCC Dig. Tech. Papers*, pp. 112-113, Feb. 2014.

[2] W. Jung *et al.*, "A 1.2mW 0.02mm^2 2GHz Current-Controlled PLL Based on a Self-Biased Voltage-to-Current Converter," *ISSCC Dig. Tech. Papers*, pp. 310-311, Feb. 2007.

[3] T.-K. Jang *et al.*, "A 0.026mm^2 5.3mW 32-to-2000MHz Digital Fractional-N Phase-Locked Loop Using a Phase-Interpolating Phase-to-Digital Converter," *ISSCC Dig. Tech. Papers*, pp. 270-271, Feb. 2013.

[4] J. Liu *et al.*, "A 0.012mm2 3.1mW Bang-Bang Digital Fractional-N PLL with a Power-Supply-Noise Cancellation Technique and a Walking-One-Phase-Selection Fractional Frequency Divider," *ISSCC Dig. Tech. Papers*, pp. 268-270, Feb. 2014.

[5] M. Song *et al.*, "A 0.009mm2 2.06mW 32-to-2000MHz 2nd-Order ΔΣ Analogous Bang-Bang Digital PLL with Feed-Forward Delay-Locked and Phase-Locked Operations in 14nm FinFET Technology," *ISSCC Dig. Tech. Papers*, pp. 226-227, Feb. 2015.

[6] Y. Li *et al.*, "A Reconfigurable Distributed All-Digital Clock Generator Core with SSC and Skew Correction in 22nm High-k Tri-Gate LP CMOS," *ISSCC Dig. Tech. Papers*, pp. 70-72, Feb. 2012.

[7] A. Elkholy *et al.*, "A 20-to-1000MHz 14ps Peak-to-Peak Jitter Reconfigurable Multi-Output All-Digital Clock Generator Using Open-Loop Fractional Dividers in 65nm CMOS," *ISSCC Dig. Tech. Papers*, pp. 272-273, Feb. 2014.

[8] A.A. Abidi, "Phase Noise and Jitter in CMOS Ring Oscillators." *IEEE J. Solid-State Circuits*, vol. 41, no. 8, pp. 1803-1816, Aug. 2006.

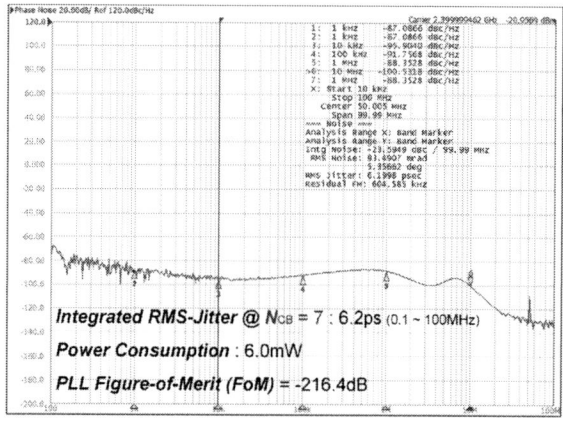

Fig. 7. Measured phase noise and jitter of 2.4 GHz at N_{CV} = 7.

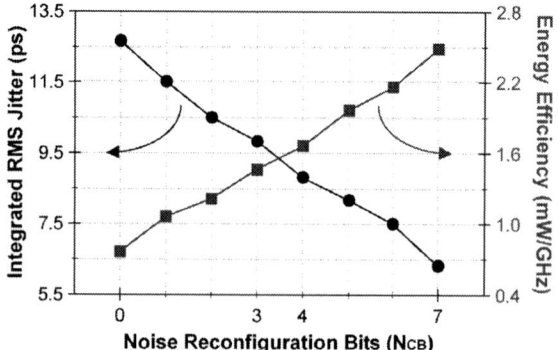

Fig. 8. Measured jitter and power reconfiguration according to N_{CB}.

7-3 (7153)

IEEE Asian Solid-State Circuits Conference
November 4 – 6, 2019
The Parisian Macao, Macao SAR, China

A 2.4-GHz 500-µW 370-fs$_{rms}$ Integrated Jitter Sub-Sampling Sub-Harmonically Injection-Locked PLL in 90-nm CMOS

Chun-Yu Lin, Yu-Ting Hung, and Tsung-Hsien Lin

Graduate Institute of Electronics Engineering and Department of Electrical Engineering, National Taiwan University, Taiwan

thlin@ntu.edu.tw

Abstract - This paper proposes a sub-sampling (SS) sub-harmonically injection-locked (SI) PLL. Both the SS and SI techniques are incorporated to realize a low-jitter low-power PLL. To ensure both techniques operate coherently, the injection timing of the SI path is calibrated through the assist of the SS loop. Fabricated in a 90-nm CMOS, the proposed 2.4-GHz PLL consumes only 500 µW from a 1-V supply. With a 40-MHz reference frequency, this SS-SIPLL achieves 370-fs$_{rms}$ integrated jitter (10 kHz to 30 MHz); the FOM is -251.6 dB.

Keywords - sub-sampling PLL, injection-locked PLL, gating operation.

I. INTRODUCTION

High performance phase-locked loop (PLL) is widely used in various applications such as wireless radios, wireline links, analog-to-digital converters, and digital systems. In these applications, low noise and low-power consumption are two major requirements. Several recent PLL research works demonstrated good performance [1-8]. However, they still face tradeoff among key performance specifications.

Sub-sampling PLL (SSPLL) is an attractive clock generator topology for high-performance applications, since eliminating the frequency divider reduces both in-band phase noise (PN) and power consumption [1]. However, the out-of-band PN of an SSPLL is still dominated by the VCO, which mandates large current for low-noise VCO design [2].

The sub-harmonically injection-locked PLL (SIPLL) provides an alternative low-noise PLL architecture. By properly injecting a short pulse at the reference frequency to the VCO, the VCO frequency is injection-locked to the N times of the reference frequency. Since the accumulated noise is eliminated by the injection pulse, the PLL output can achieve very low noise. One key advantage of an SIPLL is that the high-pass corner of the VCO noise transfer function (TF) is extended to a higher frequency; hence, the VCO noise is better suppressed. This attribute relaxes the VCO design and leads to lower power consumption. Moreover, it can also suppress in-band noise due to the CP. However, due to narrow injection-locking range, the main design challenge of an SIPLL is the injection timing. Several methods have been presented to address this problem [4]-[7]. Still, these methods suffer from increased power consumption or low reliability. In [4], the time amplifier is used to detect the REF and VCO edges. As the VCO edge travels through the divider and retimed DFF, the propagation delay of DFF results in the timing mismatch between the injection pulse and REF edges. In [6], an injection-locked with self-aligned injection window is proposed. However, the injection pulse width is only several tens of picoseconds, it is difficult to capture a correct voltage; especially in high-frequency operation. In [7], a

dual-edge phase detector (DEPD) is used for sensing REF and VCO edges, and it operates in a gated mode. In high frequency, the gated signal of DEPD may exceed one VCO cycle, which leads to a false operation.

In this work, we propose the sub-sampling sub-harmonically injection-locked PLL (SS-SIPLL), which takes advantages of both the SS and SI techniques, to realize a low-jitter low-power PLL. The key challenge of realizing the proposed PLL is to ensure the SS and SI techniques operate coherently without contaminating each other. This is solved by calibrating the SI injection timing before the SS and SI techniques work concurrently. The proposed calibration reuses the SS mode circuit; hence, it does not require extra hardware. Furthermore, gating operation is applied to some circuits for power saving. With these techniques, the FOM of this PLL is -251.6 dB.

II. ARCHITECTURE OF THE SS-SIPLL

A. Operation Concept of the Proposed SS-SIPLL

Fig. 1(a) shows the conceptual architecture of the proposed SS-SIPLL. The system is composed of an injection-locking path (including a digital-to-time converter (DTC) and a pulse generator (PG)) and a sub-sampling path (consists of a sub-sampling phase detector (SSPD), a voltage-to-current converter (VIC), and a loop filter (LF)). The DTC is adopted to facilitate a proper injection timing to ensure both paths work in harmony. DTC calibration will be discussed later.

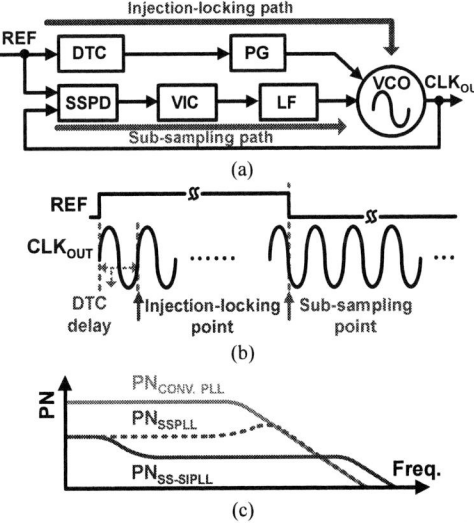

Fig. 1. (a) Conceptual block diagram of the SS-SIPLL; (b) conceptual timing operation, and (c) PN improvement.

978-1-7281-5107-6/19 $31.00 © 2019 IEEE

91

Fig. 1(b) depicts simplified operation waveforms of the SS-SIPLL. SS and SI techniques operate at different timing instants, which do not interrupt each other under proper design. Fig. 1(c) shows the conceptual output PN profiles of a conventional PLL, an SSPLL, and the SS-SIPLL. In an SSPLL, the in-band PN is improved compared to that of a conventional PLL; however, the VCO still dominates out-of-band noise. In the proposed SS-SIPLL, the in-band PN due to SSPD/VIC is further improved by the injection technique, while the out-of-band PN is improved by pushing the high-pass corner of VCO noise TF to a higher frequency.

B. SS-SIPLL Architecture

Fig. 2 shows the block diagram of the overall PLL. As discussed, this PLL exploits concurrent SS and SI operations to improve the PN. It is composed of a frequency-locked loop (FLL), an SS loop, and an SI loop. The FLL is a conventional PLL with a dead-zone PFD (DZPFD), divider, and CP. The SSPLL is realized with the sub-sampling phase detector (SSPD) and VIC. The SI loop consists of a digital-to-time converter (DTC) and a pulse generator (PG). In normal operation, V_{PULSE} is a periodic short pulse for delivering current to the LF. In the calibration mode, V_{PULSE} signal is also adopted to break the SS loop. The enable signals ($SSBUF_{EN}$ and VIC_{EN}), generated by the timing control block, gate the operation of power hungry circuits, VIC and SSBUF, for saving power.

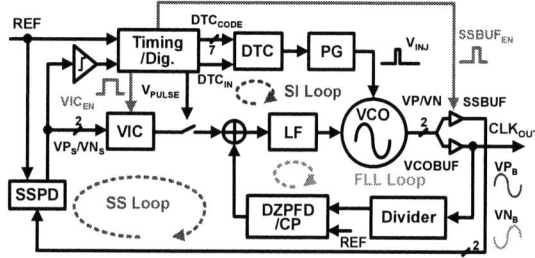

Fig. 2. Block diagram of the proposed SS-SIPLL.

C. SS-SIPLL Operation Sequence and Calibration

The simplified sequence of operation modes is depicted in Fig. 3. The whole PLL works as follows. First, the FLL operates to acquire frequency locking. Once the FLL locks, the PLL enters the SSPLL mode. The SSPD samples buffered VCO outputs and generates VP_S/VN_S. The voltage-to-current converter (VIC) is enabled to achieve phase locking, which drives the VCO to align with REF edges (falling edges in this example). The leftmost part of Fig. 3 illustrates a locked SSPLL mode.

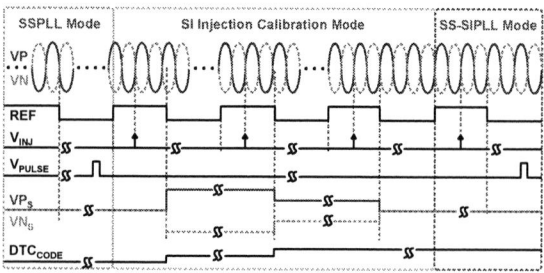

Fig. 3. Sequence of operation modes.

Next, the PLL enters the SI injection timing calibration mode (the middle part of Fig. 3). During this mode, the SS loop is disabled (via the signal V_{PULSE}). When the injection signal (V_{INJ}) is applied to the VCO, it drives the VCO outputs (VP/VN) to align with V_{INJ} forcibly. If the injection pulse timing is not proper, the phase of VP/VN will move away from the SS locked state, resulting in a phase error. The phase error is captured by the SSPD, leading to $VP_S \neq VN_S$. This information is then processed and fed to the DTC to adjust the SI injection timing. This process continues until the SSPD differential outputs return to the locked state (i.e. $VP_S = VN_S$). Finally, V_{PULSE} returns to its normal operation, and the SS and SI modes operate concurrently. The PLL enters the SS-SIPLL mode.

D. SS-SIPLL Operation Timing Diagram

Fig. 4 shows a more detailed timing diagram once the SS and SI techniques operate concurrently. Note that these two techniques operate at different timing without mutual interference. For further power reduction, the operations of SSBUF and VIC are gated, since they only operate in the timing region around the vicinity of REF falling edges.

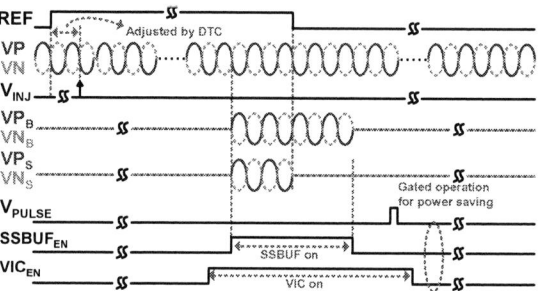

Fig. 4. Operation timing diagram of the proposed SS-SIPLL.

E. Noise Analysis of the SS-SIPLL

Fig. 5 shows the linear noise model of the proposed SS-SIPLL. Two additional TFs are identified. $T_{up}(s)$ represents the up-conversion from REF to VCO output. $T_{rl}(s)$ represents the phase realignment [9].

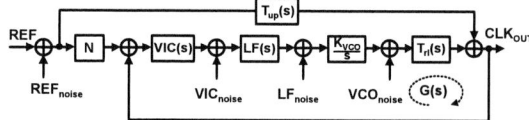

Fig. 5. Linear noise model of the proposed SS-SIPLL.

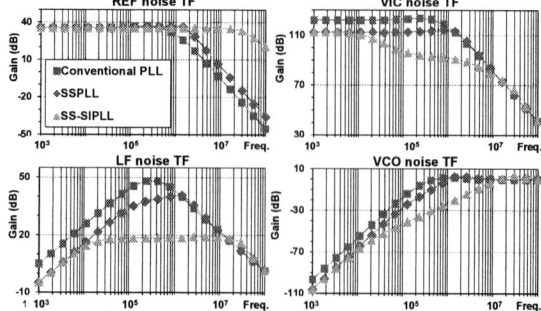

Fig. 6. Noise TFs with respect to various sources of the SS-SIPLL.

From the linear model, the TF magnitude of four major noise contributors, REF, VIC, LF, and VCO, are plotted in Fig. 6. Compared to those of an SSPLL, the VIC and LF noises in the proposed PLL are better suppressed. The high-pass corner of VCO noise TF is extended to a higher frequency due to the SI operation; hence, the VCO design can be relaxed. The REF noise sees a wider bandwidth; it determines the far-out noise in the proposed PLL.

III. CIRCUIT IMPLEMENTATION

A. DTC and Pulse Generator

Fig. 7 shows the DTC and PG circuits adopted in this work. A 7-bit binary-controlled capacitor array is used for DTC control (DTC_{CODE}). Delay range is designed to be around 250 ps (~half T_{VCO} period), for ensuring that the system can calibrate to the correct injection point. The timing difference between DTC_{IN} and V_{INJ} edges is determined by T_{DTC}. It can be expressed as:

$$T_{DTC} = T_{BASE} + DTC_{CODE} \times T_{DTC,LSB} \qquad (1)$$

where T_{BASE} is the propogation delay of DTC under $DTC_{CODE}=0$; $T_{DTC,LSB}$ represents 1 LSB of DTC dealy line, which is designed to be around 2 ps. The short injection pulse is generated by the PG, and the pulse width of V_{INJ} can be adjusted from 15 ps to 100 ps.

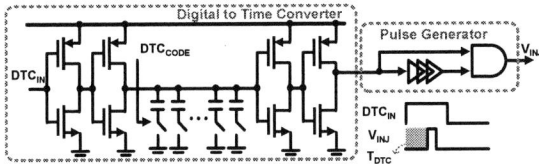

Fig. 7. DTC and PG designs.

B. VCO, Buffer, and Sub-Sampling Part

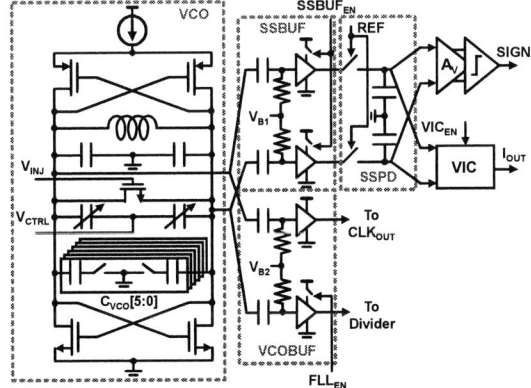

Fig. 8. VCO, SSBUF, VCOBUF, and SSPD designs.

Fig. 8 shows the VCO, SSBUF, VCOBUF, and SSPD designs. The VCO adopts dual cross-couple topology for larger gm, and 6-bit cap array ($C_{VCO}[5:0]$) for coarse tuning the VCO frequency. It covers 2.2 GHz to 2.6 GHz; the K_{VCO} is around 100 MHz/V. The VCO outputs are connected to two buffers, SSBUF and VCOBUF. The VCOBUF sends signals to CLK_{OUT} and divider of the FLL. After locking, the FLL

loop is disabled for saving power. Accordingly, the buffer that sends signal to divider is also turned off. The SSBUF, drives the SSPD, adopts an inverter-based design. The SSBUF operation is gated since its operation is only required for phase detection; hence can be turned off most of the time.

IV. MEASUREMENT RESULTS

The proposed 2.4-GHz PLL is fabricated in the 90-nm CMOS process. The chip micrograph is shown in Fig. 9. The core area of this PLL is 0.26 mm². Fig. 10 shows the power consumption pie chart. The whole SS-SIPLL consumes 500 µW from a 1-V supply. The VCO only consumes 286 µW, indicating its design is indeed relaxed under the proposed PLL architecture.

Fig. 11 shows the measured PN in 3 different PLL configurations: conventional PLL, SSPLL, and the proposed SS-SIPLL. The PN was measured using the signal source analyzer R&S FSUP. The in-band PN is improved from -94 dBc/Hz to -115 dBc/Hz (at 200-kHz offset). The integrated RMS jitter (from 10 kHz to 30 MHz) is improved from 1.7 ps to 370 fs. Fig. 12 shows the measured output spectrum under 40-MHz reference frequency. In the SS-SIPLL mode, the reference spurs are -40/-44 dBc. The output frequency is 2.4 GHz with a multiplier ratio (N) of 60. The reference spurs of the SS-SIPLL are mainly dominated by the injection-locking pulse (V_{INJ}) [4]-[5].

Fig. 13 shows the jitter versus power consumption plot. As shown in Fig. 13, this PLL achieves low jitter in the sub-mW range. Fig. 14 shows the plot of FOM versus REF frequency. It is worth noting that a PLL with a higher REF frequency typically achieves a better output PN [3]-[5]. This is benefitted from a lower N. In practice, a higher REF frequency is usually generated from a 3rd (or higher) overtone crystal oscillator (XO), which is more costly than a fundamental-tone XO. Detailed measurement results and comparison to state-of-the-art PLLs are listed in Table I. With the proposed techniques, this SS-SIPLL achieves low-jitter and low-power performance, leading to an FOM of -251.6 dB.

Fig. 9. Chip micrograph.

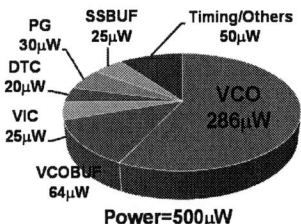

Fig. 10. Power consumption pie chart of the SS-SIPLL.

IEEE Asian Solid-State Circuits Conference
November 4 – 6, 2019
The Parisian Macao, Macao SAR, China

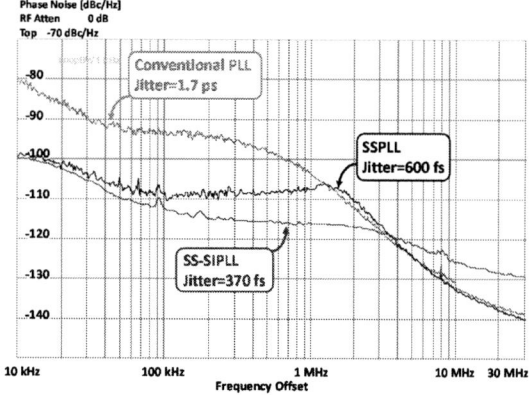

Fig. 11. Measured output phase noise

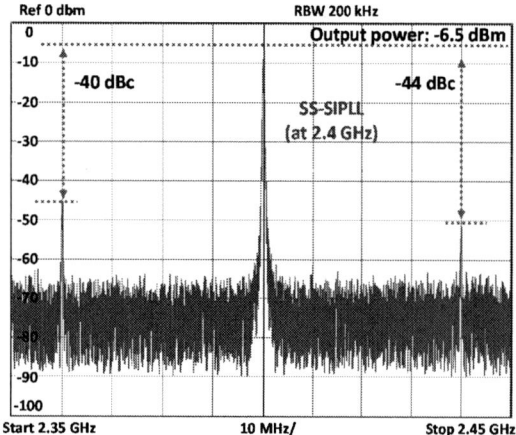

Fig. 12. Measured output spectrum.

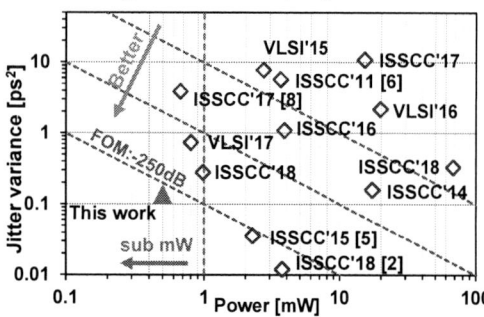

Fig. 13. Plot of jitter vs. power consumption.

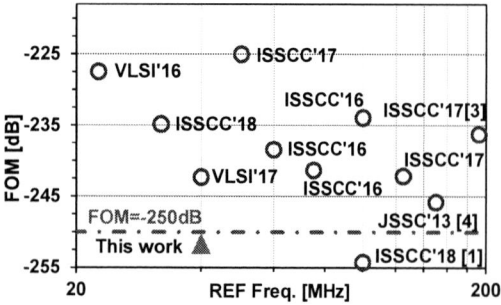

Fig. 14. Plot of FOM vs. REF frequency.

TABLE I. Performance Summary and Comparison

	This Work	ISSCC'18 [1]	ISSCC'18 [2]	JSSC'13 [4]	ISSCC'15 [5]	ISSCC'17 [8]
Process	90nm	65nm	65nm	180nm	65nm	40nm
Supply	1	0.8	1.2/0.5	1.8	0.9	0.8/1
Topology	SS-SIPLL	SSPLL	RSPLL	SIPLL	SIPLL	ADPLL
VCO Type	LC	LC	LC	LC	LC	LC
CLK$_{OUT}$(GHz)	2.4	5	2.55	2.4	6.8	1.8-2.5
REF (MHz)	40	100	50	150	106.25	N/A
Multi. Ratio (N)	60	50	51	16	64	N/A
PN (dBc/Hz) @ 200kHz	-115	-123	-121	-123	-112	-92
Jitter (fs) Int. BW (Hz)	370 10k-30M	185.3 10k-50M	110 10k-100M	145 1k-40M	190 10k-100M	1980 10k-10M
REF. Spur (dBc)	-40/-44	-64.1	-63	-40	-40	-62
Power (mW)	0.5	1.1	3.7	12.6	2.25	0.67
Area (mm²)	0.26	0.01	0.36	0.64	0.25	0.18
FOM (dB)*	-251.6	-254	-253.5	-246	-251	-236

*FOM(dB) = $10 \cdot log_{10}[(Jitter_{RMS}/1s)^2 \cdot Power/1mW]$

V. CONCLUSION

This paper presents a 2.4-GHz sub-sampling sub-harmonically injection-locked PLL (SS-SIPLL) with gating operation. The injection timing of the SI path is calibrated through the assist of the SS loop and DTC calibration scheme. With the proposed techniques, a 500-μW 370-fs$_{rms}$ integrated jitter clock source is achieved without complex circuit design. Experimental results demonstrated the effectiveness of these techniques.

ACKNOWLEDGMENT

Authors thank CIC (TSRI), Taiwan, for chip fabrication. This work is supported by Donation Grant FD105012.

REFERENCES

[1] A. Sharkia, S. Mirabbasi, and S. Shekhar, "A 0.01mm² 4.6-to-5.6GHz Sub-Sampling Type-I Frequency Synthesizer with –254dB FOM," *ISSCC Dig. Tech. Papers*, pp. 256-257, Feb. 2018.

[2] J. Sharma and H. Krishnaswamy, "A Dividerless Reference-Sampling RF PLL with -253.5dB Jitter FOM and <-67dBc Reference Spurs," *ISSCC Dig. Tech. Papers*, pp. 258-259, Feb. 2018.

[3] J. Chuang and H. Krishnaswamy, "A 0.0049mm² 2.3GHz Sub-Sampling Ring-Oscillator PLL with Time-Based Loop Filter Achieving -236.2dB Jitter-FOM," *ISSCC Dig. Tech. Papers*, pp. 328-329, Feb. 2017.

[4] Y. C. Hunag and S. I. Liu, "A 2.4-GHz Subharmonically Injection-Locked PLL With Self-Calibrated Injection Timing," *IEEE J. Solid-State Circuits*, vol. 48, no. 2, pp. 417–428, Feb. 2013.

[5] A. Elkholy, M. Talegaonkar, T. Anand, and P. K. Hanumolu, "A 6.75-to-8.25GHz 2.25mW 190fs$_{rms}$ Integrated-Jitter PVT-Insensitive Injection-Locked Clock Multiplier Using All-Digital Continuous Frequency-Tracking Loop in 65nm CMOS," *ISSCC Dig. Tech. Papers*, pp. 188-189, Feb. 2015.

[6] C. F. Liang, K. J. Hsiao, "An Injection-Locked Ring PLL with Self-Aligned Injection Window," *ISSCC Dig. Tech. Papers*, pp. 90-91, Feb. 2011.

[7] S, Choi, S. Yoo, Y. Lim, and J. Choi, "A PVT-Robust and Low-Jitter Ring-VCO-Based Injection-Locked Clock Multiplier With a Continuous Frequency-Tracking Loop Using a Replica-Delay Cell and a Dual-Edge Phase Detector," *IEEE J. Solid-State Circuits*, vol. 51, no. 8, pp. 1878–1889, Aug. 2016.

[8] Y. He, Y. H. Liu, T. Kuramochi, J. Heuvel, B. Busze, N. Markulic, C. Bachmann, and K. Philips , "A 673μW 1.8-to-2.5GHz Dividerless Fractional-N Digital PLL with an Inherent Frequency-Capture Capability and a Phase-Dithering Spur Mitigation for IoT Applications," *ISSCC Dig. Tech. Papers*, pp. 420-421, Feb. 2017.

[9] S. Ye, L. Jansson, and I. Galton, "A Multiple-Crystal Interface PLL With VCO Realignment to Reduce Phase Noise," *IEEE J. Solid-State Circuits*, vol. 37, no. 12, pp. 1795–1803, Dec. 2002.

A Sub-Sampling PLL with Robust Operation under Supply Interference and Short Re-Locking Time

Yuan Cheng Qian, Yen-Yu Chao, and Shen-Iuan Liu

Graduate Institute of Electronics Engineering & Department of Electrical Engineering
National Taiwan University, Taipei, Taiwan 10617, R.O.C.
E-mail: lsi@ntu.edu.tw

Abstract—A sub-sampling phase-locked loop (SSPLL) is presented to tolerate the supply interference and have a short re-locking time. A sampling phase detector, a voltage-to-current converter and a mini-dead zone creator are presented. The low in-band phase noise of the proposed SSPLL is kept and the power penalty is low. This SSPLL is realized in a 0.18μm CMOS process and its active area is 0.097mm². The power consumption is 13.59mW from a 1.8V supply. At the output frequency of 2.2 GHz, this SSPLL achieves an in-band phase noise of -115.33dBc/Hz at 100kHz offset frequency with a division ratio of 44. Its root-mean-square jitter integrated from 10kHz to 100MHz is 0.77 ps.

I. INTRODUCTION

PHASE-LOCKED loops (PLLs) are widely used for various clocking applications, such as frequency synthesizers, clock generators and so on. The in-band phase noise of a conventional charge-pump PLL is limited by a voltage-controlled oscillator (VCO), a charge pump and a divider. To reduce the in-band phase noise, a sub-sampling PLL (SSPLL) [1] using a sub-sampling phase detector (SSPD) is presented. While the SSPLL is locked, the divider can be turned off to reduce both the power consumption and phase noise. However, a conventional SSPLL [1] suffers from a long re-locking time [2] while the supply interference occurs. When a frequency error induced by the supply interference, it may induce a phase error which is over the phase acquisition range of the SSPD. The frequency-locked loop (FLL) cannot be active until this phase error is over the dead zone (DZ) [1]. Thus, a long re-locking time is required.

To deal with this issue, two methods have been presented. One is to combine a phase-frequency detector (PFD) and an SSPD [2]. The other is to use a soft loop switching [3] which maintains a constant loop gain to shorten the re-locking time. However, it has to compensate for the delay mismatch between the sub-sampling loop and the FLL. In addition, the off-chip components are required in [3].

In this paper, a sampling phase detector (SPD) with a mini dead zone (MDZ) creator and a voltage-to-current converter is presented to shorten the re-locking time of the SSPLL under the supply interference or divider ratio changed.

II. CIRCUIT DESCRIPTION

Fig. 1 shows the proposed SSPLL. It consists of a conventional SSPLL with a FLL, and a SPD with the MDZ creator and the second voltage-to-current converter (Gm2). The proposed circuits are discussed as follows.

A. Mini Dead Zone Creator

When the supply interference occurs or the divider ratio is switching, the frequency error will induce a phase error. To

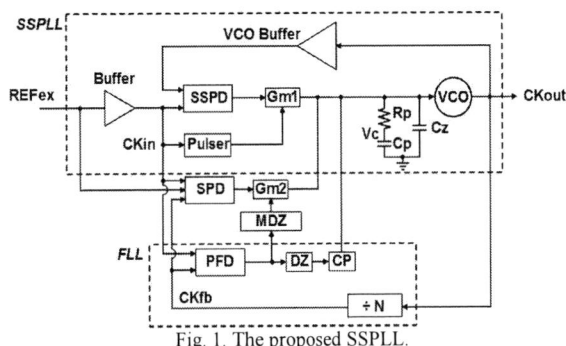

Fig. 1. The proposed SSPLL.

turn on the FLL, the corresponding phase error should be larger than the DZ range, ϕ_{DZ}, which is expressed as

$$\left| \int_0^{\Delta t} 2\pi(f_{target} - f_0)dt + \phi_0 \right| \geq \phi_{DZ} \quad (1)$$

where f_{target} is the target frequency, f_0 is the initial frequency, and ϕ_0 is the initial phase error. The re-locking time Δt is needed before the DZ creator to turn on the FLL, which is given as

$$\Delta t \geq \frac{\phi_{DZ} - \phi_0}{2\pi(f_{target} - f_0)}, \text{ if } f_0 < f_{target}$$
$$\Delta t \geq \frac{-\phi_{DZ} - \phi_0}{2\pi(f_{target} - f_0)}, \text{ if } f_0 > f_{target} \quad (2)$$

According to (2), when the difference between f_{target} and f_0 is small, the re-locking time will be long.

To shorten the re-locking time, Fig. 2(a) shows the MDZ creator which is composed of four D-flip-flops (DFFs) and an OR gate. The left-hand timing diagram in Fig. 2(b) shows that the internal reference clock CKin leads the feedback one CKfb by a timing error $t_e > T_{vco}/2$, where Tvco is the period of the VCO. Thus, the corresponding pulse width of the UP signal is also larger than Tvco/2. The falling edge of CKout samples the UP signal to enable MDZ_up1=1. Then, the rising edge of CKfb enables MDZ_up2=1. Finally, MDZ_out goes high to turn on the Gm2. Since $t_e > T_{vco}/2$, the corresponding phase error ϕ_e is given as

$$\phi_e = 2\pi \frac{t_e}{T_{CKin}} > \pi \frac{T_{VCO}}{T_{CKin}} = \frac{\pi}{N} \quad (3)$$

where T_{CKin} is the period of CKin, N is the divider ratio and $T_{CKin} = N \cdot Tvco$. On the contrary, when $t_e < Tvco/2$, the right-hand timing diagram in Fig. 2(b) shows that both the MDZ_up1 and MDZ_up2 are low. So, MDZ_out goes low and the Gm2 is turned off. In summary, when $|\phi_e| > \pi/N$, the MDZ creator will enable the Gm2 to speed up the SSPLL. When

978-1-7281-5107-6/19 $31.00 © 2019 IEEE

IEEE Asian Solid-State Circuits Conference
November 4 – 6, 2019
The Parisian Macao, Macao SAR, China

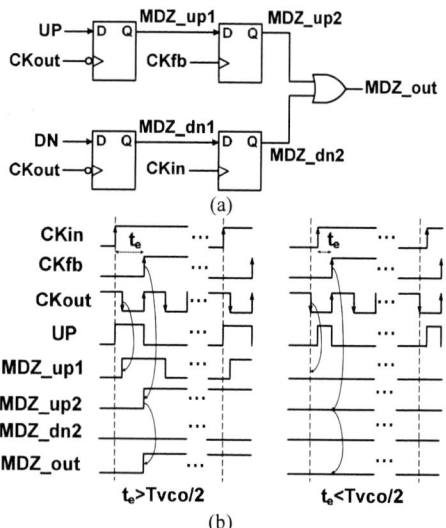

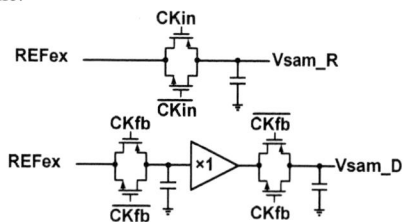

Fig. 2 (a) the MDZ creator and (b) its timing diagrams when CKin leads CKfb.

Fig. 3. A sampling phase detector.

$|\phi_e|<\pi/N$, the Gm2 will be turned off to save the power consumption.

B. Sampling PD and Gm2

Fig. 3 shows the SPD which detects the phase error ϕ_e. Note that this SPD's input has the same frequency with sampling clocks which is different from the SSPD. The sinusoidal external reference clock REFex is sampled by both the clocks CKin and CKfb. The phase difference between with REFex and CKfb is converted into a voltage Vsam_D by using two passive sample-an-hold (S/H) circuits and a unity-gain buffer. Vsam_D is kept for a whole period of CKfb. Since the phase difference between REFex and CKin is fixed, the sampled voltage Vsam_R is served as a reference voltage by using only one S/H circuit. It can save a unity-gain buffer and an S/H circuit. The gain of this sampling PD is roughly given as

$$K_{SPD} = \frac{d(Vsam\_D - Vsam\_R)}{d\phi_e} = \frac{d(Vsam\_D)}{d\phi_e} \quad (4)$$

Assumed that the voltage Vsam_D is given as

$$Vsam\_D(\phi_e) = A_{REF}\sin\phi_e \quad (5)$$

where A_{REF} is the amplitude of the external reference clock REFex. When ϕ_e is small enough, K_{SPD} can be approximated as

$$K_{SPD} = A_{REF}\cos\phi_e \approx A_{REF} \quad (6)$$

Fig. 4 shows the Gm2, which is composed of a differential pair and the current mirrors. Since Vsam_R is constant while CKin is high, the output current I_{Gm2} of the Gm2 drives the loop filter to speed up the frequency acquisition process.

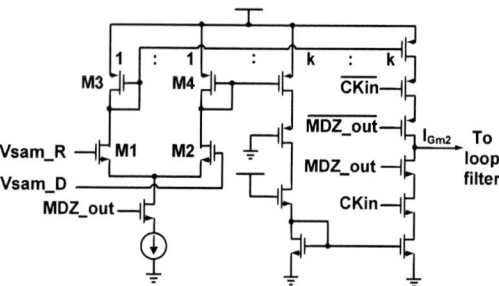

Fig. 4. The second voltage-to-current converter, Gm2.

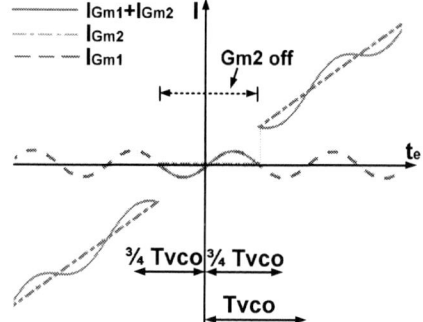

Fig. 5. The transfer curves for the currents I_{Gm1}, I_{Gm1} and $I_{Gm1}+I_{Gm2}$ versus t_e.

The gain of the Gm2 is given as

$$K_{Gm2} = g_{m2} \cdot k \cdot D_{CKin} \quad (7)$$

where g_{m2} is the transconductance of M1 and M2, k is a gain of the current mirror and D_{CKin} is the duty cycle of CKin.

Fig. 5 shows the transfer curves of the currents I_{Gm1} and I_{Gm2} versus t_e where I_{Gm1} is the output current of the first voltage-to-current converter (Gm1). When the FLL is turned off, K_{Gm2} must be carefully designed to ensure the same polarities of $I_{Gm1}+I_{Gm2}$ and t_e at the worst cases; i.e., $t_e = \pm \frac{3}{4}T_{vco}$. When $t_e > 0$, $I_{Gm1} + I_{Gm2}$ must be larger than zero at $t_e = \frac{3}{4}T_{vco}$. Similarly, when $t_e < 0$, $I_{Gm2} + I_{Gm1}$ must be smaller than zero at $t_e = -\frac{3}{4}T_{vco}$. For example, at $t_e = \frac{3}{4}T_{vco}$, the current I_{Gm2} is expressed as

$$I_{Gm2} = A_{REF}\sin\left(2\pi\frac{\frac{3}{4}T_{vco}}{T_{CKin}}\right)\cdot K_{Gm2} = A_{REF}\sin\left(\frac{3}{2N}\pi\right)\cdot K_{Gm2} \quad (8)$$

And the current I_{Gm1} is expressed as

$$I_{Gm1} = A_{vcobuf.}\sin\left(2\pi\frac{\frac{3}{4}T_{vco}}{T_{vco}}\right)\cdot K_{Gm1} = A_{vcobuf.}\sin\left(\frac{3}{2}\pi\right)\cdot K_{Gm1} \quad (9)$$

where K_{GM1} is the gain of the Gm1. Since $I_{Gm1} + I_{Gm2}$ must be larger than zero at $t_e = \frac{3}{4}T_{vco}$, the following condition must be satisfied as

$$K_{Gm2} > \frac{-A_{vcobuf.}\sin\left(\frac{3}{2}\pi\right)}{A_{REF}\sin\left(\frac{3}{2N}\pi\right)}K_{Gm1} \quad (10)$$

For example, if N=44, $A_{vcobuf.}$=0.4V, A_{REF}=0.9V, and K_{Gm1}=40uA/V, K_{Gm2} must be larger than 166uA/V. To further

978-1-7281-5107-6/19 $31.00 © 2019 IEEE

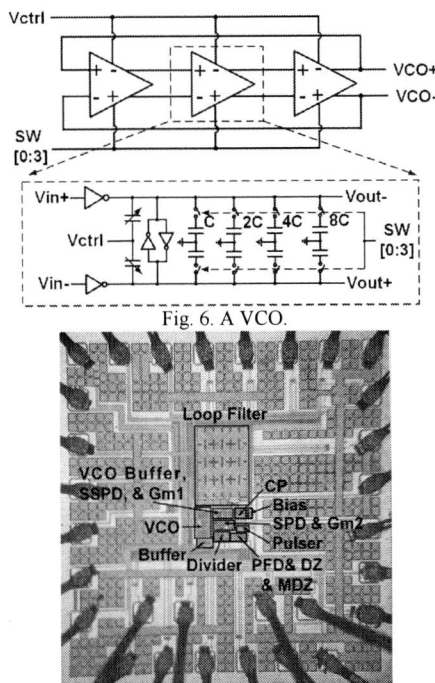

Fig. 6. A VCO.

Fig. 7. Die photo.

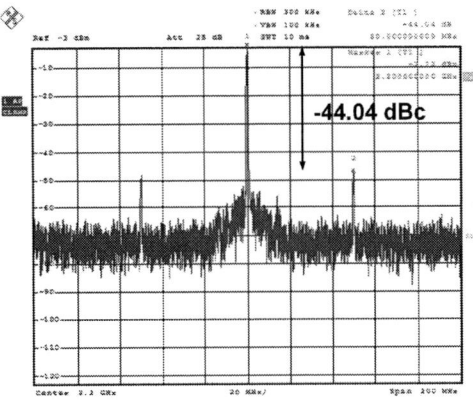

Fig. 8. Measured reference spur of the proposed SSPLL when operating at 2.2 GHz.

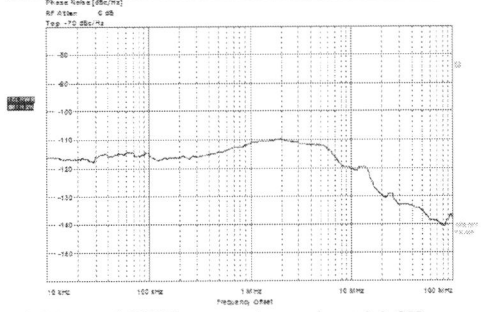

Fig. 9. Measured SSPLL output phase noise at 2.2 GHz.

reduce the re-locking time, the final K_{Gm2} is chosen as 700uA/V. The simulated re-locking time is 0.36us when the initial frequency is 2.15GHz and the target frequency is 2.2GHz.

C. VCO

A 3-stage ring oscillator and its delay stage are shown in Fig. 6. For every delay stage, its delay is controlled by a control voltage Vctrl and four digitally-controlled binary-weighted capacitors. The unit capacitance C is 8fF. A four-bit control signal SW [0:3] controls the frequency of the VCO. The simulated tuning range of the VCO covers from 1.7 GHz to 2.4 GHz and its K_{vco} is around -160 MHz/V.

III. EXPERIMENTAL RESULTS

This proposed SSPLL is fabricated in 0.18μm CMOS process. Fig. 7 shows the die photo with the active area of 0.436×0.223 mm². The frequency of REFex is 50 MHz and N=44. The power consumption of this SSPLL is 13.59mW for a supply of 1.8V at the output frequency of 2.2GHz. The VCO dissipates 9.864mW which is around 72% of the total power consumption. The power summation of the SPD, the MDZ creator and the Gm2 is 0.34mW, which is 2.5% of the total power consumption. Fig. 8 shows the measured output spectrum of the proposed SSPLL at the frequency of 2.2GHz. The measured reference spur is -44.04dBc. Fig. 9 shows the measured phase noise spectrum of the proposed SSPLL. The phase noise is -115.33dBc/Hz at the offset frequency of 100kHz.

A triggering signal is used to switch the divider ratio from 44 to 45. The measured re-locking time of the SSPLL without the MDZ creator and the Gm2 is 4.08 us to settle within the

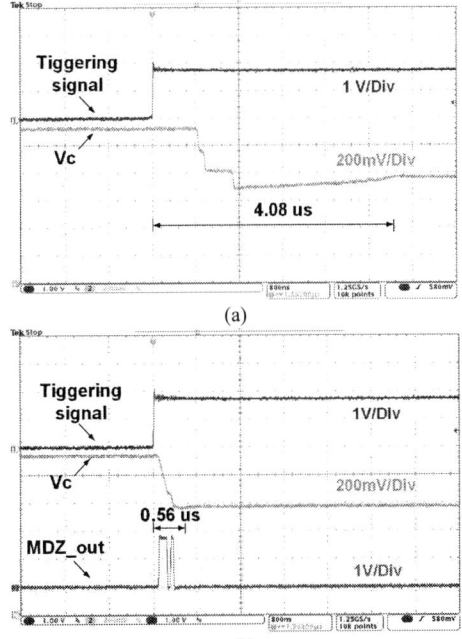

(a)

(b)

Fig. 10. Measured transient responses while the divider ratio is switched from 44 to 45 (a) with and (b) without the MDZ and the Gm2.

IEEE Asian Solid-State Circuits Conference
November 4 – 6, 2019
The Parisian Macao, Macao SAR, China

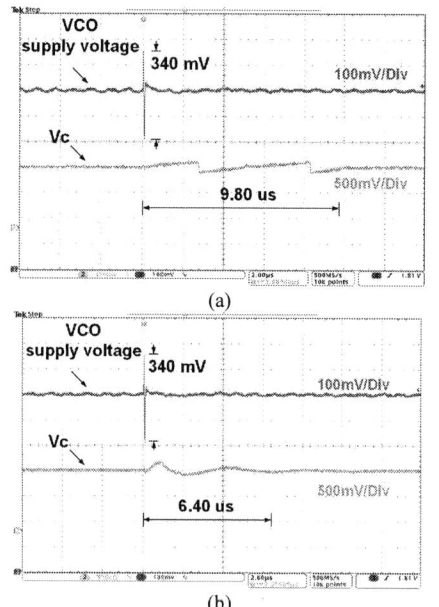

(a)

(b)

Fig. 11. Measured transient responses when a supply interference occurs (a) with and (b) without the MDZ creator and the Gm2.

frequency error of 0.27%. The measured re-locking time of the proposed SSPLL with the MDZ creator and the Gm2 is 0.56us as shown in Fig. 10(b).

To consider a peak-to-peak supply interference of 340mV occurring on the supply of the VCO, Fig. 11(a) and (b) show the measured transient responses of the proposed SSPLL without and with the MDZ and the Gm2. Without the MDZ and the Gm2, the SSPLL takes 9.8us to settle within the frequency error of 0.17%. With the MDZ and the Gm2, the proposed SSPLL takes 6.4us within the same frequency error of 0.17%. The performance summary and comparison with prior art are listed in Table I. In this work, the ring VCO is used, so it occupies less active area comparing with other two SSPLLs in [2] and [3]. Note that the proposed method can also be realized in an LC-based SSPLL.

IV. CONCLUSIONS

By using the SPD, the MDZ creator and the Gm2, the proposed SSPLL has a short re-locking time under the supply interference or divider ratio changed. The experimental results demonstrate the feasibility of the proposed technique. The low in-band phase noise of the proposed SSPL is kept and the power penalty is low.

ACKNOWLEDGMENT

The authors would like to thank National Taiwan University Donation Grant FD105012, Taiwan Semiconductor Research Institute, and Ministry of Science and Technology, Taipei, Taiwan.

REFERENCES

[1] X. Gao, E. A. M. Klumperink, M. Bohsali, and B. Nauta, "A low noise sub-sampling PLL in which divider noise is eliminated and PD/CP noise is not multiplied by N^2," IEEE J. Solid-State Circuits, vol. 44, no. 12, pp. 3253–3263, Dec. 2009.
[2] C. W. Hsu, K. Tripurari, S. A. Yu and P. R. Kinget, "A sub-sampling-assisted phase-frequency detector for low-noise PLLs with robust operation under supply interference," IEEE Trans. Circuits Syst. Part-I: Regul. Papers, vol. 62, no. 1, pp. 90-99, Jan. 2015.
[3] D. Liao, F. F. Dai, B. Nauta, and E. A. M. Klumperink, "A 2.4-GHz 16-phase sub-sampling fractional-N PLL with robust soft loop switching," IEEE J. Solid-State Circuits, vol. 53, no. 3, pp. 715–727, March 2018.
[4] J. Chuang and H. Krishnaswamy, "A 0.0049mm^2 2.3GHz sub-sampling ring-oscillator PLL with time-based loop filter achieving -236.2dB jitter- FOM," in IEEE International Solid-State Circuits Conference (ISSCC), pp. 328–329, Feb. 2017.
[5] S. S. Nagam and P. R. Kinget, "A 0.008mm^2 2.4GHz type-I sub-sampling ring-oscillator-based phase-locked loop with a −239.7dB FoM and −64dBc reference spurs," IEEE Custom Integrated Circuits Conference (CICC), pp. 1–4, April 2018.
[6] S. S. Nagam and P. R. Kinget, "A low-jitter ring oscillator phase-locked loop using feedforward noise cancellation with a sub-sampling phase detector," IEEE J. Solid-State Circuits, vol. 53, no. 3, pp. 703-714, March 2018.

TABLE I PERFORMANCE SUMMARY

	Technology (μm)	Robust @ supply interference	VCO type	Output Freq. (GHz)	Ref. Freq. (MHz)	In-band phase noise (dBc/Hz)	RMS jitter (ps) (integrated range)	Reference spur (dBc)	Supply voltage (V)	Power (mW)	Active Area (mm^2)	FOM[1] (dB)
[1]	0.18 CMOS	No	LC	2.21	55.25	-126 @200 kHz	0.15 (10k-40MHz)	-46	1.8	7.6	0.18	-247.67
[2]	0.65 CMOS	Yes	LC	2.2	50	-122 @200 kHz	0.48 (10k-40MHz)	-41.6	1.1	8.8	0.24	-236.93
[3]*	0.13 CMOS	Yes	LC	2.4	50	-116.68 @100 kHz	0.16 (10k-10MHz)	-72	1.3	21	0.43	-242.69
[4]	0.65 CMOS	No	Ring	2.3	192	-106.98 @99.9 kHz	0.72 (10k-100MHz)	-37	1.2	4.59	0.0049	-236.23
[5]	0.65 CMOS	No	Ring	2.4	100	-117.03 @100 kHz	0.42 (1k-100MHz)	-64.2	1.2	6.1	0.008	-239.68
[6]	0.65 CMOS	No	Ring	2.2	49.15	-123.5 @300 kHz	0.63 (1k-100MHz)	-55.2	0.935	5.86	0.022	-236.33
This work	0.18 CMOS	Yes	Ring	2.2	50	-115.33 @100 kHz	0.767 (10k-100MHz)	-44.04	1.8	13.59	0.097	-230.97

[1]FOM=20log$_{10}$(RMS jitter/1s)+10log$_{10}$(Power/1mW)

* [3] The data are listed while the PLL works in the integer-N mode.

AI and IoT for Social Value Creation

Yasunori Mochizuki

NEC Corporation
Tokyo, Japan
y-mochizuki@az.jp.nec.com

Abstract—The features of AI and IoT technologies leading to the realization of next smart cities/smart communities are discussed. The AI topics included are person identification AI for biometrics, white-box/explainable AI for value chain innovation, whose use cases reveal how these technologies are contributing to the broader range of new social values. IoT system also has significant implications in realizing the next smart society. Here, interoperability and openness are becoming increasingly important for scalability and cross-domain data utilization of solutions, in quest to the realization of citizen-centric and economically sustainable, thereby future-proof smart cities.

Keywords—AI; IoT; smart city; data exchange platform

I. INTRODUCTION

The idea of digital transformation is attracting global attention, which stands for leveraging digital capabilities to drive fundamental change or disruption in organizational management and/or reconstructing business process The trend has become even more prominent in the last several years as AI and IoT are rapidly applied to a broad range of business sectors under the concept of Industrie 4.0 and/or the fourth industrial revolution. The idea applies not just for private sector businesses but also to society and community. Japan formulated in 2014 the idea of Society 5.0 [1], human-centric super smart society to be achieved by digital innovation.

The topic discussed in this paper is smart city. The 'smart city' was popularized as a concept in the early 2010s to describe how advances in technology and data could allow us to plan and run our cities better. In the space of a few years the concept of a smart city shifted from a focus on technologies and systems, to a focus on citizens and services for them. We are now witnessing a challenge to the vision of a citizen-centered smart city arising from the disruption [2]

In this paper, we will discuss the important clues to leveraging AI and IoT technologies for digital disruption of society and community by looking at the social value creation activity of the author's organization.

II. AI TECHNOLOGIES FOR SOCIAL VALUE CREATION

A. Recognition AI and Biometrics solutions

NEC has decades of experience on R&D on AI technologies including fingerprint recognition and facial recognition, both of which have been proven to be with the world-leading performance through the series of global technology benchmark conducted by NIST [3] and are delivered as biometrics solutions. (Fig.1)

Biometrics solutions are applied to police investigation, law enforcement (such as passport control and citizen ID), and facility access control, thereby contributing to "Safety" and "Security" of society", sometimes combined with other analytics solutions (Fig.2)

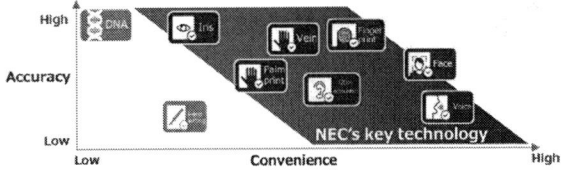

Fig. 1 NEC's biometrics solution suite

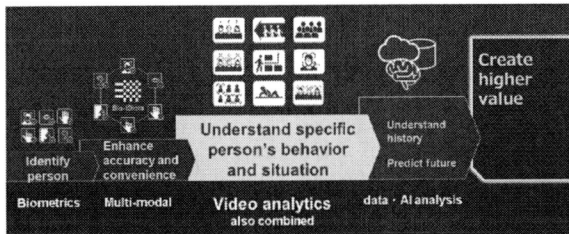

Fig. 2. Value creation by biometrics and video analytics

In 2013, NEC participated in Singapore's Safe Cities Testbed and a more sophisticated combination of recognition AI and other real-world analysis technologies for holistic city safety was demonstrated under the concept of Inter Agency Collaboration [4]. The author emphasizes the novelty of the non-technical aspect of this specific demonstration in that, for a smart society in the digital age, breaking the organizational silo and encouraging collaborations through data and information sharing are becoming crucially important and should be defined as a new management strategy.

In fact, the city of Tigre in Argentina [5], where the mayor introduced a new command control center implemented with a set of biometrics and other video analytics solutions, started to enhance the collaboration of departments and also with citizens. The outcome of such mayor's leadership was also notable that the city's car theft rate dropped by 80%. Even more amazingly, the city reported that its tourism-related industry earning was increased by 300% due to the improved brand of the city being a safe city to visit.

Biometrics is also contributing to "Efficiency" and customer experiences. NEC recently announced it will provide a facial recognition system for the new "One ID" check-in to boarding process in airport [6]. This new process will enable passengers at the airport to register facial images during initial procedures, such as check-in, and if the images match within a database of passport photos, they can advance through procedures that include baggage drop-off, passenger security screening entrance and boarding without presenting boarding passes and passports that are conventionally required. Thus, passengers are expected to be able to board more smoothly, with fewer repetitive processes and with less time in line.

Here are further use cases for which biometrics is contributing to "Equality". India operates the Aadhaar Program, which is the largest system of its kind in the world with more than one billion citizens registered. NEC provides a large-scale biometrics identification system that utilizes fingerprints, face images, and iris images for this national identification system. Based on this program, the Indian agency is promoting an effort to create a society in which the entire nation can enjoy equal access to public and financial services. Such a technology solution is particularly found helpful to the citizens living in rural regions of India, thereby improving their quality of lives [7]. For instance, the financial inclusion service allows users to perform bank transactions, including receiving subsidies from the government, by verifying their identity through biometric scans across the country without having to travel to a distant bank branch. The other example, in which biometrics is contributing to "Equality", is the fingerprint identification for the provision of legal identity to newborn children in developing countries [8]. It is becoming technically possible to identify fingerprints of newborn babies only 2-24 hours after birth. Such a feature will enable social registration of children before leaving from hospitals and will lead to a better healthcare including vaccination planning.

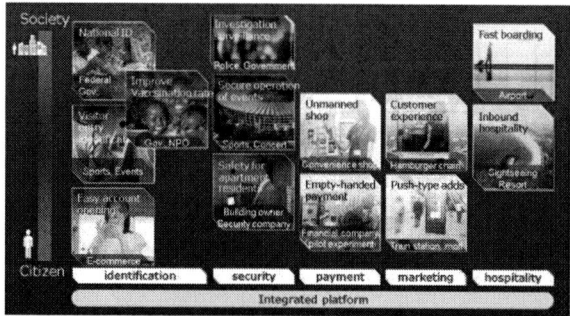

Fig. 3 Market value expansion of biometrics and video analytics solutions.

B. Explainable AI supporting high-level human decisions

Recently, the AI technology is applied to rapidly expanding applications in human society. However, some challenges are becoming also visible. One of them is the lack of

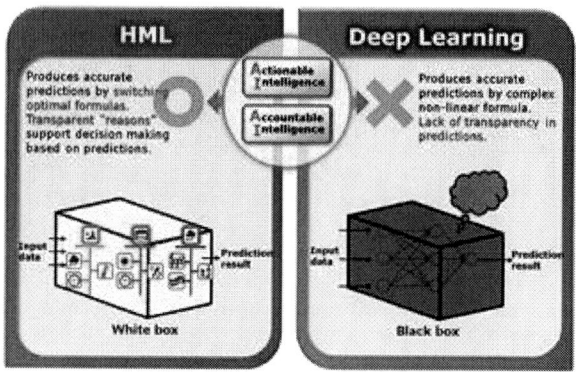

Fig. 4 Features of heterogeneous mixture learning

explainability: it is very often unclear why AI made a certain decision or gave a certain prediction.

Many of business activities, such as enterprise management, investment assessment and medical care, require decision making by authorized personnel, or a decision gives crucial impact to people, If AI does not give the reason why a certain decision was made, issues remain as the lack of fairness, consensus and persuasion. This is the background of the emerging demand for "Explainable AI" technology.

In 2012, NEC developed Heterogeneous Mixture Learning (HML) technology [9]. Previous techniques, such as the cross-validation or the Bayesian, calculated the scores (information criteria) for the model candidates and selected the model with the best score. HML, which is backed by the latest machine learning theory called "factorized asymptotic Bayesian inference", is capable of adaptive searching of the number of groups, the method of grouping and the prediction model for each group. This makes it possible to find the optimum data grouping and prediction model with high prediction accuracies without searching unpromising candidates. Furthermore, because of such an automatic discovery feature of optimum grouping and prediction models simultaneously, the predicted result obtained by HML is quantitatively explainable.

HML has been already applied to more than 160 use cases in a broad range of business sectors including retail (demand prediction), manufacturing (inventory optimization), finance (risk assessment) and energy (power demand prediction).

III. IoT and Data Exchange for Next Smart Society

A. Vision for Next Smart Societies

Smart city initiatives in the initial period have been focused on enhancing efficiency of specific service domains such as for power supply or traffic infrastructures as shown in Fig.5.

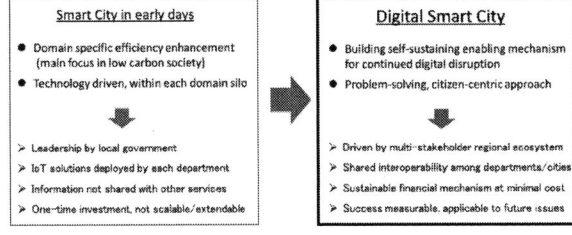

Fig. 5 Evolution of smart city vision

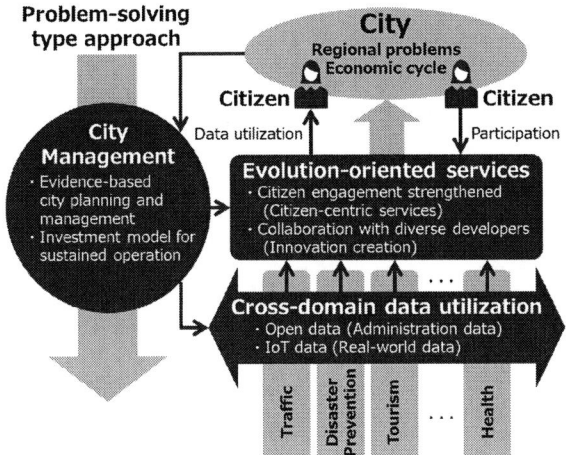

Fig. 6 Requirements for the digital smart city.

However, since the problems of cities affect each other across the domains, a more holistic approach is necessary in order to create a sustainable city. Also, early smart city projects tended to focus on introducing new technologies by working with global technology providers and the outcome of those were seriously questioned as to whether they really improved citizen's lives. There is also the issue that many projects are still in the pilot stage. Even when a smart city in an early stage is developed via government funding, it cannot be expanded unless a more sustainable financial mechanism is established that assures continuous operation. Based on the recognition of issues as described above attempts to convert smart city projects to next-generation ones are now beginning to emerge having its central focus on citizens and sustainability. In the present paper, we refer to the next-generation approach as the digital smart city (Fig.5). The objective of this approach is the digital innovation of cities by solving issues of the smart cities emerging in the initial period as described below.

Accordingly, there are set of important ingredients in order to establish digital smart cities and these are schematically listed in Fig 6 [10]. From the technological viewpoint, the smart city ICT platform should enable a cross-cutting data utilization coming from diversified sources in the form of both historical and (real time) IoT data. The data utilization systems that become the contact points with citizens are designed as citizen-centric city management design (design thinking approach) in order to encourage the engagement of citizen groups.

B. Data Exchange system for Smart Citiess

As of today, three primary barriers are preventing widespread deployment of effective, powerful smart city solutions. First is the inadequate information and knowledge transfer: custom solutions unable to exchange information with each other, and therefore, neither replicable and reusable. Second is the wide range of standards, which stands for a number of architectural standardization efforts that are underway worldwide but have not yet converged. The consequence is the poor scalability arising from insufficient interoperability and scalability of underlying Internet of Things (IoT), and Cyber-Physical Systems (CPS) technologies.

In order to overcome such a situation, interoperability and openness are attracting increasing attention as the key ingredients for the realization of "Next smart society". Open and Agile Start City, a global network of more than 140 cities, share the so-called "minimal interoperability mechanisms" (MIMs) which provide interoperability for the crucial points for data exchange among heterogeneous systems, yet allowing each cities to maintain system features to cope with local and/or custom requirements. So far, the MIMs consist of (1) standard open API, (2) federated data models and (3) ecosystem transaction management. Meanwhile, open source approach is also becoming the basic strategy of many cities to avoid vender lock-in thereby achieving the technology governance owned by the regional community stakeholders.

In Europe, the data exchange platform FIWARE has been developed starting from 2011 and has been financed under the next-generation Future Internet Public-Private Partnership (FI-PPP) program. NEC has been participating in this program from the initial phase.

The core function of FIWARE is the "context information management" that provides application systems with various IoT data at desired timings. One of its main features is that it implements this function as the open source software (OSS) of an API, conforming to the international standard (NGSI-9/10). The nature of its open architecture enables extension of the systems without relying on a single ICT vendor. In addition, the modular structure enables quick startup by freely combining the modules required for each city. The developed modules can be replicated and reused in another city, so that flexibility is very high.

When FIWARE is used as the core method for connecting various information source systems and application systems, it will be possible to utilize data efficiently by means of the cross-cutting approach (Fig. 7).

Furthermore, under the collaboration with standardization organizations such as the International Telecommunication Union (ITU) and European Telecommunications Standards Institute (ETSI), harmonization of API specifications is underway. These efforts are making the FIWARE the de-facto standard in Europe.

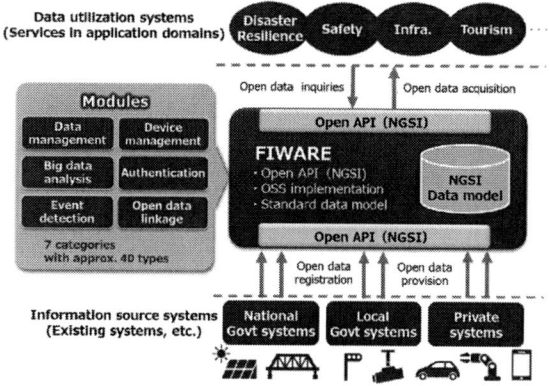

Fig.7 Features of FIWARE

With the consideration that the endeavors of the FIWARE community as described above are indispensable also for Japan to achieve its Society 5.0 policy, NEC became the first Japan-originated enterprise member of the highest-class (Platinum) of the FIWARE Foundation in March 2017. Acting as a member of the Board of Directors (BoD), NEC is leading the promotion of the enhancement and the extension of the functions of FIWARE. NEC is also working with cities to implement FIWARE-based data-exchange platform. Figure 8 is the use case developed in Takamatsu city in Japan that is aiming at both self-sustaining growth of tourism-based new economy as well as citizen-engaged disaster resiliency.

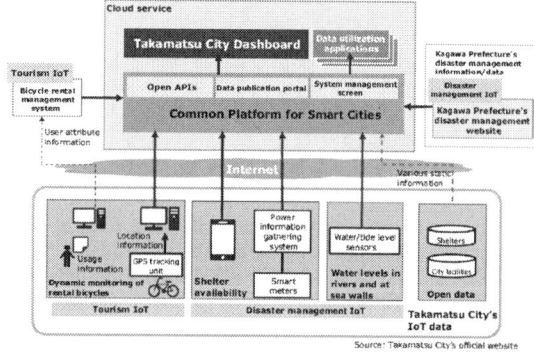

Fig. 8. FIWARE-based smart city system in Takamatsu City

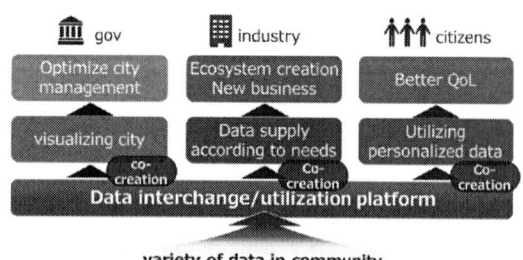

Fig. 9 Smart city leveraging data utilization platform

As a high-level picture, Fig. 9 visualizes how such a cross-domain and cross-agency data interchange will act as the core function in the next smart society. To the city government, it will deliver the capability of evidence-based planning and management, new business ecosystem to the regional industry, and better quality of life to the citizens. The scalability and openness in the smart system bring the technology governance and data governance to the hands of regional multi-stakeholder ecosystem, thereby serving as the key driving mechanism of the next smart society.

IV. CONCLUSION

The vision of next generation smart society is to leverage digital innovation and transform the society into a human-centered, livable and sustainable communities.

Recognition AI (biometrics) and explainable AI (heterogeneous mixture learning) are discussed and how the new social value creation is advancing toward safety, security, efficiency and equality leveraging these novel technologies.

The goal for the smart city has become to build a more sustainable and citizen-centric community via cross-cutting, problem-solving, and flexible approach. Such an innovation agenda calls for an open and agile nature for the IoT data exchange platform and an EU-originated open source platform called FIWARE is highlighted together with its ecosystem building and use cases.

REFERENCES

[1] Japanese Government, "Society 5.0" (in English), https://www.gov-online.go.jp/cam/s5/eng/#motiongraphicsModal

[2] Future Cities Catapult, "Smart City Strategies, A Global Review", 2017, https://futurecities.catapult.org.uk/wp-content/uploads/2017/11/GRSCS-Final-Report.pdf

[3] K. Sakurai, H.Hashimoto, Y. Morishima, A. Hayasaka, H. Imaoka, "How Face Recognition Technology and Person, Re-identification Technology Can Help Make Our, World Safer and More Secure", NEC Technical Journal, vol. 13, No.2, pp. 69-73, 2019.

[4] P. Wang, Kang W. Woo, S. K.t Kohsm, "Building a Safer City in Singapore", NEC Technical Journal, vol.9, No.1 pp. 71-74, 2014.

[5] J. Vargas, D. Bergonzelliaper, "Securing the Future in Tigre", NEC Technical Journal, Vol. 9, No. 1, pp. 75-77, 2014.

[6] NEC Release, "NEC to provide facial recognition system for new "One ID" check-in to boarding process at Narita Airport", 2019, . https://www.nec.com/en/press/201902/global_20190228_01.html

[7] NEC Release, "NEC and CSC e-Governance Services form a strategic alliance to deliver innovative digital services for rural areas in India", 2019. https://www.nec.com/en/press/201905/global_20190515_01.html

[8] NEC Release, "NGavi, NEC, and Simprints to deploy world's first scalable child fingerprint identification solution to boost immunization in developing countries", 2019, https://www.nec.com/en/press/201906/global_20190606_01.html.

[9] R. Fujimaki, S. Morinaga, "The Most Advanced Data Mining of the Big Data Era", NEC Technical Journal, vol. 7, No. 2, pp. 91-95, 2012.

[10] Y. Mochizuki, "Global Perspective for Data-Leveraged Smart City Initiatives", NEC Technical Journal, vol. 13, No. 1, pp. 14-18, 2018.

[11] K. Ishii, A. Yamanaka, "Building a Common Smart City Platform Utilizing FIWARE (Case Study of Takamatsu City)", NEC Technical Journal, vol. 13, No. 1, pp. 28-31, 2018.

9-2 (xxxx)

Millimeter-Wave System-on-Chip Applications from Space Explorations to Contactless Connectivity

Mau-Chung Frank Chang

Department of Electrical and Computer Engineering, University of California, Los Angeles
College of Electrical Engineering, National Chiao Tung University, Hsinchu, Taiwan

Abstract—This paper highlights a wide range of (sub)-mm-Wave system-on-chip (SoC) applications from space exploration to contactless connectivity. Passive and active SoCs developed at UCLA and NCTU, including heterodyne-radiometer, spectrometer processor, and multiband radio/radars, will be exemplified as cost/weight/energy-effective ways to explore astrophysics and earth/planetary sciences. Multi-Giga-bit/sec contactless connectors based on mm-Wave radios have also been successfully developed and commercialized to replace high-performance mechanical connectors/cables for smartphones, electrical cars, and data centers. Techniques to augment range and bandwidth efficiency according to PAM-4 modulations and flexible plastic waveguide are also reviewed with design and test results.

Keywords— mm-Wave system-on-chip (SoC), mm-Wave radio and radar, heterodyne radiometer, spectrometer, contactless connectivity, plastic waveguide

I. INTRODUCTION

Recent science missions such as heterodyne instrument for far-infrared (HIFI), microwave instrument for Rosetta orbiter (MIRO), and microwave limb sounder (MLS) feature a wealth of scientific knowledge that can be attained with heterodyne radiometers and spectrometers. Despite its importance being long recognized by the science community, space-borne observations remain scarce and under-sampled (summarized in [1]) primarily because space-borne micro/millimeter-wave instruments can only be flown on major missions. They often demand commitments in size, weight, and power (SWaP) with excessive cost. The infusion of Si CMOS System-on-Chip (SoC) has transformed the way in which space instruments are implemented. The ability to monolithically integrate microwave/mm-Wave, mixed-signal, and memory circuits to operate under low supply voltages (e.g. 1V) is ideally suited for payload-limited space probes. In particular, cost/performance-effective CMOS technologies can play vital roles in supporting more frequent science missions on lower cost platforms such as 1U to 6U cube satellites (CubeSat). 1U is defined as 10cm x 10cm x 10cm cubic units with maximum 1.33kg mass and 1W power consumption. For instance, fully integrated W-band heterodyne radiometer, wideband spectrometer processor, and 183GHz heterodyne spectrometer, have all been realized in 28nm CMOS for planetary science explorations. Another emerging application is the detection of water/moisture distributions for environmental sciences based on drones or unmanned vehicles (UAV). While an Earth-orbiting satellite such as soil moisture active passive (SMAP) offers global coverage, UAVs or drones could provide localized monitoring systems in particular areas.

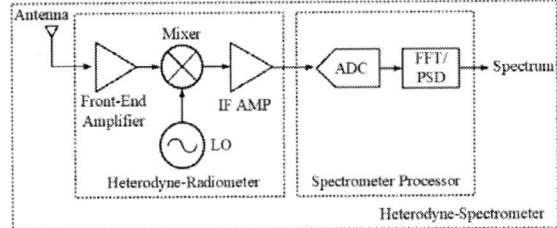

Fig. 1. Block diagram of heterodyne-radiometer/spectrometer.

In this paper, we report a Ku-band frequency modulated continuous wave (FMCW) radar developed to accurately estimate the amount of water available from mountain snow.

High-speed and short distance signal/data connectivity was accomplished by using high performance mechanical connectors and cables with serious constraints in weight/size/range/speed and configurability. Such constraints can now be overcome by using highly integrated mm-Wave contactless radios. After 10 years of research at UCLA, contactless connectors have finally been implemented and commercialized in LG's 5G smartphones to enable dual-screens and Acer's laptops to enable detachable monitors. Future research activities will focus on further extending the data rate and connectivity distance.

II. HETERODYNE RADIOMETER/SPECTROMETER

Radiometric observations on thermal radiation at (sub)mm-wavelengths are invaluable for assessing Earth's climate by profiling water vapor distribution inside storms and hurricanes. For planetary science, (sub)mm-Wave radiometers are capable of detecting water contents of comets or planetary bodies. A direct-detecting radiometer generally contains building block circuits such as the front-end amplifier, down-conversion mixer, local oscillator (LO), and intermediate frequency (IF) amplifiers. For the down-conversion, a power detector is often employed to extract the received signal's envelope to discern the atmosphere's absorption from Earth's blackbody radiation. Heterodyne radiometers, however, are created by replacing direct-detecting radiometers' mixers and LOs with square-law detectors and amplifiers. The direct-detecting radiometer may be more size/energy-efficient but lacks the capability to specify gas compositions [2]. A heterodyne radiometer can also be combined with the spectrometer processor to form a complete heterodyne spectrometer for resolving rotational emission/absorption spectra of ionized species and gas molecules.

978-1-7281-5107-6/19 $31.00 © 2019 IEEE   103

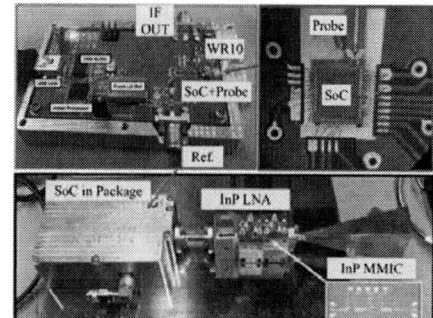

Fig. 2. Passive imager SoC with external InP LNA.

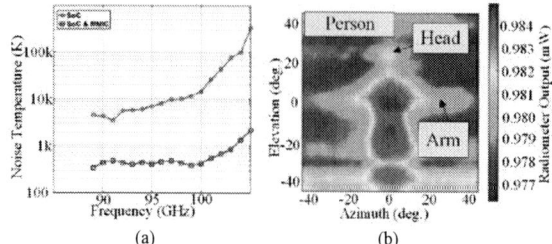

(a)   (b)

Fig. 3. (a) Measured noise temperature, and (b) captured passive image.

A spectrometer processor consists of building blocks including an analog-to-digital converter (ADC), an FFT/power-spectral-density (PSD) processor, and an accumulator.

A. W-Band CMOS/InP-Hybrid Radiometer & Passive Imager

The W-band radiometer is prototyped in 28nm CMOS and coupled with a 35nm InP HEMT low-noise amplifier (LNA) front-end to decrease the noise for meeting space exploration requirements [3]. The LO (Fig. 1) is generated by an on-chip frequency synthesizer to cover 90 to 100GHz. A transformer-coupled front-end amplifier is structured with cascoded NMOS devices. Transformers are used to isolate low frequency interferences among amplifier stages and ensure their DC stability. The entire radiometer system with on-chip frequency synthesizer is packaged in a waveguide block and consumes only 260mW. The coupling between WR10 waveguide and subsequent SoC input is achieved by using a probe and through micro-strip lines fabricated on alumina substrate. Despite 3-4dB signal attenuation, the preceding InP LNA front-end can easily compensate the loss with solid 30dB gain. A WR10 noise source (Quistar QNS10) is introduced to measure radiometer's noise temperature and is measured with/without the InP LNA. Test results from both cases are compared in Fig. 3(a), showing substantially improved noise temperature with the InP LNA front-end. The captured W-band radiation pattern, detected one meter away from a laboratory staff's body, is mapped in Fig. 3(b). This is made possible by using a 25dBi WR10 horn antenna and an angular (Az/El) raster scanner.

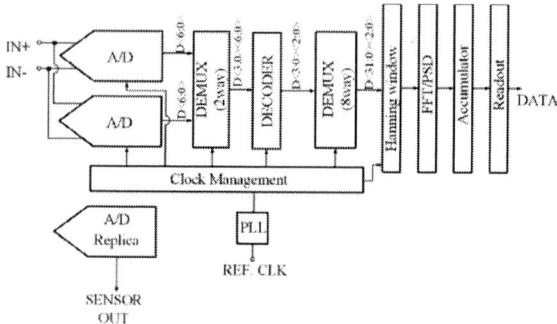

Fig. 4. Block diagram of a spectrometer processor.

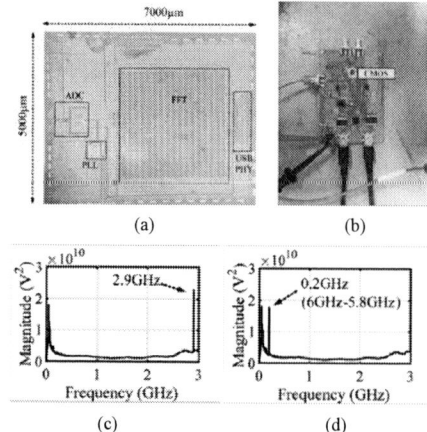

(a)   (b)

(c)   (d)

Fig. 5. (a) Fabricated spectrometer die in 65-nm CMOS, (b) Board assembled spectrometer processor, (c) 2.9-GHz input tone measurement, (d) 5.8-GHz tone input tone measurement for frequency folding.

B. 6-GS/s 4096-Point Spectrometer Processor

Conventional spectral analysis for space/planetary sciences was performed based on FPGAs under stringent SWaP challenges, especially with multiple arrays. Fig. 4 shows the building block diagram of a spectrometer processor. An interleaved-by-2 flash ADC is first used to convert differential input signals to seven-level thermometer-coded digital outputs at 6GS/s sampling rate. A replica ADC is also placed to monitor DC biases for track-and-hold amplifiers and pre-amplifiers. As analyzed in [4], a 3-bit ADC is sufficient to resolve Earth and planetary spectra since their quantization noise can be averaged down over the integration time. For FFT/PSD processor to surpass its digital timing constraints, full-speed thermo-coded data must be first de-multiplexed via 2-way (becoming 4-way due to 2-channel ADC), then decoded into 3-bit binary form, and further de-multiplexed via 8-way. Consequently, the FFT/PSD processor is designed to channel 32 parallel data streams at 187.5MHz clock speed each. Cooley-Tukey algorithm is also implemented with symmetric frequency bins removed. Accordingly, the fabricated spectrometer SoC in 65nm CMOS and its assembled module are shown in Fig. 5(a) and 5(b), respectively.

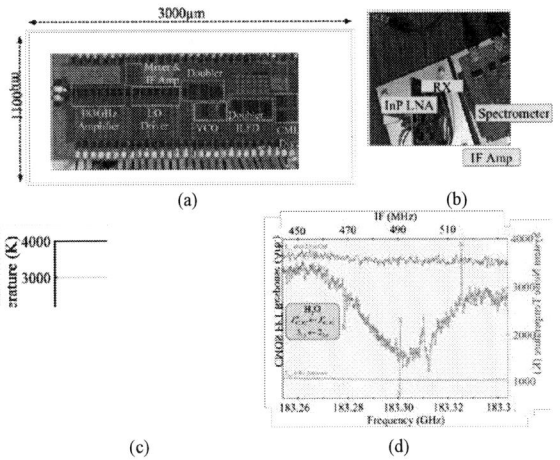

(a)   (b)

(c)   (d)

Fig. 6. (a) Fabricated 183-GHz heterodyne radiometer, (b) assembled complete heterodyne spectrometer, (c) measured noise temperature, and (d) detected water molecule.

The fabricated spectrometer consumes only 1.5W under 1V supply and clocked at 6GS/s sampling speed. To validate its full-speed operation, 2.9GHz and 5.8GHz sinusoidal signals are applied and successfully detected, as shown in Fig. 5(c) and 5(d), respectively. In Fig. 5(d), a frequency folding effect is also confirmed.

C. 183 GHz CMOS/InP-Hybrid Heterodyne-Spectrometer

A 183GHz InP/CMOS-hybrid heterodyne-radiometer is also realized in [5]. Although 65nm transistor's cut-off frequency does not exceed 150GHz, the further scaled 28nm transistor offers useful gain above 200GHz. In fact, both 183GHz front-end amplifiers and frequency synthesizer are implemented in 28nm CMOS. While sampling at 6GS/s, a 183GHz heterodyne spectrometer is developed to characterize rotational spectroscopic responses of water vapor and other volatiles including organics and other radical species. The outgassing velocity escaping from jets or plumes can be assessed by observing Doppler frequency shifts on these known spectroscopic lines; pressure profiles can be assessed by observing spectroscopic line's broadening effect. The fabricated CMOS heterodyne radiometer is shown in Fig. 6(a), and the complete heterodyne spectrometer assembly is shown in Fig. 6(b). The system's noise temperature is measured spectroscopically from the frequency bins. While the complete system is running, the 'hot' spectrum N_1 is recorded at room temperature (T_1=293K), and the 'cold' spectrum N_2 is recorded with the liquid nitrogen soaked absorber (T_2=77K) placed in front of the horn antenna. The Y-factor is then calculated by N_1/N_2 for each spectral bin. The measured noise temperature is plotted in Fig. 6(c). The entire system's noise temperature ranges from 700 to 1000K within 1GHz IF bandwidth. The 183GHz heterodyne radiometer consumes 515mW, and the entire heterodyne spectrometer consumes 3.315W. After characterizing the SoC performance, the complete system is installed in a gas cell for molecule detection. The observed response for the water vapor line at 183.310GHz is clearly seen in Fig. 6(d).

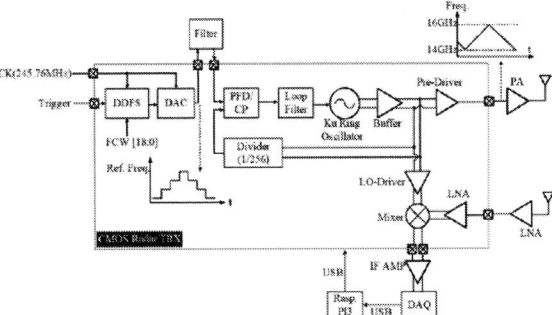

Fig. 7. Block diagram of Ku-band FMCW snow radar.

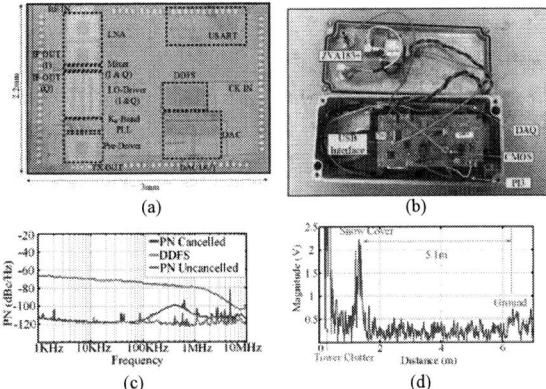

(a)   (b)

(c)   (d)

Fig. 8. (a) Fabricated CMOS FMCW radar, (b) assembled radar module, (c) measured phase noise cancellation, and (d) measured depth of snow.

With the measured system noise temperature (~1000K), the stratospheric water signal's (~100K) brightness is readily observable with a sub-second integration time. The modular CMOS design strategy suggests that each subsystem can be readily infused into another SoC to formulate a low-SWaP instrument. For instance, a Schottky diode-assisted terahertz (520 to 600GHz) heterodyne spectrometer from Jet Propulsion Laboratory now adopts CMOS spectrometer processor instead of conventional FPGAs and adopts CMOS frequency synthesizers (28 to 34 GHz) as drivers for 600GHz LOs instead of Gunn oscillators [6].

III. SNOW SENSING RADAR

While climate change has influenced a large portion of the Earth in terms of temperature and sea water level, some of the most dramatic effects have appeared in reduction of snowpack and subsequent water availability globally. As snowpack retreats, the available water from snowmelt is limited, so more vigorous water resource management is required to ensure that water demands are satisfied. The key to this planning exercise is to accurately estimate the water content in snowpack. Among the three major snow properties (cover, depth, density), a Ku-band FMCW radar is designed to measure the depth of snow [7]. As the frequency of radar operation increases beyond the Ku-band regime, an inter-grain scattering mechanism starts to dominate over the bulk absorption (Mie regime) for typical snow grain size, which is the reason the Ku-band operation is chosen for penetration.

978-1-7281-5107-6/19 $31.00 © 2019 IEEE

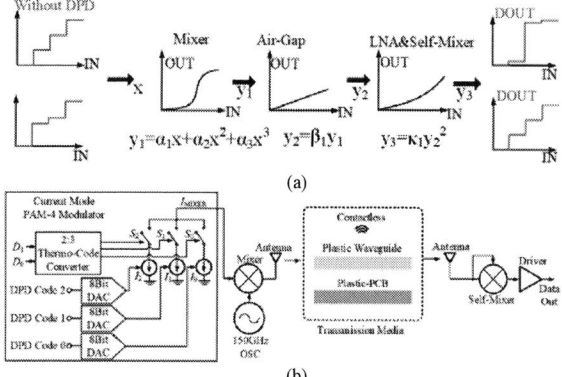

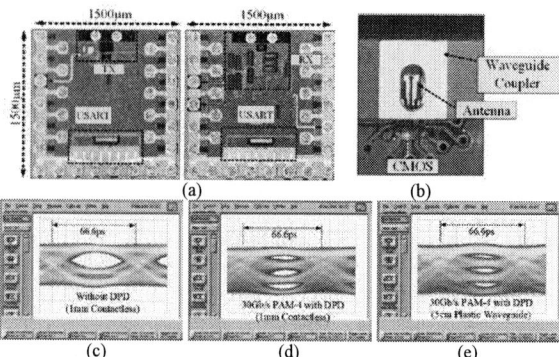

(a)

Fig. 9. (a) System concept with and without DPD, and (b) block diagram of 150 GHz CMOS PAM-4 TX/RX with current-mode DPD.

Fig. 10. (a) Fabricated chips, (b) assembled module, (c) measured without DPD, (d) measured with DPD, and (e) measured via plastic waveguide.

The radar system targets for 7.5cm axial resolution, which results in a chirp bandwidth of 2GHz. To achieve a compact transceiver while supporting such a wide chirp bandwidth, a ring oscillator-based chirp generator is adopted. The synthesizer accepts a low frequency reference chirp from the DDFS, synthesizes, and then interpolates to turn a discrete frequency staircase into a smooth frequency ramp at Ku band. The overall system block diagram is shown in Fig. 7. The fabricated radar transceiver in 65 nm CMOS, shown in Fig. 8(a), integrates DDFS, DAC, synthesizer, mixer, and amplifiers. The entire CMOS radar consumes 250mW under 1.1V supply. The assembled module is shown in Fig. 8(b). The ring oscillator's phase noise becomes problematic when there are secondary scatters around the target of interest. The phase noise can be overcome by the range correlation effect, where the phase noise can be partially compensated depending on the target distance. This is identical to that of the first-order sigma-delta modulator. The measured phase noise in Fig. 8(c) proves the effectiveness of the range correlation effect. The radar module successfully detects the depth of snow at Mammoth Mountain, as shown in Fig. 8(d).

IV. CONTACTLESS/WAVEGUIDE COMMUNICATIONS

Contactless connectivity is categorized as the extreme case where wireless connectivity distance is only a few millimeters. Although successfully deployed publically, the existing solution is lacking in terms of throughput (currently support 6 Gb/s) for the latest standard (e.g. 10 GB/s for USB 3.0) and thus requires a new direction to scale the bandwidth. Higher-depth modulations are attractive, but as illustrated in Fig. 9(a), multi-level signals may suffer system non-linearity and experience amplitude distortion. To overcome such limitations, a digital pre-distortion (DPD) is inserted to the PAM-4 modulator [8]. As shown in Fig. 9(b), the transmitter is built upon the RF-mixing-DAC, where the current-steering DAC is combined into the up-conversion mixer [9]. The PAM-4 modulator consists of a binary to 3-bit thermometer-code converter, three current sources, and three switches. The thermometer-coded data decides which current branch to flow and creates four-level current signals at the mixer's input. The 8-bit R-2R DACs control the amount of each current and enable DPD.

The 150GHz transmitter and receiver are prototyped using a 28nm CMOS process (Fig. 10(a)). The CMOS chips and antennas are assembled on the same PCB substrate, as shown in Fig. 10(b). In Fig. 10(c), without DPD, the upper-eye is completely closed and center-eye is wide open. After DPD applied, a 30Gb/s linkage over 1 mm air-gap is achieved with 2pJ/bit energy efficiency. The connectivity distance can be extended by simply inserting plastic waveguides between the modules; the measured eye-diagram is captured in Fig. 10(e).

V. CONCLUSION AND FUTURE DEVELOPMENT

CMOS technologies continue to pave the way into un-explored areas such as space/planetary sciences and niche connectivity applications. The 183GHz heterodyne spectrometer is currently being assembled in a balloon-craft at the Jet Propulsion Laboratory for launching into Earth's stratosphere. A complete passive/active snow sensor system on a UAV is under development for future deployment. Contactless connectors are explored for use in smartphones, data centers, robotics, automobiles, and industrial factories. Lastly, more exciting plastic waveguides for (sub)mm-Wave or terahertz system applications is on the horizon.

REFERENCES

1. P. H. Siegel, "THz instruments for space," *IEEE Trans. Antennas Propag.*, vol. 55, no. 11, pp. 2957-2965, Nov. 2007.
2. S. Padmanabhan, *et al.*, "Radiometer payload for the temporal experiment for storms and tropical systems technology demonstration mission," *IEEE Int. Geosci. Remote Sens. Symp.*, 2017.
3. A. Tang, *et al.*, "A W-band 65nm CMOS/InP-hybrid radiometer & passive imager," *IEEE MTT-S Int. Microw. Symp.*, 2016.
4. Y. Zhang, *et al.*, "Integrated wide-band CMOS spectrometer systems for spaceborne telescopic sensing," *IEEE Trans. Circuits Syst. I*, 2019.
5. Y. Kim, *et al.*, "A 183-GHz InP/CMOS-hybrid heterodyne-spectrometer for spaceborne atmospheric remote sensing," *IEEE Trans. THz Sci. Technolo.*, 2019.
6. G. Chattopadhyay, *et al.*, "Compact terahertz instrumets for planetary missions," *Proc. 9th Eur. Conf. Antennas Propag.*, 2015.
7. Y. Kim, *et al.*, "A Ku-band CMOS FMCW radar transceiver for snowpack remote sensing," *IEEE Trans. Microw. Theory Tech.*, 2018.
8. Y. Kim, *et al.*, "A millimeter-wave CMOS transceiver with digitally pre-distorted PAM-4 moudlation for contactless communications," *IEEE J. Solid-State Circuits*, 2019.
9. Y. Kim, et al., "30Gb/s 60.2mW 151GHz CMOS transmitter/receiver with digitally pre-distorted current mode PAM-4 modulator for plastic waveguide and contactless communications," *IEEE MTT-S Int. Microw. Symp.*, 2019.

An Energy-Efficient BJT-Based Temperature-to-Digital Converter with ±0.13°C (3σ) Inaccuracy from -40 to 125°C

Rushil K. Kumar, Hui Jiang, and Kofi A. A. Makinwa

Delft University of Technology, The Netherlands

Abstract— This paper describes an energy-efficient BJT-based temperature-to-digital converter (TDC) with a state-of-the-art resolution FoM (RFoM) of 5.4pJ·K². It consists of a BJT front-end, a capacitively-coupled instrumentation amplifier (CCIA) and a 2nd-order continuous-time (CT) ΔΣADC. The front-end and the CCIA employ bitstream-controlled dynamic element matching (DEM) and chopping to mitigate mismatch and 1/f noise without incurring quantization noise fold-back. The sensor achieves ±0.13°C (3σ) inaccuracy from -40 to 125°C and has a supply sensitivity of 8.2mK/V.

I. INTRODUCTION

BJT-based TDCs are widely used, because they only require room-temperature calibration to achieve high accuracy [1-5]. However, they usually employ switched-capacitor (SC) readout architectures, and so their resolution and energy efficiency are limited by kT/C noise. For example, a state-of-the-art TDC achieves a resolution FoM of 7.8pJ·K² [4]. Although the design in [5] is more energy-efficient, achieving 3.6pJ·K², its output is a duty-cycle modulated signal, which must still be sampled by a high-speed clock and digitized by an external counter. This paper describes a TDC that uses a continuous-time approach. A CCIA is used to boost a PTAT voltage ΔV_{be} (generated by a BJT front-end) before it is digitized by a CT ΔΣADC. As in SC circuits, the CCIA's gain is also determined by a capacitor ratio, resulting in good accuracy and stability. The TDC achieves ±0.13°C (3σ) from -40°C to 125°C, and a state-of-the-art RFoM of 5.4pJ·K².

II. PROPOSED DESIGN

Fig. 1 shows the proposed TDC. A BJT front-end generates PTAT and CTAT voltages ΔV_{be} and V_{be}. The CCIA amplifies ΔV_{be} by a factor $C_{in}/C_{fb} = \alpha/2$. Its output is then applied to a 2nd-order incremental CT ΔΣADC, which implements a charge-balancing scheme, in which $2 \cdot \alpha/2 \cdot \Delta V_{be}$ is integrated when the bitstream output (bs) is "0" and $-V_{be}$ when bs is "1". This results in a bitstream average $\mu = \alpha \Delta V_{be}/(\alpha \Delta V_{be} + V_{be})$, which is proportional to absolute temperature (PTAT) [6].

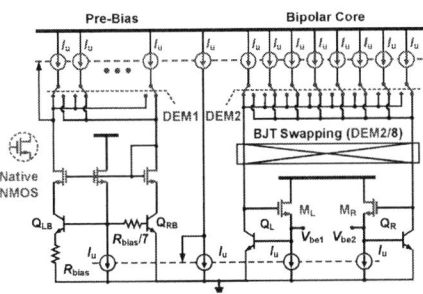

Fig. 2 Simplified schematic of the BJT front-end.

The BJT front-end consists of a bias circuit and a bipolar core (Fig. 2), built around two pairs of identical NPNs (Q_{LB}, Q_{RB}, Q_L and Q_R). Each pair is biased at a collector current ratio of 1:7, with a unit current of 53nA at room temperature. The bias circuit generates a PTAT current that is used by the NPNs of the bipolar core to generate V_{be} and ΔV_{be}. The base currents of the NPNs are provided by source followers M_L and M_R, which also drive the input resistors of the CT ΔΣADC without affecting accuracy. To minimize their mismatch, the current sources of the bipolar core and biasing circuit are dynamically matched. As in [6], this is done in a bitstream-controlled manner to avoid quantization noise fold-back due to intermodulation between the DEM residuals and the modulator's bitstream. To mitigate their mismatch, the NPNs of the sensor core are chopped at the end of every complete cycle of DEM2, avoiding the intermodulation.

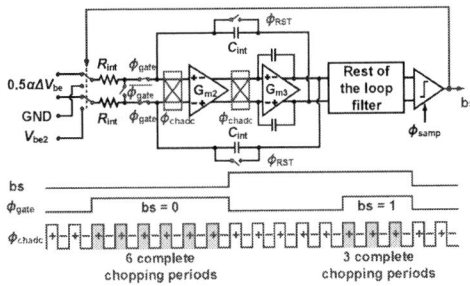

Fig. 3 Block diagram of CT ΔΣADC with the associated timing.

The CCIA consists of a 2-stage Miller-compensated amplifier (Fig. 1). It employs a current-reuse input stage for energy efficiency. Its closed-loop gain is set to $\alpha/2$ (=7), which limits its output swing to ~0.5V over the full temperature range and ensures that its loop-gain is high enough to keep its gain spread below 0.01%. Two CMFB loops (not shown) ensure

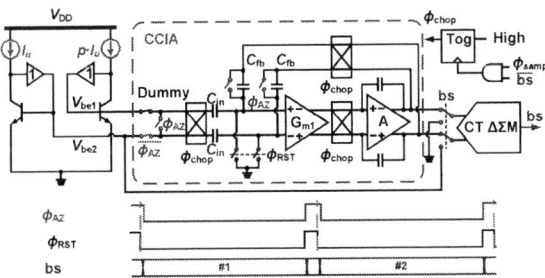

Fig. 1 a) Block diagram of the TDC and b) the associated timing.

978-1-7281-5107-6/19 $31.00 © 2019 IEEE

adequate robustness to process and temperature variations. The CCIA is also chopped in a bitstream-controlled manner.

At the start of each incremental conversion, the opamp is auto-zeroed by shorting its feedback caps C_{fb} (ϕ_{AZ}, Fig. 1) [7]. This effectively stores the opamp's offset on the input caps C_{in}, resulting in significantly reduced chopping ripple. Therefore, a ripple reduction loop is not required.

At the chopping transitions, the CCIA generates output spikes, which would be a source of error. To avoid this, the input of the 1st integrator of the $\Delta\Sigma$ADC is gated as in [8] (Fig. 3). Furthermore, the integrator is chopped at $8f_s$ to suppress its $1/f$ noise and offset. This choice ensures that the chopped offset is always averaged out despite the different lengths of the integration windows corresponding to bs=0 and bs=1 (Fig. 3). To conserve area, the 2nd integrator and feed-forward coefficient are implemented with SC techniques.

III. MEASUREMENT RESUTLS

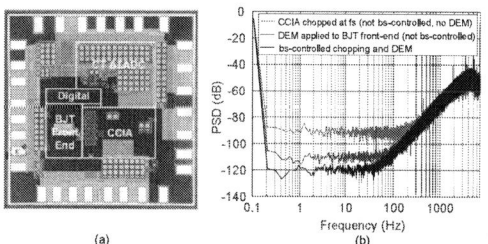

(a)   (b)

Fig. 4 a) Die micrograph of the TDC, b) PSD of the TDC's bitstream.

Fabricated in a standard 180nm CMOS process, the TDC occupies an active area of 0.35mm^2 (Fig. 4a) and consumes 5.6µA from a 1.6V supply at room temperature. The sinc^2 decimation filter is implemented off-chip. As shown in Fig. 4b, the use of bitstream-controlled chopping and DEM effectively prevents quantization noise fold-back and preserves the noise floor of the BJT front-end.

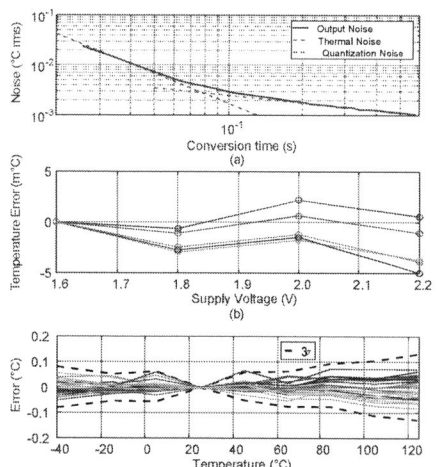

Fig. 5 a) Resolution versus conversion time, b) Temperature error versus supply voltage, and c) Inaccuracy of 25 devices.

When sampled at f_s = 15.625kHz, the TDC achieves a resolution of 1.67mK (rms) in a conversion time of 218ms (Fig. 5a). From 1.6V to 2.2V, its supply sensitivity is 8.2mK/V (Fig. 5b). Measurements were made on 25 devices in ceramic DIL packages. To compensate for V_{be} spread, each sample was calibrated at 25°C. In the same step, the CCIA gain (C_{in}/C_{fb}) of each sample is calibrated with the help of an external voltage. This information is used by the digital backend to perform a PTAT trim on each sample, resulting in an inaccuracy of ±0.25°C (3σ) from -40°C to 125°C. This improves to ±0.13°C (3σ) after a 5th order polynomial is used to remove systematic non-linearity (Fig. 5c). In Table I, the performance of the TDC is summarized and compared with previous BJT-based designs. As can be seen, this work achieves both competitive accuracy and state-of-the-art energy-efficiency.

Table I. Performance summary and comparison with the state-of-the-art.

	JSSC2013 [1]	JSSC2015 [2]	IEEE SJ 2017 [3]	JSSC2017 [4]	This work
Technology	0.16µm	0.16µm	0.18µm	0.16µm	**0.18µm**
Chip area [mm²]	0.54	0.085	0.198	0.16	**0.35**
Supply voltage [V]	1.6	1.4	1	1.5-2	**1.6-2.2**
Supply current [µA]	3.4	4.5	1.1	5	**5.6**
Supply sensitivity (m°C/V)	500	--	--	10	8.2
Temperature range	-55°C to 125°C	-40°C to 85°C	25°C to 45°C	-55°C to 125°C	**-40°C to 125°C**
3σ inaccuracy [°C] (Trimming points)	±0.2 (1)	--	±0.2 (1)	±0.06 (1)	**±0.13 (1)**
Conversion time [ms]	5.3	6	500	5	**218**
Resolution [mK]	20	25	10	15	**1.67**
Resolution FoM [pJ·K²] *	11	24	55	7.8	5.4

* Resolution FoM = Energy / (Conversion x (Resolution)²)

REFERENCES

[1] K. Souri, Y. Chae, K. A. A. Makinwa, "A CMOS temperature sensor with a voltage-calibrated inaccuracy ±0.15 °C (3σ) from −55 °C to +125 °C," *IEEE Journal of Solid-State Circuits*, vol. 48, no. 1, pp. 292–301, Jan. 2013.

[2] S. Zaliasl, et al., "A 3 ppm 1.5 × 0.8 mm2 1.0 µA 32.768 kHz MEMS-based oscillator," *IEEE Journal of Solid-State Circuits*, vol. 50, no. 1, pp. 291–302, Jan. 2015.

[3] M. Law et al., "A 1.1 µW CMOS Smart Temperature Sensor With an Inaccuracy of ±0.2 °C (3σ) for Clinical Temperature Monitoring", *IEEE Sensors Journal*, vol. 16, no. 8, Apr. 2016.

[4] B. Yousefzadeh, S.H. Shalmany, K.A.A. Makinwa, "A BJT based temperature-to-digital converter with ±60 mK (3σ) inaccuracy from −55 °C to +125 °C in 0.16 µm CMOS," *IEEE Journal of Solid-State Circuits*, vol. 52, no. 4, pp. 1044–1052, Apr. 2017.

[5] A.Heidary; G.Wang; K.A.A.Makinwa; G.Meijer, "A BJT-based CMOS temperature sensor with a 3.6pJ·K2-resolution FoM," *2014 IEEE International Solid-State Circuits Conference (ISSCC)*, pp. 224–225, 2014.

[6] M.A.P. Pertijs et al., "A CMOS smart temperature sensor with a 3σ inaccuracy of ±0.5 °C from -50 °C to +120 °C," *IEEE Journal of Solid-State Circuits*, vol. 40, no. 2, pp. 454–461, Jan. 2005.

[7] H. Wang et al., "A 19 nV/√Hz Noise 2-µV Offset 75-µA Capacitive-Gain Amplifier With Switched-Capacitor ADC Driving Capability," in *IEEE Journal of Solid-State Circuits*, vol. PP, no. 99, pp. 1-10, Aug.2017.

[8] H. Jiang, S. Nihtianov and K. A. A. Makinwa, "An Energy-Efficient 3.7-nV/√Hz Bridge Readout IC With a Stable Bridge Offset Compensation Scheme," in *IEEE Journal of Solid-State Circuits*, vol. 54, no. 3, pp. 856-864, March 2019.

A 16.1-b ENOB 0.064mm² Compact Highly-Digital Closed-Loop Single-VCO-based 1-1 SMASH Resistance-to-Digital Converter in 180nm CMOS

Elisa Sacco[1], Johan Vergauwen[2] and Georges Gielen[1]

[1]K.U. Leuven, Dept. Elektrotechniek ESAT-MICAS, Leuven, Belgium
[2]Melexis Technologies, Tessenderlo, Belgium

Abstract—This paper presents a novel highly-digital area- and energy-efficient closed-loop time-based CMOS single-ended resistive sensor-to-digital readout circuit. It achieves a high resolution of 16.1 bits utilizing a time-based implementation in an extremely small area of only 0.064mm². It employs a single VCO and a digital feedback loop for the read-out of an external single-ended resistive sensor such as a NTC thermistor. In addition to inherent 1st-order quantization noise shaping due to the oscillator, a second loop in SMASH configuration creates 2nd-order noise shaping. The fabricated prototype in a 180nm CMOS process achieves 16.1 bit of resolution for 1ms conversion time and consumes 171μW, resulting in a 2.4pJ/c.s. FOM_W, which is excellent for a resistive sensor interface. Thanks to the closed-loop architecture, it also achieves more than 13 bits of linearity.

Index Terms—resistive sensor readout circuit, VCO-based, time-based, $\Delta\Sigma$ modulator, noise shaping.

I. INTRODUCTION

Resistive sensors are used in many sensing systems, e.g. for temperature sensing. Single-ended NTC thermistors, for example, are used for most automotive applications (e.g. HVAC, TMAP). In literature, solutions have been presented for the readout of integrated NTC and/or PTC resistive sensors [1], [2]. In [3] the temperature dependency of the VCO oscillation frequency is used as sensing element. In [2] a traditional 2nd-order CT $\Delta\Sigma$ modulator is employed, thus resulting in poor process scalability due to the use of OTAs. In [1] and [3] more digital-friendly solutions are proposed, employing a VCO in a frequency-locked loop (FLL) and a VCO followed by a counter (FDC), respectively. However, these solutions require 2-point trimming and systematic error correction, resulting in an increased test time. Moreover, they suffer from poor linearity and only achieve 1st-order noise shaping. For the readout of external differential resistive sensors, [4] employs a traditional SAR architecture, while in [5] and [6] two VCOs are used in a phase-looked loop (PLL) configuration, leading to a highly digital implementation with 1st-order noise shaping. These solutions, however, all showed limited resolution.

This paper presents a novel time-based architecture for the high-resolution readout of external single-ended resistive sensors, such as NTCs. The main characteristics are:

- a maximum resolution of 16.1 bits.
- a small chip area and highly-digital technology-scalable solution due to the VCO-based implementation.

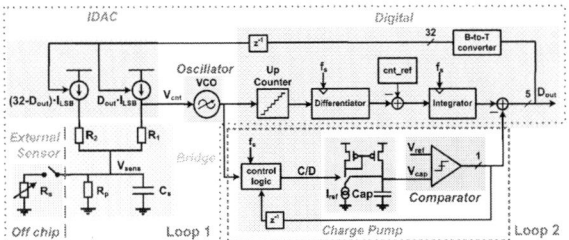

Fig. 1: Sensor readout block diagram.

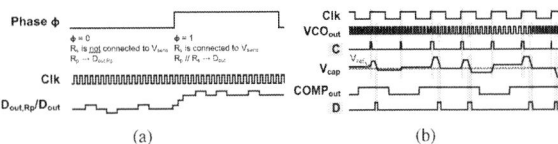

Fig. 2: (a) System time diagram and (b) *Loop 2* time diagram.

- a high linearity due to the closed-loop configuration.
- 2nd-order noise shaping by implementing a time-based 1-1 SMASH $\Delta\Sigma$ modulator.
- operation over the whole automotive temperature range (-40°–125°C), while requiring only 1-point trimming at room temperature to achieve a high temperature accuracy.

The paper is organized as follows. In Section II an overview of the sensor interface architecture is given. Section III details the circuit implementation in CMOS technology. Experimental measurement results are provided in Section IV. Finally, Section V concludes the paper.

II. SYSTEM-LEVEL OVERVIEW

Fig. 1 shows the basic building blocks of the resistive sensor readout circuit. *Loop 1* consists of a current-steering DAC (IDAC), a resistive bridge (R_1, R_2 and R_p), a VCO, and a digital part. The resistive sensor R_s is external to the chip. The digital output D_{out} results from integrating the difference between two counts: the number of VCO edges in one sampling period and a constant variable cnt_{ref}. The latter is performed to remove the common-mode signal generated by the central frequency of the VCO. Thanks to the integrating property of the VCO, the closed-loop system (*Loop 1*) shows inherent 1st-order quantization noise shaping.

The conversion consists of two phases, as shown in Fig. 2(a): in phase ϕ_0, the resistive sensor R_s is not connected to the bridge, thus the digital output D_{out,R_p} is the representation of R_p only. In phase ϕ_1, R_s is connected in parallel to R_p, resulting in the system digital output D_{out}.

Due to the closed loop, in equilibrium, the voltage V_{cnt} at the VCO input is on average equal to an operating voltage V_0 determined by the bridge. When expressing the VCO characteristic as: $f_{\mathrm{VCO}} = f_0 + K_{\mathrm{VCO}} \cdot (V_{\mathrm{cnt}} - V_0)$, this implies that the VCO frequency is on average f_0, which is such that the number of VCO edges occurring within each sampling clock period ($T_s = 1/f_s$) is on average equal to the reference count value $\mathrm{cnt}_{\mathrm{ref}} = f_0/f_s$.

By equating the value of the voltage V_{cnt} in the two phases, ϕ_0 and ϕ_1, assuming equilibrium, the input-output system characteristic equation can be expressed as follows:

$$R_p \parallel R_s = (R_1 \cdot D_{\mathrm{out},R_p})/2^N + R_p - (R_1/2^N) \cdot D_{\mathrm{out}} \quad (1)$$

where R_p is the internal parallel resistor at the input (see details in Section III), R_s is the external resistive sensor to be measured, R_1 is the bridge resistor, D_{out,R_p} is the digital output when only R_p is at the input, and D_{out} is the digital output with $R_p \parallel R_s$. Thanks to the closed-loop architecture, the transfer characteristic (1) depends only on passive resistive component values and not on sensitive IDAC and/or VCO parameters. These components are relatively temperature- and process-independent, thus requiring only one calibration point.

A second loop (*Loop 2* in Fig. 1) is added to the architecture, which consists of control logic, a charge pump and a comparator. Its 1-bit output is directly subtracted from the *Loop 1* output, thus realizing a 1-1 Sturdy MASH (SMASH) [7] architecture with 2nd-order noise shaping for the *Loop 1* quantization error. Being a 1-1 SMASH modulator, the stability of this architecture is similar to that of a 1st-order system, and thus not a problem. As shown in Fig. 2(b), during the time difference between the sampling edge and the next VCO edge, which corresponds to the quantization error in the time domain computed by the control logic of *Loop 2* (charge signal C), the capacitor Cap is charged by a constant current I_{ref}. According to the previous comparator decision, the capacitor Cap is then either discharged by the current I_{ref} for a VCO period or nothing happens (discharge signal D).

III. Circuit Design

The main features and implementation details of the building blocks in the proposed design are described in this section.

1) VCO: The schematic of the VCO in *Loop 1* is shown in Fig. 3(a). The PMOS converts the control voltage V_{cnt} into a current that drives a 7-stage current-controlled oscillator (CCO). The VCO central frequency is made tunable by activating 1 to 4 identical PMOS units. A tunable resistor R_{VCO} degenerates the PMOS, improving the linearity and lowering the $1/f$ corner frequency at the expense of some gain loss.

2) IDAC and Resistive Bridge: Since the IDAC provides both the feedback signal and the biasing of the external resistive sensor R_s, the loop can be closed with no extra power,

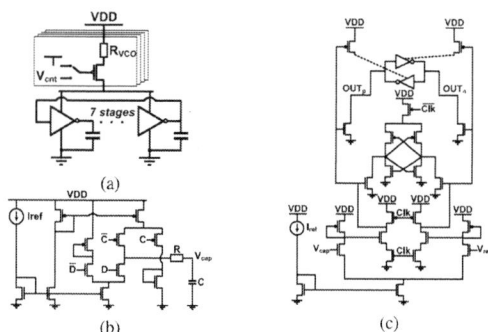

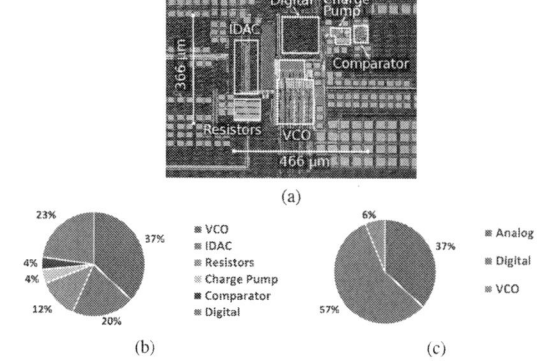

Fig. 3: Circuit schematics of (a) the VCO in *Loop 1*, and of the charge pump (b) and the comparator (c) in *Loop 2*.

Fig. 4: Chip micrograph (a), core area breakdown (b) and typical core power breakdown (c).

resulting in a high energy efficiency. To limit the voltage seen at the input for a large resistive sensor range (kΩ to MΩ), an internal resistor ($R_p \approx 20\mathrm{k}\Omega$) is placed in parallel to the external sensor R_s. The resistor R_1 is around 25kΩ to ensure proper functionality across the whole sensor input range.

3) Digital Circuitry: All digital circuitry has been synthesized with standard cells and laid out with a commercial P&R tool. The sampling clock f_s is generated by a $2\times$ clock divider from an off-chip master clock. In order to avoid errors due to the counter reset (the VCO and the sampling clock are asynchronous), a Johnson counter has been implemented, followed by a differentiator, which limits the worst-case count error to ± 1. The variable $\mathrm{cnt}_{\mathrm{ref}}$ is set externally.

4) Charge Pump and Comparator: The charge pump, shown in Fig. 3(b), has been designed to provide the same charge/discharge current using always-on current mirrors. The 1-bit quantizer has been implemented as a latch-based comparator preceded by an amplifier (Fig. 3(c)) to minimize the kickback. Since both circuits operate in *Loop 2*, the noise requirements are greatly relaxed, allowing scalability.

IV. Measurement Results

The prototype chip, shown in Fig. 4(a), has been fabricated in a standard 180nm CMOS process and occupies an active area of only 0.064mm^2. Fig. 4(b) and Fig. 4(c) show the area and power breakdown, respectively.

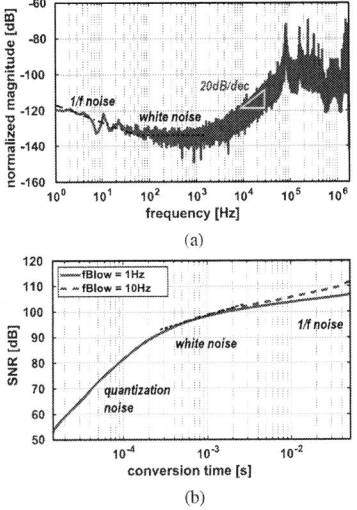

(a)

(b)

Fig. 5: Measured normalized PSD of the output bitstream (a) and SNR vs. conversion time plot (b) in fast sampling mode: $f_s = 1.8$MHz, $cnt_{ref} = 2$ and all 4 VCO PMOS units activated.

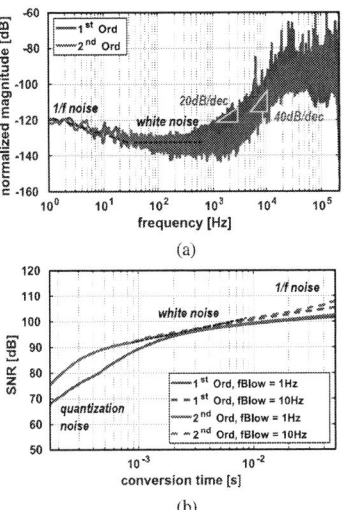

(a)

(b)

Fig. 6: Measured normalized PSD of the output bitstream (a) and SNR vs. conversion time plot (b) for 1st- and 2nd-order configuration in slow sampling mode: $f_s = 210$kHz, $cnt_{ref} = 8$ and 1 VCO PMOS unit activated.

A. Sensor Readout Characterization

Depending on the number of VCO PMOS units activated and the variable cnt_{ref}, the chip can operate in a fast and a slow sampling mode. Fig. 5(a) shows the measured output spectrum normalized to the full-scale digital output [6] of the sensor interface in the fast sampling mode. Since cnt_{ref} is set to 2 in this case, *Loop 2* cannot be activated[1], resulting in 1st-order noise shaping only. A resolution of 16.1 bits is achieved in 1ms conversion time (Fig. 5(b)), while consuming 171µW. The $1/f$ noise degrades the performance for long conversion time when the lower integration limit $f_{B_{low}}$ is small (Fig. 5(b)).

Normalized FFT plots of the measured bitstream output in slow sampling mode are shown in Fig. 6(a): when only *Loop 1* is active, 1st-order noise shaping is achieved, while 2nd-order is obtained with *Loop 2*. In the quantization noise limited region, the 2nd-order quantization noise shaping system indeed achieves a higher resolution for the same conversion time compared to the 1st-order system with the same settings, as shown in Fig. 6(b).

The prototype has been fully characterized at room temperature (RT) in order to determine the coefficients of the linear input-output characteristic (1), as shown in Fig. 7(a). The chip also achieves more than 13 bits of linearity (Fig. 7(b)).

B. NTC sensor measurements

A commercial NTC thermistor [8] has been used as external sensor to perform temperature measurements in a temperature-controlled environment. Fig. 8(a) shows the measured sensor characteristic over temperature and the corresponding digital output. The chip is functional over the whole automotive temperature range (from -40°C to 125°C). Using (1) and

---
[1]*Loop 2* requires at least eight VCO periods in each sampling period to operate properly, leading to the requirement $cnt_{ref} \geq 8$ (see Fig. 2(b)).

the RT characterization from Section IV-A, the sensor value R_s has been calculated. The maximum measured temperature inaccuracy is only 0.3°C over the whole temperature range (Fig. 8(b)), while the circuit is calibrated only at RT.

C. Performance summary

In Table I, the sensor interface performance is summarized and compared to the state of the art. The presented system compares favorably to other time-based solutions [1], [3], [5], [6] in terms of achievable resolution: it is the first to achieve 16.1 bits. This chip also achieves a better FOM_W value compared to recent state-of-the-art sensor readout circuits using a conventional amplitude approach [2], [4], while offering a more digital, therefore smaller and process scalable solution.

V. CONCLUSION

This paper has presented a novel closed-loop sensor interface. Employing only one VCO, the architecture is suitable for the readout of single-ended resistive sensors, such as NTC thermistors. It realizes an ultra-small energy-efficient time-based design with high resolution. Its highly-digital nature allows to synthesize large portion of the design. The design has been fabricated in a standard 180nm CMOS technology and occupies an active area of only 0.064mm². The chip consumes 171µW from a 1.8V supply and achieves a resolution of 16.1 bits in 1ms conversion time, resulting in a 163.5dB FOM_S. By trading resolution for bandwidth, the chip achieves 14.5 bits in 0.2ms conversion time, leading to a 1.48pJ/c.s. FOM_W. These are excellent FOM values for a resistive sensor interface.

The 2nd-order noise shaping property has been demonstrated in a closed-loop time-based system. The chip is the first to achieve 16.1 bits in time domain. The prototype is operational over the entire automotive temperature range, capable of

TABLE I: Performance Summary and Comparison with the State of the Art.

	This work	JSSC 2013 [5]	ISSCC 2018 [1]	ISSCC 2017 [3]	JSSC 2019 [6]	ISSCC 2019 [2]	ISSCC 2018 [4]
Topology (-based)	time (count)	time (PLL)	time (FLL)	time (FDC)	time (PLL)	amplitude (CTSDM)	amplitude (SAR)
Sensor type	ext, res	ext, res	int, p-res	int, VCO	ext, res	int, p-res	ext, bridge
Technology (nm)	180	130	65	180	180	180	180
Area$_{core}$ (mm^2)	0.064	0.2	0.007	0.22	0.26	0.12	1.7
Power$_{core}$ (μW)	171	125	68	0.57	3410	79	2.65
t$_{conv}$ (ms)	0.2 | 1	0.05	1	8	0.014	10	1
ENOB	14.5† | 16.1†	8.9†	13.6††	8.3††	9.9†	18.3††	7.9†
Temp. range (°C)	-40–125	-20–80	-40–85	-20–100	-40–175	-55–125	- -
Inaccuracy (°C)	0.3*,◇	0.56*,◇	±0.35**,◇◇	0.76*,◇◇	- -	±0.14**,◇◇	- -
FOM$_W$* (pJ/c.s.)	1.48 | 2.4	13.03	5.27	14.21	50	2.43	11.25
FOM$_S$+ (dB)	160.7 | 163.5	134.4	152.6	132.3	131.5	170	131.9

†from integrated noise, †† $\mathrm{SNR} = 20\log\left(\frac{T_{range}}{2\sqrt{2}T_{res}}\right)$, *min or max, **3σ, ◇1-point trimming, ◇◇2-point trimming, *FOM$_W$ = $\frac{Power \cdot t_{conv}}{2^{ENOB}}$, +FOM$_S$ = SNR+10log$\left(\frac{1}{2 \cdot Power \cdot t_{conv}}\right)$

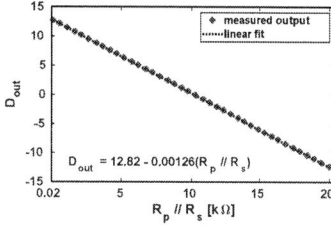

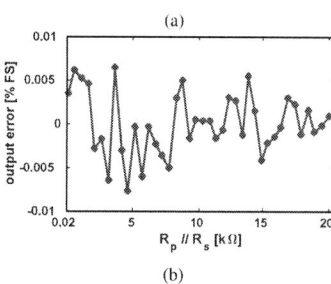

Fig. 7: Measured system input-output characteristic (a) and output error (b), as calculated using the fitted linear equation shown in (a). These measurements have been performed at room temperature.

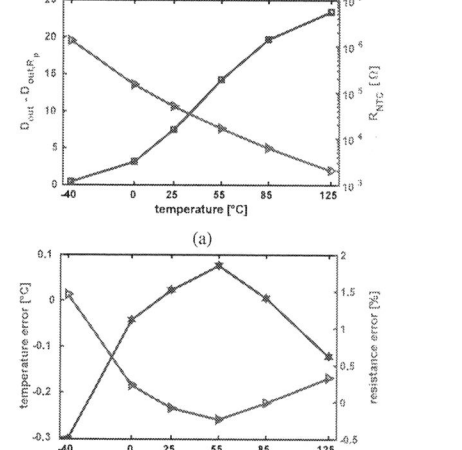

Fig. 8: Measured NTC characteristic vs. temperature and corresponding digital output (a), and relative resistor error with respect to the NTC characterization and corresponding temperature error (b). The temperature error is temperature- and sensor-dependent and it is calculated according to [8].

converting thermistor sensor values from kΩ to MΩ with a maximum inaccuracy of only 0.3°C, while requiring only room-temperature characterization. Compared to the state of the art, the proposed architecture achieves excellent sensor readout FOM values with a small and highly-digital thus scalable implementation.

Acknowledgment

The authors would like to thank VLAIO for its financial support and Melexis for its technical and financial support.

References

[1] W. Choi, Y.-T. Lee, S. Kim, S. Lee, J. Jang, J. Chun, K. A. Makinwa, and Y. Chae, "A 0.53 pJ·K^2 7000μm^2 resistor-based temperature sensor with an inaccuracy of ±0.35°C (3σ) in 65nm CMOS," in *IEEE Int. Solid-State Circuits Conference (ISSCC)*, pp. 322–324, 2018.

[2] S. Pan and K. A. Makinwa, "A Wheatstone Bridge Temperature Sensor with a Resolution FoM of 20f·K^2," in *IEEE International Solid-State Circuits Conference-(ISSCC)*, pp. 186–188, 2019.

[3] K. Yang, Q. Dong, W. Jung, Y. Zhang, M. Choi, D. Blaauw, and D. Sylvester, "A 0.6 nJ -0.22/+0.19°C Inaccuracy Temperature Sensor Using Exponential Subthreshold Oscillation Dependence," in *IEEE Int. Solid-State Circuits Conference (ISSCC)*, pp. 160–161, 2017.

[4] S. Oh, Y. Shi, G. Kim, Y. Kim, T. Kang, S. Jeong, D. Sylvester, and D. Blaauw, "A 2.5 nJ Duty-Cycled Bridge-to-Digital Converter Integrated in a 13mm^3 Pressure-Sensing System," in *IEEE Int. Solid-State Circuits Conference (ISSCC)*, pp. 328–330, 2018.

[5] J. Van Rethy, H. Danneels, V. De Smedt, W. Dehaene, and G. Gielen, "Supply-noise-resilient design of a BBPLL-based force-balanced wheatstone bridge interface in 130-nm CMOS," *IEEE Journal of Solid-State Circuits*, vol. 48, no. 11, pp. 2618–2627, 2013.

[6] J. Marin, E. Sacco, J. Vergauwen, and G. Gielen, "A Robust BBPLL-Based 0.18-μm CMOS Resistive Sensor Interface With High Drift Resilience Over a -40°C-175°C Temperature Range," *IEEE Journal of Solid-State Circuits*, 2019.

[7] N. Maghari, S. Kwon, G. Temes, and U. Moon, "Sturdy MASH ΔΣ modulator," *Electronics Letters*, vol. 42, no. 22, p. 1, 2006.

[8] TDK, NTC thermistor B57867S. Datasheet: https://www.tdk-electronics.tdk.com/inf/50/db/ntc/NTC_Mini_sensors_S867.pdf.

978-1-7281-5107-6/19 $31.00 © 2019 IEEE

10-3 (7089)

A Low-Noise Sub-Bandgap Reference with a ±0.64% Untrimmed Precision in 16nm FinFET

Matthias Eberlein
Communication and Device Group, Intel Germany
and Johannes Kepler University, Linz 4040, Austria
Email: matthias.eberlein@intel.com

Harald Pretl
Communication and Device Group, Intel Austria
and Institute for Integrated Circuits (IIC)
Johannes Kepler University, Linz 4040, Austria

Abstract— This paper presents a new concept for sub- bandgap circuits, based on direct generation of PTAT current with larger and adjustable temperature coefficient. The reference voltage is generated inside the feedback loop, which facilitates good supply rejection, low-noise and high precision performance. A 3σ - spread of only ± 0.64 % across split-lots was observed in a 16 nm FinFET process on 4400 μm², without trimming or any switching techniques. By using NPN bipolar devices, noise levels of ~130nV/sqrt(Hz) at 1 kHz can be achieved at 125 μA power drain. Different configurations are explained, which allow flexible reference levels or very low supply voltages of < 0.85 V.

Keywords— bandgap, voltage reference, NPN bipolar, low-noise

I. Introduction

Aggressive technology scaling has motivated numerous bandgap structures, often variants of the "Banba" current-mode principle [4] , which cope with a decreasing VDD (<< 1.3V). For mixed-signal SOCs there are additional requirements for precision and very low noise, across a wide frequency range. Hence this sub-bandgaps need also high PSRR, and can't use switching techniques like chopping for offset compensation. Realizations in recent FinFET process, though for 1.8V supply, achieve improved accuracy through careful layout [1], or by using a NPN bipolar core [2]. But to suppress flicker and supply noise, it is essential to generate the reference inside the main feedback loop, and provide a low impedance output, like in [5].

In [3] a different approach for a sub-bandgap without subdivision was used, by generating a down-shifted PTAT voltage from MOS short-channel effects. However, this open loop solution is sensitive to process and still requires a fairly large multiplication (k1) of the PTAT signal to balance the temperature coefficients (Tc), like the classical bandgap (Fig. 1). Here we introduce a concept to create a "strong PTAT" voltage having a higher relative Tc. In that case, as explained in Fig. 1, a smaller weighting (k2) is required to achieve flat temperature response. This results in sub-bandgap levels < 1.2V and provides more robustness towards amplifier offsets.

II. Circuit Realization

Fig. 2 presents the simplified circuitry, which includes also cascodes and a startup function (not shown). The *key features* result from *rationed NPN* devices (Q1:Q2 = N = 8) available in triple-well FinFET process, which offer superior mismatch and flicker noise performance for RF applications. The native reference output is a low-impedance net and drives a resistor string for variable Vref-tapping. While the structure is similar to the well-known "Brokaw Bandgap", resistor Rc plays a vital role to enable sub-bandgap operation: Connected in parallel to the base-emitter junction of Q1, Rc injects a CTAT current into the

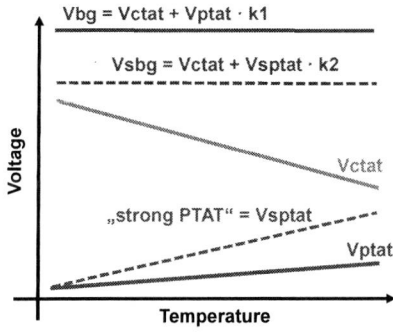

Fig. 1. Bandgap composition from exemplary CTAT and PTAT signals.

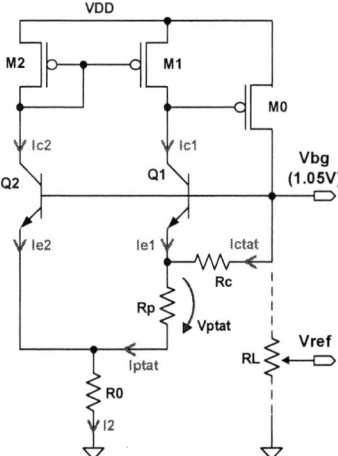

Fig. 2. Proposed sub-bandgap circuit with 1:N rationed NPN bipolar core.

978-1-7281-5107-6/19 $31.00 © 2019 IEEE

critical node, where the PTAT voltage is generated. In consequence, the feedback loop formed by the asymmetric differential amplifier and M0 is forced to create emitter currents Ie1, Ie2 with a larger Tc, to establish Vptat = Vbe2 − Vbe1 across Rc. This "strong PTAT" signal is multiplied by R0 and facilitates a sub-bandgap level of Vbg = 1.05V in our circuit.

The interaction of signals becomes clear from calculating the emitter currents in Fig. 2 (assuming a constant current gain β):

$$Ie = Ie1 = Ie2 = Iptat - Ictat = \frac{Vptat}{Rp} - \frac{Vbe1}{Rc} \quad (1)$$

where Vptat = $V_T \cdot ln(N)$, V_T is the thermal voltage and Vbe1 equals the base-emitter drop of Q1. Using typical temperature coefficients for both voltages, we can explain the branch current with a boosted Tc, resulting from a negative CTAT component:

$$\frac{dIe}{dT} = ln(N) \cdot \frac{90\mu V/°C}{Rp} + \frac{1.8mV/°C}{Rc} \quad (2)$$

Adding PTAT and CTAT voltages yields the native reference:

$$Vbg = Vptat \cdot \left(1 + \frac{2 \cdot R0}{Rp}\right) + Vbe1 \cdot \left(1 - \frac{R0}{Rc}\right) \quad (3)$$

Obviously the "strong PTAT" appears in the term like a scaling of Vctat, and for $Rc \rightarrow \infty$ Vbg turns into a classic bandgap.

III. PERFORMANCE DISCUSSION

Equation (3) helps also to understand the noise properties: The *thermal* component is dominated by Vptat and the multiplication factor (1+2·R0/Rp), which is around 10 for the classic bandgap. With inclusion of Rc, however, this factor reduces when the Tc's are balanced, and the critical resistor Rp can be smaller with respect to the same branch current (1). A *low flicker noise* is achieved mainly by using a BJT differential input instead of MOS. Contributions from the PMOS mirror were mitigated by using large device sizes and gate-overdrive. Fig. 3 shows the resulting performance, including an optional RC-filter (f_{3db} = 12 kHz) to meet the target spec at high frequencies.

Since our application is a complex mixed-signal SoC, this bandgap requires a good suppression of supply noise, too. Thanks to the closed-loop architecture, the simulated PSRR (Fig. 4) meets the required specification of > 60dB under all conditions and for low VDD > 1.1V.

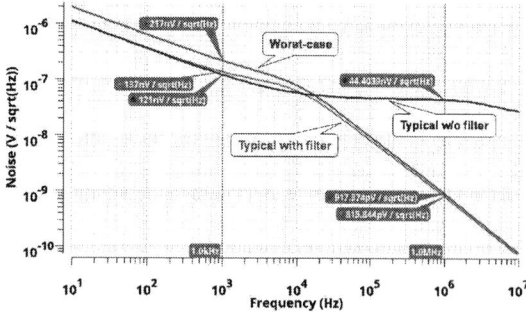

Fig. 3. Simulated noise at Vref=0.9V for typical and worst-case conditions.

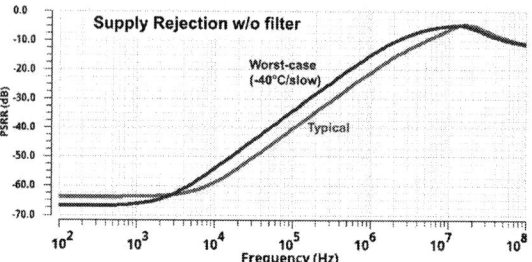

Fig. 4. Small signal PSRR of the sub-bandgap at VDD = 1.2V (simulated).

Another circuit variant for even lower supply is possible, when the reference is tapped from a high impedance node, as shown in Fig. 5. The separation of Rc into Rc1 and Rc2 yields the output voltage:

$$Vbg = Vptat \cdot \left(1 + \frac{2 \cdot R0}{Rp}\right) + Vbe1 \cdot \frac{Rc1 - R0}{Rc1 + Rc2} \quad (4)$$

The reference level is adjustable by the Rc1/Rc2 ratio, in combination with R0. For R0 = 0Ω the circuit effectively presents a reverse bandgap, with lowest supply constraint and similar noise performance. Simulations in Fig. 6 confirm a full functionality down to VDD = 0.8V across temperature.

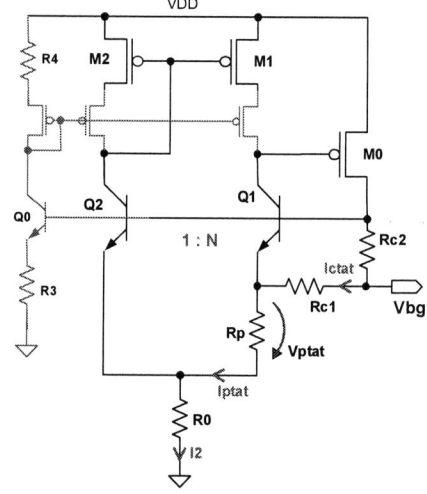

Fig. 5. Circuit variant with Vctat scaling for very low supply voltages, including proposed cascode bias.

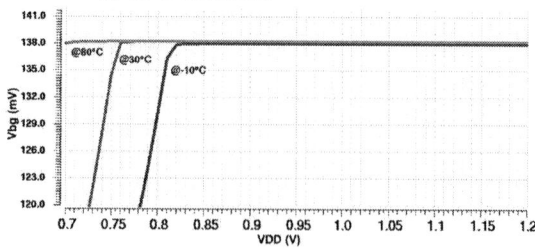

Fig. 6. Line regulation for a circuit configuration as reverse bandgap.

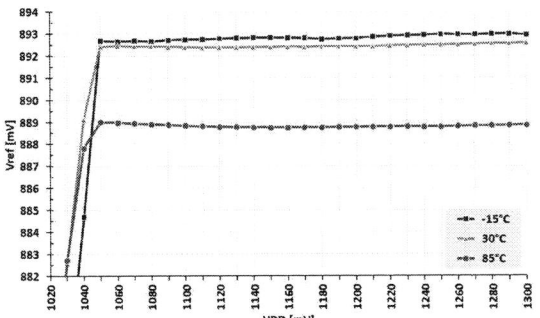

Fig. 7. Measured reference voltage versus VDD for a typical sample.

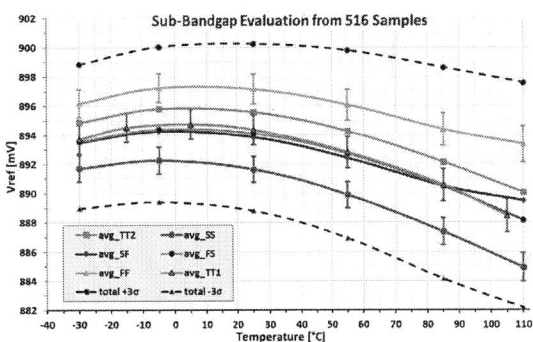

Fig. 8. Measured output voltage (at 900mV tab) with averaged values for each wafer, including ± 3-sigma limits for the full split lot (dotted lines).

IV. MEASURED SILICON RESULTS

The sub-bandgap IP according to Fig. 2 was realized as part of a wireless transceiver SoC on 16 nm FinFET, with strict requirements on reference noise. The deep system integration and test limitations did not allow a direct reading of the bandgap noise. But indirect measurements on system parameters confirmed the target specification of < 220 nV/√Hz at 1 kHz and < 1 nV/√Hz at 1MHz according to Fig. 3.

Only the 900 mV output, derived from the native reference (1.05 V), was tested. Results for supply sensitivity are plotted in Fig. 7 for few temperatures, and the PSRR was measured with ~73dB for VDD > 1.1V. Fig. 8 shows the statistical performance of the bandgap voltage, which was evaluated from a split lot with 6 skew wafers and a total of 516 samples. Individual wafers feature a variance of ~1mV, as explained by the error bars. Without any trimming, the overall 3σ - spread is ± 0.64% (at 25°C). The dominant error source is from mismatch in the PMOS mirror, additional to the process skews. A residual, negative temperature coefficient can be observed from the averaged results in Fig. 8. This deviation was traced back to the NPN devices, which were processed on all wafers close to the "lower" model corner, according to PCM data.

V. CONCLUSION

We presented a novel reference concept [6], which features low-impedance output, low-supply capabilities and effective noise suppression at once. Unlike other Sub-Bandgaps, the reference is generated *inside* the feedback loop, and by using a "strong PTAT" signal. In combination with the benefits of a NPN differential pair, this architecture yields low flicker noise and high precision, without trimming or offset compensation. The low circuit complexity produces a small silicon area of 4400μm² (Fig. 9), with only 125 μA power drain. Table I summarizes the measured performance in relation to prior work.

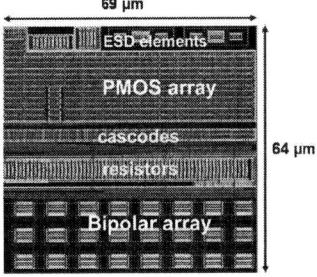

Fig. 9. Floorplan of the bandgap core circuitry in 16nm FinFET silicon.

TABLE I. PERFORMANCE OVERVIEW

	This work	[1] (Wadhwa)	[3] (Annema)	[5] (Chahardori)	[7] (Sanborn)	[8] (Chen)
Process	16nm FinFET	16nm FinFET	160nm CMOS	180nm CMOS	500nm BiCMOS	12nm FinFET
VDD	1.1 (0.85) V	1.8 V	> 1.1 V	1.2 V	1.0 V	0.7 V
Power	125 μA	190 μA	1.4 μA	112 μA	20 μA	9.8 μA
Vref	1.05~0.13 V	1.22 V	0.944 V	698 mV	191 mV	207 mV
Spread (3σ) w/o trim	± 0.64 %	± 0.67 %	± 2.4 %	± 2.2 %	± 0.57 %	± 2.35 %
DC-PSRR · Vref	73 dB	-	-	92 dB	72 dB	-
Area	0.0044 mm²	0.011 mm²	0.0025 mm²	0.027 mm²	< 0.4 mm²	0.065 mm²
# Samples	516	-	60	20	32	35
Temp. Range	–40 ~ 125°C	–40 ~ 125°C	–40 ~ 140°C	–40 ~ 140°C	–40 ~ 125°C	-20 ~ 125°C
Type	NPN BG	NPN BG	Sub-Vth BG	Current-mode BG	PNP reverse BG	Current-mode BG

978-1-7281-5107-6/19 $31.00 © 2019 IEEE

In comparison this bandgap features not only a small area, but particularly an excellent accuracy across process skews, with a high confidence level of > 500 samples measured.

Acknowledgment

The authors would like to thank especially Zdravko Georgiev for excellent support with lab measurements, and Michael Schittenhelm for data provision from ATE.

References

[1] S. K. Wadhwa and N. Chaudhry, "High accuracy, multi-output bandgap reference circuit in 16nm FinFet", VLSID 2017, pp. 259-262.

[2] C. H. Chang, J. J. Horng, A. Kundu, C. C. Chang and Y. C. Peng, "An ultra-compact, untrimmed CMOS bandgap reference with 3σ inaccuracy of +0.64% in 16nm FinFET," *2014 IEEE Asian Solid-State Circuits Conference (A-SSCC)*, KaoHsiung, 2014, pp. 165-168.

[3] A. Annema and G. Goksun, "A 0.0025mm^2 bandgap voltage reference for 1.1V supply in standard 0.16μm CMOS," *2012 IEEE International Solid-State Circuits Conference (ISSCC)*, San Francisco, CA, 2012, pp. 364-366.

[4] H. Banba *et al.*, "A CMOS bandgap reference circuit with sub-1-V operation," in *IEEE Journal of Solid-State Circuits*, vol. 34, no. 5, pp. 670-674, May 1999.

[5] M. Chahardori, M. Atarodi and M. Sharifkhani. "A sub 1 V high PSRR CMOS bandgap voltage reference," Microelectronics Journal 42 (2011), pp. 1057-1065.

[6] M. Eberlein, "Bandgap reference circuit", *Patent # US 8816756 B1. issued*, Aug. 2014.

[7] K. Sanborn, D. Ma and V. Ivanov, "A Sub-1-V Low-Noise Bandgap Voltage Reference," in *IEEE Journal of Solid-State Circuits*, vol. 42, no. 11, pp. 2466-2481, Nov. 2007.

[8] Y. Chen, J. Horng, C. Chang, A. Kundu, Y. Peng and M. Chen, "18.7 A 0.7V, 2.35% 3σ-Accuracy Bandgap Reference in 12nm CMOS," *2019 IEEE International Solid- State Circuits Conference - (ISSCC)*, San Francisco, CA, USA, 2019, pp. 306-307.

10-4 (7005)

A 1.2V 86dB SNDR 500kHz BW Linear-Exponential Multi-Bit Incremental ADC Using Positive Feedback in 65nm CMOS

Biao Wang, Sai-Weng Sin, Seng-Pan U[1], Franco Maloberti[2], R. P. Martins[3]

State-Key Laboratory of Analog and Mixed-Signal VLSI, IME/DECE/FST, University of Macau, Macao, China

E-mail: terryssw@um.edu.mo

1-Also with Synopsys Macau Ltd, 2-Also with University of Pavia, Pavia, Italy

3-On leave from Instituto Superior Técnico / Universidade de Lisboa, Portugal

Abstract— This paper presents a linear-exponential two-phase multi-bit incremental ADC (IADC). The exponential integration in the proposed IADC is generated by positively feedback the integrator output to the input, which can accumulate the signals stably due to the reset operation in IADC. To avoid the nonlinearity due to the signal-dependent charge injected from the reference, this work separates the sampling capacitor and the DAC capacitor. It will relax the requirement of reference buffer for fast-settling under a high sample rate. Then, we reconfigure the DAC capacitor to directly offer the exponential integration, resulting in saving in the usage of integration capacitor with a compact implementation. The linear-exponential two-phase scheme provides data-weighted-averaging-friendly weighting function to suppress the multi-bit DAC mismatch error. Fabricated in a 65nm CMOS under 1.2V supply and clocked at 128MHz, the ADC achieves an SNDR/DR/SFDR of 86.02/94.6/103.03dB with 500kHz BW, 20mW & 0.26mm², resulting in FoMs of 168.57dB.

Keywords—Linear-exponential, Multi-Bit Incremental ADC, Data Weighted Averaging, Positive Feedback

I. INTRODUCTION

Incremental $\Sigma\Delta$ ADCs are popular in multi-channel applications for their low-latency and simple decimation filter properties, due to the periodically resetting in memory blocks [1]. Moreover, they don't suffer from idle tones problems compared with the traditional $\Sigma\Delta$ modulator. However, when compared with $\Sigma\Delta$ modulators of the same order, the intrinsic sample-variant weightings of high-order IADCs would induce penalties on the thermal noise and the effectiveness of the data weighted averaging (DWA) technique (for multi-bit DACs). A smart-DEM algorithm from [2] was proposed to compensate for the mismatch error. The DEM and DAC in [3] have a fine-stage with large OSR=2k, thus relaxing the mismatch errors with limited bandwidth. A dynamic integrator slicing technique was presented in [4] to improve the tradeoff between the first integrator's power and signal power. Incremental ADCs with extended counting could achieve high resolution [5], but they are sensitive to inter-stage gain error in real implementations. The two-step architecture [6] reaches good performance with an extended cyclic ADC. It is susceptible to interstage gain mismatch, and the cyclic ADC does not help to suppress the kT/C noise.

A two-phase linear-exponential IADC, firstly introduced in [7], uses a noise-coupling (NC) path to achieve exponentially speed up accumulation. When the performance enhances to high-resolution with a sampling frequency of a hundred Mega-Hz, the sampling capacitor will draw a

This work was financially supported by the Science and Technology Development Fund, Macao S.A.R (FDCT)-055/2012/A2 & Research Committee of University of Macau - MYRG2017-00192-FST.

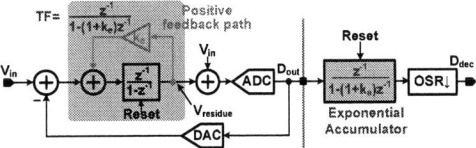

Fig. 1. Exponential incremental ADC by using positive feedback.

significant amount of signal-dependent current from the reference. It would induce nonlinearity caused by the inadequately settling. This work solves this problem by using separated sampling and DAC capacitors, with the DAC network reconfigured as a positive feedback path, providing the exponential integration without extra hardware cost and avoiding the detrimental signal-dependent DAC current on the reference. Therefore, we can eliminate the need of an accurate NC network for exponential accumulation and reduce the power dissipation by 33% from the second opamp in [7]. Without the NC network, a passive adder can be employed to replace the power-hungry active adder.

Thus, the proposed exponential IADC reuses the DAC branch as a positive feedback path to generate an exponential integrator in the loop, enhancing the SQNR. The overall IADC architecture adopted a two-phase linear-exponential scheme similar to [7], the positive feedback will be disabled in the 110 cycles of linear phase, then enabled in the following 18 cycles of the exponential integration phase. As a result, the loop finally combines the best features of thermal noise and DAC mismatch suppression from the linear-phase and the SQNR boosting capability from the following exponential phase. To verify the proposed architecture, we implemented a prototype of two-phase IADC with 500kHz BW (f_s = 128MHz) in 65nm CMOS, achieving 86/94.6dB SNDR/DR. The SFDR is 103.3dB, verifying the linearity performance of the proposed architecture.

II. THE PROPOSED ARCHITECTURE

A. Linear-Exponential Incremental ADC

In the traditional order-based incremental ADCs, the signal would be accumulated linearly to bi-quadratically from 1st to 4th order. This work presents an exponential integrator with different topology, as illustrated in Fig. 1. Instead of the equivalent exponential integrator with a noise transfer function of $(1 - (1 + k_e)z^{-1})\varepsilon_q$ in [7], the residue is positive feedback to the integrator's input, directly offering an exponential integrating function of $\frac{z^{-1}}{1-(1+k_e)\,z^{-1}}$, where k_e is the positive feedback coefficient. After resetting the system, both the analog modulator and the digital filter accumulate

978-1-7281-5107-6/19 $31.00 © 2019 IEEE

IEEE Asian Solid-State Circuits Conference
November 4 – 6, 2019
The Parisian Macao, Macao SAR, China

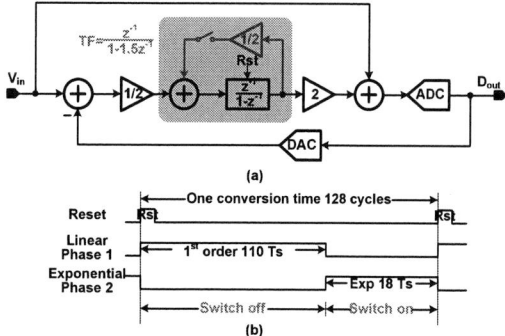

(a)

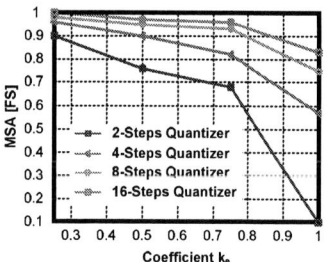

(b)

Fig. 2. (a) Block diagram of a two-phase linear-exponential IADC by using positive feedbak path and (b) its timing diagram.

Fig. 3. The relationsihp among quantizer steps, the coefficient k_e and the maximum stable amplitude.

from zero until the activation of the next reset signal. The modulator accumulates the difference between the input signal and the tracking DAC, while the digital part accumulates the bitstreams. Thus, the estimation of the moving input signal V_{in} with N clock cycles is,

$$\overline{V_{in}} = \frac{\sum_{i=1}^{N} D_{out}[i]V_{refm}(1+k_e)^{N-i}}{\sum_{i=1}^{N}(1+k_e)^{N-i}} + \frac{V_r(N)}{\sum_{i=1}^{N}(1+k_e)^{N-i}} \quad (1)$$

where V_{refm} is the LSB of the quantizer, and $V_r(N)$ is the residue voltage after N clock cycles. It implies that the resolution of the exponential IADC depends on the N (or the oversampling ratio), the coefficient k_e and the quantizer level. A multi-bit quantizer enhances the resolution by reducing the residue voltage $V_r(N)$ boundary. The larger the coefficient k_e is, the larger the accumulation slope would be.

The accumulation weightings of the exponential IADC are non-uniform, leading to a thermal noise penalty [7]. The more advanced scheme, linear-exponential IADC, resolves the thermal noise penalty of the exponential IADC, as illustrated in Fig. 2. The scheme will periodically deactivate the positive feedback in the linear phase. Thus, the oversampling will fully contribute to the reduction of the thermal noise in the linear phase (where the weightings are uniform). In the second phase, the residue feeds back to the integrator's input in a positive manner, which creates an exponential accumulation loop. To achieve the thermal-noise-limited SNR of 98dB, we select a total sampling capacitor of 8pF with 128 clock cycles under 128MHz sampling frequency, leading to a bandwidth of 500kHz. The first 110 cycles are for the linear-phase, and the remaining cycles perform the exponential-phase. Therefore, the scheme with linear-exponential integration works complementarily, combining the best features of thermal noise suppression in linear phase and SQNR boosting in exponential

phase. Besides, the uniform weightings of the DAC in the linear phase allows the DWA technique to work well averaging the mismatch error. The weights at the second phase decrease exponentially and are much lower than the weights in the linear phase, then they do not degrade the overall linearity.

B. Loop Stability and Parameters' Selection

When compared with the noise-coupling implementation in [7], the exponential integrator by using a positive feedback path would substantially influence the stability of the system as it is included in the loop filter, making the loop stability vulnerable to a larger coefficient of k_e. We start to analyze the loop stability from its continuously running $\Sigma\Delta$ equivalent without reset operation, as follows.

Considering the system of Fig. 1 under continuously running mode, the transfer function of the analog modulator becomes,

$$D_{out}(z) = V_{in}(z) + \frac{1-(1+k_e)z^{-1}}{1-k_e z^{-1}}E(z) \quad (2)$$

where the coefficient k_e must be less than 1 to ensure a stable system. Intuitively, the negative feedback DAC is not strong enough to pull back the positive feedback under a larger coefficient, leading to an unstable feedback system.

When it comes to the incremental mode, the loop suffers less from the stability issue by using exponential accumulation. The reason comes from the fact that the integrator has been reset to zero before it goes without bound. For example, with a coefficient k_e of 1, the loop could accumulate in a faster and stable way, if the time for exponential accumulation was less than 20 clock cycles under a 9-level quantizer. Generally, the coefficient value, the clock cycles for the exponential phase and the quantizer's resolution influence the loop stability. Moreover, the selection of coefficient k_e has some impact on the speed of loop accumulation and the maximum stable amplitude (MSA).

Fig. 3 shows a behavior study regarding the above aspects. Additional levels in the quantizer are beneficial for enlarging the MSA for a given coefficient k_e. Meanwhile, it could also reduce the integrator swing. However, more levels of the quantizer lead to higher power consumption and DWA complexity. Considering the speed of loop accumulation within a stable manner, the MSA and the capacitor sharing technique (details present in Section III), we chose a 9-level quantizer with a coefficient k_e of 0.5 in the proposed scheme.

III. CIRCUIT IMPLEMENTATION

A. The Switched-Capacitor Circuitry

Fig. 4 shows the simplified circuit implementation of the proposed two-phase linear-exponential IADC. We exhibit only a single-ended implementation for simplicity, but the real circuit is fully-differential. The coefficient along the accumulation path was reallocated for reducing the integrator swing, as shown in Fig. 2. Moreover, a gain of 1/2 before the integrator permits to reuse the DAC capacitor for a positive feedback path, as the positive feedback has the same coefficient k_e of 0.5. To retain the full scale, the DAC capacitor size equals to the sampling capacitor.

For the SC circuitry, we employ a bootstrapped switch BS_SW (Fig. 4) to ensure the critical input switches' linearity. Besides, we use the bottom-plate sampling to reduce the

978-1-7281-5107-6/19 $31.00 © 2019 IEEE

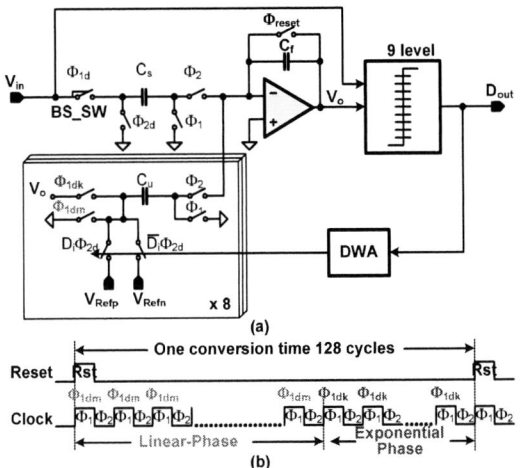

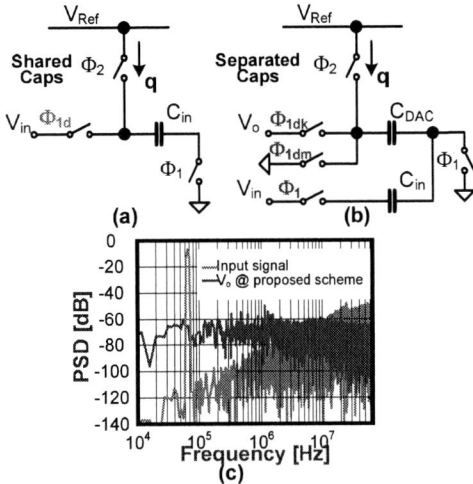

Fig. 4 (a) Circuit diagram of the proposed IADC by using traditional integrator with positive feedback path and (b) its timing diagram.

Fig. 5 (a)/(b) Circuit diagram of the signal-dependent/independent reference loading, and (c) spectrum of the integrator output.

signal-dependent charge injection. With a sampling frequency of 128MHz, it is not necessary to use the chopping technique, since the flicker noise has less impact on the overall performance. Moreover, the chopping technique increases the parasitic capacitance on the opamp's virtual ground, degrading the closed-loop bandwidth. As for the offset, it can be solved at the system-level chopping, improving the final thermal noise limited SNR by 3dB incidentally.

In the linear phase, the circuit works as a first-order IADC with a low-distortion feedforward structure. Thanks to the feedforward architecture, the quantization noise of multi-bit quantizer bounds the integrator swing, relaxing the opamp's requirements. The integrator is reset at Φ_1 in the first cycle, as illustrated in the timing diagram of Fig. 4 (b).

When the exponential phase is enabled, we need an extra branch to sample the residue from the previous cycle, as illustrated in Fig. 2 (a). Straightforwardly, we reuse the DAC capacitor to achieve the exponential integrator, since the connection of the DAC capacitor is to V_{cm} at Φ_{1dm} during the linear phase. Thus, the DAC capacitors would sample the residue information at Φ_{1dk} in the exponential-phase and then fed back the residue information to the integrator, achieving an exponential accumulation. With the reusing of DAC capacitor, the extra positive feedback branch does not degrade the feedback factor.

B. Considerations about the Reference

In the traditional switched-capacitor implementation, the capacitor C_{in} samples the input voltage at the end of the track phase Φ_1. During the integration phase, the sampling capacitor C_{in} connects the reference by the DAC action, as illustrated in Fig. 5 (a). It is helpful for low noise design, but it draws a signal-dependent current on the reference since the injected charge is equal to $C_{in}(V_{ref} - V_{in})$. It should achieve an adequate settling within half sampling period; otherwise, the nonlinearity from the reference oscillation will cause harmonic distortion. When the ADC's input is close to full scale with a sampling frequency of a hundred Mega-Hz, this reference settling requirement becomes demanding [8][9]. To solve this issue, we utilize a separate capacitor strategy, which

foresees the separation of sampling and DAC capacitors. The setting of the DAC capacitor is V_{cm} at Φ_1, while it will connect to reference at the integration phase. It relaxes the reference buffer requirement with the KT/C noise penalty so that the opamp needs more power to achieve the same SNR.

Fig. 5 (b) shows the input branches in the proposed scheme, with separated input and DAC branches. With the DAC capacitor also sampling the residue voltage, the impact on the reference due to the residue V_o should be considered. Fortunately, the residue contains mainly the quantization noise, because of the low-distortion feedforward structure. Moreover, the multi-bit quantizer turns the residue much more uncorrelated with the input, as shown in Fig. 5 (c). As a result, the injected reference charge does not induce harmonic distortion, due to its signal-independent property. It relaxes the reference buffer requirement, which would consume much power to achieve adequately fast-settling under a high sample rate.

C. Integrator Opamp, Passive Adder, and Comparator

The first integrator consumes most of the power in the switched-capacitor modulator, to satisfy the high SNR and linearity requirements. Thanks to the feedforward structure and multi-bit quantizer, we can employ an energy-efficient single-stage gain-boosted telescopic-cascode opamp under 1.2V supply.

By eliminating the noise-coupling network for the exponential integrator in work [7], we can also employ a power-efficient passive adder. Considering the signal attenuation and offset, the comparator consists of a switched-capacitor network for the passive adder, a pre-amplifier, and a strongARM latch, followed by an SR latch. With the auto-zero technique [8], the input-referred offset meets the 9-level quantizer requirement easily.

IV. EXPERIMENTAL RESULTS

The ADC has been fabricated in a 65nm CMOS with an active area of 0.26 mm², as shown in Fig. 6. Under a 1.2V supply and 128MHz sampling clock, 500kHz bandwidth, the analog modulator consumes 20mW, with the opamp being responsible for 86% of the consumed power.

978-1-7281-5107-6/19 $31.00 © 2019 IEEE

IEEE Asian Solid-State Circuits Conference
November 4 – 6, 2019
The Parisian Macao, Macao SAR, China

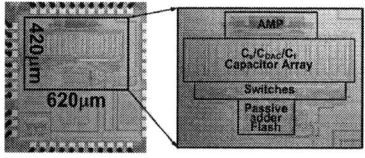

Fig. 6 Chip micrograph.

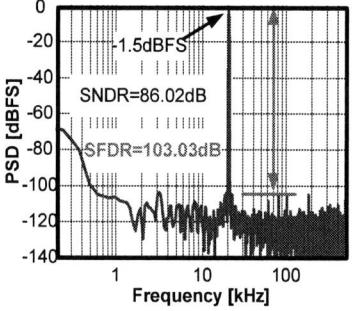

Fig. 7 Measured spectrum of the incremental ADC with DWA.

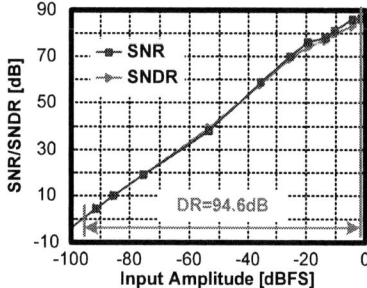

Fig. 8 Measured SNDR versus the input amplitude (DWA on).

Fig. 7 shows the spectrum of the design after the digital decimation filter with a 19.89kHz, -1.5dBFS input sinusoidal signal. The measured peak SNDR/SFDR is 86.02dB/103.0dB. It indicates that the reconfiguration of the DAC capacitor can keep high linearity. The injected residue on the reference in the exponential phase did not affect the linearity. Fig. 8 reveals the measured IADC's SNR/SNDR versus the input amplitude, and the ADC achieves a dynamic range of 94.6dB. The proposed circuit reaches high linearity IADC with a FoM$_S$ of 168.57dB. Table I lists the performance summary and the comparison with the state-of-the-art. Scaled well with decreasing advanced process with a low-supply, this work achieved a peak SNDR exceeding 85dB with a bandwidth of 500kHz only under a 1.2V supply.

V. CONCLUSIONS

This work proposed a single-loop linear-exponential IADC for high-resolution multi-channel application. It avoids the signal-dependent current drawing on the reference and provides well-balanced system-level tradeoffs on the accumulation efficiency, the thermal noise penalty and the efficiency of the DWA technique for DAC mismatch error. In the initial linear phase, the IADC works as a first-order architecture and fully utilizes the oversampling operation on thermal noise suppression. After that, the DAC array is reconfigured as a positive feedback to the integrator's input, achieving exponential accumulation. The scheme allows the DWA technique to work well in improving the linearity by rotating the multi-bit DAC elements. With a multi-bit quantizer, the integrator can minimize the integrator power

Tabel I Performance summary and comparison with the state-of-the-art.

Parameter	Wang [7]	Vogelmann [4]	Katayama [6]	Agah [10]	This Work
Architecture	Linear-Exp IADC	3rd-order IADC	IADC +Cyclic	IADC +SAR	Linear-Exp IADC
Process	65nm	180nm	180nm	180nm	65nm
Supply	1.2V	3V	3V	1.8V	1.2V
Clock Frequency	10.24MHz	30MHz	55MHz	45MHz	128MHz
Bandwidth	20kHz	100kHz	625kHz	500kHz	500kHz
C$_s$/C$_{DAC}$	Shared	Shared	Shared	Shared	Separated
Peak SNDR	100.8dB	86.6dB	96.6dB	86.3dB	86.02dB
Dynamic Range	101.8dB	91.5dB	100.1dB	90.1dB	94.6dB
Power	550μW	1.10mW	27.7mW	38.1mW	20mW
FoM$_{S,SNDR}$(dB)#	176.4	166.18	170.13	157.48	160.00
FoM$_{S,DR}$(dB)##	177.4	171.09	173.6	161.3	168.57
Area [mm²]	0.13	0.36	0.71	3.5	0.26

FoM$_{S,SNDR}$ = SNDR+10log$_{10}$(BW/Power);FoM$_{S,DR}$ = DR+10log$_{10}$(BW/Power)

dissipation by employing an energy-efficient opamp topology. More significant is the fact that the separated capacitors strategy do not inject signal-dependent charge on the reference, which is critical and will cause nonlinearity for the high-resolution, high sampling rate ADC design. Implemented in 65nm CMOS, the ADC reaches an SNDR/DR/SFDR of 86.02dB/94.6dB/103.0dB with 500kHz BW while consuming 20mW, resulting in FoM$_S$ of 168.57dB.

ACKNOWLEDGMENT

The authors would like to thank Dr. U-Fat Chio for the PCB design, and Liang Qi, Yuan Ren, Jiaji Mao, as well for useful technical discussions and measurement support.

REFERENCES

[1] J. Markus, J. Silva, and G. C. Temes, "Theory and applications of incremental delta-sigma converters," *IEEE Trans. Circuits Syst. I, Reg. Papers*, vol. 51, no. 4, pp. 678–690, Apr. 2004.

[2] Y. Liu, E. Bonizzoni, and F. Maloberti, "A 105-dB SNDR, 10 kSps multi-level second-order incremental converter with Smart-DEM consuming 280μW and 3.3-V supply," in *IEEE Proc. of European Solid-State Circuits Conference (ESSCIRC)*, pp. 371–374, Sep. 2013.

[3] J. Y. Chae, K. Souri, and K. A. Makinwa, "A 6.3 μW 20bit incremental zoom-ADC with 6 ppm INL and 1 μV offset," *IEEE J. Solid-State Circuits*, vol. 48, no. 12, pp. 3019–3027, Dec. 2013.

[4] P. Vogelmann, M. Haas, M. Ortmanns, "A 1.1mW 200kS/s Incremental ΔΣ ADC with a DR of 91.5dB Using Integrator Slicing for Dynamic Power Reduction," in *IEEE ISSCC Dig. Tech. Papers*, pp 236-237, Feb. 2018.

[5] Y. Zhang, C. H. Chen, T. He, and G. C. Temes, "A 35μW 96.8dB SNDR 1 kHz BW multi-step incremental ADC using multi-slope extended counting with a single integrator," *IEEE Proc. Symp. VLSI Circuits*, pp. 24–25, Jun. 2016.

[6] Takato Katayama, Shiko Miyashita, Kazuki Sobue, and Koichi Hamashita, "A 1.25MS/s Two-Step Incremental ADC with 100dB DR and 110dB SFDR," *in Proc. IEEE Symp. VLSI Circuits, pp. 205-206, Jun. 2018.*

[7] B. Wang, Sai-Weng Sin, Seng-Pan U, Franco Maloberti and R. P. Martins, "A 550μW 20kHz BW 100.8dB SNDR Linear-Exponential Multi-Bit Incremental ΣΔ ADC with 256-cycles in 65nm CMOS," *IEEE J. Solid-State Circuits*, vol. 54, no. 4, pp. 1161-1172, Apr. 2019.

[8] R. Zanbaghi, S. Saxena, G. C. Temes, and T. S. Fiez, "A 75-dB SNDR, 5-MHz bandwidth stage-shared 2–2 MASH modulator dissipating 16 mW power," *IEEE Trans. Circuits Syst. I, Reg. Papers*, vol. 59, no. 8, pp. 1614–1625, Aug. 2012.

[9] S. Devarajan, Larry Singer, Dan Kelly, Steven Decker, Abhishek Kamath, and Paul Wilkins, "A 16 b 125 MS/s 385 mW 78.7 dB SNR CMOS pipeline ADC," in *IEEE ISSCC Dig. Tech. Papers*, pp. 86–87, Feb. 2009.

[10] A. Agah, K. Vleugels, P. B. Griffin, M. Ronaghi, J. D. Plummer, and B. A. Wooley, "A high-resolution low-power incremental ΣΔ ADC with extended range for biosensor arrays," *IEEE J. Solid-State Circuits*, vol. 45, no. 6, pp. 1099-1110, Jun. 2010.

978-1-7281-5107-6/19 $31.00 © 2019 IEEE

A Multi-Slice VCO-based Quantizer for On-Chip Power Supply Noise Analysis Achieving 0.11 (mV)^2/sqrt(MHz) Noise Floor

Pengfei Zhai, Xiong Zhou, Yan Cai, Zheng Zhu, Fan Zhang, Zixiao Lin, Qiang Li

Institute of Integrated Circuits and Systems, University of Electronic Science and Technology of China, Chengdu, China

Abstract—This paper presents a multi-slice VCO-based quantizer (MSVQ) for high-resolution power supply noise analysis. To relax the trade-off between accuracy and measurement time, a multi-slice quantizer at a relatively higher sampling rate is proposed, and the auto-covariance replaces the autocorrelation to obtain PSD while immunes the spurs introduced by the quantizers itself and further reduces noise floor. 0.11 (mV)^2/sqrt(MHz) noise floor is achieved with a significantly reduced measurement time of 336s.

I. INTRODUCTION

Power supply noise (PSN) has ultimate impact on the performance of ICs, especially for high-performance SoCs and systems with intensive interconnection around sensitive analog blocks. On-chip PSN analyzers provide in-vivo monitoring of power supplies.

In principle, an ultra-wideband spectrum could be brutally captured by a high-speed ADC. However, this approach suffers from limited bandwidth, aliasing, and high power consumption. The spectrum of a periodic stationary PSN can be recovered from the autocorrelation (R_{acor}) or autocovariance (R_{acov}) via two lower-speed ADCs [1]. Low-resolution averaging-based VCO quantizers avoids the need of a S&H circuit and reaches higher bandwidth [2], while the sampling clocks uncorrelated with noise are often required to measure the PSN accurately. [3] has reported a PSN analyzer which avoids using uncorrelated sampling clocks through accumulating the phase noise directed by a PN sequence. Moreover, the efficiency of noise floor reduction is still limited by a simple averaging, since noise floor decreases with the square root of the number of averaging points. In addition, it is worth mentioning that the results of the two quantizers need to be multiplied for autocovariance operation, therefore, the unit of autocovariance should be V^2, corresponding the unit of noise floor is $V^2/\sqrt{Hz}$, rather than $V/\sqrt{Hz}$.

In this work, to achieve lower quantization noise, a multi-slice VCO quantizer (MSVQ) is proposed. Compared with autocorrelation, theoretical analysis shows that autocovariance permits clearer spectrum with less self-induced spurs. The prototype exhibits significant improvement on the measurement time and noise floor simultaneously.

II. CIRCUITS AND DESIGN CONSIDERATIONS

Fig. 1 illustrates the relationship between the periodic stationary signal and its autocorrelation or autocovariance. The autocorrelation or autocovariance reaches its maximum value

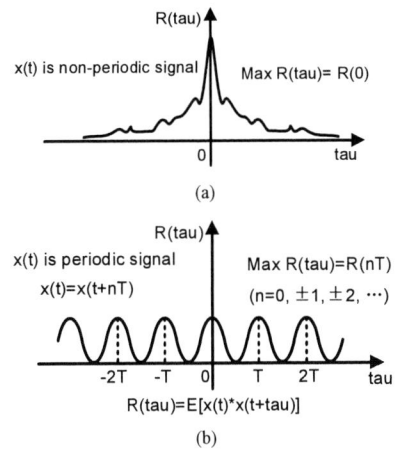

Fig. 1. Illustration of autocorrelation or autocovariance for (a) a non-periodic signal and (b) a periodic signal.

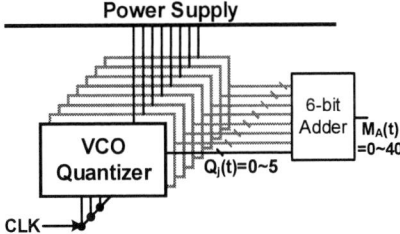

Fig. 2. The proposed multi-slice VCO quantizer for on-die power supply noise spectrum analyzer.

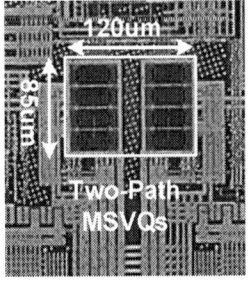

Fig. 3. Die photo of the proposed MSVQ.

when *tau* reaches any integer times of the period of the signal x(t). Information of x(t) is contained in the autocorrelation as well as the autocovariance of x(t). Therefore, by scanning *tau* with a fixed delay step Δtau, the spectrum of the PSN can be obtained.

Fig. 2 shows the architecture of proposed MSVQ which is used to sample the PSN. It consists of multiple slices of identical VCO-based quantizers. Each VCO quantizer samples the PSN simultaneously. Different from a single VCO quantizer, the proposed MSVQ takes advantage of uncorrelated phase noise across the VCOs, increasing the independence of the output (Q_j) of each VCO. The variance of M_A is n times lower than the variance of nQ_j, as given in

$$\sigma^2\{n[Q_j(t)]\} = n^2\sigma^2[Q_j(t)] \tag{1}$$

$$\sigma^2[M_A(t)] = n\sigma^2[Q_j(t)] = (1/n)\sigma^2\{n[Q_j(t)]\} \tag{2}$$

where n is the number of slices of the MSVQ. The variance is the main contributor of the noise floor of the PSN spectrum [2]. Thus, the MSVQ creates an additional dimension in lowering the noise floor.

Moreover, theoretical analysis has suggested two advantages of utilizing autocovariance versus autocorrelation. Firstly, the spurs introduced by the quantizers can be reduced drastically through autocovariance operation, since the spurs correlated to the sampling clock are periodic stationary if the *tau* is unchanged, the discrete value after sampled can be equivalent to a DC component. Secondly, autocovariance allows a lower noise floor when the mean value of the quantizer output is not zero. In the process of autocorrelation, the quantization noise of one-path quantizers is amplified by the DC component of the outputs of the-other-path quantizers, resulting in an increase in the noise floor of the measured PSD of PSN. Fortunately, this part of the amplified noise can be effectively reduced during the autocovariance process.

III. MEASUREMENT RESUTLS

The prototype was fabricated in a 40nm CMOS process, with the die photo given in Fig. 3. The area of the two-path MSVQs is 0.01mm². Fig. 4 gives the measured power spectrum under adjacent inputs with minimum frequency space. It shows that the measurable scale of amplitude can be as large as 100mV. Fig. 5 shows the noise floor is effectively reduced with the increased number of slices. Fig. 6 shows the measured power spectrum from autocovariance is free of spurs as compared to autocorrelation based measurement, and has a lower noise floor than that from autocorrelation. In the power spectrum from autocorrelation, the largest measured tone is almost submerged in the spurs which are produced by measurement system. In conclusion, the proposed multi-slice VCO-quantizer has improved the noise performance and measurement speed simultaneously.

ACKNOWLEDGMENT

This work was supported by the National Natural Science Foundation of China (61534002, 61761136015).

IV. REFERENCES

[1] E. Alon, et al, *JSSC*, pp.820-828, April 2005.

[2] E. Alon, et al, *TAP*, pp. 248-259, May 2009.

[3] T.-C. Hsueh, et al, *JSSC*, pp. 1711-1721, July 2015.

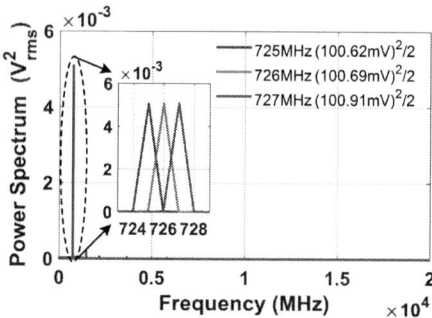

Fig. 4. Multiple measurements under adjacent inputs with minimum frequency space.

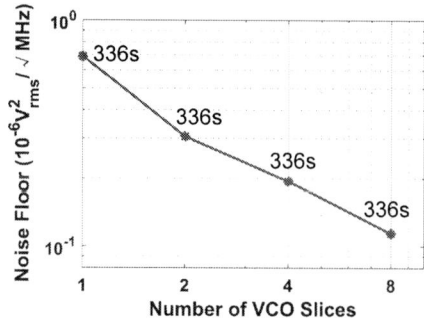

Fig. 5. Measured RMS noise floor versus the number of VCO slices under fixed measurement time.

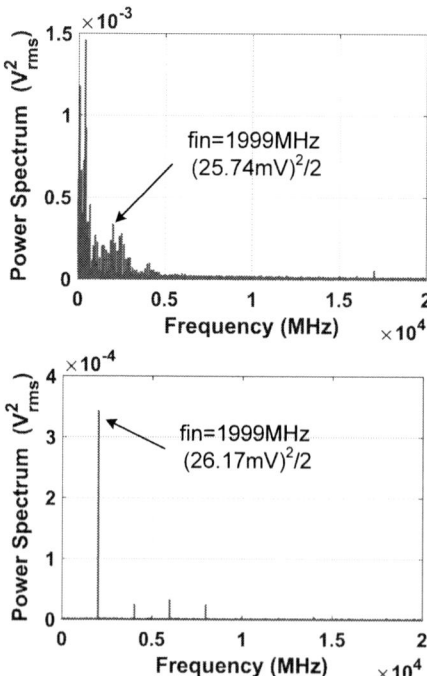

Fig. 6. Measured PSD from (a) autocorrelation and (b) autocovariance under the same conditions.

10-6 (7086)

A 265μW Continuous-Time 1-2 MASH ADC Achieving 100.6 dB SNDR in a 24 kHz Bandwidth

Sujith Billa, Suhas Dixit and Shanthi Pavan
Indian Institute of Technology, Madras

Abstract—**We present a high-performance audio CT$\Delta\Sigma$M that uses an FIR-filtered 1-2 MASH architecture. The modulator uses a singe-bit first-order input stage that achieves a maximum stable amplitude close to full-scale in spite of using a 1-bit quantizer. We show that FIR feedback can be used to advantage in a MASH converter by effectively filtering the first-stage error that is coupled to the second stage. Our design achieves 100.6 dB peak SNDR in a 24 kHz bandwidth, resulting in a Schreier FoM in excess of 180 dB.**

I. INTRODUCTION

High-performance audio CT$\Delta\Sigma$Ms have typically been realized using single-loop techniques. Further, many of the designs that achieve state-of-the-art performance have employed single-bit (rather than multibit) quantizers [1], [2], [3]. This makes sense, since a multibit ADC results in increased power dissipation in the clock distribution network. Also, a multibit feedback DAC needs to be linearized, which usually complicates design, and increases power dissipation. In this work, we describe the design of a cascaded continuous-time modulator based on single-bit ADCs. The paper is organized as follows. Section II describes the architecture and unique features of our continuous-time 1-2 MASH ADC. We show how FIR feedback, apart from its usual benefits, can improve the performance of a MASH converter. Measurement results from a prototype chip, fabricated in a 180 nm CMOS process, are given in Section III. The CT$\Delta\Sigma$M achieves a peak SNDR of 100.6 dB, and a dynamic range of 104 dB in a 24 kHz bandwidth. It consumes only 265μW from a 1.8 V supply, resulting in a Schreier (SNDR) FoM of 180.2 dB. Section IV concludes the paper.

II. 1-2 CT-MASH ARCHITECTURE AND CIRCUIT DESIGN

One problem with the use of a 1-bit ADC in a high-order single-loop $\Delta\Sigma$ converter is that it restricts the maximum stable amplitude (MSA) to about -3 dBFS. If the MSA could somehow be increased to full-scale (while still using a single-bit quantizer), the power-efficiency of the CT$\Delta\Sigma$M (quantified by the Schreier FoM) would increase by 3 dB. A way of doing this is to use a 1-X MASH architecture, whose input stage is realized as a single-bit first-order CT$\Delta\Sigma$M. This is the principle behind our design, whose simplified architecture is shown in Fig. 1(a). The input stage (MOD1) is a first-order loop. ADC$_1$ is a single-bit design, resulting in a power efficient quantizer. The main feedback path of MOD1 is a twelve-tap FIR DAC with transfer function $F(z)$, implemented using semi-digital techniques. The DAC is resistive, chosen for low-noise. The compensation DAC ($F_c(z)$) reuses the flip-flops of the main DAC. The gain α of the input

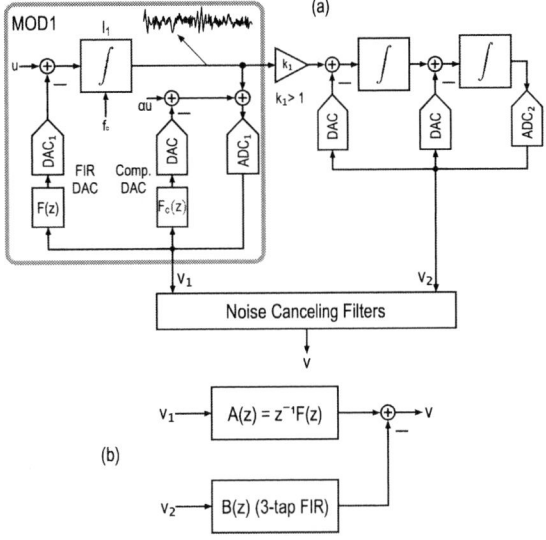

Fig. 1. (a) Simplified architecture of the 1-2 MASH CT$\Delta\Sigma$M and (b) Reconstruction of v from v_1 and v_2.

feedforward path is chosen so that the output of I_1 only consists of quantization noise. Thanks to the FIR filter in the main feedback path, the output of I_1 is a *filtered* version of the quantization noise introduced by ADC$_1$ – this results in a lower swing than in a design without FIR feedback. This means that the output of I_1 can be amplified before being coupled to the second stage, thereby increasing its effective resolution when referred to MOD1's input. In our design, which operates with OSR=128, the second-stage is a second order, single-bit CIFB CT$\Delta\Sigma$M. The OTA in the active-RC integrator I_1 is chopped at a frequency f_c to reduce flicker noise. By choosing $f_c = f_s/24$, folding of shaped noise (due to chopping) into the signal band is mitigated [2]. The use of FIR feedback, therefore, not only results in low jitter sensitivity, improved integrator linearity, and negligible chopping artifacts, but also improves the performance of a MASH converter by filtering its error.

Fig. 1(b) shows the structure of the digital reconstruction filters. v_1 is filtered by an FIR filter with transfer function $z^{-1}F(z)$. The output of this filter consists of the input (which is largely unaffected by $F(z)$) and the filtered quantization noise of MOD1.

978-1-7281-5107-6/19 $31.00 © 2019 IEEE

v_2 consists of a digitized version of the first-stage output and its own quantization noise, which is second-order noise shaped. v_2 is filtered by a 3-tap FIR filter before being subtracted from the output of the $A(z)$ filter. The taps of $B(z)$ are estimated using the LMS algorithm (implemented offline), and the same weights are used for all measurements.

III. MEASUREMENT RESULTS

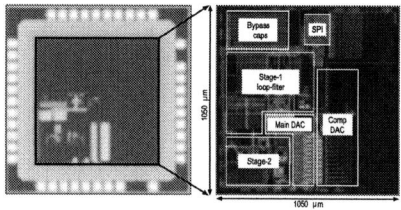

Fig. 2. Chip photograph and layout snapshot.

The 1-2 MASII CT$\Delta\Sigma$M was fabricated in a $0.18\,\mu$m CMOS process through Europractice. Fig. 2 shows the die photograph, and the layout of the active area, which measures about $0.65\,\text{mm}^2$.

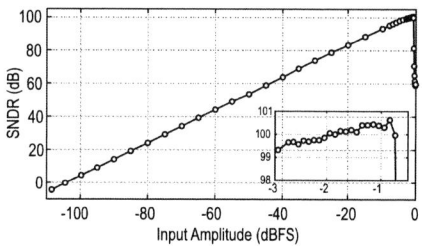

Fig. 3. Measured SNDR as a function of input amplitude.

Fig. 3 shows the measured SNDR as a function of input amplitude. The peak SNDR is $100.5\,$dB, and the dynamic range is $104\,$dB. Thanks to the first-order first stage, the maximum stable amplitude is as large as -0.7 dBFS, even with a single-bit quantizer. Fig. 4 shows the measured PSD of the MASH CT$\Delta\Sigma$M for a -0.7 dBFS input. The SNDR is $100.6\,$dB, with second and third harmonic distortion being $109\,$dB and $119\,$dB respectively.

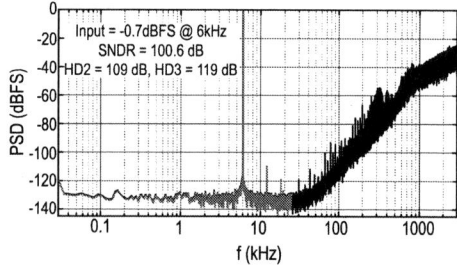

Fig. 4. Measured output PSD for a -0.7 dBFS input. A 2^{18} point Blackman-Harris window is used for spectral estimation.

Table I compares the performance of our design with state-of-the art audio modulators reported in the literature. We see that our design achieves an excellent Schreier FoM and SFDR, even in a 180 nm process. When compared to [2], the Schreier FoM is about 2.4 dB higher, mostly on account of the increased stable amplitude offered by the 1-X MASH architecture.

Reference	[4]	[5]	[6]	[7]	[1]	[2]	[3]	This work
BW (kHz)	24	20	25	20	24	24	24	24
Tech. (nm)	65	160	65	65	180	180	65	180
Vdd (V)	1.0	1.8	1.0	1.2	1.8	1.8	1.2	1.8
Power (μW)	94	1120	175	550	280	280	68	265
Peak SNDR(dB)	91.2	103	94.6	100.8	98.2	98.5	94.1	100.6
Peak SNR(dB)	91.9	106	96.1	-	99	99.3	94.8	101.7
DR (dB)	93	109	98.5	101.8	103	103.6	98.2	104
SFDR (dB)	104.8	-	100.3	121	106	107.6	107	109
Area (mm^2)	0.11	0.16	0.39	0.13	1	1	0.27	0.64
Chopping	yes	no	no	no	no	yes	no	yes
FoM$_{Schreier}$	175.3	175.5	176.2	176.4	177.5	177.8	179.6	180.2

$$\text{FoM}_{\text{Schreier}} = \text{SNDR}_{max} + 10 \log_{10}\left(\frac{\text{BW}}{\text{Power}}\right).$$

TABLE I
PERFORMANCE COMPARISON WITH STATE-OF-THE-ART $\Delta\Sigma$ CONVERTERS.

IV. CONCLUSION

We described the advantages of a 1-X MASH $\Delta\Sigma$ architecture, and showed how FIR feedback can be used to improve the performance of such a design. These claims are borne out by measurement results from a continuous-time 1-2 MASH audio ADC that achieves a peak SNDR of 100.6 dB in a 24 kHz bandwidth while consuming only 265μW from a 1.8 V supply, resulting in a Schreier FoM of 180.2 dB. This appears to be the highest FoM CT-MASH ADC to be reported.

REFERENCES

[1] A. Sukumaran and S. Pavan, "Design techniques for audio single-bit continuous-time delta sigma modulators with FIR feedback," *IEEE Journal of Solid State Circuits*, vol. 49, no. 11, pp. 2515–2525, Nov 2014.

[2] S. Billa, A. Sukumaran, and S. Pavan, "Analysis and design of continuous-time delta-sigma converters incorporating chopping," *IEEE Journal of Solid State Circuits*, vol. 52, no. 9, pp. 2350–2361, Sep 2017.

[3] M. Jang, C. Lee, and Y. Chae, "Analysis and design of low-power continuous-time delta-sigma modulator using negative-R assisted integrator," *IEEE Journal of Solid-State Circuits*, vol. 54, no. 1, pp. 277–287, Jan 2019.

[4] B. Zhang, R. Dou, L. Liu, and N. Wu, "A 91.2 dB SNDR 66.2 fJ/conv. dynamic amplifier based 24kHz $\Delta\Sigma$ modulator," in *Proceedings of the 2016 IEEE Asian Solid-State Circuits Conference (A-SSCC)*. IEEE, 2016, pp. 317–320.

[5] B. Gönen, F. Sebastiano, R. Quan, R. van Veldhoven, and K. A. A. Makinwa, "A dynamic zoom ADC with 109-dB DR for audio applications," *IEEE Journal of Solid-State Circuits*, vol. 52, no. 6, pp. 1542–1550, June 2017.

[6] S. Liao and J. Wu, "A 1 V 175 μW 94.6 dB SNDR 25 kHz bandwidth delta-sigma modulator using segmented integration techniques," in *Proceedings of the 2018 IEEE Custom Integrated Circuits Conference (CICC)*. IEEE, 2018, pp. 1–4.

[7] B. Wang, S. Sin, S. U, F. Maloberti, and R. P. Martins, "A 550-μW 20-kHz BW 100.8-dB SNDR linear-exponential multi-bit incremental $\Sigma\Delta$ ADC with 256 clock cycles in 65-nm CMOS," *IEEE Journal of Solid-State Circuits*, vol. 54, no. 4, pp. 1161–1172, April 2019.

Drop-In Energy-Performance Range Extension in Microcontrollers Beyond VDD Scaling

Saurabh Jain, Longyang Lin, Massimo Alioto
Department of Electrical and Computer Engineering
National University of Singapore
Singapore
elesau@nus.edu.sg

Abstract— This work introduces reconfigurable thread count augmentation for existing microcontroller architectures, and row aggregation for their dedicated SRAM memory, to extend their energy-performance tradeoff beyond traditional voltage scaling, while at minimal design effort ("drop-in"). The proposed techniques are architecture-agnostic as the added reconfigure-ability does not modify the original instruction execution down to the cycle level. Reconfiguration permits to occasionally boost the throughput of simple architectures that were originally not conceived to allow multi-thread operation, while allowing the original single-thread operation in less performance-critical tasks. From a design viewpoint, thread count augmentation is fully automated and directly manipulates the gate-level netlist of an existing single-thread processor, allowing its application to commercial Intellectual Property cores (even if obfuscated by the IP vendor). Similarly, SRAM row aggregation can be applied on commercially compiled 6T SRAM arrays with minor modification in the row decoder. A 40nm ARM Cortex-M0 testchip shows 1.8X (1.4X) core (memory) performance boost beyond a baseline at nominal voltage, 1.4X lower minimum energy point at only 16% (4%) area (timing) overhead, and lowest energy/cycle to date.

Keywords— Energy efficiency, beyond-voltage scaling energy-performance tradeoff, processor, microarchitecture, SRAM.

I. INTRODUCTION

Microcontrollers with wide energy-performance scalability are being demanded in energy-autonomous platforms, so that energy can be reduced in the common case, while providing extra peak performance when needed [1]-[2] (e.g., to respond to events). Energy-performance scaling is routinely achieved via wide dynamic voltage scaling [1]-[3], [8]-[10]. Recently, beyond-voltage scaling extension of the energy-performance tradeoff has been demonstrated via microarchitectural reconfiguration down to the pipestage level in application-specific accelerators [3]. However, microarchitectural reconfiguration [3] is difficult to adopt in microcontrollers, as repipelining of existing architectures requires a full redesign of the control flow. Compared to the original architecture, this poses legacy issues, can severely degrade the per-thread throughput, and prohibits design automation. Also, such speedup translates into higher performance only if the dedicated memory is sped up as well.

In this work, reconfiguration is introduced both at the microcontroller and the SRAM level to dynamically improve their performance or energy beyond the range and the capabilities of conventional voltage scaling, as shown in Fig. 1. At the processor level, thread interleaving is introduced to boost

the thread count from the value allowed by the baseline architecture (P mode) to a larger number (P+ mode), when needed. This reconfiguration is introduced directly on the gate-level netlist without the need for specific knowledge of the processor architecture, allowing its application to commercial Intellectual Property (IP) cores (even if encrypted or obfuscated by the IP vendor). This technique is shown to be deployable in the form of fully automated digital design flows, thus entailing low design effort. At the memory level, row aggregation is selectively enabled to boost up the read access rate when needed (M+ mode), compared to the baseline array organization (M mode). Row aggregation is achieved by simply manipulating the MSB of the address, thus requiring a minor modification of a commercial 6T-bitcell SRAM macro generated from memory compiler, as opposed to prior art (e.g., [4], [5]). The above joint processor/memory reconfiguration was demonstrated in an ARM Cortex-M0 core testchip in 40nm, where processor/SRAM reconfiguration is jointly adjusted with V_{DD} to either boost performance beyond the original architecture at nominal voltage, or to further reduce the minimum energy when scaling down V_{DD} as in Fig. 1.

The paper is organized as follows. The proposed processor and memory reconfiguration techniques are introduced in Sections II. Section III discusses the testchip design and measurements results. Conclusions are drawn in Section IV.

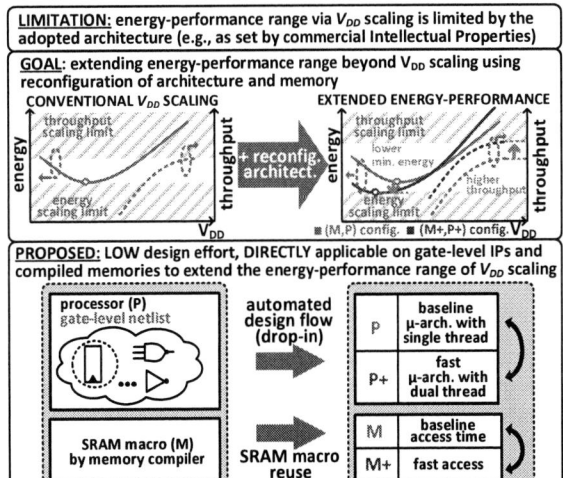

Fig. 1. Low-overhead drop-in processor/memory reconfiguration to extend the energy-performance tradeoff beyond conventional voltage scaling (applicable to both proprietary designs and commercial IPs).

II. PROCESSOR AND MEMORY RECONFIGURATION

In this section, it is assumed that a conventional design of a processor baseline architecture is available in the form of behavioral description (e.g., Verilog) or gate-level netlist, as available from synthesis. This also includes the case of encrypted or obfuscated netlists offered by commercial IP vendors. Also, the associated SRAM memory is assumed to be available, as generated with a conventional memory compiler.

A. Processor Energy-Performance Reconfiguration

The extension of the processor energy-performance tradeoff by re-architecting its pipeline structure is generally unfeasible. Indeed, this disrupts the control flow, can severely degrade the instructions/cycle, and is not supported by design automation. Accordingly, such performance enhancement needs to be performed by keeping the same instruction execution, and hence be moved up to the thread level, introducing the ability to occasionally increase the number of concurrent threads. This is achieved by introducing time-interleaved operation through the manipulation of the available gate-level netlist, as described by the proposed design flow in Fig. 2. Indeed, time-interleaved execution of each thread is provably guaranteed to be equivalent to the cycle-level execution of the original netlist [6], [7], when subsampling signals every n cycles (where n is the amount of time interleaving, which is assumed to be 2 for simplicity). This maintains the instruction execution unaltered, while delivering a larger number of simultaneous threads and hence higher performance (i.e., as if multiple cores were being available).

In the proposed design flow in Fig. 2, the netlist is parsed to identify the flip-flops at step 1. At step 2, each flip-flop is replaced by two cascaded flip-flops. This enables time interleaving of two completely independent input and instruction streams (i.e., two separate threads, as desired). At any given cycle, odd-numbered stages process one thread, whereas even-numbered process the other. This is fundamentally different from [3], which requires explicit pipeline insertion. At this step, no throughput improvement is achieved compared to the original design, since the clock frequency remains the same as the baseline (since the combinational logic delay in each pipestage has not changed). To truly improve throughput, retiming is applied at step 3 to balance the delays among pipestages, thus approximately halving the clock cycle in P+ mode. The retimed netlist is now able to run at nearly doubled frequency since the logic depth in each pipestage is now halved compared to the original design. This preserves nearly the same per-thread throughput, while doubling the thread count and hence the overall throughput. Such throughput increase would be unfeasible via simple repipelining, as the whole processor architecture would need to be completely revisited in terms of control flow (i.e., the original architecture would not be preserved, and very substantial architectural verification effort would be required). To add the capability of dynamically reconfiguring the thread count, at step 4 in Fig. 2 the flip-flops belonging in even-numbered register levels are replaced with their bypassable version, where a multiplexer selects either the register output or its input (i.e., normal register operation or bypass).

When single-thread operation is required (P mode, DT=0),

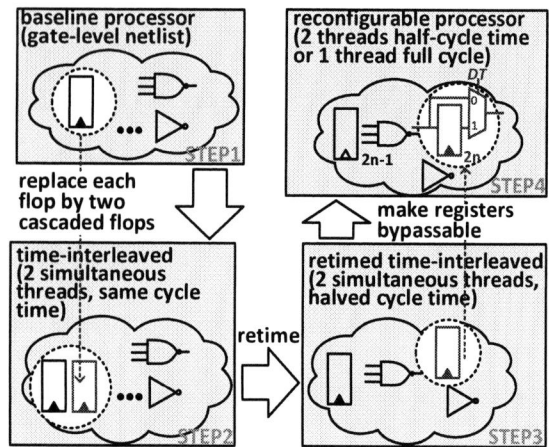

Fig. 2. Automated design flow to introduce thread interleaving in the gate-level netlist of the baseline processor.

even-numbered register levels are bypassed to assure the same operation as the baseline. In this case, the bypassable registers are clock gated to suppress their energy consumption since they are not utilized, and all registers share the same clock network. Instead, when DT=1 (P+ mode), no register is bypassed and time interleaving takes place, leading to dual-thread operation as desired. The concept can be generalized to more than two threads, although at the cost of higher timing and area overhead due to multiplexers (see Fig. 2, step 4).

Fig. 3 shows the architecture and the instruction flow in the processor pipeline for both configurations, as exemplified for an ARM Cortex-M0 core. In P mode, the processor runs instructions every cycle, with each instruction being completed in three cycles (provided there are no stalls). In dual-thread (P+) mode, the instructions from each thread are time interleaved and completed in a doubled number of cycles, although at nearly doubled clock frequency.

In Fig. 3, the first (second) thread is executed in odd (even) numbered cycles, and its control flow is inherently independent and equal to the original architecture [6], [7]. Hence, when a

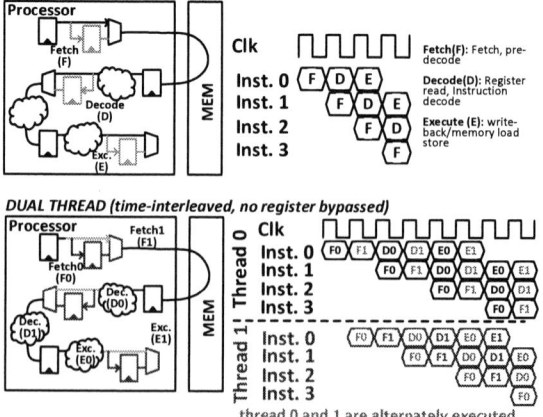

Fig. 3. Instruction execution in single-thread mode (P) and dual-thread mode (P+). For illustration purposes, the ARM Cortex-M0 architecture is assumed.

978-1-7281-5107-6/19 $31.00 © 2019 IEEE

thread is stalled, the other one continues to run independently as guaranteed by time-interleaved operation. This improves the throughput without changing the pipeline structure (i.e., hazards and the instructions per cycle) seen in each thread. Accordingly, the proposed reconfigurable approach can dynamically double the effective number of cores and throughput, while doubling only the registers (i.e., at reduced area cost), and maintaining exactly the same per-thread control flow (i.e., no change in the software stack or the original architecture is needed). The approach is architecture-agnostic, and can be performed in a fully automated manner, using the above design flow.

Regarding the register file, it needs to be split (i.e., halved per-core size, nearly at same area) or replicated (i.e., same per-core size, doubled register file area), to assure the necessary architectural state separation between the two threads. Although both approaches are equally applicable, the latter is here adopted. Indeed, replication preserves the register file size across single- and dual-thread mode, so that no change is needed in the entire software stack from compiler to application (i.e., even the very same binary executable code runs correctly in both modes).

In regard to the memory mapping across threads, the easiest option is to divide the address space in two independent halves as in the following subsection, so that the very same binary executable code can be used independently for the two threads (i.e., no change is needed in the software stack). Other options are available at the additional effort to modify the software stack, such as sharing the same full address space across threads.

B. Memory Energy-Performance Reconfiguration

To speed up read access, memory row aggregation was introduced to double the effective 6-transistor bitcell read current being drawn from the bitline capacitance, and hence accelerate its discharge as shown in Fig. 4. The simultaneous activation of two rows and wordlines is achieved by simply replacing the conventionally inverted address MSB in the second half bank (addresses with MSB=1) by its XNOR with the *ROWAGG* signal (i.e., conventional inverter if *ROWAGG*=0, or buffer if *ROWAGG*=1 to replicate the same MSB in both half banks). The wordline pulsewidth is adjusted to shorten the allotted bitline discharge time, when two rows are simultaneously activated. These slight modifications do not alter the compiled-memory array, or any other part of the periphery (Fig. 4). As an additional benefits (Fig. 5), the bitline discharge time variability is reduced by ~25%, as expected by the averaging of the currents delivered by the two simultaneously activated bitcells. This translates into 3X improvement in the bitline discharge time value margined by 6σ, in M+ mode.

The adoption of row aggregation does not require any change in the software stack, as correct memory mapping simply requires compiler execution under the available address space.

III. TESTCHIP AND MEASUREMENT RESULTS

The proposed reconfigurable core was demonstrated by implementing an ARM Cortex-M0 processor with two 8-kB SRAM arrays for program and data memory in 40nm, using a commercial 6-transistor bitcell (Fig. 6a). The total core area of 0.3mm^2 includes two SRAMs (0.032mm^2 each), and the testing harness (0.15mm^2). From Fig. 6b, the processor-memory

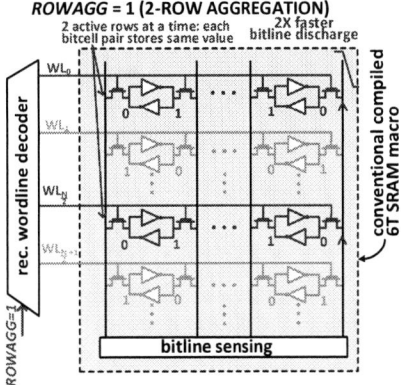

Fig. 4. SRAM under *ROWAGG = 1* simultaneously activates two wordlines.

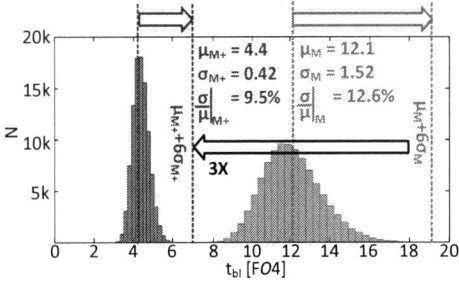

Fig. 5. Monte Carlo simulation in 40nm at V_{DD}=1.1V shows 3X speed-up in the bitline discharge time (evaluated at 6-σ design margin, 100,000 runs).

interaction is controlled by the AHB bus and the thread translator logic. The latter handles the memory mapping of threads onto memory. As the two threads are alternately executed under *DT*=1, the translator logic toggles the MSB of the address every cycle, reserving each memory half to the corresponding thread (when *DT*=0, no toggling is needed).

Row aggregation (M+ mode) in SRAM allows 1.4-1.85X access time speed-up at 0.6-1.1V (Fig. 7), at 1-4% energy penalty (Fig. 8), compared to single active row. The speed-up in M+ over M mode improves at lower V_{DD}, thanks to the above-mentioned reduction in the bitline discharge time variability (larger at low V_{DD}). From Fig. 8, write is faster than read, hence it does not limit speed in M+ mode.

For the entire core including processor and SRAM, the

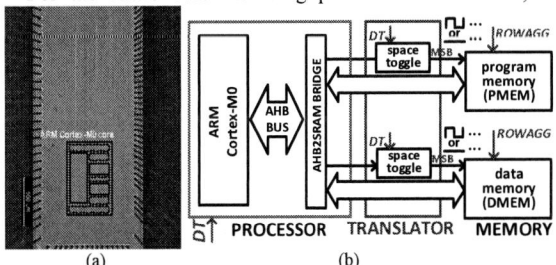

Fig. 6. a) Die micrograph, b) proposed reconfigurable microarchitecture applied to ARM Cortex-M0 core with adjustable number of threads (controlled by signal *DT*), and adjustable memory read access time via row aggregation (controlled by signal *ROWAGG*).

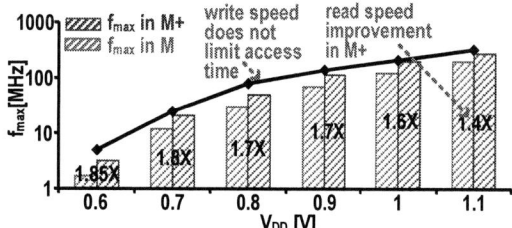

Fig. 7. Maximum SRAM frequency vs V_{DD}, and speed-up of M+ w.r.t. M mode.

(M+,P+) mode enhances the maximum throughput by 1.8X (Fig. 9) and reduces the minimum energy by 1.4X (Fig. 10), using matrix multiplication as benchmark in both threads.

This comes at the cost of halved memory capacity, compared to the baseline (M,P) mode. Compared to prior ARM Cortex-M0 research prototypes (Fig. 11), the proposed core has the lowest minimum energy of 8.64 pJ/cycle in (M,P) (7.4 pJ/cycle in (M+,P+) mode), and the highest maximum throughput, thanks to reconfiguration. The (M+,P+) configuration boosts performance and lowers energy at MEP, whereas (M, P) has lower energy from nominal voltage down to 0.6 V.

Compared to a traditional single core, the above advantages come at the cost of a slight performance degradation due to 6% (4%) processor (memory) timing overhead. Interestingly, reconfiguration enables the execution of up to two threads and 1.8X throughput increase at only 16.4% larger area, compared to a traditional single core.

IV. CONCLUSION

In this paper, simultaneous processor/memory reconfiguration was proposed for low-end processor cores to extend the energy (performance) range by 1.4X (1.8X), beyond allowed by V_{DD} scaling. This is achieved at only 16.4% area

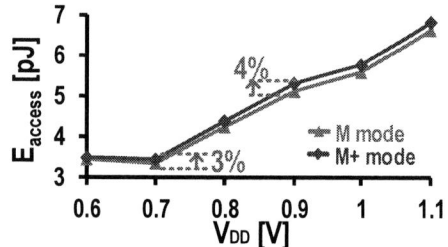

Fig. 8. SRAM energy per access vs V_{DD} in M, M+ mode is nearly the same.

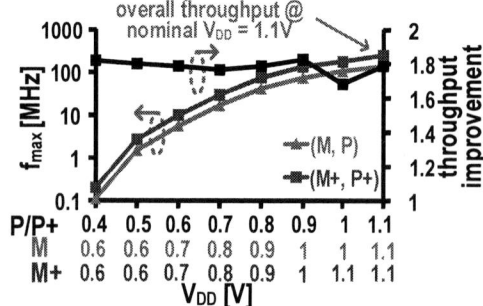

Fig. 9. Maximum processor frequency vs V_{DD} (ARM Cortex-M0 with SRAM).

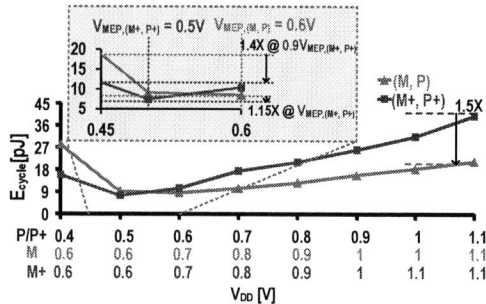

Fig. 10. Energy/thread at various configurations (ARM Cortex-M0 with SRAM, designed with 10% voltage margin).

	tech.	processor	area (area/F^2)	freq. range	min. E_{cycle} (frequency)	max. throughput improvement beyond V_{DD} scaling	min. energy improvement beyond V_{DD} scaling
(M+,P+) this work (M,P)	40nm	ARM Cortex-M0	0.15mm² (10⁸F^2)	200kHz-250MHz	7.4 pJ @0.5V (2.7MHz)	1.8X	1.15X*, 1.4X**
				100kHz-140MHz	8.64pJ @0.6V (5.56MHz)	1X	1X
[8] VLSI '17	65nm	ARM Cortex-M0+	3.76mm² (57·10⁶F^2)	20kHz-50MHz	12.9pJ @0.35V (174kHz)	1X	1X
[9] JSSC '17	40nm	ARM Cortex-M0	0.39mm² (2.4·10⁸F^2)	0.8MHz-50MHz	43.2pJ @0.37V (15MHz)	1X	1X
[10] ISSCC '12	180nm	ARM Cortex-M0	1.7mm² (5.3·10⁷F^2)	160kHz-330kHz	17.2pJ @0.26V (160kHz)	1X	1X

*1.15X @ exact MEP, **1.4X @ 10% V_{DD} margin

Fig. 11. Comparison with the state of the art (ARM Cortex-M0 with SRAM).

increase, and 6% throughput degradation, compared to a single traditional core. Reconfiguration makes the core more adaptable to a wider range of applications and operating conditions, while preserving the same per-thread control flow of the original architecture (i.e., no further architectural verification), allowing fully automated processor design based on commercial tools, maintaining the same software stack, and requiring minor change in the compiled SRAM macro.

ACKNOWLEDGEMENT

We acknowledge the support of Intel, Singapore NRF grant (NRF-CRP20-2017-0003), and TSMC for fabrication.

REFERENCES

[1] M. Alioto (Ed.), *Enabling the Internet of Things,* Springer 2017.

[2] D. Rossi, A. Pullini, I.Loi, M. Gautschi, Frank K. Gurkaynak, A. Teman, et. al., "Energy-efficient near-threshold parallel computing: The PULPv2 cluster," *IEEE Micro*, vol. 37, no. 5, pp. 20-31, 2017.

[3] S. Jain, L. Lin., M. Alioto, "Dynamically adaptable pipeline for energy-efficient microarchitectures under wide voltage scaling," *JSSC,* 2018.

[4] H. Fujiwara, S. Okumura, Y. Iguchi, H. Noguchi, H. Kawaguchi and M. Yoshimoto, "A 7T/14T dependable SRAM and its array structure to avoid half selection," *22nd Int. Conf. VLSI Design*, 2009, pp. 295-300.

[5] S. Okumura, S. Yoshimoto, K. Yamaguchi, Y. Nakata, H. Kawaguchi and M. Yoshimoto, "7T SRAM enabling low-energy simultaneous block copy", *CICC*, 2010

[6] D. Markovic, R. Brodersen, *DSP Architecture Design Essentials*, Springer, 2011.

[7] N. Weaver, Y. Markovskiy, Y. Patel, J. Wawrzynek, "Post-placement C-slow retiming for the Xilinx Virtex FPGA," in *FPGA*, 2003.

[8] J. Myers, A. Savanth, P. Prabhat, S. Yang, R. Gaddh, S. O. Toh, et al., "A 12.4pJ/cycle sub-threshold, 16pJ/cycle near-threshold ARM Cortex-M0+ with autonomous SRPG/DVFS and temperature tracking clocks," *VLSI Symp.* 2017.

[9] H. Reyserhove, W. Dehaene, "A differential transmission gate design flow for minimum energy sub-10-pJ/cycle ARM cortex-M0 MCUs," *JSSC,* 2017.

[10] Y. Lee, G. Kim, S. Bang, Y. Kim, I. Lee, P. Dutta, et al., "A modular 1mm³ die-stacked sensing platform with optical communication and multi-modal harvesting," *ISSCC*, 2012.

978-1-7281-5107-6/19 $31.00 © 2019 IEEE

A 28nm fully digital voltage monitor with 16.5uV/°C accuracy and 0.8mV quantized error from -40 to 160°C for ISO26262 ASIL-D capable MCU

Toshifumi Uemura, Yuko Kitaji, Kazuki Fukuoka

Renesas Electronics Corporation, Tokyo, Japan

{toshifumi.uemura.ra, kazuki.fukuoka.xn}@renesas.com

Abstract—A fully digital voltage monitoring system has been developed for automotive ISO26262 ASIL-D capable MCUs to detect internal under-voltage fault at IR-drop hotspots and to mitigate large Vmin margin due to voltage detection variation. Proposed oscillators employing temperature coefficient tuners enable digitized voltage detection with small temperature drift and quantized error. A digital voltage monitor fabricated in 28nm shows temperature drift of 16.5uV/°C and 0.8mV quantized error. The total voltage detection variation of 4.1mV is the smallest and the temperature range from -40 to 160°C is the largest compared with previous works

I. INTRODUCTION

Vehicle electric/electrical systems have been rapidly evolving to realize CASE (Connected, Autonomous, Shared and Electric) technology. Especially MCUs require both higher performance and safer operation over a wide temperature range such as Tj = -40 to 160°C to control multiple functions of integrated electronic control units (ECUs). Functional safety (FuSa) assures safe and sustainable operation of ECUs. Thus critical control applications on MCU need the highest level FuSa defined as ISO26262 automotive safety integrity level-D (ASIL-D) [1]. Power supply delivery is one of major single point faults of MCUs. A voltage monitor (VMON) is generally employed as a safety mechanism of ASIL-D capable MCUs. If the VMON detects under-voltage of power supply, the MCU should transition into a safe state such as hardware reset before the system operation fails.

Fig. 1 shows a conventional analog voltage monitor (AVM), where a reference voltage Vref generally has offset error and temperature drift. Although post-silicon trimming compensates offset errors of Vref and a comparator, it is difficult to make them smaller than quantized error corresponding to one-tap resolution Vtap. Besides that, wide temperature range for automotive MCUs makes it difficult to mitigate temperature drift. As a result, the AVM for automotive MCU has large voltage detection variation.

Fig. 2 shows a voltage monitoring system of a typical MCU. A core voltage VDD is monitored by an AVM. In order to assure a safe state transition, all modules should perform correctly until the AVM detects under-voltage fault. The detection voltage variation of AVM requires large Vmin margin and that results in performance degradation. On the other hand, the AVM should be placed at a small noisy peripheral region of a chip to avoid digital noise interference.

While the AVM detects power supply fault at PCB and PKG, it is difficult to detect excessive IR-drop such as higher resistance fault of power mesh at hotspots.

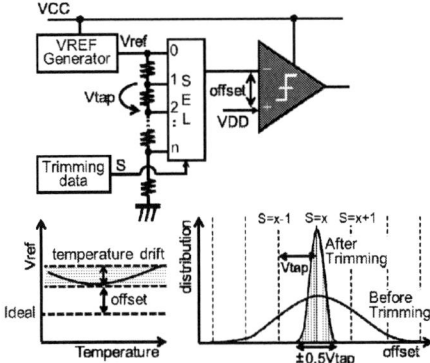

Fig. 1. Conventional analog voltage monitor

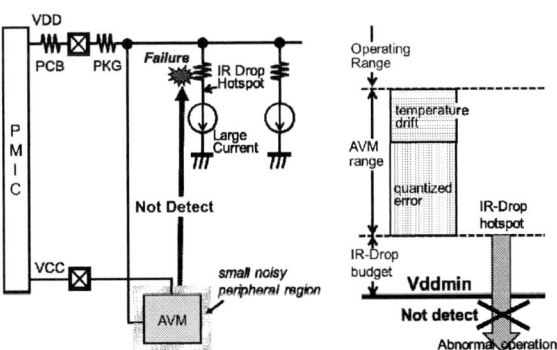

Fig. 2. Voltage monitoring system of a typical MCU

In order to overcome those issues, we propose a novel voltage monitoring system. The design highlights are as follows:

(1) Reference-voltage-less fully digital voltage monitor to enable distributed placement and to detect internal under-voltage faults

(2) Robust voltage monitoring with a barrier mechanism against invalid signals and soft errors

(3) Small temperature drift and quantized error to reduce Vmin margin and to achieve higher performance

(4) Wide temperature range of Tj = -40 to 160°C

II. PROPOSED FULLY DIGITAL VOLTAGE MONITORING SYSTEM

A. Distributed voltage monitoring

Fig. 3 shows a proposed voltage monitoring system. Digital voltage monitors (DVMs) are placed at both of hot and non-hot IR-drop spots. A DVM CNTL controls all DVMs. Both the DVM and the DVM CNTL are regarded as voltage safety islands which are assured lower voltage operation until a coarse VMON detects excessive under-voltage. Non-safety region excluding them can operate until the DVMs detect under-voltage. The detection level is programmed by IR-drop design budget of each point. If at least one spot causes under-voltage fault, the DVM issues voltage error and all non-safety regions transition to a safe state.

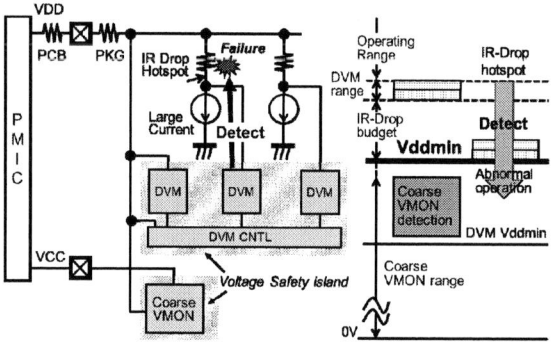

Fig. 3. Proposed distributed voltage monitoring system with DVMs

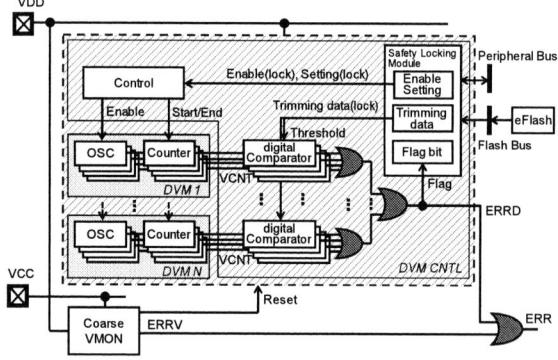

Fig. 4. Block diagram of DVMs and a DVM CNTL

Fig. 4 shows a block diagram of a DVM CNTL and DVMs. Each DVM consists of oscillators (OSCs) and counters and digitizes voltage level as count numbers (VCNT). VCNT has 12-bit width to realize small quantized error when trimming. Digital comparators compare the VCNT with threshold values.

If at least one of VCNT is less than the threshold, the DVM CNTL issues under-voltage error 'ERRD'. The threshold values are transmitted initially as trimming data from eFlash. Before the DVMs become uncontrollable due to excessive under-voltage, a coarse VMON issues a reset to the DVM CNTL and holds the under-voltage error by 'ERRV'.

B. Barrier mechanism for robust voltage monitoring

It is necessary to sustain correct voltage monitoring operation from ASIL-D points of view even though non-safety region induces faults or soft errors are injected to the safety islands. A safety locking module shown in Fig. 5 works as a barrier against invalid signals and soft errors. Once an enable bit is issued by a peripheral bus access, it locks all flip-flops for enabling, setting and trimming data. Fig. 6 shows a timing chart of safety locking. After the locking operation, i.e. the beginning of voltage monitoring, all data cannot be overwritten if invalid register writing or flash operation are occurred due to faults at non-safety regions. Triple module redundancy (TMR) is also employed for all locked flip-flops. Even if a soft error is occurred, a majority voter outputs correct data of locked FFs and recovers data from the error.

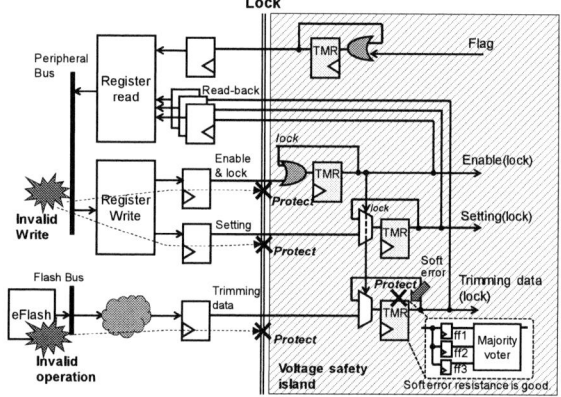

Fig. 5. Barrier structure for invalid access and soft error injection

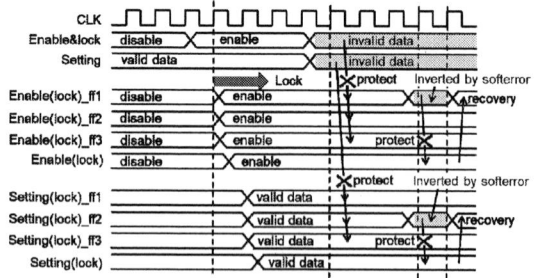

Fig. 6. Timing chart of safety locking for an enable bit and setting data

III. HIGH ACCURACY VOLTAGE MONITOR

Fig. 7 shows a schematic of proposed DVM. The DVM has four OSCs and their counters. Each node of the OSC has a temperature coefficient tuner (TCT) which consists of serial connected tunable PMOS resistance and MOS capacitance. A native ring OSC without TCTs gains larger counts at higher temperature because I_{ds} becomes larger at high temperature in 28nm. On the other hand, the TCT performs as larger capacitance load at higher temperature because larger I_{ds} of PMOS makes its on-resistance (Ron) lower. That means that TCT can add opposite temperature coefficient to the OSCs following the amount of Ron and C. As a result, four different temperature coefficient OSCs are equipped in the DVM.

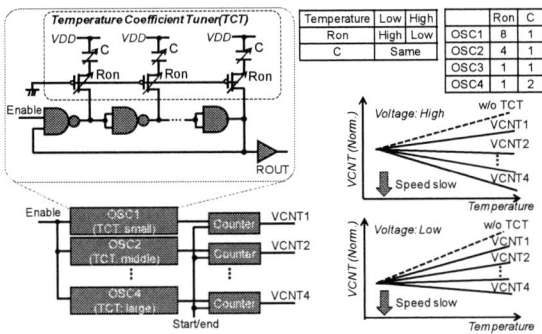

Fig. 7. Schematic of DVM with temperature coefficient tuners

Fig. 8 shows a post-silicon trimming and a mechanism of small temperature drift voltage detection. Trimming tests at each target detection voltage are performed at -40 and 160°C and smaller VCNT of each OSC is stored as the threshold (a). Although each OSC works as a digital voltage monitor after the trimming, the temperature drift is still large (b). VCNT comparison of four OSCs with each thresholds simultaneously composes small temperature drift detection voltage (c). Redundant four different temperature coefficient composing realizes robustness of small temperature drift over required voltage detection range even the coefficient shift is occurred due to change of Vth and mobility effect to I_{ds}. In accordance with the coefficient decided by the target voltage and process variation, two out of four OSCs work as the voltage monitor. Disabling OSCs except them achieves smaller power consumption while keeping small temperature drift. However, four OSCs are always operating in order to simplify the trimming flow.

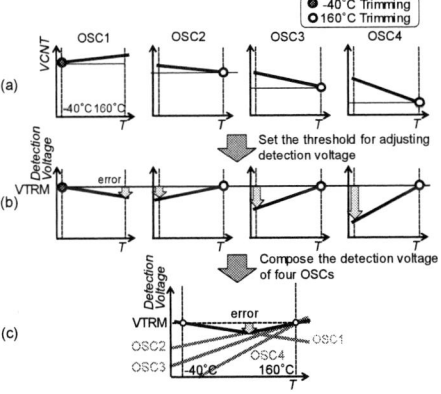

Fig. 8. Post-silicon trimming and voltage detection with small temperature drift scheme

IV. MEASUREMENT RESULTS

Fig. 9 shows a test chip micrograph fabricated in 28nm process with eFlash technology. Area of a DVM is 150um by 50um. We have measured DVMs of 10 chips.

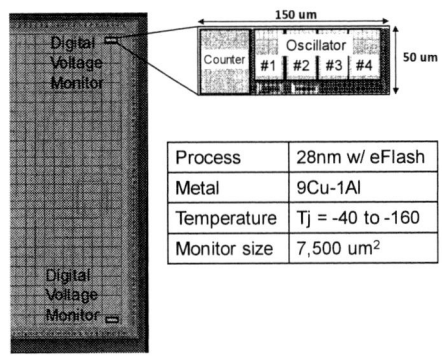

Process	28nm w/ eFlash
Metal	9Cu-1Al
Temperature	Tj = -40 to -160
Monitor size	7,500 um²

Fig. 9. Test chipmicrograph and features

Fig. 10 (a) shows measured temperature dependence of VCNT at 1.00V. Each VCNT of OSC1-4 shows different temperature coefficient by TCT effect. A solid line shown in fig. 10 (b) shows detection voltage after composing of four OSCs. We have confirmed that the composing enables small temperature drift of voltage detection.

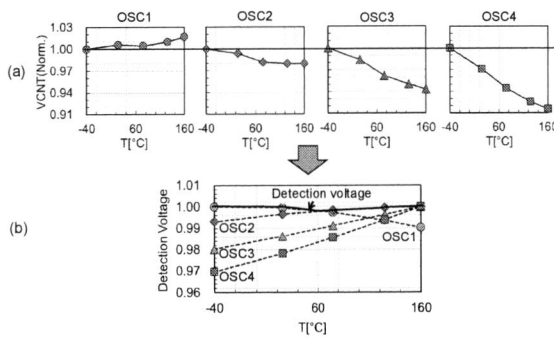

Fig. 10. Measured temperature dependence of OSCs and composing effect

Fig. 11 (a) shows measured detection voltage from -40 to 160°C. Trimming is performed at 1.000V, 0.960V and 0.920V, respectively. Fig. 11 (b) plots the worst error of each DVM normalized by I_{ds}. The results show that the worst voltage detection error is 3.3mV, i.e. 16.5uV/°C temperature drift even though process variation varies I_{ds}.

Quantized error of a DVM is determined by the worst VCNT resolution of four OSCs. Fig. 12 shows measured VCNT resolution of the worst OSC at -40, 25 and 160°C, respectively. Counting period of the DVM is 3.2usec. The results show that the quantized error is 0.8mV from 0.9 to 1.0V.

978-1-7281-5107-6/19 $31.00 © 2019 IEEE

IEEE Asian Solid-State Circuits Conference
November 4 – 6, 2019
The Parisian Macao, Macao SAR, China

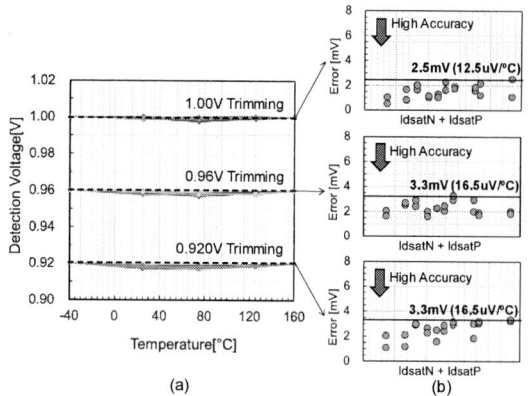

(a)   (b)

Fig. 11. Measured detection voltage with temperature dependence

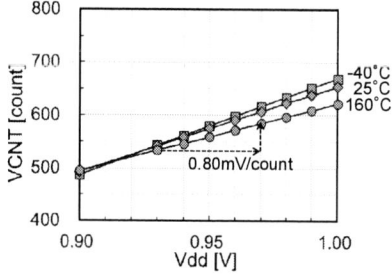

Fig. 12. Measured VCNT resolution of the worst OSC

Fig. 13 shows a shmoo plot of minimum voltage where a DVM can issue a voltage error. The minimum voltage of 0.70V is lower by 220mV compared with the detection voltage of 0.92V. That means we can use a 200mV variation voltage monitor as a coarse VMON for this system.

minimum voltage [V]					
0.80	Pass	Pass	Pass	Pass	Pass
0.75	Pass	Pass	Pass	Pass	Pass
0.70	Pass	Pass	Pass	Pass	Pass
0.65	Fail	Pass	Pass	Pass	Pass
0.60	Fail	Fail	Pass	Pass	Pass
0.55	Fail	Fail	Fail	Pass	Pass
0.50	Fail	Fail	Fail	Fail	Fail
	-40	25	75	125	160
			Temperature[°C]		

Fig. 13. Shmoo plot of DVM minimum voltage

TABLE I. shows a comparison result with previous works [2], [3]. This work has the smallest variation as a voltage

monitor and the largest temperature range even though reference-voltage-less digital configuration. Voltage detection latency of 3.2usec is a key to realize safe operation. While the OSCs oscillating at over several hundred MHz achieves that, power consumption of 1mW is larger than others. However, it is almost negligible in high performance MCU with several watts. The Area becomes larger due to inserting many redundant TCT in each OSCs for adjusting temperature coefficient over a wide range.

TABLE I.   COMPARISON WITH PREVIOUS VOLTAGE MONITORS

	[2] ESSCIRC'15	[3] ASSCC'16	This work
Process	250nm	250nm	28nm
Topology	Analog	Analog	Digital
Reference voltage	Need	Need	No need
Detection level	0.52-0.85V	1.88-4.67V	0.92-1.00V
Resolution/Quantized error	<49mV	50mV	<0.8mV
Temperature drift [uV/°C]	110	170	16.5
Temperature range [°C]	-20 to 80	-20 to 80	-40 to 160
Total detection variation [mV]	60	67	4.1
Area [um2]	9120	41400	7500
Power	248pW	13nW	1mW
Detection Latency	n/a	3.2ms	3.2us

V.   CONCLUSIONS

We have proposed a fully digital voltage monitoring system to enable distributed monitor placement and to mitigate large Vmin design margin due to voltage detection variation. Composing four different temperature coefficient OSCs achieves small temperature drift of a voltage monitor. TCT is a key to control temperature coefficient of OSCs. The measurement results of test chips fabricated in 28nm show temperature drift of 16.5uV/°C and 0.8mV quantized error. The total voltage detection variation of 4.1mV is the smallest and the temperature range from -40 to 160°C is the largest compared with previous works.

REFERENCES

[1] Sugako Otani, Norimasa Otsuki, Yasufumi Suzuki, Naoto Okumura, Shohei Maeda, Tomonori Yanagita, Takao Koike, Yasuhisa Shimazaki, Masao Ito, Minoru Uemura, Toshihiro Hattori, Tadaaki Yamauchi, and Hiroyuki Kondo, "A 28nm 600MHz automotive flash microcontroller with virtualization-assisted processor for next-generation automotive architecture complying with ISO26262 ASIL-D," ISSCC, pp.54-55, Feb. 2019.

[2] Teruki Someya, Hiroshi Fuketa, Kenichi Matsunaga, Hiroki Morimura, Takayasu Sakurai, and Makoto Takamiya, "246pW, 0.11mV/C glitch-free programmable voltage detector with multiple voltage duplicator for energy harvesting," ESSCIRC, pp.249-252, Sep. 2015.

[3] Teruki Someya, Kenichi Matsunaga, Hiroki Morimura, Takayasu Sakurai, and Makoto Takamiya, "56-level programmable voltage detector in steps of 50mV for battery management," ASSCC, pp.49-52, Nov. 2016.

978-1-7281-5107-6/19 $31.00 © 2019 IEEE

HyCUBE: A 0.9V 26.4 MOPS/mW, 290 pJ/op, Power Efficient Accelerator for IoT Applications

Bo Wang, Manupa Karunarathne, Aditi Kulkarni, Tulika Mitra, Li-Shiuan Peh

School of Computing, National University of Singapore

Abstract—IoT devices use ultra-low-power micro-controllers that cannot handle the performance demands of emerging compute-intensive applications. Accelerators can be added to improve system power-performance efficiency. We present Hy-CUBE, a Coarse-Grained Reconfigurable Array (CGRA) accelerator chip that realizes 127× improvement in power efficiency compared to TI Sensortag IoT platform. HyCUBE has a bufferless Network-on-Chip (NoC), enabling single-cycle data traversal to boost throughput and a software-scheduled architecture, automatically extracting application parallelism. Our 40nm test chip delivers peak efficiency of 26.4 MOPS/mW with 290 pJ/operation, realizing a power efficiency improvement of 28.6× and 26.5× compared to Xilinx Zynq FPGA and ARM Cortex-A7 core.

Keywords— CGRA; accelerator; IoT; power efficient

I. INTRODUCTION

Internet of Things (IoT) place enormous performance-per-mW demands on the devices that have to run diverse IoT applications ranging from face recognition [1] to healthcare diagnostics [2]. Among IoT applications, high parallelism and loop-intensive processing are increasingly prominent. However, the ultra-low-power processors in IoT devices have difficulty keeping up with the applications' demands. For instance, Fig. 1 shows a heart rate monitoring application, TROIKA [3], running on a wearable platform, TI Sensortag with ARM Cortex-M3 core. As profiled, TROIKA is a loop intensive application where the innermost loops from the two kernels consume 37.9% and 37.5% of the total execution time, respectively. However our measurements show that the Sensortag can only support up to 48MHz frequency while consuming 350mW when running TROIKA, thus delivering a peak efficiency of 0.21 MOPS/mW, which do not suffice to support such a throughput critical application.

CGRAs are promising accelerators for IoT devices. Unlike FPGAs, where bit-level reconfiguration leads to high area and power overheads, CGRAs support much more efficient configuration at word level. Prior CGRA architectures have demonstrated good power efficiency. For instance, Samsung's SRP CGRA [4] achieves 22.56 MOPS/mW efficiency in post-layout simulation. The works in [5], [6], [7] rely on large number of PEs for acceleration, which is not practical for IoT devices which are area constrained. Besides, this leads to low PE utilization and poor power efficiency.

Enhancing PE utilization and throughput is critical for realizing power-efficient CGRAs. This paper presents the HyCUBE CGRA accelerator chip targeting IoT applications. It has two key features: (1) A low-power NoC that offloads communications from the ALUs, delivering data across the

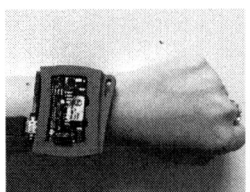

(a)

TROIKA kernel	Loop Iteration*	Share of run time
MPInverse	8000K	37.9%
FOCUSS	16000K	37.8%

*Iteration is calculated based on the innermost loop

(b)

Fig. 1. (a) TI Sensortag IoT platform running TROIKA and (b) profiling of the innermost loops in two kernels.

chip within a single cycle, improving PE utilization and throughput, (2) A software-scheduled architecture extracting parallelism and orchestrating the cycle-by-cycle schedule of the PEs and NoC, enabling lightweight hardware design. Our 40nm test chip shows a power efficiency of 26.4 MOPS/mW with energy consumption of 290 pJ/operation, which is on average 127.5×, 28.6× and 26.5× higher than Sensortag, Xilinx Zynq FPGA and ARM A7 core, respectively.

II. HyCUBE ACCELERATOR DESIGN

A. Execution Model

Fig. 2(a) shows the architecture of HyCUBE, which consists of a 4 × 4 PE array, 2 dual-port data memories (DMs), 16 configuration memories (CMs) and an SPI interface for communication with host. HyCUBE is statically scheduled, thus the compiler [8] determines the PE and the cycle where each operation is scheduled. The instructions are encoded with routing information that can be used for communication, concurrently to the computations. Inter-PE communication is realized by crossbar switch circuits and links between the PEs, which forms a bufferless NoC. Moreover, HyCUBE's unique ability to communicate to distant PEs in a single cycle is enabled through the NoC that can bypass intermediate nodes in multi-hop path. The compiler thus exploits such connections to increase instruction-level parallelism (ILP) in kernels.

Initially, the compiler analyzes the control and data dependencies of the kernels and constructs a control-data dependency graph (CDFG). Then compilation models architecture

978-1-7281-5107-6/19 $31.00 © 2019 IEEE

IEEE Asian Solid-State Circuits Conference
November 4 – 6, 2019
The Parisian Macao, Macao SAR, China

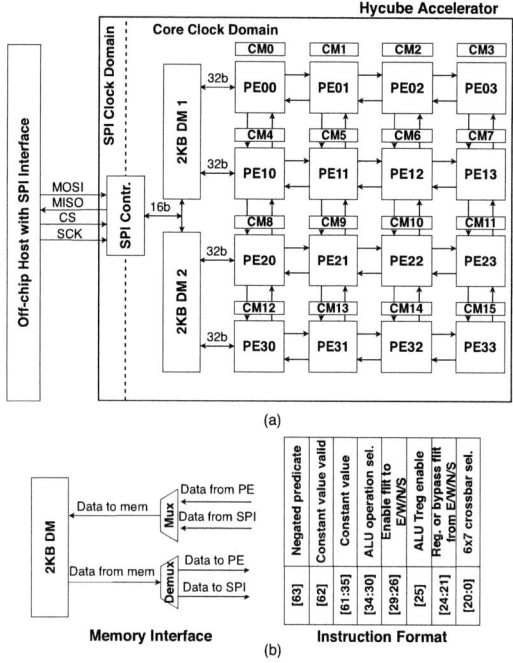

(a)

(b)

Fig. 2. (a) Architecture of HyCUBE accelerator and (b) its memory interface with instruction format

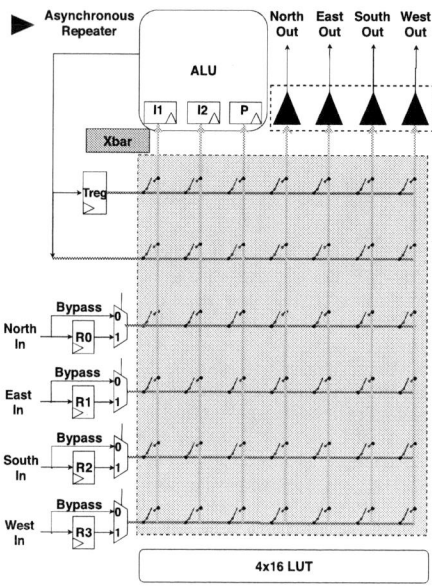

Fig. 3. Circuit diagram of HyCUBE tile

as a time-extended resource graph – MRRG[9], that includes single-cycle multi-hop connectivity. Finally, the compiler maps the CDFG to the MRRG exploiting higher degree of ILP and generates instruction streams for configuration.

Once the instruction and data are ready, they are loaded via SPI to memories. Thereafter, each PE will read its CM and execute the instruction. After the execution, the host reads back the computed data, again using the SPI.

B. Detailed Architecture of HyCUBE Accelerator

As Fig. 2(b) shows, the memory interfaces with both SPI and PEs so initial data and configuration loaded from SPI can be propagated to all PEs. HyCUBE has a custom 64b instruction format where crossbar selection, bypass enabling and ALU operations can be configured (Fig. 2(b)). The accelerator has two clock domains, a core clock domain including PEs and memory, and an SPI clock domain synchronized by the host. When crossing the domains, the 16b data from memory is buffered in a FIFO before being accessed by SPI and vice versa. The 2KB DM bank is organized as 512×32 where each 32b word can be partially updated by an 8b data chunk to accommodate 1B/2B/4B data sizes. Each 192B CM is able to store 24 instructions to orchestrate PE execution. A PE either executes computation or helps to bypass the received data to a distant PE. When computation completes, the DMs can be updated by the leftmost PEs where all other PEs have to transfer the output to one of them at first.

C. Tile design

To minimize critical path, each PE is grouped with its NoC crossbar switch to form a tile. A tile propagates flits to its neighbours for communication. A flit word incorporates 32b data and a predication bit P for control divergence. The circuit diagram of tile is depicted in Fig. 3. It comprises 4 input registers (R0~R3), a 6×7 crossbar circuit (Xbar), an ALU and a lookup table (LUT).

The registers cache an incoming flit from N/S/E/W direction if it is locally consumed. Otherwise the flit bypasses the register and is directly routed to a neighbour PE without buffering. The crossbar is able to select incoming data from neighbours and eject it to N/S/E/W/PE direction. It can eject 2 operands ($I1$, $I2$) and the predication bit (P) to ALU for computation. The ALU then sends the result to Xbar or a single register $Treg$. A tiny LUT maintains the start and end locations of loops so the loop pointer can access easily for innermost loop acceleration. The output channel is connected with asynchronous repeaters (buffers). Conventional NoCs employ output registers so that flits can be buffered for each hop. As HyCUBE offloads flow control to the compiler, buffers and output registers are no longer necessary. Hence, the data path from ALU to output is no longer re-timed and the propagation is not terminated until the flit is latched by a tile. This facilitates traversal across multiple hops within a single cycle [10], which will be elaborated in Section D.

Fig. 4 plots the simulated power breakdown of the tile circuit. The result shows the NoC crossbar and the ALU dissipating 36.3% and 14.7% of total power, respectively.

978-1-7281-5107-6/19 $31.00 © 2019 IEEE   134

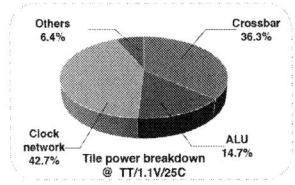

Fig. 4. Power breakdown of a tile circuit

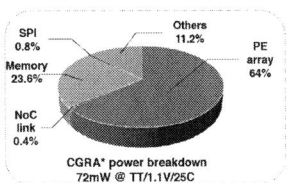

Fig. 7. Power breakdown of HyCUBE CGRA circuit

D. Software-scheduled NoC for single cycle traversal

The statically-scheduled HyCUBE chip comes with a compiler tool chain which takes C programs as input, runs profiling to identify frequently occurring kernels, then automatically extracts parallelism as CDFGs and schedules them onto the chip, storing the cycle-by-cycle scheduling into the 16 CMs. The NoC is then orchestrated cycle-by-cycle by the CMs. Each cycle, HyCUBE PE reads a 64b instruction which encodes the configuration for ALU and crossbar. This static scheduling obviates the power and area overheads consumed by scheduling, route computations and flow control.

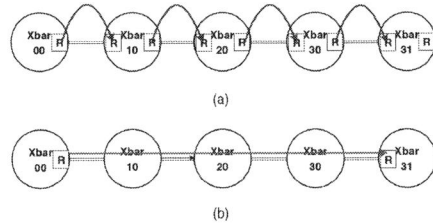

Fig. 5. 4-hop paths in (a) conventional NoCs and (b) HyCUBE NoC

In conventional CGRAs, it takes multiple cycles for data to travel between two distant PEs as data is registered every time when reaching a PE. Fig. 5(a) shows a 4-hop data path between PE00 and PE31. The intermediate nodes, PE10~PE30 are merely used for communication rather than computing, leading to low PE utilization and throughput. Fig. 5(b) shows the output from PE00 is directly sent to PE31 in a single cycle in HyCUBE, forming a critical path through the intermediate hops. As HyCUBE NoC cancels data re-timing along PE10~PE30, it can use all the nodes for compiler mapping after one cycle. Fig. 6 compares the critical path (4-hop latency) in HyCUBE against that in conventional

TABLE I
COMPARISON OF SIMULATED POWER EFFICIENCY ON VARIOUS CGRAs

	SRP[4]	REMUS[5]	RAA[6]	BiiRC[7]	DRRA[11]	This work
Technology	40nm	65nm	180nm	90nm	65nm	40nm
VDD (V)	1.1	1.2	1.8	1.0/1.2	1.2	1.1
PE count	9	256	64	39	12	16
FFT benchmark	256-pt	256-pt	256-pt	1024-pt	2048-pt	256-pt
Freq. (MHz)	100	200	70	492	450	853
Throughput (MOPS)	467[1]	NA	NA	8757.6	~5395	6482.8
Power (mW)	20.7	NA	~185	NA	~135.2	72
Power efficiency (MOPS/mW)	22.56	NA	NA	NA	39.9	90
Norm. efficiency[2] (MOPS/mW)	22.56	NA	NA	NA	64.8	90
Core area (mm²)	NA	21.6	NA	3.67	NA	2.87

1. 467 MOPS in SRP work is estimated from published 467 MIPS. Real MOPS can be less
2. Norm. efficiency is calculated as performance scaling up to 40nm

NoCs. While it prolongs the cycle time, it shortens the 4-hop latency by 47.5%. As 4-hop is able to cover the chip edge, HyCUBE constrains the maximum number of hops to 4 as trade-off between throughput and frequency. The timing loops occurring at place-and-route can be broken up with appropriate configuration of backend tools. As prior CGRAs [4][5][6][7][11] only published simulation results, we compare them with HyCUBE's post-layout results at nominal voltages based on the FFT benchmark in Table I. HyCUBE throughput (MOPS) is calculated by multiplying operations per cycle by frequency, while the efficiency (MOPS/mW) is obtained by dividing throughput by power. HyCUBE achieves a throughput of 6482.8MOPS and power efficiency of 90 MOPS/mW at 1.1V, 4× and 1.4× as high as [4] and [11], respectively. Fig. 7 shows the simulated power breakdown of the CGRA circuit. The PE array and memory consume most of the power (64% and 23.6%, respectively) while the link (sans crossbar) only dissipates 0.4% of the total power.

III. MEASUREMENT RESULTS

The HyCUBE test chip was fabricated in 40nm CMOS technology with core area of 2.86 mm². Fig. 8 shows the

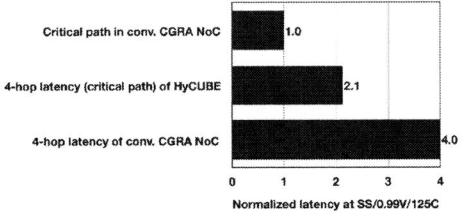

Fig. 6. Comparison of 4-hop latency in conventional NoC and HyCUBE.

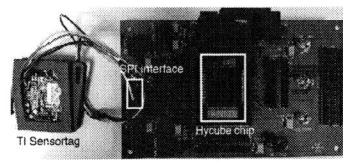

Fig. 8. Prototype of HyCUBE chip interfacing with TI Sensortag

978-1-7281-5107-6/19 $31.00 © 2019 IEEE

TABLE II
CHIP MEASUREMENT COMPARISON OF PERFORMANCE-PER-MW

Benchmark	Xilinx Zynq FPGA				ARM Cortex-A7				This work[2]			
	Iteration exe. cycles	Freq (MHz)	Power (mW)	Norm. MOPS/mW	Iteration exe. cycles	Freq (MHz)	Power (mW)	Norm.[1] MOPS/mW	Iteration exe. cycles	Freq (MHz)	Power[3] (mW)	Norm. MOPS/mW
FFT	2	222	1105	1	27	1400	440	1.2	5	488	140	6.9
GEMM	8	226	1224	1	23	1400	323	8.2	1.5	488	139	101.2
DCT	7	274	255	1	77	1400	434	0.3	11	488	141	2.0
TROIKA	1.8	274	629	1	125	1400	474	0.1	1.75	488	140	8.2

1. All benchmarks in HyCUBE utilize single-cycle 4-hop path
2. HyCUBE measured power @ 0.9V includes power from on-chip clock generator and SPI controller due to single power rail in the design. Cortex-A7 and FPGA are measured power numbers including only that of all the cores

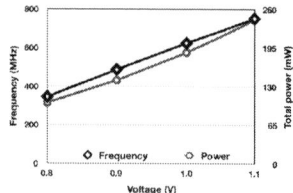

Fig. 9. Measured frequency and power of HyCUBE chip

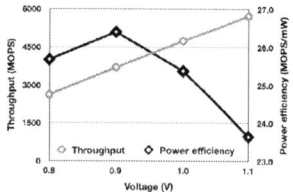

Fig. 10. Measured throughput of HyCUBE chip

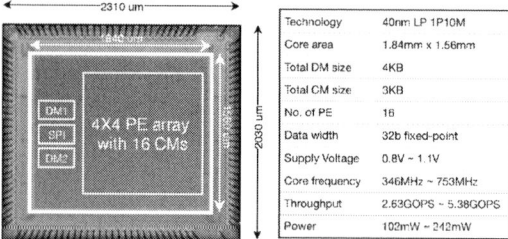

Fig. 11. Die photo of the HyCUBE chip with summary

prototype of the HyCUBE accelerator interfacing with the host, Sensortag via SPI.

Fig. 9 depicts the operating frequency with voltage scaling. The frequency varies from 753MHz to 346MHz when the supply goes from 1.1V to 0.8V. Fig. 10 plots the measured throughput and power efficiency. The HyCUBE chip achieves the maximum throughput of 5380MOPS at 1.1V and the minimum of 2630MOPS at 0.8V. The peak power efficiency is observed at 0.9V, as high as 26.4MOPS/mW with an energy of 290 pJ/op based on the FFT benchmark. For the TROIKA application, the chip (24.8MOPS/mW at 0.9V) outperforms 117×, 7.2× and 81.7× compared to TI Sensortag (0.21MOPS/mW), Xilinx Zynq FPGA (3.02MOPS/mW) and ARM Cortex-A7 core (0.3MOPS/mW), respectively. We choose ARM Cortex-A7 core for comparison as it was developed for wearable and IoT devices. Further peak power efficiency comparison (Table II) is made among 4 benchmarks. From the table, HyCUBE improves power efficiency by 28.6× and 26.5× on average compared to Zynq FPGA and Cortex-A7 core, respectively. Note that the power of FPGA and Cortex-A7 are underestimated as they merely include core power whereas HyCUBE includes entire on-chip power. Fig. 11 shows the die photo with summary.

IV. CONCLUSION

We propose HyCUBE, a novel CGRA chip with 16 PEs, where communication is handled by a bufferless NoC that is scheduled and controlled by the accompanying compiler, leading to high PE utilization at low power. Chip measurements show the HyCUBE chip has power efficiency improvement of 127.5×, 28.6× and 26.5× compared to the commercial IoT platform TI Sensortag, Xilinx Zynq FPGA and ARM A7 core, respectively.

REFERENCES

[1] S. Kodali et al., "Applications of deep neural networks for ultra low power IoT," in ICCD, pp. 589–592, Nov 2017.

[2] A. Limaye et al., "HERMIT: A benchmark suite for the internet of medical things," IEEE Internet of Things Journal, vol. 5, no. 5, pp. 4212–4222, 2018.

[3] Z. Zhang et al., "TROIKA: A general framework for heart rate monitoring using wrist-type photoplethysmographic signals during intensive physical exercise," IEEE Trans. on Bio. Engineering, vol. 62, no. 2, pp. 522–531, 2015.

[4] C. Kim et al., "ULP-SRP: Ultra low power Samsung Reconfigurable Processor for biomedical applications," in FPT, IEEE, 2012.

[5] J. Wei et al., "An efficient implementation of FFT based on CGRA," ICCSNT, vol. 2018-Janua, pp. 493–497, 2018.

[6] Y. Kim et al., "Hierarchical reconfigurable computing arrays for efficient CGRA-based embedded systems," DAC, 2009.

[7] O. Atak et al., "BilRC: An execution triggered coarse grained reconfigurable architecture," IEEE Trans. on VLSI Sys., vol. 21 (7), 2013.

[8] M. Karunaratne et al., "Hycube: A CGRA with reconfigurable single-cycle multi-hop interconnect," in 2017 54th ACM/EDAC/IEEE Design Automation Conference (DAC), pp. 1–6, IEEE, 2017.

[9] B. Mei et al., "DRESC: A retargetable compiler for coarse-grained reconfigurable architectures," in 2002 IEEE International Conference on Field-Programmable Technology, 2002.(FPT). Proceedings., pp. 166–173, IEEE, 2002.

[10] T. Krishna et al., "Breaking the on-chip latency barrier using SMART," in 2013 IEEE 19th International Symposium on High Performance Computer Architecture (HPCA), pp. 378–389, Feb 2013.

[11] N. Farahini et al., "39.9 GOPS/watt multi-mode cgra accelerator for a multi-standard basestation," in ISCAS, pp. 1448–1451, IEEE, 2013.

A 54% Power-Saving Static Fully-Interruptible Single-Phase-Clocked Shared-Keeper Flip-Flop in 14nm CMOS

Amit Agarwal, Steven Hsu, Monodeep Kar, Mark Anders, Himanshu Kaul, Raghavan Kumar,
Vikram Suresh, Sanu Mathew, Ram Krishnamurthy, Vivek De

Circuits Research Lab, Intel Labs
Intel Corporation,
Hillsboro, OR 97124
amit1.agarwal@intel.com

Abstract—A low clock power, static, fully-interruptible, single-phase-clocked, shared-keeper flip-flop without local clock inverters and no write-back failure reduces the clock transistor count to 6 instead of 12 in conventional transmission-gate flip-flop, achieving 54% reduction in total cell level power and 100mV improved V_{MIN}. An experimental microcontroller with shared-keeper sequentials, fabricated in 14nm CMOS, shows 6.5% lower measured chip level power at iso-frequency compared to the previously published single-phase-clocked AOI sequentials at 0.75V, 25°C.

I. INTRODUCTION

In modern synchronous CPU/GPU/SoC for edge, client and data center platforms, a large percentage of the overall power dissipation (>30%) is in the clock grid and final sequential circuits, limiting power, performance, and cost. Designing digital sequential circuits for low power consumption and low voltage operation is critical to reduce overall power for these platforms. Since the majority of sequentials have a very low data activity (typically 5-10%), clock power dominates their overall dynamic power. Also, a large percentage of timing paths in these circuit blocks have positive slack (setup, clk-q margin). These non-critical paths typically use the lowest drive strength sequentials with min-sized clock transistors to reduce power. These cells cannot be downsized any further to trade timing slack for power reduction, since the sizing depends on the min-sized transistors allowed by the process technology or V_{MIN} requirements (lowest operating voltage).

Previous work has shown a significant amount of research in the development of energy-efficient sequential circuits [1-2]. Pulsed latches have shown promise in reducing the power and delay; however, they suffer from hold time degradation, significant increases in min-delay buffer insertion, and pulse width evaporation issues at low supply voltages under process variations [3-4]. Internally gated flip-flops reduce power consumption based on data activity factor; however, for min-sized flip-flops the additional circuitry may consume more power than saved, resulting in no overall power benefit [5]. Previously reported single-phase-clocked flip-flops suffer from glitching, charge-sharing, and internal power switching, even when input data is constant [6-7]. Usage of multi-bit flip-flops to reduce power has become the industry standard and has been integrated into many synthesis/APR flows [8]; however it restricts and limits the physical placement of the flip-flops and degrades low voltage operation due to high fan-out loads on internal clock inverters. The previously published fully static AOI flip-flop [9] with single-phase-clocking reduces the

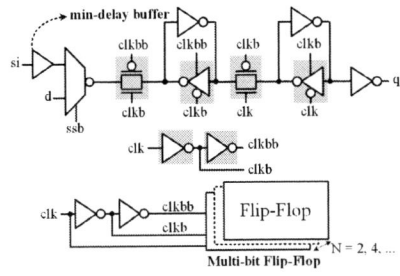

Fig. 1. Conventional transmission-gate (TG) flip-flop with 12 clock transistors and multi-bit flip-flop.

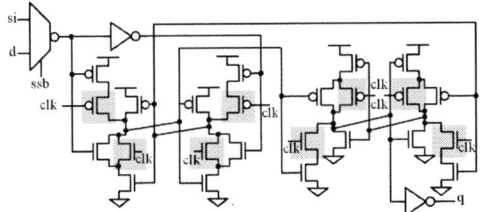

Fig. 2. Previously published single-phase-clocked AOI flip-flop (swapped stack) with 8 clock transistors [9].

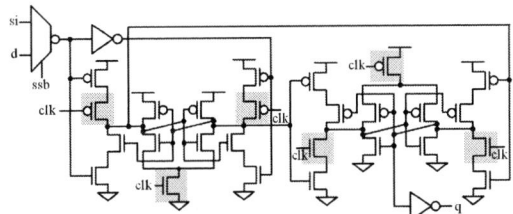

Fig. 3. Proposed single-phase-clocked SK flip-flop with 6 clock transistors.

number of clock transistors compared to a conventional transmission-gate (TG) flip-flop [2] (Fig. 1) for lower clock power (Fig. 2). In this paper, a family of static, fully-interruptible, min-sized shared-keeper (SK) flip-flop circuits are proposed, which takes advantage of timing slack in non-critical paths to reduce the number of clocked transistors further than AOI, resulting in lower clock pin-cap and power with improved hold V_{MIN} for reduced supply voltage operation.

II. SHARED-KEEPER (SK) FLIP-FLOP

Fig. 3 shows the proposed SK flip-flop topology with only 6 clock transistors vs. 8 and 12 in AOI and TG flip-flop, respectively. An SK flip-flop is derived from combining low

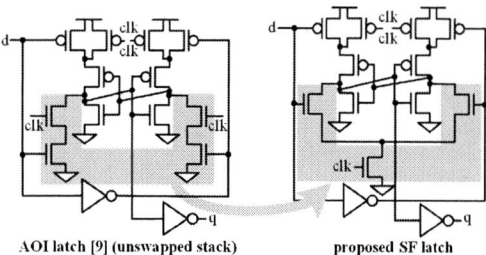

AOI latch [9] (unswapped stack)   proposed SF latch

Fig. 4. Shared-footer (SF) latch with 3 clock transistors.

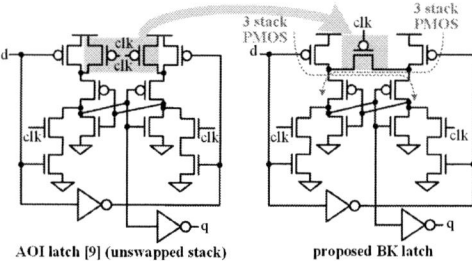

AOI latch [9] (unswapped stack)   proposed BK latch

Fig. 5. Bridge-keeper (BK) latch with 3 clock transistors.

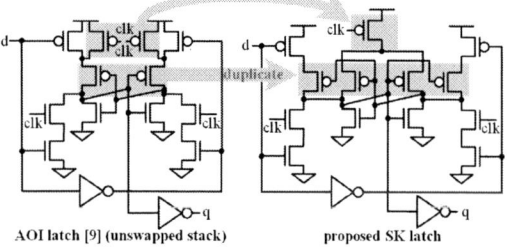

AOI latch [9] (unswapped stack)   proposed SK latch

Fig. 6. Shared-keeper (SK) latch with 3 clock transistors.

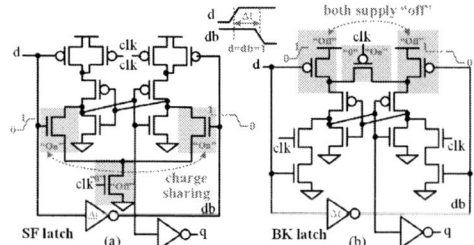

Fig. 7. (a) Charge sharing condition in SF and (b) "off" supply condition in BK latch topologies when data is latched.

and high phase latches with only 3 clock transistors. Three clock optimized latch topologies are proposed which reduce the total number of clock transistors even further to 3 compared to 4 in previously published AOI latch. The first clock optimization is the shared-footer (SF) topology (Fig. 4), which swaps the data and clock NMOS stack, allowing the footer clock transistor to be shared. Although, this circuit reduces the clock transistor count, there exists a charge sharing condition after data is latched and NMOS clock footer is "off". When input d switches from "0" to "1", for one inverter delay (Δt) both data pull-down NMOSs are "on", creating a charge sharing path between both state nodes, which corrupts the data under worst case systematic and random process variation (Fig. 7a). In the bridge-keeper (BK) topology (Fig 5), two clocked PMOS keepers in cross-coupled AOI gates are replaced with a single bridged PMOS keeper. This topology reduces the clock transistor count but increases the keeper stack to three. Moreover, just like shared-footer in bridge-keeper, when data is latched and input d switches from "0" to "1", both PMOSs connected to supply are turned "off" for one inverter delay (Δt) (Fig. 7b). During this period, the state retaining cross-coupled inverter is without power supply and may lose stored data.

The proposed SK topology duplicates the cross-coupled PMOS devices, allowing a single clocked keeper PMOS device to be shared (Fig. 6). This topology maintains a keeper stack of two; however, increases the state node capacitances because of the duplicated PMOS devices. Compared to SF and BK, this latch does not suffer from any charge sharing or supply off condition, making it the robust choice to reduce clock power. SK latch reduces the number of clock transistors to 3 vs. 4 and 6 in AOI and TG, respectively (Table I). This results in 16% vs. AOI and 46% vs. TG, total cell level power savings (data activity factor $\alpha_d = 10\%$), while maintaining fully static and interruptible (contention-free) design, good for low voltage

TABLE I. MIN-SIZED LATCH COMPARISON

Latch Type	Avg. T_{DQ} (Norm. Fo4)	Clock Trans. (Norm. Width)	Clock Pin-cap (Norm. Width)	Trans. Count	Power* (α_d= 10%)	Area
TG	1.9	6	4	14	1.0	1.0
AOI [9]	2.9	4	4	16	0.64	1.0
proposed SK	3.0	3	3	17	0.54	1.2

TABLE II. MIN-SIZED SCAN FLIP-FLOP COMPARISON

FF Type	TG	AOI[9]	proposed SK
Avg. T_{Setup} (Norm. Fo4)	2.0	3.5	3.6
Avg. $T_{Clock-Q}$ (Norm. Fo4)	2.5	2.8	2.9
Avg. T_{Hold} (Norm. Fo4)	-0.1	-2.1	-2.2
Clock Trans. (Norm. Width)	12	8	6
Clock Pin-cap (Norm. Width)	4	8	6
Trans. Count (w/ Scan)	38	34	36
Power* (α_d=10%)	1.0	0.57	0.46
Area	1.0	0.88	0.94
Scan Min-delay Buffer	Yes	No	No

* power includes upsized local clock buffer

operation. SK latch improves the clock pin-cap to 3 vs. 4 in AOI and TG, which lowers the back propagated load on clock grid buffers and clock gates, reducing the clock power further. One extra transistor in a SK latch (vs. AOI latch) results in unused empty layout space and 20% latch area overhead. The empty layout space can be utilized by extra circuitry in more complex standard cells such as scan flip-flops/integrated clock gates to minimize area overhead. The min-sized SK latch with higher delay offers a lower power standard cell beyond the lowest drive strength TG for non-critical timing paths.

A true single-phase-clocked SK flip-flop (derived from SK latch) without local clock inverters reduces the clock transistor

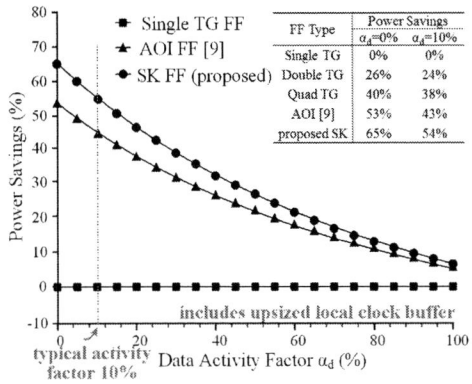

Fig. 8. Flip-flop power vs. data activity factor.

FF Type	Power Savings	
	α_d=0%	α_d=10%
Single TG	0%	0%
Double TG	26%	24%
Quad TG	40%	38%
AOI [9]	53%	43%
proposed SK	65%	54%

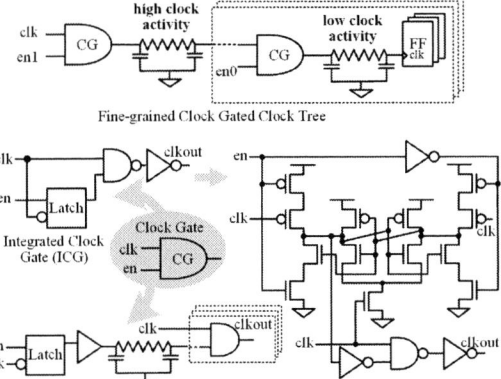

Fig. 9. Multi-level fine-grained clock gating using integrated clock gate or shared latch and proposed SK integrated clock gate.

TABLE III. INTEGRATED CLOCK GATE COMPARISON

ICG Type	Avg. T_{Setup} (Norm. Fo4)	Switched Clock Trans. (Norm. Width)		Clock Pin-cap (Norm. Width)	Power* p_{en}= 20%	Area
		en = 0	en = 1			
TG	2.0	12	18	10	1.0	1.0
AOI [9]	3.5	10	16	10	0.77	1.0
proposed SK	3.6	9	15	9	0.70	1.0

count to 6 vs. 8 and 12 in AOI and TG (Table II), respectively, achieving 19% vs. AOI and 54% vs. TG, total cell level power reduction (α_d=10%, includes upsized local clock buffer due to clock pin-cap increase, Fig. 8). The single-bit SK flip-flop achieves better power savings than a multi-bit TG flip-flop, eliminating the physical design overhead of clustering and routing congestion (Fig. 8). Improved hold time of SK flip-flop reduces the number of min-delay buffers in data-path after placement and clock tree synthesis, reducing congestion. It also eliminates the need of a scan min-delay buffer in flip-flop standard cell, resulting in 6% cell area reduction. Higher setup and clock-q delay of SK flip-flop can be absorbed by inserting them in non-critical paths and logic re-sizing/re-optimization without affecting block level frequency for lower power.

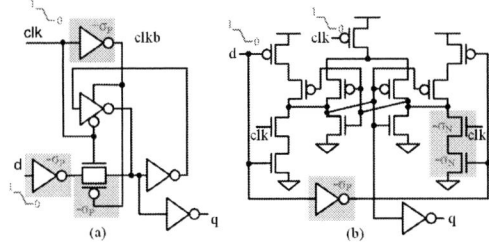

Fig. 10. Worst case hold time V_{MIN} condition a) TG and b) proposed SK latch.

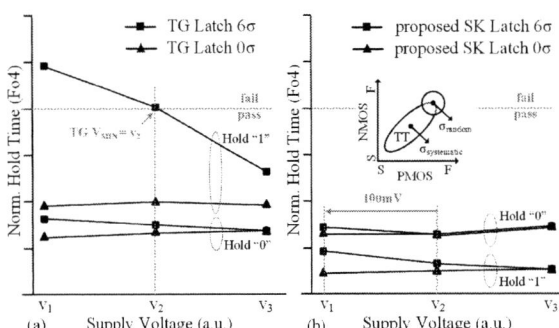

Fig. 11. Hold time V_{MIN} simulation with aging a) TG b) proposed SK latch.

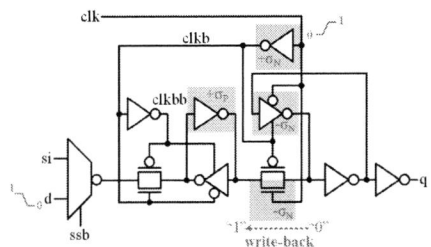

Fig. 12. Worst case TG flip-flop hold time V_{MIN} condition with write-back.

III. INTEGRATED CLOCK GATE (ICG)

In current synchronous circuits, fine-grained clock gating using integrated clock gates (ICGs) or a single latch driving multiple AND gates, are implemented at all levels of the clock tree to reduce the dynamic power (Fig. 9). However, these clock gates add additional clock load onto the high-activity un-gated portion of the clock grid, in order to reduce both logic and clock activity downstream. Integrating the SK latch into an ICG results in 11% vs. AOI and 30% vs. TG cell level power savings (enable probability p_{en} = 20%, Table III). These clock gate power savings along with lower clock pin-cap are on high clock activity nodes, which reduces significant block level clock power, even though there are fewer ICG instances compared to flip-flops.

IV. HOLD TIME V_{MIN} ANALYSIS

Hold time V_{MIN} statistical simulations for latching "0" and "1" were performed on both TG and SK latches (Fig. 10). Latch is first aged (at high temperature/high voltage) and then simulated across fast/slow systematic and 6σ random

IEEE Asian Solid-State Circuits Conference
November 4 – 6, 2019
The Parisian Macao, Macao SAR, China

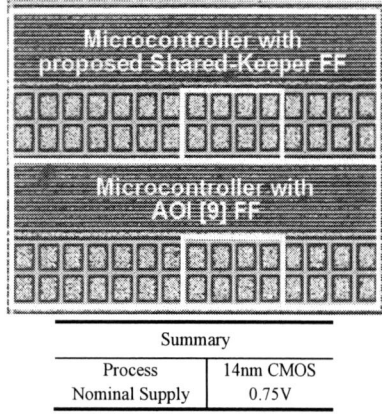

Summary	
Process	14nm CMOS
Nominal Supply	0.75V

Fig. 13. 14nm CMOS die micrograph.

parameter variations, and 0˚C-85˚C (Fig. 11). A weaker local min-sized clock inverter in TG latch under worst case process variation (exacerbated with aging) delays the latch closing, resulting in significant hold time degradation at low supply voltages with 6σ vs. 0σ random variations (worst case Hold "1"). Both single-phase-clocked SK and AOI latch do not have a local clock inverter, variation can only make the forward path fast under worst case hold time condition (Fig. 10). Simulations show that the SK latch results in comparable hold time between 0σ and 6σ for both latching "0" and "1" (Fig. 11), improving hold time V_{MIN} by 100mV vs. TG. Similar to the latch, the TG flip-flop results in degraded hold time under 6σ compared to 0σ due to weakening of local clock inverters under process variation/aging (worst case Hold "1", Fig. 12). Moreover, SK flip-flop does not suffer from same write-back failure as the TG flip-flop (worst at low supply voltages under variation), resulting in improved hold V_{MIN}.

V. 14NM CMOS MEASUREMENT RESULTS

To estimate power savings achieved by SK sequentials, an experimental low-frequency microcontroller was fabricated in 14nm CMOS [10] (Fig. 13). This experiment consists of pipelined random/data-path logic with clock gating and memories for data and instruction storage. Two side-by-side chips designed with the same target frequency using proposed SK and previously published AOI sequentials (flip-flops and integrated clock gates) were implemented for comparison. The power consumed by both chips is measured across 0.45V to 0.75V (nominal supply) at 25˚C while running a workload with varying flip-flop data activity (Fig. 14). The experimental microcontroller with SK sequentials consumes 6.5% lower measured chip level power compared to AOI for average flip-flop data activity factor $α_d = 10\%$ at 0.75V, 25˚C. Since SK flip-flops have larger data capacitance compared to AOI, power savings decreases to 5.2% at higher data activity. Reducing the supply voltage lowers the power savings, since leakage portion of the total power increases at low supply voltages. The SK-based microcontroller measurements show 6% and 4.2% total power savings at 0.6V and 0.45V, respectively.

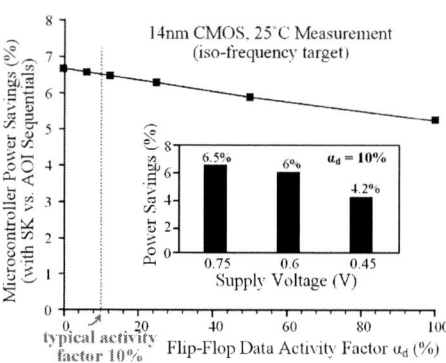

Fig. 14. Microcontroller power savings (proposed SK vs. AOI).

VI. CONCLUSIONS

A low clock power, static, fully-interruptible, single-phase-clocked, shared-keeper flip-flop is shown with 54% lower total cell level power and 100mV better V_{MIN} compared to the conventional TG flip-flop. Measurements of an experimental test-chip in 14nm CMOS show 6.5% better chip level power using SK sequentials compared to AOI at 0.75V, 25˚C.

ACKNOWLEDGMENT

The authors thank S. Realov, B. Kommandur, I. Rajwani, S. Wijeratne, J. Tschanz, and M. Haycock for their encouragement and discussions.

REFERENCES

[1] M. Alioto, E. Consoli, and G. Plaumbo, "Analysis and comparison in the energy-delay-area domain of nanometer CMOS flip-flops: part I—methodology and design strategies," *in IEEE Transaction on Very Large Scale Integration (VLSI) Systems, 2011,* 19.5, pp. 725-736.

[2] J. Tschanz et al., "Comparative delay and energy of single edge-triggered and dual edge-triggered pulsed flip-flops for high-performance microprocessors," in *IEEE International Symposium on Low Power Electronics and Design ISLPED, Aug. 2001,* pp. 147-152.

[3] B. Stackhouse et al., "A 65 nm 2-billion transistor quad-core Itanium processor," *IEEE Journal of Solid-State Circuits 2009,* 44.1, pp. 18-31.

[4] E. Consoli, M. Alioto, G. Palumbo, and J. Rabaey, "Conditional push-pull pulsed latches with 726fJ·ps energy-delay product in 65nm CMOS," *IEEE International Solid-State Circuits Conference ISSCC, Feb. 2012,* pp. 482-484.

[5] T. Fischer et al., "Design solutions for the Bulldozer 32nm SOI 2-core processor module in an 8-core CPU," in *IEEE International Solid-State Circuits Conference ISSCC, Feb. 2011,* pp. 78-80.

[6] Y. Kim et al., "A static contention-free single-phase-clocked 24T flip-flop in 45nm for low-power applications," in *IEEE International Solid-State Circuits Conference ISSCC, Feb. 2014,* pp. 466-467.

[7] C. K. Teh, T. Fujita, H. Hara, and M. Hamada, "A 77% energy-saving 22-transistor single-phase-clocking D-flip-flop with adaptive-coupling configuration in 40nm CMOS," in *IEEE International Solid-State Circuits Conference ISSCC, Feb. 2011,* pp. 338-340.

[8] I. H.-R. Jiang, C.-L. Chang, and Y.-M. Yang, "INTEGRA: fast multibit flip-flop clustering for clock power saving," *IEEE Transaction on Computer Aided Design of Integrated Circuits and Systems, 2012, vol. 31.2,* pp. 192–204.

[9] S. Hsu et al., "A microwatt-class always-on sensor fusion engine featuring ultra-low-power AOI clocked circuits in 14nm CMOS," *in IEEE Symposia on VLSI Circuits, June 2019.*

[10] S. Nataranjan et al., "A 14nm logic technology featuring 2nd-generation FinFET, air-gapped interconnects, self-aligned double patterning and a 0.0588 μm2 SRAM cell size," *in IEEE International Electron Devices Meetings IEDM, 2014,* pp. 3.7.1-3.

11-5 (7113)

IEEE Asian Solid-State Circuits Conference
November 4 – 6, 2019
The Parisian Macao, Macao SAR, China

0.54 pJ/bit, 15Mb/s True Random Number Generator Using Probabilistic Delay Cell for Edge Computing Applications

Fei Li, Ming Ming Wong, Aarthy Mani, Vishnu Paramasivam, and Anh Tuan Do
Institute of Microelectronics, Agency for Science, Technology and Research (A*STAR), Singapore
email: doat@ime.a-star.edu.sg

Abstract—This paper presents a true random number generator utilizing the probabilistic delay of the cross-coupled latch during start up as an entropy source. The start-up signal is used as a slow and jittered clock to sample a fast ring oscillator to harvest random bit stream. The tunable jitter delay range and all-digital design ensure high quality random bits generation across a wide range of supply voltages. The test chip consumes 0.54 pJ/bit with a bit rate of 15Mb/s at 0.5 V supply voltage while occupying only 18% area of the state of the art. The design's compact area, simple interface, and high energy efficiency make it a well suited choice for edge computing applications.

I. INTRODUCTION

A true random number generator (TRNG) that is robust and low in power consumption plays a vital role in ubiquitous edge applications such as cryptography, stochastic computing, probabilistic neural computing, and Bayesian inference [1].

Essentially, TRNG is evaluated using the concept of entropy, which is measured from its stochastic model. While quantum processes appear as ideal sources of entropy, its generation circuitry is impractical in hardware implementation [2]. Subsequently, CMOS-based TRNGs have often been designed using alternative entropy sources such as the noise function in transistors [3]. Such an analog approach is bulky and difficult to scale with nano-scale digital circuits. More recently, digital-like implementation based on cross-coupled pair with meta-stability [4] as well as jitter noise in ring oscillators [5] were also reported. However, they required either complex on-chip compensation circuit, high performance time-to-digital converter (TDC) or bit-trimming for high quality output stream. This leads to area and power overhead, which is not desirable for light-weight on-edge devices.

We proposed an energy efficient TRNG which offers output stream with high randomness quality with a tiny footprint. It has a very simple interface with the host and does not require complex on-chip calibration (for transistor matching) or TDC for accurate time counting. The contributions of this work are two-folds. First, as opposed to the conventional designs, our design exploits the stochasticity of meta-stability resolution time of a cross-coupled inverter to produce a highly jittered delay version of a clock signal that can be used to sample a fast running ring oscillator (RO).

This research is supported by Programmatic grant no. A1687b0033 from the Singapore government's Research, Innovation and Enterprise 2020 plan (Advanced Manufacturing and Engineering domain)

Second, a series of in-depth on-chip testing are conducted to obtain a comprehensive characterization of the generated output stream. The results prove that the proposed TRNG design generates high-quality random bit streams which pass all random tests from NIST [7] without any post processing or trimming.

II. PROPOSED DESIGN

A. Cross-coupled latch as a probabilistic delay cell (CCPD)

The proposed CCPD is constructed based on a cross-coupled inverter pair, as shown in Fig. 1. The sources of the PMOS transistors (i.e. node V_x) of the pair are not connected to the VDD rail directly but via a PMOS switch P1. In addition, V_X and the internal nodes (i.e. Q and QB) can also be discharged to ground via pull down transistors, N1-N3, which are also shown in Fig. 1. The CCPD works in two distinct modes, namely *Idle* and *Evaluation*.

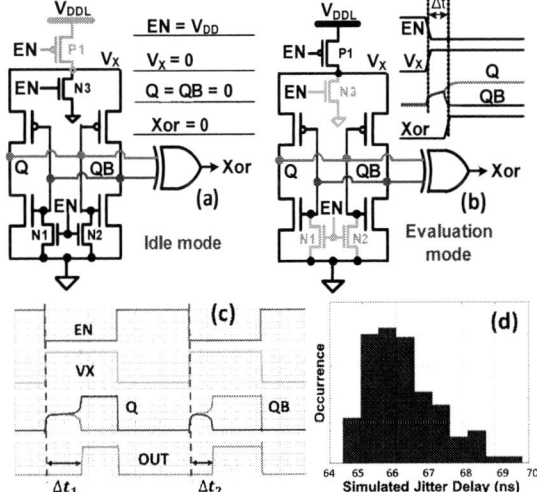

Fig. 1 Schematic diagram of the cross-coupled inverter (a) During idle mode (b) During evaluation mode. (c) Simulation waveform (d) Jitter delay distribution of the cross-coupled pair based delay cell at 0.5 V.

During *Idle* mode, EN signal is kept at VDD, turning on N1-N3 while keeping P1 turned off. Hence, V_X, Q and QB are all discharged to ground (see the timing diagram in Fig. 1a). As a result, the XOR gate outputs a '0'. In *Evaluation* mode, EN is driven to ground. Thus, N1, N2, and N3 are turned off to

978-1-7281-5107-6/19 $31.00 © 2019 IEEE   141

release V_X, Q and QB. At the same time, P1 is turned on to pull V_X to VDDL. Because both Q and QB were initially at ground, the PMOS pair in the CCPD conduct and gradually charge Q and QB at similar rates (i.e. $V_Q \approx V_{QB}$). The CCPD is temporarily maintained at its meta-stability (i.e. $V_Q \approx V_{QB}$) during this transition, due to the very small voltage gap between Q and QB and the initial small loop gain.

Eventually, a large enough voltage gap is developed across Q and QB due to thermal noise, and the cross-coupled inverter will latch to one of its stable states, as shown in Fig. 1b. Since Q and QB are always complementary in stable states, the XOR gate will eventually output a '1'.

The circuit structure in Fig. 1 thus behaves like an inverted delay cell with unique characteristics: Its delay is largely decided by how much time the cross-coupled structure stays in meta-stability. In addition, it is noise dependent and thus non-deterministic. This characteristic can be exploited as a high-quality entropy source to harvest random bit streams. Fig. 1c shows the simulation waveform of the CCPD under the condition of transient noise. It can be observed that the states of Q and QB exhibit non-deterministic behavior and the cell's delay varies in two operation cycles. Its delay distribution is obtained from 200 runs of noise simulation at 0.5 V as shown in Fig. 1d. The profile follows a chi-square distribution.

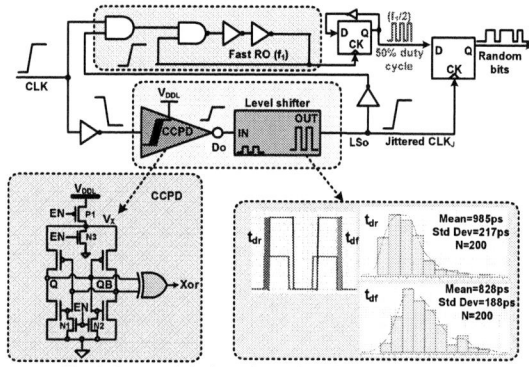

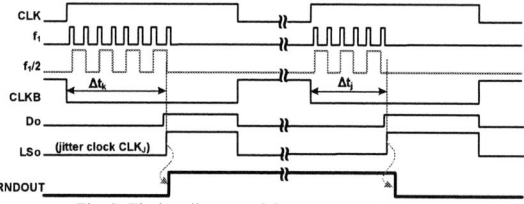

Fig. 2 Proposed design schematic and operating principle.

Fig. 3 Timing diagram of the proposed design.

B. True RNG using CCPD

The proposed TRNG (Fig. 2) consists of only 5 key components: (1) a 3-stage inverter-based ring oscillator to provide very fast oscillation frequency f_1, (2) a frequency divider to obtain near--50% duty cycle signal, (3) the proposed CCPD, (4) a robust, wide operating range level shifter and (5) a DFF to sample random data. Its operation timing diagram is shown in Fig. 3 and details are explained as follows:

When the input signal CLK is low the CCPD is in *Idle* mode producing a '0' at its output as discussed in Section A; while the AND produces a '0' which in turn stabilizes the output of the RO to '1' and keeps it turned off. When CLK goes high the CCPD enters *Evaluation* mode. Its output remains at '0' during

the meta-stable state, the output of the AND gate changes to '1' and with the feedback branch remaining at '1' the RO starts to oscillate at the frequency of f_1. The first self-feedback DFF converts the oscillation waveform to 50% duty cycle clock and halves the oscillation frequency ($f_1/2$). When the CCPD leaves the meta-stable state, a deterministic '1' is produced at its output which is level-shifted from V_{DDL} to V_{DD} level by the level shifter. The AND produces a '0' that turns off the RO. The fast, balanced clock is sampled by the slower, jittered clock CLK_J at the second DFF and this generates a random bit. The input CLK then goes to low and the CCPD enters *Idle* mode. The above is repeated every input clock cycle and this generates one random bit each time.

Power analysis shows that the RO consumes more than 80% of the dynamic power due to its fast switching speed. Therefore our RO is gated by the combination of both input CLK and the output of the CCPD cell to reduce the unnecessary power consumption after a random bit has been sampled. Simulation shows that 55% power reduction can be achieved by turning it off when necessary as shown in Fig. 4.

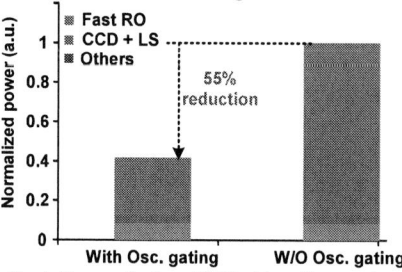

Fig. 4 Power reduction of 55% with oscillator gating.

It is essential that the output of the RO is sampled by the self-feedback DFF to achieve a 50% duty cycle clock. It helps to reduce the bias towards a particular logic value during the generation of random bit streams. Fig. 5 shows the simulation results of two random bit streams of 100 bits with and without clock balancing. Without clock balancing, the RO output has a duty cycle of 60% and it can be clearly observed that the resulting random bit stream is biased towards '1' with a mean value of 0.58. After duty cycle balancing, the mean value is reduced to 0.496.

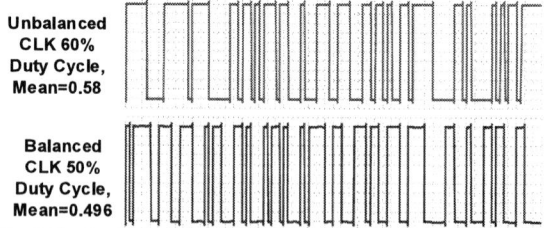

Fig. 5 Simulated waveform and mean value of output bit stream with/without 50% duty cycle.

A high-speed, low-leakage level shifter is needed to shift the output voltage level from V_{DDL} to V_{DD}. Implemented in 40 nm CMOS, this design is robust enough to convert signals of all the RVT transistors from near or subthreshold (i.e. from 200 mV) to nominal supply voltage (i.e. 1.1 V). Fig. 2 shows a typical conversion waveform and the rise and fall delays, t_{dr} and t_{df}, of the level shifter. Monte-Carlo simulation is performed to obtain the distribution profiles of the two delays.

Note that, this delay is insignificant when compared to the delay of the latch-cell in Section A.

Two key conditions need to be satisfied for the proper operation of the TRNG and the generation of high-quality random numbers. First, the cycle time of the balanced fast clock need to be smaller than the minimum time required for the CCPD to exit meta-stability. Second, the jitter of the CCPD should not be smaller than the period of the balanced fast clock $2/f_1$. We can ensure these two conditions are met through proper tuning of both V_{DD} and V_{DDL} and by giving a wide design space for the TRNG to function as required.

III. CHARACTERIZATION AND MEASUREMENT RESULTS

Our TRNG (chip microphotograph as depicted in Fig. 6) is fabricated in 40 nm CMOS. The core RNG circuit occupies only 550 μm^2 while the total area including the testing and monitoring circuits occupies 2500 μm^2.

Characterization with on-chip test circuits are performed to obtain the jitter delay corresponding to various V_{DDL} values, and to measure the oscillation frequency of the fast clock across a range of V_{DD} values. For each V_{DD} (and each oscillation frequency of $f_1/2$), there exists a maximum V_{DDL} and its associated minimum jitter delay range under which the output bit stream with high entropy is obtained. With incremental V_{DD} values (e.g. from 0.5V to 1.1 V), V_{DDL} is swept from low (e.g. 0.3 V) to high (e.g. 0.6 V) using the characterized jitter delay and oscillation frequency.

Under every test condition, an output stream of 3 Mb data is collected and checked using NIST test (except for RandomExcursions and RandomExcursionsVariant tests). The entropies for the 3 Mb data (corresponds to each V_{DD} and maximum V_{DDL}) are plotted in Fig. 7. The results prove that our design is robust in generating high entropy bits across the voltage range and the worst-case entropy is 0.999999615/bit at 0.7 V.

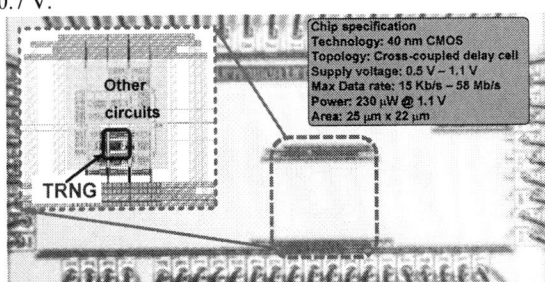

Fig. 6 Test chip (40nm CMOS) microphotograph and the chip's specification.

Furthermore, the 3 Mb of data from each of the seven V_{DD} values are also concatenated into a single ASCII file and checked with NIST test. The results (tabulated in Table I) showing all passes in 15 tests indicate that our design is capable to generate high quality random bits across the tested voltage range.

In order to testify the quality of the random bit streams, Auto Correlation Function (ACF) is computed on the 1 Mb data (obtained from each test conditions) to examine the obvious repeated patterns. The derived result, shown in Fig. 8, reflects very low autocorrelation factor in the collected data. In addition, the frequency of all the 8-bit patterns (from "00000000" to "11111111") occurred in the bit stream across three V_{DD} values is counted and plotted in Fig. 9. The result shows even distribution of all the patterns across the bit stream.

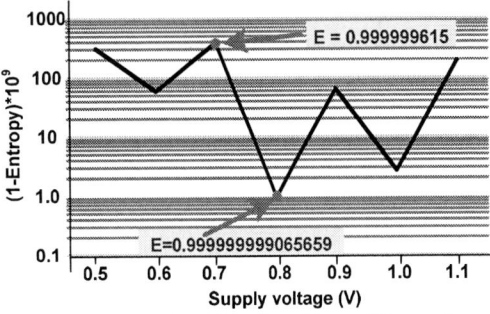

Fig. 7 Measured entropies of the collected data at different voltages.

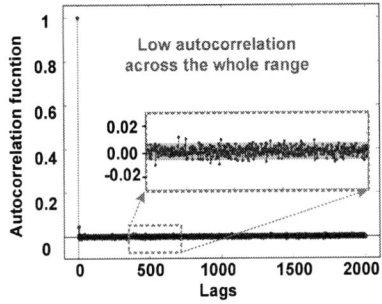

Fig. 8 Autocorrelation plot of the collected data at 1.1 V.

TABLE I. NIST TEST RESULTS

Test Name	P-Value	Pass Rate	Passed?
Frequency	0.35	100%	Yes
BlockFrequency	0.485	95%	Yes
CumulativeSums	0.466	95%	Yes
Runs	0.789	99%	Yes
LongestRun	0.719	100%	Yes
Rank	0.243	100%	Yes
FFT	0.689	100%	Yes
NonOver. Temp.	0.497	95%	Yes
Over. Temp.	0.469	100%	Yes
Universal	0.521	100%	Yes
Approximate Entropy	0.178	100%	Yes
Rand. Excursion	0.278	100%	Yes
Rand. Exc. Variant	0.244	100%	Yes
Serial	0.367	98%	Yes
LinearComplexity	0.585	100%	Yes

To provide a visual insight into the NIST Runs test results of our bit stream, the Run of '1' plot across the seven V_{DD} values is presented (refer to Fig. 10). The plot shows that the number of runs decreases exponentially with the matching Run length and the oscillation frequency of '1' and '0' is maintained within the acceptable range.

Meanwhile, the 1 Mb of data is folded into a 1000x1000 2D matrix. With bit '1' represented in pixel white and bit '0' in pixel black, its scatter plot is depicted in Fig. 11a. No significant pattern or bias is observed in the data. Fig. 11b shows the probability density function plot of the output stream, whose distribution is well modelled as a Gaussian function with a mean value of 0.50008 and standard deviation of 0.005.

IEEE Asian Solid-State Circuits Conference
November 4 – 6, 2019
The Parisian Macao, Macao SAR, China

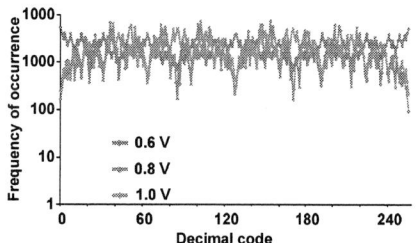

Fig. 9. Occurrence frequency of all the possible 8-bit patterns.

The obtained waveform at 0.5 V, together with the energy (per bit) and the highest attainable data rate measured across various V_{DD} is as depicted in Fig. 12. Our chip is well functioned across the voltage range (0.5 to 1.1 V) dissipating 2.82 pJ per bit at a data rate of 58 Mb/s at 1.1 V and 0.54 pJ per bit at a data rate of 15 Mb/s at 0.5 V. The maximum data rate is capped at 58 Mb/s after exceeding the V_{DD} of 0.9 V due to the limitation of our test board. The energy consumption in our design is comparable to the state of the art design [4] but occupies less than 20% of its silicon area. Table II tabulates the performance comparison of our test chip with the existing designs.

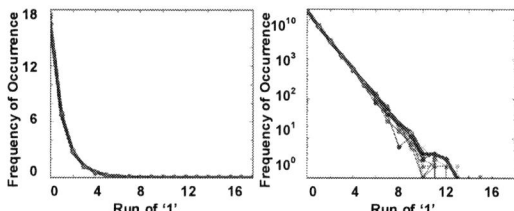

Fig. 10 Frequency of occurrence plot of various run of '1' lengths.

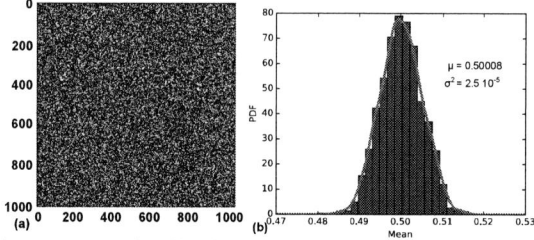

Fig. 11 (a) Scatter plot of the bit stream folded into a 2D matrix (b) Probability density function plot of the measured bit stream at 1.1 V.

IV. CONCLUSION

A TRNG based on a probabilistic delay cell realized by a cross-coupled latch has been implemented in 40 nm CMOS technology. Measurement is conducted across supply voltages of 0.5 V to 1.1 V and the measured data have passed all the tests under the NIST randomness test suit. Our TRNG chip is able to produce high quality random bit streams with very low energy consumption of 0.54 pJ/bit at a maximum data rate of 15 Mb/s at 0.5 V while achieving an 82% area reduction than the state of the art design.

ACKNOWLEDGMENT

The authors acknowledge Yun Kwan, Nursuria, Mun Wei and Kim Seng for layout, PCB and chip assembly support.

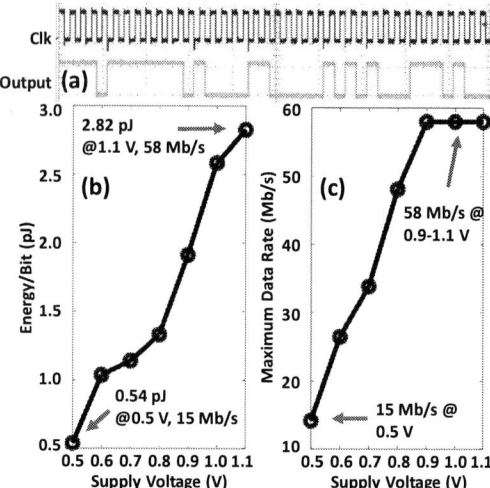

Fig. 12 (a) Input clock and the obtained random bit stream at 0.5 V, (b) Energy/bit measurement and (c) Highest data rate at various voltage values.

TABLE II. BENCHMARK COMPARISON WITH THE EXISTING WORKS

	VLSI Symp.'10 [4]	ISSCC'14 [5]	L-SSC' 18 [6]	This work
Technology	45 nm	28 nm	65 nm	40 nm
Design	Meta-stability	Jitter in 3-edged RO	Meta-stability	Probabilistic delay cell
V_{DD} (V)	0.28-1.35	1	0.53-1.0	0.5-1.1
Bit rate (Mb/s)	15-2400*	23.16	3.2 (0.33V)-86 (1.0V)	15 (0.5V)-58 (1.1V)***
Area (F²)	1.9×10^6	478×10^3	2.3×10^6	344×10^3
NIST pass	15/15	15/15	15/15	15/15
Calibration	Required, on-chip	-	Required, on-chip	Not required
Efficiency (pJ/bit)	0.3@0.28V** 3@1V**	23	2.58@0.53V 6.08@1.0V	0.54@0.5V 2.83@1.1V

* Extracted from Fig. 5 in [4]   **Estimated from Fig. 5 in [4]
*** Highest bit rate (at 1.1 V) is restricted with test board setup

REFERENCES

[1] S. N. Dhanuskodi, A. Vijayakumar and S. Kundu, "A chaotic ring oscillator based random number generator," IEEE International Symposium on Hardware-Oriented Security and Trust (HOST), 2014, pp. 160-165.

[2] N. Massari et al., "16.3 A 16x16 pixels SPAD-based 128-Mb/s quantum random number generator with -74dB light rejection ratio and -6.7ppm/C bias sensitivity on temperature," Digest of Technical Papers, IEEE International Solid-State Circuits Conference, 2016, pp. 292-293.

[3] M. Matsumoto, et. al., "1200 µm2 physical random-number generators based on SiN MOSFET for secure smart-card application," Digest of Technical Papers, IEEE International Solid-State Circuits Conference, 2008, pp. 414-624

[4] S. Srinivasan et. al., "2.4GHz 7mW all-digital PVT-variation tolerant True random number generator in 45nm CMOS," VLSI Circuits Symp., Honolulu, 2010, pp. 203-204.

[5] K. Yang, et. al., "16.3 A 23Mb/s 23pJ/b fully synthesized true-random-number generator in 28nm and 65nm CMOS," Digest of Technical Papers, IEEE International Solid-State Circuits Conference, 2014, pp. 280-281.

[6] V. R. Pamula, et. al, "A 65-nm CMOS 3.2-to-86 Mb/s 2.58 pJ/bit highly digital true-random-number generator with integrated de-correlation and bias correction," in IEEE Solid-State Circuits Letters, vol. 1, no. 12, pp. 237-240, Dec. 2018.

[7] Csrc.nist.gov. (2019). NIST SP 800-22: documentation and software - random bit generation | csrc. [online] Available at: http://csrc.nist.gov/groups/ST/toolkit/rng/documentation_software.html [Accessed 5 June. 2019].

11-6 (7205)

A Smart Hardware Security Engine Combining Entropy Sources of ECG, HRV and SRAM PUF for Authentication and Secret Key Generation

Sai Kiran Cherupally[*], Shihui Yin[*], Deepak Kadetotad[*], Chisung Bae[†], Sang Joon Kim[†], and Jae-sun Seo[*]

[*]School of Electrical, Computer and Energy Engineering, Arizona State University, Tempe, AZ, USA
[†]Samsung Advanced Institute of Technology, Suwon, Gyeonggi-do, South Korea
Email: scherupa@asu.edu

Abstract—We present a smart hardware security engine that combines three different sources of entropy, electrocardiogram (ECG), heart rate variability (HRV) and SRAM-based physical unclonable function (PUF), to perform real-time authentication and generate unique and random signatures. Such hybrid signatures vary person-to-person, device-to-device, and over time, and hence can be used for personal device authentication as well as secret random key generation, significantly reducing the scope of an attack. The prototype chip fabricated in 65nm LP CMOS consumes 4.04 μW at 0.6 V for real-time authentication. Compared to ECG-only authentication, the equal error rate of multi-source authentication is reduced by 18.9X down to 0.09% for an in-house ECG database. 256-bit secret keys generated by optimally combining ECG, HRV and PUF values pass NIST randomness tests with 100% pass rate.

I. INTRODUCTION

Traditional hardware designs for device authentication and secret key generation typically employ PUF based on SRAM, delay, or analog circuit elements [1]–[4]. While silicon PUFs can be highly stable, they do not represent liveliness. On the other hand, biometric authentication using fingerprint or iris for smartphones and IoT devices has become standardized, but effective spoofing attacks also have been reported. Our physiological signals such as ECGs provide liveliness proof, and have emerged as a new modality for authentication [5] and secret key generation [6], but lack satisfactory error rate and randomness. Another challenge for ECG-based security is that ECG manifests in an identical way on multiple devices for the same user, which exposes a larger attack surface and cannot be revoked once leaked. In this paper, we present a smart security engine that combines multiple entropy sources including individually-unique ECG features, heart rate variability (HRV), and device-unique SRAM-based PUF, to perform enhanced authentication and secret key generation.

This approach has three key advantages. First, by combining the entropies from multiple independent sources, we enable an efficient multi-factor authentication where both possession of silicon asset and biometric match is accounted for simultaneously in the authentication process. Second, biometric information is never used in its raw form. Hence, the scope of an attack is substantially reduced by making it imperative for an attacker to know both the silicon PUF as well as the user's ECG/HRV features to uncover the root of trust. A unique

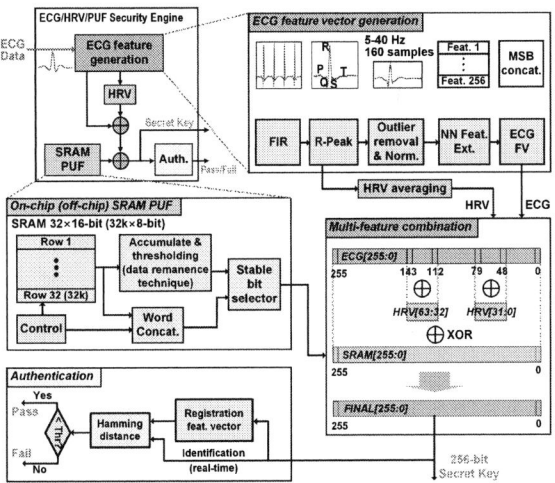

Fig. 1. Illustration of the proposed multi-source hardware security engine. Features based ECG, HRV and SRAM PUF are extracted and combined to form a 256-bit secret key, which is also used for authentication.

silicon PUF will have different IDs on different devices, and the same chip can be provisioned to authenticate multiple users due to their own unique IDs owing to unique ECG/HRV features. Third, authentication can still be performed when one or two of these entropy sources are compromised.

We fabricated the prototype chip for the multi-entropy-source hardware security engine in 65nm LP CMOS. The prototype chip consumes 4.04 μW at 0.6 V for real-time authentication. Compared to ECG-only authentication scheme [5], the equal-error-rate (EER) has been improved by 18.9X for a 741-subject in-house database. In addition, we also evaluated the effectiveness of combining multiple entropy sources for random secret key generation using NIST randomness tests [7], by analyzing various trade-offs and optimizations in aggregating ECG and HRV features with SRAM PUF values.

II. HARDWARE SECURITY ENGINE DESIGN

The proposed hardware security engine can perform two main tasks: authentication and secret key generation. For both tasks, each subject's 256-bit ECG feature vector (FV), average HRV value, and each device's PUF (from 64-bit to 256-bit)

978-1-7281-5107-6/19 $31.00 © 2019 IEEE

IEEE Asian Solid-State Circuits Conference
November 4 – 6, 2019
The Parisian Macao, Macao SAR, China

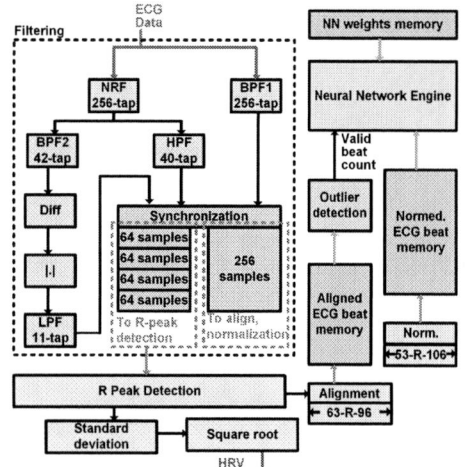

Fig. 2. The ECG pre-processing flow is adopted from [5], but in this work, the number of BPFs is reduced by 2X and memory used for alignment and normalization are reduced by 4X.

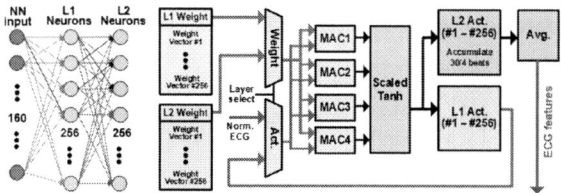

Fig. 3. The neural network has two hidden layers and no output layer as only the MSBs of the second hidden layer's neurons are used for feature extraction. Four neurons are evaluated per clock cycle, taking a total of 128 clock cycles to obtain the final 256 features at the second hidden layer output.

are combined using bitwise XOR to generate a 256-bit secret key (Fig. 1). The variations in ECG patterns from different subjects, heart rates over time, and SRAM power-on states of different chips enables the resultant bit-vector to be used as a secret random key as well as for device authentication.

A. Pre-processing and Neural Network

As shown in Fig. 2, raw digitized ECG data goes through filtering, peak detection, alignment and normalization. While ECG pre-processing and neural networks (NNs) are built upon our prior work on ECG-only authentication [5], a number of improvements have been made in this work. [5] has a large number of band-pass filters (BPFs) and four parallel single-layer NNs that require four data channels for alignment/normalization. In this work, we reduced the number of BPFs by 2X, only employed one two-layer NN with a single data channel, reducing the memory for alignment/normalization by 4X. In addition, while the NNs in [5] used the conventional one-hot labels for training, the NN in this work was trained with the authentication-specific cost function in [6], which leads to a large reduction in EER.

After aligning the ECG beats around the R-peak with 63/96 samples before/after the R-peak, the extracted ECG beats are passed through outlier removal and normalization

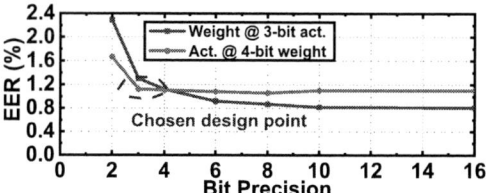

Fig. 4. The design point of 3-bit activation and 4-bit weight precision was chosen at the inflection point of the EER vs. precision curve.

modules. The precision of FIR filtering, R-peak detection, outlier detection and normalization were reduced to 13-b, 13-b, 11-b and 12-b, respectively, with minimal error rate degradation compared to software simulations.

These pre-processed ECG beats are fed to a single fully-connected NN with 160 input neurons and 2 hidden layers containing 256 neurons per layer (Fig. 3). The NN is trained using an in-house 741-subject ECG database with the objective to maximize separation between intra-subject and inter-subject distributions [6]. 256 ECG features are extracted from the output of the 2nd hidden layer, and the security engine can be configured to extract features from averaging either 30 or 4 ECG consecutive heart beats of the subject, for more stable or faster feature extraction, respectively. Finally, each user's representative 256-bit ECG feature vector is obtained by taking the MSBs from these average values of 256 features (Fig. 3).

Furthermore, the power and area were optimized by using a low precision NN and time-multiplexed operation. ~0.3% EER was traded-off for reducing the NN precision to 3-bit activation and 4-bit weights, substantially reducing the NN area and computation (Fig. 4).

B. Heart Rate Variability and SRAM PUF

In addition to the ECG features, our ECG processing engine also produces a continuous 10-bit HRV value. HRV is a measure of variation in instantaneous heart rates, which is computed on-chip as the inverse of the time interval between successive R-R peaks. These HRV values are collected for the duration of the ECG data for each subject and averaged externally to obtain a 64-bit HRV value using floating point arithmetic. The averaged HRVs are bitwise XORed with bits [143:112] and [79:48] of the 256-bit ECG FV to enhance randomness. Based on extensive software simulations, these

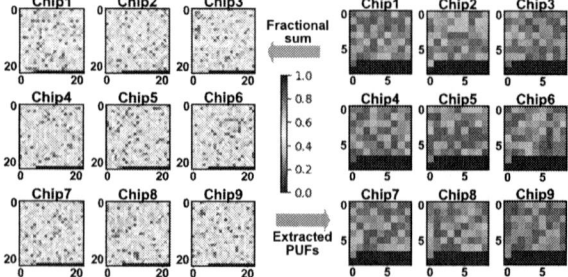

Fig. 5. Accumulated fractional sums of on-chip SRAM power-on states of 9 different chips across 100 power-on events (left) and their corresponding extracted PUFs with stability thresholds set to 0.65 and 0.35 (right).

978-1-7281-5107-6/19 $31.00 © 2019 IEEE

IEEE Asian Solid-State Circuits Conference
November 4 – 6, 2019
The Parisian Macao, Macao SAR, China

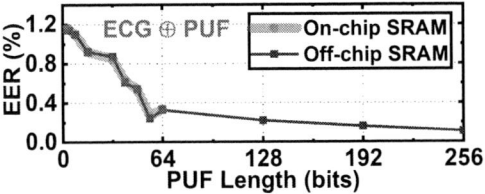

Fig. 6. EER dependence on the length of on-chip and off-chip SRAM PUF.

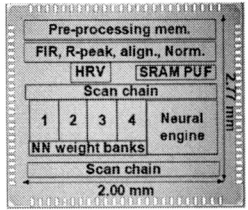

Fig. 7. 65nm LP CMOS prototype chip micrograph.

particular bit positions were selected where the average bit activity was closer to 0.5 than other positions of the ECG FV.

For on-chip SRAM PUF, a small on-chip 32×16 SRAM is employed to generate a 64-bit PUF response, which acts as a unique device signature. Initially, a 512-bit mask indicating the stable bit positions of the SRAM state is generated by accumulating the SRAM power-on state ('1' or '0') across 100 power-on instances. We considered a SRAM bit to be stable if the bit's accumulation result is higher than 65 or less than 35, to obtain a 64-bit PUF response. If the stability threshold values are more aggressive, the PUF stability will increase, but less than 64 PUF bits will be obtained from the same on-chip SRAM.

We compare the SRAM states against these stability threshold values and prune out unstable bits (Fig. 5).

In order to extract a larger number of PUF bits for different users with ECG features, we also employed an off-the-shelf SRAM chip (Alliance Memory AS6C62256-55SIN) and generated one 256-bit PUF vector from each 8K×8-bit SRAM memory block, using the data remanence approach proposed in [8]. In total, four 32K×8-bit off-the-shelf SRAM chips were used to generate 16 256-bit stable PUF vectors, which are bitwise XORed with 16 256-bit ECG FVs of 16 randomly selected subjects from the 741-subject database. The optimal remanence period was found to be between 60 ms and 100 ms for the four chips, which resulted in selection of 100% stable cells that are either strongly biased towards '0' or '1', all passing the NIST randomness tests.

C. Authentication and EER

To perform authentication (Fig. 1), the chip first operates in registration mode, where a 256-bit FV corresponding to the user (ECG/HRV) and device (SRAM PUF) is generated and stored. The 256-bit registration FV is obtained by bitwise XORing the 256-bit NN output average over 30 ECG beats of the user, the 64-bit average HRV value, and the 256-bit off-chip SRAM PUF (or 64-bit on-chip SRAM PUF) vector. During identification mode, using the same process, a 256-bit identification FV is obtained. Then, the normalized Hamming distance between registration FV and identification FV is computed and compared to a pre-defined threshold value to determine whether the user/device can be authenticated or not. To evaluate the effectiveness of this method, we have studied the distribution of Hamming distances for feature vectors obtained from the same and different users' raw ECG data, both of which were acquired over multiple times and days.

With the objective of minimizing EER, parameters such as SRAM PUF length, HRV bit-width, NN precision were

optimized through experimentation. For example, Fig. 6 shows the SRAM PUF length experimental results for the ECG⊕PUF scheme, where the EER decreased from 0.336% to 0.11% when the SRAM PUF vector size is increased from 64-bit (on-chip SRAM) to 256-bit (off-chip SRAM).

III. MEASUREMENT RESULTS

A prototype chip for the proposed hardware security engine was implemented in 65nm LP CMOS (Fig. 7). Clock-gated and time-multiplexed datapaths are used to optimize the latency and throughput of computing functions such as mean, variance, inverse square root, and weighted sum over time. The total on-chip memory is 64 kB, where 52 kB is used for NN weights. The chip occupies a total area of 7.54 mm^2 with 98 digital I/O pads for external communication.

Due to the limited number of 64-bit/256-bit PUF vectors (16) that we measured from on-/off-chip SRAMs, we used the ECG/HRV features of the 16 randomly selected subjects from the in-house 741-subject ECG database. 30 beats per subject were averaged to obtain one ECG FV. Four different starting points were used per subject to evaluate the time variance in the ECG signals. Fig. 8 shows the measured intra-class (intra-user/-device) and inter-class Hamming distance distributions, where the ECG⊕PUF⊕HRV scheme achieves 6.4X separation of μ. Fig. 9 shows that EER values (when false acceptance rate

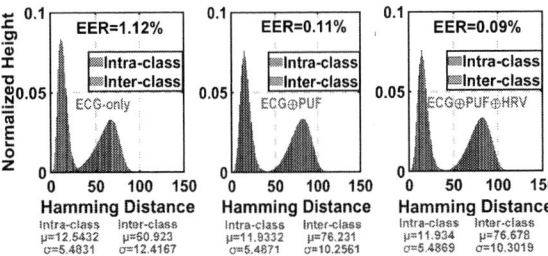

Fig. 8. The intra-class and inter-class Hamming distance distributions are shown for ECG-only, ECG⊕PUF and ECG⊕PUF⊕HRV schemes.

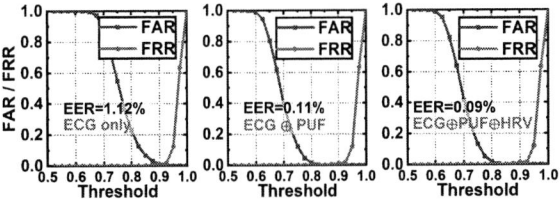

Fig. 9. FAR/FRR plots and EER (@FAR=FRR) values are shown for ECG-only, ECG⊕PUF and ECG⊕PUF⊕HRV schemes.

978-1-7281-5107-6/19 $31.00 © 2019 IEEE

IEEE Asian Solid-State Circuits Conference
November 4 – 6, 2019
The Parisian Macao, Macao SAR, China

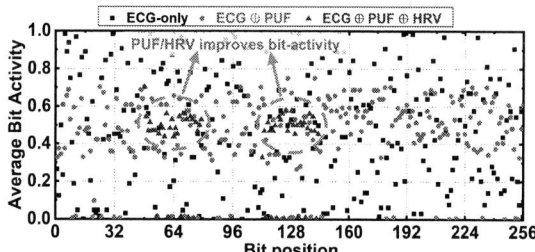

Fig. 10. The average bit activity is tightened to be ∼0.5 in the regions where SRAM PUF and HRV vectors are bitwise XORed with ECG FV.

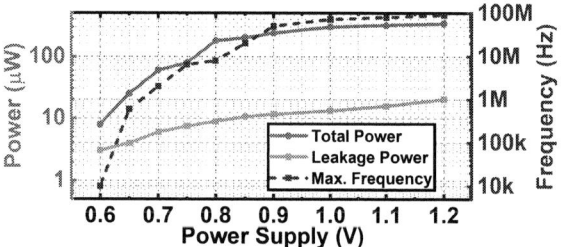

Fig. 11. Chip power consumption with voltage and frequency scaling.

(FAR) equals false rejection rate (FRR)) progressively improve from 1.12% (ECG-only) to 0.11% (ECG⊕PUF) to 0.09% (ECG⊕PUF⊕HRV), as we combine more entropy sources.

The average bit activity calculated from the feature vectors obtained from 16 randomly selected subjects with four different starting points of ECG input data is shown in Fig. 10 for each bit position. It is evident that the bit activity factor is improved to ∼0.5 (equal number of 0s and 1s) by introducing additional entropy sources such as SRAM PUF and HRV.

At 0.6V supply, the total processor power is 8.01 μW for maximum 10 kHz frequency (Fig. 11). For real-time ECG authentication operation at 0.6V, the frequency can be reduced down to 2.2 kHz, at which the total power is 4.04 μW.

Using the obtained secret key values with ECG-only, ECG⊕PUF and ECG⊕PUF⊕HRV schemes, we performed six NIST randomness tests [7]. For the selected 16 subjects, the NIST test results in Fig. 12 shows randomness improvement after combining ECG FV with SRAM PUF and HRV. It is also shown how different number of SRAM PUF bits from 64 to 256 affects the NIST test results. With 256-bit PUF length, ECG⊕PUF⊕HRV scheme passes all six NIST tests.

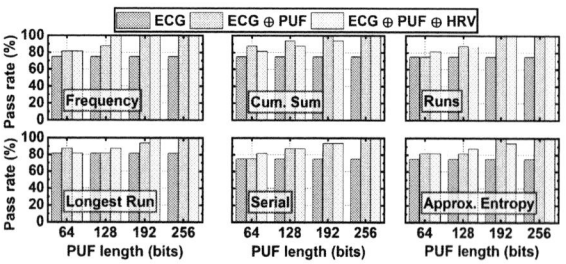

Fig. 12. NIST test pass rates for 16 randomly selected subjects. Employing multiple entropy sources, NIST test pass rates were improved to 100%.

Table I shows the comparison with prior works. Due to the slow rate of real-time ECG acquisition and pre-processing/NN, ECG schemes operate at a lower frequency and consume more area compared to conventional security engines [2]–[4], but conventional schemes do not exhibit biometric features. Compared to ECG-only work [5], the EER of our multi-factor authentication is 18.9X lower (0.09%) by trading off higher power, but the power is still sufficiently low even for wearable devices. To the best of our knowledge, our work is the first to integrate individually-unique biometric features (ECG, HRV) and device-unique PUF (SRAM states), substantially reducing the scope of attacks and enhancing the security of IoT devices.

TABLE I
COMPARISON WITH PRIOR WORKS.

	[2]	[3]	[4]	[5]	This work
Technology	65 nm	22 nm	65 nm	65 nm	65 nm
Supply Voltage	0.7-1.4	0.9 V	1 V	0.55 V	0.6 V
Area (mm²)	0.018	24	1	5.94	7.54
Power	N/A (15-26 fJ/bit)	25 μW	N/A (15-163 fJ/bit)	1.06 μW	4.04 μW
Clock Freq.	N/A	2 GHz	N/A	2 kHz	2.2 kHz
Memory Size	N/A	N/A	N/A	19.5 kB	64 kB
Entropy source	Inv. amp.	Delay/ SRAM	Sense amp.	ECG	ECG, HRV, SRAM
Silicon PUF	Yes	Yes	Yes	No	Yes
Biometric features	No	No	No	Yes	Yes
EER	0%	0%	0%	1.7%	0.09%

IV. CONCLUSION

This paper presents a new multi-factor authentication and secret key generation scheme based on three entropy sources of ECG features, HRV, and SRAM PUF values. The 65nm prototype chip consumes 4.04 μW at 0.6V for real-time user/device authentication. Optimally combining extracted features from individually-unique (ECG, HRV) and device-unique (SRAM PUF) entropy sources, a low equal error rate of 0.09% is achieved. For 16 randomly selected subjects, the generated 256-bit secret keys fully pass the NIST randomness tests.

REFERENCES

[1] D. E. Holcomb *et al.*, "Power-up SRAM state as an identifying fingerprint and source of true random numbers," *IEEE Transactions on Computers*, vol. 58, no. 9, pp. 1198–1210, 2009.

[2] D. Li and K. Yang, "A 562F² physically unclonable function with a zero-overhead stabilization scheme," in *IEEE ISSCC*, 2019.

[3] S. K. Mathew *et al.*, "A 0.19 pJ/b PVT-variation-tolerant hybrid physically unclonable function circuit for 100% stable secure key generation in 22nm CMOS," in *IEEE ISSCC*, 2014.

[4] A. Alvarez *et al.*, "15fJ/b static physically unclonable functions for secure chip identification with <2% native bit instability and 140× inter/intra PUF hamming distance separation in 65nm," in *IEEE ISSCC*, 2015.

[5] S. Yin *et al.*, "A 1.06 μW smart ECG processor in 65 nm CMOS for real-time biometric authentication and personal cardiac monitoring," in *IEEE Symp. on VLSI Circuits*, 2017.

[6] S. Yin *et al.*, "Designing ECG-based physical unclonable function for security of wearable devices," in *IEEE EMBC*, 2017.

[7] A. Rukhin *et al.*, "A statistical test suite for random and pseudorandom number generators for cryptographic applications," *National Institute of Standards and Technology*, 800-22 Rev 1a, 2010.

[8] M. Liu *et al.*, "A data remanence based approach to generate 100% stable keys from an SRAM physical unclonable function," in *IEEE/ACM ISLPED*, 2017.

978-1-7281-5107-6/19 $31.00 © 2019 IEEE

A Packaged Fully Digital 390GHz Harmonic Outphasing Transmitter in 28nm CMOS

Alexander Standaert, Patrick Reynaert
KU Leuven ESAT-MICAS
Kasteelpark Arenberg 10, 3001 Leuven, Belgium

Abstract—This paper presents a fully digital 390 GHz harmonic outphasing transmitter in a 28nm CMOS technology. The proposed architecture features the combination of phase predistortion and outphasing at the third harmonic enabling higher-order non constant envelope modulation schemes which is inherently difficult in multiplier-last architectures used in conventional THz data transmitters. The transmitter includes a 5-bit outphasing symbol generator at 15 GHz which can generate the necessary symbols for a variety of modulation schemes from OOK to STAR16 QAM. The transmitter is fully packaged with a micromachined waveguide flange such that it provides a standard waveguide output. It produces a peak output power of -16 dBm with a 3 dB IF bandwidth of 16.5 GHz and has a maximum measured datarate of 6 Gbps.

Index Terms—THz, Outphasing, Transmitter

I. INTRODUCTION

The need of high speed communication links requires an increasingly large bandwidth. The unallocated frequency spectrum above 275 GHz can provide this bandwidth but poses many challenges for the CMOS chips that form the transceivers for the channel. Conventional transmitter architectures that have an power amplifier(PA) as output stage don't work anymore as the lower Ft/Fmax of the CMOS devices prevent any power gain. Multiplier/mixer last architectures have been proposed as alternative . Due to their simplicity, multiplier last architectures are quite popular in literature[1][2]. They can provide descent output power but due to the inherent phase multiplication, most constellation schemes are distorted. For this reason, only modulation schemes such as OOK, BPSK or QPSK in combination with triplers are found in literature as their phases form the same constellation scheme with some symbol perturbation. Additionally, the multiplication also causes amplitude distortion due to the fact the the Pin Pout curve doesn't follow a slope of 1dB/dB. [3][4] resolves this by making use of cubic/ square mixer last architecture which in which the IF operates linearly, but in order to achieve the linearity requirements, power amplifiers are operating at back-off conditions which results in lower output power or the need for massive power combining.

This is not the only way to modulate amplitude efficiently however, in this design the authors propose to combine two phase modulated tripled signals and rely on the harmonic outphasing angle to efficiently perform the amplitude modulation. This way no power back-off is needed and the multipliers can be used to provide high output power and form higher-order modulation schemes.

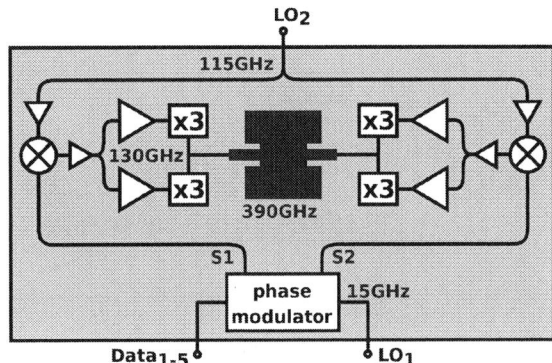

Fig. 1: High level system architecture

II. TRANSMITTER DESIGN

Figure 1 shows the system level architecture of the design. Two predistorted outphasing signals are generated from a 15 GHz LO signal using a phase modulator that can accept 5 parallel data streams. The outphasing signals are each up converted with a 115 GHz LO to a 130 GHz RF signals. These are combined back together after being amplified and tripled.

A. Phase Modulator

Given the importance of accurate phase in the architecture, the phase modulator is arguably the most imported building block of the design. An overview of its architecture can be seen in figure 2 and it can be functionally subdivided into two sections: First the different phases of the constellation diagram are generated, after which small permutations on these phases are applied to generate the outphasing angle for the amplitude modulation.

The phases of the constellation diagram are generated through a combination of a DLL and interpolator. A 15 GHz LO signal is feed to a 12-stage DLL giving a $30°$ phase shift per stage. Of those 12 phases, 5 ($0°$, $30°$, $60°$, $90°$, $120°$) are used to interpolate to 8 phases ($0°$, $15°$, $30°$, $45°$, $60°$, $75°$, $90°$, $105°$). Each of these phase can be selected by a 3-bit transmission gate multiplexer and after tripling, these phases are able to form a 8PSK signal. Due to mismatch between the different DLL stages and the limited accuracy of the phase detector, the phases of the DLL will naturally exhibits some errors which will be amplified after tripling. There for each phase passes through an 8-bit programmable phase shifter before going to

978-1-7281-5107-6/19 $31.00 © 2019 IEEE

IEEE Asian Solid-State Circuits Conference
November 4 – 6, 2019
The Parisian Macao, Macao SAR, China

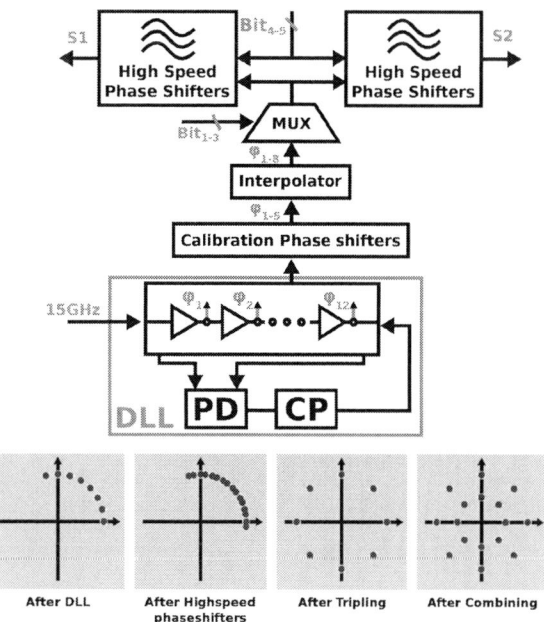

Fig. 2: Block diagram of the phase modulator and build up of constellation points

the interpolator. The phase shifter has a range of 10° at 15 GHz and is implemented as varactor controlled by an R2R DAC. After the interpolator, each of the phases can be selected using a high speed 3-bit multiplexer. The output of the multiplexer is split in three paths: one signal is brought off-chip for calibration and debugging purposes. The other two are each send to a high speed switchable phase shifters to generate the outphasing angle. The high speed phase shifts are implemented as delay lines with 8 switchable capacitors and can provide a total phase shift of 61° at 15 GHz. Because this phase shift gets tripled, it is sufficient to achieve full outphasing angles. The 8 capacitors are switched by two data bits and the ON/OFF state of the capacitors for each of the two bits is programmable. This allows for flexibility in the necessary outphasing angle for different modulation schemes: the modulation index for OOK is e.g different than that of PAM8.

B. RF/Packaging

The two phase modulated signals are upconverted with Gilbert mixers to 130 GHz using a 115 GHz LO signal which is provided from an off-chip source. The LO signal is brought on chip using a waveguide to bondwire transition similar as demonstrated in [5]. The upconverted signal is amplified and then tripled using two-way power combining (Figure 1). After tripling, the amplified signals are recombined and converted from a differential signal to a microstrip line (Figure 3). Finally, the two outphasing signals are recombined in the output coupler which can be seen in Figure 3. The output coupler is a differential patch which excites the fundamental mode in a

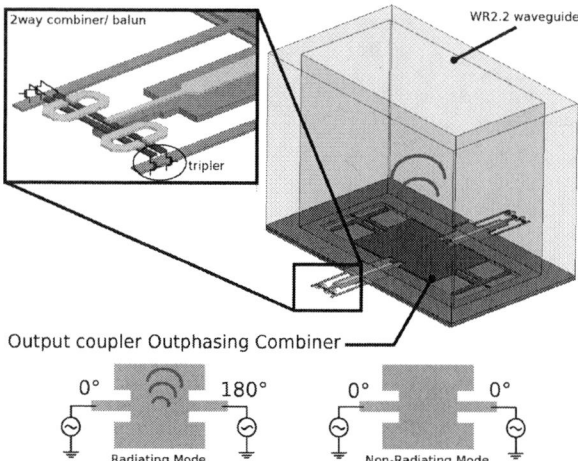

Fig. 3: Layout of the power combiner of the the tripler and the outphasing combiner/output coupler

WR2.2 waveguide. When this patch is excited differentially, it provides maximum output power. When an in-phase signal is applied, no power is coupled to the waveguide. In this design, the two outphasing paths are perfectly symmetrical, the 180° phase shift needed to excite the output coupler correctly is implemented in the input transformer/splitter of the 115 GHz LO signal. Due to the presence of bondwires and interference to the RF circuits, a WR2.2 waveguide flange can't practically be screwed against the chip. The chip is therefore packaged with a micromachined waveguide flange which provides a cavity where the bondwire can reside in (Figure 4).

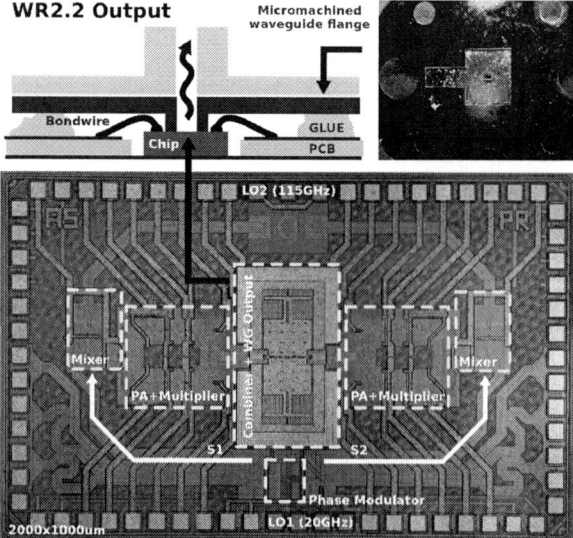

Fig. 4: Chip picture and Packaging of the output coupler

978-1-7281-5107-6/19 $31.00 © 2019 IEEE   150

IEEE Asian Solid-State Circuits Conference
November 4 – 6, 2019
The Parisian Macao, Macao SAR, China

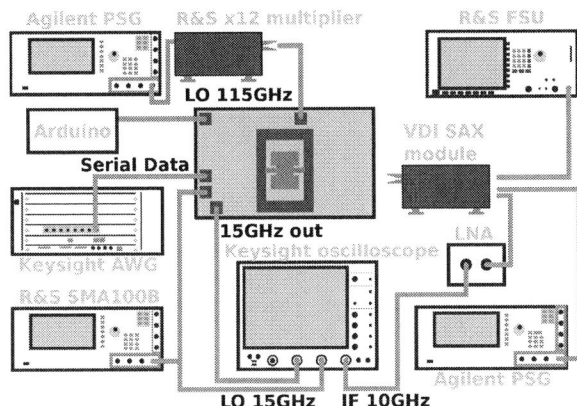

Fig. 5: Measurement Setup

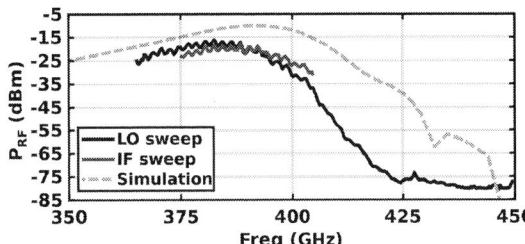

Fig. 6: Output power in function of LO frequency (IF = 15 GHz) and IF frequency (LO = 115 GHz)

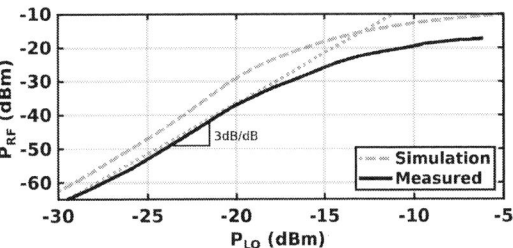

Fig. 7: Output power at 390 GHz in function of the 115 GHz LO input power

III. MEASUREMENTS

The outphasing transmitter was fabricated in a 28nm CMOS technology and has an area of 2 by 1mm (Figure 4). The measurement setup and a diagram of its configuration is shown in Figure 5.

A. Continuous Wave Measurements

Figure 6 shows the output power in function of the frequency for both the 115 GHz LO and the 15 GHz IF signal. The power is measured with the VDI SAX configured as a 2port mixer and connected to a spectrum analyser. This setup is calibrated with a VDI SGX and Erikson power meter. The Peak measured output power is −16 dBm. The IF has a 3 dB bandwidth of 16.5 GHz and the 3 dB LO bandwidth is 20 GHz. The output power at 390 GHz versus the input power of the 115 GHz LO is depicted in Figure7. The LO power is varied with a tuneable D-band attenuator. Simulated data is used to deembbed the waveguide to (bondwire) chip transition to get the LO power at the input of the chip. An under estimation of this loss might very well be the cause of the discrepancy between simulated and measured Pin in Pout curves. The measured output power clearly shows the the sloop of 3dB/dB which is to be expected of a tripler last system. The measurements of Figures 6 and 7 are shown for an outphasing angle of 0°.

In Figure 8 the outphasing performance is evaluated by measuring the phase and amplitude of the transmitter in function of outphasing angle. The amplitude can easily be measured with the spectrum analyzer. The phase however, is not very trivial to measure. The conventional way to measure the phase (Figure 8b) would be to down convert the signal and compare it to a common reference. Because of drift and phase noise between all of the different pieces of measurement equipment this technique proves to be very inaccurate. The measurement is therefore done by switching the outphasing phase shifter of the transmitter between a reference angle and a outphasing angle (at very low data rates) and demodulating

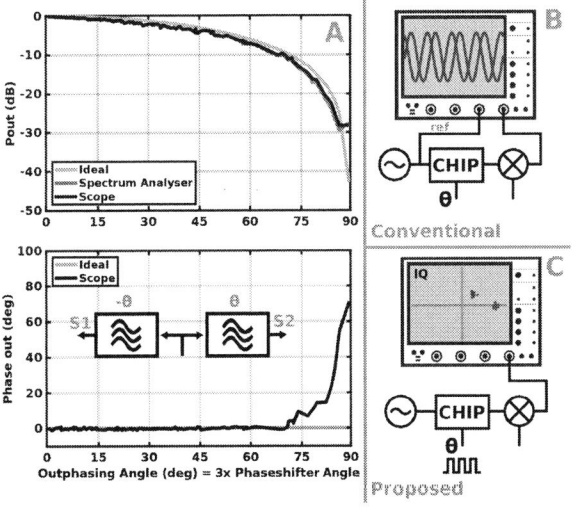

Fig. 8: Amplitude and phase at 390 GHz in function of the simulated outphasing angle of high speed phase shifters

the signal in the IQ domain (Figure 8c). The amplitude and phase can be determined compared to the reference much more reliable and accurate. The transmitter follows the ideal curve very well both in amplitude and phase. For outphasing angles greater then 80° the error in phase increases rapidly which is probably due to a slight difference in output power between the two outphasing paths.

978-1-7281-5107-6/19 $31.00 © 2019 IEEE

IEEE Asian Solid-State Circuits Conference
November 4 – 6, 2019
The Parisian Macao, Macao SAR, China

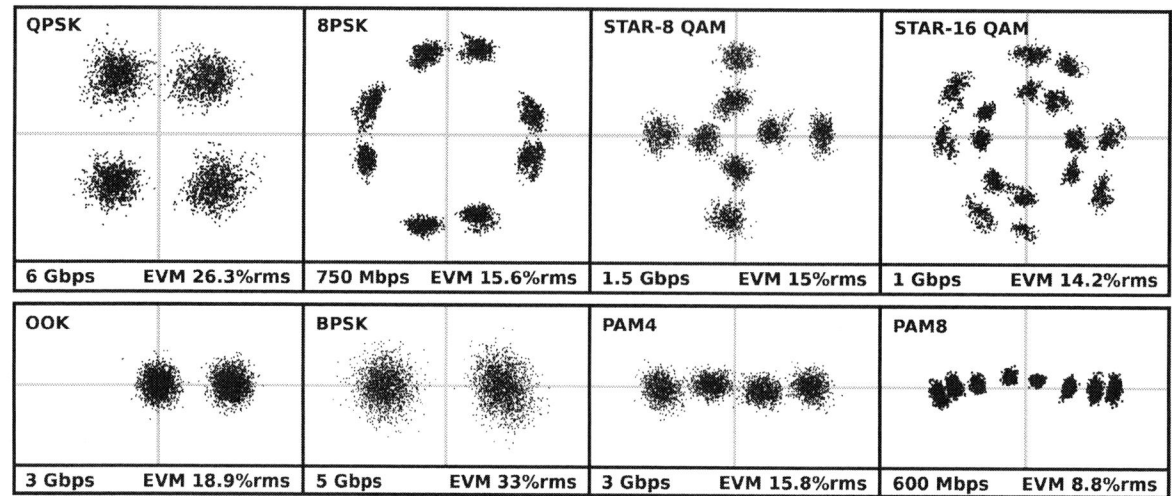

Fig. 9: Various measured constellation diagrams

TABLE I: STATE-OF-THE-ART IN TRANSMITTERS AROUND 400GHz

	This work	[3]	[4]	[2]	[1]
Technology	28nm CMOS	40nm CMOS	40nm CMOS	0.13um SiGe	65nm CMOS
Freq (GHz)	390	275-305	300	434	240
Pout (dBm)	-16	-14.5	-5.5	-18.5	0
Pdc (W)	1.1 (250mW BB, 850mW RF)	1.4	1.4	0.12	0.22
Data rate(Gb/s)	6	17.5x6	105	10	16
Integration	Baseband, RF, Packaging	RF	RF	ASK modulator, RF	QPSK modulator, RF
RF Output	Waveguide flange	GSG pad	GSG pad	antenna	antenna

B. Data Measurements

To perform the data measurements, the VDI SAX was configured as a 3port mixer and the 390 GHz RF signal from the chip was down converted to 10 GHz. This IF signal was amplified and feed to the scope where it could be demodulated using VSA software. Figure 9 shows the constellation diagrams and corresponding error vector magnitude for various modulation schemes. The results each contain 4096 symbols and no equalization was used. An issue in the interpolator causes the constellation points of 8PSK and STAR-16 QAM to be a bit skewed. The maximum measured datarate was 6 Gbps using a QPSK modulation scheme.

IV. CONCLUSION

A 390 GHz harmonic outphasing transmitter was presented as alternative way to generate complex modulated signals above fmax/ft while still generating adequate output power. Comparing to other transmitters that can produce higher order modulation schemes [3][4], this work can achieve similar output powers at a lower DC power consumption even though the frequency is quite a bit higher and the chip is fully packaged and not probed.
A 5-bit outphasing symbol generator was included onchip capable of generating a variety of modulation schemes although

its implementation is limiting the data performance compared to literature.

ACKNOWLEDGMENT

The authors would like to thank Mike Keaveney and Analog Devices, Limerick, Ireland for supporting this project. This work was also supported by the CHIST-ERA project "WISDOM" under FWO grant G0H5816N.

REFERENCES

[1] S. Kang, S. V. Thyagarajan, and A. M. Niknejad, "A 240ghz wideband qpsk transmitter in 65nm cmos," in *2014 IEEE Radio Frequency Integrated Circuits Symposium*, June 2014, pp. 353–356.

[2] S. Hu, Y. Xiong, B. Zhang, L. Wang, T. Lim, M. Je, and M. Madihian, "A sige bicmos transmitter/receiver chipset with on-chip siw antennas for terahertz applications," *IEEE Journal of Solid-State Circuits*, vol. 47, no. 11, pp. 2654–2664, Nov 2012.

[3] K. Katayama, K. Takano, S. Amakawa, S. Hara, A. Kasamatsu, K. Mizuno, K. Takahashi, T. Yoshida, and M. Fujishima, "20.1 a 300ghz 40nm cmos transmitter with 32-qam 17.5gb/s/ch capability over 6 channels," in *2016 IEEE International Solid-State Circuits Conference (ISSCC)*, Jan 2016, pp. 342–343.

[4] K. Takano, S. Amakawa, K. Katayama, S. Hara, R. Dong, A. Kasamatsu, I. Hosako, K. Mizuno, K. Takahashi, T. Yoshida, and M. Fujishima, "17.9 a 105gb/s 300ghz cmos transmitter," in *2017 IEEE International Solid-State Circuits Conference (ISSCC)*, Feb 2017, pp. 308–309.

[5] K. Guo, Y. Zhang, and P. Reynaert, "A 0.53-thz subharmonic injection-locked phased array with 63- μ w radiated power in 40-nm cmos," *IEEE Journal of Solid-State Circuits*, vol. 54, no. 2, pp. 380–391, Feb 2019.

978-1-7281-5107-6/19 $31.00 © 2019 IEEE

A Fully Integrated 27.5-30.5 GHz 8-Element Phased-Array Transmit Front-end Module in 65 nm CMOS

An'an Li[#*], Yingtao Ding[#], Zipeng Chen[*], Wei Wang[*], Sijia Jiang[*], Shiyan Sun[#*], Zhiming Chen[#], and Baoyong Chi[*]

[#]School of Information and Electronics, Beijing Institute of Technology, 100081, China
[*]Institute of Microelectronics, Tsinghua University Beijing, 100084, China
Email: czm@bit.edu.cn, chibylxc@tsinghua.edu.cn

Abstract—A fully integrated 27.5-30.5 GHz 8-element phased-array transmit front-end module in 65 nm CMOS is presented. The module is based on the all-RF phased-array architecture and consists of 8 phased-array elements. One 1:8 power splitter divides the input power into 8 elements after the driver stages. Each element consists of one 5-bit digital attenuator, one 5-bit phase shifter, one extra 0/180° switching amplifier, one power amplifier and some buffer stages between the above stages to compensate the gain loss. A 5-bit digital attenuator with amplitude and phase calibration is presented to provide 15.5 dB attenuation coverage with 0.5 dB step. A 180° reflection-type phase shifter, along with an 0/180° switching amplifier is proposed to realize 360° phase shifting coverage with 5.625° step. Amplifiers, including power amplifier, 0/180° switching amplifier and buffer stages, adopt wideband inter-stage matching technique to ensure 27.5-30.5 GHz bandwidth coverage. All of the features are digitally controllable. The measurements show that the module achieves <-10 dB input return loss, about 36 dB gain across 27.5-30.5 GHz and delivers about 12.71 dBm output-referred 1-dB compression point (OP1dB) power at 29 GHz. The module achieves 2.61° RMS (root mean square) phase error with 1.26 dB RMS gain variation across 27.5-30.5 GHz from the phase shifter and 0/180° switching amplifier. The attenuator achieves 0.26 dB RMS amplitude error with 2.79° RMS phase variation. The chip occupies $6.95 \times 3.17\ mm^2$ area and consumes 1.66 A from the 1.2 V power supplies.

Keywords— Phased-array, phase shifter, attenuator, power amplifier, CMOS, transmit front-end module.

I. INTRODUCTION

Phased arrays system contains thousands of transmit/receive (T/R) front-end modules to achieve fast beam-forming and avoid the interference from other directions, hence increasing signal-to-noise ratio (SNR) and/or improving channel communication capacity [1]. Therefore, phased arrays are widely used in satellite communication and radar applications. Matured GaAs technology, together with silicon-based digital control block, is looked as a competitive scheme [2]. However, it is extremely costly. As the silicon technologies are getting developed, silicon-based T/R front-end modules have been introduced as a lower cost solution for such systems due to its high integration density. This paper focus on the implement of the Si-based transmit front-end module.

Transmit front-end module consists of three basic functions, including signal amplification, phase shifting, and amplitude/gain controlling. Precise amplitude and phase controlling are critical in millimetre-wave phased array system. [3] proposed a fully X-band transceiver chip in 0.18-μm CMOS. It adopted the switched PI/T topology and switch-embedded passive-type to realize 6-bit attenuator and 6-bit phase shifter, respectively. [4] presents a X-band phased-array T/R chipset in 0.13-μm CMOS. Switch-embedded T-type topology and high pass/low-pass network based on the SPDT and DPDT switches are utilized to realize 5-bit attenuator and 6-bit phase shifter. Due to the switch parasitic, the attenuators in these work achieve RMS amplitude error of 0.25-0.3 dB with large RMS phase variation (>7°) in X-band. The challenge in amplitude/phase controlling is increased when the operation frequency is further improved. [5] introduced a Ka-band CMOS chipset. It employs high/low-pass network and PI/T topology based on the SPDT/DPDT switch to realize the phase shifter and attenuator, respectively, while only achieving the RMS phase error of 7° and RMS amplitude error of 0.5 dB across 27-29 GHz.

This paper presents a fully integrated 27.5-30.5 GHz 8-element phased array transmit front-end module in 65 nm CMOS. In order to improve the accuracy of amplitude control, the amplitude/phase calibration technique is proposed in the attenuator. The 0/180° switching amplifier relaxes the phase shifting coverage requirement of the passive reflection type phase shifter (RTPS), result in higher resolution. The amplifiers, including power amplifier (PA), 0/180° switching amplifier and buffer stages, utilize the wideband inter-stage matching to ensure the bandwidth. The measurements show that the proposed transmit front-end module achieve competitive performance.

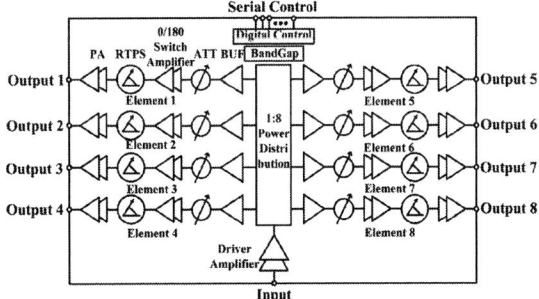

Fig. 1. Block diagram of proposed 27.5-30.5 GHz 8-element phased-array transmit front-end module.

II. CIRCUIT DESCRIPTIONS

A. System Architecture

Fig. 1 shows the block diagram of the proposed transmit front-end module. The module mainly consists of a 2-stage driver amplifier (DA), a 1:8 power splitter network and 8 phased-array elements. The on-chip bandgap-based biasing circuit provides the dc bias for various blocks. Digital control units, including the SPI interface and integrated lookup tables (LUTS) for configuration, generate the needed digital control bits. Each element is composed of one 5-bit digital attenuator, one 5-bit RTPS phase shifter, one extra 0/180° switching amplifier, one power amplifier and some buffer stages between the above stages to compensate the gain loss. All active blocks (mainly amplifiers) adopt differential structure and interface with 50-Ω input and output ports so as to allow for individual design and ease the cascaded connection between each other. All passive components, such as the transformer for impedance matching, RTPS, and the key routing path in the attenuator, are simulated with a full-wave electromagnetic tool to ensure the design accuracy.

B. Amplifiers

All the amplifiers adopt differential common-source amplifier stages to provide sufficient gain and reject the common-mode interference. However, the poor reverse isolation caused by the parasitic gate-drain capacitance C_{gd} of the transistors induces the stability risk for the amplifier. The neutralization capacitors are introduced to address this issue and eliminate the influence as much as possible, resulting in improved circuit stability and gain performance. Fig. 2 shows the schematic of the proposed driver amplifier and power amplifier. Two amplifier stages are required in these two blocks. 2-bit switched-capacitor array are inserted in the input and output impedance matching network to provide the frequency tuning capability and overcome the process variation.

In the DA, the neutralization capacitor uses the parasitic capacitance of a transistor. The transistor size is 90% of the input device to avoid over-compensation. A big source degeneration resistor makes its trans-conductance negligible. Both the input and output impedance matching network use the conjugation matching method based on the transformer to realize maximum gain and single-ended-to-differential

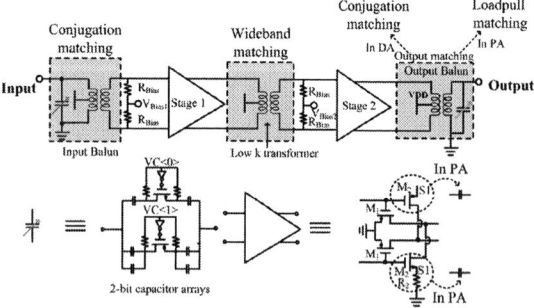

Fig. 2. Schematic of the driver amplifier and power amplifier.

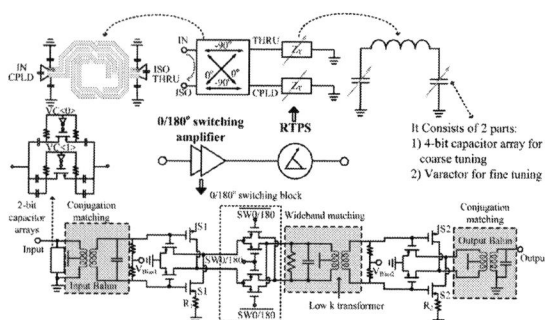

Fig. 3. Schematic of the 0/180° switching amplifier and the RTPS phase shifter.

conversion, while the inter stage matching adopts the wideband matching method based on low k transformer to ensure >5 GHz 1 dB bandwidth in the design. The DA consumes 51.7 mA current from one 1.2 V power supply and provides a simulated gain of 17.5 dB and 1dB bandwidth of 27-32 GHz.

The PA has the same topology as the DA, but with different component values. Load-pull matching method is used in the output to provide the maximum output power. MIM capacitors are introduced to replace the transistor parasitic capacitor as the neutralization capacitor, since large neutralization transistor brings large trans-conductance and introduces the compensation risk. The PA consumes 120 mA DC current from a 1.2 V power supply and provides 21.3 dB gain with >5 GHz 1 dB bandwidth while achieving >12dBm output-referred 1-dB compression point output power.

C. 0/180° Switching Amplifier and RTPS Phase Shifter

The phase shifter is composed of one 0/180° switching amplifier and one RTPS phase shifter. Fig. 3 shows the schematic of the 0/180° switching amplifier and the RTPS phase shifter. A 0/180° switching block is inserted after the first stage to realize 180° phase shifting. The 0/180° switching amplifier adopts the same amplifier topology and matching method as the DA, as described in Subsection II.B. It consumes 22.38 mA current and provides 12.03 dB gain.

The 0/180° switching amplifier relaxes the phase shifting coverage requirement for the RTPS phase shifter and provides some gain to compensate the gain loss. At this case, the RTPS phase shifter only need provide 180° phase shifting range. As shown in Fig. 3, the RTPS phase shifter consists of one 3dB quadrature coupler and two PI-type loads. As introduced in [6], the targeted phase shifting is realized by adjusting the termination impedance at the through port and coupled port, since the reflection coefficient variation of the through port and coupled-port leads to varied phase shifting between the input port and the isolated port in the 3dB quadrature coupler. In this work, the impedance load adjustment is realized by 2 parts, including the 4-bit MOM switched-capacitor array for coarse tuning and varactor for fine tuning. The control voltage of the varactor comes from one 3-bit DAC. The RTPS phase

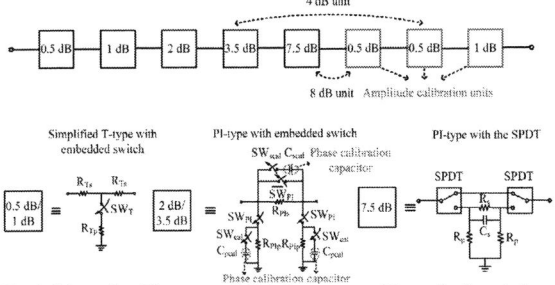

Fig. 4. Schematic of the proposed digital attenuator with amplitude and phase calibration.

shifter can cover above 200° phase shifting range in the simulation. 7-bit control bit is sufficient to achieve 6-bit resolution for the RTPS phase shifter, since the lower phase shifting coverage requirement due to the 0/180° switching amplifier is beneficial to the phase shifting resolution. An integrated LUTS is utilized to simplify the control bit generation for the RTPS phase shifter.

D. 5-Bit Digital Attenuator

Fig. 4 shows the schematic of the proposed digital attenuator with amplitude and phase calibration. It uses three kind of attenuator topologies, including simplified T-type unit with embedded switch for 0.5 dB and 1 dB attenuation, PI-type unit with embedded switch for 2 dB and 3.5 dB attenuation and PI-type unit with the SPDT switch for 7.5 dB attenuation. The attenuator employs one PI-type unit with the SPDT switch to replace the conventional PI-type unit with embedded switch to realize 7.5dB attenuation, since the SPDT can improve the cascade matching performance and alleviate high inserted loss issue while maintaining compact chip area. A different unit cascade strategy is proposed, consisting of 0.5 dB, 1 dB, 2 dB, 3.5 dB and 7.5 dB attenuation units, together with three amplitude compensation units to realize the amplitude compensation. The extra calibration switched-capacitors are inserted in the 2 dB and 3.5 dB attenuation units to form the high-pass or low-pass network and compensate the phase variation. This amplitude/phase calibration technique improves the circuit robustness against the process variation. The simulation at the 0 dB reference state of the attenuator achieves 12.62 dB insertion loss at 29 GHz, which is improved by about 1.8 dB by using the PI-type unit with the SPDT switch, compared with the convention embedded switch

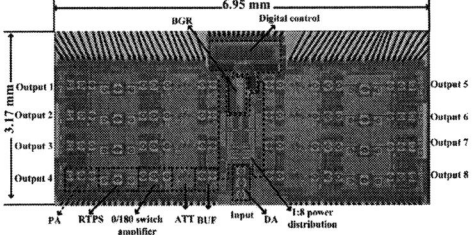

Fig. 5. Microphotograph of the proposed 8-element phased array transmit front-end module.

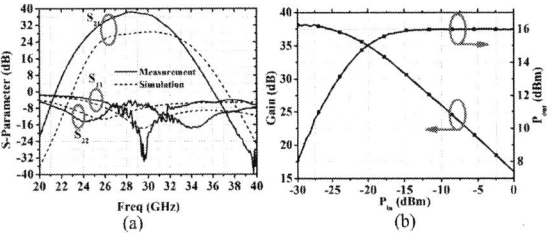

Fig. 6. Simulated and measured s-parameter and measured OP1dB of the proposed transmit front-end module for Element #1 at reference attenuation state. (a) s-parameter (b) OP1dB.

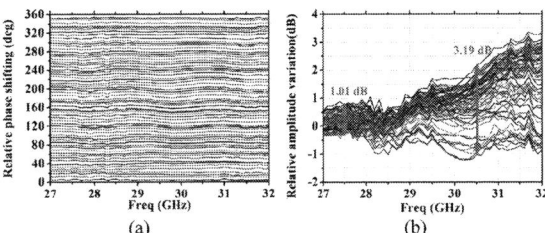

Fig. 7. Measured relative phase shifting and amplitude variation at 64 states of Element #1 for the RTPS phase shifter and 0/180° switching amplifier when the attenuator is set at the reference state.

topology. Thanks to the amplitude and phase calibration technique, the RMS amplitude error and phase variation of the attenuator are smaller than 0.4 dB and 2.55° over the 27.5 GHz-30.5 GHz, respectively.

III. MEASUREMENT RESULTS

The Ka-band 8-element phased array transmit front-end module has been implemented in 65 nm CMOS and the microphotograph is shown in Fig. 5. The chip size is 6.95 × 3.17 mm^2 including the pads. The transmit front-end module is measured via on-wafer probing by using Agilent N5245A network analyzer. It consumes 1.66 A current from 1.2 V power supplies.

Fig. 6 shows the simulated and measured s-parameter and OP1dB of Element #1. Most of the measurements agree well with the simulations. As shown in Fig. 6 (a), input return loss of Element #1 is better than 10 dB over 27.5-30.5 GHz. The output return loss is better than -5.55 dB since the output matching network of the PA adopts the load-pull design method. About 36 dB gain is achieved across 27.5-30.5 GHz. As shown Fig. 6 (b), the transmit front-end module delivers about 12.71 dBm OP1dB power at 29GHz, where the simulated OP1dB of the PA is about 12.84 dBm.

Fig. 7 shows the measured relative phase shifting and amplitude variation at 64 states of Element #1 for the RTPS phase shifter and 0/180° switching amplifier when the attenuator is set at the reference state. The transmit front-end module realizes the 360° phase shifting coverage with the step of 5.625°. The relative amplitude variation is 1.01 dB@27.5 GHz and 3.19 dB@ 30.5 GHz.

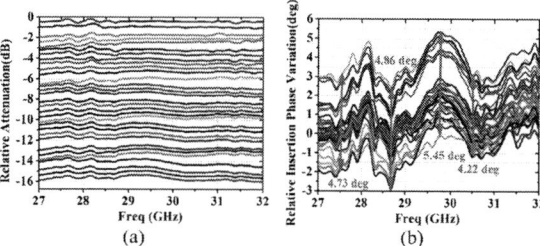

Fig. 8. Measured relative attenuation and insertion phase variation of Element #1 for the digital attenuator when the phase shifter is set at the reference state.

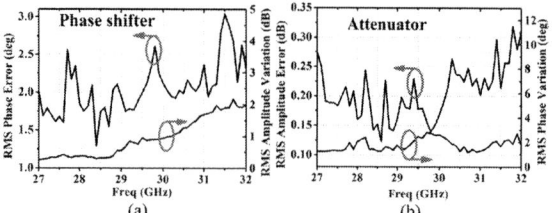

Fig. 9. Measured RMS amplitude error/variation and RMS phase variation/error as varying (a) the phase states and (b) the attenuation states.

Fig. 8 shows the measured relative attenuation and relative insertion phase variation of Element #1 for the digital attenuator when the phase shifter is set at the reference state. The attenuator achieves 15.5 dB attenuation range with 0.5 dB step. The insertion phase variation is less than 5.45° across 27.5-30.5 GHz.

Fig. 9 (a) shows the measured relative phase error and amplitude variation of Element #1 when the attenuator is set at the reference state. The phase shifter achieves less than 2.61° RMS phase error and <1.26 dB RMS amplitude variation across 27.5-30.5 GHz, respectively. Fig. 9 (b) shows the measured RMS amplitude error and RMS phase variation of Element #1 when the phase shifter is set at the 0° reference state. The attenuator achieves the RMS amplitude error of less than 0.26 dB and less than 2.79° RMS phase variation across 27.5-30.5 GHz. The performance in other elements is similar to the performance in Element #1 in this work.

Table 1 compares the measured performance of the proposed phased-array transmit front-end module with the-state-of-the-art. It shows that the proposed transmit front-end module shows the competitive RMS amplitude error, RMS phase error and the output 1dB compression point compared with the others.

IV. CONCLUSION

This paper presents a 27.5-30.5 GHz 8-element phased array transmit front-end module. The transmit front-end module has been implemented in 65 nm CMOS. It consumes $6.95 \times 3.17~mm^2$ die area and 1.99 W power consumption. The measurement shows good input return loss. The module achieves about 36 dB gain across 27.5-30.5GHz and about 12.71 dBm OP1dB output power. The transmit front-end module shows the RMS amplitude error of 0.26 dB with 2.79°

Table 1 Performance summary and comparison.

Ref.	This work*	[4]	[5]	[7]
Technology	65 nm CMOS	0.13 μm CMOS	65 nm CMOS	SiGe BiCMOS
Frequency (GHz)	27.5-30.5	8.5-10.5	26-32	28-32
Gain (dB)	36	3.5	>10	12
Output P1dB (dBm)	12.71	6.5	-2	10.5
Phase shifting range/Step (°)	360/5.625	360/5.625	348/11.25	360/5.625
RMS Phase Error/ RMS Amplitude Variation (Phase Shifter)	2.61°/1.26 dB	4.3°/0.8 dB	7°/0.3 dB	6°/NA
Attenuation range/Step (dB)	15.5/0.5	31/1	31/1	14/1**
RMS Amplitude Error/RMS Phase Variation (Attenuator)	0.26 dB/2.79°	0.33 dB/7.4°	0.5 dB/5°	0.8/NA
Chip Area Per element (mm^2)	1.28	1.19	0.68**	1.29**
Power Consumption Per element (mW)	249***	150	237.5***	200

*The Element #1. **Estimated from the paper. ***Total power consumption/Number of elements.

phase variation from the attenuator. The proposed RTPS phase shifter, together with the 0/180° switching amplifier, realizes the RMS phase error of 2.61° and RMS amplitude variation of 1.26 dB across the band. The module shows competitive performance compared with the-state-of-the-art.

REFERENCES

[1] H. L. Van Trees, *Optimum Array Processing. Part IV of Detection, Estimation, and Modulation Theory*. New York: Wiley, 2002.

[2] S. Hong, S.-G. Kim, C. T. Rodenbeck, and K. Chang, "A multiband, compact, and full-duplex beam scanning antenna transceiver system operating from 10 to 35 GHz," *IEEE Trans. Antennas Propag.*, vol. 54, no. 2, pp. 359–367, Feb. 2006.

[3] K. Gharibdoust, N. Mousavi, M. Kalantari, M. Moezzi, and A. Medi, "A fully integrated 0.18-μm CMOS transceiver chip for X-band phased-array systems," *IEEE Transactions on Microwave Theory and Techniques*, vol. 60, no. 7, pp. 2192–2202, July 2012.

[4] S. Sim, L. Jeon, and J. Kim, "A compact X-band bi-directional phased-array T/R chipset in 0.13μm CMOS technology," *IEEE Transactions on Microwave Theory and Techniques*, vol. 61, no. 1, pp. 562–569, Jan 2013.

[5] J. Han, J. Kim, J. Park, and J. Kim, "A Ka-band 4-ch bi-directional CMOS T/R chipset for 5G beamforming system," in *2017 IEEE Radio Frequency Integrated Circuits Symposium (RFIC)*, June 2017, pp. 41–44.

[6] R. Garg and A. S. Natarajan, "A 28-Ghz low-power phased-array receiver front-end with 360° RTPS phase shift range," *IEEE Transactions on Microwave Theory and Techniques*, vol. 65, no. 11, pp. 4703–4714, Nov 2017.

[7] K. Kibaroglu, M. Sayginer, and G. M. Rebeiz, "A quad-core 28–32 GHz transmit/receive 5G phased-array IC with flip-chip packaging in SiGe BiCMOS," in *2017 IEEE MTT-S International Microwave Symposium (IMS)*, June 2017, pp. 1892–1894.

12-3 (7199)

IEEE Asian Solid-State Circuits Conference
November 4 – 6, 2019
The Parisian Macao, Macao SAR, China

A 2Mbps sub-100μW Crystal-less RF Transmitter with Energy Harvesting for Multi-Channel Neural Signal Acquisition

Heng Huang*, Milin Zhang*, Guolin Li*, Zhihua Wang[†]

*Department of Electronic Engineering, [†]Institute of Microelectronics, Tsinghua University, China
Corresponding author email: zhangmilin@tsinghua.edu.cn

Abstract—This paper presents a 427MHz OOK modulated transmitter with energy harvesting for multi-channel neural acquisition. A power efficient OOK modulator is proposed integrated a 3-input AND gate array to drive the power transistors. The requirement on the driving capability of the VCO output is greatly reduced. A 9-stage ring oscillator based VCO is designed with PLL and edge-combining technology. A crystal based 6.78MHz oscillator is combined with the wireless power transmitter coil to power up the entire transmitter design, as well as to generate a reference frequency to the transmitter, resulting a crystal-less transmitter design. The proposed design was fabricated in 180nm technology, occupying a silicon area of only 0.26mm². Experimental results show a power consumption of 96μW with a -19dBm output. A 48pJ/bit efficiency is achieved under a data rate of 2Mbps.

Index Terms—Energy havering, wireless power transmission, transmitter, neural signal acquisition, ultra-low-power, ring oscillator, edge-combining, phase locked loop.

I. INTRODUCTION

WIRELESS multiple channel analog signal acquisition devices are strongly required in in-vivo neural signal acquisition experiments using freely-moving animal subjects. The sampling rate for the acquisition of the deep brain signals, i.e. action potentials (APs),or local field potentials (LFPs), is typically up to several tens of kS/s per channel with a resolution of 10 to 12 bits. A total throughput up to Mbps is required while tens of channels are applied in the experiments. In addition, the total load can be carried by the subjects are usually limited to several grams while small animals, i.e. mice, rat, are applied. The requirement of small volume size and long batter life becomes a paradox in the design.

Energy harvesting (EH) has been widely used as a promising solution to ultra-low-power data communication with limited volume size restriction. Various energy sources have been utilized in literature for energy harvesting, such as RF power [1, 2], thermal [3], and coil coupling power [4]. The RF-powered EH and EH using thermoelectric are widely used in wireless sensing interface designs and sensor network applications. The total output energy is usually at the level of Micro-watts, which limits its potential application scenarios. On the other hand, wireless power transmission (WPT) integrated inductive coupling coils, enabling an output energy at the level of Milli-Watts, is usually applied to application with a higher power budget requirement, i.e. multiple channel neural signal

acquisition. In addition, the output voltage and efficiency of the WPT is higher than RF powered EH.

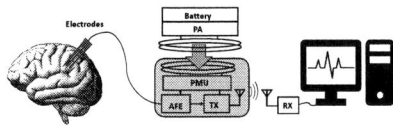

Fig. 1. A block diagram of a wireless neural signal acquisition system with WPT. The implantable circuit integrated an analog front end (AFE), a power management unit (PMU) and a transmitter (TX). A battery powered energy source with a power amplifier (PA) is used to power up the implantable module. A receiver (RX) is used to collect data sent out from the TX.

In the application multiple channel neural signal acquisition, wireless data transmission capability is usually required for experiments operated on small freely-moving animal subjects. The power consumption of the transmitter is curial since the total amount of data to be transmitted is as high as Mbps. Fig.1 illustrated a block diagram of a wireless neural signal acquisition system with WPT. The implantable circuit integrated an analog front end (AFE), a power management unit (PMU) and a transmitter (TX). A battery powered energy source with a power amplifier (PA) is used to power up the implantable module. A receiver (RX) is used to collect data sent out from the TX.

In order to reduce the power consumption of the TX, simple modulation schemes, such as On-Off Keying (OOK), or Binary Phase Shift Keying (BFSK), have been widely used in literature [1–3]. [2] is a traditional PLL based design with a low power supply voltage. However, two ring oscillators are required causing a relative high power consumption. [3, 5] proposed to reduce the power consumption by implementing TX without a phase locked loops (PLL), but with an injection-locked oscillator operated in a lower frequency with edge-combing technology. However, an off-chip crystal is required and an extra circuit is used to generate reference. [1] adopt low voltage and injection-locked divider and multiplier. two locked ring increase power consumption.

This paper proposed a Ultra-low-power crystal-less transmitter with energy harvesting for multi-channel neural signal acquisition. it is based on PLL and frequency multiplication to reduce power. The rest of the paper is organized as follows. Section II introduces the architecture of the proposed

978-1-7281-5107-6/19 $31.00 © 2019 IEEE   157

IEEE Asian Solid-State Circuits Conference
November 4 – 6, 2019
The Parisian Macao, Macao SAR, China

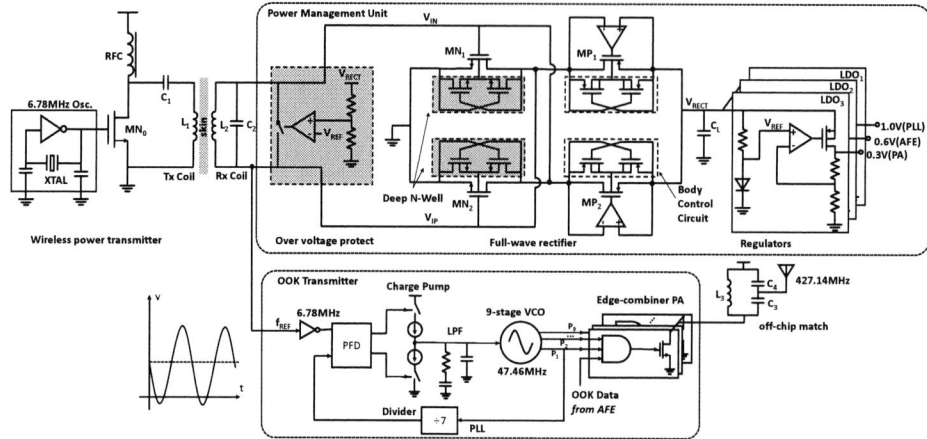

Fig. 2. Architecture of the proposed power management unit (PMU) and transmitter(TX). A rectifier is designed for energy harvesting from the receiver coil. Three LDOs are implemented to generate three different supply voltage for the PLL, AFE and PA, respectively.the proposed OOK modulated transmitter consists of a phase frequency detector(PFD), a charge pump(CP), a low pass filter(LPF), a frequency divider, a VCO and edge-combining power amplifier(PA)

transceiver, as well as a detailed analysis on the key modules in the design. Section III illustrates the experimental results, while Section IV concludes the entire work.

II. ARCHITECTURE AND PROPOSED SYSTEM

Fig.2 illustrated system overview of the proposed design. It consists of an OOK modulated transmitter and a power management unit (PMU). An crystal based 6.78MHz oscillator is combined with the wireless power transmitter coil to power up the entire transmitter design, as well as to generate a reference frequency to the transmitter, resulting a crystal-less transmitter design. The off-chip components for the transmitter are only two parts: 1) the RX coil for energy harvesting, and 2) the off-chip matching circuit to suppress the out-of-band spurs for data transmission at 427.14MHz.

A. Design of the Power Management Unit

The PMU integrated a full-wave CMOS rectifier and three regulators as illustrated in the top-right of Fig.2. The full-wave rectifier consists of four power transistors, a body voltage control circuit and two comparators. In order to avoid the latch-up effect, a cross gate control circuit, as shadowed in Fig.2 is employed to select either the highest voltage form the PMOS body or the lowest voltage form the NMOS body. The NMOS in the cross gate control circuit is implemented in the deep-N-well. In order to further improve the efficiency of the power converter, the comparator is designed to control PMOS switching. The simulated maximum efficiency of the rectifier was around 87%. Three LDOs are implemented with three different supply voltages outputs, 1.0V for the phase locked loop (PLL), 0.6V for the AFE and 0.3V for the power amplifier (PA) module.

B. Design of the OOK modulated Transmitter

A 9-stage ring oscillator based VCO is designed with PLL and edge-combining technology for the proposed transmitter.

The PLL consists of a phase frequency detector(PFD), a charge pump, a low pass filter(LPF), a frequency divider and a voltage controlled oscillator(VCO). A 6.78MHz sinusoidal outputted from the EH RX coil is introduced into the PLL as a reference frequency.

Fig.3 illustrated the architecture of the PFD, the charge pump and the LPF. In order to improve the robustness, two static D-flip-flops are integrated to detect the phase difference between the reference clock and divider output. No extra delay compensate is required, due to the two-phase output scheme of the D-flip-flop. The output of the PFD is fed into the charge pump to charge or discharge the LPF. to reduce the influence of UPN and DW signal on V_{ctrl}. the switch was placed far from V_{ctrl}. In order to improve the stability of the system, a fully integrated second-order passive filter is used with a bandwidth around 100kHz and phase margin around 56°.

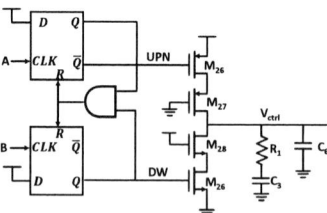

Fig. 3. The schematic of the phase frequency detector(PFD), the charge pump, and the low pass filter(LPF).

A 9-stage current starved ring-oscillator is used in the VCO design, is illustrated in Fig.4, which features lower power consumption while comparing with the LC-based VCO design [6]. Although the noise performance of the ring-based VCO

978-1-7281-5107-6/19 $31.00 © 2019 IEEE   158

is not as good as the LC-based structure, it is acceptable for OOK modulation [4]. A inverter chain is designed generating nine outputs, $P_1 \sim P_9$, with different phase wave. The power consumption of the ring oscillator can be evaluated as

$$P = NCV^2 f \qquad (1)$$

Where N stands for stages of oscillator, C is node capacitance, V is supply voltage, and f is frequency. A 66.7% reduction on power consumption can be expected by increasing the total stages of the ring oscillator from the traditional 3 stages to 9 stages as used in the proposed work.

$M_{19} \sim M_{25}$ compose a voltage control current. M_{25} provide an extra current bias avoiding too low frequency.

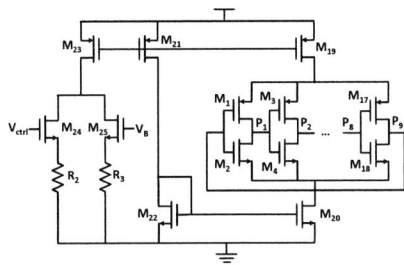

Fig. 4. The schematic of the 9-stage current starved ring-oscillator based VCO.

Fig.5 illustrated the architecture of the power amplifier(PA). A power efficient OOK modulator is proposed integrated a 3-input AND gate array to drive the power transistors. The requirement on the driving capability of the VCO output is greatly reduced. The output of the ring oscillator, $P_1 \sim P_9$, are fed into the 3-input AND gate array to drive the switch control transistors for modulation. The switch control transistors are also used as power transistor of the PA. A simultaneous edge combination and OOK modulation are realized. Compare to traditional three-transistor cascode modulator scheme, a higher power efficiency can be expected due to the low conduction impedance.

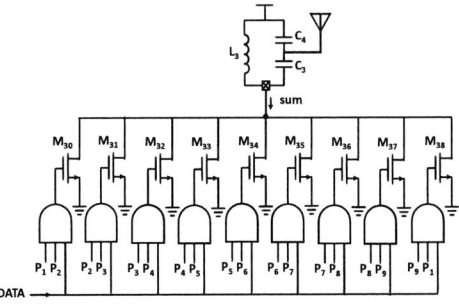

Fig. 5. The schematic of edge-combining PA and OOK modulation

Fig.6 demonstrates the principle of the edge-combining[5]. $P_1 \sim P_9$ are the output waveforms from VCO. A delay of T/18

is applied between two neighboring output of the VCO, where T the period of the VCO at 47.46MHz When $DATA$ is high, the output of the VCO,

$$\begin{aligned} sum = P_1 \cdot P_2 + P_3 \cdot P_4 + P_5 \cdot P_6 + P_7 \cdot P_8 \\ + P_9 \cdot P_1 + P_2 \cdot P_3 + P_4 \cdot P_5 + P_6 \cdot P_7 + P_8 \cdot P_9 \end{aligned} \qquad (2)$$

The frequency of sum is 427.15MHz.

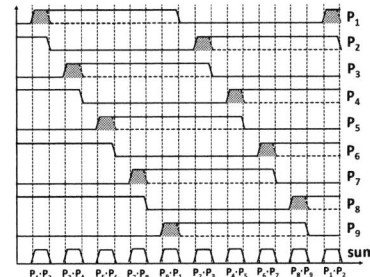

Fig. 6. The principle of edge-combining

III. EXPERIMENTAL RESULTS

The proposed work was fabricated in TSMC 180nm CMOS technology, occupying a silicon area of 0.26mm^2. Extra pins are inserted for testing and debugging purpose as shown in Fig.7(a). Bench test was performed on the proposed design as shown in Fig.7(b). The PMU test board is connected to a receiving coil to gather energy from the wireless power board, as well as to generate supply voltage and reference frequency for the TX board. Baseband output is directly fed into the PA for testing. RF signal is received by a SDR (Adalm-Pluto, ADI).

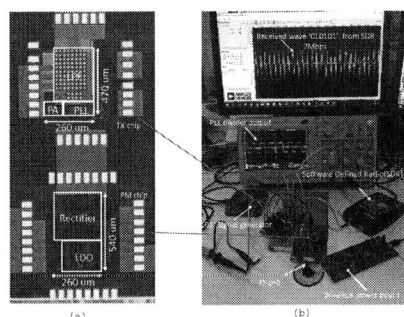

Fig. 7. Die photograph and bench test

Fig.8 and Fig.9 show the PA output spectrum and phase noise, where the peak power is -19.13dBm at 427.18MHz, the spur level is -30dB at 6.78MHz frequency offset. the phase noise is -89.79dBc/Hz at 1MHz frequency offsets. The power consumption of transmitter is 96μW, where VCO consumes 46μW, PA consumes 36μW, Thus, a PA efficiency of around 34% is achieved.

Fig.10 shows that the free VCO phase noise and locked PLL. The phase noise of VCO is suppressed in the loop

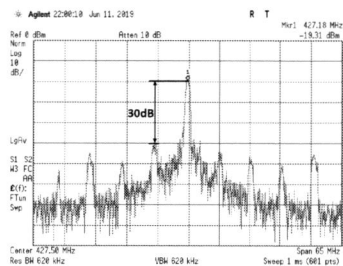

Fig. 8. The spectrum of the PA output measured by Agilent E4440A spectrum analyzer

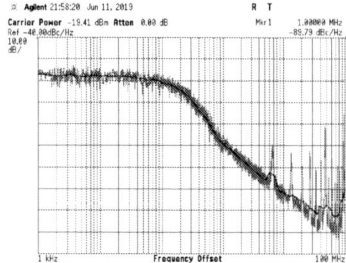

Fig. 9. The phase noise of the PA output

bandwidth which is around 100kHz. The phase noise of the PLL is about -109dBc/Hz at 1MHz offset. Compare to Fig.9, the phase noise of the PA output is -89.79dBc/Hz which is higher than the phase noise of the PA output by $20log_{10}(9) \approx 19dB$ as expected.

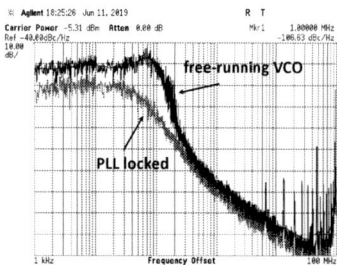

Fig. 10. The phase noise of the VCO output

TABLE I summaries the performance of proposed design. A comparison is performed between the proposed work and state-of-the-art designs reported in literature. A lowest power consumption performance was achieved.

IV. CONCLUSION

In this work, a 427MHz OOK modulated transmitter was proposed featuring a power consumption of $96\mu W$ under a 2Mbps data rate. The proposed design integrated a 9-stage ring oscillator based VCO is designed with PLL and edge-combining technology. Instead of traditional three-transistor

TABLE I
PERFORMANCE COMPARISON OF THE LOW POWER TRANSMITTER

Reference	This work	JSSC'14 [1]	JSSC'14 [2]	JSSC'13 [3]
Technology	180nm	65nm	90nm	130nm
Architecture	PLL	Injection-locked	PLL	Injection-locked
Energy Harvesting	coil-coupling	RF	RF	Thermal,RF
Modulation	OOK	OOK	OOK	BFSK
Supply Voltage	1.0V/0.3V	0.56V	1V	1.0V/0.5V
Power	96μW	215μW	380μW	160μW
TX Pout	-19dBm	-18dBm ~-16dBm	-12.5dBm	-18.5dBm
TX Frequency	427MHz	402MHz	2.44GHz	400/433MHz
Data Rate	2Mbps	250kbps	5Mbps	200kbps
TX efficiency	48pJ/bit	860pJ/bit	76pJ/bit	800pJ/bit
Chip Area	0.26mm^2	2mm^2	1.54mm^2	0.86mm^2 *

* calculate from the die photo illustrated in the paper

cascode structure, a 3-input AND gate array is proposed for OOK modulation, which greatly reduce the requirement on the driving capability of the VCO output. The proposed design reduce the off-chip components by using a crystal based 6.78MHz oscillator with the wireless power transmitter coil as the energy source for energy harvesting. Both the supply voltage and the reference clock are generated from the energy harvesting circuit. The proposed design was fabricated in 180nm technology, occupying a silicon area of only 0.26mm^2.

ACKNOWLEDGMENT

This work is supported in part by National Natural Science Foundation of China through grant 61674095, and Thousand Youth Talents Plan.

REFERENCE

[1] L. Xia, J. Cheng, N. E. Glover, and P. Chiang, "0.56 V, -20 dBm RF-Powered, Multi-Node Wireless Body Area Network System-on-a-Chip With Harvesting-Efficiency Tracking Loop," *IEEE Journal of Solid-State Circuits*, vol. 49, no. 6, pp. 1345–1355, 2014.

[2] G. Papotto, F. Carrara, A. Finocchiaro, and G. Palmisano, "A 90-nm CMOS 5-Mbps Crystal-Less RF-Powered Transceiver for Wireless Sensor Network Nodes," *IEEE Journal of Solid-State Circuits*, vol. 49, no. 2, pp. 335–346, 2014.

[3] Y. Zhang, F. Zhang, Y. Shakhsheer, J. D. Silver, A. Klinefelter, M. Nagaraju, J. Boley, J. Pandey, A. Shrivastava, E. J. Carlson, A. Wood, B. H. Calhoun, and B. P. Otis, "A Batteryless 19 μW MICS/ISM-Band Energy Harvesting Body Sensor Node SoC for ExG Applications," *IEEE Journal of Solid-State Circuits*, vol. 48, no. 1, pp. 199–213, 2013.

[4] Y. Liao, W. Chen, and C. Wu, "A CMOS MedRadio-band low-power integer-N cascaded phase-locked loop for implantable medical SOCs," in *2013 IEEE Biomedical Circuits and Systems Conference (BioCAS)*, Conference Proceedings, pp. 286–289.

[5] J. Pandey and B. P. Otis, "A Sub-100μW MICS/ISM Band Transmitter Based on Injection-Locking and Frequency Multiplication," *IEEE Journal of Solid-State Circuits*, vol. 46, no. 5, pp. 1049–1058, 2011.

[6] W. Chen, W. Loke, and B. Jung, "A 0.5-V, 440-μW frequency synthesizer for implantable medical devices," *IEEE Journal of Solid-State Circuits*, vol. 47, no. 8, pp. 1896–1907, 2012.

12-4 (7031)

IEEE Asian Solid-State Circuits Conference
November 4 – 6, 2019
The Parisian Macao, Macao SAR, China

Direct-Conversion Receiver Front-End for 180 GHz with 80 GHz Bandwidth in 130 nm SiGe

Paul Stärke, Andres Seidel, Corrado Carta, Frank Ellinger
Chair for Circuit Design and Network Theory
Technische Universität Dresden
01069 Dresden, Germany
{paul.staerke; andres.seidel; corrado.carta; frank.ellinger}@tu-dresden.de

Abstract—This work presents an integrated direct-conversion in-phase receiver front-end for carrier frequencies around 180 GHz with an RF bandwidth above 80 GHz. The circuit consists of a double-balanced active mixer core with baseband output buffer, local oscillator (LO) driver and low noise amplifier. Passive baluns are used for the single-ended to differential conversion. The design is optimized for ultra-wideband communication systems with good linearity to support higher order modulation schemes. The saturated conversion gain is 14 dB with a double-sideband noise figure of 12 dB. The optimal LO power is −10 dBm and the RF input referred 1-dB compression point occurs at −15 dBm. The LO driver allows a usable carrier range of 160 GHz to 200 GHz. The total power consumption is 100 mW and the final chip occupies an area of 0.75 mm². It is fabricated in a 130 nm SiGe BiCMOS process with a maximum oscillation frequency of 450 GHz. Compared to the state-of-the-art this circuit offers one of the highest RF bandwidth of any receiver with fixed carrier, making it a viable option for wireless transmission systems achieving data rates beyond 100 Gbit/s.

Index Terms—receiver, direct conversion, mixer, amplifier, balun, communication system, G band, mm-wave, SiGe, BiCMOS

I. INTRODUCTION

Transceiver systems operating in the mm-wave region allow wireless communication with data rates above 100 Gbit/s, otherwise only achievable with wired or optical links. For such data rates a direct-conversion architecture with binary modulation scheme offers the advantage of digital in- and output streams, without the need for complex high-speed mixed-signal components. The lower spectral efficiency is compensated by the higher carrier frequency and makes the system also more resilient. Another possible scenario are parallel multichannel links with individual data rates below 10 Gbit/s. The lower necessary bandwidths improve the signal-to-noise ratio and make higher order modulation schemes, like 16-QAM, more viable and easier to realize.

In any case, a large absolute bandwidth is required to achieve symbol rates beyond 50 GBd wirelessly. In contrast to radar and imaging applications, this bandwidth must be achieved not only at RF, but also in the baseband. Therefore, a mm-wave receiver is proposed, intended to support ultra-wideband single-channel, as well as narrowband multichannel, communication systems. For this purpose the design is focused on high linearity and large bandwidth. It is part of a 180 GHz transceiver system, with the matching transmitter presented in [1]. The proposed design is implemented in a commercially available 130 nm

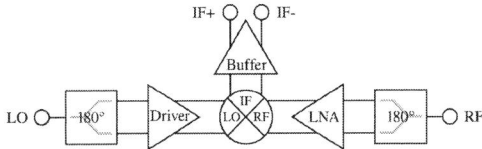

Fig. 1. Differential receiver architecture with 180° baluns , LO driver, low noise amplifier (LNA), active mixer and IF output buffer

SiGe BiCMOS process with a maximum oscillation frequency f_{max} of 450 GHz.

II. DESIGN

A. Architecture

The receiver is based on a direct-conversion architecture consisting of a differential active mixer, a low noise amplifier (LNA), a local oscillator (LO) driver and a baseband (IF) buffer. The complete circuit is intended to be interconnected with other system components, such as a LO frequency multiplier and an IF limiting amplifier. For this reason passive, wideband baluns are used to realize single-ended RF and LO inputs, which simplify chip-to-chip wire bonding and also enable probing for on-chip characterization. A block diagram of the receiver concept is given in Fig. 1.

The active mixer core offers positive conversion gain and allows for larger RF and IF bandwidths, compared to passive approaches. The LO driver reduces the necessary input power and also limits the maximum amplitude level at the mixer, since switching signals too large would degrade the linearity by putting the transistors into saturation. Usually, the LNA is placed before the balun to achieve the lowest noise figure (NF). However, the possible improvement is small and the chosen configuration allows the reuse of the balun for the LO driver, as well as improving the common mode suppression. The output buffer is intended to drive arbitrary differential loads with low impedance and is dc-coupled, because a blocking capacitor in the range of 1 nF, required for long bit sequences, is impractical to integrate.

B. Implementation

For the mixer core a double-balanced topology with common collector (CC) output buffer is chosen. The simplified schematic is given in Fig. 2. Bias voltages are generated on-chip from

978-1-7281-5107-6/19 $31.00 © 2019 IEEE   161

IEEE Asian Solid-State Circuits Conference
November 4 – 6, 2019
The Parisian Macao, Macao SAR, China

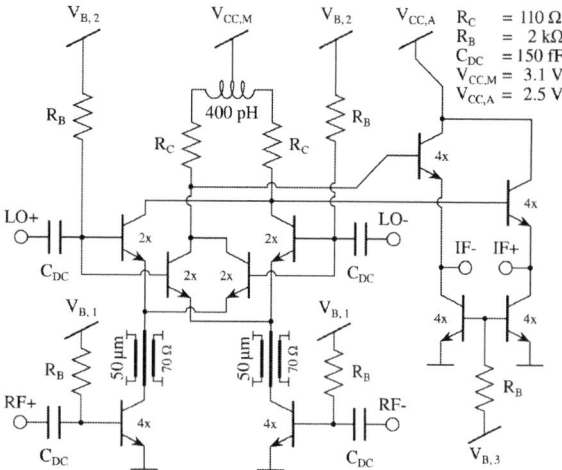

Fig. 2. Simplified schematic of the double-balanced active mixer core with inductive load and dc-coupled common collector output buffer

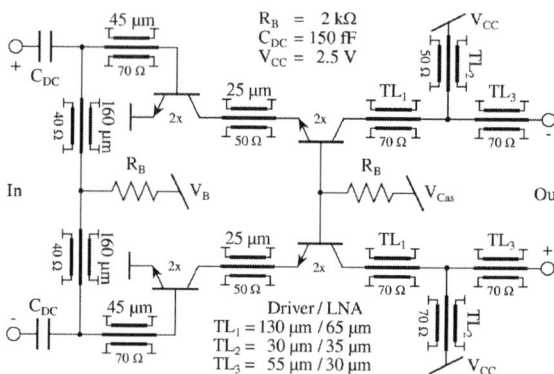

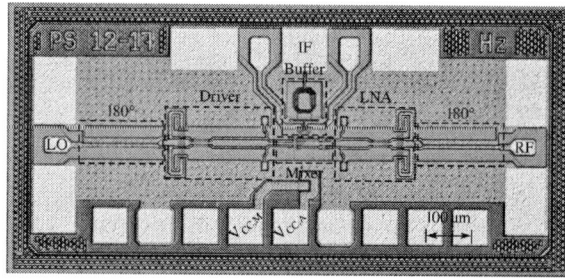

Fig. 3. Simplified schematic of the LO driver and the LNA, differing only in operating point and output matching network

the supplies and all RF-relevant dc nodes are blocked by arrays of large capacitors. The active approach allows a higher bandwidth, due to the lower load impedance at the input of the CC stage, which is further enhanced by inductive peaking with a differential 400 pH inductor. Larger transistors of size 2 and 4, relative to a single element with $0.07\,\mu m \times 0.9\,\mu m$ emitter area, are chosen to improve the linearity. They are biased closed to f_{max}, with a nominal collector current of $1.5\,mA$ per unit. An LO level of around $-3\,dB$ is required to reach the optimal operating point, in regard of conversion gain and linearity. DC blocking capacitors are used to decouple the LO and RF inputs, while the IF output stays dc-coupled.

The same circuit topology is used for the LO driver and the LNA, since they serve a similar purpose of amplifying an external signal. The shared, simplified schematic is given in Fig. 3. For both a single-ended input, with good matching, needs to be converted to a differential output. While the LNA should exhibit a low NF over a large bandwidth, the LO driver must have a higher output power narrowband. A single-stage differential cascode provides a maximum gain of around $9\,dB$ and is sufficient to reduce the necessary LO power to levels, more easily generated externally. The achievable gain / NF of the cascode and the mixer are around $9\,dB\,/\,9.5\,dB$ and $5\,dB\,/\,16\,dB$. This results in a cascaded gain of $14\,dB$ and NF of $11.4\,dB$, which is only $1.9\,dB$ higher than the minimum. Therefore, additional LNA stages would not improve the NF significantly, but impede a wideband realization.

The LNA input matching network is optimized together with the balun for a low NF. It is also used to provide the base current of the transistors over a common resistor connected in between the parallel lines. This approach reduces the impact of component mismatch and improves the common mode suppression, because the virtual ground at the resistor is only valid for differential inputs. The LO driver and the LNA differ in the operating point and the sizing of the output matching network. The LNA is biased with $1.4\,mA$ and the driver, slightly

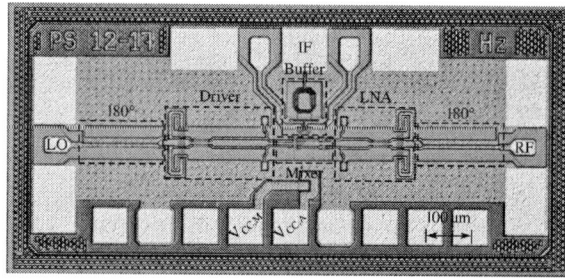

Fig. 4. Image of the fabricated chip ($1.25\,mm \times 0.6\,mm = 0.75\,mm^2$) with functional blocks and pad assignment

higher for more output power, with $1.9\,mA$ per transistor unit. The drivers output is matched around the center frequency of $180\,GHz$, while the LNA is matched at $205\,GHz$ to enhance the bandwidth by compensating the loss at higher frequencies.

The balun design is based on three coupled lines of slightly less than $\lambda/4$ length and offers a compact solution with low losses and good symmetry between the outputs. A similar implementation is demonstrated and characterized in [2]. The introduced loss is around $0.7\,dB$, increasing the total NF by the same amount. The balun is also part of the wideband input matching network, achieving a good compromise of small-signal and noise matching. The balun and all other passive structures, including the transistor connections, are analyzed and optimized by full-3D electromagnetic simulations.

III. Experimental Results

The fabricated chip has a size of $1.25\,mm \times 0.6\,mm = 0.75\,mm^2$, with a pad pitch of $100\,\mu m$. A chip photograph, with functional blocks and pad assignment labeled, is given in Fig. 4. Additional pads are connected to internal dc nodes and can be used for verification and limited correction of the different operating points. The total power consumption is $100\,mW$, with a current of $12\,mA$ from the $3.1\,V$ mixer supply and $25\,mA$ from the $2.5\,V$ amplifier supply.

The circuit is characterized in a wafer prober environment with a $67\,GHz$ 4-port vector network analyzer (VNA), with two $140\,GHz$ to $220\,GHz$ range extension modules and a $67\,GHz$

978-1-7281-5107-6/19 $31.00 © 2019 IEEE

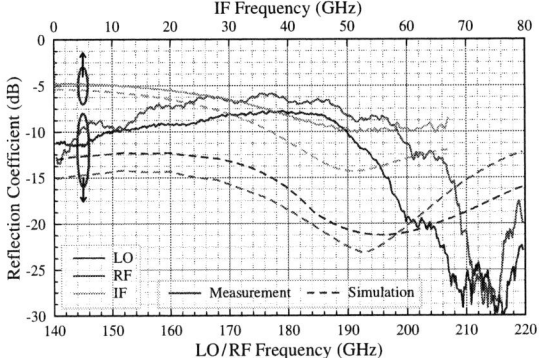

Fig. 5. Simulated and measured small-signal reflection coefficients of the single-ended (RF, LO) and differential (IF) receiver ports

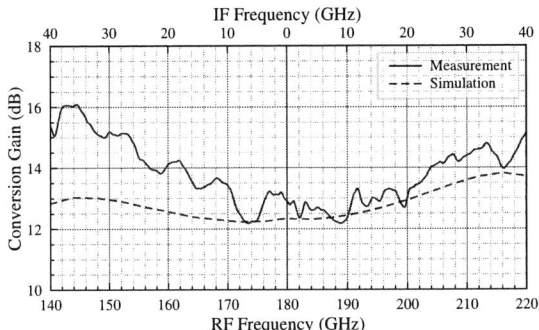

Fig. 6. Simulated and measured differential conversion gain versus RF frequency, with resulting IF component, for a fixed LO frequency of 180 GHz.

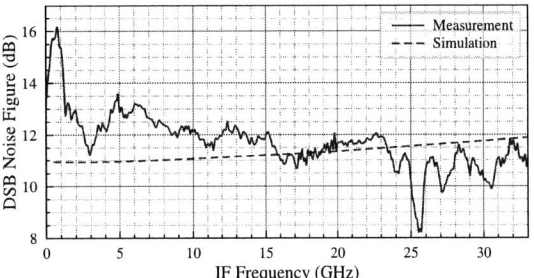

Fig. 7. Simulated and measured differential DSB NF versus IF frequency

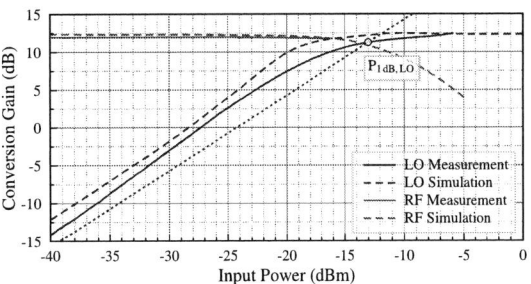

Fig. 8. Simulated and measured differential conversion gain versus LO and RF input power; linear reference and $IP_{1\,dB}$ included for the LO sweep; LO level is −10 dBm for the RF sweep

spectrum analyzer, with integrated low noise preamplifier. Calorimetric power meters, one for dc to 67 GHz and one for sub-THz frequencies above 100 GHz, are used to calibrate the absolute RF and IF power levels. A waveguide frequency 4× multiplier, with up to −5 dBm output power at the chip interface, is used for the LO input. The RF signal is generated by a converter module, which can only provide between −20 dBm and −15 dBm, including probe loss. The VNA simultaneously generates the necessary input signals for the multiplier and the converter modules and measures the resulting IF component. Losses introduced by the probes and cables are calibrated afterwards by individually measured S-parameters.

Small signal simulations and measurements of the receivers reflection coefficients are given in Fig. 5. While in the simulations the RF and LO matching is always below −10 dB the measurements show values up to −6 dB. However, the simulations include the contact pads capacitance, which is in the order of 15 fF and is partly removed by the S-parameter calibration process. The results for the IF port are in good agreement with the simulations.

The simulated and measured conversion gain at the nominal carrier frequency of 180 GHz, with an LO level of −10 dBm, is shown in Fig. 6. The RF converter module allows a characterization only between 140 GHz and 220 GHz, but extrapolating based on the simulations indicate a usable RF bandwidth of

approximately 130 GHz to 230 GHz. Here, the measurements for different LO frequencies confirm an IF bandwidth of at least 40 GHz. Compared to the simulations the conversion gain is slightly higher below 160 GHz. However, a lower actual gain towards higher frequencies was anticipated and taken into account by intentionally introducing an imbalance of +1 dB in the simulations. The resulting imbalance between the lower and upper sideband (LSB / USB) is therefore below 1.5 dB over the full bandwidth.

Simulations and measurements of the double-sideband (DSB) NF are given in Fig. 7. The measurements are performed differentially with a 33 GHz balun to remove low-frequency common mode noise introduced by the resistive bias circuitry. The NF is calculated from noise floor measurements at the IF port, because no suitable noise source was available. Since the receiver has positive conversion gain with a comparable large NF value, the increased output noise can be detected by the spectrum analyzer. The RF probe, with the converter module connected, is used as input termination. The DSB NF then can be approximated by removing the input thermal noise ($k_B T \approx -173.8\,dBm/Hz$ at 300 K), the conversion gain and the balun and cable losses. The obtained results are in good agreement with the simulations, with an average NF of 12 dB.

For assessing the linearity of the receiver, the conversion gain is simulated and measured versus the LO and RF input power. The gathered results are shown in Fig. 8. The available RF power is not sufficient, to clearly determine the input referred 1-dB compression points ($IP_{1\,dB}$), but the results indicate a slightly better performance than simulated, with values estimated around −15 dBm. The behavior of the LO

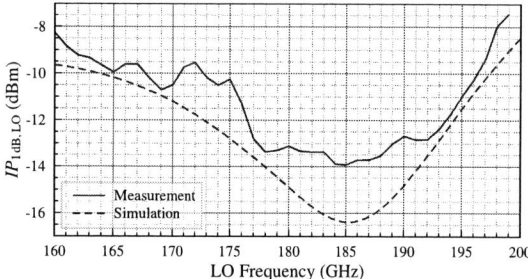

Fig. 9. Simulated and measured $IP_{1\,dB,LO}$ versus LO frequency; simulations indicate optimal input levels approximately 3 dB above

TABLE I
COMPARISON OF RECENT MM-WAVE RECEIVER DESIGNS

Reference	This	[3]	[4]	[5]	[6]	[7]	[8]
Technology	SiGe	SiGe	SiGe	SiGe	InP	InGaAs	CMOS
Node Size (nm)	130	130	130	130	250	35	65
f_{max} (GHz)	450	450	450	550	650	1000	300
I-Q Demodulation	no	no	yes	yes	no	no	yes
Carrier (GHz)	180	190	240	240	300	240	240
LO Multiplier	1	1	16	16	1	4	18
RF BW (GHz)	80	35	18	24	20	32	14[a]
Conv. Gain (dB)	14.0	47.0	11.0	7.8	26.0	16.0	25.0
DSB NF (dB)	12.0	10.7	16.0	8.7	14.0	6.3[a]	12.0
P_{LO} (dBm)	-10.0	-20.0	0.0	0.0	-[b]	-2.0	0.0
$IP_{1\,dB,RF}$ (dBm)	-15.0[a]	-45.0	-18.0	n.a.	-30.0	n.a.	n.a.
DC Power (mW)	100	122	866	916	482	n.a.	260
Area (mm^2)	0.75	1.24	1.57	1.26	1.32	3.25	2.00

[a] estimated   [b] VCO integrated

sweep shows more deviation from the simulation, with an earlier and less pronounced expansion of the conversion gain. The $IP_{1\,dB,LO}$ occurs at −13 dBm and is used as orientation to evaluate the optimum LO power, since adequate two-tone measurements are not possible with the used setup. Simulations of the RF-to-IF 3rd order intermodulation intercept point (IIP3) show a maximum of 8 dBm at approximately 3 dB above the $IP_{1\,dB,LO}$. Overdriving the LO input further has a negative impact on the linearity, with a slight degradation of the IIP3 to 3 dBm at 0 dBm power. The $IP_{1\,dB,LO}$ values for different carrier frequencies between 160 GHz and 200 GHz are shown in Fig. 9. In general, the measurements are slightly above the simulations, but the characteristic of the LO driver can be seen with its maximum between 175 GHz and 195 GHz. With higher input levels are also frequencies between 160 GHz and 200 GHz viable. Beyond this range the LO driver does not saturate the mixer anymore.

IV. CONCLUSION

This work presented an integrated mm-wave receiver for fixed carrier frequencies between 160 GHz and 200 GHz. It was optimized for 180 GHz and exhibits an RF bandwidth

of 80 GHz, with an average NF of 12 dB. An LO level of −10 dBm is sufficient for an optimal operation, in regard of conversion gain and linearity, while even comparably large RF signals up to −15 dBm can be handled. The high linearity is also a requirement for higher-order modulation schemes and the main application is ultra-wideband short-range communication, where a large absolute RF bandwidth is required. However, the wide LO range of 40 GHz also allows the use in multichannel systems as well. An in-phase transmitter, which matches the large bandwidth, is also available and presented in [1].

An overview of recent mm-wave receiver designs is given in Table I. To the authors knowledge, the proposed design offers one of the largest available absolute RF bandwidth of any receiver with fixed LO. It also shows a low dc power consumption, but a direct comparison is difficult, since most of the designs implement other necessary components, like frequency multipliers, oscillators or baseband amplifiers as well. The noise performance is a compromise of complexity and bandwidth, showing a good result for a SiGe implementation, but in general much lower values are achievable with III-V technologies, like InP or GaAs.

ACKNOWLEDGMENT

This work was supported by the German Federal Ministry of Education and Research (BMBF), within the project "Agent3D Hertz" and the German Research Foundation (DFG), within the frame of the Collaborative Research Center 912 "Highly Adaptive Energy-Efficient Computing" (HAEC), subproject A01. The authors also thank the Center for Information Services and High Performance Computing (ZIH) at Technische Universität Dresden for the allocations of compute resources.

REFERENCES

[1] P. Stärke, X. Xu, C. Carta, and F. Ellinger, "Direct-Conversion I-Q Transmitter Front-End for 180 GHz with 80 GHz Bandwidth in 130 nm SiGe," in *European Solid-State Circuits Conference (accepted)*, Sep. 2019.
[2] P. Stärke, C. Carta, and F. Ellinger, "A Deembedding Method for Reciprocal Three-Port Devices Demonstrated with 200-GHz Baluns," *IEEE Microw. Wirel. Compon. Lett.*, 2018.
[3] D. Fritsche, G. Tretter, P. Stärke, C. Carta, and F. Ellinger, "A Low-Power SiGe BiCMOS 190-GHz Receiver With 47-dB Conversion Gain and 11-dB Noise Figure for Ultralarge-Bandwidth Applications," *IEEE Trans. Microw. Theory Tech.*, vol. 65, no. 10, pp. 4002–4013, Oct. 2017.
[4] N. Sarmah *et al.*, "A Fully Integrated 240-GHz Direct-Conversion Quadrature Transmitter and Receiver Chipset in SiGe Technology," *IEEE Trans. Microw. Theory Tech.*, vol. 64, no. 2, pp. 562–574, Feb. 2016.
[5] P. R. Vazquez, J. Grzyb, N. Sarmah, B. Heinemann, and U. R. Pfeiffer, "A 219–266 GHz fully-integrated direct-conversion IQ receiver module in a SiGe HBT technology," in *12th European Microwave Integrated Circuits Conference (EuMIC)*, Oct. 2017, pp. 261–264.
[6] S. Kim *et al.*, "300 GHz Integrated Heterodyne Receiver and Transmitter With On-Chip Fundamental Local Oscillator and Mixers," *IEEE Trans. Terahertz Sci. Technol.*, vol. 5, no. 1, pp. 92–101, Jan. 2015.
[7] C. Grötsch, A. Tessmann, A. Leuther, and I. Kallfass, "Ultra-wideband quadrature receiver-MMIC for 240 GHz high data rate communication," in *42nd International Conference on Infrared, Millimeter, and Terahertz Waves (IRMMW-THz)*, Aug. 2017.
[8] S. V. Thyagarajan, S. Kang, and A. M. Niknejad, "A 240 GHz Fully Integrated Wideband QPSK Receiver in 65 nm CMOS," *IEEE J. Solid-State Circuits*, vol. 50, no. 10, pp. 2268–2280, Oct. 2015.

A Blocker-Tolerant Direct Sampling Receiver for Wireless Multi-Channel Communication in 14nm FinFET CMOS

Barosaim Sung[1], Chilun Lo, Jaehoon Lee, Sangdon Jung, Seungjin Kim, Jaehong Jung, Seungyong Bae,
Youngsea Cho, Yong Lim, Dooseok Choi, Myeongcheol Shin, Soonwoo Choi, Byungki Han,
Seunghyun Oh and Jongwoo Lee
Samsung Electronics, Hwasung, Korea
[1]brs.sung@samsung.com

Abstract— **This paper presents a low power and area efficient wireless direct sampling receiver (DSR) implemented in 14nm FinFET process for frequency modulation (FM) receiver. By employing digital mixer for channel selection, the inductor based local oscillator can be removed, and I/Q matching requirements in analog front-end are vastly relaxed. Implemented in a 14nm FinFET process, the proposed FM-DSR with FPGA based demodulation achieves 31 dB SNR with –90 dBm sensitivity level and 71 dB SNR with -47 dBm input power while consuming 11.74 mW.**

Keywords— **Frequency modulation, receiver, software radio.**

I. INTRODUCTION

As wireless communication technologies develop, there are increasing needs for a low-power, compact-size receiver in portable devices. Recently, FM radio receiver is re-emerging to be integrated on cellular phone for disaster broadcast in many countries. Conventional multi-channel receivers such as FM receiver are based on analog pre-signal processing for channel selection in Analog Front-End (AFE) which increases the portion of analog hardware including multiple-stage filters, ADC and Local Oscillator (LO) as Fig 1. Besides, in order to deal with I/Q analog signals, two analog paths, having cascade Channel Selection Filters (CSF), Anti-Aliasing Filter (AAF), Received Signal Strength Indication (RSSI) and ADC, should be utilized as well as I/Q mismatch calibrations. These are the bottleneck of taking advanced technology power and area benefits. To improve the hardware efficiency, a Direct Sampling Receiver (DSR) having digital channel selection was developed to take advantages of digital processing [1]. However, applications of the previous DSR have been limited to wire-line systems because of power-hungry high speed ADC to deal with multi-channels.

In this work, thanks to the development of the current re-use LNA and the digital-assisted wide dynamic range ADC, a blocker-tolerant DSR is proposed for wireless FM mobile radio. This paper is organized as follows. The proposed architecture is addressed in II, circuit implementation is explained in section III, and the measurement results and conclusion are described in IV and V.

II. PROPOSED ARCHITECTURE

A. FM Direct Sampling Receiver (DSR) architecture

The proposed wireless DSR consists of AFE and Digital Front-End (DFE) as Fig 2. The AFE has Low Noise Amplifier

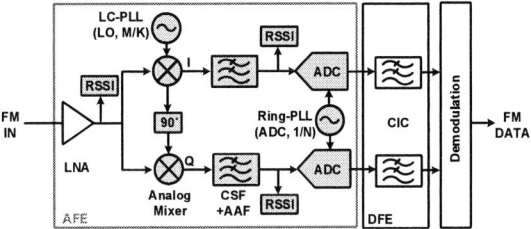

Fig. 1 Block diagram of the conventional FM receiver

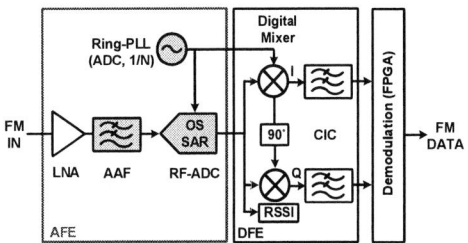

Fig. 2 Block diagram of the proposed FM-DSR

(LNA), AAF, PLL, and RF-ADC, and the DFE is composed of digital mixer and digital filters for digital channel selection, which has avoided expensive analog channel selection due to fractional type LO clock jitter requirement and huge analog filter area for narrow bandwidth.

Since there is no analog filter except AAF for aliased signals from A/D conversion, the proposed RF-ADC is designed to have wide dynamic range (DR) with oversampling to keep low interested signal bandwidth noise floor even with large adjacent channels and blockers. Thanks to the digital channel selection, inductor based LO PLL is removed, and a ring-VCO-based PLL is used for ADC clock generation.

All analog RSSI blocks, which detect each stage RF input power level for automatic gain control (AGC) of FM receiver, are eliminated by using the RF-ADC and AAF having wider DR than that of LNA, so the AGC only uses ADC output to detect LNA saturation.

In addition, only single channel path of AFE is utilized to deal with modulated I/Q signals because channel selection and I/Q signal separation are processed in DFE. Thus, there is no mismatch between I/Q channels in AFE, and this makes FM-DSR have the enhanced power consumption and area compared to those of the conventional approach by removing the one side of receiver path and I/Q calibration in AFE.

B. FM DSR system link budget

The Fig.3 shows link budget analysis of sensitivity with maximum gain case. The 12 bit 240 MS/s RF-ADC and AAF have 7 dBm full-scale and LNA has -2 dBm full-scale with 42 dB maximum gain to remove analog RSSI circuit, and 55 dB ADC SNR is estimated based on kT/C and comparator noise power. The additional 27 dB DR of ADC is achieved by using 240 MHz sampling rate for 250 kHz bandwidth input signal after digital filtering as (1) and (2) for over-sampling.

$$ADC\ DR = 6.02N + 1.76 + 10 \log 10\ OSR \quad (1)$$

$$OSR = f_S/(2*f_{BW}) \quad (2)$$

where N is the number of bits, OSR is the oversampling ratio, f_S is ADC sampling rate, and f_{BW} is input signal bandwidth. Thus, effective 82 dB DR of the ADC is achieved. The -110 dBm sensitivity level from antenna is amplified by 42 dB LNA gain to -68 dBm signal, and thermal noise and LNA noise figure with the maximum gain show -72 dBm. Hence, 4 dB modulated SNR is achieved without any analog channel selection filters.

In maximum SNR case as analysed in Fig. 4, from -47 dBm input power, signal power has -5 dBm with maximum gain and total harmonic distortion and noise show -70.8 dBm, so the receiver achieves 65.8 dB SNR which is higher than required 60 dB audio SNR.

In -93dBm wanted signal with -33dBm adjacent channel signal condition, ACS SNR requirement is satisfied as Fig. 5, and in sensitivity case which makes receiver chain have maximum gain, even though adjacent channels exist, the RF-ADC is not saturated because adjacent channel power is amplified as -8 dBm from -50 dBm at ADC input.

C. Sampling clock jitter requirement for FM DSR

ADC sampling clock phase noise can be interpreted as noise added on input signal which determines the input SNR based on (3).

$$SNR_{jitter} = -20*\log_{10}(2\pi*fin*t_{jitter}) \quad (3)$$

where, fin is the highest input signal frequency, and t_{jitter} is sampling jitter. Thus, the limited SNR due to sampling clock jitter is 55dB with maximum input frequency 108MHz when sampling jitter is 2.5ps which is the same as LO-PLL jitter specification for conventional FM receiver. In addition, in this FM receiver, a reciprocal mixing of blockers with PLL phase noise is also critical noise source because blockers located at ±200 kHz and ± 400 kHz have 50 dB and 60 dB higher signal power than input signal, so a reciprocal mixing degrades its system SNR. Meanwhile, FM receiver specifies adjacent channel selectivity (ACS) as table 1.

Table 1 FM adjacent channel selectivity (ACS)

Item	Condition		Value
ACS	30dB SNR at the audio output (8dB for ADC SNR)	±200 KHz	50 dB
	Wanted Signal: -93dBm(14.1uV)	±400 KHz	60 dB

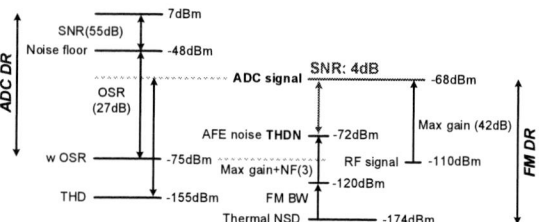

Fig.3 Sensitivity link budget (Required SNR: 3dB)

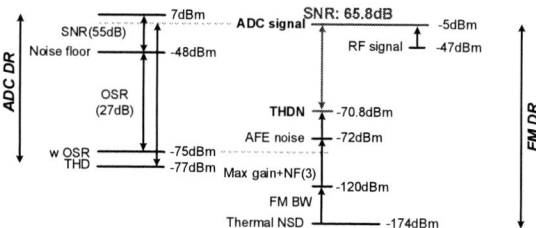

Fig.4 Maximum SNR link budget (Required SNR: 60dB)

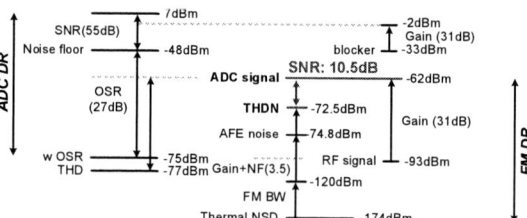

Fig.5 ACS SNR link budget (Required SNR: 8dB)

Because the largest adjacent channel power is defined as 60 dB stronger power than -93 dBm wanted signal at ± 400 kHz offset frequency, a ADC needs more than 68 dB DR, and it is covered by the RF-ADC having 82 dB DR.

Table 2 Noise density requirement of PLL for FM-DSR

Blocker power	Offset [Hz]	BW [Hz]	Target SNR	Phase noise [dBc/Hz]
Pin + 50dB	200k	250k	8	-107
Pin + 60dB	400k	250k	8	-115

Although this PLL for RF-ADC has the same phase noise requirement as conventional LO-PLL, it can be implemented with inductor-less ring type PLL to get the same phase noise performance because this PLL is not required to have fractional output for ADC sampling clock generation, which is not channel selection as LO-PLL.

III. CIRCUIT IMPLEMENTATION

A. LNA and anti-aliasing filter (AAF)

The Fig. 6 shows LNA and AAF of FM-DSR. It supports single-end input signal with working frequency tuneable characteristic by on-chip CDAC input matching cap array with on-board inductor. The LNA consists of two branches; a main gain path and attenuation path as noted in Fig. 6. This structure

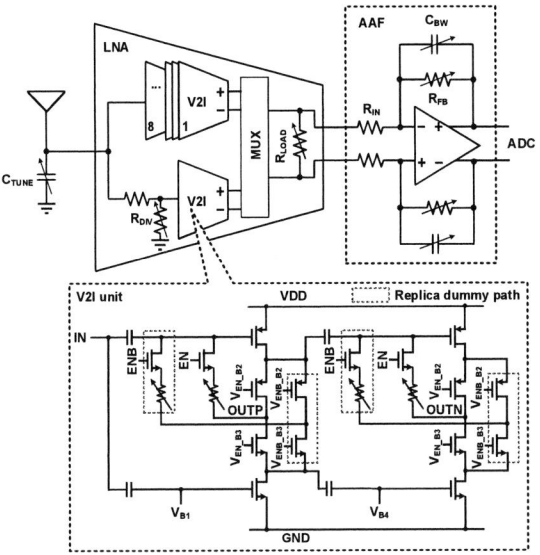

Fig.6 LNA and AAF block diagram and LNA gm-cell unit (V2I)
(Bias circuits are not shown.)

$$Av_{LNA} \cong Gm*(R_{LOAD}*R_{IN})/(R_{LOAD}+R_{IN}) \qquad (4)$$

$$Av \cong Gm*(R_{LOAD}*R_{FB})/(R_{LOAD}+R_{IN}) \qquad (5)$$

was chosen due to the target of 42 dB gain range with the maximum input signal of -2 dBm. In the main gain path, eight basic trans-conductance cells are used for gain control, and the resistive divider with 3 bit control (R_{DIV}) and one trans-conductance cell, used for better isolation between the main gain and attenuation branches, is implemented for the attenuation path. For normal gain path, the gain at LNA output and AAF output could be roughly calculated as (4), (5). Traditionally the LNA output connected to another gm stage which decouples the next stage load with LNA R_{LOAD} design. The R_{LOAD} directly impacts the gain, linearity and input matching at the same time. On the contrary, in the proposed topology, to do input impedance matching and achieve required gain and linearity of LNA Gm cell, the LNA output resistor (R_{LOAD}), AAF input (R_{IN}), and feedback resistor (R_{FB}) are carefully chosen to satisfy gain, linearity, and matching requirements without additional gm stage power and area penalty. To reduce LNA current consumption, a cascode current re-used inverter topology is chosen to implement due to high gain, low noise performance, and high input-output isolation characteristics. Besides, this topology also provides single-ended input to differential output conversion function. The LNA gain control is utilized by paralleling eight cells of gm unit. By turn on and off different numbers of gm unit, effective gain is achieved but this topology potentially would change the input impedance of LNA which makes input matching be difficult. When gm cell is turn off, to keep input impedance similar as gain changing, the additional dummy path is added.

The designed AAF is composed of Miller compensated

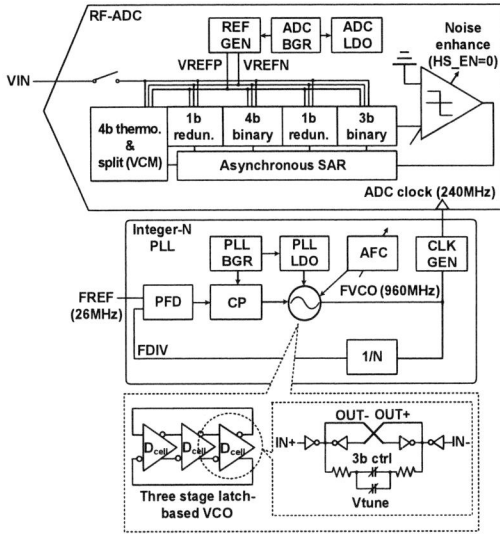

Fig.7 RF-ADC and integer-N PLL block diagram
(ADC is simplified as single-ended.)

operational amplifier and passive components. (R and C) As mentioned in LNA section, LNA swing and linearity are affected by R_{IN} value, so AAF gain is mainly controlled by R_{FB} with 3 dB per step, and the bandwidth of AAF is designed by C_{BW} for attenuating aliased signals in higher than 132MHz which is aliased zone to attack FM in-band signals.

B. RF-ADC

For low power high speed ADC as RF-ADC, digital friendly SAR type ADC is implemented. Fig 7 shows a 12 bit 240 MS/s charge-redistribution binary SAR with 2 bit redundancy. This redundancy helps alleviate settling error [2] and helps in reducing the size of reference capacitor. And a LSB of 0.125 fF customized metal-oxide-metal (MOM) unit capacitor of CDAC is used based on total CDAC kT/C noise for 67dB SNR budget for wide bandwidth of ADC input sampling network. In this design, we have 4bit thermometer code on MSB to improve INL and DNL and we split the 2 MSB capacitor arrays in two halves to generate common mode voltage (VCM) and to drive with complimentary output of SAR during conversion. The other bits use monotonic switching method as [3].

C. Ring-VCO-based PLL

The ring-voltage controlled oscillator (VCO)-based PLL is implemented in FM-DSR PLL for a compact size and a low cost compared with LC-VCO-based PLL. The output frequency of this PLL is 960MHz to satisfy the integrated RMS jitter requirement, and its output is divided by 4 to generated 240MHz ADC sampling clock. The ring-VCO is designed based on latch-type 3 stage delay cells as shown in Fig. 7. As the process technology is advanced from CMOS to FinFET, the noise characteristics of ring-VCO become vulnerable due to the lower supply headroom and especially flicker noise. However, by taking the high trans-conductance and improved switching

978-1-7281-5107-6/19 $31.00 © 2019 IEEE

speed of the process technology as well as by using the additional passive resistor in delay-cell, the designed VCO has the improved phase noise performance, especially flicker noise, without utilizing inductor. This VCO is designed to cover the frequency range from 700 MHz to 1400 MHz, and taps the supply voltage from a low noise low drop-out (LDO) regulator, which improves a power supply rejection ratio (PSRR) characteristic.

IV. Measurement Results

The prototype FM-DSR is implemented in a 14nm FinFET CMOS. The required clock, bias currents, internal LDOs for LNA and ADC supply, and references are generated on chip, and DFE and demodulator are implemented in external FPGA. Chip photograph of FM-DSR is shown in Fig. 8. Total 0.23 mm² area is occupied for FM-DSR.

From FM signal source generator, wired FM modulated signal having 22.5 kHz frequency deviation with 95.1 MHz carrier is applied to FM-DSR. RF-ADC outputs are demodulated by on-board FPGA which has FM-DSR DFE and demodulator. The RF-ADC outputs are decimated as 120 MHz output from 240MHz to reduce FPGA driving noise because FPGA driving noise is the dominant noise source to degrade SNR in sensitivity of this prototype FM-DSR, even though there is 3dB ADC SNR loss from 1/2 decimation. This noise issue can be smaller if DFE and modulator are integrated in FM-DSR chip. Fig. 9 shows FM-DSR demodulated SNR measurement results with 95.1 MHz input frequency. The FM-DSR achieves 71.2 dB demodulated SNR of R and L sides with -47 dBm input power which meets stereo 60dB audio SNR requirement, and 31 dB demodulated SNR of R and L with -90 dBm input power for sensitivity. From 65 MHz to 108 MHz input frequency, demodulated SNR results maintain more than 64 dB while consuming the power consumption of 11.74mW as shown in Fig 9. Because the matching network including external inductor is optimized to 94 MHz input frequency, the maximum demodulated SNR is measured at 95.1 MHz.

V. Conclusion

A wireless-DSR has been proposed and verified about the functionality and performance, and the efficiency of the architecture for area and power reduction and design consideration are also analysed. Due to digital intensive design in back-end signal processing, the analog circuit burden has been minimized and relaxed. It is particularly useful in advance CMOS process by taking advantage of the low power and small area digital circuit implementation. With advanced 14nm

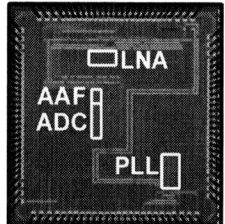

Fig. 8 Chip photo

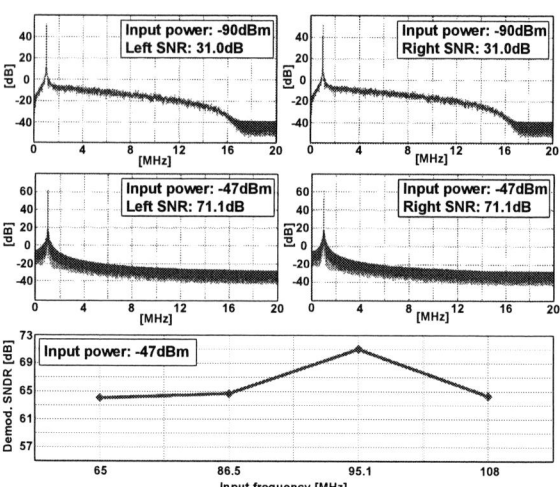

Fig. 9 Measured demodulated SNR of FM DSR

Table 3 FM-DSR power and area comparison

FM receiver		This work	[4]	[5]	[6]
Process [nm]		14	55	60	65
Active Area [mm²]		0.23		0.8	
Receiver architecture		DSR	Low-IF	Low-IF	Low-IF
Input range [MHz]		65-108	76-108	76-108	76-108
Supply [V]	VDD1	1	2.8	1.35	1.2
	VDD2	1.2	-	3.3	-
Core Power [mW]		11.74	*23.8	19.59	12.96
Maximum SNR [dB]		71.1	-	-	60
Channel selection		Digital	Analog	Analog	Analog
I/Q calibration		Not required	Required	Required	O
Analog wideband RSSI		Not required	Required	Required	O
VCO type		Ring	LC	LC	LC
Channel scan		Digital only	Analog+digital	Analog+digital	Analog+digital

*Including digital power

CMOS FinFet process, the proto-type wireless FM-DSR is implemented with the total 11.74mW power consumption and 0.23mm² area which are competitive performance among published FM receivers. Therefore, the proposed wireless-FM DSR is a remarkable architecture as future wireless receiver standards as well as mobile wireless receiver application.

References

[1] Jiangfeng We, et al., "A 2.7mW/Channel 48-1000 MHz Direct Sampling Full-Band Cable Receiver," in IEEE Journal of Solid-State Circuits, vol. 51, no. 4, pp. 845-859, Apr. 2016

[2] Sang-Hyun Cho, et al., "A 550uW 10-b 40MS/s SAR ADC with Multistep Addition-only Digital Error Correction," IEEE Journal of Solid State Circuits, pp. 1881-1892, Aug. 2011

[3] C. Liu, S. Chang, G. Huang and Y. Lin, "A 10-bit 50-MS/s SAR ADC With a Monotonic Capacitor Switching Procedure," in IEEE Journal of Solid-State Circuits, vol. 45, no. 4, pp. 731-740, Apr. 2010

[4] Jing-Hong Conan Zhan, et al., "A 55nm CMOS 4-in-1 (11b/g/n, BT, FM, and GPS) Radio-In-a-Package with IPD Front-End Components Directly Connected to Antenna," RFIC, pp 209-212, Jun. 2014

[5] Yuan-Hung Chung, et al., "4-in-1 (WiFi/BT/FM/GPS) Connectivity SoC with Enhanced Co-Existence Performance in 65nm CMOS," ISSCC Dig. Tech. Papers, pp 172-174, Feb. 2012

[6] Chungyeol Paul Lee, et al., "A MultiStandard, Multiband SoC with Integrated BT, FM, WLAN Radios and Integrated Power Amplifier," ISSCC Dig. Tech. Papers, pp 460-461, Feb. 2010

12-6 (7109)

A Sub-0.6V, 330 μW, 0.15 mm² Receiver Front-End for Bluetooth Low Energy (BLE) in 22 nm FD-SOI with Zero External Components

Ehsan Kargaran[1], Carl Bryant[2], Danilo Manstretta[1], Jon Strange[2], Rinaldo Castello[1]

[1] University of Pavia, Italy, [2] MediaTek, Kent, United Kingdom

Abstract— A 2.4 GHz highly power efficient receiver front-end for Bluetooth Low Energy (BLE) is presented in this paper. Thanks to the extensive current reuse scheme combined with the passive source-boosting topology and exploiting forward back gate biasing (FBB) of active devices, the low-noise transconductance amplifier (LNTA) input impedance is matched to 50 Ω with 18 times less current compared to a single common-gate (CG) LNTA. Moreover, a high swing Class-C VCO operating at the twice the desired frequency drives the frequency divider by 2 to create 4 phases with 25% duty cycle. Implemented in 22 nm FD-SOI technology, the proposed receiver front-end consumes only 330 μW from 0.55 V and occupies an active area of 0.15 mm². The measured conversion gain and NF are 32.3 dB and 9.4 dB respectively. The proposed receiver consumes the least power among all previous published papers while its performance is largely superior to the requirement of intended applications.

Keywords— ultra low-power (ULP), current reuse, gm-boosting, Internet of thing (IoT), LNTA, Class-C VCO.

I. INTRODUCTION

The quick proliferation of internet-of-Things (IoT) and wireless sensor networks (WSNs) demands ultra-low power (ULP) radios with drastically reduced cost, complexity and IC area. Advanced CMOS technology offers both high f_t and low V_{th}, enabling sub-1V operation and low power. Aggressively reduced supply voltage (down to 0.3 V in [1] and 0.18 V in [2]) can strongly lower power dissipation but requires bulky on-chip inductors and boost supply converters for proper device biasing. This, however, significantly enlarges chip area, 1.65 mm² in [2] and 2.5 mm² in [1]. Moreover, with such a low supply voltage, flip-flop-based quadrature frequency dividers cannot be used. This problem was addressed inserting an I/Q splitter in the signal path while running the voltage-controlled oscillator (VCO) at the carrier frequency [2]. Eliminating the dividers further lowers the power but at the increased risk of VCO pulling. On the other hand, recycling the bias current in different blocks can effectively minimize the power consumption as well. In [3], stacking the low-noise amplifier (LNA), voltage-controlled oscillator (VCO), the mixer, and the input stage of baseband leads to drawing only 530 μA from a 0.8 V supply. However, stacking more blocks performing different functions can degrade the overall performance and

may also be more prone to VCO pulling [3]. Following a similar approach, [8] reports partially stacked LNA-baseband stage receiver front-end for SoC coexistence which leads to a notable OOB-IIP3 of 6 dBm. Even though the same bias current is shared between LNA and the first stage of both I and Q baseband transimpedance amplifiers (TIAs), power consumption exceeds 1 mW. In [5] a receiver front-end was reported with a NF below 6.5 dB while consuming only 350 μW, excluding the VCO. The LNTA makes use of transformer-based passive gain boosting and an efficient current reuse scheme, while the baseband adopts a low-power common-gate stage, at the price of reduced linearity at small offset frequencies. Targeting a single sub-0.6 V supply allows to lower the power consumption by exploiting stacked devices while, unlike [1,2], avoiding inductive loads, thus saving considerable chip area. We propose an ULP receiver front-end for Bluetooth Low-Energy (BLE) applications operating at 2.4 GHz. Its performance is largely superior to the requirements of the envisioned applications and it consumes only 330 μW.

II. IMPACT OF BODY BIASING

A 22 nm Fully Depleted Silicon-On-Insulator (FD-SOI) process enables very low voltage supply design using super low-V_{TH} (SLV_{th}) transistors. SLV_{th} transistors utilize flip-well: nFET devices are located on n-type back gates and pFET devices are located on p-type back gates [6]. This configuration is suitable for forward body biasing (FBB), helping to increasing performance. Here, to explore the FBB effect in 22 nm FDSOI, the charactcristic of a 10μm/22nm nFET transistor is initially extracted and shown in Fig.1. The typical V_{TH} of a SLV_{th} nFET is around 180 mV and it can be lowered thanks to FBB, i.e. a positive body-source voltage (V_{BS}). Fig.1 is extracted by varying V_{gs} from 0 to 400 mV, while biasing V_{BS} at either 0 or 0.4V (dashed and solid line respectively). Once V_{BS} is at 0.4V, V_{TH} decreases, and smaller voltage is required at the gate node to form the channel. The intrinsic gain of nFET devices stays almost the same when FBB is employed at low V_{GS} but is slightly reduced at higher V_{GS}. FBB can also increase f_t of active devices. By contrast, for a given V_{GS}, g_m/I_d is slightly reduced. Biasing transistors in weak inversion results in maximum current efficiency, g_m/I_d, but results in a relatively

978-1-7281-5107-6/19 $31.00 © 2019 IEEE

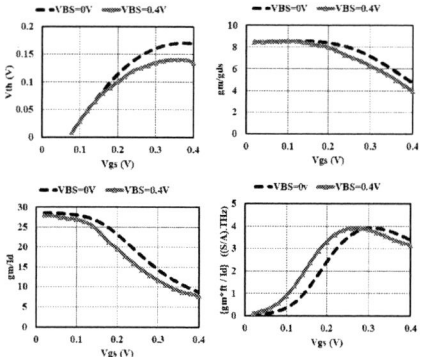

Fig.1. Simulated plots of (a) Vth, (b) intrinsic gain (gm/gds), (c) gm/Id, and (d) FoM of nFET device versus Vgs with and without FBB (*permission obtained from Global Foundry (GF)*)

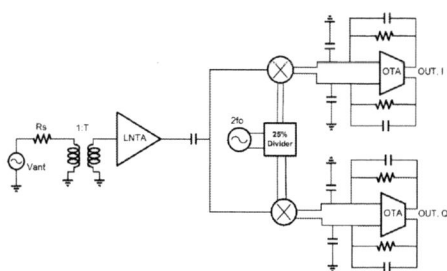

Fig.2. Proposed frontend architecture

poor frequency response. Therefore, the product of the current efficiency and cut-off frequency can be considered as the proper Figure-of-Merit (FoM) for RF design. FoM versus V_{GS} is depicted in Fig.1(d). The peak of the FoM is shifted toward lower V_{GS}, from 320 mV to 260 mV, thanks to FBB. This effect is even more pronounced at very low V_{GS}. At V_{GS}=100mV, FoM is 90 (S/A)THz versus 35 (S/A)THz for V_{BS}=0V and V_{BS}=0.4V respectively, demonstrating the benefits of FBB.

III. RECEIVER FRONTEND ARCHITECTURE

The schematic of the proposed RF front-end is shown in Fig.2. A single-ended 2.4 GHz RF input voltage signal is converted to current by LNTA, which is followed by a single-balanced passive mixer and transimpedance amplifiers (TIAs).

A. LNTA Design

CG amplifiers have moderate low noise and wide bandwidths, but their current consumption is constrained by the device g_m required for input impedance matching. Current consumption can be lowered by a factor T^2 using a 1:T step-up transformer in front of the LNTA. Passive gain boosting across the gate to source of the input transistor (source boosting) is an effective method to lower the required device g_m for input impedance matching. Combining both techniques together with a current-reuse scheme that can be operated below 0.6 V is the basic idea

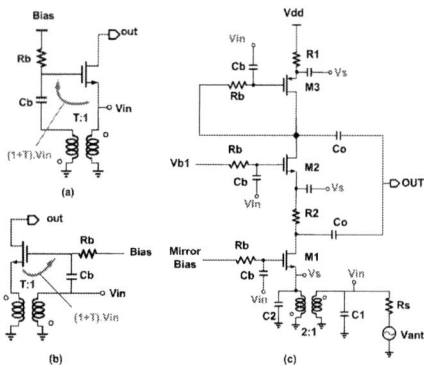

Fig.3. a) Conventional passive gate boosting, b) Power efficient passive source boosting, and c) Proposed source boosting LNTA

in the proposed LNTA. As shown in Fig.3b, for an ideal transformer with k=1, the source voltage (v_s) is T times the input voltage (v_{in}), whilst the gate-source voltage is (1+T) v_{in}. Hence, the drain current is i_d= g_m(1+T)v_{in}, the input current is i_{in}=T·i_d and the input impedance is Z_{in}=1/{T(1+T) g_m}.

The main benefit of source boosting compared with gate boosting, shown in Fig 3a, is a current reduction by a factor of T. For T=2 input power matching is achieved with a device g_m equal to 1/(6R_s) and the transconductance gain (G_m) is 1/2 that of a conventional CG amplifier. To further reduce power dissipation, more than one device can be stacked to improve voltage efficiency by re-using the bias current. A previous design [4] shows 4x bias current reuse. Even operating all devices in deep sub-threshold, i.e. using extremely large devices, the stacking configuration in [4] cannot function below 0.8 V. Below 0.6 V conventional solutions can only achieve 2x current reuse. In the proposed LNTA in Fig. 3c, three devices are stacked, and the gate and source of each transistor are ac coupled to the primary and secondary of the transformer respectively. Assuming that all devices have the same g_m, the total transconductance is tripled and the required device g_m to perform input matching is only 1/(18R_s). In the proposed LNTA, shown in Fig.3c, two NMOS and one PMOS share the same bias current, have the same gate-source applied voltage signal and are all biased in weak inversion to maximize g_m/I_d with an acceptable f_t. Signal isolation between the two stacked NMOS devices is achieved by interposing a large resistor (R_2) and the PMOS device is biased with a large resistor (R_1) toward the supply. A capacitive combiner adds the output currents.

B. Baseband TIA Design

The baseband TIAs following the mixer convert the signal current to voltage and act as 1st order filter for the subsequent stage. In order to guarantee linear mixer operation, the active-RC TIA stage should provide low impedance at its input. Good power efficiency and low noise can be achieved using self-biased inverter-based amplifiers. Simple inverters are highly

978-1-7281-5107-6/19 $31.00 © 2019 IEEE

IEEE Asian Solid-State Circuits Conference
November 4 – 6, 2019
The Parisian Macao, Macao SAR, China

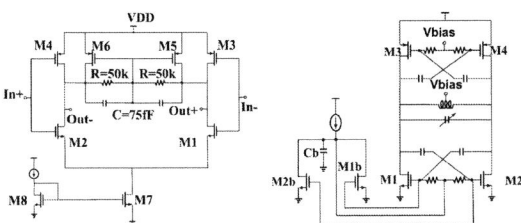

Fig.4. (a) Inverted based OTA amplifier, (b) high swing Class-C VCO

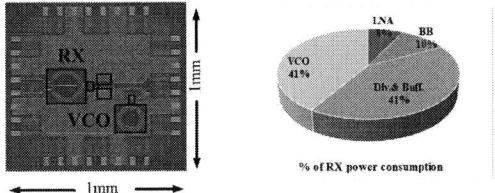

Fig.5. (a) Chip micrograph, (b) RX Power break-down

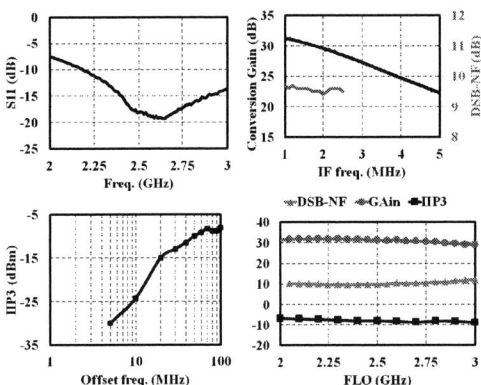

Fig.6. (a) S11, (b) RX gain, DSB-NF vs IF, (c) IIP3 vs offset, (d) RX performance vs LO freq.

sensitive to supply voltage variations and, to decrease the common mode (CM) gain, an extra parallel CM feedback (CMFB) path is required, adding extra power and noise. To have a good compromise between power, noise, reduced sensitivity to supply variations and good CM rejection, the amplifier used in our TIA was designed as shown in Fig.4a. The tail current source reduces the sensitivity to supply variations and the CM gain of the NMOS devices. The PMOS devices are made of two identical halves: one half operates as input stage and the other half as a diode connected load for CM signal and as a current source for differential signals. Hence, a differential gain of $(g_{mn2}+g_{mp4})R$ and a CM gain of g_{mp4}/g_{mp6} are obtained respectively. Considering the finite transistors output impedance, the CM gain is below unity, ensuring CM stability. Capacitors C at the amplifier output improve the CM rejection at high frequency. Finally, placing an RC around the amplifier performs 1st order filtering.

C. VCO Design

In order to save power, ultra-low power VCOs are typically operated at the lowest possible output voltage swing required by the frequency dividers. This leads to poor voltage efficiency and degraded phase noise for a given power. To counter this, a class-C VCO topology has been adopted in this design. Class-C VCOs have the highest current efficiency, hence they require the lowest current consumption for a given resonance tank impedance and voltage swing. Using a complementary p-n topology a factor of two current efficiency improvement is further achieved. This allows to achieve good phase noise and low current consumption under a low supply. Although using a transformer to set the bias current [7] the tail current source could be removed, in our design a capacitive cross coupled connection is used instead. This is compatible with the reduced swing required but has a smaller area. The schematic of the VCO is shown in Fig. 4b. The two cross-coupled pairs M1/M2 and M3/M4 operate in Class-C and provide negative resistance to restore the energy losses in the resonant load. At start-up, the VCO bias current is set to N times I_{bias} by current mirror composed by M1/M2 and M1b/M2b and ensures a robust start-up. As the VCO oscillation builds up, M1b/M2b start to operate in class-AB and their current tends to grow, forcing the voltage V_{gateN} to be reduced and leading the NMOS pair to operate in Class-C. The center tap of the VCO load inductor is connected to the gates of the M3/M4 for biasing purpose. Since the

common mode voltage of the VCO is slightly above $V_{dd}/2$, the PMOS devices are pushed into the triode region in some portion of period. The LC-VCO combines a 2-bit capacitor bank and a varactor to cover the range from 4.7 to 5.13 GHz. The VCO consumes 137 µW and has a figure of merit (FoM) of 184.3.

IV. EXPERIMENTAL RESULTS

The fabricated receiver in 22 nm FD-SOI technology operates from 2 to 3 GHz, is supplied with 0.55 V and consumes only 330 µW. To reduce the threshold voltage of NMOS and PMOS devices, back gates are forward biased to 0.55 V and 0V respectively. The chip micrograph is shown in Fig.5 and it has an active area of 0.15 mm². The measured RX conversion gain and DSB-NF are 32.3 dB, 9.4 dB respectively, as shown in Fig. 6. RX shows wideband impedance matching and s_{11} is below -10 dB from 2.2 to above 3 GHz with zero external components. The -3dB bandwidth is 2.1 MHz and it allows RX BW of 1MHz centered at IF of 1.5MHz. A good image rejection (IR) of 27.7 dB is achieved at the cost of reduced filtering and linearity. At low frequency offset, i.e., placing two tones at 5 and 8 MHz, IIP3 is -30 dBm, limited by the baseband stage, and IIP3, with two tones at [LO+60MHz] and [LO+118MHz], is -8 dBm, limited by the LNTA. Additionally, IIP2 is 12 dBm when placing two tones at 100 and 102MHz away from LO. The VCO is tested separately and a replica of the divider is cascaded to the VCO. It has a phase noise of -107.7 dBc/Hz at 1MHz offset from the carrier. The VCO tuning range is about 11% and can be tuned from 2.35 to 2.65 GHz. Benchmarking with state-of-the-art, our RX features the lowest power dissipation, the smallest active area and NF<10dB. In contrast to [2] which represents the second least power-hungry RX, our

978-1-7281-5107-6/19 $31.00 © 2019 IEEE   171

TABLE I: PERFORMANCES SUMMARY AND COMPARISON WITH STATE-OF-THE-ART RECEIVERs

	This work	JSSC'13 [1]	JSSC18 [2]	'JSSC'15 [3]	JSSC'18 [9]
Technology (nm)	22	65	28	130	65
External components	Zero	1 inductor+ 2 caps	Zero	1 inductor + 1 cap	FABR resonator + FBAR filter
# on chip inductors	2	4	4	1	0
S11< -10dB Bandwidth (MHz)	>1000 (2.2 to > 3 GHz)	>600 (2 to 2.6 GHz)	100 (2.45 to 2.55 GHz)	250 (2.3 to 2.55 GHz)	100 (2.4 to 2.5 GHz)
Vdd(V)	0.55	0.3	0.18	0.8	1
Power (µW)	330	1600	382	600	1800
Gain (dB)	32.3	83	34.5	55.5	57.8
NF (dB)	9.4	6.1	11.3	15.8	15.7
IIP3 IB/OOB (dBm)	-30/-8	-21.5/NA	-12.5/NA^	-16.8/NA	-45/-16.7
VCO PN (dBc/Hz) @1MHz	-107.7	-112.3	-106*	-105*	-141
IRR (dB)	27.7	Without IQ	26.2	30.5	>30
LO-to-RF leakage (dBm)	<-80	NA	NA	NA	NA
Active area (mm2)	0.15	2.5	1.6	0.25	0.45

* Estimated from plot, ^ LNA IIP3 is -17.4 dBm

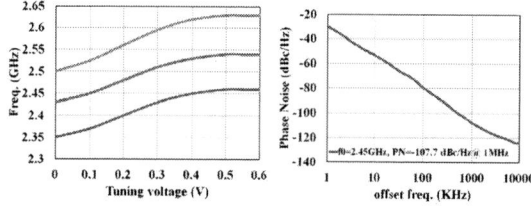

Fig.7. (a) VCO tuning, (b) VCO phase noise

proposed RX shows 2 dB better NF with 15% less power dissipation. Additionally, our solution occupies significantly smaller chip area (0.15 mm² versus in 1.6 mm² in [2]) leading to substantially reduced cost. Even though [1] has 3 dB lower NF than our design, it consumes much higher power and occupies considerably larger chip area.

V. CONCLUSION

A receiver frontend for Bluetooth Low Energy in 22nm FD-SOI with 0.15mm² active area was presented in this paper. It includes LNTA, passive mixers, BB filtering TIAs, class-C VCO and clock divider. By boosting the driving impedance and the input signal with a 2:1 transformer and stacking 3 devices between the rails the LNTA current is 40 µA. The VCO consumes 137 µW. The proposed receiver consumes only 330 µW from 0.55 V. The measured RX gain, DSB-NF and IRR at 2.4 GHz are 32.3 dB, 9.4 dB and 27.7dB respectively. Out of Band IIP3 is -8dBm. The performance of the proposed

receiver is highly competitive with the-state-of-art while consumes the least power and has smallest active area among all the prior works.

ACKNOWLEDGMENT

Special thanks to Global Foundry (GF) for shuttle tapeout.

REFERENCES

[1] F. Zhang et al.," Design of a 300-mV 2.4-GHz Receiver Using Transformer-Coupled Techniques," IEEE JSSC, vol. 48, no. 12, pp. 3190–3205, Dec. 2013.

[2] W. H. Yu et al., "A 0.18V 382µW Bluetooth Low-Energy (BLE) Receiver with 1.33nW Sleep Power for Energy-Harvesting Applications in 28nm CMOS," IEEE JSSC, vol. 53, issue 6, pp. 1618-1627, June 2018.

[3] A. Selvakumar et al., " Sub-mW Current Re-Use Receiver Front-End for Wireless Sensor Network Applications", *IEEE* JSSC, vol. 50, no. 12, pp. 2965–2974, Dec. 2015.

[4] E. Kargaran, et al.," Design and Analysis of 2.4 GHz 30µW CMOS LNAs for Wearable WSN Applications," IEEE TCAS-I, vol.65, no.3, 2018.

[5] E. Kargaran, et al., "A Sub-1V, 350 µW, 6.5 dB integrated NF Low-IF Receiver Front-End for IoT in 28nm CMOS," to be appear in IEEE Solid State Circuit Letter (SSC-L), 2019.

[6] R.Carter et al., "22nm FDSOI Technology for Emerging Mobile, Internet-of-Things, and RF Applications," in proc. IEEE IEDM, 2016.

[7] M. Tohidian, et al., "High-swing Class-C VCO," in Proc. ESSCIRC, pp. 495–498, Sept. 20117.

[8] M. Ramella et al., "A SAW-Less 2.4 GHz Receiver Front-End with 2.4mA Battery Current for SoC Coexistence", IEEE JSSC, vol. 52, issue 9, pp. 2292-2305, September 2017.

[9] K. Wang, et al., "A 1.8 mW PLL-free channelized 2.4GHz ZigBee receiver utilizing fixed-LO temperature-compensated FBAR resonator," IEEE JSSC, Vol. 53, No. 6, June 2018.

A 4-Mbps 41-pJ/bit On-off Keying Transceiver for Body-channel Communication with Enhanced Auto Loss Compensation Technique

Jian Zhao[1,2], Jingna Mao[2,3], Wenyu Sun[2], Yuxuan Huang[2], Yixiong Yang[2], Huazhong Yang[2] and Yongpan Liu[2]

[1]Department of Micro/Nano Electronics Shanghai Jiao Tong University, Shanghai, China 200240
[2]Department of Electronic Engineering Tsinghua University, Beijing, China 100084
[3]The Institute of Automation, Chinese Academy of Science, Beijing, China 100190
E-mail: zhaojianycc@sjtu.edu.cn, ypliu@tsinghua.edu.cn

Abstract—This paper presents a body-channel communication (BCC) transceiver for wireless body sensor network applications. The transceiver employs on-off keying (OOK) modulation and operates in the quasi-static band. It achieves high energy efficiency and stability through two innovative methods: 1) an auto-loss compensation (ALC) technique attenuates the time-variant path loss in the wearable scenario; 2) a capacitive interface in the receiver strengthens the path loss compensation. The transceiver is designed and fabricated in 180-nm CMOS process, it achieves 4-Mbps data rate with a 41-pJ/bit energy efficiency over 1.5-m distance. It also exhibits resiliency to the ground variations and achieves 4.1× reduction in path loss.

Keywords—wireless-body-area network; body-channel communication; auto-loss compensation; transceiver; energy efficiency.

I. INTRODUCTION

Body channel communication (BCC) uses human body as the transmission medium, which results in a lower channel loss and makes it one of the best candidates in wireless-body-area network (WBAN) applications. BCC technology can be classified into three categories based on their coupling mechanisms, which are capacitive coupling, galvanic coupling and magnetic coupling. Among the three categories, capacitive coupling BCC (CC-BCC) is regarded to be more superior than the other two competitions due to the following advantages: (1) high data rate and energy efficiency, (2) high safety level for human beings, (3) lower cost as it requires no antennas and coils. [1].

BCC transceiver is critical as it determines the performance and efficiency of a BCC system. The BCC transceiver usually occupies < 200 MHz frequency band, which can be further separated into two regions. One is far-field region ($f \geq 30$ MHz), the other is quasi-static region ($f < 30$ MHz), considering the communication distance on human body. The former region has a lower path loss, however it suffers from the multi-path effect and exhibits very large variations [2]. The quasi-static region relatively tends to be more stable with deterministic features [3]. As such, IEEE 802.15.6 standard select 21 MHz in the quasi-static band as the carrier frequency.

As the backward path of CC-BCC is formed by the coupling capacitance between two ground electrodes of the

This work was supported by the National Natural Science Foundation of China (Grant Nos. 61904104, 61934005, 61674094).

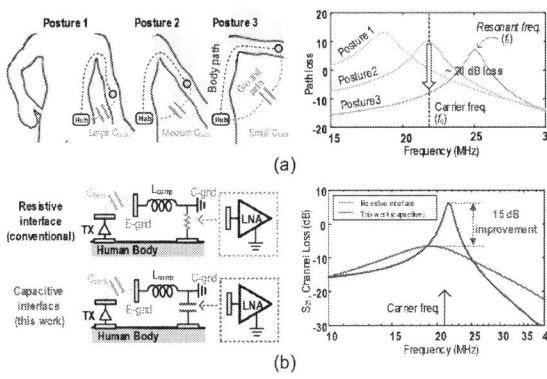

(a)

(b)

Fig. 1: Two main challenges in resonance-matching BCC-transceivers: (a) Path loss increase due to human-posture changes, (b) Resonant-matching compensation weaken due to a resistive interface of the BCC receiver.

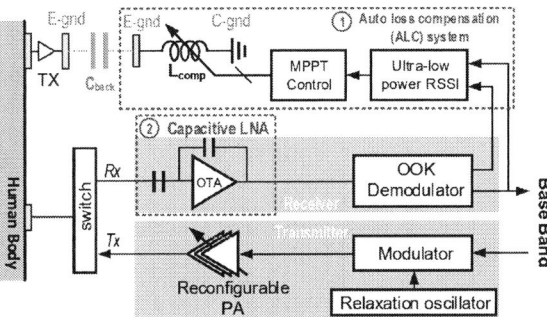

Fig. 2: Concise system block diagram of the proposed BCC transceiver with enhanced ALC technique.

transmitter and receiver, the channel impedance increases when the frequency gets lower. The signal in the quasi-static region normally has relatively lower frequency and thus suffers from higher loss, leading to lower energy efficiency of the transceiver [4], [5]. The resonance-matching method renders an efficient approach to compensate the capacitance induced path loss through resonating the backward capacitance (C_{back}) with a compensation inductor (L_{comp}) [6].

IEEE Asian Solid-State Circuits Conference
November 4 – 6, 2019
The Parisian Macao, Macao SAR, China

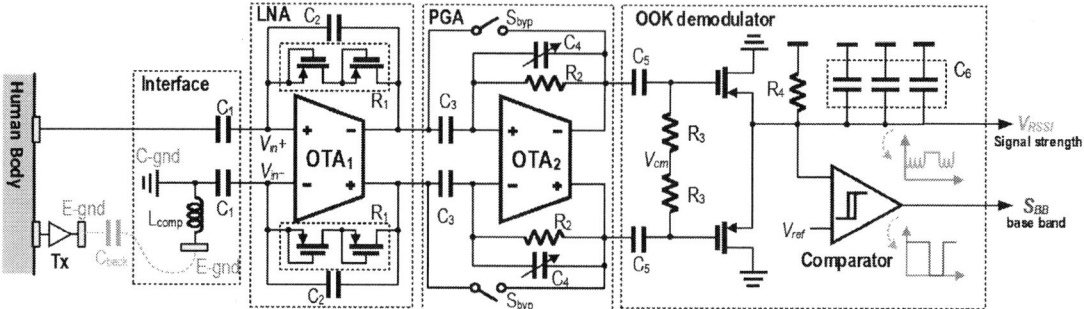

Fig. 3: Schematic of the BCC receiver with a capacitive interface.

However, the BCC transceiver with resonance-matching technique still faces with two main problems. Firstly, since backward capacitance varies with the environment and human posture changes, neither fixed nor manually tuned compensation can properly compensate for the path loss in practical scenario. Secondly, the quality-factor (Q-factor) of the compensation inductor (L_{comp}) and thus the compensation strength will be degraded due to the resistor interface posed to the conventional BCC receivers. As a consequence, the resonance-matching technique can only operate in quiescent and low input impedance scenarios.

II. System Description

This work proposes an energy-efficient BCC transceiver, which employs auto-loss-compensation (ALC) technique to attenuate the backward coupling capacitance induced path loss in CC-BCC, and thus reduces the required power consumption for data transmission. The BCC receiver also employs capacitive interface to increase the input impedance of the receiver without degrading the compensation inductor, As a result, the loss compensation strength and the energy efficiency could be further improved.

Fig. 1 shows the proposed BCC transceiver, which consists of three blocks, namely, BCC receiver, BCC transmitter and the ALC system. The BCC receiver employs an LNA with a purely capacitive interface so that the Q-factor of the compensation inductor (L_{comp}) will not be degraded, and thus leading to an enhancement of the compensation strength. The ALC system consists of an ultra-low power received signal strength indicator (RSSI) to monitor the received signal power, and a maximum power point tracking (MPPT) block that can automatically keep the communication system in the minimum loss situation. Moreover, the transceiver employs an OOK modulation scheme to further reduce the power consumption.

III. Circuit Design

A. BCC Receiver with a Capacitive Interface

Fig. 3 depicts the schematic of the proposed BCC receiver. To resonant with the ground capacitance (C_{back}) and increase the channel gain, a tunable compensation inductor (L_{comp}) is used to bridge circuit ground (C-gnd) and electrode ground (E-gnd) on the receiver end. The low-noise amplifier (LNA) is a trans-capacitance amplifier with 20 dB gain (10×) and an equivalent 1 kΩ input impedance at the carrier frequency,

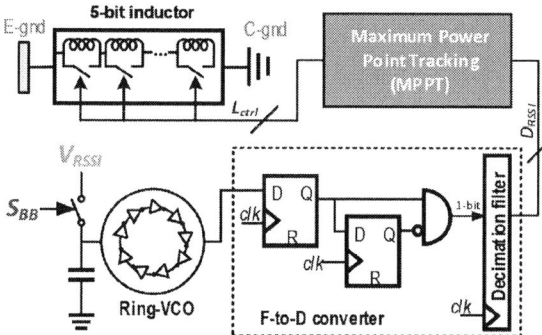

Fig. 4: Schematic of the auto-loss-compensation circuit.

it has a capacitive interface that will not degrade the Q-factor of L_{comp}. An additional 15 dB gain enhancement can be achieved compared to conventional resistive interface according to simulation results as shown in Fig. 1(b). The gain of the subsequent programmable gain amplifier (PGA) can be tuned from 1× to 10×, when the required gain is less than 5×, the tail current of the operational amplifier (OTA$_2$) in the PGA will be cut in half to save power. The fully differential topology attenuates the second harmonics and thus increases the data rate as well as the energy efficiency. It is followed by a hysteresis comparator with enhanced decision robustness.

B. Auto-loss-compensation Circuit

As shown in Fig. 1(a), for the purpose of minimizing the channel loss, the basic concept of the ALC technique is to automatically match the channel resonant frequency (f_r) with the carrier frequency (f_c) by tuning the compensation inductor (L_{comp}), thus minimizing the channel loss.

The proposed ALC system consists of 3 key blocks, including an ultra-low-power RSSI, a digital controlled inductor and an MPPT block. The conventional logarithmic RSSI consumes too much power, which impede the power benefit from the ALC system. To solve the power consumption issue, an ultra-low-power RSSI is proposed as shown in Fig. 4. It achieves low power with sufficient accuracy based on 3 key innovations: (1) extracting the signal strength (V_{RSSI}) by reusing the existing OOK demodulator in Fig. 3 to save power; (2) resampling the rectified OOK signal with the received base-band

978-1-7281-5107-6/19 $31.00 © 2019 IEEE

IEEE Asian Solid-State Circuits Conference
November 4 – 6, 2019
The Parisian Macao, Macao SAR, China

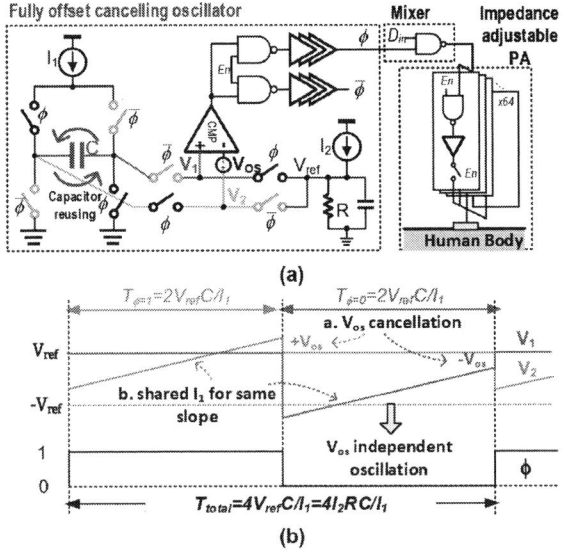

(a)

(b)

Fig. 5: Energy-efficient low-offset BCC receiver, (a) the schematic and (b) the timing diagram.

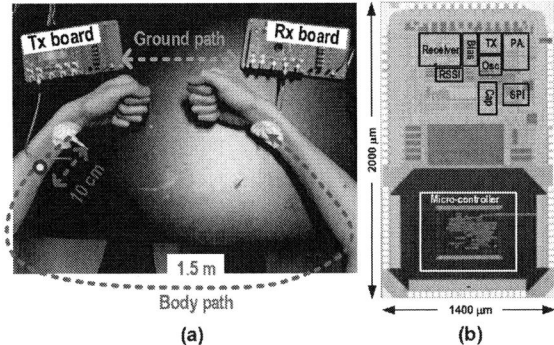

(a)   **(b)**

Fig. 6: (a) The test setup and (b) die photo of the proposed BCC transceiver.

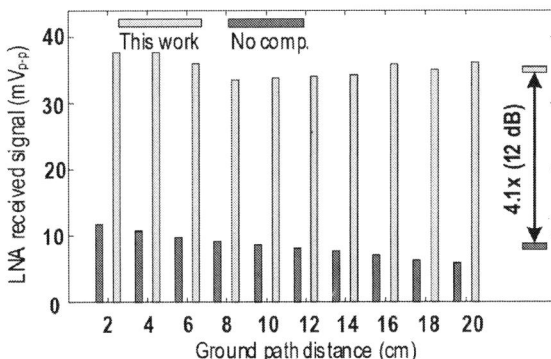

Fig. 7: Measurement result of the proposed ALC system.

signal to reduce the sampling rate. In this way, the detection accuracy can also be improved by eliminating the base-band data dependency of the OOK signal. (3) The time-domain ADC quantizes the sampled signal strength in digital domain as well as to provide a 1st order noise shaping. With the above mentioned methods, the resolution of RSSI (2 mV_{pp} @ 100 Hz sampling rate) is sufficiently high for ALC control whereas the power consumption (30 μW) is much lower than the conventional logarithmic RSSI.

The digital controlled inductor is implemented with a 5-bit binary-weighted serial inductor array with a resolution of 0.5 μH, covering the possible variation range of the ground capacitance (C_{back}). The MPPT block is implemented to converge the system to the minimum loss point (or the maximum received signal power point) by tuning the digitally controlled compensation inductor (L_{comp}) at the receiver side without measuring the exact value of C_{back} [3]. In addition, both the digital-controlled inductor and the MPPT block are implemented with off-chip devices .

C. Energy-efficient BCC transmitter

The schematic of the transmitter is shown in Fig. 5(a), which consists of a relaxation oscillator, a mixer and an impedance adjustable power amplifier (PA). In the oscillator, the integration capacitor will be flipped and reused in the positive and negative half cycles. In this way, the offset of the comparator can be effectively canceled thus leading to significant power saving for the comparator. The measurement shows the oscillator only consumes 26-μW power when operating at a frequency of 21 MHz, equivalent to 1.2-nW/kHz energy efficiency. In order to match the human impedance as well as to optimize the energy efficiency, an impedance adjustable power amplifier (PA) is designed. Thanks to these approaches, the entire transmitter consumes only 120 μW power.

IV. MEASUREMENT RESULTS

The proposed transceiver was fabricated using TSMC 180-nm CMOS process and occupies a total area of 1.4×2 mm^2. Fig. 6 shows the test setup and the die photos. The chips were mounted on the PCB prototypes through chip-on-board (COB) technology. Two prototypes were configured as transmitter and receiver, respectively. During the testing, the prototypes are attached to human bodies through 2.5 cm-radius-size Ag/AgCl medical electrodes. The E-gnd of each prototype is formed by the copper of the PCB whose area is 7×9 mm^2. Both prototypes are battery powered to avoid building-ground return path.

The first experiment validates the functionality of the ALC system. The ground path distance was changed from 2 cm to 20 cm, and the received signal magnitudes of the LNA were monitored when the ALC system was either turned ON or OFF. Fig. 7 shows the measured results. When the ALC system is turned on, the channel loss almost stays constant against the variation of ground path distance, and the magnitude of the received signal can be improved by around 12 dB compared to the "OFF" case.

The second experiment measures the data rate of the proposed transceiver on human body under two different cases: long-distance case (body path distance = 150 cm, ground path distance = 50 cm), and short-distance case (body path distance

978-1-7281-5107-6/19 $31.00 © 2019 IEEE   175

TABLE I: Performance comparison with the previously reported CC-BCC transceivers. (*active area, **entire endoscope SoC)

ParameternPub.	ISSCC'18 [1]	JSSC'15 [4]	JSSC'16 [5] (ET-mode)	JSSC'16 [5] (LP-mode)	JSSC'17 [7]	JSSC'19 [8]	This work
Frequency (MHz)	40, 160	21	40, 160	13.56	20-120	21	21
Process(nm)	65	130	65	65	65	65	180
Supply (V)	1.2	1.2	1.2	0.8	1.1	1.2	0.8
Data Rate (Mbps)	80	1.3125	80	0.1	2	1.3125	4
TX Power (μW)	800-1700	1400	1700-2600	21	870	3520	120
RX Power (μW)	8000	5000	6300	42.5	1100	620	160
RX Bit energy (pJ/bit)	100	3960	79	430	550	472	41
Modulation	BPSK, QPSK	FSDT	BPSK	OOK	P-OFDM	FSDT	OOK
Bit Err. Rate	$< 10^{-5}$	N.A.	$< 10^{-5}$	$< 10^{-5}$	$< 10^{-7}$	$< 10^{-7}$	$< 10^{-5}$
Area (mm^2)	16**	N.A.	5.8*	0.17*	0.54*	0.67	2.8
ALC control	No	No	No	No	No	No	Yes

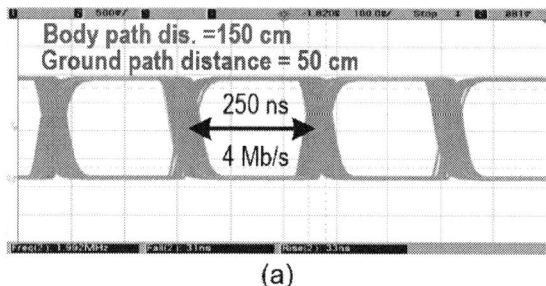

(a)

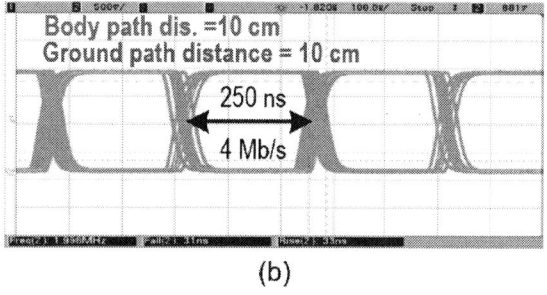

(b)

Fig. 8: Measured eye diagrams of the proposed BCC transceiver in (a) the long-distance scenario and (b) the short-distance scenario.

= 10 cm, ground path distance = 10 cm). The transmitter is configured to send repetitive patterns. Fig. 8 shows the measured eye diagrams under two cases from the receiver side. The data rate of this transceiver achieves 4 Mbps over 1.5-m body path distance and the sensitivity is -48 dBm, which is sufficient for most wearable applications. The transceiver operates under a 0.8-V single supply, and the measured power consumption of the transmitter and receiver are 120 μW and 164 μW, respectively, with an energy efficiency of 41 pJ/bit. Table. I shows the comparison with the existing CC-BCC transceivers, this work achieves the highest power efficiency thanks to the enhanced ALC techniques.

V. CONCLUSION

This paper proposes a BCC transceiver with an enhanced ALC technique, it can automatically adjust the compensation

inductor for resonance-matching technique and thus minimize the path loss. The receiver is designed with a capacitive interface to enhance the compensation strength of the ALC system. In which an RSSI with an extremely low power consumption of 30 μW is designed to monitor the received signal power in real-time by utilizing base-band re-sampling and time-domain data conversion techniques. The measurement result shows the ALC system can lower the path loss by 12 dB compared to that without compensation. And the transceiver achieves 4-Mbps data rate with a 41-pJ/bit efficiency over 1.5-m distance, it shows great prospect in low power wireless body-area-network applications.

REFERENCES

[1] J. Jang, J. Lee, K. Lee, J. Lee, M. Kim, Y. Lee, J. Bae, and H. Yoo, "4-camera vga-resolution capsule endoscope with 80mb/s body-channel communication transceiver and sub-cm range capsule localization," in *2018 IEEE International Solid - State Circuits Conference - (ISSCC)*, Feb 2018, pp. 282–284.

[2] J. Lee, K. Kim, M. Choi, J. Sim, H. Park, and B. Kim, "A 16.6-pj/b 150-mb/s body channel communication transceiver with decision feedback equalization improving >200 % area efficiency," in *2017 Symposium on VLSI Circuits*, June 2017, pp. C62–C63.

[3] J. Zhao, J. Mao, W. Sun, Y. Huang, B. Zhao, H. Yang, and Y. Liu, "An auto loss compensation system for capacitive coupled body channel communication," *IEEE Transactions on Biomedical Circuits and Systems*, pp. 1–1, 2019.

[4] H. Cho, H. Lee, J. Bae, and H.-J. Yoo, "A 5.2 mw ieee 802.15. 6 hbc standard compatible transceiver with power efficient delay-locked-loop based bpsk demodulator," *IEEE journal of solid-state circuits*, vol. 50, no. 11, pp. 2549–2559, 2015.

[5] H. Cho, H. Kim, M. Kim, J. Jang, Y. Lee, K. J. Lee, J. Bae, and H. Yoo, "A 79 pj/b 80 mb/s full-duplex transceiver and a 42.5μw 100 kb/s super-regenerative transceiver for body channel communication," *IEEE Journal of Solid-State Circuits*, vol. 51, no. 1, pp. 310–317, Jan 2016.

[6] J. Bae, K. Song, H. Lee, H. Cho, and H. Yoo, "A 0.24-nj/b wireless body-area-network transceiver with scalable double-fsk modulation," *IEEE Journal of Solid-State Circuits*, vol. 47, no. 1, pp. 310–322, Jan 2012.

[7] W. Saadeh, M. A. B. Altaf, H. Alsuradi, and J. Yoo, "A 1.1-mw ground effect-resilient body-coupled communication transceiver with pseudo ofdm for head and body area network," *IEEE Journal of Solid-State Circuits*, vol. 52, no. 10, pp. 2690–2702, Oct 2017.

[8] B. Zhao, Y. Lian, A. M. Niknejad, and C. H. Heng, "A low-power compact ieee 802.15. 6 compatible human body communication transceiver with digital sigma–delta iir mask shaping," *IEEE Journal of Solid-State Circuits*, vol. 54, no. 2, pp. 346–357, 2019.

A battery-less 31 µW HBC receiver with RF energy harvester for implantable devices

Jihee Lee, Jaeeun Jang, Jaehyuk Lee, and Hoi-Jun Yoo
School of Electrical Engineering
KAIST (Korean Advanced Institute of Science and Technology), Daejeon, Korea
Email: jihee.lee@kaist.ac.kr

Abstract—A power-efficient human body channel (HBC) RX IC which operates without an external battery is implemented in 65 nm CMOS mixed mode process with 0.6 V supply voltage. The HBC RX IC is designed to receive intermittent data from an external device while generating power by itself. The HBC RX IC integrates an ambient RF energy harvester (EH), a wireless power transfer (WPT)-based wake-up receiver (WuRX), and a low-power main HBC RX. The ambient RF EH harvests stand-by power of the proposed HBC IC with dynamic input range. The WPT-based WuRX receives incoming wake-up signal and generates power for main HBC RX with 68% power conversion efficiency (PCE) at the same time. The main HBC RX demodulates both ASK and FSK information with the same oscillation with 31 µW low-power consumption. As a result, the proposed HBC RX IC works stand-alone while performing 0.97 µW standby power and 31 µW active power with 0.13nJ/b energy efficiency.

Keywords—*battery-less, body area network, body channel communication, energy harvester, human body communication, implantable device, low power, receiver, wake-up, wireless power transfer*

I. INTRODUCTION

In recent years, low-power implantable devices have drawn significant attention for wireless physiological monitoring applications such as a pacemaker [1] and a neuro-stimulator [2]. Recent smart implantable devices integrate the communication block inside to transmit physiological signal and receive an order to decide the operation of devices [3]. However, this communication system suffers from highly-constrained power and area restriction as the volume of the implantable devices has been shrinking continuously [4]. The communication system including an RF antenna, an IC, and a battery should occupy a small area to affect patients' life minimally while performing the long-term operation.

Energy harvesting (EH) and wireless power transfer (WPT) have been widely adopted in the implantable devices to generate power for the sensor node and extend battery life [5]. An ambient RF EH technology has several advantages that it can be always available and it does not require other large external components except an antenna to capture the RF signal. However, the harvested power density by ambient RF signal is under 1µW/cm², which is too limited to operate the sensor node [6]. On the other hand, the WPT can generate enough power to operate the sensor node, but it only receives

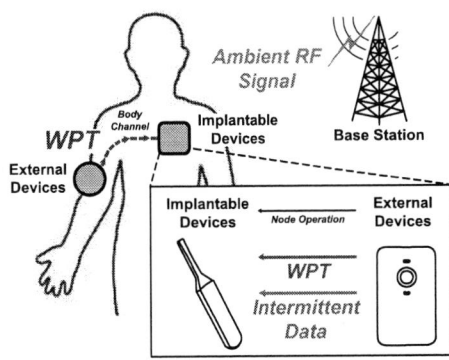

Fig. 1. Data communication and power transfer/harvesting for proposed implantable devices.

power when the external device transmits power [7]. Therefore, utilizing both EH and WPT can harvest a small amount of power all the time, and generate enough power when it requires.

The communication system should consume low power to perform the long-term operation with restricted power. The human body communication (HBC) which uses the human body as the communication channel [8] can be utilized for low-power operation as it has shown energy-efficient performance compared to air channel communication. Moreover, it can be integrated into a smaller area since it receives an RF signal through electrodes, not an antenna [9]. To reduce more power consumption, the use of auxiliary and always-on wake-up receiver (WuRX) is one of the most promising solutions because the communication of implantable devices is usually intermittent, not operating all the time. The always-on low-power WuRX defines the sleep mode of the sensor node by an incoming wake-up signal with the heavily duty-cycled operation which can significantly reduce the unnecessary communication power of the sensor nodes. [10] Thus, the hardware development of the HBC system with a WuRX can satisfy the tight power and cost requirement of the implantable system.

In this paper, we present an HBC receiver system that can be operated without an external battery (Fig. 1). The ambient RF EH harvests power all the time for operation of the always-on WuRX. When the external device transfers power wirelessly with the modulated wake-up signal, a WPT-based WuRX decodes an incoming signal and turns on the sensor

978-1-7281-5107-6/19 $31.00 © 2019 IEEE

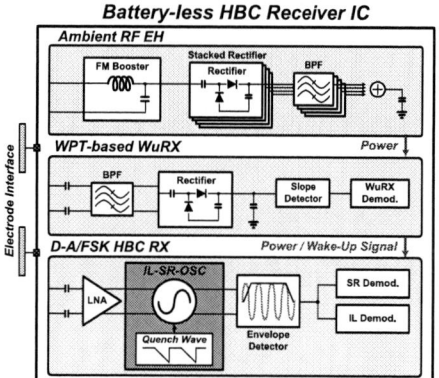

Fig. 2. Overall architecture of proposed battery-less HBC receiver IC.

node while generating power to perform the main RX. Finally, the main HBC RX receives the RF signal with low-power consumption.

The rest of this paper is organized as follows. In Section II, the proposed battery-less HBC receiver system will be introduced. Section III discusses the key building block of the proposed RX including: 1) dual amplitude-shift-keying / frequency-shift-keying modulated HBC receiver (D-A/FSK RX); 2) WPT-based WuRX; 3) wide-sensitivity ambient RF EH. Section IV shows the implementation and measurement results, and finally, Section V concludes this paper.

II. PROPOSED BATTERY-LESS HBC RECEIVER SYSTEM

Fig. 2 shows the overall block diagram of the proposed battery-less HBC receiver system. The HBC receiver IC consists of an ambient RF EH, a WPT-based WuRX, and a D-A/FSK RX. An ambient RF EH harvests and stores power to operate the WuRX continuously. The resonance-based RF booster is used to increase the sensitivity and a multi-stage stacked rectifier is adapted to guarantee high power conversion efficiency (PCE) whereas input signal strength is different. An always-on WuRX turns on the main receiver from the sleep mode when wake-up signal incomes to the HBC RX IC. Also, the WuRX generates power to operate the main receiver from the input RF wake-up signal. The main receiver is composed of an LNA, an injection-locking super-regenerative oscillator (IL-SR-OSC), an envelope detector, and demodulators. The IL-SR-OSC plays an important role in making the receiver power-efficient because it gives both amplitude-to-envelope and frequency-to-envelope conversion for ASK and FSK demodulation, respectively.

To operate HBC RX IC without battery, the building blocks should work organically. An ambient RF EH harvests a small amount of power for the WuRX all the time, and the WuRX always turns on and listens wake-up signal. The incoming wake-up signal should last at least 3 seconds with a minimum -15 dBm input to generates enough power to operate the main HBC RX for 1 second.

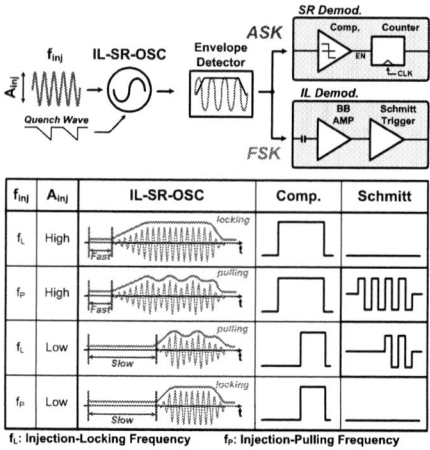

Fig. 3. Concept diagram of the proposed IL-SR-OSC and its demodulation.

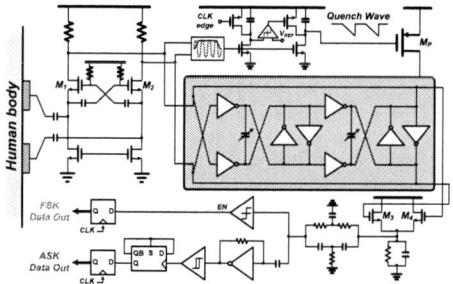

Fig. 4. Schematic of the dual ASK/FSK modulated HBC receiver.

III. CIRCUIT IMPLEMENTATION

A. Dual ASK/FSK modulated HBC receiver (D-A/FSK RX)

Fig. 3 describes the overall HBC RX architecture with the conceptual diagram of the IL-SR-OSC. The IL-SR-OSC receives the input data which is amplified by the LNA and the output waveform of the IL-SR-OSC differs by the input signal data. The input sine wave is modulated by both ASK and FSK with an amplitude of low and high, and its frequency of 40-MHz f_L (injection-locking frequency) and 32-MHz f_P (injection-pulling frequency), respectively. The ASK modulated data change the starting time of the oscillation of an SR-OSC [9] whereas the FSK modulated data defines the envelope shape of the IL-OSC [11]. The SR-OSC requires more time to start oscillation if the amplitude of received input is lower, corresponding to input data 0. On the other hand, the IL-OSC differs its envelope shape according to the frequency of the received input data. The proposed HBC receiver demodulates both ASK and FSK with a common IL-SR-OSC. Nothing that both demodulations share the same oscillator which is the most power-hungry block of the receiver, the total power consumption of the IL-SR-OSC is comparable to the previous IL-OSC or SR-OSC with twice modulation index.

In Fig. 4. a detailed schematic of the D-A/FSK HBC RX is shown. The common gate LNA with the gm-boosting cross-coupled NMOS M_1, M_2 amplifies the input signal. The amplified signal is injected into a power-efficient current-

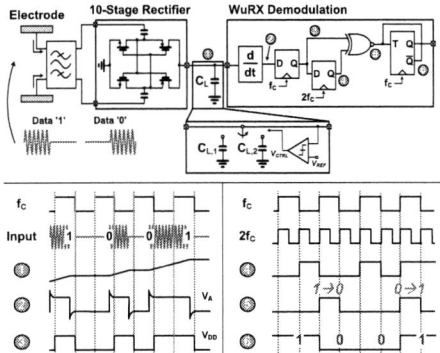

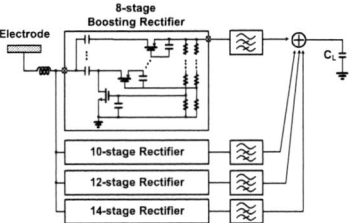

Fig. 5. Demodulation of proposed WPT-based wake-up receiver.

steering ring oscillator whose free-running frequency is 40-MHz. The ring oscillator shows better power and area performance compared to the common R-C and L-C oscillator. The oscillation frequency can be tuned by capacitors placed on the intermediate nodes. However, the time difference of the ring-based SR-OSC in accordance with the input amplitude is less than other oscillators. To control the time difference more dramatically, the input signal is injected into not only the oscillator but also the quench wave generator after the envelope detector. The envelope information decides the fall timing of the quench wave which also affects oscillation timing. The quench wave controls the IL-SR-OSC through the power MOS M_P. The output of the oscillator is converted into the envelope information by an envelope detector which adopts common drain differential NMOS M_3, M_4 with a low-pass filter load. The envelope information is transferred to 2-path: one is for the ASK demodulation with a comparator and a counter; the other is for the FSK demodulation with an AC-coupled inverter, a Schmitt trigger, an SR flip-flop, and a D flip-flop. The off-chip 1-MHz crystal oscillator is used to generate clock source for D flip-flops.

B. Wireless power transfer-based wake-up receiver (WPT-based WuRX)

The proposed WPT-based WuRX receives incoming wake-up signal while harvesting power to operate the main HBC RX. (Fig. 5) The WPT-based WuRX is composed of a band-pass filter (BPF), a 10-stage V_{th}-cancellation rectifier [12], two 60-μF super-capacitors, and a demodulator. The input is pulse position modulated to transfer the same amount of energy whereas its transmitting data is different. Incident RF power at least -15 dBm is provided wirelessly for 3 seconds and rectified through an RF rectifier front-end consisting of a 10-stage charge pump. The voltage of the charging node increases progressively and the timing of the slope of the charged voltage is different by transmitted data. A differentiator extracts the slope of the charged voltage and it shows high-to-low transition and low-to-high transition for data 1 and 0, respectively. As the charge is accumulated into the capacitor, the voltage is saturated and the slope of the charging capacitor becomes gentle. Two capacitors are used alternatively based on the ping-pong technique to prevent bit-lose while storing charge continuously. The result of the differentiator is sampled by a D flip-flop and the other D flip-flop which is operating

with twice clock generates a 180°-phase-shifted signal. An XNOR gate that operates with the original signal and shifted signal as input shows 'high' when the transition of the input signal occurred. Finally, a T flip-flop extracts the transmitted signal by the result of the XNOR gate. Consequently, the WPT-based WuRX can demodulate the input signal while accumulating energy into super-capacitor. The total WPT-based WuRX only consumes 0.97 μW power, as the WuRX is composed of passive building blocks and a demodulator with a simple differentiator and digital gates.

C. Multi-stage stacked ambient RF energy harvester

The human body under electromagnetic fields behaves as an antenna with its resonance frequency determined by the wavelength equal to twice of the human height. The human body operates as a wideband antenna in 40-400 MHz frequency range and it receives up to -30 dBm of 88-108 MHz FM radio broadcasting signal [13].

The ambient FM signal strength differs by the distance between user and radio broadcasting tower and the location of the user. As the input sensitivity can be varying dramatically, the rectifier should be designed to harvest wide-dynamic-range of the input RF signal. Fig. 6 shows the proposed ambient RF energy harvester. The received RF energy is boosted to increase the sensitivity of the rectifier with off-chip 0.2 mH inductor [14] and it is converted into DC voltage by several rectifiers. The optimized number of the stages of the rectifiers is different according to input sensitivity and power conversion efficiency (PCE) [15]. The proposed ambient RF rectifier stacks 4 multi-stage rectifiers in parallel to guarantee high PCE for wide input range. The number of stages of 4 parallel rectifiers is 8, 10, 12, and 14. Each rectifier is composed of the single-ended voltage doubler with 3.3 nF stage capacitor to reduce the area. For each branch, the RF signal is blocked by a low-pass filter to pass only the DC component to the output load and it is summed into the 5-μF charging capacitor.

IV. MEASUREMENT RESULTS

The proposed HBC receiver IC in 65-nm mixed-mode CMOS process while occupying an area of 1.1×0.9 mm². The performance summary with the die micrograph and the comparison with the state-of-the-art HBC RX are illustrated in Fig. 7 and Table I. The proposed HBC RX consumes 31 μW with a data rate of 240 kb/s with 0.13 nJ/b energy efficiency. The proposed D-A/FSK RX achieves the lowest power consumption with a comparable data rate.

978-1-7281-5107-6/19 $31.00 © 2019 IEEE

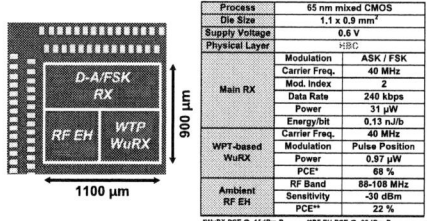

Process	65 nm mixed CMOS		
Die Size	1.1 x 0.9 mm²		
Supply Voltage	0.6 V		
Physical Layer	HBC		
Main RX	Modulation	ASK / FSK	
	Carrier Freq.	40 MHz	
	Mod. Index	2	
	Data Rate	240 kbps	
	Power	31 µW	
	Energy/bit	0.13 nJ/b	
WPT-based WuRX	Carrier Freq.	40 MHz	
	Modulation	Pulse Position	
	Power	0.97 µW	
	PCE*	68 %	
Ambient RF EH	RF Band	88-108 MHz	
	Sensitivity	-30 dBm	
	PCE**	22 %	

*WuRX PCE @ -15dBm P_in   **RF EH PCE @ -30dBm P_in

Fig. 7. Chip micrograph and performance summary

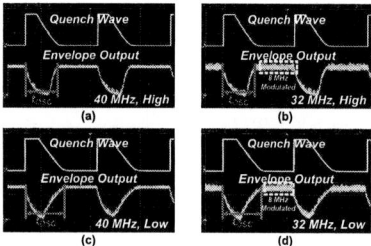

Fig. 8. Measured transient waveform of D-A/FSK RX

Fig. 8 shows the measured transient waveforms of the D-A/FSK RX with 4 different inputs at the 240 kbs/ data rate. A distance between TX and RX is 20-cm while transferring 10-dBm input signal at the TX. The ASK and FSK information can be distinguished by the starting time of the oscillation and the envelope shape, respectively.

Fig. 9 illustrates the transient waveform of the WPT-based WuRX. The voltage of the charging node increases as input comes and the demodulator extracts modulated input based on the slope of the charging node. The sensitivity of the WPT-based WuRX is -15 dBm to guarantee the slope of the charging node.

The PCE of the ambient RF EH with wide dynamic input range is shown in Fig. 10. The measured PCE of the proposed multi-stage stacked rectifier with -30 dBm input is 22% which is 4.2% higher than the PCE of the single 10-stage rectifier.

V. CONCLUSION

In this paper, a battery-less HBC RX IC for implantable devices with low-power and high-PCE is proposed. For low-power energy-efficient HBC RX, a D-A/FSK RX with an IL-SR-OSC is presented. It demodulates both ASK and FSK data with the same oscillator which is the most power-hungry block among receiver components. Also, a WPT-based WuRX is adapted to receive incoming wake-up signal while harvesting

TABLE I. COMPARISON OF THE STATE-OF-THE-ART HBC RX

	[9]	[11]	[15]	**This Work**
Process	180 nm	65 nm	130 nm	**65 nm**
Supply	0.8 V	0.7 V	0.7 V	**0.6 V**
Frequency	13.56 MHz	80 MHz	45 MHz	**40 MHz**
Modulation	OOK	FSK	FSK	**ASK/FSK**
Demodulation	SR*	IL**	IL **	**SR* / IL****
Modulation Index	1	1	1	**2**
Power	42.5 µW	45 µW	37.5 µW	31 µW
Data rate	100 kb/s	312 kb/s	200 kb/s	**240 kb/s**
Energy	0.43 nJ/b	0.14 nJ/b	0.19 nJ/b	0.13 nJ/b

*SR: Super-Regenerative   **IL: Injection-Locking

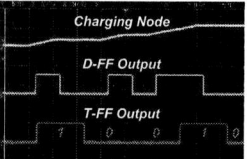

Fig. 9. Measured transient waveform of the WPT-based WuRX.

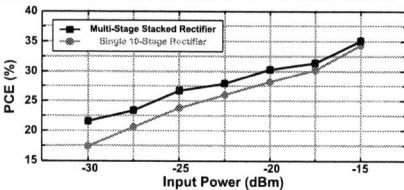

Fig. 10. PCE of the ambient RF EH depending on the input power.

power for main RX with 68% PCE. An ambient RF EH with multi-stage stacked rectifier is proposed to perform high PCE with wide dynamic range input. As a result, the proposed RX IC consumes only 31 µW power and it can be operated without the external battery.

REFERENCES

[1] L. S. Y. Wong et al., "A very low-power CMOS mixed-signal IC for implantable pacemaker applications," *IEEE Journal of Solid-State Circuits*, vol. 39, no. 12, pp. 2446-2456, Dec. 2004

[2] G. J. Suaning et al., "CMOS neurostimulation ASIC with 100 channels, scaleable output, and bidirectional radio-frequency telemetry," *IEEE Trans. on Biomedical Engineering*, vol. 48, no. 2, pp. 248-260, Feb. 2001.

[3] Biomonitor 2, "Technical manual Biomonitor 2-AF/-S," Aug. 2017. [Online]. Available: https://www.biotronik.com/en-us/product

[4] Y. Shi et al., "A 10 mm³ inductive coupling radio for syringe-implantable smart sensor nodes," *IEEE Journal of Solid-State Circuits*, vol. 51, no. 11, pp. 2570-2583, Nov. 2016.

[5] J. Olivo et al., "Energy harvesting and remote powering for implantable biosensors," *IEEE Sensors Journal*, vol. 11, no. 7, pp. 1573-1586, Jul. 2011.

[6] S. Kim et al., "Ambient RF energy-harvesting technologies for self-sustainable standalone wireless sensor platforms", *Proceedings of the IEEE*, vol. 102, no. 11, pp. 1649-1666, Nov. 2014.

[7] A. K. RamRakhyani et al., "Design and optimization of resonance-based efficient wireless power delivery systems for biomedical implants," *IEEE Trans. on Biomedical Circuits and Systems*, vol. 5, no. 1, pp. 48-63, Feb. 2011.

[8] N. Cho et al., "The human body characteristics as a signal transmission medium for intrabody communication," *IEEE Trans. on Microwave Theory and Techniques*, vol. 55, no. 5, pp. 1080-1086, May 2007.

[9] H. Cho et al., "A 79 pJ/b 80 Mb/s full-duplex transceiver and a 42.5 µW 100 kb/s super-regenerative transceiver for body channel communication," *IEEE Journal of Solid-State Circuits*, vol. 51, no. 1, pp. 310-317, Jan. 2016.

[10] I. Demirkol and C. Ersoy, "Wake-up receivers for wireless sensor networks: Benefits and challenges," *IEEE Wireless Communications*, pp. 88-96, Aug. 2009.

[11] J. Bae and H. Yoo, "A 45 µW injection-locked FSK wake-up receiver with frequency-to-envelope conversion for crystal-less wireless body area network," *IEEE Journal of Solid-State Circuits*, vol. 50, no. 6, pp. 1351-1360, Jun. 2015.

[12] K. Kotani et al., "High-efficiency differential-drive CMOS rectifier for UHF RFIDs," *IEEE Journal of Solid-State Circuits*, vol. 44, no. 11, pp. 3011-3018, Nov. 2009.

[13] N. Cho et al., "A 60 kb/s-10 Mb/s adaptive frequency hopping transceiver for interference-resilient body channel communication," *IEEE Journal of Solid-State Circuits*, vol. 44, no. 3, pp. 708-717, Mar. 2009

[14] L. G. Carli et al., "Maximizing the power conversion efficiency of ultra-low-voltage CMOS multi-stage rectifiers," *IEEE Transactions on Circuits and Systems I: Regular Papers*, vol. 62, no. 4, pp. 967-975, Apr. 2015.

[15] H. Cho et al., "A 37.5 µW body channel communication wake-up receiver with injection-locking ring oscillator for wireless body area network", *IEEE Transactions on Circuits and Systems I: Regular Papers*, vol. 60, no. 5, pp. 1200-1208, May 2013.

A Piezoelectric Energy Harvesting Interface for Irregular High Voltage Input with Partial Electric Charge Extraction with 3.9× Extraction Improvement

Muhammad Bilawal Khan, Hassan Saif, and Yoonmyung Lee

Department of Electrical and Computer Engineering, Sungkyunkwan Unversity, Suwon, South Korea

bilawal786@skku.edu, yoonmyung@skku.edu

Abstract—A novel energy harvesting technique for piezoelectric transducers (PZTs) is proposed to enable energy harvesting from irregular high voltage inputs. With the proposed partial electrical charge extraction (PECE) scheme, a small portion of the charges generated by the PZT is repeatedly transferred to the battery until the input excitation is finished. With PECE, the PZT output voltage is automatically regulated at the maximum voltage level allowed by technology (V_{MAX}) while maximizing energy extraction from the PZT. This allows energy harvesting from irregular PZT excitations that can generate open circuit voltage (V_{OC}) higher than V_{MAX}. The measurement results with a test chip fabricated in 350nm process show that the proposed harvester successfully harvests energy from the excitation with V_{OC} of up to 60V, higher than V_{MAX} of 30V, with up to 3.9× improvement compared with a conventional full bridge rectifier (FBR).

Keywords—*piezoelectric energy harvester, PZT, PECE, partial electric charge extraction, partial harvesting*

I. INTRODUCTION

Energy harvesting from the ambient environment has been drawing attention for its potential to realize self-powered IoT systems. If sufficient energy can be harvested, such systems can potentially achieve perpetual operation without battery replacement or recharging, which can significantly reduce maintenance costs for remotely dispatched IoT systems or implanted diagnostic systems.

One of the most widely investigated harvesting schemes in the research community is kinetic energy harvesting with piezoelectric transducers (PZTs). A PZT can be electrically modelled as a current source (I_{PZ}) and an internal capacitance (C_{PZ}) as shown in Fig. 1. As mechanical force is applied on a PZT, I_{PZ} is generated and develops output voltage by charging C_{PZ}. There have been many previous harvesting interface circuits proposed [1-5] to efficiently collect charge and energy generated by the PZT.

One of the most popular PZT energy harvesting schemes is synchronous switch harvesting on inductor (SSHI) [1-2], which can be simplified as shown in Fig. 1(a). In this scheme, energy generated by the PZT is transferred to a large load capacitor, C_L, which acts as a temporary storage, allowing DC-DC converter-like harvesting circuit implementation in a later stage. For efficient operation, bias flipping (Fig. 1(b)) is often implemented (with L_1 and S_L in Fig. 1(a)) to handle alternating input from the PZT. However, if the input is not regular and periodic (e.g. harvesting from sporadic human motion or vehicle movement), maintaining efficient harvesting operation become significantly challenging due to the difficulty in

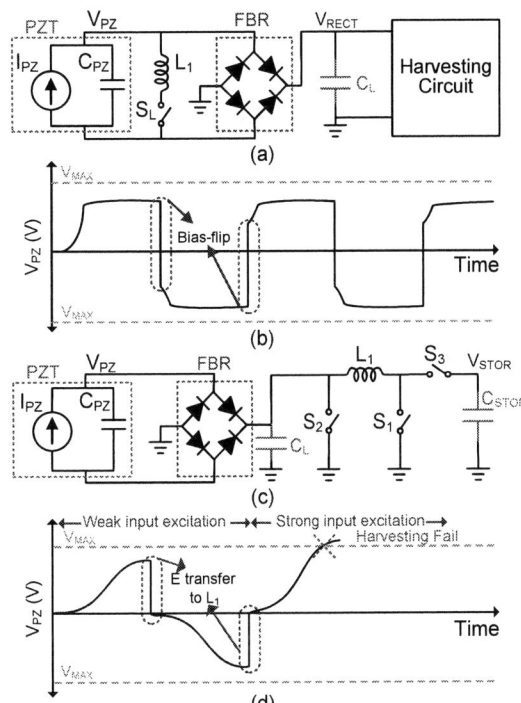

Fig. 1. (a) SSHI harvesting interface circuit and (b) its operation waveform (c) SECE harvesting interface circuit and (d) its operation waveform with different input excitation strength

tracking continuously a changing maximum power point and bias flipping timing.

Synchronous electric charge extraction (SECE) is another popular scheme for PZT energy harvesting [3-4], which can be simplified as shown in Fig. 1(c). In this scheme, sporadic input can be handled since the generated energy is collected from the PZT when the output voltage (V_{PZ}) is peaked as shown in Fig. 1(d). The peak voltage is proportional to the strength of excitation. Therefore, with strong excitation, peak V_{PZ} can exceed the maximum voltage allowed by technology (V_{MAX}) and damage the integrated circuit (IC). A simple solution to keep the peak V_{PZ} below V_{MAX}, is to use a larger C_L. However, increasing C_L significantly reduces the extractable energy from PZT [5] for the following reason: When the PZT is deformed, current I_{PZ} is generated and charges C_{PZ}. The amount of generated charge (Q_{PZ}) is proportional to the amount of

IEEE Asian Solid-State Circuits Conference
November 4 – 6, 2019
The Parisian Macao, Macao SAR, China

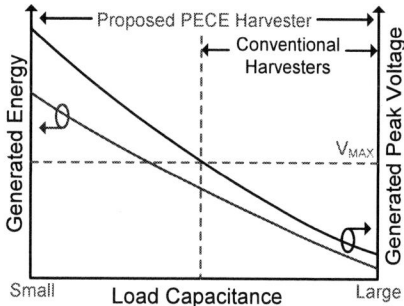

Fig. 2. Comparison between conventional harvesters and proposed PECE harvester for different load capacitances

physical deformation the PZT experienced. Therefore, for a given amount of deformation, Q_{PZ} does not change even if the load capacitance seen by the PZT changes, and the PZT generated energy (E_{PZ}) can be written as

$$E_{PZ} = \frac{1}{2}(C_{PZ} + C_L)V_{PZ}^2 = \frac{1}{2}\frac{Q_{PZ}^2}{C_{PZ}+C_L} \qquad (1)$$

where V_{PZ} refers to the final voltage at the PZT. Therefore, for a given deformation and Q_{PZ}, E_{PZ} is maximized with the smallest C_L. This is because a smaller C_L allows faster V_{PZ} development, and higher V_{PZ} on the PZT makes its damping force stronger [6]. Therefore, to apply same amount of deformation, more energy is required, hence more mechanical energy is converted to electrical energy. This relation is depicted by the plot shown in Fig. 2. As the load capacitance (C_L+C_{PZ}) is reduced, more energy can be generated but the resulting peak voltage also increases. Therefore, to maximize energy extraction, the load capacitance should be minimized while avoiding exceeding V_{MAX}. With a conventional harvesting scheme, V_{PZ} cannot be increased beyond V_{MAX}, limiting harvestable energy.

To address this challenge, a novel harvesting scheme, termed as partial electric charge extraction (PECE), is presented in this paper. With PECE, charges are partially extracted from the PZT whenever the output voltage approaches V_{MAX}. Therefore, this approach allows operation with $C_L=0$ to maximize energy extraction while limiting output voltage within a safe range. This allows the circuit to handle strong input excitation that can generate open circuit peak voltage (V_{OC}) higher than V_{MAX} and therefore harvest energy from PZT materials with very high output voltages. Our measurement with a prototype IC fabricated with 350nm technology confirmed that the proposed harvesting interface successfully harvests energy from stronger excitations with higher V_{OC} up to 60V with up to 3.9× improvement compared with a conventional full bridge rectifier (FBR).

II. PROPOSED PZT HARVESTING INTERFACE WITH PECE

The top level implementation details of the proposed harvesting interface are shown in Fig. 3, and the conceptual operation waveform is shown in Fig. 4. A passive FBR is used to rectify V_{PZ} to V_{RECT}, and an inductor-based converter structure is used to transfer energy from the PZT to the battery. The operation sequence is as follows: As the PZT is deformed, charges are accumulated on C_{PZ}, and V_{PZ} is increased. If V_{PZ} approaches V_{MAX} without a peaking event, a high-voltage tracking comparator detects this condition, and partial charge extraction is performed, as shown in Fig. 4(b), to transfer a small amount of PZT-generated charge to a storage element immediately. Therefore, V_{PZ} is decreased by ΔV, and the PZT

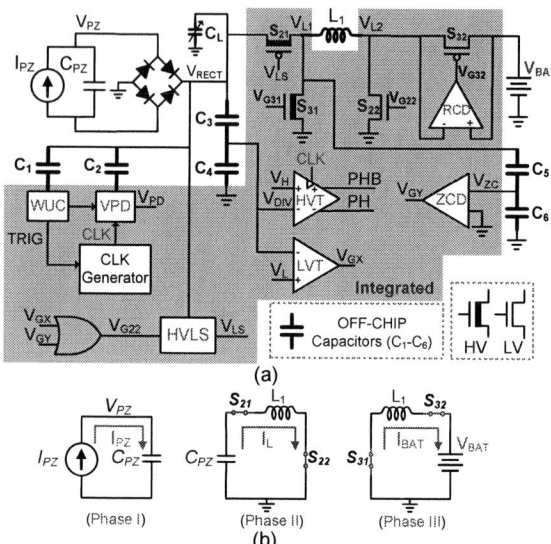

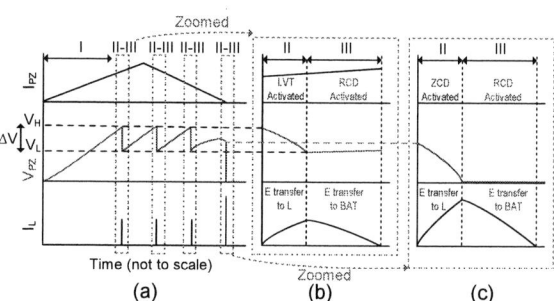

Fig. 3. (a) Top level implementation of proposed harvesting circuit (b) summary of operational phases

Fig. 4. Conceptual diagram of harvesting phases of proposed harvesting circuit (a) overall harvesting waveform (b) partial charge extraction (c) full charge extraction

continues to accumulate V_{PZ}. Whenever V_{PZ} approaches V_{MAX} again, a partial charge extraction is repeated until a peaking event of V_{PZ} (or negative slope) is detected. Then a full charge extraction, as shown in Fig. 4(c), is executed to collect all residual energy generated by the PZT.

To realize this PECE operation, many circuits with dedicated functions are required. Firstly, since the PZT output can be irregular, a wake-up controller (WUC) is employed to detect when energy is harvested by the PZT and activate other circuits. The integrated WUC [8] is interfaced with the PZT using an off-chip capacitor C_1, as shown in Fig. 3(a), which is necessary to deal with high V_{PZ} (V_{RECT}). When energy input is detected, the increasing V_{RECT} needs to be monitored and limited under V_{MAX}. Therefore, a clocked high voltage tracker (HVT) and a voltage peak detector (VPD) [8] are enabled. For the HVT, discrete C_3 and C_4 are used to generate capacitively divided V_{RECT} (V_{DIV}), which is fed to the HVT. An external reference voltage V_H, which serves as the equivalent of V_{MAX}, is compared with V_{DIV} at every rising edge of CLK, and a PH signal is triggered whenever $V_{DIV}>V_H$ (or $V_{RECT}>V_{MAX}$). This activates partial charge extraction and a low voltage tracker (LVT).

During a PECE cycle, S_{21} and S_{22} are turned on for a short duration to transfer energy from C_{PZ} to L_1 until V_{PZ} decreases by ΔV (Phase II in Fig. 3(b)). For this cycle, lower limit V_{PZ} is mapped with the V_L, which is compared with V_{DIV} by the LVT.

978-1-7281-5107-6/19 $31.00 © 2019 IEEE   182

IEEE Asian Solid-State Circuits Conference
November 4 – 6, 2019
The Parisian Macao, Macao SAR, China

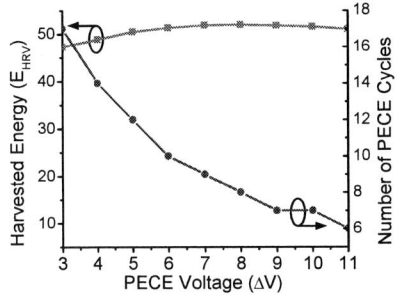

Fig. 5. ΔV variation simulation results (@V_{OC}=100V)

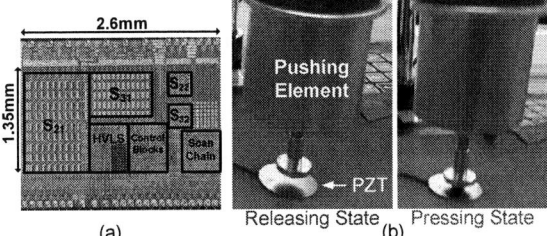

Fig. 6. (a) Chip micrograph (b) PZT pressing setup

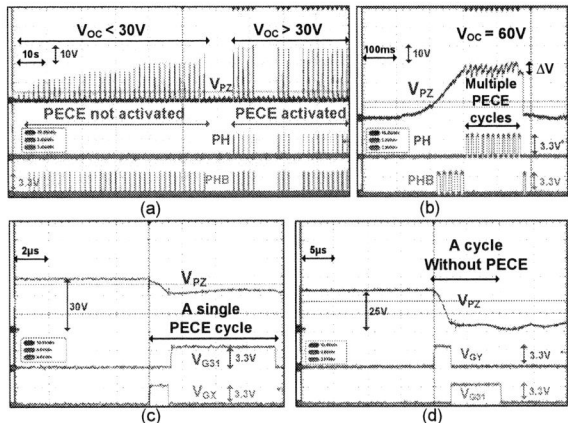

Fig. 7. Measured waveforms with the proposed harvesting circuit (C_L=0) (a) PECE activation with gradual increase of V_{OC}>30 V (b) harvesting @60 V V_{OC} with PECE activation (c) a single harvesting cycle w/ PECE (d) a single harvesting cycle w/o PECE

As C_{PZ} is discharged partially, V_{RECT} decreases as well as V_{DIV}, and harvesting is stopped when V_{DIV}<V_L is detected by the LVT. At this point, S_{21} and S_{22} are turned off, and the energy stored on the inductor is transferred to the battery by turning on S_{31} and S_{32} (Phase III). A reverse current detector (RCD) is used to monitor V_{L2} and V_{BAT} to prevent reverse current flow to the battery, finishing Phase III. At the end of a single PECE cycle, HVT and VPD are activated again simultaneously to monitor V_{PZ}. If V_{RECT} approaches V_{MAX} again, another PECE cycle will be activated, whereas if a peaking event of V_{RECT} is detected, full charge extraction is performed. In the latter case, S_{21} and S_{22} are turned on, and all the energy on the PZT is transferred to L_1 (full harvesting). A zero-crossing detector (ZCD) is used to detect when the inductor current peaks, which can be detected when V_{L1} becomes 0 V. As ZCD is integrated, V_{L1} cannot be directly monitored, and another discrete capacitive divider (C_5, C_6) is used to generate a divided version of V_{L1}, i.e. V_{ZC}. When the inductor current peaks, S_{21} and S_{22} are turned off, and S_{31} and S_{32} are turned on to transfer energy from L_1 to the battery. The RCD is again used to prevent reverse current flow to the battery.

At the end of Phase II during the partial charge extraction cycle, S_{21} needs to be completely turned off. Since S_{21} is a PMOS transistor, it requires a high voltage level shifter (HVLS) that can pull down the gate terminal of S_{21} (V_{LS}) to GND when S_{21} is turned on, and pull it up to V_{RECT} when $_{21}$ is turned off. The details of the HVLS and a few other sub-blocks are omitted here due to space constraints.

For PECE, finding the optimal ΔV is critical for extracting the maximum energy from the PZT. If ΔV is too small, PECE will be triggered too many times before peak detection, incurring large switching and control losses. In contrast, if ΔV is too large, V_{PZ} would be dropped to low voltage, lowering the damping force on the PZT and lowering the amount of extracted energy for a given excitation. To find the proper ΔV, a set of simulations are performed to sweep ΔV for a given PZT excitation input that would generate V_{OC} of 100V, and the net harvested energy is estimated as shown in Fig. 5. It shows that for ΔV lower than 7V, the number of PECE activation cycles increases as ΔV is lowered, and hence the net harvested

energy decreases due to the increasing harvesting overhead. For ΔV higher than 9V, the number of PECE activation cycles decreases as ΔV is increased. However, a higher ΔV also results in a lower damping force at the end of a partial harvesting cycle and hence lower net harvested energy from the PZT. Therefore, ΔV of 8 V is used as the optimal design point for maximum energy extraction in this work.

III. MEASUREMENT RESULTS

A test chip for the proposed harvesting interface circuit was fabricated in 350nm BCD process as shown in Fig. 6(a). A commercially available disc-type piezo element (ABT-441-RC) with 20nF C_{PZ} and a pushing tester (NWT1001V10) were used for testing as shown in Fig. 6(b), and V_{MAX} of 30V is used for measurement experiments.

Fig. 7 shows the measured waveforms with the proposed harvesting interface. Fig. 7(a) shows harvesting operation with gradually increased V_{OC}, i.e. stronger excitation. For V_{OC}<30V, only PHB is activated, indicating full harvesting operations. As V_{OC} is increased beyond 30 V, PH is triggered, indicating PECE activation. Fig. 7(b) shows a zoomed in waveform for PECE operation (11 cycles) for V_{OC}=60V. Fig. 7(c) shows a zoomed in view of a single PECE operation cycle where only 8V of V_{PZ} is decreased with partial harvesting. In contrast, Fig. 7(d) shows a full harvesting operation where V_{PZ} is fully discharged to 0V.

The performance of the proposed harvesting interface is evaluated in Fig. 8 with varying excitation conditions characterized across a range of V_{OC} values. Higher V_{OC} represents the condition with stronger excitation. Fig. 8(a) shows the amount of harvested energy (E_{HRV}) under various excitation conditions. To verify the effectiveness of the proposed PECE technique, harvesting is performed with different load capacitances (C_L). The black curve at the top represents the amount of energy harvested when no load capacitance is used (C_L=0). As V_{OC} exceeds 30V, PECE is required to avoid damaging the IC. Even with very strong excitation (V_{OC}=60V), harvesting can be performed since PECE limits the maximum voltage seen by the IC to 30V. To perform harvesting without PECE, load capacitance needs to be added to lower the maximum V_{PZ} seen by the IC for excitations with V_{OC}>30V. When 10nF C_L is added (pink in Fig. 8(a)), harvesting can be performed with excitation with V_{OC} up to 35V. However, E_{HRV} significantly drops due to inefficient

978-1-7281-5107-6/19 $31.00 © 2019 IEEE

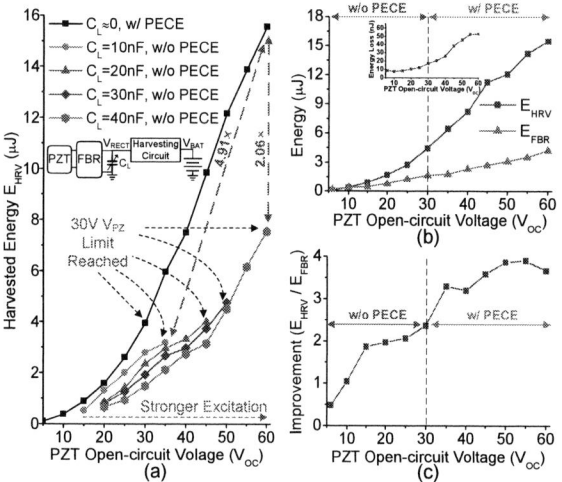

Fig. 8. Measurement results with a range of excitation conditions (a) E_{HRV} w/ different C_L used to keep peak voltage below V_{MAX} (b) E_{HRV} w/ PECE and its comparison with energy harvested using FBR (E_{FBR}) (c) energy extraction improvement compared to FBR-based harvesting

energy extraction with larger load capacitance seen by the PZT source. Similarly, with C_L of 20nF, 30nF, and 40nF, harvesting can be performed for excitations with V_{OC} of 45V, 50V, and 60V, respectively, with significantly lower energy harvested than when PECE is applied with $C_L \approx 0$. This indicates that PECE allows energy harvesting for excitations with V_{OCS} larger than the technology limit (30V) without additional C_L, which makes energy extraction more effective (Eq. (1)). For example, to harvest energy for excitation with V_{OC} of 60V without PECE, 40nF C_L is required and 7.54μJ can be harvested. However, with PECE, no C_L is required, which makes energy extraction more efficient, allowing 15.56μJ of energy to be harvested under the same condition, which is a 2.06× improvement compared with that obtained without PECE.

Fig. 8(b) compares the measured energy harvested using the proposed circuit (E_{HRV}) and using FBR only with C_L=0 (E_{FBR}). E_{HRV} is significantly higher than E_{FBR} for the entire range V_{OC} thanks to the efficient energy extraction achieved by the proposed harvesting interface. The inset diagram in Fig. 8(b) shows the energy consumed during a single harvesting operation (E_{LOSS}), which stays within a reasonable range, on the order of nJs. In addition, the standby power of the proposed

harvesting circuit is 346pW, making it a suitable harvesting interface for harvesters with sporadic energy input. Fig. 8(c) shows the energy extraction improvement compared to a conventional FBR; the maximum improvement ratio was 3.90×. Table I compares the proposed PECE harvester with prior-arts.

IV. CONCLUSIONS

A novel PZT energy harvesting interface with PECE is proposed in this paper. The proposed single-inductor-based harvesting interface is capable of harvesting energy from irregular energy input with very high open-circuit voltages generated by the PZT. Once the PZT voltage exceeds the maximum allowed voltage, the PECE scheme is enabled to harvest maximum energy from the PZT. A prototype circuit fabricated in 350nm technology demonstrated 3.9× energy extraction improvement compared to a conventional FBR.

ACKNOWLEDGEMENT

This research was supported by the National Research Council of Science & Technology (NST) grant by the Korea government (MSIP) (No. CAP-17-04-KRISS). The EDA tool was supported by the IC Design Education Center (IDEC), Korea.

REFERENCES

[1] S. Javvaji, et al., "Multip-step Bias-flip Rectification for Piezoelectric Energy Harvesting," *IEEE ESSCIRC*, pp. 42-45, Sep. 2018.

[2] Y. Ramadass, et al,, "An Efficient Piezoelectric Energy Harvesting Interface Circuit using a Bias-flip Rectifier and Shared Inductor," *IEEE J. Solid State Circuits*, vol. 45, no. 1, pp. 189-204, Jan. 2010.

[3] M. Meng, et al., "Multi-beam Shared-inductor Rconfigurable Voltage/SECE-mode Piezoelectric Energy Harvesting of Multi-axial human motion," *IEEE ISSCC*, pp. 426-427, Feb. 2019.

[4] A. Quelen, et al., "A 30nA Quiescent 80nW-to-14mW Power-Range Shock-Optimized SECE-based Piezoelectric Harvesting Interface with 420% Harvested Energy Improvement," *IEEE ISSCC* , Feb 2018.

[5] M. B. Khan, et al., "Performance Improvement of Flexible Piezoelectric Energy Harvester for Irregular Human Motion with Energy Extraction Enhancement Circuit", *Nano energy*, vol. 58, pp. 211-219, April 2019.

[6] D. Kwon, et al., "A Single-inductor 0.35μm Energy-investing Piezoelectric Harvester," *IEEE ISSCC*, pp. 78-79, Feb. 2013.

[7] J. Yang, et al., "A 2.5-V, 160-μJ-Output Piezoelectric Energy Harvester and Power Management IC for Batteryless Wireless Switch (BWS) Applications," *IEEE VLSI*, Jun 2015, pp. C282-C283.

[8] H. Saif, et al., "A High-Voltage Energy-Harvesting Interface for Irregular Kinetic Energy Harvesting in IoT Systems with 1365% Improvement using All-NMOS Power Switches and Ultra-low Quiscent Current Controller," *Sensors* 2019, 19, 3685.

TABLE I: COMPARISON WITH PRIOR-ARTS

	This Work	[1] 2018 ESSCIRC	[2] 2010 JSSC	[3] 2019 ISSCC	[4] 2018 ISSCC	[7] 2015 VLSI
Process	350nm (High Voltage)	130nm (Standard Voltage)	350nm (Standard Voltage)	350nm (Standard Voltage)	40nm (High Voltage)	250nm (High Voltage)
Die Active Area (mm²)	2.6×1.35	1.9×0.28	4.25	1.4×2.9	1.1×0.5	2×1.2
Harvesting Technique	PECE	SSHI	SSHI	Reconfigurable VM-SECE	SECE	Series-Parallel SC
Piezoelectric Material	Piezo Element ABT-441-RC	PPA1022	MIDE V22B	Custom PZT/Nickel/PZT	MIDE PPA1011	PMN-PT On disc
Piezoelectric Internal Capacitance	20nF (Released state) ~40nF (Pressed state)	14nF	12nF	17nF-49nF	43nF	150nF
Inductor (Off-Chip)	220μH	47μH	0.82mH	2.2mH	2.2mH	470μH
Excitation Type	Irregular Pulse	Periodic	Periodic	Periodic & Shock	Periodic & Shock	Irregular Pulse
Input Voltage	5V-60V	2.5V	2.4V	<5V*	<6V*	35V
Output Voltage	3V-4V	-	1.8V	-	1.5V, 2.8V	2.5V Regulated
Max. Improvement**	3.90×	3.85×	4×	5.11×	4.2×	-
Max. E_{HRV} (Per Pulse)	>15.56μJ	0.122μJ*	0.15μJ*	-	0.55μJ*	160μJ

* Estimated from paper   ** E_{HRV}/E_{FBR}

978-1-7281-5107-6/19 $31.00 © 2019 IEEE

A 100-pA Adaptive-FOCV MPPT Circuit with >99.6% Tracking Efficiency for Indoor Light Energy Harvesting

Peng-Chang Huang and Tai-Haur Kuo
Department of Electrical Engineering
National Cheng Kung University, Tainan, Taiwan
Email: {pchuang_msic, thkuo}@ee.ncku.edu.tw

Abstract—To harvest indoor light energy via photovoltaics (PVs), this paper proposes an adaptive-fractional-open-circuit-voltage (AFOCV) maximum power point tracking (MPPT) circuit to automatically adjust the FOCV fraction under different illuminances for amorphous PV modules. Designed in 0.5-μm CMOS process, the AFOCV circuit consumes an ultra-low average current of 100 pA. Over 100–1000 lux illuminance, the measured MPPT efficiency with AFOCV is >99.6%, which outperforms that of other state-of-the-art energy harvesting (EH) ICs with conventional FOCV methods.

I. INTRODUCTION

Light energy harvesting (EH) is an attractive solution for low-power autonomous devices, such as wireless sensor nodes, to reduce maintenance costs associated with frequent battery replacement. In such applications, amorphous photovoltaic (PV) modules are widely adopted for indoor PV EH [1]. Because the harvested power from small-sized amorphous PV modules under indoor light is low, it is common to adopt a very low-power consuming fractional-open-circuit-voltage (FOCV) method [1]-[3], instead of the power-hungry perturb & observe (P&O) method [4] for maximum power point tracking (MPPT). According to our measured P-V curves, this paper first observes that FOCV fraction k, which is the maximum power point (MPP) voltage V_{MP} divided by the open-circuit voltage V_{OC}, decreases linearly as illuminance increases in several commercial amorphous PV modules, including those from two globally renown manufacturers [5], [6]. Fig. 1(a) shows the measured results from [5] under indoor LED light from 100 to 1000 lux, in which the circles are the MPPs and the triangles are the tracking power point voltages V_{FOCV} in conventional FOCV method. However, as can be seen in Fig. 1(b), the conventional FOCV method with fixed k can not maintain high MPPT efficiency η_{MPPT} over a wide range of illuminance, where k is usually designed for 1000 lux illuminance because a low η_{MPPT} at higher illuminance leads to higher power loss. To reduce the difference between V_{MP} and V_{FOCV} without complicating the operation, this paper proposes an enhancement to conventional FOCV called the adaptive-FOCV (AFOCV) method. Further, this paper realizes a dual-path three-switch (2P3S) converter [1], which has high efficiency and fewer power switches for managing power between the PV, load, and battery, to demonstrate the AFOCV capability.

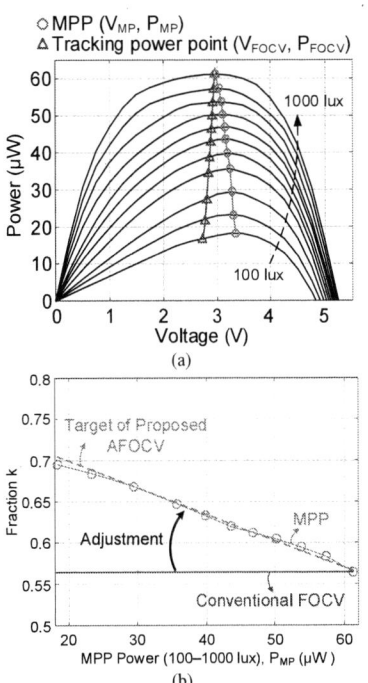

Fig. 1. (a) Measured P-V curves of a commercial amorphous PV module and (b) concept of adaptive-FOCV (AFOCV).

II. PROPOSED AFOCV EH IC

Fig. 2(a) shows the proposed IC with the AFOCV MPPT, including the three power switches and controller for realizing the 2P3S converter. For low power consumption, the converter operates in discontinuous-conduction mode with constant-on-time pulse-skipping and pulse-frequency modulation, while the system clock clk, generated from the oscillator, decreases its frequency from 100 kHz to 10 kHz at stable loads. The AFOCV circuit generates a reference voltage V_{Pref} for regulating the PV voltage V_P at the MPP. The power monitor monitors V_P and load voltage (V_L) for regulation. The pulse generator generates corresponding on-time pulses and off-time

IEEE Asian Solid-State Circuits Conference
November 4 – 6, 2019
The Parisian Macao, Macao SAR, China

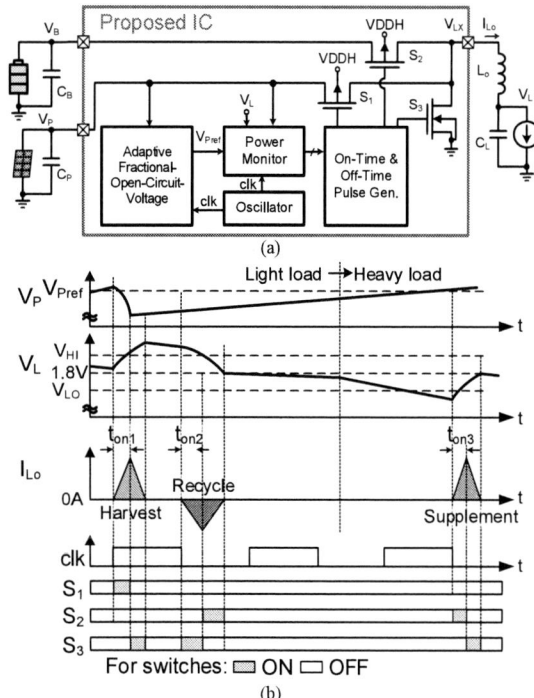

(a)

(b)

Fig. 2. (a) Block diagram of the proposed IC and (b) operating waveforms for the 2P3S converter.

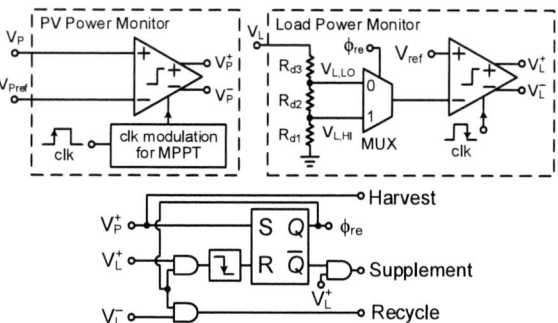

Fig. 3. Power monitor.

pulses to control the power switches. V_{DDH} is connected to V_P or V_{DD}, whichever is higher.

A. Operating Principle

Fig. 2(b) shows the operating waveforms, in which the energy delivery via the PV→load, load→battery and battery→load paths are called the harvest, recycle, and supplement operations, respectively. At the clk rising edge, if $V_P > V_{Pref}$, the converter performs the harvest operation; at the clk falling edge, if $V_L < V_{LO}$, the converter performs the supplement operation; and, at the clk falling edge, if the previous clk rising edge triggers the harvest operation and $V_L > V_{HI}$, the converter performs the recycle operation.

B. Adaptive-Fractional-Open-Circuit-Voltage Method

In Fig. 1(a), at the MPP locus, the MPP voltage V_{MP} linearly decreases with increasing MPP power P_{MP}, and thus fraction k linearly decreases with increasing P_{MP}, as shown in Fig. 1(b). For the conventional FOCV method, fraction k is fixed and the $V_{FOCV} = k \cdot V_{OC}$. If k is designed as V_{MP}/V_{OC} under 1000 lux, the difference between V_{MP} and V_{FOCV} linearly increases with decreasing P_{MP}, as shown in Fig. 1(a), thereby decreasing η_{MPPT}. To reduce the difference without complicating the operation, this paper proposes an adaptive-FOCV (AFOCV) method, which is an enhancement of the conventional FOCV method. In the measured MPP locus in Fig. 1(b), the value of k under 1000 lux is obtained, and the slope of

k under 100–1000 lux is fitted. The AFOCV method initially operates at the k value under 1000 lux, estimates the decrease in the current P_{MP} from that under 1000 lux, and linearly increases k accordingly. Thus, k is adaptive to the current P_{MP} for achieving high η_{MPPT}. For circuit implementation, V_{Pref} is initially set to the V_{FOCV}. The decrease in the current P_{MP} from that under 1000 lux is estimated from the period between two consecutive harvest operations, which is proportional to P_P. In this manner, k is automatically adjusted by charging a capacitor with a compensation current in the period to increase V_{Pref} from the V_{FOCV} toward the V_{MP} for the current P_{MP}. Further, if the AFOCV method is applied to different PV modules, the compensation current can be trimmed for different slopes of k or turned off, leaving the conventional FOCV method. The circuit implementation is described in detailed in Section III.

III. CIRCUIT IMPLEMENTATION

In this section, the circuit implementations of the proposed IC is illustrated.

A. Power Monitor

Fig. 3 shows the power monitor. The PV power monitor regulates V_P around V_{Pref} while the load power monitor regulates a scaled V_L around V_{ref}. To reduce power consumption, the dynamic comparator for the PV power monitor is enabled at the clk rising edge; and while in the MPPT compensation period, the clk frequency of 100 kHz is divided into 1.56 kHz through clk modulation. The dynamic comparator for the load power monitor is enabled at the clk falling edge. A mode selection signal (ϕ_{re}), which is generated from reconfigurable controller, and an analog multiplexer (MUX) are used to select either $V_{L,LO}$ or $V_{L,HI}$ to be compared with V_{ref} to save one comparator [1]. At the clk rising edge, the V_P^+ triggers the harvest operation and sets the SR latch to pull ϕ_{re} high for comparing $V_{L,HI}$ with V_{ref} and deciding whether to recycle the surplus energy in C_L at the coming clk falling edge. At the clk falling edge, if ϕ_{re} is high, the V_L^- triggers the recycle operation and resets the SR latch to pull ϕ_{re} low. In contrast, if ϕ_{re} is low, the V_L^+ triggers the supplement operation. In addition, the V_P^+, V_P^- signals are also sent to the AFOCV circuit to realize the AFOCV operation.

978-1-7281-5107-6/19 $31.00 © 2019 IEEE   186

IEEE Asian Solid-State Circuits Conference
November 4 – 6, 2019
The Parisian Macao, Macao SAR, China

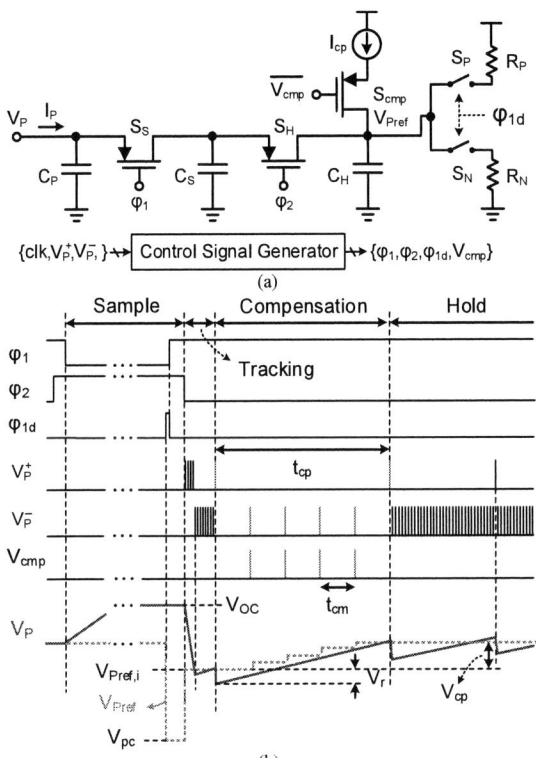

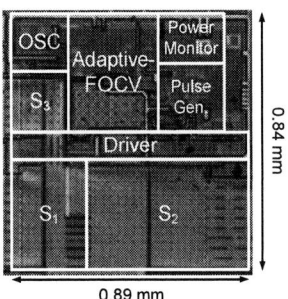

Fig. 6. Chip micrograph.

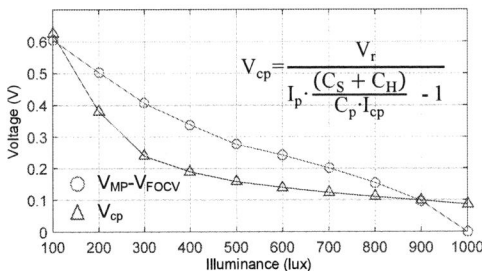

Fig. 4. (a) Proposed AFOCV MPPT circuit and (b) its operating waveforms.

$$V_{cp} = \frac{V_r}{\dfrac{I_p \cdot \dfrac{(C_S + C_H)}{C_P \cdot I_{cp}}} - 1}$$

Fig. 5. Compensation voltage (V_{cp}) and voltage difference between V_{MP} and V_{FOCV}.

B. Adaptive-Fractional-Open-Circuit-Voltage Circuit

Fig. 4 presents the AFOCV circuit and its operating waveforms, where the AFOCV circuit has 4 operating modes: sample, tracking, compensation, and hold. In the sample mode, φ_1 turns low to turn on S_S for charging C_S to V_{OC}. When φ_1 turns high, φ_{1d} turns on transmission gates S_P and S_N for a short period to pre-charge C_H to the pre-charge level of $V_{pc} = V_{DD} \cdot R_N/(R_P+R_N)$. When entering the tracking mode, φ_2 turns low to connect C_S and C_H and generate the initial V_{Pref}, $V_{Pref,i} = 0.5 \cdot (V_{OC} + V_{pc})$, where $C_S = C_H = 25$ pF. To ensure $V_{Pref,i}$ equals V_{MP} at 1000 lux, V_{pc} is designed by adjusting the off-chip R_P and R_N for different PV modules, where $V_{Pref,i}$ is nearly the

same as V_{FOCV} (locus with delta marker) in Fig. 1(a) from 100 to 1000 lux. In the tracking mode, the clk operates at 100 kHz to guarantee V_P (blue solid line) immediately tracks $V_{Pref,i}$ by delivering Harvesting energy packets. After $V_P < V_{Pref,i}$, V_P^- is triggered at every clk rising edge until V_P reaches $V_{Pref,i}$, at which point V_P^+ is triggered and MPPT enters the compensation mode, where V_P^+ and V_P^- are signals from the power monitor. In the compensation mode, I_{cp} charges C_S and C_H, and V_{Pref} (red dash line) is increased. While V_P reaches V_{Pref}, at which point V_P^+ is triggered and I_{cp} stops charging C_S and C_H, then MPPT enters the hold mode. With a constant I_{cp} charging C_S and C_H in compensation mode, different illuminances lead to different values of I_P charging C_P, which causes different increasing rates of V_P, resulting in different compensation periods t_{cp}, compensation voltages V_{cp}, and V_{Pref} in the hold mode. Fig. 5 shows V_{cp} under different illuminances with I_{cp} designed as 1 nA and $C_P = 4.7$ μF, where $V_r = 0.25$ V is the voltage drop on V_P due to a Harvest operation. Although V_{cp} does not perfectly fit the difference between V_{FOCV} and V_{MP}, with a proper design, V_{Pref} can be compensated toward V_{MP} and has higher η_{MPPT} compared with FOCV. Further, due to the flatness of the P-V curve around V_{MP} shown in Fig. 1(a), the harvested power is near the maximum and the tolerance range of V_{cp} on η_{MPPT} is wide. To simplify the circuit design, the compensation of V_{Pref} is made stepwise instead of continuous by a duty-cycled current source I_{cp}. The V_P^- pulse, used as V_{cmp}, controls I_{cp}, where each I_{cp} pulse width is around 11μs, while the period t_{cm} is 1/(1.56 kHz). Therefore, I_{cp} is implemented to be about 50 nA, and can be trimmed for variations in P-V curves or different PV modules; and, I_{cp} can also be turned off for realizing FOCV with fixed k. Additionally, I_{cp} and its bias are enabled only during t_{cp} in a roughly 12s sample-hold period; thus, due to this duty-cycled operation, the resulting average current of I_{cp} and its bias are <100 pA.

IV. MEASUREMENT RESULT

Figure 6 shows the chip micrograph of the AFOCV EH IC. Fig. 7 shows the measured waveforms of AFOCV MPPT operation at 100 and 400 lux illuminances, in which the P_{MPS} are 18 μW and 40 μW, respectively. At 100 lux, after V_P reaches V_{FOCV}, the 2.78V V_{Pref} is compensated to 3.36 V. And at 400 lux, after V_P reaches V_{FOCV}, the 2.86V V_{Pref} is compensated to 3.23V. Accordingly, the compensated V_{Pref} at 100 and 400 lux are near V_{MP}. Fig. 7(d) shows the measured

978-1-7281-5107-6/19 $31.00 © 2019 IEEE

Table I: Performance Comparison with Other State of the Arts.

	This Work	[1] JSSC 2016	[2] VLSI 2018	[3] ISSCC 2017	[4] ISSCC 2013
Process	0.5μm	0.5μm	0.18μm	0.35μm	0.5μm
MPPT method	Adaptive-FOCV	FOCV	FOCV	FOCV	P&O
MPPT efficiency	>99.6% (100lux–1000lux)	>90%	>90%	>95%*	>99%
MPPT circuit current	100 pA	N/A	N/A	240 nA	35 μA
Controller current	0.13μA	0.85μA	N/A	65μA	100μA
Load power range	0μW–18mW	0μW–120mW	100μW–120mW	N/A	N/A
Load voltage range	1V–3.3V	1V–3.3V	0.5V–5V	N/A	N/A
Peak conversion efficiency	95.1%	95%	81.7%	92.6%	94%
Inductor value	4.7μH	4.7μH	9.9μH	10μH	22μH

*The lowest illuminance in the measurement is not mentioned.

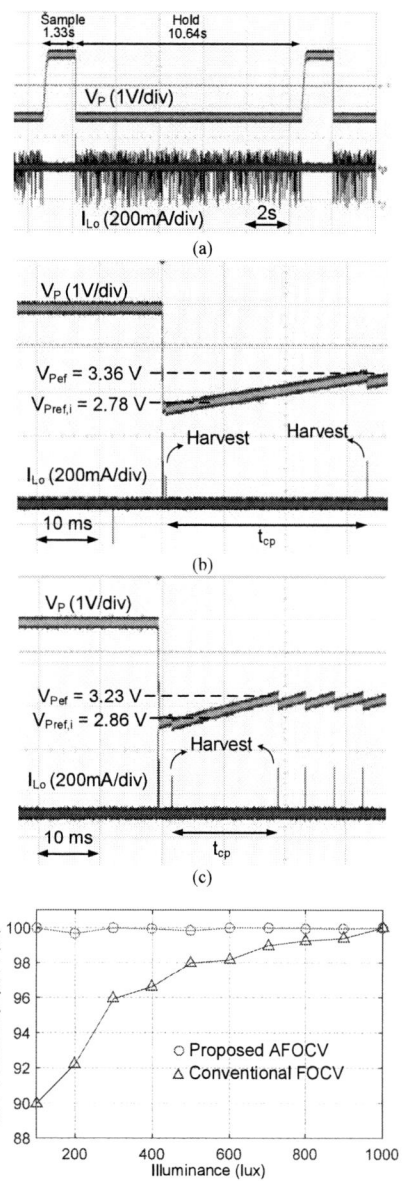

Fig. 7. Measured waveforms of (a) AFOCV and operations of compensation under (b)100 lux and (c) 400 lux. (d) Measured MPPT results

the highest power conversion efficiency. Most importantly, this work has higher η_{MPPT} than conventional FOCV approaches [1]-[3], and has lower MPPT circuit current than [3]. Moreover, compared with the P&O MPPT [4], this work achieves higher η_{MPPT} while consuming much lower current.

V. CONCLUSION

This paper first observes that FOCV fraction decreases linearly as illuminance increases and proposes an AFOCV MPPT to automatically adjust the fraction under different illuminances for specific amorphous PV modules. The AFOCV circuit consumes an ultra-low average current of 100 pA. Over 100–1000 lux illuminance, the measured MPPT efficiency with AFOCV is >99.6%, which outperforms that of other state-of-the-art EH ICs with conventional FOCV methods.

ACKNOWLEDGMENT

The authors would like to acknowledgement chip fabrication support provided by Taiwan Semiconductor Research Institute (TSRI), Taiwan.

REFERENCES

[1] Y.-H. Wang, Y.-W. Huang, P.-C. Huang, H.-J. Chen, and T.-H. Kuo, "A single-inductor dual-path three-switch converter with energy-recycling technique for light energy harvesting," in *IEEE J. Solid-State Circuits*, vol. 51, no. 11, pp. 2716-2728, Nov. 2016.

[2] S. Kim, V. Vaidya, C. Schaef, A. Lines, H. Krishnamurthy, S. Weng, X. Liu, D. Kurian, T. Karnik, K. Ravichandran, and J. Tschanz, V. De, "A single-stage, single-inductor, 6-input 9-output multi-modal energy harvesting power management IC for 100μW-120mW battery-powered IoT edge nodes," in *IEEE Symp. VLSI Circuits Dig. Tech. Papers*, Jun. 2018, pp. 195–196.

[3] S. Uprety and H. Lee, "A 93%-power-efficiency photovoltaic energy harvester with irradiance-aware auto-reconfigurable MPPT scheme achieving >95% MPPT efficiency across 650μW to 1W and 2.9ms FOCV MPPT transient time," *in IEEE ISSCC Dig. Tech. Papers*, Feb. 2017, pp. 378–379.

[4] W.-C. Liu, Y.-H. Wang, and T.-H. Kuo, "An adaptive load-line tuning IC for photovoltaic module integrated mobile device with 470 μs transient time, over 99% steady-state accuracy and 94% power conversion efficiency," in *IEEE ISSCC Dig. Tech. Papers*, Feb. 2013, pp. 70–71.

[5] Panasonic Eco Solution Amorton, Japan, "Amorphous silicon solar-cells," Jan. 2016.

[6] EnOcean, Germany, "Solar cells ECS 300 and ECS 310", Oct. 2013.

MPPT performance. With only 100 pA average-current overhead, the proposed AFOCV MPPT circuit improves the FOCV η_{MPPT} from 90% to 99.6% and increases the harvested power. In addition, although PVT variations affect both AFOCV and FOCV, the AFOCV still has higher η_{MPPT} and can be further calibrated by adjusting I_{cp}. Table I lists the performance comparison. Compared with other state-of-the-arts, the AFOCV EH IC has the lowest quiescent current and

14-1 (7133)

IEEE Asian Solid-State Circuits Conference
November 4 – 6, 2019
The Parisian Macao, Macao SAR, China

A Single-Supply Buffer-Embedding SAR ADC with Skip-Reset having Inherent Chopping Capability

Min-Jae Seo[1,3], Dong-Hwan Jin[1], Ye-Dam Kim[1], Jong-Pal Kim[2], Dong-Jin Chang[1], Won-Mook Lim[1],
Jae-Hyun Chung[1], Chang-Un Park[1], Eun-Ji An[1] and Seung-Tak Ryu[1]

[1]School of Electrical Engineering, KAIST, Daejeon, Korea
[2]Samsung Advanced Institute of Technology (SAIT), Suwon, Korea
[3]now with Samsung Electronics, Hwasung, Korea
Email: seom0429@kaist.ac.kr

Abstract—**This paper presents a power-efficient buffer-embedding successive approximation register (SAR) analog-to-digital converter (ADC) that utilizes a core power supply for the source-follower buffer, having a rail-to-rail signal swing owing to the capacitive level shifting bias scheme. In conjunction with 8x oversampling and the power-saving skip-reset technique that has the inherent chopping capability, the prototype 180nm CMOS 12b ADC operating at a 5.12 MS/s sampling rate achieved a 74.8 dB SNDR under a 1.5V supply voltage.**

Keywords—low power, analog-to-digtal converter (ADC), successive approximation register (SAR), loop-embedded input buffer, push-pull source follower, capacitive level-shifting.

I. INTRODUCTION

As the interests in implantable and wearable devices for biomedical applications grows explosively, the importance of low power and small size chip implementation for a battery-powered system is also growing. In terms of ADC implementation, owing to the scaled CMOS process, a charge-redistribution SAR ADC has been a very attractive architecture for the last decade for energy-efficient systems [1]. However, one of the major drawbacks with conventional high-resolution SAR ADCs is the large capacitance of the capacitive DAC (CDAC) required to satisfy the matching accuracy for linearity as well as the kT/C noise specification. This large input capacitance requires high power consuming input driving buffer. The relatively short sample time period due to the many decision cycles makes this problem more serious in SAR ADCs, which makes the power consumption by the SAR ADC core to be overwhelmed by the power consumption of the driving buffer. In [2] and [3], input buffer is embedded in the SAR loop to reduce the linearity requirement and to increase the driving capability: as the nonlinearity via the input path is cancelled exactly by the identical DAC path, a simple two-PMOS source follower could be utilized as the input buffer. However, the current DAC (IDAC) consuming considerable static power under a dedicated high supply voltage makes the structure in [3] unsuitable for low power applications. Recently, a buffer-embedding SAR ADC with a CDAC have been reported in [4], but the high supply voltage required for the buffer still becomes the bottleneck in achieving further power saving.

In this paper, we propose a buffer-embedding CDAC-based SAR ADC that utilizes a low-power push-pull source follower

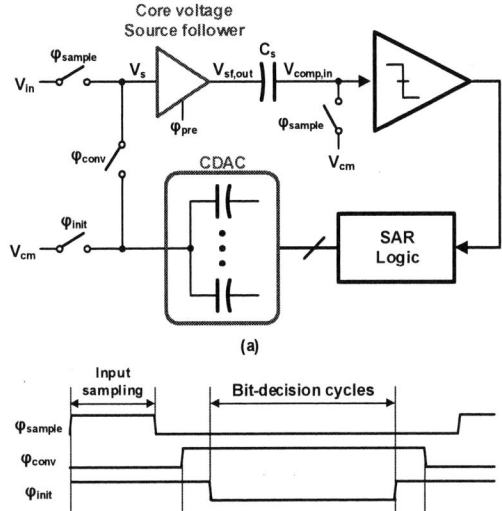

Fig. 1. (a) Block diagram of the proposed low power SAR ADC with integrated input buffer and (b) timing diagram of the control switches.

with a capacitive level-shifting bias under a core supply voltage. In addition, the power saving skip-reset technique reported in [5] is improved here to provide an inherent flicker noise chopping capability.

II. SINGL-SUPPLY SAR ADC WITH EMBEDDED INPUT BUFFER

A. Buffer-embedding SAR Loop with a Capacitive DAC

The current-DAC in [3] occupies a large area for the matching property and consumes considerable power, about 55% of the total ADC power consumption. Thus, using a static current consuming DAC is very burdensome in low power applications. As shown in Fig. 1(a), the proposed buffer-embedding architecture generates decision reference voltages by using a CDAC without static current. Considering kT/C noise, the input sampling capacitor C_S is designed with 1 pF, and the total capacitance of CDAC (C_{total}) is designed to be 5.4 pF considering the matching requirement. The conversion

978-1-7281-5107-6/19 $31.00 © 2019 IEEE

189

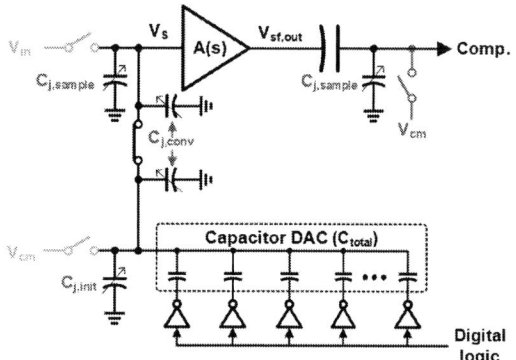

Fig. 2. Parasitic capacitance on conversion phase.

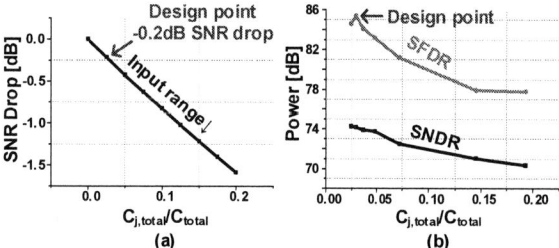

Fig. 3. (a) SNR drop depending on the junction capacitance and (b) Estimated performance depending on the junction capacitance.

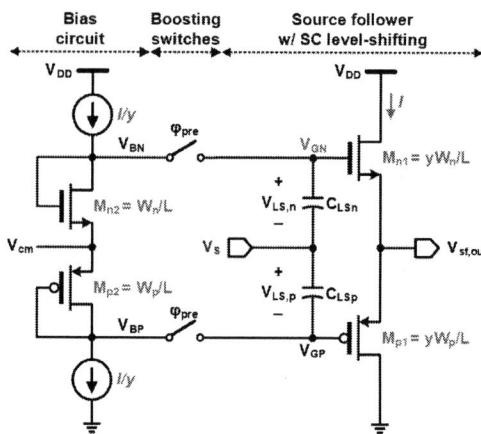

Fig. 4. Core-voltage push-pull source follower.

process of the proposed ADC is shown in Fig. 1(b). At first, the input signal is sampled on C_S through the buffer at the falling edge of φ_{samp}. After this bottom-plate sampling is done, the input common level V_{cm} is connected to the input node of the buffer by φ_{conv} and φ_{init} in order not only to transfer the sampled input signal to the input of the comparator, but also to reset any signal-dependent charge at the summing node (V_S). The bit decision process begins from the falling edge of φ_{init} as in a typical SAR ADC. After A/D conversion is done, φ_{pre} sets the bias condition of the buffer (will be discussed later).

Although the signal-dependent charge remaining at V_S node is reset after sampling, the nonlinearity of the parasitic capacitance on that node should be also considered as it affects the DAC nonlinearity: as Fig. 2 illustrates, the DAC output at V_S node appears as $V_{REF} \times kC/(C_{j,total}(V_S) + C_{total})$, where $C_{j,total}$ is the total junction capacitance at node V_S, C_{total} is the total capacitance of the CDAC, and k is the number of capacitors connected to the reference according to the CDAC input code. Owing to the CMOS switch design, the capacitance nonlinearity by each of the NMOS and PMOS switch could be largely compensated by each other. In order not to degrade the target performance by the residual $C_{j,total}$ nonlinearity, the switch sizes were carefully designed by sweeping them as shown in Fig. 3. In our design, $C_{j,total}/C_{total}$ is about 2.8%.

B. Push-Pull Source Follower with Switched-capacitor (SC) Level-Shifting Bias

Owing to the buffer-embedding architecture, the power and linearity burden imposed on the input buffer could be greatly alleviated. However, due to the V_{GS} voltage drop in the source follower, a dedicated high-voltage supply is required for the buffer to widen the signal swing range [3], which means increased power consumption. In order to avoid the additional supply problem, we propose a rail-to-rail swing push-pull source follower working under a core supply voltage. Shown in Fig. 4, the proposed source follower consists of a bias circuit, switches, and a push-pull source follower with a capacitive level shifter. The source follower operates with two phases: First, the source follower charges the level-shifting capacitors ($C_{LS,n}$ and $C_{LS,p}$) through the bias circuit and the switches for class-AB operation. Note that the bias circuit is a replica of the source follower and the bias circuit has been scaled down with the ratio of y, similar to [6]. Thus, the output DC level of the source follower ($V_{amp,out}$) is set to V_{cm}. In this pre-charge phase, the input of the source follower (V_S) is connected to V_{cm} ($V_{cm} = V_{DD}/2$) according to Fig. 1. After the bias condition is set up, the switches are turned off ($\varphi_{pre} = 0$) for input tracking and SAR operation, referring to Fig. 1(b). Owing to the DC level shifting by $C_{LS,n}$ and $C_{LS,p}$, the two gate voltages of the source follower (V_{GN}, V_{GP}) follows the input signal on their own DC levels.

However, these DC level-shifted V_{GN} and V_{GP} nodes raise the leakage current problem through the switches when the signal is close to VDD and GND. For example, if the input level is close to GND, V_{GP} will be lower than V_{SS}. Then a simple NMOS switch which is supposed to be turned off with GND at its gate will be turned on through the body diode and thus the level-shifting capacitor $C_{LS,p}$ cannot preserve its voltage. To eliminate this leakage problem, the switches are driven by clock boosting circuits. Note that if the NMOS switch connected to V_{GP} is turned off with -V_{DD} at its gate and body, it can be fully turned off with no leakage issue. In order to generate the negative boosting voltage (-V_{DD}) for the NMOS switch connected to V_{GP}, the negative clock boosting circuit is proposed as in Fig. 5. When φ_{preb} is = 0, the bottom node of C_{boost} samples V_{DD} through M_{p1}. At the same time, $V_{body,n} = V_{SS}$, and $V_{boost} = V_{DD}$ through M_{p3} and M_{p4}. Thus, in this phase, the NMOS boosting switch is on to sample the bias level (V_{BP}). When $\varphi_{preb} = 1$, GND is connected to the bottom plate of C_{boost} through M_{n1}. Eventually, $V_{body,n}$ is down-boosted to -V_{DD}. Owing to $V_{body,n} = -V_{DD}$, although the gate voltage of M_{dn2}

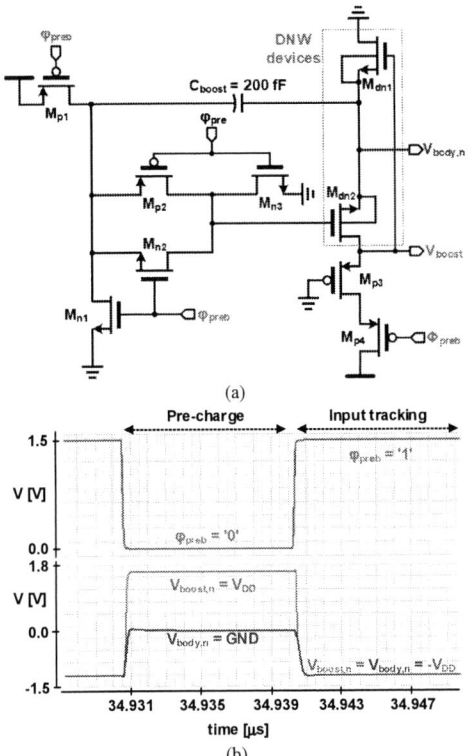

(a)

(b)

Fig. 5. (a) Schematic of the negative clock boosting circuit and (b) the transient waveform.

drops to GND via M_{n2}, $-V_{DD}$ is transferred to V_{boost}. Note that M_{dn1} and M_{dn2} are implemented using deep-nwell devices to allow non-GND body potential. Also note that the voltage stress of all the transistors are within V_{DD}, with no reliability issue. The switch for V_{GN} connection is designed with a PMOS, and its clock boosting circuit is the upside-down version of Fig. 5(a).

III. PROTOTYPE OVERSAMPLED SAR ADC WITH INTEGRATED INPUT BUFFER

Fig. 6 shows the block diagram of the prototype SAR ADC. The CDAC has a 12b resolution and the input sampling path has an alternating switch for implementation of the skip-reset technique proposed in [5] that saves the reset power during the input sampling phase. Unlike [5] where the swapped input is re-swapped at the comparator input, in this design, the swapped input is directly connected to the comparator so that the skip reset implements the chopping capability inherently. As illustrated in Fig. 7, the inherent chopping operation by the skip reset sampling upmodulates the even harmonics as well as the DC offset and flicker (1/f) noise. The chopper is implemented by the analog input switch (MX_{in}) and digital output multiplexer. In addition to the skip reset technique based on the delta conversion principle, the MSB rounding technique and the tri-level switching techniques utilized in [5] have also been applied in this design. The capacitors of 2^5C and 2^4C are for the MSB rounding technique and the error compensation required for the delta-conversion based skip-reset operation.

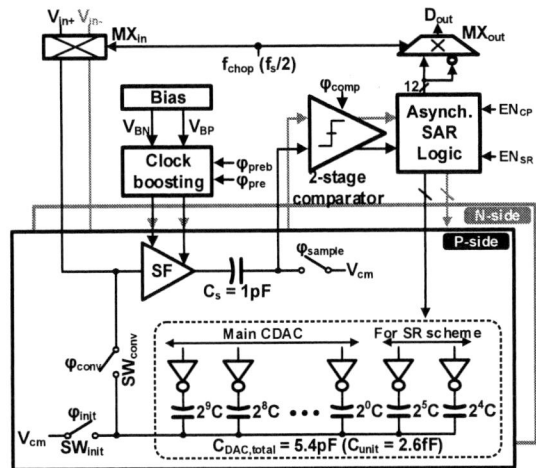

Fig. 6. The block diagram of the prototype ADC.

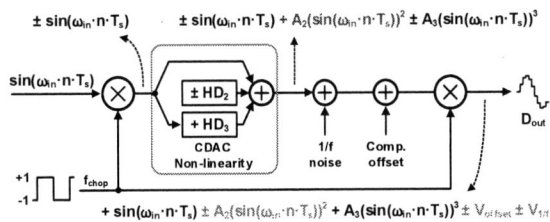

Fig. 7. Signal flow of the hybrid chopping technique.

Depending on the control bits (EN_{CP}, EN_{SR}), the prototype ADC has three operational modes. When $EN_{CP} = 0$ and $EN_{SR} = 0$, the ADC operates in a conventional SAR conversion mode (Normal mode) with no skip reset and chopping. When $EN_{CP} = 1$ and $EN_{SR} = 0$, the chopper with the chopping frequency $f_{chop} = f_s/2$ is enabled (Chopping mode with no skip reset). When $EN_{SR} = 1$, the ADC operates in delta-readout mode with the skip-reset scheme (SR mode with inherent chopping). In the normal mode and the chopping mode, the ADC performs every bit decision and the CDAC is switched back to the initial state after the conversion is completed. In the SR mode, the residue is predicted by utilizing the previous MSB codes, meaning no MSBs (6 bit) switching from the previous decision for input sampling. For code decision, only the remaining 6 bit LSB codes are decided to determine the final output, resulting in power saving especially at low frequency input or oversampled operation.

IV. MEASUREMENT RESULTS AND COMPARISION

The prototype ADC with a 12b resolution was fabricated in a 180 nm CMOS process (Fig. 8), and it works under a 1.5V supply. In order to check the basic performance of the ADC and its power breakdown, the ADC was run in the normal mode first with no chopping or skip reset techniques: The FFT spectrum is taken at 5.12 MS/s with a 50kHz low-frequency input (Fig. 9). The measured SNDR and the SFDR are 65.6 dB (10.6 ENOB) and 77.6 dB, respectively. The measured DNL and INL from 2^{21} samples are shown in Fig. 10. The peaks of the DNLs and INLs are +1.1/−0.80 LSB and +1.0/−1.0 LSB,

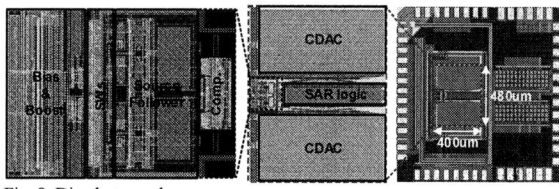

Fig. 8. Die photograph.

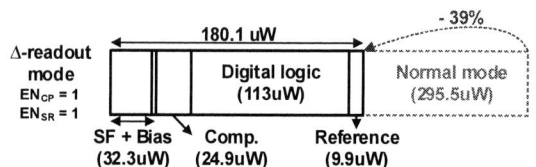

Fig. 12. Power breakdown.

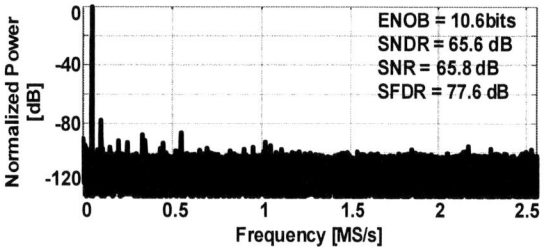

Fig. 9. (a) Measured spectra and (b) Power breakdown.

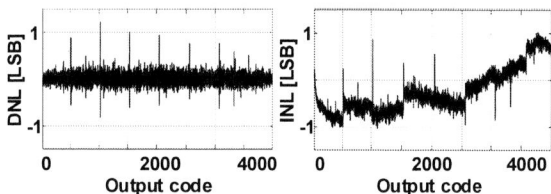

Fig. 10. Static performance

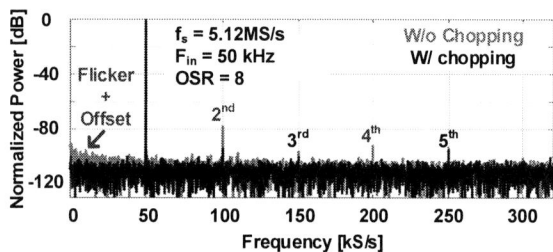

Fig. 11. Measured spectrum with and without chopping.

respectively. The measured total power consumption is 295.5 µW, and the proposed input buffer consumes only 32.3 µW, corresponding to 10% of the entire power consumption.

The effect of the chopping is shown in Fig. 11 with the measured FFT spectra, with a 50-kHz input signal and 8x oversampling. When the input chopper is disabled (red), the SNDR in signal band of 320kHz is 70.8 dB with noticeable flicker noise and even harmonics. When the input chopper is enabled (black), the SNDR is improved by 5 dB, from 70.8 dB to 75.8 dB owing to the shifted flicker noise even harmonics to high frequency.

When the skip-reset is enabled with 8x OSR, the prototype ADC could reduce the power consumption by 39% at a low frequency input (Fig. 12) and achieves a 74.8dB SNDR; this condition is set assuming that the prototype ADC is utilized for multi-channel bio signal readout applications (particularly, 16 channel readout with signal bandwidth of 20 kHz). Table I summarizes the measured performances in different operational

TABLE I
PERFORMANCE SUMMARY OF PROTOTYPE ADC

Architecture	12b buffer-embedding SAR ADC at 5.12MS/s		
Operation Mode	Normal	w/ Chopping	w/ Skip-reset
OSR	1	8	8
SNDR [dB]	65.6	75.8	74.8 (@ 80 Hz)
SFDR [dB]	77.6	85.6	86.1 (@ 80 Hz)
Power [µW]	295.5		180.1
FoM$_S$ [fJ/c.s.]	165.0	166.2	167.3

TABLE II
COMPARISON WITH PREVIOUS LOOP-EMBEDDED WORKS

	This work	[4]	[3]	[2]
Tech. [nm]	180	65	40	65
Reference DAC	CDAC	CDAC	IDAC	IDAC
Input capacitance	10 fF	200 fF	200 fF	-
SF's Gain	0.94	-	0.94	0.64
Core voltage	O	X (2.1V)	X (2.5V)	O (swing↓)
Power portion of (buffer+DAC)	31% (10% + 21%)	72% (59% + 13%)	78% (23% + 55%)	--

modes. Note that the skip-reset mode shows the best FoM$_S$ owing to the reset power saving as well as to the up-modulated flicker noise and even harmonics. Table II shows a brief comparison with the previous buffer-embedding SAR ADCs. Previous work utilizing an IDAC has considerable power consumption [3]. The designs with a dedicated high supply voltage for the buffer also show considerable power consumption [3-4]. Note that the proposed design could save power consumption owing to the proposed low-supply buffer and CDAC. In addition, the proposed source follower has a very low input capacitance of 10 fF, which greatly reduces the driving burden of the preceding stage.

V. CONCLUSION

This paper demonstrates a low-power SAR ADC with an embedded input buffer. The capacitive level-shifting bias enables a low power input buffer under a single power supply with rail-to-rail signal swing capability. The inherent chopping function of the skip reset technique enhances the SNDR with reduced power consumption, achieving SNDR around 75 dB SNDR with 8x OSR.

REFERENCES

[1] P. Harpe, *IEEE Solid-State Circuits Mag.*, pp. 64–73, 2016.
[2] K. Doris *et al.*, *IEEE ISSCC*, pp. 180–182, Feb. 2011.
[3] M. Kramer *et al.*, *IEEE ISSCC*, pp. 284–286, Feb. 2015.
[4] T. Kim *et al.*, *IEEE CICC*, pp. 1-4, Apr. 2019.
[5] M.-J. Seo *et al.*, *IEEE TCAS-I*, pp. 3617-3627, Nov. 2018.
[6] C.-K. Lee et al., IEEE TCAS-II, pp. 557-561, Sep. 2013.

14-2 (7055)

A 68 dB SNDR Compiled Noise-Shaping SAR ADC With On-Chip CDAC Calibration

Harald Garvik*, Carsten Wulff† and Trond Ytterdal*

Email: harald.garvik@ntnu.no

*Dept. of Electronic Systems, Norwegian University of Science and Technology (NTNU), Trondheim, Norway

†Nordic Semiconductor, Trondheim, Norway

Abstract—This paper[1] presents a noise-shaping SAR ADC with an on-chip, foreground capacitive DAC (CDAC) calibration system. At start-up, the ADC uses the LSBs in the CDAC to measure and digitize the errors of the MSBs. A synthesized digital module accumulates the noise-shaped measurements, computes calibration coefficients, and corrects ADC codes at run-time. The loop filter implements two optimal zeros and two poles, and achieves 27.8 dB in-band attenuation at an oversampling rate (OSR) of 4. The prototype is implemented in 28 nm FDSOI, and achieves 68.2 dB SNDR at 5 MHz bandwidth, while consuming 108.7 µW from a 0.8 V supply. The Walden FOM is 5.2 fJ/conv.-step. The layout of the ADC is compiled from a netlist, a rule file, and an object definition file.

Index Terms—Noise-shaping, SAR, ADC, Noise-shaping SAR, CDAC calibration, analog layout synthesis.

I. INTRODUCTION

Noise-shaping SAR ADCs (NS-SARs) have a loop filter that shapes quantization noise and comparator noise [1]–[5]. This enables energy-efficient operation at 10-12 bits ENOB, but mismatch in the CDAC usually limits the overall accuracy. This work introduces an on-chip CDAC calibration system for NS-SARs. At start-up, the ADC is run in a low-range mode with only the LSB part of the DAC, and the errors of the MSBs are measured. The measurements are repeated with the loop filter engaged, and the results are averaged in a synthesized digital module, which then corrects ADC codes at run-time.

The presented ADC has an active, OTA-based loop filter that provides high noise suppression at low OSR. This is an interesting contrast to the current trend of using simpler loop filters [2]–[4] that only consume dynamic power. Higher OSR is needed for the same performance in many of these filters.

II. NOISE-SHAPING SAR ARCHITECTURE

The proposed NS-SAR architecture is shown in Fig. 1, and consists of a 9-bit MOM CDAC, a bootstrapped sampling switch, a dual input strong-arm comparator, a loop filter, and multiple digital modules: The asynchronous SAR logic and calibration logic control the CDAC through the multiplexer DAC driver. The synthesized code correction module runs at the sample clock. The CDAC is divided in an MSB part which uses split monotonic switching, and an LSB part that uses monotonic switching. The actual ADC implementation is differential.

[1]This work has been supported by the Centre for Innovative Ultrasound Solutions (CIUS), a Norwegian Research Council appointed Centre for Research Based Innovation. Research Council of Norway grant number 237887.

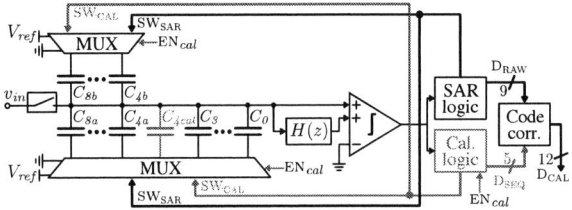

Fig. 1. Proposed noise-shaping SAR architecture. Blue blocks and paths are only active in calibration mode.

A. Calibration system

In calibration mode, the 5-bit LSB NS-SAR formed with C_{3-0} and C_{4cal} is used to measure the errors of the five MSB capacitors C_{8-4} (a–b capacitor pairs are merged in calibration mode). The code correction module runs the main calibration state machine, and asserts EN_{cal} at start-up. The calibration logic now controls the CDAC, and produces measurements of the MSB error ΔC_8 at the digital bus D_{SEQ}. When a specified number of measurements (2048 in this paper) have been accumulated by the correction module, the operation is switched to measure ΔC_7, and so forth. The switching algorithm used by the calibration logic is similar to the classical SAR calibration algorithm [6]. For instance, to measure ΔC_6, C_6 is first connected to ground while C_{5-0} are connected to V_{ref}. The ADC then samples common mode, before C_6 is flipped to V_{ref}, and C_{5-0} to ground (C_{8-7} are held at V_{ref} all the time). ΔC_6 is now at the top plates, and digitized with C_{3-0} and C_{4cal}. This extra capacitor enables bidirectional DAC operation, and is normally connected to V_{ref}. If the first SAR comparison is positive, it is switched to ground.

B. Loop filter

The positive branch of the loop filter is shown in Fig. 2, together with the implemented noise transfer function (NTF). The filter realizes two optimally placed zeros and two poles, through the use of two integrators, a passive summer, and a local feedback path. The coefficient g is set by the local feedback, and a_{1-2} by the summer together with the gain of the first integrator. The current mirror OTAs only need to transfer charge and provide valid output in ϕ_1, and are therefore shut down in ϕ_2 to save power. The loop filter input transfers SAR residue charge directly from the CDAC to the

978-1-7281-5107-6/19 $31.00 © 2019 IEEE

IEEE Asian Solid-State Circuits Conference
November 4 – 6, 2019
The Parisian Macao, Macao SAR, China

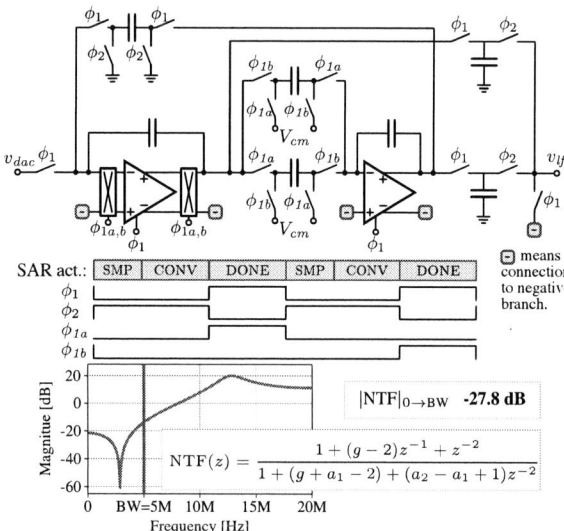

Fig. 2. Loop filter implementation and noise transfer function.

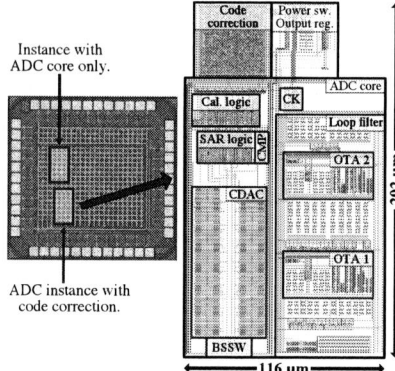

Fig. 3. Die photo and ADC layout.

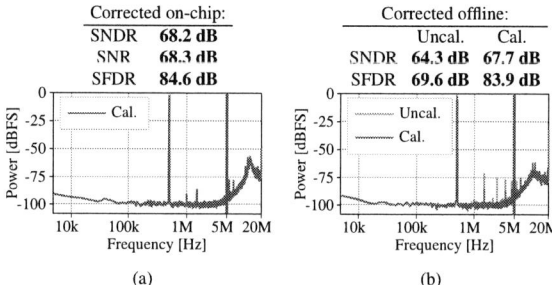

Fig. 4. Measured results and power spectrums: (a): On-chip correction, (b): offline correction. The power spectra have 2^{12} bins from DC to $f_s/2$.

first integrator [5]. This is a noise-effective solution because the residue is neither attenuated, buffered or resampled.

III. MEASUREMENT RESULTS

The prototype is implemented in 28 nm FDSOI, and Fig. 3 shows the die photo, layout and dimensions. The entire region marked "ADC core" is compiled from a netlist, rule file, and object definition file using the layout compiler presented in [7]. Fig. 4a shows a measured power spectrum with on-chip code correction. Because the correction module cannot be disabled, uncorrected and offline corrected spectrums from another ADC instance are shown in Fig. 4b.[2] On eight measured chips, mean uncalibrated/calibrated SNDR and SFDR are 65.6 dB/67.3 dB, and 74.4 dB/83.1 dB, respectively. The ADC is compared to prior NS-SARs in Table I. To the best of the authors' knowledge, the Walden FOM of 5.2 fJ/conv.step is currently the lowest reported for a noise-shaping SAR.

REFERENCES

[1] J. Fredenburg and M. Flynn, "A 90-MS/s 11-MHz-Bandwidth 62-dB SNDR Noise-Shaping SAR ADC," *IEEE Journal of Solid-State Circuits*, vol. 47, no. 12, pp. 2898–2904, Dec. 2012.

[2] C. Liu and M. Huang, "28.1 A 0.46mW 5MHz-BW 79.7dB-SNDR noise-shaping SAR ADC with dynamic-amplifier-based FIR-IIR filter," in *2017 IEEE International Solid-State Circuits Conference (ISSCC)*, Feb. 2017, pp. 466–467.

[3] S. Li, B. Qiao, M. Gandara, D. Z. Pan, and N. Sun, "A 13-ENOB Second-Order Noise-Shaping SAR ADC Realizing Optimized NTF Zeros Using the Error-Feedback Structure," *IEEE Journal of Solid-State Circuits*, vol. 53, no. 12, pp. 3484–3496, Dec. 2018.

[4] Y. Lin, C. Lin, S. Tsou, C. Tsai, and C. Lu, "20.2 A 40MHz-BW 320MS/s Passive Noise-Shaping SAR ADC With Passive Signal-Residue Summation in 14nm FinFET," in *2019 IEEE International Solid-State Circuits Conference - (ISSCC)*, Feb. 2019, pp. 330–332.

TABLE I
COMPARISON TO PRIOR STATE-OF-THE-ART NOISE-SHAPING SARs.

	[1]	[2]	[3]	[4]	**This work**
CDAC correction	None	DWA	Off-chip cal	None	On-chip cal
NTF type	1z, 2p	1z, 2p	2z opt	1z, 1p	2z opt, 2p
OSR	4	13.2	8	4	4
Technology (nm)	65	28	40	14	28
Area (mm²)	0.03	0.0049	0.024	0.0021	0.0234
Supply (V)	1.2	1	1.1	0.9	0.8
Bandwidth (MHz)	11	5	0.625	40	5
SNDR (dB)	62.0	79.7	79.0	66.6	68.2
SFDR (dB)	72.5	92.6	89.0	77.4	84.6
Power (μW)	806 .0	460.0	84.0	1250.0	108.7
FOM$_w$ (fJ/c.step)	35.8	5.8	9.2	8.9	5.2
FOM$_s$ (dB)	163.3	180.1	177.7	171.7	174.8

$\text{FOM}_w = \text{P}/(2^{(\text{SNDR}-1.76)/6.02} \cdot 2\,\text{BW})$, $\text{FOM}_s = \text{SNDR} + 10\log(\text{BW}/\text{P})$

[5] K. Obata, K. Matsukawa, T. Miki, Y. Tsukamoto, and K. Sushihara, "A 97.99 db sndr, 2 khz bw, 37.1 μw noise-shaping sar adc with dynamic element matching and modulation dither effect," in *2016 IEEE Symposium on VLSI Circuits (VLSI-Circuits)*, Jun. 2016, pp. 1–2.

[6] H. S. Lee, D. A. Hodges, and P. R. Gray, "A self-calibrating 15 bit CMOS A/D converter," *IEEE Journal of Solid-State Circuits*, vol. 19, no. 6, pp. 813–819, Dec. 1984.

[7] C. Wulff and T. Ytterdal, "A Compiled 9-bit 20-MS/s 3.5-fJ/conv.step SAR ADC in 28-nm FDSOI for Bluetooth Low Energy Receivers," *IEEE Journal of Solid-State Circuits*, vol. 52, no. 7, pp. 1915–1926, Jul. 2017.

[2]The ADC instance only contains the ADC core, and is marked in Fig. 3. Accumulation/calculation of capacitor measurements and code correction are performed on a PC using the same algorithms as in the correction module.

A Digitally-Calibrated 70.98dB-SNDR 625kHz-Bandwidth Temperature-Tolerant 2nd-order Noise-Shaping SAR ADC in 65nm CMOS

Jae Sik Yoon, Jiyoon Hong, and Jintae Kim

Department of Electrical and Electronics Engineering, Konkuk University, Seoul, Korea

Email : js.yoon@msel.konkuk.ac.kr

Abstract

This paper presents a temperature-tolerant 2nd-order noise-shaping (NS) SAR ADC employing a cascade of a temperature-compensated dynamic amplifier and a ring amplifier to realize a low-power/low-noise feedback path gain. A new mismatch calibration technique optimized for a noise-shaping SAR ADC is also presented. Fabricated in 65nm CMOS, the prototype ADC shows peak SNDR of 70.98dB with a signal bandwidth of 625kHz. Over 70-degree of the temperature range, it shows only 2dB SNDR drop. With a total power consumption of 130 µW, it achieves Walden FoM of FoM$_W$=35.9f and Schreier FoM of FoM$_S$=168dB, respectively.

Introduction

To maintain the desired noise-shaping, the feedback path in NS SAR ADCs must achieve precise and temperature-tolerant signal gain with low-power. Previous works [1]-[3] have demonstrated that using a dynamic amplifier in the feedback path leads to very power efficient NS SAR ADCs, but the gain variation over temperature and the nonideality in CDAC due to capacitor mismatch or parasitic capacitance still remain as design issues, requiring dedicated solutions such as digital-domain background temperature calibration [1] or CDAC Dynamic Element Matching (DEM) [3].

This work presents two approaches that can advance the design of NS SAR ADCs. First, a feedback-path design consisting of a cascade of a temperature-compensated dynamic amplifier and a ring amplifier realizes a low-power and temperature-tolerant amplifier. Our approach enables temperature-tolerant operation without a digital feedback loop, simplifying the overall design. Our prototype chip maintains SNDR>68dB over 5°C to 75°C with maximum 2dB SNDR degradation. Second, this paper presents a new calibration algorithm and accompanying correction calibration circuit that can remove both random and systematic CDAC mismatch in noise-shaping SAR ADC. Unlike the traditional calibration method for a standard SAR ADC [4], our calibration algorithm can find optimal bit weights of an NS SAR ADC under the presence of shaped quantization noise, leading to 10dB higher SNDR compared to conventional least-squares based calibration algorithm.

Noise-Shaping SAR ADC with Temperature-Tolerant Dynamic Amplifier

Fig. 1 shows overall circuit architecture. The core ADC is a bottom-plate-sampled 9-bit SAR ADC with binary capacitor scaling. In the feedback path, the residue at the end of 9-bit SAR conversion is first amplified by a gain-of-4.3 dynamic amplifier whose output is subsequently sampled and amplified by the following gain-of-7 ring amplifier with switched capacitor. We use a fully-differential and PVT-insensitive ring amplifier [5]. The total feedback path gain G$_{total}$=30 is chosen to achieve optimal SQNR in conjunction with a passive FIR filter, which creates zero in the NTF. The feedback is accomplished by charge sharing summation as demonstrated in [1]. Due to the up-front gain of 4.3 in the dynamic amplifier, the noise contribution from the ring

amplifier and the FIR filters are negligible and can be designed with low power. Dynamic amplifiers, while having superior power efficiency, are known to have poor gain accuracy over temperature changes. To maintain the gain accuracy, we introduce a new temperature-compensated dynamic amplifier. The key idea is to proportionally adjust the amplifier activation pulse-width with temperature change by utilizing PTAT current and local low-dropout (LDO) regulator. Fig. 2 shows a detailed schematic of the pulse generator and the feedback path implementation. The PTAT current is generated by an on-chip bandgap reference and is mirrored to generate a voltage that decreases with temperature as V$_{REF\_BUF}$=V$_{DDLDO}$ − I$_{PTAT}$×R$_{EXT}$, where R$_{EXT}$ is an external resistor with a low temperature coefficient. On-chip LDO takes the V$_{REF\_BUF}$ as a reference of the negative feedback to generate a local buffer supply voltage V$_{DD\_BUF}$ such that V$_{DD\_BUF}$≈V$_{REF\_BUF}$. The core of the pulse generator for the dynamic amplifier is powered under this V$_{DD\_BUF}$. Specifically, the delay buffers consisting of I$_1$ and I$_2$ under V$_{DD\_BUF}$ determine the pulse-width T$_{pulse}$ for the dynamic amplifier. Therefore, T$_{pulse}$ increases as temperature rises and V$_{DD\_BUF}$ decreases. By choosing proper R$_{EXT}$ value and PTAT current, the gain can be maintained reasonably well over a wide temperature range. As shown in Fig. 2, our simulation indicates that the gain of the amplifier stays with ±1.3% error bound over -45°C ~ 125°C, while the gain without proposed technique varies from -5% to +13%.

Digital Calibration Algorithm for NS-SAR

The performance of NS SAR is also critically impacted by the random and systematic mismatch in CDAC capacitor array. Finding optimal bit weights for a CDAC with mismatch is straightforward in traditional SAR ADC via foreground least-squares method [4], but reusing the same method in NS SAR ADC leads to sub-optimal bit weights due to the presence of shaped quantization noise. To exclude the impact of shaped quantization noise in the calibration process, finding optimal bit weights is cast as a convex optimization problem with 1+z^{-1} filter being applied to the error calculation. The overall calibration procedure as well as the digital calibration logic is illustrated in Fig. 3-(a). A notable difference from the standard least-squares method is that the object function is ‖err[1:N-1]+err[2:N]‖$_2$ where err[k] is the error between k$_{th}$ ADC output $\sum_{i=0}^{8} d_{out}[k][i] \cdot w[i]$ and the best-estimate of input during the calibration mode. Here, $d_{out}[k][i]$ is k$_{th}$ ADC output at i$_{th}$ bit position out of N total record length, and w[i] is the optimal bit weight for i$_{th}$ bit position. The 1+z^{-1} filter on the error de-emphasizes high-frequency quantization noise when finding optimal bit weights. Numerical experiments using fully-extracted simulations show that the calibrated ADC output achieves SNDR of 83dB when our presented algorithm is used to find optimal bit weights (Fig. 3-(c)). On the other hand, SNDR of only 73dB is achieved when the standard least-squares based calibration is applied for finding bit weights (Fig. 3-(b)). The synthesized calibration logic in Fig. 3-(a) consumes 4uW with

10MHz clock, which is only 3% of the total ADC power.

Measurement Results and Conclusion

Shown in Fig. 4, the prototype chip is fabricated in 65nm CMOS process and occupies 0.072 mm². Total power dissipation is 130 μW under 1V/1.8V where 9μW is for 1.8V LDO supply. Synthesized digital calibration logic is not fabricated on-chip but its power is included in the total power budget in Fig. 4. The total input capacitance of the ADC is 2.56pF and the input full scale is $1.8V_{ppdiff}$. Fig. 5-(a) shows the measured spectrum of the ADC when f_{sig}=197kHz which shows SNDR of 70.98 dB and 2nd-order noise shaping. Shown in Fig. 5-(b), the measured dynamic range is 74dB. Fig. 5-(c) shows measured results of temperature-tolerance testing. For this test, the entire ADC board is placed inside the temperature-cycling oven while varying the temperature from 5°C to 75°C. The measurement shows that our ADC maintains the SNDR within 2dB. On the other hand, when we disable the temperature-aware pulse generation, the performance drops as much as 5dB at 75°C, demonstrating the effectiveness of the temperature-compensated dynamic amplifier. Also, as shown in Fig. 5-(d), the bit-weight calibration using the optimal bit improves the SNDR by 13.6dB when compared to using binary-scaled bit weights. Table I compares the performance of our ADC with previously published NS SAR ADCs mostly in 65nm for fair comparison. Our work demonstrates the performance stability over a wide temperature range by utilizing a temperature-tolerant dynamic amplifier with state-of-the-art power efficiency.

Acknowledgments

This work was supported by Samsung Future Interconnect Technology Research and NRF, funded by the Ministry of Science ICT and Future Planning (201 6M3A7B4909668).

References

[1] S. Li, et al., ISSCC, 2018
[2] C.-C. Liu, et al., ISSCC,2017
[3] M.Miyahara et al., CICC, 2017
[4] H. Garvik, et al., CICC, 2017
[5] Y. Lim et al., ISSCC, 2015
[6] Z. Chen , et al., ASSCC, 2016
[7] J. Fredenburg, et al.,ISSCC, 2012
[8] Y.H . Hwang , et al., ASSCC, 2018

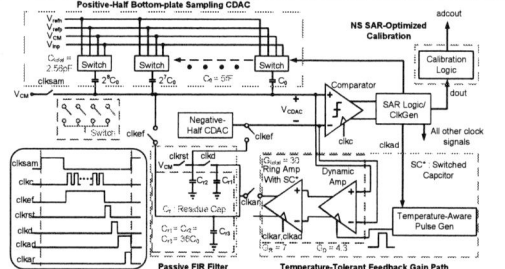

Fig. 1 Overall architecture of proposed temperature-tolerant NS-SAR ADC.

Fig. 2. Temperature-aware pulse generator and adaptive supply voltage generator that enable the temperature-compensated dynamic amplification.

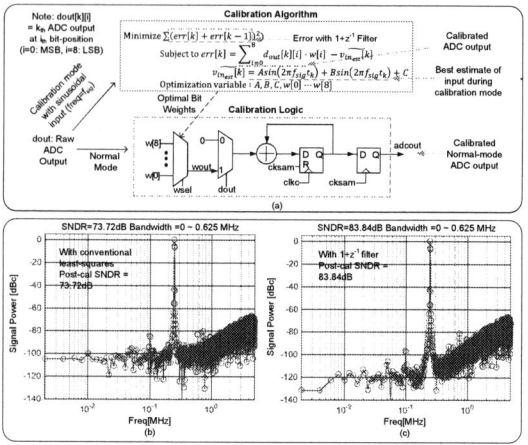

Fig. 3. (a) Calibration algorithm and correction logic. (b) FFT with conventional calibration algorithm (c) FFT with Proposed NS-optimized calibration algorithm.

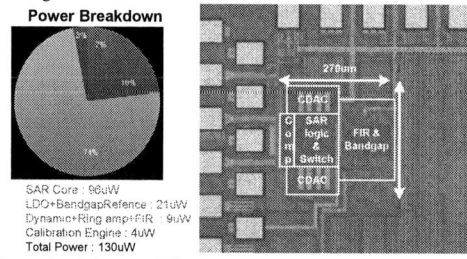

Fig.4 . Power breakdown and die photograph.

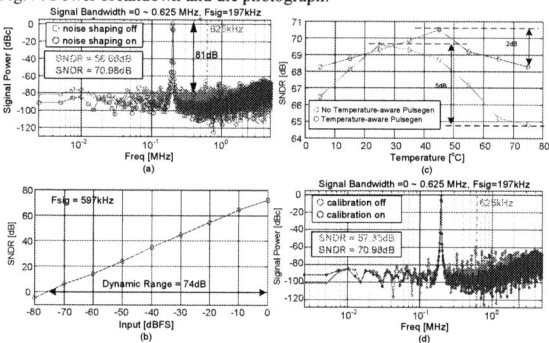

Fig. 5. Measured (a) FFT with/without noise shaping, (b) dynamic range, (c) SNDR vs temperature with/without temperature-aware pulse generator, and (d) FFT with/without the proposed calibration method.

Table I. Performance Summary and Comparison

	This work	[6] ASSCC16	[3] CICC17	[7] ISSCC12	[8] ASSCC18
Process [nm]	65	65	65	65	28
Type	EF[1]	CIFF[2]	CIFF[2]	CIFF[2]	CIFF[2]
Area [mm²]	0.072	0.0129	0.08	0.03	0.0575
VDD [V]	1	1	1	1.2	1
Amplifier Type	Dynamic +Ring	Passive	Dynamic	Opamp	Passive
Temperature Tolerance	O	O	X	O	O
Temperature Calibration	Not Required	Not Required	Digital Control	Not Required	Not Required
DAC Error Correction	NS-Opt Calibration	N/A	DEM	N/A	DEM
OSR	8	4	20	4	16
Power [uW]	130	253	630	806	118
Bandwidth [kHz]	625	8000	625	11000	100
SNDR [dB]	70.98	64.9	80.4	62	69.3
FoM_W [fJ/step]	35.9	10.9	58.9	36	251
FoMs [dB]	168	170	170	163	159

[1]EF: Error-Feedback, [2]CIFF: Cascaded Integrator Feed-Forward

8.6fJ/step VCO-Based CT 2nd-Order $\Delta\Sigma$ ADC

Akshay Jayaraj*, Abhijit Das[†], Srinivas Arcot* and Arindam Sanyal*

*Electrical Engineering, University at Buffalo, Buffalo, NY 14260, USA.

[†]Texas Instruments, Dallas, TX 75243, USA.

Email: akshayja@buffalo.edu, arindams@buffalo.edu.

Abstract—A purely VCO-based continuous-time (CT), single-loop second-order $\Delta\Sigma$ ADC is proposed in this work. Two ring oscillators are used as integrators to perform second-order quantization noise shaping. The proposed CT ADC does not require additional circuit for excess loop delay compensation. A current-reuse DAC architecture is proposed to simultaneously reduce ADC noise and power consumption. A 65nm prototype consumes 105μW from 1V supply at sampling frequency of 32.6MHz, and achieves a walden FoM of 8.6fJ/step over 2.3MHz bandwidth, which is the best among current CT $\Delta\Sigma$ ADCs.

Index Terms—voltage-controlled oscillator, analog-to-digital converter, delta-sigma, continuous-time ADC

I. INTRODUCTION

Voltage-controlled oscillator (VCO) based analog-to-digital converters (ADCs) are a popular choice for data conversion in advanced CMOS technologies. This is because VCOs are highly digital in nature and can perform simultaneous integration and multi-bit quantization with first-order noise shaping. While VCO-ADC comes with lot of inherent advantages, VCO is highly nonlinear and sensitive to variations in process, voltage and temperature (PVT). Previous attempts to linearize VCO-ADC have embedded the VCO inside a loop with high gain op-amp based loop filter, employed digital calibration of open-loop VCO [1] or used a two-stage architecture [2]. Recent continuous-time (CT) purely VCO-ADCs have used single-loop $\Delta\Sigma$ architecture to suppress VCO nonlinearity and reduce susceptibility to PVT variation [3], [4]. The higher-order VCO-ADCs in [5], [6] have used an open-loop VCO followed by a second-order VCO-based single-loop $\Delta\Sigma$ to achieve third-order quantization noise shaping.

This work presents the first CT $\Delta\Sigma$ VCO-ADC to exhibit *sub-10fJ/step walden FoM*. The second-order ADC consists of two VCO integrators in a feedback loop. While the ADC is based on the modified DPLL architecture reported in [7]–[9], a current-reuse DAC and optimized design methodology is used in this work to reduce walden FoM by $17\times$ compared to our previous prototype [9], and $4\times$ compared to the second-order VCO-ADC in [10]. The rest of this paper is organized as follows: Section II presents the proposed ADC architecture and design optimization methodology, measurement results are presented in Section III and the conclusion is brought up in Section IV.

II. PROPOSED ARCHITECTURE

A. ADC Circuit and Model

Fig. 1(a) shows the circuit schematic of the proposed ADC. Analog voltage input, V_{IN}, is converted to current, I_{IN}, through two off-chip resistors, R, before entering the ADC. A tri-state phase/frequency detector (PFD) extracts phase difference of the input current-controlled oscillators (CCOs). The PFD provides two 1-b outputs 'UP' and 'DN' such that the difference in widths of 'UP' and 'DN' pulses encodes the phase difference of the input CCOs. The 1-b 'UP' and 'DN' pulses drive the second CCO integrators such that they switch between only 2 frequencies f_H and f_L corresponding to currents I_H and I_L (see Fig. 1(a)). Hence, the second CCO integrators act as switched ring oscillators (SROs). Since the SROs switch between 2 frequencies, they have very high linearity. Both CCO (first integrator) and SRO are built using a chain of 19 pseudo-differential inverters as shown in Fig. 1(a). Use of dual CCO architecture in combination with pseudo-differential inverter stages reduce even-order distortion and improve common-mode rejection and power-supply rejection [9]. The SRO output is digitally differentiated using XOR gates and unit delay, which implements $1 - z^{-1}$, and fedback to the CCO input using a multi-element current steering non-return-to-zero (NRZ) DAC. Digital differentiation using XOR naturally scrambles the element selection pattern of the DAC such that its static mismatch is first-order shaped.

Fig. 1(b) shows the mathematical model of the proposed ADC. A single-ended model is shown for sake of simplicity. In order to mathematically analyze the ADC, we use pulse-frequency modulation (PFM) [9] model for CCO+PFD which operate in continuous-time. In the PFM model, the CCO acts as a pulse frequency encoder which encodes phase information in rising edges of the CCO output. The PFD integrates the dirac-delta impulses which correspond to the timing instants when CCO phase crosses 2π, and converts the PFM output into a pulse-width modulated output [9]. As shown in [9], the PFM signal contains i) a dc term proportional to CCO center frequency, f_{cco} ii) the input signal multiplied by CCO tuning gain, k_{cco} and iii) distortion terms with modulation sidebands centered around harmonics of f_{cco}. The PFM distortion terms is denoted by q_1 in Fig. 1(b). Since sampling happens immediately after the SRO, the SRO is modeled as a phase domain integrator rather than PFM encoder [9]. The SRO has a transfer function of $2\pi k_{sro}/s$. The feedback NRZ DAC has

978-1-7281-5107-6/19 $31.00 © 2019 IEEE

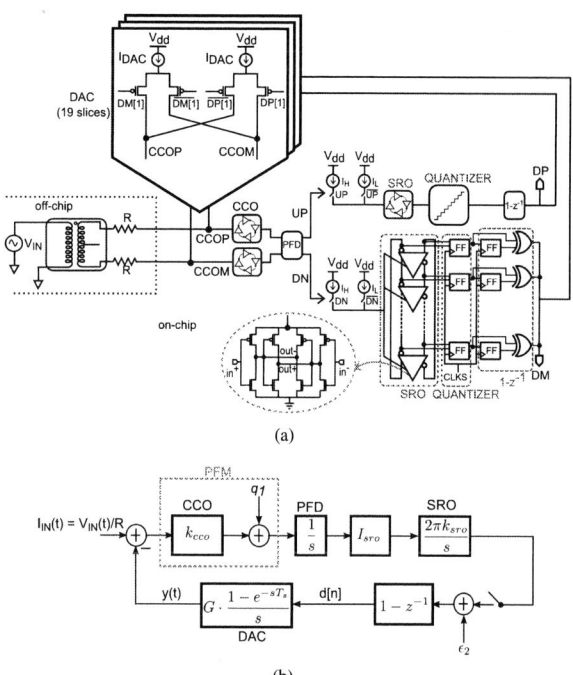

(a)

(b)

Fig. 1: (a) Circuit schematic of the proposed ADC (b) mathematical model

a gain of G and $I_{sro}=I_H - I_L$. SRO quantization noise is modeled by ϵ_2. The ADC output can then be written as

$$
\begin{aligned}
d &= \left[I_{IN} H_1(s) \frac{(1 - e^{-sT_s})^2}{(sT_s)^2} \right]^* + \epsilon_2 NTF_2(z) \\
&+ \left[q_1 \frac{H_1(s)}{k_{cco}} \frac{(1 - e^{-sT_s})^2}{(sT_s)^2} \right]^*
\end{aligned}
\tag{1}
$$

where $H = 2\pi k_{cco} k_{sro} I_{sro}$, $[]^*$ denotes sampling operation and T_s is the sampling period. $H_1(s)$ and NTF_2 are given by

$$
H_1(s) = \frac{2HT_s^2}{2 + (GHT_s^2 - 2)\, e^{-sT_s} + (GHT_s^2)e^{-2sT_s}}
\tag{2}
$$

$$
NTF_2(z) = \frac{2(1 - z^{-1})^2}{2 + (GHT_s^2 - 2)\, z^{-1} + (GHT_s^2)z^{-2}}
\tag{3}
$$

It can be seen from (1) that the ADC input and PFM tones are second-order sinc-filtered before sampling. Since the PFM tones are centered around f_{cco}, they can be adequately suppressed by setting $f_{cco}=f_s$ ($f_s=1/T_s$). However, without background calibration, it is hard to ensure $f_{cco}=f_s$ across PVT variations and small drift in f_{cco} will alias PFM tones into signal-band and degrade ADC SNDR [9]. We set f_{cco} to $1.25f_s$ instead such that PFM tones alias mostly out-of-band and does not degrade ADC SNDR.

B. Design Optimization

The design trade-offs between noise, power and bandwidth (BW) for the proposed ADC are tightly coupled through the

parameters N, I_{DAC} and T_s which are optimized for maximum energy efficiency. In order to reduce energy consumption, current in the feedback DAC is re-used to bias the first CCO integrator as shown in Fig. 1(a). The PMOS DAC supplies the differential CCOs with currents given by $\{I_{DAC}(N + DP[n] - DM[n])\}$ and $\{I_{DAC}(N + DM[n] - DP[n])\}$ in the n-th cycle. Thus, center frequency of the CCOs, f_{cco}, is set by $N \cdot I_{DAC}$. The proposed current-reuse DAC architecture results in both lower noise and power consumption than the ADCs which use a PMOS current source to bias the CCO and an NMOS DAC for feedback [4], [9], [10]. Since the PMOS current source in [9], [10] contributes to a significant fraction (28%) of overall thermal noise, its absence in the proposed architecture reduces noise. Also, re-using the DAC current for CCO bias reduces current consumption by $N \cdot I_{DAC}$. Thus, the proposed architecture can achieve higher energy efficiency compared to [9], [10].

In order to further increase energy efficiency, we need to reduce thermal noise from CCO and DAC, as well as reduce overall power consumption. Thermal noise from SRO is first order high-pass shaped and is not a significant contributor to overall noise.

Input referred thermal noise due to the DAC is given by

$$
\sqrt{\overline{i_{dac,n}^2}} = \sqrt{2} \cdot \sqrt{N \cdot 4kT\gamma g_m \cdot f_B}
\tag{4}
$$

where g_m denotes transconductance of unit DAC current source, f_B is the ADC bandwidth and the factor of $\sqrt{2}$ in (4) accounts for the differential DACs. g_m is directly proportional to I_{DAC} and the input referred DAC noise increases linearly with $\sqrt{N \cdot I_{DAC}}$.

The input-referred thermal noise due to CCO is given by

$$
\sqrt{\overline{i_{cco,n}^2}} = \sqrt{2} \cdot \frac{\sqrt{2DT_s}}{2\pi k_{cco} T_s} \cdot \frac{1}{\sqrt{OSR}}
\tag{5}
$$

where D is phase diffusion constant given by $D=\mathcal{L}(\Delta\omega) \cdot (\Delta\omega)^2/2$ where $\mathcal{L}(\Delta\omega)$ is the phase noise at an offset frequency of $\Delta\omega$. Input-referred CCO thermal noise is proportional to f_{cco}, and hence, $N \cdot I_{DAC}$.

The input-referred quantization noise is given by

$$
\sqrt{\overline{i_{q,n}^2}} = \frac{I_{DAC}}{\sqrt{12}} \cdot \frac{\pi}{\sqrt{5}} \cdot (OSR)^{-5/2}
\tag{6}
$$

From (4) and (5), it can be seen that reducing $N \cdot I_{DAC}$ reduces ADC input-referred thermal noise. However, reducing $N \cdot I_{DAC}$ alone does not improve SNR since input swing is also reduced proportionally. In addition, reducing $N \cdot I_{DAC}$ reduces the DAC gain, G, which can in turn reduce ADC SNR [9]. Thus, to keep the ADC open-loop gain unchanged, reduction in $N \cdot I_{DAC}$ is accompanied by increase in T_s which further reduces input referred CCO thermal noise (see (5)) as well as ADC power consumption. Hence, reduction in $N \cdot I_{DAC}$ and simultaneous increase in T_s increases ADC energy efficiency. However, for a given ADC bandwidth, T_s cannot be set too high as quantization noise will limit ADC SNR and energy efficiency. To find an optimum ADC sampling frequency, we sweep f_s and scale $N \cdot I_{DAC}$, I_{sro} and input swing by the same factor as f_s. Fig. 2 shows the ADC input referred thermal noise, quantization noise and SNR as function of f_s for BW of 1.5MHz. Thermal noise is calculated from SPICE noise simulations on DAC and CCO. At low f_s, quantization noise limits ADC SNR, while at high f_s, thermal noise limits ADC SNR. At f_s=32MHz, thermal noise and quantization noise are almost equal. Hence, f_s is set to 32MHz for maximum energy efficiency (walden FoM), and the corresponding $N \cdot I_{DAC}$ is 15μA.

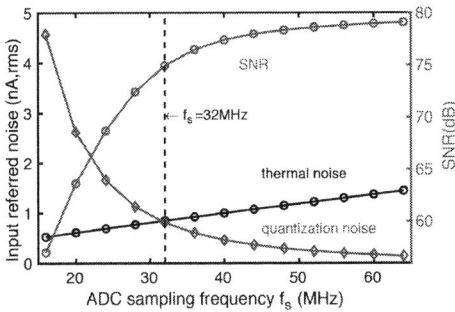

Fig. 2: ADC noise and SNR versus f_s

For a given $N \cdot I_{DAC}$, optimum value of N is decided by quantization noise. If N is very small, in-band quantization noise dominates in-band thermal noise, while for very large N, in-band quantization noise is much smaller than in-band thermal noise. Fig. 3 shows the ADC input referred thermal noise, quantization noise and SNR as function of N for BW of 1.5MHz, f_s=32MHz and input current amplitude of 18.5μA (pk-pk). The product of $N \cdot I_{DAC}$ is kept constant as N is swept. Fig. 3 shows that for N=18, quantization noise and thermal noise are almost equal. We choose N=19 for this design for an SNR of 74.9dB. The current-reuse DAC architecture as well as the optimum choice of N, I_{DAC} and f_s results in 17$\times$ better energy efficiency than our first prototype [9].

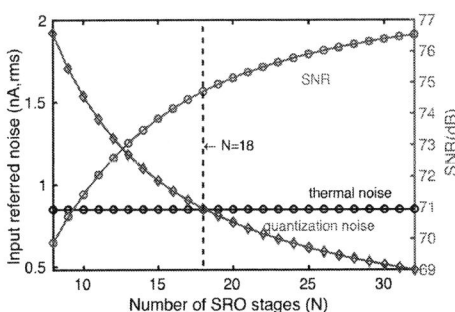

Fig. 3: ADC noise and SNR versus N

Excess loop delay (ELD) in CT $\Delta\Sigma$ ADC can de-stabilize the system by introducing additional poles to ADC transfer function. Typically, ELD is compensated by adding another DAC around the quantizer. For the proposed ADC, ELD is compensated without additional DAC by judiciously selecting the CCO and SRO gains during the design phase. As shown

in [9], the CCO and SRO gains are set to ensure that the ADC can tolerate an ELD of upto 1 sampling period even if CCO, SRO gains vary by $\pm$10% due to PVT variations.

III. MEASUREMENT RESULTS

A prototype ADC is fabricated in 65nm CMOS process and Fig. 4(a) shows the die microphotograph. The ADC core occupies an area of 0.06mm^2 and runs from a power supply of 1V. The ADC output is converted from thermometer-to-binary (T/B) code before being brought off-chip. I_{DAC} is set to 0.8μA for this design. The ADC consumes 105μW power at f_s=32.6MHz. Fig. 4(b) shows the measured SNR and SNDR versus input amplitude at a BW=1.5MHz. The ADC has a measured dynamic range of 74dB.

Fig. 5 shows the measured ADC spectrum for an input frequency of 50kHz and amplitude of -5dBFS. Analog voltage input is converted into current input through off-chip 6kΩ resistors. The ADC has an SNDR of 72.7dB and SFDR of 81dB for BW=1.5MHz without any calibration.

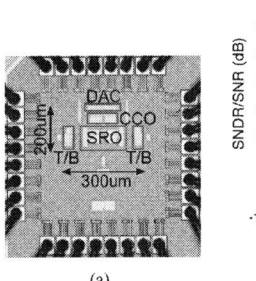
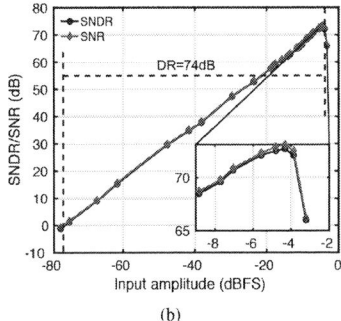

(a)   (b)

Fig. 4: (a) Die micro-photo (b) ADC dynamic range plot

Fig. 6 shows the measured schreier and walden FoM versus BW for the ADC. The ADC achieves 174.2dB schreier FoM at 1.5MHz BW and 8.6fJ/step walden FoM at 2.3MHz BW Fig. 7 shows the measured SNDR of the ADC versus input frequency at BW=1.5MHz. The ADC SNDR varies between 72-73dB as input frequency is varied from 10-100kHz. Fig. 8 shows the measured SNDR for five chips at BW of 1.5MHz across power supply of 0.9-1.1V and temperatures from 0-50C. The ADC maintains high SNDR ($>$ 70dB) across VT corners without calibration.

Table I compares the fabricated ADC with state-of-the-art CT $\Delta\Sigma$ VCO-ADCs with similar bandwidths. The proposed ADC achieves the lowest walden FoM of 8.6fJ/step at BW=2.3MHz and lowest power consumption of 105μW. Thanks to the optimized design procedure, the proposed ADC has high SNDR even at small OSR. The figure next to Table I compares walden FoM of the proposed ADC with state-of-the-art CT $\Delta\Sigma$ ADCs. The proposed ADC has the lowest walden FoM among reported CT $\Delta\Sigma$ ADCs.

978-1-7281-5107-6/19 $31.00 © 2019 IEEE

TABLE I: Comparison with state-of-the-art CT $\Delta\Sigma$ VCO-ADCs.

	Process (nm)	Area (mm^2)	Fs (MHz)	Power (mW)	BW (MHz)	SNDR (dB)	FoM$_w$ (fJ/step)[1]
JSSC'10 [1]	65	0.075	1300	11.5	5.08	75	246
ESSCIRC'16 [3]	130	0.13	250	1.05	3	70.2	66.2
JSSC'17 [5]	65	0.01	1000	1.5	10	55.1	158
VLSIC'17 [11]	40	0.028	330	0.5	6	68.6	19.8
ASSCC'18 [10]	40	0.086	260	0.91	5.2	69.6	34.7
TCAS–I'19 [9]	65	0.06	205	1	2.5	64.2	150.9
This Work	**65**	**0.06**	**32.6**	0.1	**1.5**	**72.7**	9.9
					2.3	**70.2**	8.6

[1]FoM$_w$ = Power/(2^{ENOB}x2BW)

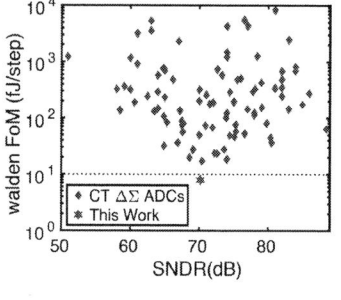

Comparison with state-of-the-art CT $\Delta\Sigma$ ADCs

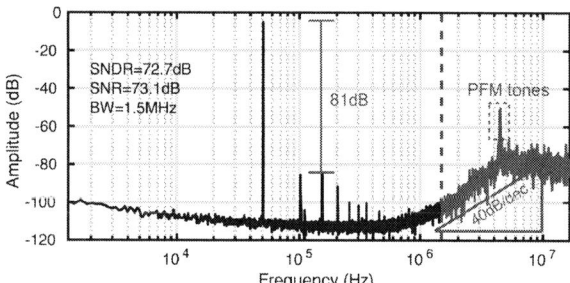

Fig. 5: Measured FFT for 50kHz input

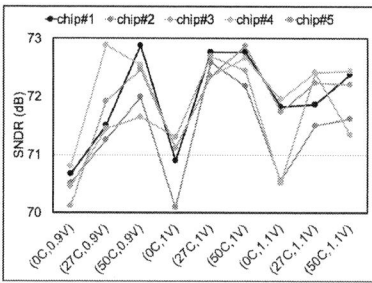

Fig. 8: Measured ADC SNDR across VT corners for 5 chips

at Dallas' Texas Analog Center of Excellence (TxACE).

REFERENCES

[1] G. Taylor and I. Galton, "A mostly-digital variable-rate continuous-time delta-sigma modulator ADC," *IEEE Journal of Solid-State Circuits*, vol. 45, no. 12, pp. 2634–2646, 2010.

[2] A. Sanyal and N. Sun, "A 18.5-fJ/step VCO-based 0-1 MASH Delta-Sigma ADC with digital background calibration," in *IEEE Symposium on VLSI Circuits*, 2016, pp. 26–27.

[3] S. Li and N. Sun, "A 174.3 dB FoM VCO-based CT $\Delta\Sigma$ modulator with a fully digital phase extended quantizer and tri-level resistor DAC in 130nm CMOS," in *IEEE European Solid-State Circuits Conference*, 2016, pp. 241–244.

[4] K. Lee, Y. Yoon, and N. Sun, "A Scaling-Friendly Low-Power Small-Area $\Delta\Sigma$ ADC With VCO-Based Integrator and Intrinsic Mismatch Shaping Capability," *IEEE Journal on Emerging and Selected Topics in Circuits and Systems*, vol. 5, no. 4, pp. 561–573, 2015.

[5] A. Babaie-Fishani and P. Rombouts, "A Mostly Digital VCO-Based CT-SDM With Third-Order Noise Shaping," *IEEE Journal of Solid-State Circuits*, vol. 52, no. 8, pp. 2141–2153, 2017.

[6] F. Cardes, E. Gutierrez, A. Quintero, C. Buffa, A. Wiesbauer, and L. Hernandez, "0.04-mm^2 103-dB-A Dynamic Range Second-Order VCO-Based Audio ADC in 0.13-μm CMOS," *IEEE Journal of Solid-State Circuits*, vol. 53, no. 6, p. 1731, 2018.

[7] A. Sanyal and N. Sun, "Second-order VCO-based $\Delta\Sigma$ ADC using a modified DPLL," *Electronics Letters*, vol. 52, no. 14, pp. 1204–1205, 2016.

[8] V. Prathap, S. T. Chandrasekaran, and A. Sanyal, "2nd-Order VCO-based CT $\Delta\Sigma$ ADC architecture," in *IEEE International MidWest Symposium on Circuits and Systems*, 2017, pp. 687–690.

[9] A. Jayaraj, M. Danesh, S. T. Chandrasekaran, and A. Sanyal, "Highly Digital Second-Order $\Delta\Sigma$ VCO ADC," *IEEE Transactions on Circuits and Systems I: Regular Papers*, 2019.

[10] Y. Zhong, S. Li, A. Sanyal, X. Tang, L. Shen, S. Wu, and N. Sun, "A Second-Order Purely VCO-Based CT $\Delta\Sigma$ ADC Using a Modified DPLL in 40-nm CMOS," in *IEEE Asian Solid-State Circuits Conference (A-SSCC)*, 2018, pp. 93–94.

[11] S. Li and N. Sun, "A 0.028mm^2 19.8 fJ/step 2nd-order VCO-based CT $\Delta\Sigma$ modulator using an inherent passive integrator and capacitive feedback in 40nm CMOS," in *IEEE Symposium on VLSI Circuits*, 2017, pp. C36–C37.

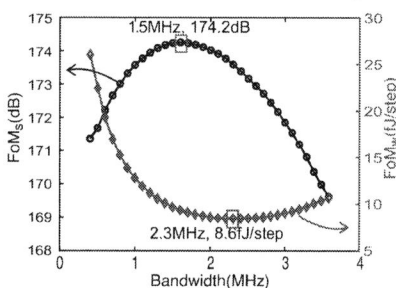

Fig. 6: Schreier and walden FoM vs BW

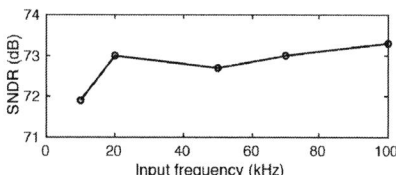

Fig. 7: Measured SNDR vs input frequency

IV. CONCLUSION

A second-order purely VCO-based CT $\Delta\Sigma$ ADC is presented in this work. A 65nm prototype achieves a walden FoM of 8.6fJ/step which is the lowest among state-of-the-art CT $\Delta\Sigma$ ADCs. Energy efficiency of the ADC is expected to improve further with CMOS technology scaling.

ACKNOWLEDGMENT

This work is supported by Semiconductor Research Corporation (SRC) task # 2712.020 through The University of Texas

15-1 (7206)

A 110.3-bits/min 8-Ch SSVEP-based Brain-Computer Interface SoC with 87.9% Accuracy

Wooseok Byun
Dept. of Electronics Engineering
Chungnam National University
Daejeon, Republic of Korea

Dokyun Kim[1] and Sung Yeon Kim[2]
Dept. of Electrical Information
Engineering[1] / Electronic Engineering[2]
Seoul National University of Science
and Technology
Seoul, Republic of Korea

Ji-Hoon Kim
Dept. of Electronic and Electrical
Engineering
Ewha Womans University
Seoul, Republic of Korea
jihoonkim@ewha.ac.kr

Abstract— A wearable brain-computer interface (BCI) based on steady-state visual evoked potential (SSVEP) has been widely studied to enable paralyzed patients to communicate with others. However, target identification accuracy and information transfer rate (ITR), which are general performance indicators of SSVEP-based BCI system, still need to be further improved in wearable devices. This paper proposes 8-channel SSVEP-based visual target identification system-on-chip (SoC) to improve the ITR of low-cost wearable BCI device while dramatically reducing the computational complexity without accuracy degradation. The proposed target identification algorithm, CCA-CR, includes algorithmic optimizations and candidate reduction (CR) method that reduce signal processing load by at least 75% without degrading target identification accuracy and ITR. This paper also proposes a matrix decomposition processor (MDP) that calculates complex matrix arithmetic operations through systolic array based CCA-CR engines. Compared to the state-of-the-art CCA-based algorithm, the proposed SoC implemented in FPGA exhibits 63% better ITR with 33% reduction of recording time without accuracy degradation.

Keywords—Brain-computer interface (BCI); system-on-chip; canonical correlation analysis (CCA); matrix decomposition; steady-state visual evoked potential (SSVEP); target identification

I. INTRODUCTION

Brain-computer interface (BCI) can translate brain signals into interpretable information that reflects the user's intentions [1]. Steady-state visual evoked potential (SVEP) has been widely used in paralyzed patients due to high information transfer rates (ITRs) and little user training [2]. The behavior of the SSVEP-based wearable BCI system is illustrated in Fig. 1(a). When the user is looking at a specific flickering character on the display (1), SSVEP, the brain response of the same frequency as the flickering target, is generated. The SSVEP can be measured at the occipital region of the scalp (2). The target identification algorithm is then performed to identify the target being focused on (3). The results of the target identification algorithm are wirelessly transmitted to the display (4). After single typing, the user continues typing by moving his or her eyes to the next visual target (5). In a previous study, in-ear SSVEP-based BCI system was reported [3]. However, in this system, the wearable in-ear device only wirelessly transmits

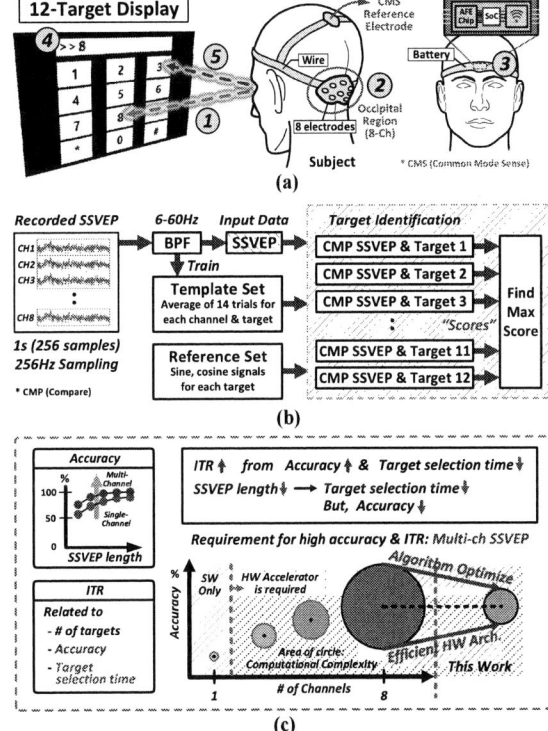

Fig. 1. (a) Behavior of SSVEP-based BCI system (b) target identification processing flow (c) motivation of this study.

measured raw SSVEP data with high power consumption [4]. Embedded BCI systems performing canonical correlation analysis (CCA) on a low-cost MCU were proposed [5][8]. However, these wearable platforms still require further improvement of performance, such as target identification accuracy and ITR. As described in Fig. 1(c), multi-channel SSVEP is required to improve target identification accuracy and ITR. For wearable devices, high computational complexity due to multi-channel SSVEP should be reduced. In this paper,

978-1-7281-5107-6/19 $31.00 © 2019 IEEE

IEEE Asian Solid-State Circuits Conference
November 4 – 6, 2019
The Parisian Macao, Macao SAR, China

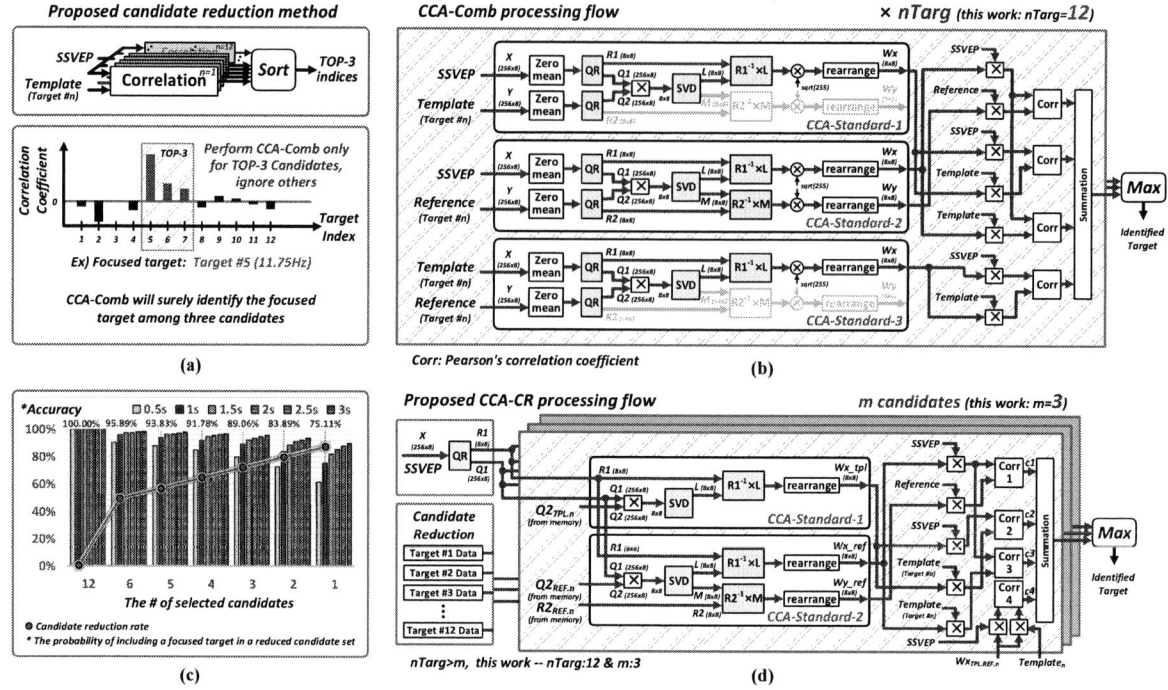

Fig. 2. (a) Proposed candidate reduction method shows (c) high candidate reduction accuracy and candidate reduction rate. The proposed method can be applied to (b) conventional CCA-Comb algorithm which has huge computational complexity. (d) Algorithmic optimizations for CCA-Comb can drastically reduce computational complexity with the proposed candidate reduction method. (the ratio of the number of operations of proposed CCA-CR to CCA-Comb: QR decomposition (QRD) 1.38%, singular value decomposition (SVD) 18.75%, back substitution 18.75%)

we propose CCA-CR, efficient target identification algorithm, that dramatically reduced computational complexity while maintaining target identification accuracy and improving ITR. we also propose an energy efficient hardware architecture that supports multiple complex matrix arithmetic operations required by CCA-CR. Through the proposed algorithm and hardware architecture, we identified the feasibility of high-performance BCI systems in low-cost wearable devices. This work used 8-channel SSVEP data for 12 visual targets in [6] along with the target identification process shown in Fig. 1(b).

II. PROPOSED ALGORITHM

A. Candidate Reduction

The CCA-based target identification algorithms typically compare the SSVEP to all target frequencies. This affects the computational complexity in proportion to the number of visual targets. In this paper, the baseline algorithm is CCA-Comb, a combination of CCA-Standard and individual template-based CCA (*IT-CCA*) [6], which detects temporal features of SSVEP signals using CCA-Standard between input data and individual template data that can be obtained by averaging multiple training tests for specific subject. As indicated in Fig. 2(b), there are three CCA-Standard operations in a single CCA-Comb calculation. The number of targets, *nTarg*, increases the computational complexity. If the BCI system can detect targets that are not related to input SSVEP early on, it is not necessary

to perform the CCA-Comb on them, which can reduce the computational complexity. We propose candidate reduction (CR) method using Pearson's correlation coefficient between SSVEP and individual template data to reduce the targets to be used in CCA-Comb calculations. As described in Fig. 2(a), the SSVEP is used to 12 correlation calculations with different templates. Then, the CR treats the Top-N target indices of correlation coefficient as the most relevant targets and ignores the rest. The feasibility of CR method is shown in Fig. 2(c). This graph represents the accuracy, which is the probability of including a focused target in a reduced candidate set, according to the number of selected candidates, N. The red circle shows the candidate reduction rate. The fewer selected candidates, the less computational complexity, but the less accurate. The number of candidates selected in this paper is fixed as 3.

B. CCA-CR: approx. CCA-Comb with Candidate Reduction

The processing flow of CCA-Comb described in Fig. 2(b) can be optimized with the CR and algorithm approximations. First, CR reduces the number of CCA-Comb operations by 75%. In addition, if a complex matrix operation is not related to the current input SSVEP, it can be precomputed and stored in memory to reduce the amount of computation. It increases memory, but reduces the amount of computation and memory access, which is beneficial for system performance. This algorithm is called CCA-CR, the approximated CCA-Comb with candidate reduction.

IEEE Asian Solid-State Circuits Conference
November 4 – 6, 2019
The Parisian Macao, Macao SAR, China

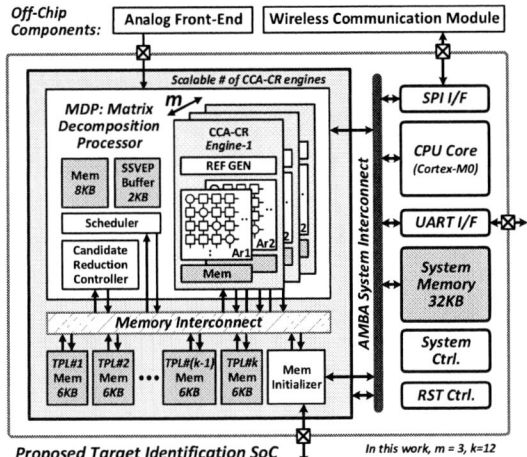

Fig. 3. Proposed target identification SoC architecture.

III. PROPOSED ARCHITECTURE

Fig. 3 shows the target identification SoC (system-on-chip) architecture. The main component of the SoC is the matrix decomposition processor (MDP). The memory interconnect between the 12 template memories and the MDP is controlled by CR controller and CCA-CR engine scheduler in the MDP.

A. Matrix Decomposition Processor (MDP)

Target identification algorithm CCA-CR can be performed on MDP with highly reusable CCA-CR engines. As indicated in Fig. 3, the number of CCA-CR engines m and the number of targets k are scalable while designing the SoC. Since the CR always selects 3 candidates and the dataset was recorded for the 12 targets, this paper chose m as 3 and k as 12. All CCA-CR engines are managed by scheduler in the MDP for parallel processing. When the SSVEP is written to the SSVEP buffer, the correlation operation of CR is performed four times on each of the three CCA-CR engines. The required template data is loaded from TPL memories via memory interconnect.

In Fig. 4(a), the MDP scheduler managing three CCA-CR engines is controlled by the Cortex-M0. Large-scale operations such as QRD and SVD, and control-intensive operations are assigned by software. Instructions for these software-assigned operations are listed in the array configuration registers. All Ar1 and Ar2 of the CCA-CR engine can share the data required for CCA-CR, such as SSVEP buffer, intermediate results of QRD of SVD. The CCA-CR default operation assignments for the two systolic arrays Ar1 and Ar2 are illustrated in Fig. 4(b). All operations after the QRD of SSVEP should wait until the QRD ends, except for the CR and the 'Corr 4' in Fig. 2(d).

B. CCA-CR Engine

In Fig. 5, The CCA-CR engine includes two 8x8 systolic arrays with a processing element (PE) controller, which allows the PE to perform various matrix arithmetic operations such as QRD, SVD, back substitution, matrix multiplication, Pearson's correlation, and system-defined operations.

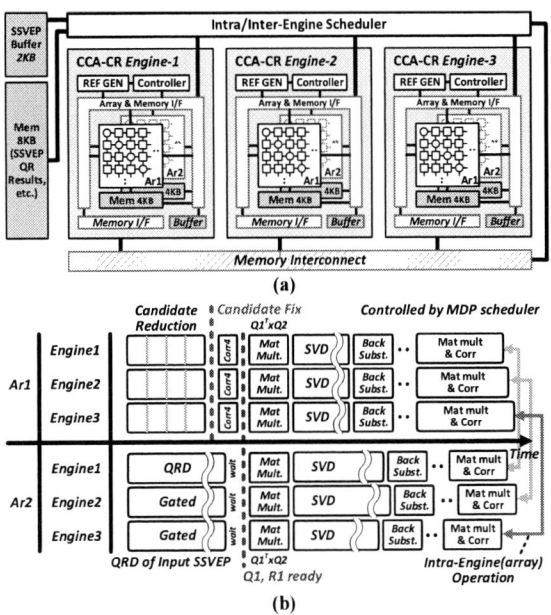

Fig. 4. (a) Three CCA-CR engines (b) the CCA-CR operation assignment of two systolic arrays in each CCA-CR engine

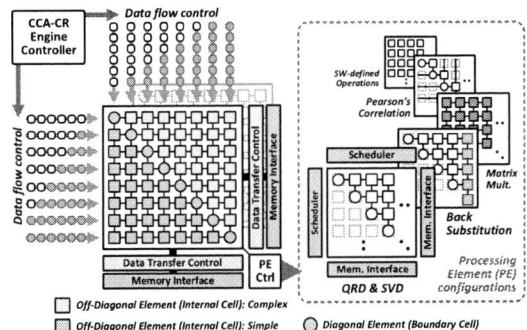

Fig. 5. Configurable systolic array (Ar1 & Ar2)

IV. IMPLEMENTATION AND EXPERIMENTAL RESULTS

The proposed SoC is implemented in Intel Altera Cyclone IV EP4CE115 with an operating frequency of 50MHz. This SoC consists of 1113k gates (2-input NAND) in a 65nm CMOS library with 138KB on-chip SRAM. Fig. 6 shows the target identification accuracy and ITR performance from RTL simulation results for 1-s length SSVEP and the hardware modeled simulation results for remaining length of SSVEP. (The results of the RTL and 1-s hardware model are the same.) The ITR is evaluated using the following equation:

$$ITR = \log_2 \left(N_f \cdot P^P \cdot \left[\frac{1-P}{N_f - 1} \right]^{1-P} \right) \times \left(\frac{60}{T} \right) \quad (1)$$

where N_f is the number of target frequencies (in this work, 12), P is the target identification accuracy, and T is the average target selection time (in this work, 1.5s = recording 1.0s + gaze

978-1-7281-5107-6/19 $31.00 © 2019 IEEE

IEEE Asian Solid-State Circuits Conference
November 4 – 6, 2019
The Parisian Macao, Macao SAR, China

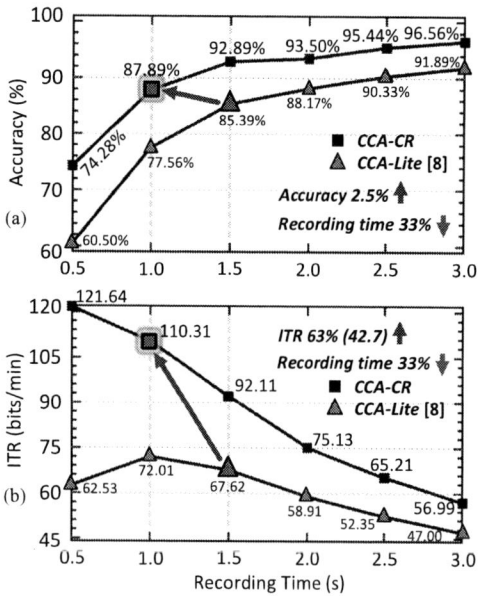

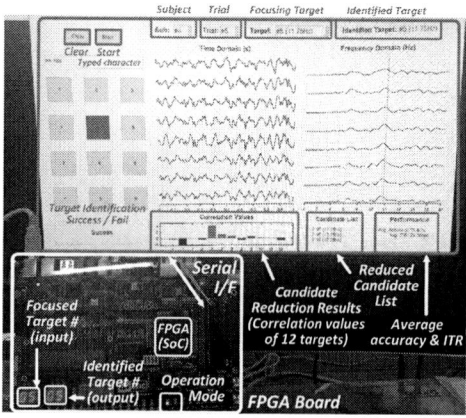

Fig. 6. Hardware modeled simulation performance of (a) target identification accuracy and (b) ITR

Fig. 7. The experimental setup

TABLE I. COMPARISON TABLE

	PLoS ONE'15 [6]	EMBC'18 [7]	Access'19 [8]	**This Work**
Identification Algorithm	CCA-Comb	CCA-Comb w/ FBA (Filter Bank Analysis)	CCA-Lite	**CCA-CR**
Dataset	Dataset [6]	Dataset [7]	Dataset [6]	**Dataset [6]**
Accuracy (%)	92.78	91.43	85.39	**87.89**
ITR with std.^{c} (bits/minute)	91.68 ± 20.32	89.04 ± 13.17	67.62 ± 26.95	**110.31 ± 33.43**
Record / Gaze-Shift Time (s)	1.0 / 1.0	1.2 / 0.5	1.5 / 0.5	**1.0 / 0.5**
# of Targets	12	9	12	**3 (12^a)**
# of Channels	8	9	1	**8**
Comput. Load^b	86.0	97.2	18.0	**24.0**
Processing Platform	PC (Intel Zeon 3.7G)	PC (Detail Info N/A)	MCU (Arm Cortex-M3 72MHz)	**SoC in FPGA (50MHz)**

^{a}. The proposed candidate reduction algorithm reduces the number of targets from 12 to 3.
^{b}. Computational Load = <Recording Time> × <Number of Targets> × <Number of Channels>
^{c}. std. : standard deviation

V. CONCLUSION

This work proposes 8-channel SSVEP target identification SoC using CCA-CR, which reduces CCA-Comb calculations by 75% without degrading target identification accuracy and ITR. Thanks to multi-channel SSVEP, recording time is short, which significantly reduces target identification latency and increases ITR by 63%. The CCA-CR is performed efficiently by matrix decomposition processor with three CCA-CR engines. The reconfigurable systolic array of the CCA-CR engine dramatically reduces the number of memory accesses and allows the parallelism of PEs to efficiently perform complex matrix arithmetic operations.

REFERENCES

[1] N. Birbaumer et al., ''A spelling device for the paralysed,'' *Nature*, vol. 398, pp. 297–298, Mar. 1999

[2] H. Cecotti, ''Spelling with non-invasive brain–computer interfaces–current and future trends,'' *J. Physiol.-Paris*, vol. 105, pp. 106–114, Jun. 2011.

[3] J. W. Ahn, Y. Ku, D. Y. Kim, J. Sohn, J.-H. Kim, and H. C. Kim, ''Wearable in-the-ear EEG system for SSVEP-based brain–computer interface,'' *Electron. Lett.*, vol. 54, no. 7, pp. 413–414, Mar. 2018.

[4] N. Verma, A. Shoeb, J. Bohorquez, J. Dawson, J. Guttag, and A. P. Chandrakasan, ''A micro-power EEG acquisition SoC with integrated feature extraction processor for a chronic seizure detection system,'' *IEEE J. Solid-State Circuits*, vol. 45, no. 4, pp. 804–816, Apr. 2010.

[5] M. Salvaro, S. Benatti, V. J. Kartsch, M. Guermandi, and L. Benini, ''A minimally invasive low-power platform for real-time brain computer interaction based on canonical correlation analysis,'' *IEEE Internet Things J.*, vol. 6, no. 1, pp. 967–977, Feb. 2019.

[6] M. Nakanishi, Y. Wang, Y.-T. Wang, and T.-P. Jung, ''A comparison study of canonical correlation analysis based methods for detecting steadystate visual evoked potentials,'' *PLoS ONE*, vol. 10, no. 10, Oct. 2015, Art. no. e0140703.

[7] X. Chen, Y. Wang, S. Zhang, X. Gao, "Enhancing Detection of SSVEPs with Intermodulation Frequencies Using Individual Calibration Data," *IEEE 40th Ann. Int. Conf. Eng. Med. Biol. Soc. (EMBC)*, July. 2018.

[8] D. Kim, W. Byun, Y. Ku, J.-H. Kim, "High-Speed Visual Target Identification for Low-Cost Wearable Brain-Computer Interfaces," *IEEE Access*, vol. 7, pp. 55169-55179, April. 2019.

shift 0.5s). As indicated in Fig. 6(a), the accuracy of the CCA-CR is increased by 2.5% even though the recording time is reduced by 33% compared to the state-of-the-art CCA-Lite. Fig. 6(b) shows a 63% increase in ITR. These results account for two main sources of performance improvement, multi-channel SSVEP and CR. Multi-channel SSVEP basically helps to improve performance. The proposed CR accurately reduces the candidates that will be found not to be a focused target. Fig. 7 shows the experimental setup. Table 1 shows that the *Computational Load* of CCA-CR is similar to that of CCA-Lite, but can be compared with other studies in terms of accuracy and is much better in terms of ITR. The result from [6] are from the same dataset as ours, but we have a much higher ITR and a slightly lower accuracy.

FPGA-Based Sparsity-Aware CNN Accelerator for Noise-Resilient Edge-Level Image Recognition

Seungsik Moon, Hyunhoon Lee, Younghoon Byun, Jongmin Park,
Junseo Joe, Seokha Hwang, Sunggu Lee, and Youngjoo Lee
Department of Electrical Engineering
Pohang University of Science and Technology (POSTECH), Pohang, Korea
Email: youngjoolee@postech.ac.kr

Abstract—**This paper presents a novel sparsity-aware CNN accelerator supporting the edge-level image recognition even for noisy images. In the proposed accelerator, we characterize the class of input noises by utilizing the FFT-based on-the-fly noise classifier. The proposed convolution engine then accesses the external memory to load the dedicated network that provides the accurate inference processing for the detected noise class. To save the energy consumed by the external DRAM accesses, in addition, we present the filter-level pruning algorithm with the memory-reduced indexing scheme, which can reduces the processing latency by utilizing the indexing method. To verify the effectiveness of the proposed methods, the proposed CNN accelerator is implemented in the commercialized FPGA-based platform, achieving the processing rate of 57.6GOPS at the speed of 100MHz. Utilizing the dedicated network for each noise type, the prototype accelerator reduces the energy consumption by 62% compared to the conventional network with a similar recognition accuracy, which is suitable for the intelligent edge-level devices subjected to the various noises in practice.**

Keywords—**CNN accelerator, FPGA, Image recognition, Noise-resilient processing.**

I. INTRODUCTION

In the past few years, the convolutional neural network (CNN) has been gaining huge popularity because of its attractive performance especially for recognizing the images [1]. However, the recent CNNs are in general based on the numerous processing layers, consuming a huge amount of energy by the frequent DRAM accesses [2]. In addition, the input raw images are in general affected by lots of noise sources, which requires additional CNN layers as well as pre-processing steps [3]. Hence, it is hard to directly apply the advanced CNN algorithms to on-device edge-level recognition engines whose energy budgets are tightly limited [4].

In this work, we present the novel sparsity-aware CNN accelerator for edge devices that provides the efficient processing steps based on our advanced noise-resilient CNN algorithm in [5]. In contrast that the previous works adopt energy-starving operations for reducing the effects of noises, the proposed algorithm simply uses the same CNN network structure as the original one to provide the noise-resilient recognition by only changing the weight kernels, which is dedicated to the current noise types determined by the FFT-based on-the-fly noise

This work was supported by the Samsung Research Funding & Incubation Center of Samsung Electronics under Project Number SRFC-TB1703-07.

classifier. To reduce the energy consumption, furthermore, we introduce the hardware-friendly pruning scheme that eliminates less important filters. Without sacrificing the recognition accuracy, the proposed scheme provides the structured sparsity in filter level, which reduces the memory demands with the simple indexing scheme and also enhances the energy efficiency by highly utilizing hardware units. The proposed CNN accelerator is implemented by using the commercialized verification platform with Intel's Stratix-V FPGA solution. Based on the proposed algorithm-level innovations, the prototype design provides the energy efficiency of 28.4GOPS/W while providing the noise-resilient image recognition, which is happily acceptable for the practical edge-level devices.

II. NOISE-RESILIENT CNN ALGORITHM

To design the noise-resilient CNN, as described in [5], we first mathematically model two practical noise sources in edge devices, i.e., Gaussian-type and blur-type noises. More precisely, the pixel values of two-dimensional (2D) image suffered from the Gaussian noise can be expressed as $P_G(x, y) = P_C(x, y) + N_G(x, y)$, where $P_C(x, y)$ represents the pixel values of clean image and $N_G(x, y)$ is value from the normal distribution of $(0, \sigma_G)$. Similarly the noisy image with the blur-type source is represented as $P_B = P_C(x, y) * N_B(x, y)$, where the operator $*$ is the 2D convolution and $N_B(x, y)$ is the parameter defined with the blur parameter σ_B as detailed in [5]. By increasing σ_G (or σ_B), conceptually, we can add more Gaussian (or blur) noises to the clean image. As two types of noise sources affect the clean images in frequency domain differently, i.e., the Gaussian type increases the high-frequency signals where the blur type lowers them, it is possible to categorize the type and amount of noises by checking the input image in frequency domain. When we define 7 noise classes, i.e., three Gaussian-type classes with different σ_G, three blur-type classes by changing σ_B, and the clean one, it is reported that the FFT-based classifier can categorize more than 80% of noise class of input images [5].

To develop the noise-resilient CNN-based recognition system, we train the dedicated filters for each noise class with the same network structure. After checking the input noise class with the FFT-based classifier, therefore, it is possible to provide the dedicated CNN operation by loading the corre-

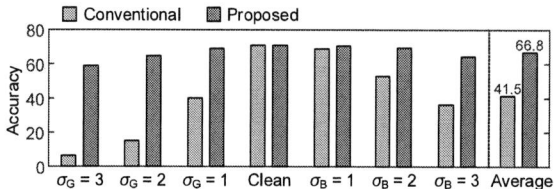

Fig. 1. Image recognition performances of VGG16 networks for CIFAR-100 data having different input noises.

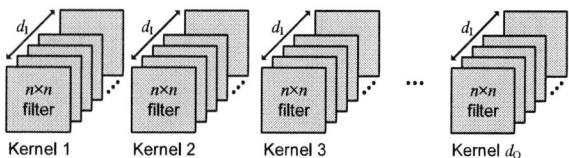

Fig. 2. Structure of a CNN layer using terminologies in this work.

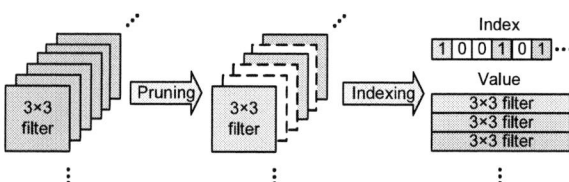

Fig. 3. The proposed filter-level pruning with the hardware-friendly on-off indexing scheme.

TABLE I. CNN PERFORMANCES WITH DIFFERENT PRUNING AND INDEXING METHODS

Type	Pruning	Indexing	Accuracy[*]	Memory
Unpruned	No	No	71.7%	14.0MB
Conventional	Element-level	SME [4]	71.5%	6.16MB
Proposed	Filter-level	On-off	70.3%	3.86MB

[*] For clean images

sponding filters. As all the networks for different noise classes are trained for the same CNN structure, we can perform the different noise-aware recognition processes at the same accelerator. Training the VGG16 network for CIFAR-100 data [6], Fig. 1 illustrates the performances of different networks under the various noises. Note that the conventional network cannot be acceptable for the practical solution due to the poor accuracy for noisy inputs. Regarding to the input noise classes, on the other hand, the proposed noise-resilient CNN always provides the attractive accuracy as shown in Fig. 1.

III. FILTER-LEVEL PRUNING SCHEME

Although the proposed CNN algorithm is effective to provide the accurate recognition for the practical noisy inputs, it is still hard to be used for the edge-level processing due to the huge amount of processing energy especially for the frequent external DRAM accesses [4]. In order to reduce the number of coefficients stored in the external memories, therefore, it is necessary to apply the aggressive pruning scheme that eliminates less important connections of the original network [7]. As the conventional pruning removes the small coefficients in order, the sparsity of the previous pruned network is random in general. Due to the unstructured sparsity, the amount of indexing data, which indicates the position of survived coefficients, sometimes dominates the size of quantized coefficients, even increasing the memory accessing energy.

It this work, a new hardware-friendly filter-level pruning that results the structured sparsity is presented. For the sake of simplicity, we define a CNN layer by four parameters; input depth, output depth, filter, and kernel. More precisely, as shown in Fig. 2, a CNN layer has d_O kernels each of which consists of d_I $n \times n$ filters, where d_I and d_O represent the values of input and output depths, respectively. Applying the conventional pruning, it is sometimes observed that a number of consecutive positions in a filter are eliminated, and only one or two coefficients are survived among n^2 positions in a filter. These barely-survived filters tend to occur more frequently when we increase the pruning factor to eliminate more coefficients for increasing the sparsity of the original network. Based on this observation, we perform the filter-level pruning that eliminates a whole coefficient of a filter by examining average values of filters [8]. Fig. 3 conceptually illustrates the proposed filter-level pruning. After applying the proposed method, it is important that we can obtained the structured sparsity. As shown in Fig. 3, therefore, the complexity of indexing can be relaxed remarkably as we

just denote the survived filters by using the filter-level on-off code rather than the previous sparse matrix encoding (SME) associated with the number of consecutive zeros between two survived coefficients [4]. Similar to the work in [9], in addition, we only use 3×3 filters that relaxes the mapping overheads at the accelerator architecture. Note that filters with different size can be handled by decomposing the filters.

Table I shows the effectiveness of the proposed sparsity-aware pruning scheme for VGG16 network and CIFAR-100 dataset. For the same 75% pruning factor, note that the proposed method still provides the acceptable recognition accuracy compared to the conventional pruning while reducing the network size. The proposed pruning scheme sacrifices the accuracy about 1% while relaxing the memory overheads by more than 72%. Hence, the proposed method is definitely suitable for the intelligent edge devices by reducing the number of DRAM accesses for loading the target network, accordingly saving the energy consumption. Thanks to the structured sparsity, moreover, we can easily improve the hardware utilization of the accelerator operation without applying the complicated skipping techniques, further enhancing the overall energy efficiency of the image recognition process.

IV. ACCELERATOR ARCHITECTURE

Fig. 4 depicts the conceptual architecture of the proposed noise-resilient CNN accelerator for recognizing CIFAR-100 data, which consists of the FFT-based classifier, the filter loader, on-chip memories for storing pruned networks and input/output feature maps, the non-linear operator including the activation function and the pooling operation, and the

IEEE Asian Solid-State Circuits Conference
November 4 – 6, 2019
The Parisian Macao, Macao SAR, China

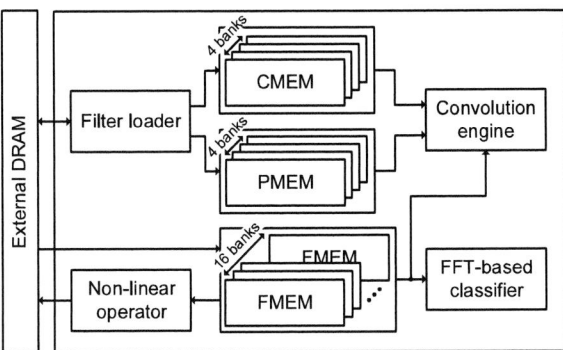

Fig. 4. The proposed noise-resilient CNN accelerator architecture.

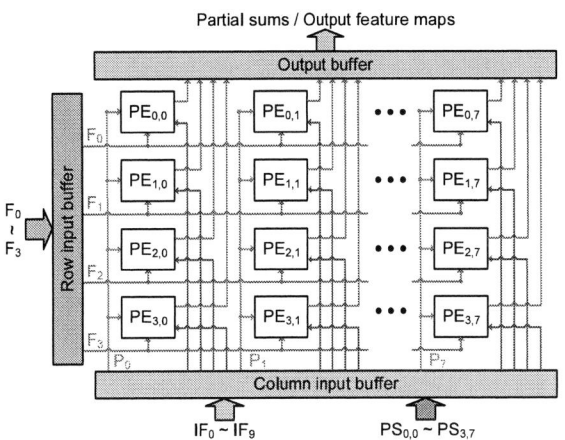

Fig. 5. Detailed architecture of the convolution engine.

convolution engine with multiple processing elements (PEs). When the input image is issued to the proposed accelerator, the 32×32 13b-quantized pixel values are firstly stored to the on-chip buffer for feature maps, i.e., FMEM, as they are regarded as the initial feature maps. To provide the sufficient bandwidths, we utilize 16 512-entry SRAM banks for implementing the FMEM. Note that the proposed architecture can be scaled to the different data sets by only changing the size of initial feature maps. After storing the initial feature map, we first pass the 2D pixel values to the FFT-based noise classifier, which is based on the 1D FFT module realizing Cooley-Tukey algorithm [10]. Performing 1D FFT twice in opposite directions, we can get the frequency-domain signals of input image, and the input noise class can be determined by accumulating the high-frequency parts as described in [5]. The recognition process is then activated by loading the CNN dedicated to the detected noise class. In the external DRAM memory, note that we simply prepare all the pruned networks for different noise types and only detected network is selected according to the output of the FFT-based classifier, providing the attractive recognition accuracy even for the noisy inputs.

In order to process each CNN layer, the filter loader in Fig. 4 accesses the external memory to serially get the pruned filters and their indexing information, which are based on the proposed filter-level pruning scheme. Note that the filter loader identifies the exact position of each survived filter based on the filter-level indexing data, and then stores the nine 8b-quantized 3×3 coefficients and the 12b-encoded filter-position vector at each row of on-chip memories, which are denoted as CMEM and PMEM, respectively. In the previous CNN accelerators associated with the conventional coefficient-level pruning method [7], it is hard to increase the hardware utilization at the following convolution engine due to the unstructured sparsity, limiting the processing efficiency. On the other hand, the proposed accelerator architecture simply enhances the ratio of active PEs by skipping the operation of zero weights since CMEM and PMEM already contain only the survived filters. In order to further improve the processing efficiency, as shown in Fig. 4, both PMEM and CMEM include four 512-entry SRAM banks, providing four 3×3 filters to the convolution engine at the same time.

Fig. 5 shows the detailed architecture of the proposed convolution engine including total 32 PEs. Note that the row input buffer (RIB) and the column input buffer (CIB) receive the elements of survived filters and input feature maps from CMEM and FMEM, respectively. Accepting the four filters simultaneously from RIB, which are denoted as F_i ($0 \leq i \leq 3$), eight PEs in horizontal direction share a 3×3 filter as depicted in Fig. 5. Similarly, the proposed PE array accepts eight 3×3 patches denoted as P_j ($0 \leq j \leq 7$) from CIB, and a P_j is shared by four PEs in vertical directions, i.e., $PE_{i,j}$ in Fig. 5. To construct the P_j, we access 10 banks of FMEM at the same time where each bank provides an 1×4 element vector IF_k ($0 \leq k \leq 9$) for supporting the seamless layer processing. As the IF_k contains four consecutive elements in vertical direction having the same input depth of the current input feature map, it is possible to generate eight 3×3 patches by combining three IF_k as shown in Fig. 5. Accepting two sets of 3×3 elements, i.e., F_i and P_j, $PE_{i,j}$ performs the convolution operation by utilizing 3×3 parallel multipliers followed by one accumulator. As each PE manages the different element of output feature maps, corresponding 32 partial sums for the current processing, denoted as $PS_{i,j}$, are also loaded to the corresponding PEs by accessing FMEM again, consequently computing 32 elements of partial sum or output feature maps in parallel as depicted in Fig. 5. The output buffer temporally holds the results of the 32 convolution operations, and arrange the storing format to FMEM. After completing the convolution operations of each layer, the results are then stored in FMEM and the non-linear operator in Fig. 4 starts to generate the output feature maps to be stored at the external memory.

Basically, the proposed accelerator reuses the filters as many as possible by changing the elements of input feature maps. The processing latency per CNN layer can be reduced as we handle only necessary calculations by totally eliminating the pruned filters at the filter loader. Note that 32 PEs in the convolution engine are fully activated except for the last processing step of each CNN layer that performs the remained computations, remarkably improving the hardware efficiency.

978-1-7281-5107-6/19 $31.00 © 2019 IEEE

207

IEEE Asian Solid-State Circuits Conference
November 4 – 6, 2019
The Parisian Macao, Macao SAR, China

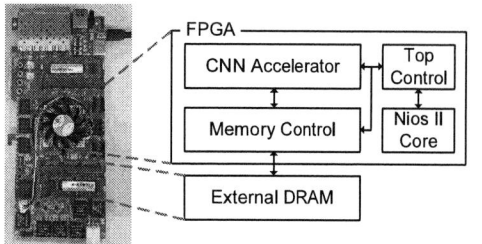

Fig. 6. The FPGA-based testing platform for the proposed accelerator.

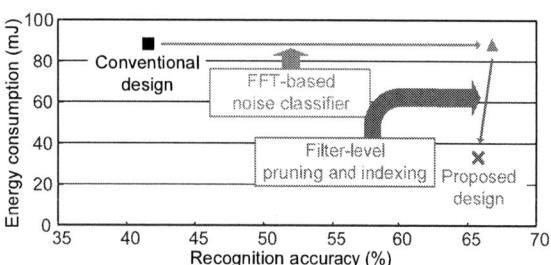

Fig. 7. Performance improvements by the proposed schemes.

V. IMPLEMENTATION RESULTS

Based on the proposed schemes, we implemented the CNN accelerator for VGG16 network in the FPGA-based verification platform including Intels Stratix-V FPGA. Fig. 6 shows our FPGA-based testing platform, which includes the proposed accelerator connected to the external DRAM with the memory controller. Note that additional embedded processor manages the whole system and checks the demonstration results. The prototype system is operated at the speed of 100MHz, providing the peak processing throughput of 57.6GOPS while consuming only 2.03W. Fig. 7 shows how the proposed schemes provide the energy-efficient noise-resilient CNN processing. By applying the FFT-based classifier, as shown in the figure, the average accuracy for recognizing the practical noisy inputs can be improved significantly. Preserving the recognition accuracy, in addition, the filter-level pruning drastically reduces the number of external DRAM access and offers the efficient convolution processing with high hardware utilization, which results in reduction of total energy consumption by 62%.

Compared with the previous FPGA-based CNN accelerator designs, as summarized in Table II, the proposed CNN accelerator achieves a similar power efficiency, which can be acceptable for the embedded edge solutions. Moreover, our design is the only option that provides noise-resilient algorithm considering the practical processing scenarios. Note that the fully-utilized convolution processing also offers the acceptable recognition speed even supporting the frame rate of video inputs in real time. Therefore, the proposed CNN accelerator is a promising solution for the practical edge-level image recognition due to its energy-efficient and noise-resilient processing.

TABLE II. COMPARISON OF FPGA-BASED CNN ACCELERATORS

	This work	[11]	[12]
Network	VGG16	AlexNet	VGG16
Frequency (MHz)	100	100	150
Bit precision	13b	16b	16b
Throughput (GOPS)	57.6	229.5	829.8
Power (W)	2.03	9.4	31.2
Efficiency (GOPS/W)	28.4	24.42	26.6
Frame per second	64.7*	N/A	N/A
Noise-resilient	Yes	No	No

* For 75% pruned network

VI. CONCLUSION

In this paper, we have presented a novel FPGA-based CNN accelerator that shows high recognition accuracy even for noisy images. By classifying the type of noise from input image by the proposed FFT-based classifier, we can use dedicated network for the detected noise type. In addition, the hardware-friendly pruning method is introduced to relax the number of external DRAM access as well as the processing latency. Implementation results show that our design provides the comparable performance with state-of-the-art FPGA-based designs while supporting the noise-resilient recognition, which is essential for the practical edge-level devices.

REFERENCES

[1] M. Tan *et al.*, "Mnasnet: Platform-aware neural architecture search for mobile," in *Proc. IEEE Conference on Computer Vision and Pattern Recognition (CVPR)*, June 2019.

[2] V. Sze, Y.-H. Chen, T.-J. Yang, and J. S. Emer, "Efficient processing of deep neural networks: A tutorial and survey," *Proceedings of the IEEE*, vol. 105, no. 12, pp. 2295–2329, 2017.

[3] S. Diamond *et al.*, "Dirty pixels: Optimizing image classification architectures for raw sensor data," *arXiv preprint arXiv:1701.06487*, 2017.

[4] S. Han, H. Mao, and W. J. Dally, "Deep compression: Compressing deep neural networks with pruning, trained quantization and huffman coding," *arXiv preprint arXiv:1510.00149*, 2015.

[5] Y. Byun, M. Ha, J. Kim, S. Lee, and Y. Lee, "Low-complexity dynamic channel scaling of noise-resilient cnn for intelligent edge devices," in *Proc. Design, Automation Test in Europe Conference Exhibition (DATE)*, Mar. 2019, pp. 114–119.

[6] A. Krizhevsky and G. Hinton, "Learning multiple layers of features from tiny images," Citeseer, Tech. Rep., 2009.

[7] S. Han, J. Pool, J. Tran, and W. Dally, "Learning both weights and connections for efficient neural network," in *Proc. Advances in Neural Information Processing Systems*, 2015, pp. 1135–1143.

[8] J. Park, S. Moon, Y. Byun, S. Lee, and Y. Lee, "Multi-level weight indexing scheme for memory-reduced convolutional neural network," in *Proc. International Conference on Artifical Intelligence Circuit and Systems (AICAS)*, 2019, pp. 284–287.

[9] L. Du *et al.*, "A reconfigurable streaming deep convolutional neural network accelerator for internet of things," *IEEE Transactions on Circuits and Systems I: Regular Papers*, vol. 65, no. 1, pp. 198–208, 2017.

[10] J. W. Cooley and J. W. Tukey, "An algorithm for the machine calculation of complex fourier series," *Mathematics of computation*, vol. 19, no. 90, pp. 297–301, 1965.

[11] Q. Xiao *et al.*, "Exploring heterogeneous algorithms for accelerating deep convolutional neural networks on FPGAs," in *Proc. ACM/EDAC/IEEE Design Automation Conference (DAC)*, 2017, pp. 1–6.

[12] L. Gong, C. Wang, X. Li, H. Chen, and X. Zhou, "MALOC: A fully pipelined FPGA accelerator for convolutional neural networks with all layers mapped on chip," *IEEE Transactions on Computer-Aided Design of Integrated Circuits and Systems*, vol. 37, no. 11, pp. 2601–2612, 2018.

978-1-7281-5107-6/19 $31.00 © 2019 IEEE

Flexible Low Power CNN Accelerator for Edge Computing with Weight Tuning

Miaorong Wang, Anantha P. Chandrakasan
Department of Electrical Engineering and Computer Science
Massachusetts Institute of Technology
Cambridge, Massachusetts 02139
Email: {miaorong, anantha}@mit.edu

Abstract—To support various edge applications, a neural network accelerator needs to achieve high flexibility and classification accuracy within a limited power budget. This paper proposes a weight tuning algorithm to improve the energy efficiency by lowering the switching activity. A flexible and runtime-reconfigurable CNN accelerator is co-designed with the algorithm and demonstrated with a feature extraction processor on an FPGA. The system is fully self-contained for small CNNs and speech keyword spotting is shown as an example. A fully integrated custom ASIC is also being fabricated for this system. Based on post place-and-route simulation of the ASIC, the weight tuning algorithm reduces the energy consumption of weight delivery and computation by 1.70x and 1.20x respectively with little loss in accuracy.

I. INTRODUCTION

Smart edge devices that support efficient neural network (NN) processing have recently gained public attention. With algorithm development, previous work has proposed small-footprint NNs achieving high performance in various medium complexity tasks, e.g. speech keyword spotting (KWS), human activity recognition (HAR), etc. Among them, convolutional NNs (CNNs) perform well [1], which gives rise to the deployment of CNNs on edge devices. A hardware platform for edge devices should be (1) flexible to support various NN structures optimized for different applications; (2) energy efficient to operate within the power budget; (3) achieving high accuracy to minimize spurious triggering of power-hungry downstream processing, since it is often part of a large system.

Both algorithms, such as quantization and model compression, and accelerator designs for energy efficient processing of CNNs have been proposed. Quantization reduces the bit precision. But some experiments show that quantizing NNs to extremely low bitwidth, e.g. 1 bit, does not necessarily lead to model size reduction, because the model structure needs to be modified to retain the accuracy [2]. Model compression algorithms focus on minimizing the model size with little loss in accuracy. However, pruning-based algorithms mainly need specialized hardware to exploit the resulting sparse tensors for energy reduction. Previous work has proposed CNN accelerators targeting edge computing. However, many of them support limited flexibility for the NN shapes, are designed only for a specific task or sacrifice the accuracy [3]–[5].

To address the challenges in flexibility, energy efficiency and accuracy in CNN accelerator design, this work takes

an algorithm-and-hardware co-design approach. The key contributions of this paper are highlighted as follows: (1) a weight tuning algorithm that reduces the energy consumption associated with weight delivery and computation by lowering the toggle count of weight sequence; (2) the co-design of a CNN accelerator that supports the proposed algorithm and is flexible for a wide range of NN model structures; (3) the demonstration of speech keyword spotting (KWS) as an example on an FPGA and the design of a fully integrated ASIC (being fabricated) with the proposed CNN accelerator and a feature extraction processor[1].

II. WEIGHT TUNING ALGORITHM

The weight tuning algorithm reduces the energy consumption of the CNN accelerator with little loss in accuracy by tuning the bit representation of weights. Fundamentally different from quantization and model compression algorithms, the proposed algorithm is based on the theory that the dynamic power of a CMOS gate is linearly proportional to its switching activity, which is influenced by the toggle count (TC) of its input sequence. In a NN accelerator, weights are read from the memory, delivered through network-on-chip (NoC) and then multiplied with input activations (IAs) following a sequence set by the designer. The TC of this weight sequence affects the switching activity of weight buses and the multipliers. And for a memory that can do conditional pre-charge based on previously read data, e.g. [7], it also affects the pre-charge activity of bit-lines. Therefore, minimizing the TC of weight sequence can reduce the power consumption of those components. To gain those benefits, we propose a weight tuning algorithm that contains 3 sequential steps: (1) tensor decomposition with retraining; (2) quantization and sign-magnitude representation; (3) weight scaling and bit perturbation with retraining.

A. Tensor Decomposition (TD) with Retraining

TD with retraining [8] is a model compression method that breaks a convolutional layer into 3 layers without activation function in between and retrains to maintain accuracy as shown in Fig. 1. The total parameters and computation of the resulting layers are less than those of the original layer. Thus it is favorable for reducing the energy per inference.

---

[1]This unit was first designed by M. Price [6] and then modified by S. Lauwereins from Prof. Marian Verhelst's group at MICAS – KU Leuven.

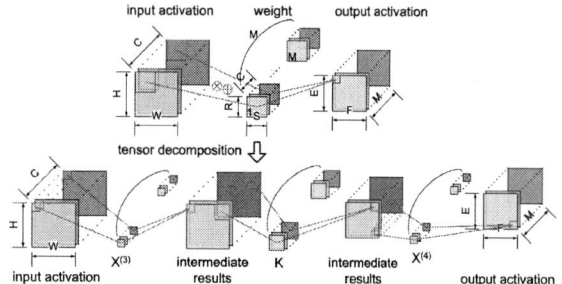

Fig. 1. An illustration of TD for CNN [8] and the shape parameters used in this paper. The decomposed NN is then retrained as proposed in [8].

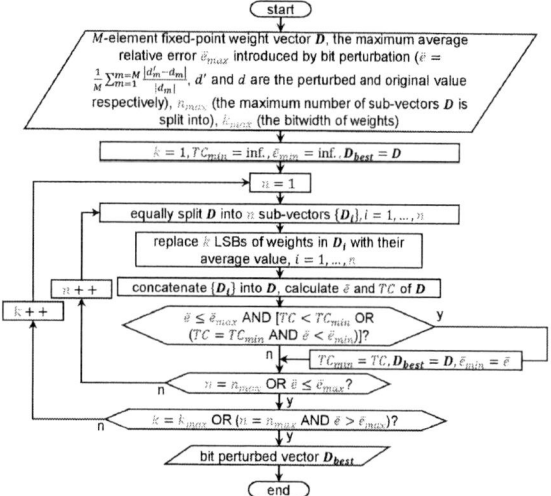

Fig. 2. The flowchart of the bit perturbation algorithm. Relative error $e = \frac{|d'-d|}{|d|}$, where d' and d is the perturbed and original value respectively, of weight is used to evaluate the deviation caused by bit perturbation.

B. Quantization and Sign-Magnitude (QSM) Representation

We then convert the model to 8 bit fixed-point numbers with linear quantization as used in many NN accelerators. Instead of 2's complement (2C) format, we use sign-magnitude (SM) representation of weights to reduce weight sequence TC [9], [10]. The overhead of that is the implementation of adder. Section III discusses in detail how we minimize the overhead.

C. Weight Scaling and Bit Perturbation (WSBP) with Retraining

The 4-D weight tensor of every layer is flattened to a 1-D vector following the sequence that weights are read, delivered and calculated in the NN accelerator. We sequentially apply weight scaling and bit perturbation to it to further reduce the TC and incorporate them with retraining to maintain accuracy.

1) Weight Scaling: Weight Scaling is to scale the weights and biases uniformly in every layer to lower the TC of weight sequence, which is inspired by the coefficient scaling for FIR

filters [11]. The scaling factor K_l of layer l is determined by $K_l = argmin_{k_l} \texttt{ToggleCount}(k_l \mathbf{W_1}), k_l \in (a, b), a \geq 0$, where $\mathbf{W_1}$ is the weight tensor, $\texttt{ToggleCount}$ is the function to calculate the TC and a, b are the user defined bounds of K_l. K_l is determined by an exhaustive search in the given range with a predefined step size s. It can be applied to layers using ReLU as the activation function without impact on the classification accuracy, given that $ReLU(K_l\mathbf{x}_l) = K_l ReLU(\mathbf{x}_l)$.

2) Bit Perturbation: Inspired by the coefficient perturbation for FIR filters proposed in [11], we apply the bit perturbation to weight sequence as shown in Fig. 2 to reduce the TC. It equally splits the weight vector $\mathbf{D}$ into n sub-vectors $\{\mathbf{D_i}\}$ and then replaces k LSBs of weights in every sub-vector $\mathbf{D_i}$ with their average value. It loops through all possible n and k searching for a bit tuned low-TC weight sequence with small deviation from the original weights.

3) Retraining: Retraining is applied to restore the accuracy loss caused by WSBP. The proposed QSM and WSBP are applied to the pretrained floating point NN as a wrapper function of parameters in the forward pass. During back-propagation, the straight-through estimator (STE) [12] is adopted, which passes the gradients through the wrapper function as-is.

III. SYSTEM ARCHITECTURE

As shown in Fig. 3, we implemented a standalone system with 80kB model parameter memory and 1kB configuration buffer for storage and configuration of the entire NN with up to 12 layers during the setup phase. All the data buffering is done on-chip using a 2kB circular input buffer and a 48kB activation memory without the need of off-chip DRAM.

The processing element (PE) array level dataflow is shown in line 4 – 8 of Fig. 4. The PE array is logically treated as having $C1$ columns and $M1$ rows as shown in line 7–8. Reconfigurable NoCs are used to support that flexibility with NoC controller adopted from Eyeriss [13]. Following this logical mapping, weights are unicast to PEs, and thus tree-structured NoC with a depth of 2 as shown in Fig. 3 is used for weight delivery. The ID of every controller at each level is unique and fixed. Activations are multicast across multiple logical rows. To support that, only level 0 NoC controllers are used. Their IDs, determined by $M1s$ and $C1$, are configured at runtime before the execution of every layer. The delivery of bias, partial sum and output activations are similar except that not every PE needs data. Thus the unused controllers and FIFOs can be gated. Different from Eyeriss which implements row stationary (RS) dataflow, the proposed design follows weight stationary (WS) dataflow in the PE array level considering that 1×1 or $1 \times x$ convolutional layers resulting from TD take up a large part of the CNNs. Fully-connected layers can be organized as a special case of convolutional layers where $W = R, H = S, E = 1, F = 1$

The PE level dataflow is shown in line 9 – 10 of Fig. 4 and the PE structure is illustrated in Fig. 5. PE has local storage to hold $M0 \times C0$ weights and $C0$ IAs to achieve temporal reuse of $E \times F$ and $M0$ times respectively as shown in line 6 and 9 of Fig. 4. $M0$ and $C0$ are runtime configurable to balance the

IEEE Asian Solid-State Circuits Conference
November 4 – 6, 2019
The Parisian Macao, Macao SAR, China

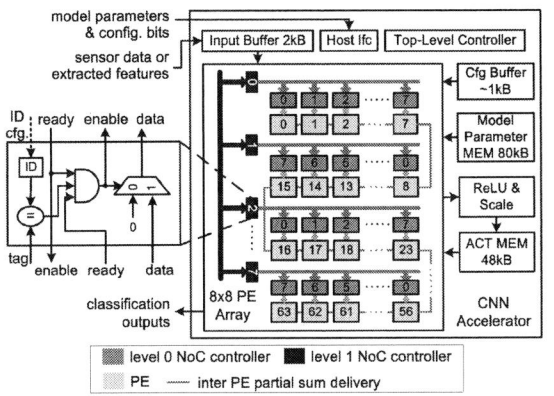

Fig. 3. System architecture of the NN accelerator and the micro-architecture of the NoC controller.

```
1)   int i[C][W][H];      // Input activations, C = C1*C0 <= 256
2)   int w[M][C][R][S];   // Filter weights, M = M1t*M1s*M0 <= 256
3)   int o[M][E][F];      // Output activations, E*F*M <= 48k

     // PE array level -- temporal
4)   for (r1=0; r1<R; r1++) { for (s1=0; s1<S; s1++) {
5)     for (m1t=0; m1t<M1t; m1t++) { // R*S*M1t <= 2^16
6)       for (e1=0; e1<E; e1++) { for (f1=0; f1<F; f1++) {
         // PE array level -- spatial
7)         parallel_for (c1=0; c1<C1; c1++) {
8)           parallel_for (m1s=0; m1s<M1s; m1s++) { // C1 * M1s <= 64
             // PE level -- temporal
9)             for (m0=0; m0<M0; m0++) { // M0 = [0, 1, 2, 3]
10)              for (c0=0; c0<C0; c0++) { // C0 = [3, 4]
11)                c = c0 + c1*C0;
12)                m = m0 + m1s*M0 + m1t*M0M1s;
13)                r = r1; s = s1; e = e1; f = f1;
14)                w = U*e1 + r; h = V*f1 + s; // U, V: filter stride in width and height
15)                o[m][e][f] += i[c][w][h] * w[m][c][r][s]; // U,V <= 1024
} ... }
```

Fig. 4. The dataflow illustrated by loop nests. The tensor dimension parameters are illustrated in Fig. 1. Bias is ignored for simplicity and batch size = 1 for real-time application. For the first layer, where $C = 1$, we replace C with R and remove the original R loop to increase the PE array utilization. The read, delivery and computation sequence of weight is as shown. PEs are filled up with weights in sequence. The limits on supported NN shapes are annotated.

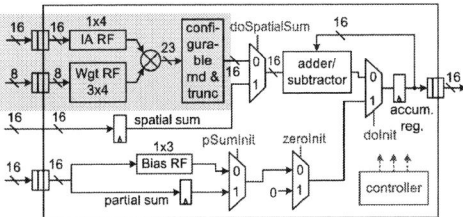

Fig. 5. The PE structure. The shaded part is the SM domain. The rest is in the 2C domain.

workload between PEs given layer shapes. All PEs are chained in a sequence for the inter-PE partial sum delivery as shown in Fig. 3. That supports spatial sum with flexible configuration of logical rows and columns, since partial sums can be spatially accumulated across an arbitrary number of PEs.

As discussed in Section II-B, weights are represented in SM format to reduce TC. Exploiting the fixed calculation pattern in CNN, we implement a mixed-representation datapath. Weights and IAs are multiplied in SM format using an unsigned multiplier and an XOR gate that generates the sign bit. An adder-subtractor is used to convert the SM product to 2C representation with the sign bit of the product as the carry bit, and do accumulation. The 2C outputs are delivered to other PEs for spatial sum or buffered in memory for temporal accumulation to generate the output activations. Only after all the computation of this layer is finished, do we need to convert them back to SM format for the next layer. Thus the energy overhead of the conversion is mitigated.

IV. IMPLEMENTATION RESULTS

The CNN accelerator with a feature extraction processor is implemented on Xilinx XC7K410T. The CNN accelerator operates at 50MHz and consumes 68mW based on Vivado power estimation. The FPGA demo for the KWS task and the post place-and-route (P&R) resource utilization is shown in Fig. 6.

The proposed system is also being fabricated using TSMC 40nm LP process with a core area of 2.16 mm^2. Based on the ASIC design, we evaluate the weight tuning algorithm on several CNNs designed for KWS on speech command dataset [14], including CNNs under 80kB (referred to as CNN80) and 200kB memory constraints in [1] and the fstride-4 model in [15]. TD with retraining is applied to most layers except ones that largely impact the accuracy, e.g. the last layer. The resulting models in 2C format serve as the baseline. QSM and WSBP with retraining is applied to the decomposed layers with a step size s of 0.05, the scaling factor bounds $a = 0.8, b = 1.8, e_{max} = 0.15$ and n_{max} equal to half of the

vector length. It reduces weight sequence TC by 1.79x–2.56x with less than 0.75% accuracy loss on the testing set for those cases. As shown in Fig. 7, the Hamming distance between successive weights are reduced. To analyze energy reduction, we implement a baseline CNN accelerator that uses 2C MACs following the same procedures for synthesis and P&R as the proposed accelerator using the same technology. Parasitics, SDF and SAIF obtained after P&R are annotated during power analysis. We simulate the execution of a tensor decomposed CNN80 on the dataset. Table I summarizes the total energy consumption of different components during the execution, and compares the accuracy and TC. $E_{mult.}$ and $E_{add.}$ are the total energy consumption of all the multipliers and adders in the PEs respectively. E_{MAC} is the sum of them. E_{wgtBus} is the energy of weight buses between the weight memory and PEs. It is obtained by summing up the internal and switching energy of buffers inserted in between. As shown, the weight tuning algorithm with the mixed-representation MAC reduces the computation energy by 1.20x compared to the 2C baseline. The energy of weight buses is reduced by 1.70x. Although the energy consumption of memory and activation delivery

978-1-7281-5107-6/19 $31.00 © 2019 IEEE
211

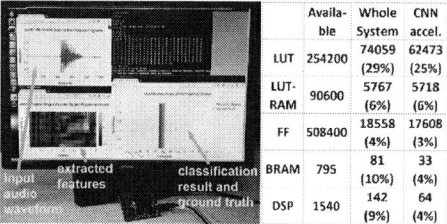

	Available	Whole System	CNN accel.
LUT	254200	74059 (29%)	62473 (25%)
LUT-RAM	90600	5767 (6%)	5718 (6%)
FF	508400	18558 (4%)	17608 (3%)
BRAM	795	81 (10%)	33 (4%)
DSP	1540	142 (9%)	64 (4%)

Fig. 6. The FPGA demo of the CNN accelerator with a feature extraction processor on KWS, and a summary of FPGA post-P&R resource utilization.

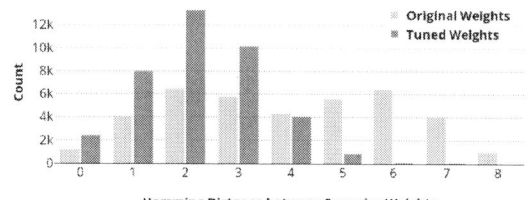

Fig. 7. The weight tuning algorithm reshapes the distribution of Hamming distance between weights with tensor-decomposed CNN80 as an example.

TABLE I
THE EFFECT OF THE WEIGHT TUNING ALGORITHM ON ACCURACY AND ENERGY CONSUMPTION BASED ON POST-P&R SIMULATION

	original 2C	bit tuned SM	loss/reduction
acc. (%)	89.3	88.8	0.5%
toggle count	154k	86k	1.79x
$E_{mult.}$	1.58uJ	1.11uJ	1.42x
$E_{add.}$	0.55uJ	0.66uJ	0.83x
E_{MAC}	2.13uJ	1.77uJ	1.20x
E_{wgtBus}	96.91nJ	56.89nJ	1.70x
$E_{totalSwitch}$	8.94uJ	7.68uJ	1.16x

is not affected by the algorithm, 1.16x reduction in the total switching energy of the entire system $E_{totalSwitch}$ is observed.

Compared with other digital ASICs for KWS, e.g. [3]–[5], our design is flexible, accurate and achieves comparable energy efficiency. The proposed architecture supports flexible shape and stride of inputs and weights for up to 12 layers. However, [5] is restricted to a fixed structure, [3] supports up to 2 layers with up to 64 nodes/layer for LSTMs, and [4] is designed for a fixed input stride and 3x3 1-bit convolution. Our supported CNNs can achieve 91.6% accuracy with 12 output classes on the public available dataset [1], while [3]–[5] only report accuracy on custom designed dataset and [3] only shows binary classification. Based on the post-P&R power analysis, the proposed design achieves 24.38 pJ/MAC with 8bit weights and 16bit activations at the worst case corner of 0.99V with a latency of 10ms for real-time KWS processing.

V. CONCLUSION

We co-designed a weight tuning algorithm and a CNN accelerator to improve energy efficiency with little loss in accuracy. Furthermore, the accelerator features high flexibility and runtime reconfigurability to support various applications.

It is demonstrated on an FPGA with a feature extraction processor for KWS. Moreover, a fully integrated ASIC is being fabricated. The post-P&R power analysis of the ASIC shows the proposed algorithm reduces the energy consumption of weight delivery and computation by 1.70x and 1.20x respectively. For future work, the weight tuning algorithm can be applied to other NNs, such as LSTM, co-designed with other hardware architectures and dataflows, e.g. output stationary dataflow, and exploited by data-dependent memory [7].

ACKNOWLEDGMENT

The authors would like to thank Foxconn Technology Group for supporting this project and the TSMC University Shuttle Plan for chip fabrication.

REFERENCES

[1] Y. Zhang, N. Suda, L. Lai, and V. Chandra, "Hello Edge: Keyword spotting on microcontrollers," *arXiv:1711.07128 [cs, eess]*, Nov. 2017. [Online]. Available: http://arxiv.org/abs/1711.07128
[2] G. Gobieski, B. Lucia, and N. Beckmann, "Intelligence Beyond the Edge: Inference on intermittent embedded systems," in *Proceedings of International Conference on Architectural Support for Programming Languages and Operating Systems (ASPLOS)*. ACM, 2019, pp. 199–213.
[3] J. Giraldo and M. Verhelst, "Laika: A 5uW programmable LSTM accelerator for always-on keyword spotting in 65nm CMOS," in *Proceedings of IEEE European Solid State Circuits Conference (ESSCIRC)*, 2018, pp. 166–169.
[4] S. Yin, P. Ouyang, S. Zheng, D. Song, X. Li, L. Liu, and S. Wei, "A 141 uW, 2.46 pJ/neuron binarized convolutional neural network based self-learning speech recognition processor in 28nm CMOS," in *IEEE Symposium on VLSI Circuits*, 2018, pp. 139–140.
[5] M. Shah, J. Wang, D. Blaauw, D. Sylvester, H.-S. Kim, and C. Chakrabarti, "A fixed-point neural network for keyword detection on resource constrained hardware," in *IEEE Workshop on Signal Processing Systems (SiPS)*. IEEE, 2015, pp. 1–6.
[6] M. Price, J. Glass, and A. P. Chandrakasan, "14.4 A scalable speech recognizer with deep-neural-network acoustic models and voice-activated power gating," in *IEEE International Solid-State Circuits Conference (ISSCC) Digest of Technical Papers*, 2017, pp. 244–245.
[7] C. Duan, A. Gotterba, M. E. Sinangil, and A. P. Chandrakasan, "Reconfigurable, conditional pre-charge SRAM: Lowering read power by leveraging data statistics," in *Proceedings of IEEE Asian Solid-State Circuits Conference (A-SSCC)*, Nov. 2016, pp. 177–180.
[8] Y.-D. Kim, E. Park, S. Yoo, T. Choi, L. Yang, and D. Shin, "Compression of deep convolutional neural networks for fast and low power mobile applications," *arXiv:1511.06530 [cs]*, Nov. 2015. [Online]. Available: http://arxiv.org/abs/1511.06530
[9] A. P. Chandrakasan and R. W. Brodersen, "Minimizing power consumption in digital CMOS circuits," *Proceedings of the IEEE*, vol. 83, no. 4, pp. 498–523, 1995.
[10] P. N. Whatmough, S. K. Lee, D. Brooks, and G.-Y. Wei, "DNN engine: A 28-nm timing-error tolerant sparse deep neural network processor for iot applications," *IEEE Journal of Solid-State Circuits*, vol. 53, no. 9, pp. 2722–2731, 2018.
[11] N. Kasturi, "Power reducing algorithms in FIR filters," Master's thesis, Massachusetts Institute of Technology, 1997.
[12] C. Zhu, S. Han, H. Mao, and W. J. Dally, "Trained ternary quantization," in *Proceedings of International Conference on Learning Representations (ICLR)*, 2016.
[13] Y. H. Chen, T. Krishna, J. S. Emer, and V. Sze, "Eyeriss: An energy-efficient reconfigurable accelerator for deep convolutional neural networks," *IEEE Journal of Solid-State Circuits*, vol. 52, no. 1, pp. 127–138, Jan. 2017.
[14] P. Warden. (2017) Speech command: A public dataset for single-word speech recognition. [Online]. Available: download.tensorflow.org/data/speech_commands_v0.01.tar.gz
[15] T. Sainath and C. Parada, "Convolutional neural networks for small-footprint keyword spotting," in *Proceedings of Interspeech*, 2015.

An Asynchronous Reconfigurable SNN Accelerator With Event-Driven Time Step Update

Jilin Zhang*, Hui Wu*, Jinsong Wei†, Shaojun Wei*‡, Hong Chen*‡

*Institute of Microelectronics, Tsinghua University,
†School of microelectronics, University of Science and Technology of China, Hefei, China
‡Beijing National Research Center for Information Science and Technology, Beijing, China
Email: hongchen@tsinghua.edu.cn

Abstract—In this paper, we put forward an asynchronous spiking neural network (SNN) accelerator with 1024 neurons and 1 million synapses, which is reconfigurable in terms of network connection and neuron parameters. Bundled data asynchronous circuits are adopted to design the neuromorphic computation core and mesh network. Multicast communication is used to transmit packet among and within each core for less packet transmission and better energy efficiency. A novel time step update mechanism, which updates neurons in an event-driven manner without considering the chip-wide activity of other unrelated neurons, is proposed to improve the performance of speed. The SNN accelerator is verified by classifying MNIST handwritten digit with Xilinx VC707 FPGA. The results show that the accelerator achieves 98% accuracy with MINST database, and more than 1 GIPS/W energy efficiency which is 32 times better than previous work.

Keywords—SNN; asynchronous circuit; FPGA; neuromorphic computation

I. INTRODUCTION

Spiking neural network (SNN), getting inspiration from brain, approaches the energy efficiency of biology by using temporally sparse connection, reduced precision, and approximate computation. However, SNN is inherently parallel in architecture whereas today's von Neumann CPU architectures and GPU variants are serial processing architecture. As a result, it is difficult for CPU and GPU to realize SNN to solve complex problem effectively. When implemented on parallel hardware, like FPGA, SNN can take full advantage of its inherent parallelism and run orders of magnitude faster than software simulations; thus, becoming appropriate for real-time applications [1].

Several SNN hardware accelerators on FPGA have been proposed in the scientific literature. NeuroFlow [2] represented the state of art in the field of FPGA architectures for SNN. It can simulate up to 600,000 neurons by six FPGA boards with 30-40W power consumption per FPGA board. Authors in [3] implemented a 1,440 neurons network with power consumption up to 8.5W when executing emulation with a 100 MHz clock on the FPGA. Researcher in [4] proposed Minitaur, which is an event-based implementation of SNN with 65,000 neurons with the power consumption of 1.5W.

All these accelerators aim to realize large-scale neural network. However, they consume too much power to be used in embedded applications. As we know, asynchronous circuits have the advantages of power efficiency because of its event-driven nature. That is, asynchronous circuits only work when data needs to be processed, otherwise it will remain idle and consume little power. This property makes asynchronous circuits suitable for the implementation of SNN because the spikes in SNN are sparse in terms of space and time. In this paper, we propose an asynchronous reconfigurable SNN accelerator with event-driven time step update mechanism to achieve flexibility, high performance and energy efficiency.

II. LINK-JOINT CIRCUITS

We adopt link-joint-based asynchronous bundled data circuits to design our SNN accelerator. The link-joint is a model proposed by asynchronous research center [5] which concludes all the self-timed asynchronous controller with a concise and comprehensive structure. The asynchronous controller with Click element [6] using link-joint model is shown in Fig. 1, in which every link has four handshake signals: *empty, full, fill,* and *drain*. A *full* signal indicates that the link's data signals are stable and valid. The link presents *full* signal at its output end to let a receiver know whether or not it is full. An *empty* signal indicates that the link's data have been released and may be replaced by fresh data. The link presents an *empty* signal at its input end to let a sender know whether or not it is empty. A joint may act when some or all of its input links are full and some or all of its output links are empty. When it acts, it computes results from data it gets from its full input links, and passes these results to some or all of its empty output links. It then fills selected empty output links, and drains selected full input links, destroying the conditions that enabled its action. Links that it fills or drains begin their data transport actions to carry results away or fetch fresh input data [5].

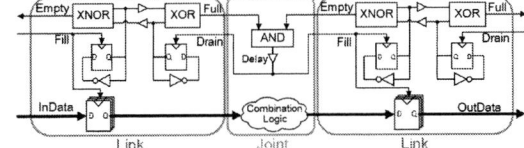

Fig. 1. Click based link-joint circuit.

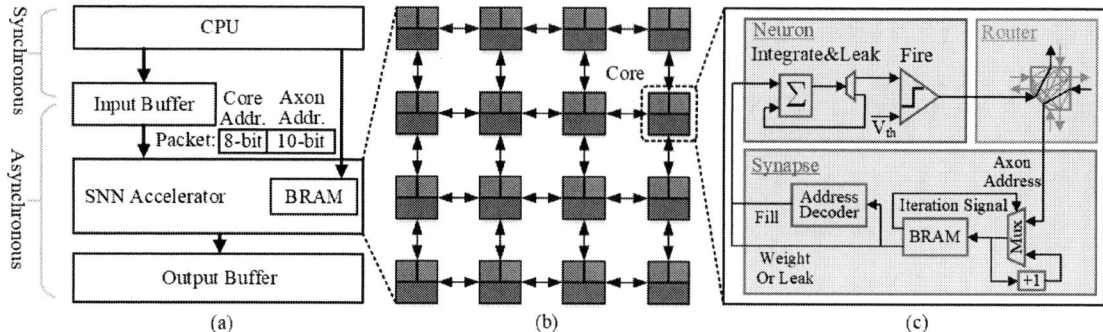

Fig. 2. (a) The System architecture of our SNN accelerator, (b) mesh network (16 cores, 1024 neurons, 1 million synapses), (c) neuromorphic core.

III. DESIGN OF THE SNN ACCELERATOR

A. System Overview

The system architecture of the SNN accelerator is illustrated in Fig. 2 (a). The address information of a spike is stored in a spike packet, including 8-bit-dual-rail address of its target core and 10-bit address of its axon, the input spike packets of the mesh network are first stored in the input buffer. The configuration information from CPU will be stored in the block RAM (BRAM) in each core. The SNN accelerator is comprised of 1024 neurons and 1 million synapses, which are divided into 16 neuromorphic cores, and each core is connected by a mesh network (shown in Fig. 2(b)). The spike packet from input buffer is multicast throughout the mesh network, and each core is responsible for receiving and sending the spike packet to neuron for the integration operation, as shown in Fig. 2(c). The final results will be stored in the output buffer.

The operation of our accelerator can be divided into three phases. The first one is configuration phase, in which the network connection, synapses weight, neuron leak and threshold voltage will be determined according to the configuration information stored in BRAM. The numbers of layer, core, synapse, and the topology of the SNN are reconfigurable for different applications. The second phase is integration phase. The spike packets from input buffer are sent to the BRAM of its target core, and weight and the address of its target neuron will be read from BRAM based on the axon address in the spike packets. The target neuron will receive the weight and perform integration operation by adding the weight into its membrane potential. The third phase is called time step update (TSU) phase, in which a TSU packet is used to update each core's time step in an event-driven manner. In this phase, the leak values and addresses of the neuron whose time step needs to be updated will be read form BRAM. The value of the leak will be subtracted from the neuron's membrane potential, and the new membrane potential will be compared to the threshold voltage. If it exceeds threshold voltage, a spike will be generated and sent to the next neurons using address event representation (AER) circuit [7].

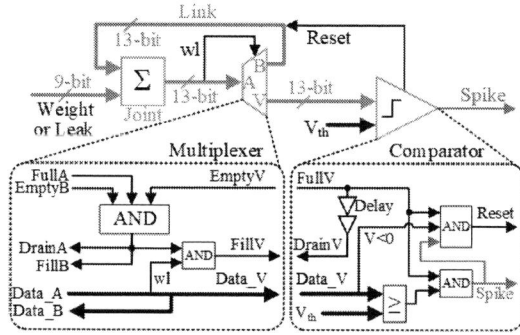

Fig. 3. Asynchronous neuron deisgn.

B. Design of Neurons

The Leak-Integrate-and-Fire (LIF) model is adopted to design the neuron circuit. The operation of LIF neuron is described in equation (1):

$$u(t+1) = u(t) + \sum sw - L \qquad (1)$$

where $u(t)$ denotes the membrane potential at time step t, s is an input spike, w is the weight of that spike, L is the leak. LIF neuron integrates the weight of all the spike into its membrane potential across time step along with leak. When its membrane potential exceeds the threshold voltage, the neuron will generate spike and pass it to the next neurons.

The structure of the LIF neuron is depicted in Fig. 3, in which the blue line and orange polygon represent link and joint respectively. Here we regard the leak of neuron as an inhibitory synapse (i.e. synapse with negative weight). In integration phase, 8-bit weight and 1-bit wl signal used to distinguish weight (when wl is logic '0') and leak (when wl is logic '1'), will be the input of target neuron. Here we choose 8-bit weight to allow a relatively high accuracy and low memory footprint. The weight value will be integrated into neuron's membrane potential and the updated membrane potential will not be sent into the comparator in Fig. 3 in this phase because wl is now logic '0'. 13-bit is chose to represent membrane potential so that the sum of spike's weight in one time step will not overflow. In TSU phase, the leak will be

978-1-7281-5107-6/19 $31.00 © 2019 IEEE

sent and integrated to membrane potential, and the updated membrane potential will be sent into the comparator in Fig. 3. If it exceeds the threshold voltage, this neuron will generate a spike and reset its membrane potential to 0. If the membrane potential is small than 0, which can be caused by inhibitory synapse or leak, neuron will reset its membrane potential into 0 without producing any spike.

C. Design of Routers

The schematic of the router in each core is shown in Fig. 4, in which five input and output channels are local, north, south, east, and west channels respectively. The local channel is connected to the core, and other four channels are connected to other adjacent routers in the cores on its north, south, east, and west sides respectively. As discussed above, the spike packet contains 8-bit destination core address and 10-bit axon address. We adopt dual rail encoding to encode the core address in order to multicast packet among cores. Each bit in the address for a set of destination cores is represented by a 2-bit symbol. Each symbol can be 0, 1, or '*'. With '*' indicating that cores with either 0 or 1 at that bit location in their addresses are destinations. When processing the '*' symbol in a router, packets branch to multiple output ports and continue to different destination cores [8].

When the router receives a packet from any input channel, the Routing Logic selects its output channel according to the destination core address and then generates a request signal along with the output channel information. Allocator will grant access for its input channel based on the status of requested output channels, which are represented by the *Empty* signals. When one input channel has been granted, the packet will be sent to the corresponding output channel after Mux.

D. Design of Synapses

As shown in Fig. 2 (c), the synapse circuit contains BRAM, which stores the neuron address, synapse weight, and 1-bit iteration signal to implement multicast operation within a core. An address decoder is used to change 6-bit neuron address from BRAM into 64-bit *fill* signals. To multicast packets within core, BRAM iterates over a list of target neurons for an axon's fan-out distribution. For example, if an axon is connected with multiply neurons in the core, after the first neuron's address and synapse weight is read from BRAM, an iteration signal will make BRAM continue to read the next address of neuron and its synapse weight.

IV. EVENT-DRIVEN TIME STEP UPDATE MECHNISM

There are two families of neuron update mechanisms for the implementation of spiking neural networks. One is time step (or clock-driven) update, in which the operation of neurons has been divided into several discrete time step and updated simultaneously at every tick of a clock. The other one is asynchronous (or event-driven) update, in which the neurons are updated only when they receive or emit a spike [9]. The time step update has the drawback of system performance proportional to network size, and the event-driven update is not as widely used as clock-driven for its complicated computation process. In order to take the advantage of both event-driven and clock-driven update, we put forward a new event-driven time step update mechanism, which improves the system performance and enables the system support universal discrete time step update.

As mentioned in section III, the TSU packet is used to update the time step of each core separately as shown in Fig. 5. Once all the spike packets are read out from the input buffer within a time step, a TSU packet from input buffer will be sent into mesh network. The TSU packet will appear in the first layer of neural network, which indicates the spike packets in current time step are all integrated. Then a TSU packet with new target core address appears in the second layer core. Similarly, when core in the second layer receives all the TSU packet from first layer core, it will send a new TSU packet target in the third layer cores, and go into the next time step. Finally, every neuron's time step can be updated asynchronously in an event-driven manner, which avoids the worst-case chip-wide activity of other unrelated neurons.

However, in order to ensure all the neurons are updated at an appropriate time, the TSU packet needs to stay between the adjacent spike packets with time step t and $t+1$ respectively. This requires the routing method of mesh network to be deterministic, which means the routing path between any source and destination core needs to be fixed and not affected by the network congest. Indeterministic routing will adjust the routing path based on the congest level of network. It may result in fewer hops than time step t packet to reach target neuron and early updates for TSU packet.

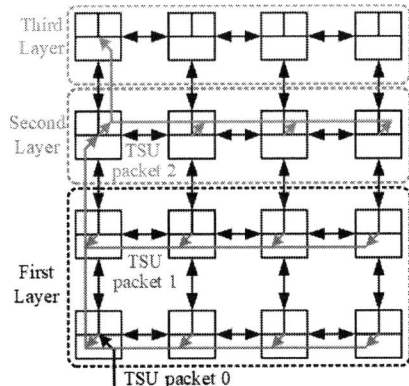

Fig. 5. Time step update example.

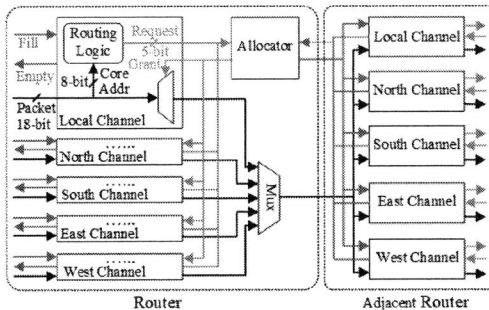

Fig. 4. The schematic of the router in mesh network.

978-1-7281-5107-6/19 $31.00 © 2019 IEEE

TABLE I. Experiment Results with Comparison of Previou Work

	TVLSI 2014[4]	ISCAS 2016[10]	IJCNN 2015[11]	This work
Platform	Spartan-6 LX150	Spartan-6 LX150	SpiNNaker (ASIC)	XC7VX6 90T
Neuron number	65K	65K	2400	1024
Synapse number (million)	16	33	48	1
Weight precision	16-bit fix point	16-bit fix point	22-bit fix point	8-bit fix point
Clock rate (MHz)	75	132	150	-
MINST accuracy	92%	94%	95%	98%
MINST classification latencies (ms)	9.2	2	20	1.1
Peak Performance (MIPS)	19	53.5	16	726
Power(idle/active, W)	1.1/1.5	-	-/0.50	0.36/0.70
Energy efficiency (GIPS/W)	0.013	-	0.032	1.037

V. EXPERIMENTAL RESULTS

We implemented our design with Xilinx FPGA VC707 board, and a four-layer feedforward neural network is adopted to perform handwritten digital recognition from MINST database. This 784-512-384-10 network has two hidden layer with neurons fully connected between layers. This network is trained with a feedforward backpropagation method on PC with the method discussed in [12]. The experimental platform is shown in Fig. 6, and the results show that accelerator has the final error rate of 2%, achieving 98% accuracy. The comparison results between our accelerator and previous works are illustrated in table I, from which we find that our accelerator is 2 times faster than that in [10], and the energy efficiency is 32 times better than that in [11].

The Demo link is https://vimeo.com/342519869 (password: asscc2019zjl).

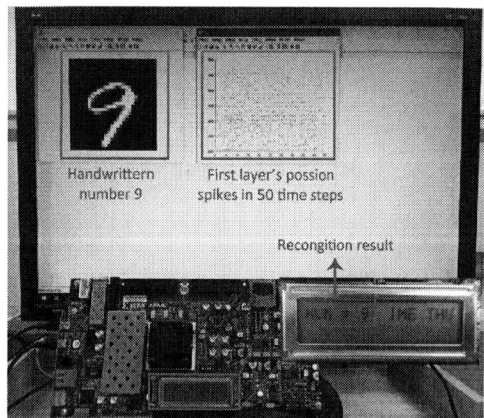

Fig. 6. The experimental platform.

VI. CONSULSION

In this paper, we implemented an asynchronous reconfigurable SNN accelerator, and a novel time step update mechanism is proposed to improve the speed. Packet is multicast among and within cores to improve energy-efficiency. The SNN accelerator has been verified on Xilinx VC707 board, and the results show that the accelerator achieves 98% accuracy at more than 1 GIPS/W of energy-efficiency, which is 32 times better than other works. The performance will be further improved in the future.

ACKNOWLEDGMENT

This work is supported by National Natural Science Foundation of China (No. 61674090).

REFERENCES

[1] L. P. Maguire et al. "Challenges for large-scale implementations of spiking neural networks on FPGAs," in Neurocomput. 71, 1-3 (December 2007), 13-29.

[2] Cheung Kit, Schultz Simon R., Luk Wayne, "NeuroFlow: A General Purpose Spiking Neural Network Simulation Platform using Customizable Processors," Frontiers in Neuroscience, vol. 9, p.516, 2016.

[3] Pani Danilo, Meloni Paolo, Tuveri Giuseppe, et al., "An FPGA Platform for Real-Time Simulation of Spiking Neuronal Networks," Frontiers in Neuroscience, vol. 11, p. 90, 2017.

[4] D. Neil and S. Liu, "Minitaur, an Event-Driven FPGA-Based Spiking Network Accelerator," in IEEE Transactions on Very Large Scale Integration (VLSI) Systems, vol. 22, no. 12, pp. 2621-2628, Dec. 2014.

[5] M. Roncken, I. Sutherland, C. Chen, Y. Hei, et al., "How to think about self-timed systems," 2017 51st Asilomar Conference on Signals, Systems, and Computers, Pacific Grove, CA, 2017, pp. 1597-1604.

[6] A. Peeters, F. t. Beest, M. d. Wit and W. Mallon, "Click Elements: An Implementation Style for Data-Driven Compilation," 2010 IEEE Symposium on Asynchronous Circuits and Systems, Grenoble, 2010, pp. 3-14.

[7] K. A. Boahen, "Point-to-point connectivity between neuromorphic chips using address events," in IEEE Transactions on Circuits and Systems II: Analog and Digital Signal Processing, vol. 47, no. 5, pp. 416-434, May 2000.

[8] G. K. Chen, R. Kumar, H. E. Sumbul, P. C. Knag and R. K. Krishnamurthy, "A 4096-Neuron 1M-Synapse 3.8-pJ/SOP Spiking Neural Network With On-Chip STDP Learning and Sparse Weights in 10-nm FinFET CMOS," in IEEE Journal of Solid-State Circuits, vol. 54, no. 4, pp. 992-1002, April 2019.

[9] R. Brette, M. Rudolph, T. Carnevale, M. Hines, D. Beeman, J. M. Bower, et al., "Simulation of networks of spiking neurons: A review of tools and strategies," J. Comput. Neurosci., vol. 23, no. 3, pp. 349–398, 2007.

[10] I. Kiselev, D. Neil and S. Liu, "Event-driven deep neural network hardware system for sensor fusion," 2016 IEEE International Symposium on Circuits and Systems (ISCAS), Montreal, QC, 2016, pp. 2495-2498.

[11] E. Stromatias, D. Neil, F. Galluppi, M. Pfeiffer, S. Liu and S. Furber, "Scalable energy-efficient, low-latency implementations of trained spiking Deep Belief Networks on SpiNNaker," 2015 International Joint Conference on Neural Networks (IJCNN), Killarney, 2015, pp. 1-8.

[12] Diehl, Peter & Neil, Dan & Binas, Jonathan & Cook, Matthew & Liu, Shih-Chii & Pfeiffer, Michael. (2015). Fast-Classifying, High-Accuracy Spiking Deep Networks Through Weight and Threshold Balancing. International Joint Conference on Neural Networks, IJCNN. 10.1109/IJCNN.2015.7280696.

16-1 (7160)

A 55nm 1-to-8 bit Configurable 6T SRAM based Computing-in-Memory Unit-Macro for CNN-based AI Edge Processors

Zhixiao Zhang[1,2], Jia-Jing Chen[1], Xin Si[1], Yung-Ning Tu[1], Jian-Wei Su[1,3], Wei-Hsing Huang[1], Jing-Hong Wang[1], Wei-Chen Wei[1], Yen-Cheng Chiu[1], Je-Min Hong[1], Shyh-Shyuan Sheu[3], Sih-Han Li[3], Ren-Shuo Liu[1], Chih-Cheng Hsieh[1], Kea-Tiong Tang[1], and Meng-Fan Chang[1]

[1]National Tsing Hua University, Hsinchu, Taiwan,
[2]Fuzhou University, Fuzhou, China, [3]Industrial Technology Research Institute, Hsinchu, Taiwan
E-mail: mfchang@ee.nthu.edu.tw

Abstract—This work presents a 1-to-8 bit configurable SRAM CIM unit-macro using (1) a hybrid structure combining 6T-SRAM in-memory binary product-sum (IMBPS) with digital near-memory-computing multi-bit-PS accumulation (DNMPSA) to achieve high read accuracy and compact area, (2) a column-based place-value-grouped weight mapping (CPVGWM) and serial-bit input mapping (SBIN) scheme to facilitate reconfiguration and increase array efficiency for various input/weight precision. A 4Kb configurable 6T-SRAM CIM unit-macro was fabricated using a 55nm CMOS process and foundry 6T cells. The resulting macro achieved 3.5ns access time with 3b PS outputs and gained a 3.91-50x improvement in figure-of-merit, compared to previous SRAM-CIMs.

Keywords—CIM, SRAM, CNN, AI edge processor.

I. INTRODUCTION

Computing-in-memory (CIM) structure is a promising candidate to overcome the memory bottleneck and improve the energy efficiency of CNN-based AI edge processors. The precision configuration requirements of these processors, including inputs (IN), weights (W), and product-sum (PS) output (PSout) can be adjusted to optimize the accuracy-energy tradeoff across various AI applications. Recent computing-in-memory (CIM) macros use in-memory-computing (IMC) and near-memory-computing (NMC) structures [1]-[6] to improve the energy efficiency of product-sum (PS) operations for AI processors. Previous analog IMC [1] macro can support only binary inputs/weights. The fully digital NMC approach consumes large area in case of high bit precision. Hybrid-structure use large-area 10T-SRAM IMC cell arrays for binary-weight product operations and analog NMC for summations [2], or 6T-SRAM IMC for weight combination and analog NMC for multi-bit product-sum (PS) operations. However, these come at the cost of large area overhead and long latency due to the use of high-precision for inputs and weights.

This work proposed a 1-to-8 bit configurable SRAM CIM unit-macro for CNN operations with various accuracy-energy tradeoffs. A hybrid foundry 6T-SRAM based CIM structure was developed combining in-memory binary product-sum (IMBPS) with digital NMC multi-bit-PS accumulation (DNMPSA) to achieve higher signal margin, better sensing yield and compact area. In addition, a column-based place-

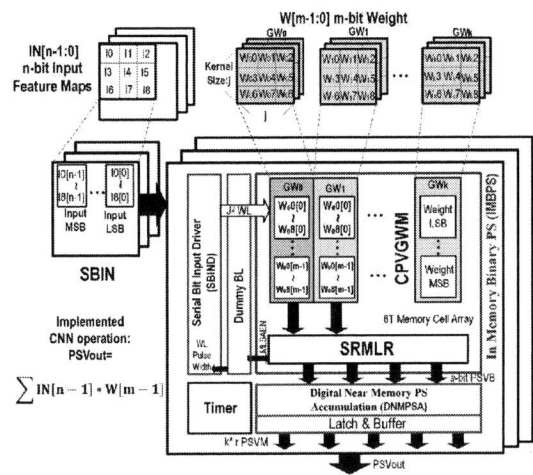

Fig. 1. Proposed configurable 6T-SRAM based SRAM-CIM structure

value-grouped weight mapping (CPVGWM) and serial-bit input mapping (SBIN) scheme was proposed to facilitate reconfiguration and improve array efficiency for various input/weight precisions.

II. PROPOSED CONFIGURABLE 6T SRAM BASED COMPUTING-IN-MEMORY UNIT-MACRO

Fig. 1 presents the proposed configurable 6T-SRAM CIM structure, comprising IMBPS, DNMPSA, CPVGWM, SBIN drivers (SBIND), and self-reference multilevel readers (SRMLR). This work computed the product-sum (PS) of j^2 n-bit inputs and j^2 m-bit weights using nxm serial cycles, where j, n, and m is the kernel size, input bit precision, and weight bit precision. The j^2 n-bit inputs are converted into n cycles of j^2 parallel-on wordline (WL) signals. In IMBPS, each 6T SRAM cell in a selected j^2-cell group on the same bitline (BL) performs the binary input-weight-product (IWP=IN×W). When a cell (IWP=1 cell) with WL=on (IN=1) and W=1 (Q=1, QB=0), it provides a discharge cell current (I_{DIS}) on its BLB. Thus, the total BLB discharge current (I_{BLB}) is ($N_{IWP=1}xI_{DIS}$), where $N_{IWP=1}$ is the number of SRAM cells with IWP=1. $I_{BL}=(N_{WL=1} - N_{IWP=1})xI_{DIS}$, where $N_{WL=1}$ is the number of cells

978-1-7281-5107-6/19 $31.00 © 2019 IEEE   217

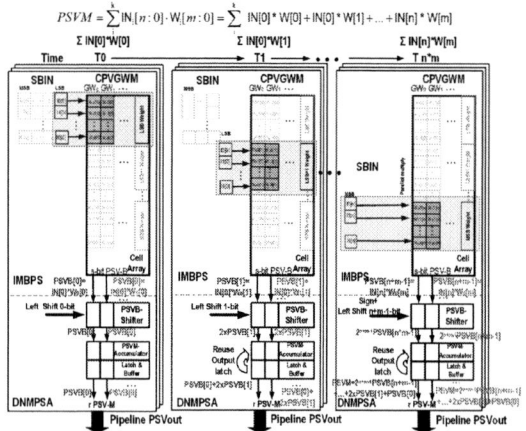

Fig. 2. Proposed column-based place-value-grouped weight mapping (CPVGWM) and serial-bit input mapping (SBIN) scheme.

with WL(IN)=1. For a given WL pulse width (T_{WL}), the BL voltage (V_{BL}) is VDD- (T_{WL} x I_{BL})/C_{BL} The resulting V_{BL} is the sum of j^2 IWP values (PS) for the p-th input bit (IN[p]) and q-th weight bit (W[q]). An SRMLR is used to detect V_{BL} and output an s-bit PS-value for a given binary input/weight (PSVB). DNMPSA then reuses the output-latch to accumulate PSVB for various combinations of IN[p] and W[q]. After mxn cycles, each DNMPSA unit (or IO) outputs the r-bit PSV for multiple-bit input/weight (PSVM). For a CIM with k columns and column-mux=1 (k IO/DNMPSA units), the CIM can output k PSVM at the same time. The hybrid structure of IMBPS with DNMPSA means that the analog read margin between PSV is determined by the kernel size (j^2) in IMBPS and the precision (s-bit) of PSVB, but is independent of the precision of inputs/weights. Thus, the proposed CIM has a constant read margin and CIM-accuracy under various precision configurations of inputs and weights. From the perspective of systems, the high-volume sequential streaming of input data is commonly used for deep CNN channels in each layer. This makes it easier to schedule the PSVout of multiple SRAM-CIM macros occurring in different cycles to hide the mxn-cycle latency. This is similar to a pipeline approach.

Fig. 2 shows the operations of proposed CPVGWM, SBIN, and DNMPSA. In CPVGWM, the m-bit weights of each jxj kernel are stored in m consecutive groups of weights (GW) in the same column/bitline (BL). Each GW includes the same place-value bits of all j^2 weights for a given kernel. Each n-bit input is applied to SBIND to output n WL pulses across n cycles. In each cycle, every BL generates the PSVB of the given input bit and GW (Σ(IN[p]xGW[q])). Every DNMPSA module includes a PSVB-shifter and a PSVM-accumulator.

The PSVB-shifter shifts PSVB by ($p+q$) bits according to the place-value of its input (p) and weight (q). The PSVM-accumulator adds/accumulates the current PSVB with the PSVM stored in the output-latch (generated in previous cycles). The new PSVMs are stored at the output-latch. After mxn cycles, the final PSVM at the output-latch is equal to Σ(IN[n-1:0]xW[m-1:0]). For CNN operations with 3x3 kernels, 8b-weigh, and 8b-input, the GW for all LSB bits (GW[0]) includes W_0[0]~W_8[0]. In the 1st cycle, PSVB[0] for Σ(IN[0]xW[0]) is

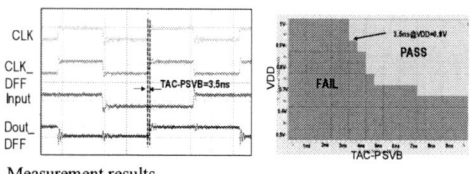

Fig. 3. Measurement results.

CHIP SUMMARY	
Technology	55nm
Unit-Macro Size	4Kb
Measured Accuracy (CIFAR-10)	85.97% (1b IN W)
	91.20% (2b IN W)
	91.78% (4b IN W)
	91.93% (8b IN W)
Access Time (Per Cycle)	3.5ns
Measured Energy (pJ)	28.6
Energy efficiency (TOPS/W)	40.2 (1b IN W)
	10.1 (2b IN W)
	2.5 (4b IN W)
	0.6 (8b IN W)

Fig. 4. Die photo and summary table.

generated at the BL and stored in the output-latch. In the 2nd cycle, PSVB[1] (for Σ(IN[0]xW[1])) is generated at the BL. DNMPSA then shifts PSVB[1] by 1-bit and adds it to PSVB[0] to generate PSVM=2xPSVB[1]+PSVB[0] at the output-latch. After 64 cycles, PSVM=Σ(2^ixPSVB[i]), i=0~15.

III. PERFORMANCE AND EXPERIMENT RESULTS

Owing to the hybrid structure combining IMBPS and DNMPSA, this work increased the sensing margin for 3x3 CNN kernels by 160x at PSVout=11b, compared to pure analog CIM schemes. This work also achieved 1.1%+ higher inference accuracy for CIFAR-10 dataset.

Fig. 3 presents the measurement results from a 55nm 4Kb SRAM-CIM unit-macro fabricated using foundry 6T cells. The captured waveforms and shmoo plot using 3x3 kernels indicate PSVB access times of 3.5ns in each cycle ($T_{AC-PSVB}$) at VDD=0.9V excluding the path-delay. A demonstration system was implemented using an SRAM-CIM testchip and FPGA-based host. Measurement data indicates that this work increased the FoM [(energy efficiency x throughput x precision)/(SRAM size x periphery overhead x cell area)] by 3.91x~50x, compared to previous silicon verified SRAM-CIM macros. Fig. 4 presents the die photo and chip summary.

ACKNOWLEDGMENT

The authors gratefully acknowledge support from CIC, TSMC-JDP, MTK-JDP, and MOST-Taiwan.

REFERENCES

[1] W. Khwa *et al.*, "A 65nm 4Kb algorithm-dependent computing-in-memory SRAM unit-macro with 2.3ns and 55.8TOPS/W fully parallel product-sum operation for binary DNN edge processors" *ISSCC*, pp. 496-498, 2018

[2] A. Biswas *et al.*, "Conv-RAM: An energy-efficient SRAM with embedded convolution computation for low-power CNN-based machine learning applications" *ISSCC*, pp. 488-490, 2018

[3] S. K. Gonugondla *et al.*, "A 42pJ/decision 3.12TOPS/W robust in-memory machine learning classifier with on-chip training" *ISSCC*, pp. 490-492, 2018

[4] J. Yang *et al.*, "Sandwich-RAM: An Energy-Efficient In-Memory BWN Architecture with Pulse-Width Modulation" *ISSCC*, pp. 394-396, 2019

[5] Xin Si *et al.*, "A Twin-8T SRAM Computation-in-Memory Macro for Multiple-bit CNN-Based Machine Learning", ISSCC, pp. 396-398,2019

[6] R. Guo et al., "A 5.1pJ/Neuron 127.3us/Inference RNN-based Speech Recognition Processor using 16 Computing-in-Memory SRAM Macros in 65nm CMOS," IEEE Symp. VLSI Circuits, 2019, pp. C120-C121.

A 24 kb Single-Well Mixed 3T Gain-Cell eDRAM with Body-Bias in 28 nm FD-SOI for Refresh-Free DSP Applications

Jonathan Narinx, Robert Giterman, Andrea Bonetti, Nicolas Frigerio, Cosimo Aprile,
Andreas Burg, and Yusuf Leblebici

Ecole Polytechnique Fédérale de Lausanne (EPFL), Lausanne, Switzerland

Abstract—Logic-compatible gain-cell embedded DRAM (GC-eDRAM) is an emerging alternative to conventional SRAM for memory-dominated system-on-chip (SoC) designs due to its high-density, low-power, and two-ported operation. Although GCs have a limited data retention time (DRT) at deeply scaled technology nodes, there are many DSP applications which only require short-term data storage and can therefore avoid refresh. In this paper, we present a novel single-well mixed 3T GC implementation in 28 nm FD-SOI technology. The proposed GC is supplied with body-bias control to improve the DRT by suppressing the leakage through the write port, and extend the maximum operating frequency by forward body-biasing the read port. A 24 kbit GC-eDRAM macro implementing the proposed 3T GC was fabricated in 28 nm FD-SOI technology, resulting in the highest density logic-compatible embedded memory fabricated in any 28 nm process with over 2× higher density compared to a 6T SRAM cell, over 4× higher DRT compared to a conventional 3T GC, and 38×–47× lower static power compared to conventional single-ported and two-ported SRAMs.

I. INTRODUCTION

Embedded memories are a key building block of digital signal processing (DSP) systems and often account for the majority of silicon real-estate and power consumption [1]–[3]. Embedded memories are used to store the ever-growing amounts of data close to the processing units, thereby reducing off-chip memory access which is costly in terms of energy and latency and has limited bandwidth. The mainstream solution for embedded memories is the 6-transistor (6T) SRAM macrocell due to its high performance and reliability under nominal supply voltages. However, it incurs high area and static power overheads, and access is limited to a single write or read operation per cycle, while pipelined DSP datapaths often require concurrent read and write operations.

Gain-cell embedded DRAM (GC-eDRAM) is an emerging alternative to the traditional 6T SRAM due to its high-density, low leakage power, and inherent two-port operation [1], [4]–[8]. Compared to conventional 1T-1C eDRAM, it is fully logic-compatible, hence it can be integrated in a standard CMOS process without additional manufacturing steps or cost. However, GC-eDRAM requires periodic power-hungry refresh cycles to reliably retain data over long periods of time, since data is stored as charge on parasitic MOSFET capacitances, and deteriorates over time due to leakage. This drawback is even more prominent in advanced technology nodes (i.e., sub-65 nm) due to a reduced storage capacitance and increased leakage currents, which lead to significantly lower DRTs [1], [5], [6]. In order to maintain a sufficiently high DRT at sub-65 nm nodes, alternative gain cell (GC) structures with additional transistors were proposed [7], [8]. However, these solutions diminish the high-density advantage of GC-eDRAMs, and incur a performance penalty due to a degraded storage node (SN) voltage, resulting in longer read access times. However, in a large variety of DSP applications

embedded memories only need to hold data for a short period of time. For example, an FFT processor implemented in [3] requires only N/2 cycles to retain data, where N is number of rows in a memory array, while an LDPC decoder reported in [1] requires only 56 ns of data retention.

In this paper, we present the highest density GC-eDRAM implementation in 28 nm technology targeted at refresh-free DSP applications. The proposed GC is composed of a novel single-well mixed PMOS/NMOS structure, equipped with body-bias control for significantly improved DRT and read access speed, and potential post-silicon calibration based on the process corner and target operating frequency. A 24 kbit array implemented in 28 nm FD-SOI technology demonstrates over 4× longer DRT compared to a conventional 3T GC, 2× smaller bitcell area and 38× less static power compared to a 6T SRAM cell.

Contributions: The main contributions of this paper are summarized as follows:

1) This work presents the highest-density logic-compatible embedded memory in 28 nm technology.
2) The proposed memory is based on a novel single-well mixed PMOS/NMOS 3T GC with integrated body bias, which allows post-silicon calibration according to the process corner and operating conditions.
3) A 24 kbit manufactured array based on the proposed two-ported GC demonstrates over 4× longer DRT compared to a conventional 3T GC, 2×–2.6× higher density and 38×–47× lower static power compared to 6T–8T SRAM cells in 28 nm technology.

Outline: The rest of this paper is organized as follows: Section II describes the proposed single-well mixed 3T gain-cell; Section III presents the complete 24 kb GC-eDRAM macro architecture and implementation; Section IV presents the test-chip integration and measurement results; Section V concludes the paper.

II. PROPOSED SINGLE-WELL MIXED 3T GAIN CELL

A conventional GC-eDRAM based on an all-NMOS 3T GC is shown in Fig. 1a [4]. The 3T GC is composed of a *write port*, including a write transistor (NW), a write bit line (WBL), and a write word line (WWL); and a *read port*, including a storage transistor (NS), a read transistor (NR), a read bit line (RBL) and a read word line (RWL). The SN capacitance (C_{SN}) is composed of the parasitic MOSFET capacitances and layout parasitic components. A major drawback of the conventional 3T GC is the low DRT at deeply-scaled (i.e., below-65 nm) technology nodes. This can be attributed to the increased leakage currents which deteriorate the SN voltage during standby. In particular, sub-threshold conduction through NW and gate leakages through NW and NS limit the storage of '1' down-to only a few hundreds of nanoseconds [9] and

IEEE Asian Solid-State Circuits Conference
November 4 – 6, 2019
The Parisian Macao, Macao SAR, China

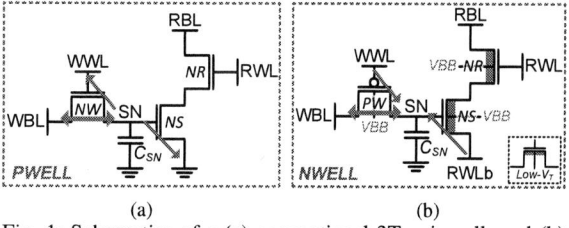

Fig. 1: Schematics of a (a) conventional 3T gain-cell, and (b) the proposed single-well mixed 3T gain-cell with their main leakage mechanisms.

significantly degrade the RBL discharge delay due to a reduced overdrive ($V_{GS} - V_{TH}$) of NS during read access.

In order to reduce the leakage currents which degrade the SN voltage and improve the read access speed while maintaining the high-density advantage of the 3T GC, we propose a single-well mixed PMOS/NMOS 3T GC structure, depicted in Fig. 1b. The proposed GC is composed of a regular-V_{TH} (RVT) PMOS write transistor (PW), and two low–V_{TH} (LVT) NMOS read transistors (NS, NR) implemented in a flipped-well configuration, which enables RVT PMOS and LVT NMOS transistors to share a common N-Well biased at VBB, without requiring an in-bit well separation that would lead to extra area. The shared body-bias applied to the cell has two main advantages. First, increasing VBB above VDD puts PW in a reversed body-bias regime, resulting in an increased V_{TH} and exponentially reduced sub-threshold leakage to SN. Second, it puts the read transistor into a forward body-bias state, resulting in reduced V_{TH} and a higher NS overdrive during read cycles, which reduces read access delay. In order to further improve the DRT and read access delay, the source diffusion of NS is connected to RWLb (the complementary of RWL), introducing *preferential coupling* that strengthens the now-critical '0' level during read access [1], [10]. In addition, compared to the conventional 3T configuration, it reduces the sub-threshold leakage path through RBL, which is pre-charged to VDD during standby.

The operating mechanism of the single-well mixed 3T GC is depicted in Fig. 2. A write operation is performed by discharging WWL to a negative voltage (V_{NEG}) in order to pass a strong '0' level from WBL to SN. As WWL is charged to VDD at the end of the write operation, coupling from WWL to SN causes the SN voltage to rise above GND for '0' and above VDD for '1', resulting in a degraded DRT of '0'. Read is performed by charging RWL to VDD (and discharging RWLb to GND), conditionally discharging the pre-charged RBL when the cell holds a '1'. The discharge of RWLb to GND in the beginning of a read cycle results in a decrease of the SN voltage, compensating for the coupling caused by WWL at the end of the write cycle. Finally, the RBL voltage is sensed by a dedicated sense-amplifier to determine the stored data.

Fig. 3 depicts the estimated DRT distribution of the single-well 3T GC under different VBB voltages. The distributions were extracted from 1000 Monte-Carlo simulations including both global and local variations under worst case biasing conditions (with the WBL biased to the opposite voltage stored in the cell). The DRT, defined at the point where the voltage difference between stored '1' and '0' deteriorated beneath

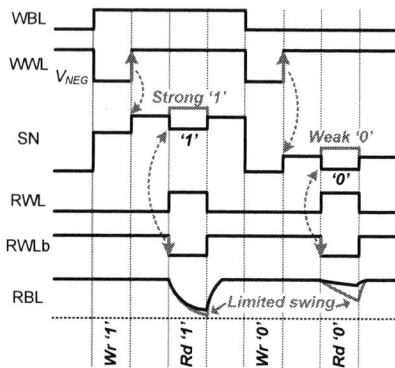

Fig. 2: Read and write operations waveforms of the proposed single-well mixed 3T GC with (black) and without (grey) preferential coupling.

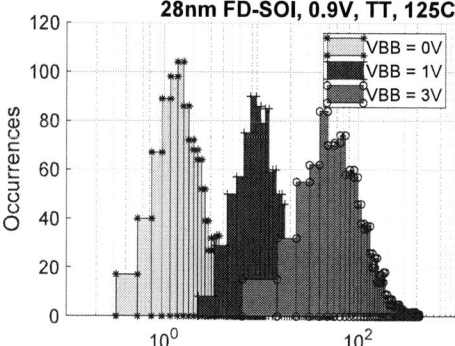

Fig. 3: Estimated DRT distributions at different body bias voltages.

VDD/2, demonstrates an almost two orders of magnitude improvement with VBB increasing from 0 V to 3 V due to a significant reduction of SN leakage.

III. MEMORY MACRO ARCHITECTURE AND IMPLEMENTATION

The single-well mixed 3T GC with body biasing was integrated into a 24 kbit memory macro, as in Fig. 4.

The memory write port includes a write address decoder, a global level shifter (LS), WWL shifted drivers, and WBL drivers. Upon a write request, the write enable signal (WE) is asserted, driving the WEb_SHIFT signal to V_{NEG} using a global level shifter. The WEb_SHIFT signal is shared across the WWL drivers column to cut-off the pull-down network of the WWL driver, driving the selected WWL to V_{NEG}. In parallel, the WE signal enables the transmission of the data signal (DI) to the WBL in order to write the data to the SNs of the selected row.

The memory read port consists of a read address decoder, RWL drivers, a differential sense amplifier and two programmable delay lines (PDLs), which are used to set the delay of the RBL pre-charge (PCH) and sense amplifier enable (SAE) signals. In order to support the large current consumption through the RWLb drivers, which are used to discharge the large RBL capacitance when reading a '1',

978-1-7281-5107-6/19 $31.00 © 2019 IEEE

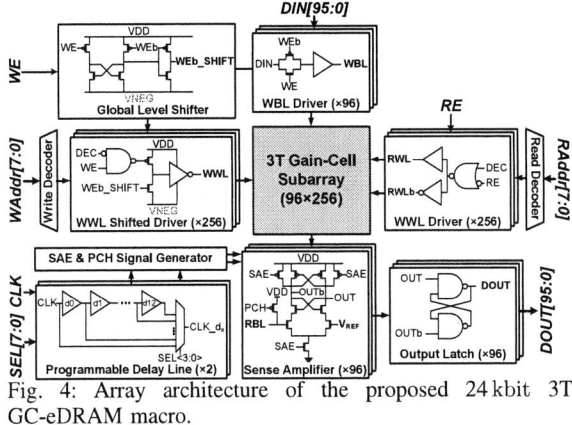

Fig. 4: Array architecture of the proposed 24 kbit 3T GC-eDRAM macro.

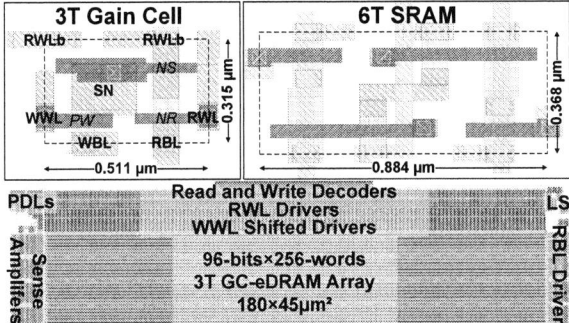

Fig. 5: 3T GC-eDRAM array layout and comparison in bit-cell layouts in 28 nm FD-SOI technology.

pull down devices were inserted into the GC array every 24 columns. Upon a read request, the read enable signal (RE) is asserted, cutting off the RBL pre-charge operation after a delay set by the PDL. Concurrently, the RWL driver of the selected read address is enabled, driving the RWL to VDD and RWLb to GND. Then, the RBLs are driven by the GCs of the selected row, conditionally discharging when the SN holds a '1'. A second PDL sets the SAE delay, asserting SAE in order to enable the sense amplifier and latch the output value, which is set according to the difference between the RBL voltage and a reference voltage (V_{REF}) which is supplied externally.

The single-well mixed 3T GC macro was designed and implemented in a 28 nm FD-SOI technology. Fig. 5 depicts the layout views of the 3T GC, as compared to a 6T SRAM using the same design rules, and the layout of the 24 kbit memory macro. The 3T GC, measured at $0.16\,\mu m^2$, was designed with minimum sized NS and NR transistors, and a 43% up-sized length for PW in order to further reduce the sub-threshold leakage through SN. Moreover, the RWLb signal, routed with M1, was horizontally extended to increase the preferential coupling with SN. The two-ported 3T GC provides over 2× higher density compared to a conventional single-ported 6T SRAM bit-cell, redrawn with standard design rules and measured at $0.325\,\mu m^2$. The entire 24 kbit memory macro has a silicon footprint of $8100\,\mu m^2$, with the GC array constituting 60% of the total macro area. In comparison, similar sized single-ported 6T and two-ported 8T compiled SRAM macros, drawn

with "pushed" design-rules, were measured at $10373\,\mu m^2$ and $20407\,\mu m^2$, respectively, which is 22%–60% larger than the proposed 3T GC memory macro.

IV. TEST-CHIP AND MEASUREMENT RESULTS

A test-chip including the 24 kbit single-well 3T GC-eDRAM was fabricated in 28 nm FD-SOI technology. The test-chip implements a test setup including an on-chip memory built-in self-test (BIST), a clock generator, and a serial interface. An external body-bias supply is delivered to the GC-eDRAM array through an analog I/O pad. The chip micrograph and its key features are shown in Fig. 6. The fabricated 3T GC-eDRAMs was successfully operated across the voltage range of 0.7 V–1.0 V and across the frequency range of 100 MHz–400 MHz.

Fig. 7 depicts the measured DRT maps of the 3T GC array at different VBB voltages under a 0.9 V memory VDD at an operating frequency of 400 MHz at 27°C, with V_{REF} and V_{NEG} set to 0.45 V and −0.3 V, respectively. The DRT maps were measured by applying '0' and '1' write patterns to each bit in the array, with the DRT representing the worst-case write-to-read delay at which the data could no longer be read out correctly. All measurements were performed under worst-case biasing conditions, which occur when the WBL is biased to the opposite voltage compared to the SN. The results clearly demonstrate the effectiveness of increasing VBB, which reduces the leakage through SN and provides a faster RBL discharge delay for a stored '1', ultimately resulting in a higher DRT. The average DRT was measured at 295 μs, 563 μs, and 743 μs with VBB set to 0.9 V, 1.35 V, and 1.7 V, respectively, providing an average DRT improvement of almost 2.5×. However, note that increasing VBB above an optimal value would result in a lower DRT due to the faster discharge of the RBL, leading to an incorrect readout of a deteriorated '0'. Therefore, the optimal choice of VBB strongly depends on the operating frequency, which determines the sampling time of the RBL.

Fig. 8 depicts the optimal VBB voltage under an average DRT criterion across a 200 MHz–400 MHz frequency range, as measured across 12 chips at 27°C. As previously noted, increasing VBB decreases the RBL discharge delay by putting the GC read port in a forward body-bias regime. As a result, the optimal VBB voltage increases with operating frequency, ranging from average values of 1.2 V –1.9 V across 200 MHz–400 MHz, and resulting in an average DRT range of 692 μs–761 μs.

Fig. 6: Die micrograph and key features of the test chip.

Technology	28nm FD-SOI
Process Corner	Fast
Die area	1.7 mm²
Array type	3T Single-Well Mixed GC-eDRAM
Array size	24 kb (96 BLs x 256 WLs)
Macro area	8100 μm²
Cell area	0.16 μm² (49% 6T SRAM)
VDD range	0.7 V-1.0 V
Max Frequency	400 MHz

IEEE Asian Solid-State Circuits Conference
November 4 – 6, 2019
The Parisian Macao, Macao SAR, China

TABLE I: Comparison between the proposed design and other silicon-proven logic-compatible embedded memories in 28 nm technology.

Name	6T SRAM	8T SRAM	4T all-NMOS GC [11]	4T Mixed-Vt IFGC [7]	**Proposed 3T Single-Well Mixed GC**
Application Target	Generic	Generic	Low-Power	Approx. Computing	**Refresh-Free DSP**
Technology Node	28 nm FD-SOI	28 nm FD-SOI	28 nm FD-SOI	28 nm Bulk	**28 nm FD-SOI**
No. of Ports	1P R/W	2P R+W	2P R+W	2P R+W	**2P R+W**
Array Size	–	–	4 kbit	8 kbit	**24 kbit**
Max Frequency	–	–	175 MHz	800 MHz	**400 MHz**
Cell Size*	$0.325\,\mu m^2$ (1×)	$0.42\,\mu m^2$ (1.29×)	$0.23\,\mu m^2$ (0.71×)	$0.25\,\mu m^2$ (0.73×)	**$0.16\,\mu m^2$ (0.49×)**
24 kb Macro Size	$10373\,\mu m^2$ (1×)**	$20407\,\mu m^2$ (1.97×)**	–	–	**$8100\,\mu m^2$ (0.78×)***
Retention Time***	Static	Static	– $820\,\mu s$ (100% yield)	5.8 μs (99% yield) 1.1 μs (100% yield)	**28 μs (99% yield) 2.64 μs (100% yield)**
Static Power	472.1 pW/bit	582 pW/bit	–	–	**12.3 pW/bit**

All results are at 0.9 V @27°C   *Logic design rule   ** "Pushed" design rule   ***Measured at max frequency

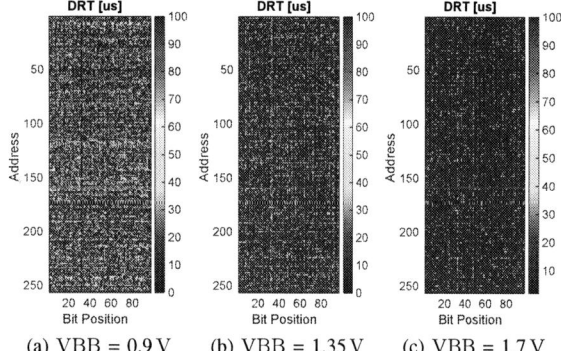

(a) VBB = 0.9 V   (b) VBB = 1.35 V   (c) VBB = 1.7 V
Fig. 7: DRT maps at 400 MHz under different VBB voltages.

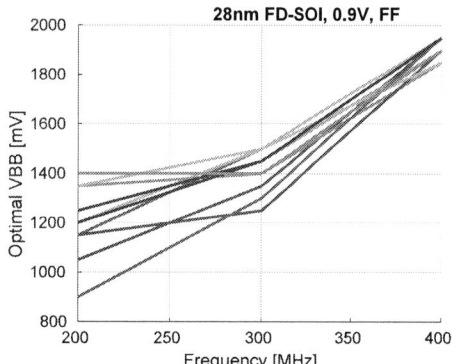

Fig. 8: Optimal VBB under an average DRT criterion at different frequencies measured across 12 chips.

Table I shows a comparison between the proposed single-well mixed 3T GC and other silicon-proven embedded memory options. The single-well 3T GC has the smallest silicon footprint among all fabricated logic-compatible embedded memories in 28 nm technology, with 2×–2.6× higher cell density compared to SRAM, and 30% smaller size compared to 4T GC-eDRAM. Moreover, the 24 kbit 3T memory macro is 22%–60% smaller than similar sized 6T and 8T SRAM macros, implemented with "pushed" design rules. In addition, the proposed two-ported 3T GC has over 4× higher minimum DRT compared to the simulated DRT at 0.67 μs of a similar-sized conventional 3T GC [9], which makes the 24 kbit 3T GC-eDRAM macro suitable for many refresh-free DSP

applications [1]–[3]. Finally, the static power consumption of the proposed 3T cell is only 12.3 pW/bit, which is 38×–47× lower than single-ported 6T and two-ported 8T SRAM cells.

V. CONCLUSION

This paper presents a single-well mixed PMOS/NMOS 3T GC with integrated body-bias in 28 nm FD-SOI technology. A 24 kbit GC-eDRAM array based on the proposed GC is the highest density logic-compatible embedded memory demonstrated in 28 nm technology. Using its unique body-biasing characteristics, the fabricated memory provides over 4× higher DRT compared to similar-sized conventional 3T GCs, and 38×–47× lower static power compared to single-ported and two-ported SRAM bitcells in 28 nm technology.

REFERENCES

[1] Y. S. Park et al., "Low-Power High-Throughput LDPC Decoder Using Non-Refresh Embedded DRAM," *IEEE Journal of Solid-State Circuits*, vol. 49, no. 3, pp. 783–794, Mar. 2014.

[2] W. Choi, G. Kang, and J. Park, "A Refresh-Less eDRAM Macro With Embedded Voltage Reference and Selective Read for an Area and Power Efficient Viterbi Decoder," *IEEE Journal of Solid-State Circuits*, vol. 50, no. 10, pp. 2451–2462, Oct. 2015.

[3] G. Kang, W. Choi, and J. Park, "Embedded DRAM-Based Memory Customization for Low-Cost FFT Processor Design," *IEEE Transactions on Very Large Scale Integration (VLSI) Systems*, vol. 25, no. 12, pp. 3484–3494, Dec. 2017.

[4] P. Meinerzhagen et al., *Gain-Cell Embedded DRAMs for Low-Power VLSI Systems-on-Chip*. Springer International Publishing, 2018.

[5] D. Somasekhar et al., "2 GHz 2 Mb 2t Gain Cell Memory Macro With 128 GBytes/sec Bandwidth in a 65 nm Logic Process Technology," *IEEE Journal of Solid-State Circuits*, vol. 44, no. 1, pp. 174–185, Jan. 2009.

[6] K. C. Chun et al., "A 667 MHz Logic-Compatible Embedded DRAM Featuring an Asymmetric 2t Gain Cell for High Speed On-Die Caches," *IEEE Journal of Solid-State Circuits*, vol. 47, no. 2, pp. 547–559, Feb. 2012.

[7] R. Giterman et al., "An 800-MHz Mixed-VT 4t IFGC Embedded DRAM in 28-nm CMOS Bulk Process for Approximate Storage Applications," *IEEE Journal of Solid-State Circuits*, vol. 53, no. 7, pp. 2136–2148, Jul. 2018.

[8] R. Giterman et al., "A 4-Transistor nMOS-Only Logic-Compatible Gain-Cell Embedded DRAM With Over 1.6-ms Retention Time at 700 mV in 28-nm FD-SOI," *IEEE Transactions on Circuits and Systems I: Regular Papers*, vol. 65, no. 4, pp. 1245–1256, Apr. 2018.

[9] E. V. Bravo, A. Bonetti, and A. Burg, "Data-Retention-Time Characterization of Gain-Cell eDRAMs Across the Design and Variations Space," in *2019 IEEE International Symposium on Circuits and Systems (ISCAS)*, May 2019, pp. 1–5.

[10] K. C. Chun et al., "A 3t Gain Cell Embedded DRAM Utilizing Preferential Boosting for High Density and Low Power On-Die Caches," *IEEE Journal of Solid-State Circuits*, vol. 46, no. 6, pp. 1495–1505, Jun. 2011.

[11] R. Giterman et al., "GC-eDRAM with Body-Bias Compensated Readout and Error Detection in 28nm FD-SOI," *IEEE Transactions on Circuits and Systems II: Express Briefs*, pp. 1–1, 2019.

978-1-7281-5107-6/19 $31.00 © 2019 IEEE

16-3 (7016)

Configurable BCAM/TCAM Based on 6T SRAM Bit Cell and Enhanced Match Line Clamping

Jongeun Koo, Eunhwan Kim, Seunghyun Yoo, Taesu Kim, Sungju Ryu, and Jae-Joon Kim

Pohang University of Science and Technology (POSTECH), Pohang, Korea

Email: {jongeun.koo, eunhwan, dysh1017, taesukim, sungju.ryu, jaejoon}@postech.ac.kr

Abstract—We present a configurable 6T binary content-addressable memory (BCAM) and 12T ternary CAM (TCAM) which is based on conventional 6T SRAM bit cell. We also propose a match line (ML) voltage/current clamping scheme to reduce ML discharge delay considerably while preventing data corruption during CAM search operation compared to the previous configurable BCAM/TCAM using 6T SRAM bit cell. Experimental data from 4Kb (64×64b) CAM testchips in a 28nm CMOS technology shows that the proposed ML clamping scheme achieves 53.7% reduction in ML discharge delay compared to the previous word line underdrive scheme.

I. INTRODUCTION

Content-addressable memory (CAM) is a high-speed search engine that performs fully-parallel comparisons between input search data and a table of storage data in a single clock cycle [1]. CAM has been widely used in applications requiring fast search operation including translation look-aside buffers [2], tag directories in fully associative caches [3], network routers [4], database accelerators and image processing [5].

However, the high search speed of a CAM comes at the cost of large area and high power consumption. As the density of CAM grows, the area and power consumption overhead increases faster. For this reason, most of recent researches on CAM have focused on the reduction of area and power consumption [6]–[8].

A conventional NOR-type CAM cell (Fig. 1a) uses a built-in XNOR circuit to compare the data (Q and Qb) stored in a 6T SRAM bit cell and search word on the search line (SL and SLb) for high-speed pattern search [6], [7]. However, the built-in XNOR circuit consisting of 4 transistors increases area in both 10T binary CAM (BCAM) and 16T ternary CAM (TCAM) cells (Fig. 1a).

Recently, a configurable BCAM/TCAM using a 6T SRAM bit cell with split word line has been proposed [8]. The combination of split WL (WL and WLb) and cell data (Q and Qb) values provides a XOR function for pattern search, thus considerable reduction of CAM area could be achieved by removing the built-in XNOR circuit in the conventional CAM cell (Fig. 1b). However, the split WL still requires redesign of the SRAM bit cell. In addition, activation of multiple SLs (WLs) for fully-parallel comparisons incurs cell data stability issues during CAM search operation so that WL-underdrive scheme, in which V_{WL} is much lower than VDD, has been used. However, it increases the match line (ML) discharge delay significantly [8].

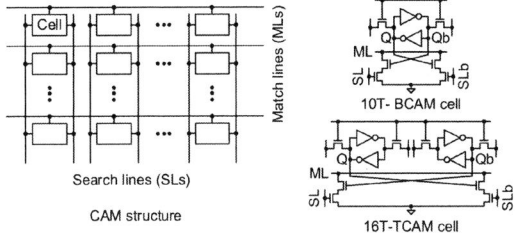

(a) Conventional CAM structure and NOR-type CAM cells.

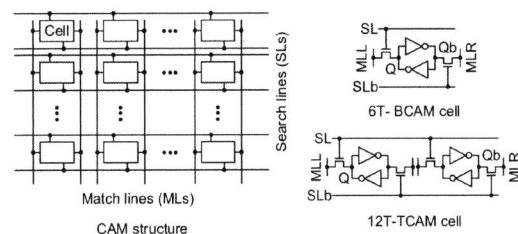

(b) Previous CAM using 6T SRAM bit cell with split WL [8].

Fig. 1. Previous CAM structures and cells.

In this paper, we present a new configurable BCAM/TCAM using an unmodified 6T SRAM bit cell. We also propose a ML voltage/current clamping scheme for considerable reduction of the ML delay overhead during CAM search operation.

II. PROPOSED CONFIGURABLE BCAM/TCAM

The proposed 6T-BCAM cell uses a single SRAM bit cell. (Fig. 2a). Similar to the 6T-BCAM cell in [8] (Fig. 1b), the proposed 6T-BCAM has the ML which consists of MLL (BL) and MLR (BLb). However, unlike [8], the proposed BCAM cell does not use split WL and the complementary search data SL and SLb are time-multiplexed on a single WL. For this reason, the search operation in the proposed BCAM is different from the search in [8] while the column write operation is exactly same.

In the first cycle of the search operation in the proposed 6T-BCAM (Fig. 2a and Table I), if SL[i]=Q[i]=1 for some i's and rest of the SL's = 0 for the cells attached to an MLL, the MLL voltage remains high and decision for pattern-match is deferred to the second cycle. At the same time, the MLR in the same column is discharged but this does not affect the BCAM output because the sense amplifier (SA) for the MLR

978-1-7281-5107-6/19 $31.00 © 2019 IEEE

223

TABLE I
MODE CONFIGURATIONS OF THE PROPOSED BCAM/TCAM.

	6T-BCAM				12T-TCAM			6T-SRAM	
	Column Write Cycle		Search Cycle		Column Write Cycle		Search	Write	Read
	1	2	1	2	1	2			
Clamp	Target Col.: OFF The Others: ON		ON		Target Col.: OFF The Others: ON		ON	OFF	
VDDC[c]	Target Col.: VDDL The Others: VDD		VDD		Target Col.: VDDL The Others: VDD		VDD	VDD	
V_{WL}	VDDL							VDD	
WL[r+1]	D[r+1]	Db[r+1]	SL[r+1]	SLb[r+1]	D[r/2] or 1 for *	Db[r/2]) or 0 for *	SL[r/2]	Row Dec. Output	
WL[r]	D[r]	Db[r]	SL[r]	SLb[r]	Db[r/2] or 1 for *	D[r/2] or 0 for *	SLb[r/2]		
BL[c]	1	0	MLL[c]	-	1	0	ML[c]	D[c]	-
BLb[c]	0	1	-	MLR[c]	0	1	-	Db[c]	-
Sense Amp.	-	-	1-Single Ended	1-Single Ended	-	-	1-Single Ended	-	Diff. Ended
Output	-	-	-	$SA_{BL}[c]$ & $SA_{BLb}[c]$	-	-	$SA_{BL}[c]$	-	$SA_{BL}[c]$

1) r: even row index (0, 2, 4, ...)   2) c: column index (0, 1, 2, ...)   3) *: don't care bit   4) 0 < VDDL < VDD

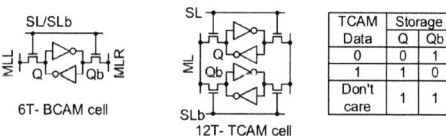

(a) Proposed 6T-BCAM and 12T-TCAM cells.

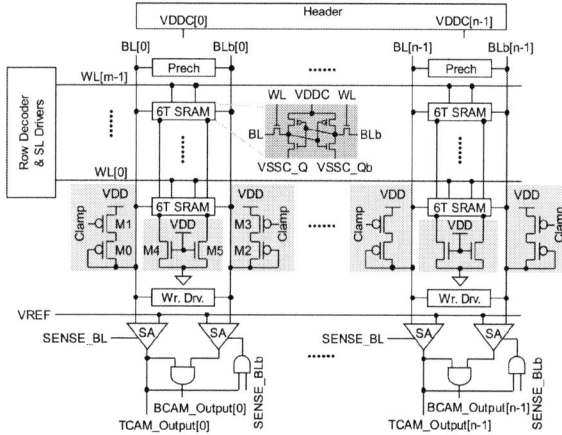

(b) Overall structure of the proposed BCAM/TCAM.

Fig. 2. Proposed CAM cells and structure.

A trade-off for this time-multiplexed approach is that it takes 2 cycles to finish a BCAM search operation (Table I).

The proposed 12T-TCAM cell is configured using two SRAM bit cells in consecutive rows while the 12T-TCAM cell in [8] uses two SRAM bit cells in consecutive columns (Fig. 2a and 1b). As a result, the number of SLs is reduced to the half of BCAM configuration in the proposed CAM. In contrast, the number of MLs is reduced to the half of BCAM configuration in [8]. The advantage of the proposed TCAM configuration against the configuration in [8] is the reduction of the number of cycles for a column write operation. While it takes 3 cycles to write 1s, 0s, and don't care bits in column-wise manner in [8], the proposed TCAM requires 2 cycles (Table I). In the first cycle of the column write operation for the proposed TCAM, input data bits for the target column are assigned to corresponding odd-numbered WLs and bit-wise inverted data are assigned to corresponding even-numbered WLs. For each don't care bit, both corresponding odd-numbered and even-numbered WLs are set to 1. At the same time, V_{BL} and V_{BLb} for the target column are set to 1 and 0 respectively by the corresponding write driver. In the second cycle, all WLs and BL/BLb values for the target column are inverted. During TCAM column write operation, V_{WL} should be set to VDDL (< VDD) and the supply voltages for all columns except the target column should be set to VDD to prevent data corruption. The supply voltage for the target column should be set to VDDL to overwrite the stored data with the input data.

For TCAM search operation, SL and SLb values are compared with Q and Qb values in parallel and the match results are evaluated using single-ended sensing of ML voltages (Fig. 2a and Table I). Thus, it takes only 1 cycle for the TCAM search operation similar to the design in [8] (Table I).

Meanwhile, in the SRAM-array style CAM structures like [8] and our design (Fig. 1b and 2b), cell data can be corrupted when multiple WLs are enabled at the same time because the lowered ML voltage can falsely write 0 to another cell which

is not evaluated in this time window. If some of the SLs are 1 (V_{SL}=VDDL) and the corresponding Q values are 0 for a MLL, the MLL is discharged and a mismatch is detected by SA for the MLL.

In the second cycle of the search operation (Fig. 2a and Table I), SLb values (either VDDL or 0 for each SLb) are applied to the WLs. In our circuit, when the mismatch is detected for MLL of a certain column in the first cycle, the SA for the MLR of the column is not evaluated in the second cycle for power saving. The two comparison results (MLLs and MLRs) are logically ANDed for BCAM output (Fig. 2b).

978-1-7281-5107-6/19 $31.00 © 2019 IEEE

IEEE Asian Solid-State Circuits Conference
November 4 – 6, 2019
The Parisian Macao, Macao SAR, China

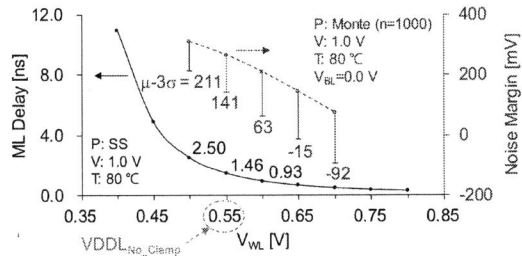

Fig. 3. Simulation results for 6T-BCAM to decide V_{WL}. (64 cells/ML, No ML clamping)

has stored value 1. As a solution, WL-underdrive scheme was used in [8]. We performed circuit simulations for 64 BCAM cells in a column to decide the optimal V_{WL} for the WL-underdrive scheme. We measured the read noise margin while reading a cell in a column when $V_{BL}= 0.0$ V. The worst ML discharge delay condition occurs when we read a single cell which stores 0. Simulation results show that the read noise margin is smaller than 100 mV when $V_{WL} > 0.6$ V while the ML discharge delay increases rapidly when $V_{WL} < 0.55$ V in our design case (Fig. 3). Note that probability of false write during search operation becomes smaller if the read noise margin of the 6T-BCAM cell when $V_{BL}= 0.0$ V becomes larger. Therefore, V_{WL} needs to be set near 0.55 V to balance the stability and the ML discharge delay.

To avoid data upset without lowering WL voltage too much, we propose a ML voltage/current clamping scheme. In the proposed BCAM/TCAM structure (Fig. 2b), the diode-connected PMOS transistors (M0 and M2) are activated to keep V_{BL} and V_{BLb} higher than half VDD during all CAM operations except for the column write operation (Table I).

While the voltage clamping prevents the false write-0, large ML discharge current can flow through multiple cells and incur large power consumption (Fig. 4a). We throttle the ML discharge current by adding footer transistors at the bottom of each column (M4 and M5 in Fig. 2b and 4b). As the

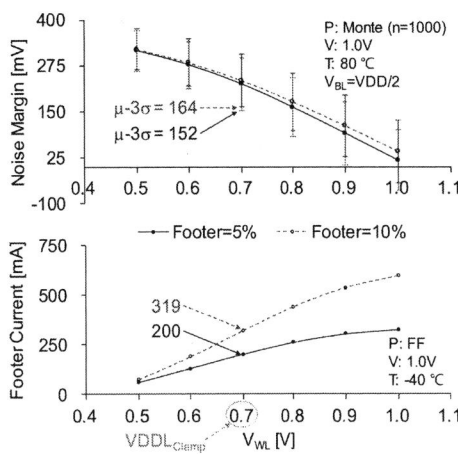

Fig. 5. Simulation results for 6T-BCAM to decide V_{WL} and footer size. (64/cells/ML, ML clamping)

footer size becomes smaller, the ML discharge current can be limited more effectively but cell read noise margin decreases due to increased voltage at the drain node of the footer. In other words, both the read noise margin and the ML current increase as footer size becomes larger.

We simulated read noise margin and footer current vs. V_{SL} for various footer sizes while reading all the cells in a column. In the results for cases of 5% and 10% of overall widths of pull-down transistors in a ML (Fig. 5), noise margin for the 10% footer case is 7.9% higher than that of 5% footer case at $V_{SL}= 0.7$ V but BL current becomes 59.5% higher at the same time. As a result, $V_{SL}= 0.7$ V with 5% footer was decided for design point because noise margin for the 5% footer case is higher than 150 mV at the SL voltage and the footer can save 59.5% of power consumption compared to the 10% footer case.

III. MEASUREMENT RESULTS

A 4Kb (64x64b) BCAM/TCAM in a 28nm CMOS technology (Fig. 6) was implemented to validate the proposed design. A programmable delay cell that adjusts WL-to-SENSE delay from 0.6 ns to 2.4 ns in 0.3 ns resolution was embedded to control unit to compare ML delay against the WL-underdrive scheme. SRAM cell was designed using logic layout rules.

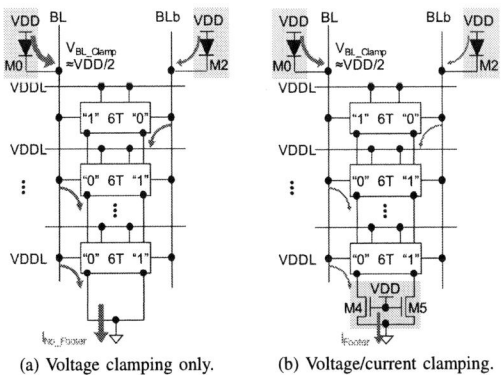

(a) Voltage clamping only. (b) Voltage/current clamping.

Fig. 4. Proposed ML voltage/current clamping scheme.

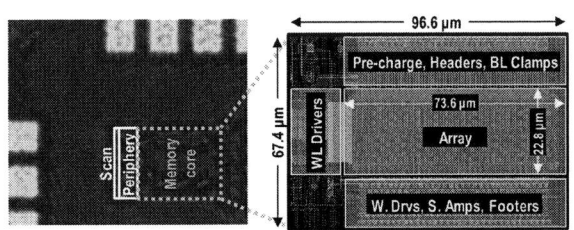

Fig. 6. Die photograph and layout.

978-1-7281-5107-6/19 $31.00 © 2019 IEEE   225

TABLE II
COMPARISON WITH PREVIOUS WORKS.

	[2]	[6]	[8]	[7]	This Work
Technology	32nm	65nm	28nm FDSOI	14nm FinFET	28nm Bulk Planar
Supply Voltage [V]	1.0	1.2	1.0	0.8	1.0
Cell Type	11T	10T	Configurable 6T-BCAM/12T-TCAM	10T	Configurable 6T-BCAM/12T-TCAM
Area/Cell [mm2]	n.a.	3.300	2) 0.152	1)1.471 (2)0.549)	1)0.362 (2)0.120)
Energy/Search/bit [fJ]	1.07	0.77	0.6	0.38 (10 MHz)	1)1.62 (10 MHz)
Search Delay [ns]	0.145	1.070	n.a.	0.560	3)2.708 (BCAM) 3)1.335 (TCAM)
Array Size	64×64b	128×128b	64×64b	128×64b	64×64b
ML Scheme	Wide AND	NOR	2-Single Ended SA	Half-and-Half Compare	2-Single Ended SA (BCAM) 1-Single Ended SA (TCAM)
Memory Modes	BCAM	BCAM	BCAM/TCAM/SRAM	BCAM	BCAM/TCAM/SRAM

1) Logic rule based layout.
2) SRAM rule based layout. In the same technology, [8] and this work will have exactly same cell area.
3) Precharge (Simulation @slow corner) + ML discharge delay (Avg. testchip measurement) + Sense enable (Simulation @slow corner).

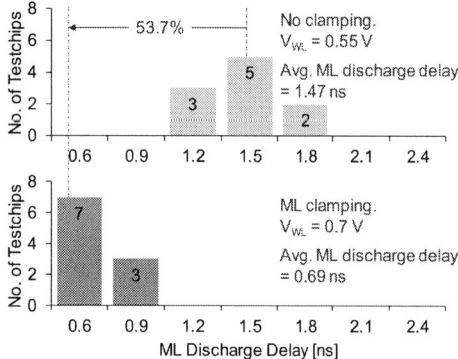

Fig. 7. Measurement results for ML discharge delay.

Single bit mismatch in MLLs (or MLRs) was used as a test pattern to measure the worst ML discharge delay and 10 chips were measured. Measurement results show that the delay in the proposed design was 53.7% lower than the delay in WL-underdrive case on average (0.69 ns vs. 1.47 ns) (Fig. 7). The testchip consumed 1.62 fJ/search/bit at 1.0 V in TCAM mode. The chip consumed 1.08 fJ/search/bit at 1.0 V when it was measured using WL-underdrive scheme without ML clamping. For comparison, the corresponding value for the design in [8] was 0.6 fJ/search/bit (Table II). The difference between 1.08 and 0.6 occurs because our design used the logic layout rule for cell layout and it increased BL loading. Among all the works in Table 2, our design is the only one to use unmodified SRAM cell and has the smallest cell area along with [8]. Compared to [8], our design has smaller ML discharge delay (Fig. 7) thanks to the increased WL voltage with the proposed ML clamping schemes.

IV. CONCLUSION

We presented a configurable BCAM/TCAM using a conventional 6T SRAM bit cell with no redesign effort. To the best of our knowledge, this is the first design to demonstrate CAM operation with conventional SRAM array structure. We also proposed a match line (ML) voltage/current clamping scheme that allows higher WL voltage while preventing data corruption during CAM search operation. Thanks to the proposed ML clamping scheme, the worst ML discharge delay was considerably reduced compared to the previous configurable BCAM/TCAM using 6T SRAM bit cell. Measurement results from 4Kb (64×64b) CAM testchips in a 28nm CMOS technology showed that the proposed ML clamping scheme achieves 53.7% reduction in ML discharge delay compared to the previous word line underdrive scheme.

ACKNOWLEDGMENT

This research was supported by Samsung Research Funding Center (SRFC-TC1603-04) and the MSIT (Ministry of Science and ICT), Korea, under the ICT Consilience Creative program (IITP-2018-2011-1-00783) supervised by the IITP.

REFERENCES

[1] K. Pagiamtzis and A. Sheikholeslami, "Content-addressable memory (CAM) circuits and architectures: a tutorial and survey," *IEEE Journal of Solid-State Circuits*, vol. 41, no. 3, pp. 712–727, March 2006.

[2] A. Agarwal, S. Hsu, S. Mathew, M. Anders, H. Kaul, F. Sheikh, and R. Krishnamurthy, "A 128×128b high-speed wide-AND match-line content addressable memory in 32nm CMOS," in *2011 Proceedings of the ESSCIRC (ESSCIRC)*, Sep. 2011, pp. 83–86.

[3] A. Hurson and S. Pakzad, "Modular scheme for designing special purpose associative memories and beyond," *VLSI Design*, vol. 2, no. 3, pp. 267–286, 1994.

[4] T.-B. Pei and C. Zukowski, "Putting routing tables in silicon," *IEEE Network*, vol. 6, no. 1, pp. 42–50, 1992.

[5] T. Ogura, M. Nakanishi, T. Baba, Y. Nakabayashi, and R. Kasai, "A 336-kbit content addressable memory for highly parallel image processing," in *Proceedings of Custom Integrated Circuits Conference*. IEEE, 1996, pp. 273–276.

[6] A. T. Do, C. Yin, K. S. Yeo, and T. T. Kim, "Design of a power-efficient CAM using automated background checking scheme for small match line swing," in *2013 Proceedings of the ESSCIRC (ESSCIRC)*, Sep. 2013, pp. 209–212.

[7] W. Choi, J. Park, H. Kim, C. Park, and T. Song, "Half-and-half compare content addressable memory with charge-sharing based selective match-line precharge scheme," in *2018 IEEE Symposium on VLSI Circuits*, June 2018, pp. 17–18.

[8] S. Jeloka, N. Akesh, D. Sylvester, and D. Blaauw, "A configurable TCAM/BCAM/SRAM using 28nm push-rule 6t bit cell," in *2015 Symposium on VLSI Circuits (VLSI Circuits)*, June 2015, pp. C272–C273.

978-1-7281-5107-6/19 $31.00 © 2019 IEEE

Sub-ns Access Sub-mW/GHz 32 Kb SRAM with 0.45 V Cross-Point-5T Cell and Built-in Y_ Line

C.Y. He[1], K. H. Tang[1], T. S. Chen[2], K. Y. Chang[1], C. H. Lin[1], K. Sato[1], S. J. Jou[1], P. H. Chen[1], H. M. Chen[1]
B. D. Rong[2], K. Itoh[1]
Institute of Electronics National Chiao Tung University[1], Etron Technology Inc.[2], Hsinchu, Taiwan

Abstract—A 0.45 V 28-nm 32-Kb SRAM with multi-power-supply low-power circuits, such as a cross-point 5T with built-in Y_line, gate-boosted drivers and adaptive tracking circuits, demonstrates a sub-ns access time and sub mW/GHz power dissipation. The 5T circuits are feasible to reduce the power of a 6T 32-Kb core to about 30% with quite the same sub-ns access time. The performance evaluation also indicates the new bit cell and array architecture open the door to the sub-ns access time and sub mW/GHz in sub-0.5 V multi-Mb era.

Keywords—Sub-0.5 V SRAM, 5T bit (memory) cell, gate-boosting driver, low-power array.

I. INTRODUCTION

Low-power SRAM cores focusing on sub-0.5 V low-power array and peripheral circuits are strongly required to low down the power consumption of an embedded SRAM module inside a SOC. There are many works proposed more than 6T bit cell and assistant circuitries to have the arrays work at below 0.5 V, for example [1-4]. These sub-0.5 V embedded SRAM arrays try to use different bit structure, asymmetric device/Vth, on-chip boosted read-bitline/wordline, etc. to enhance the reliability of read and write under low supply voltage. The achievements are able to work at near subthreshold operation with several KHz to MHz range with very low energy/operation. However, the penalty is an enlarged bit cell area. To reduce the area of the bit cell, [5] proposes a 5T SRAM bit cell with asymmetric sizing to improve read stability. However, comparable writability and readability can only down to 0.7 V and 0.5 V, respectively by using 45-nm bulk CMOS process. With the mature of high energy efficiency off-chip power management unit (PMU), multiple VDDs are available for SOC or SIP to optimize power consumption and timing for digital and analog modules. Therefore, recently, multi-VDDs are used to further improve the stability of the memory array operation and in the same time keep VDDs low to reduce the dynamic and leakage power. For examples, [6, 7] use 5T/8T to have VDD at 0.6/0.26 V to work at 100 MHz/60 KHz to have 103 fJ/31aJ per bit.

Due to the demand of increased number of sensors and computation required in IOT and AI edge device, Multi-Mb cores and GHz operation range are also needed for wider SRAM applications. For low-power array, a design concept of cross-point (CP) 5T with multi-VDDs that activates only one selected bit-line (BL) is proposed in [8] to study the feasibility for around 1 ns cycle time and least array leakage power

consumption. In this work, a pulsed CP-5T bit cell with built-in Y direction control line (YL), hierarchical memory array architecture and several multi-power-supply peripheral circuits, a sub-0.5 V sub-ns access time CP-5T SRAM is proposed. We design and implement a 28-nm 32-Kb SRAM and verify the performance through post-layout simulations and chip measurement. Based on the results, a sub-0.5 V 5T multi-Mb-core era is further explored and evaluated to confirm the low-power capability.

II. PULSED CROSS-POINT (CP) 5T CELL

Fig. 1 (a) depicts the proposed CP-5T bit cell. Here, major supplies VDD, VDH (VDD + δ) and VSL (VSS - δ) are set at 0.45 V, 0.75 V and -0.3 V, respectively, and are provided by off-chip PMU. The consideration to use both +δ and -δ for higher positive supply and negative power supply is to ease the design of PMU and also make positive and negative boosting to have symmetric capability. Word line voltage VWL (1 V) and BL pre-charged Vref (VDD/2) are generated internally. When a word line WL and a column line YL are selected, the voltages at HL and SL nodes of the cell is self-tracking to switch to the boosted VDH and -VSL by an adaptive read/write supply voltage controller (A_RW). A_RW tracks final stage amplifier trigger timing to switch the supply voltages (VDH/VSL) back to VDD/VSS to reduce the DC current of array without speed penalty. Due to only 5T with single bit line in a cell, we can embed a Y direction selection line, Y-Line in a cell. We use switches for different columns of memory array to connect to different voltage domain. To optimize read and write performance of proposed CP-5T cells, the voltage pulse of DHL and SLL is controlled by tracking circuit to ensure better performance and timing correctness. Fig. 1 (b) shows the time chart of read and write operation. In read operation, the boosting come at the very early beginning and the read signal on the BL is enhanced with 2δ. Moreover, a slow rising WL technique is used to improve the margin of the read operation. On the contrary, in write operation, the boosting occurs at the end of the write cycle to improve write ability. A fast write is achieved by a high-speed feedback during 2δ boosting. To minimize area overhead, we can use the same circuits used in the read operation for write operation to control A_RW action. Post-layout simulations verify that the read and write operation works for all 5 process corner (SS, FS, TT, SF and FF). For the half-select cells along the WL, however, their BLs are almost quiet with negligible read signals at a cycle time, due to non-YL activation (δ=0), which

978-1-7281-5107-6/19 $31.00 © 2019 IEEE

is an origin ultra-low-power array. The non-destructive read out (NDRO) is ensured for all the cells without a sense amplifier on each BL for active restore.

Fig. 1 (c) and (d) shows the post-simulation of read speed and the NDRO performance for different δ. Fig. 2 is the layout of a CP-5T bit cell. The cell size is 0.64 um by 0.27 um which is roughly the same as that of traditional single-port 6T bit cell (SP-6T) due to extra HL, SL power lines together with built-in YL in column direction. In summary, we use 4 power supply levels (VDD, VSS, VDH and VSL), one WL, one BL and one kind of Vth in a bit cell. For the other 5T [6] and 8T [7] bit cells using multi-VDDs, [6] uses different 5T bit cell structure and also 4 power voltage levels and one word and bit lines, [7] uses 5 power supply levels, 4 word/bit lines. Both of them use two kinds of Vths to optimize performance. Therefore, by using the same process technology, our proposed CP-5T bit cell shall have the least cell area.

Fig. 3 shows a read and write path comprising an open-BLs, local IOs (LIO) which are both precharged to Vref with equalizers (EQs), and global I/Os (GIO). An n-MOS current-source differential amp (n-LA) and a dummy column (DMY) enable fast and stable sensing through tracking the on-timing of a main amplifier (MA) on GIO pair. Here, the array power is for bit cells array, BL, LIO, GIO, MA, and Do buffer. When the SRAM array is idle, the cell is maintained with VDD (0.45 V) and 0 V so the overall cell array leakage power is only 1.02uW. If a single fixed power supply with 0.75 V is used, the cell array leakage power is 1.89 uW. By lower VDD in idle mode, we can get about 46% leakage power reduction which is very important factor for IOT and AI edge devices.

III. MULTI-POWER SUPPLY PERIPHERAL CIRCUITS

Fig. 4 compares an ideal gate-boosted (GB) driver (DRV) with a conventional single-VDD DRV (CV DRV). Here, each DRV consists of three tapered-inverters with a taper W-ratio of 3, and a low Vt (LVT = 50 mV) for the final-stage GB inverter while others are a regular Vt (RVT = 350 mV). Thus, the input capacitance Ci, area and leakage current are the same. The GB DRV achieves a high speed even at low VDD with increased gate-over drive of low-Vt MOSFET enabled by the input voltage swing larger than the output voltage swing. Fig. 5 compares the speed and power at δ = 0.3 V and the output capacitance Co of 20fF. Note that even at VDD = 0.1 - 0.2 V, the GB DRV works at a competitive speed with the help of the δ.

The total power (PT) increases with VDD, dominated by VDD-power at the final stage since VDH/VSL-powers are minor due to a small load capacitance at each stage. Note that the GB DRV achieves the same speed (trf ≈ 50 ps) even at 0.12 V VDD as that of CV DRV at 0.9 V VDD, reducing the CV-power to 1/12 (=19/1.6). Such a large power-reduction of GB justifies need for multi-power supplies. Fig. 6 shows various inverters (IVs), at the final stage in Fig. 4. GB-1 to GB-3 are GB-IVs, while NGB-1 to NGB-3 are non-GB-IVs. For a large Co, GB-1 is the best choice due to the lowest power, despite

need for a low-input receiver to detect the small voltage swing ΔVo at the load. If the receiver is not available, use higher-ΔVo (0.75 V) IVs as the second best choice, such as GB-2, GB-3, NGB-2 and NGB-3, with more power consumption.

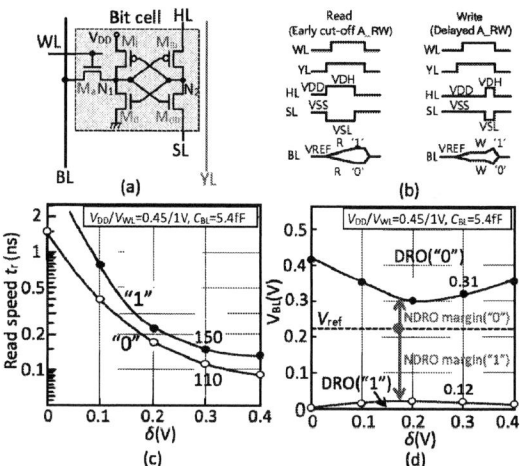

Fig. 1. (a) Pulsed CP-5T bit cell with VDH = VDD+δ = 0.75 V, VSL = VSS-δ = -0.3 V, (b) Read timing diagram and write timing diagram, (c) read speed and (d) NDRO of 1/0.

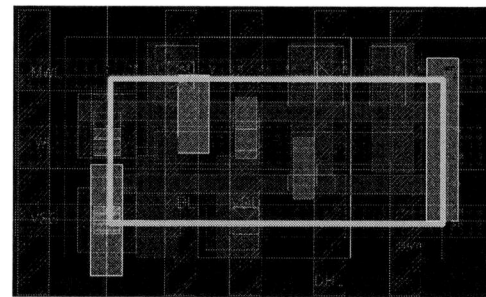

Fig. 2. CP-5T bit cell layout.

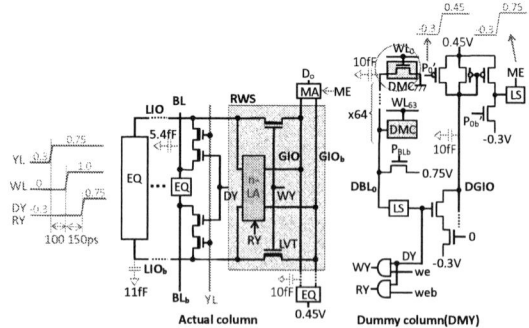

Fig. 3. Read and write path.

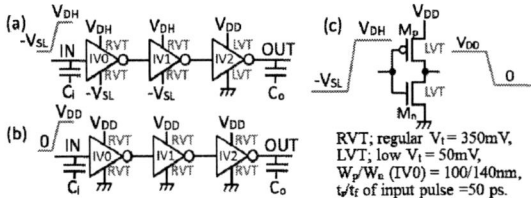

Fig. 4. (a) Gate-boosting driver (GB DRV), (b) conventional single-VDD DRV (CV DRV), and (c) concept of GB DRV.

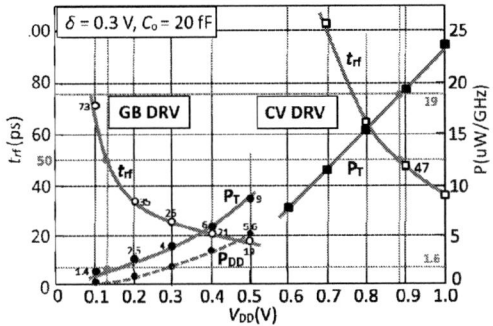

Fig. 5. Power and speed of GB DRV and CV DRV for desirable Vts. PT; total power, PDD; VDD-power (PDD), trf = (tr+tf)/2.

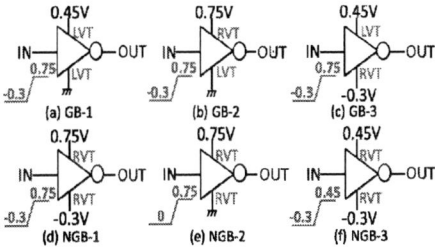

Fig. 6. Various GB and NGB inverters.

IV. 32-KB CORE CHIP ARCHITECTURE AND EVALUATION

Fig. 7 depicts a 32-Kb core composed of four 8-Kb sub-arrays, each consisting of eight 1-Kb arrays. The 1-Kb array consists of hierarchical 16 main WLs (MWL), 32 sub-WLs (SWLs), and 32 BLs. Decoded BL/LIO precharging is adopted for low power consumption. Fig. 8 shows stable post-layout simulation results of the 32-Kb core with rise/fall time of power suppliers. It can have 0.65 ns read access time and consume only 0.87 mW with 1 GHz clock at TT process corner. Moreover, the read and write operations are correct under TT, FF, SS, SF and FS process corners.

The performance of this work and 32-Kb core with single 0.9 V 6T bit cell with 1 GHz clock are shown in Fig. 8. CP-5T 32-Kb SRAM consumes roughly 30% of a 6T 32-Kb SRAM when both with 1 GHz clock. Also, using single 0.45 V (only

works at TT corner) which can only work at 55 ns cycle time is also listed for reference.

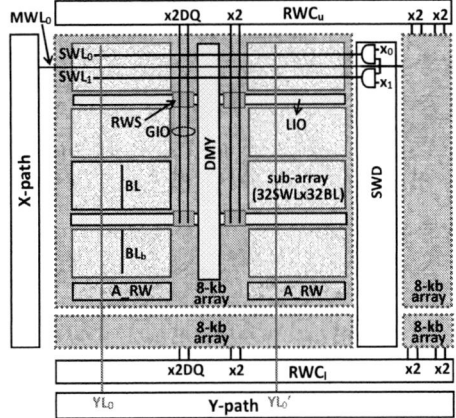

Fig. 7. 32-Kb core array

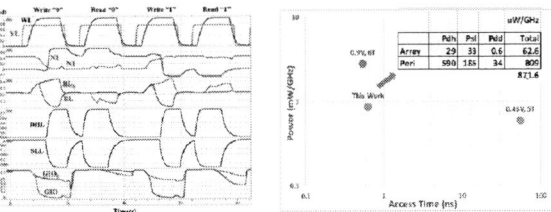

Fig. 8. Post-layout simulated performance of 32-Kb (x4D/DQ) core.

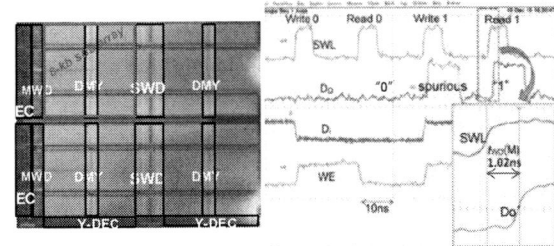

Fig. 9. 32-Kb core and measured waveforms.

Fig. 9 illustrates a chip microphotograph and measured memory operations of a key SWL-Do path. The measured delay of the path, tWD(M), is 1.02 ns. The calibrated delay, tWD(C), was 310 ps with some ambiguities, taking a slow (0.5 ns) rise time (cf. a 100 ps for designed internal pulse) of an externally-applied SWL from pulse generators, and pico-probe loading effects into account. Eventually, the estimated access time, tAC, is about 880 ps, since the calibrated Pe-SWL delay, tPW, is about 570 ps. Fig. 10 (a) depicts speed vs. VDD, experimentally verified, and the summary. Based on the measurement results, we calibrate the MOS behavior working at around subthreshold and capacitor loading of main critical path. Then, we make further enhancements of peripheral circuits, tracking circuits and sense amplifier. The new post-layout simulations show that tAC can be down to 650 ps.

978-1-7281-5107-6/19 $31.00 © 2019 IEEE

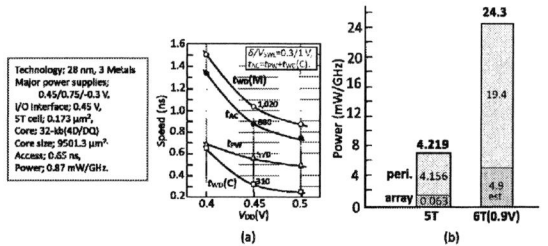

Fig. 10 (a) 32-Kb chip performance summary and (b) power of 1-Mb (x64D/DQ) core for 5T and 6T.

V. PERFORMANCE ENHANCEMENT AND MULTI_MB EVALUATION

In this section, based on the measurement results and post-layout simulation, the sub-0.5 V multi-Mb-core era is explored. Note that the low-power advantage of 5T stems from the small number of active BLs (N_active) due to CP-5Twith built-in Y_line which is independently of the array size, and smaller BL-capacitance CBL due to multi-divided BL. For 6T, N_active increases with the array size, due to activating all memory cell (MCs) on a WL and thus all BLs. Hence, the low-power advantage of 5T is more prominent with the array size.

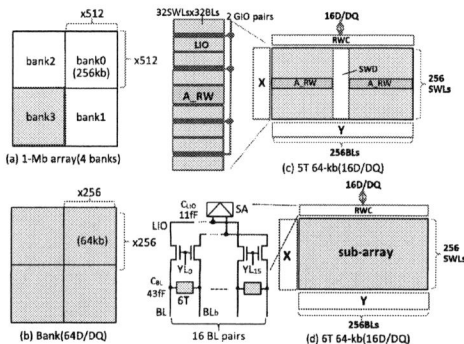

Fig. 11. (a) 1-Mb core, (b) 256-Kb array (a bank),(c) 5T 64-Kb sub-array and (d) 6T 64-Kb sub-array. n(5T)/n(6T) = 16/256. CBL(5T)/CBL(6T) = 5.4/43 fF.

For 28-nm 1-Mb core, Fig. 11 depicts a 5T 1-Mb core and a 6T counterpart. They consist of 4 banks with a 256-Kb (x64D/DQ) each. By decoding the banks, only one bank is activated with four 64-Kb sub-arrays operating in parallel. The total 1-Mb power (= array power + peripheral powers) is thus 4x64-Kb (x16D/DQ) power. For 5T, a GIO pair connects 256 MCs, unlike 128MCs for the 32-Kb core, and 256 MCs through SWD. Here, the 64-Kb(x16D/DQ) array power from the activated data path (BL, LIO, GIO, MA, and Do buffer) is almost equal to the 4x32-Kb (x4D/DQ) array power (Fig. 8), and thus 0.316 mW/GHz. This is due to the increased number of data paths from 4 to 16 and a negligible additional power by the 256 MCs connected to a GIO-pair. The peripheral power is estimated by adding 0.23 mW/GHz to the 32-Kb power (0.809 mW/GHz, Fig. 8) and is 1.039 mW/GHz. Here, the additional power is from estimated power of the 32-Kb

core and long address and control lines resulting from a larger sub-array from 32-Kb to 64-Kb. The 1-Mb total power is thus 4.22 mW/GHz from all power supplies (Fig. 10 (b)). For a 0.9 V 6T 1-Mb commercial reference design, the total power consumption is 24.3 mW/GHz. The array power is estimated to be 4.9 mW/GHz with n=256 and CBL=43 fF. Since the 6T total power is 24.3 mW/GHz, the 5T power is 17% of the 6T power.

VI. CONCLUSION

A 0.65 ns-access and 0.87 mW/GHz 28-nm 32-Kb SRAM with new pulsed CP-5T and multi-power supply circuits was demonstrated through post-layout simulations and measurements. Post-layout simulation results also verify that the read and write operation works for all 5 process corner (SS, FS, TT, SF and FF) with A_RW tracking circuits. Based on the result, the sub-0.5 V multi-Mb-core era was explored. The 5T circuits are feasible to reduce the power of a 28-nm 6T 1-Mb core to about 17%, opening the door to sub-ns access time multi-Mb sub-0.5 V era with reduced power consumption for IOT and AI edge applications.

ACKNOWLEDGMENT

This work is supported by Etron Technology, Ministry of Science and Technology and Taiwan Semiconductor Research Institute (TSRI), Taiwan. Authors also thank Untether AI, Canada and TSMC university shuttle program for their supports.

REFERENCES

[1] N. Verma and A. P. Chandrakasan, "A 256 kb 65 nm 8T subthreshold SRAM employing sense-amplifier redundancy," IEEE J. Solid-State Circuits, vol. 43, no. 1, pp. 141–149, Jan. 2008.

[2] M. Tu, J. Lin, M. Tsai, S. Jou and C. Chuang, "Single-Ended Subthreshold SRAM With Asymmetrical Write/Read-Assist," IEEE Trans. Circuits Syst. I, Reg. Papers, vol. 57, no. 12, pp. 3039-3047, Dec. 2010.

[3] M. Tu et al., "A Single-Ended Disturb-Free 9T Subthreshold SRAM With Cross-Point Data-Aware Write Word-Line Structure, Negative Bit-Line, and Adaptive Read Operation Timing Tracing," IEEE J. Solid-State Circuits, vol. 47, no. 6, pp. 1469-1482, June 2012.

[4] M. Chang et al., "A sub-0.3 V area-efficient L-shaped 7T SRAM with read bitline swing expansion schemes based on boosted read-bitline, asymmetric-VTH read-port, and offset cell VDD biasing techniques," IEEE J. Solid-State Circuits, vol. 48, no. 10, pp. 2558–2569, Oct. 2013.

[5] S. Nalam and B. H. Calhoun, "5T SRAM With Asymmetric Sizing for Improved Read Stability," IEEE J. Solid-State Circuits, vol. 46, no. 10, pp. 2431-2442, Oct. 2011.

[6] D. Jeon et al., "A 23-mW Face Recognition Processor with Mostly-Read 5T Memory in 40-nm CMOS," IEEE J. Solid-State Circuits, vol. 52, no. 6, pp. 1628-1642, June 2017.

[7] L. Lu, T. Yoo, L. Van Loi and T. T. Kim, "An Ultra-low Power 8T SRAM with Vertical Read Word Line and Data Aware Write Assist," IEEE Asian Solid-State Circuits Conference (A-SSCC), Tainan, pp. 1-2, Nov. 2018,.

[8] K. Itoh, K. A. Shaik and A. Amara, "0.5-V sub-ns open-BL SRAM array with mid-point-sensing multi-power 5T cell," IEEE International Symposium on Circuits and Systems (ISCAS), Lisbon, pp. 2892-2895, June 2015.

Privacy-Aware Data-Lifetime Control NAND Flash System for Right to be Forgotten with In-3D Vertical Cell Processing

Shun Suzuki, Kyoji Mizoguchi, Hikaru Watanabe, Toshiki Nakamura,
Yoshiaki Deguchi, Keita Mizushina and Ken Takeuchi
Department of Electrical, Electronic, and Communication Engineering, Chuo University, Tokyo, Japan
shun.suzuki@takeuchi-lab.org

Abstract— **This paper proposes Privacy-aware Data-Lifetime Control NAND Flash System (PDLCS), which changes the data-lifetime flexibly for the right to be forgotten. This system realizes both short and long term data-lifetime by In-3D Vertical Cell Processing. Furthermore, data-lifetime can be extended or shortened at any time. This system secures privacy data and thus is a universal solution for future secure storage, which is not resolved by software techniques.**

Index Terms— **3D NAND flash memories, Security, Privacy data, Lifetime control, Vertical charge de-trap, Lateral charge migration.**

I. INTRODUCTION

Recently, protection of privacy data such as human genome information becomes important. However, any encryption may be decrypted in the future advanced technologies, that is, "compromise" of privacy data. Thus, even if privacy data are protected with high encryption strength, there is a risk where encrypted data is deciphered in the future. To solve this problem, privacy data should be collapsed at the Si-chip level after the predetermined time that data is needed. This is the best way to comply with the right to be forgotten, which is stipulated as General Data Protection Regulation (GDPR) [1].

For the right to be forgotten, conventional works [2][3] achieve to delete data automatically on NAND flash storage as shown in Fig. 1. However, the conventional systems can only schedule to reduce data-lifetime because previous work just injects errors into write data. Thus, it's not suitable to store privacy data for long periods of time. The data-lifetime is sometimes defined by the law. Thus, the change of the law may extend or reduce the data-lifetime. For example, in the US hospitals, it becomes mandatory to reserve the personal medical data until 2 years after the patient's death. This means that it becomes necessary to extend the data-lifetime to 80-100 years. On the other hand, because of the right to be forgotten such as GDPR, the privacy data should be collapsed at the pre-determined time which is defined by the owner of the data.

This paper proposes Privacy-aware Data-Lifetime Control NAND Flash System (PDLCS) to realize not only data-lifetime reduction but also data-lifetime extension by on-chip hardware control. Proposed PDLCS is implemented in NAND flash controller and consists of four proposals with In-3D Vertical Cell Processing to achieve six data modes for data-lifetime controls. When privacy data are written, proposed Inverse Huffman-Coding V_{TH} Modulation (IHVM) sets short or long-term retention as mode 1 and 2, furthermore IHVM realizes long-term archive as mode 3. In addition, proposed Immediate Delete Command (IDC) collapses privacy data at the chip-level immediately as mode 4. Moreover, proposed Adaptive Lifetime Extension (ALE) and Adaptive Lifetime Reduction (ALR) realize to extend or reduce data-lifetime after writing privacy data as mode 5 and 6, respectively. Therefore, proposed PDLCS controls data-lifetime of privacy data flexibly by four proposals.

This work is supported by JST CREST Grant Number JPMJCR1532, Japan.

II. CHARACTERISTICS OF 3D-NAND FLASH MEMORIES

Fig. 2 shows the structures of 2D and 3D NAND flash. To achieve the flexible data-lifetime controls, proposed PDLCS utilizes reliability problems of charge-trap 3D NAND flash such as lateral charge migration and vertical charge de-trap. Data-Lifetime Controllers (DLCs), where neighboring word-lines (WLs) of the target WL storing privacy data, are used for this system as In-3D Vertical Cell Processing.

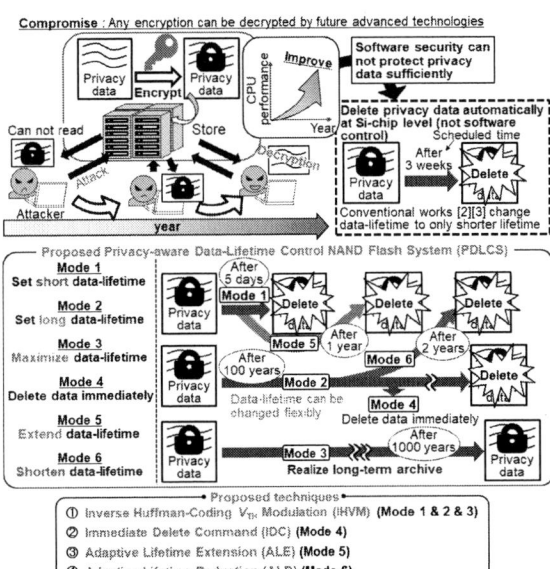

Fig. 1. Concept of this work. Proposed Privacy-aware Data-Lifetime Control NAND Flash System (PDLCS) realizes both short and long term data-lifetime. Furthermore, data-lifetime can be extended or shortened flexibly by In-3D Vertical Cell Processing.

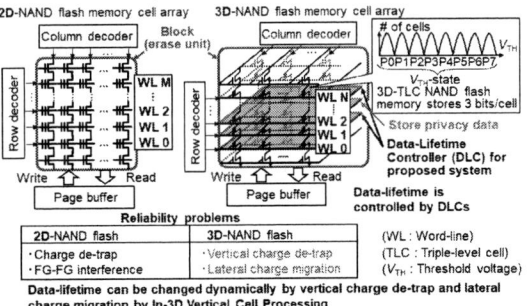

Fig. 2. Structure of NAND flash. Lateral charge migration and vertical charge de-trap control privacy data-lifetime dynamically by Data-Lifetime Controller (DLC).

Fig. 3 shows the measured error characteristics of charge-trap 3D Triple-Level Cell (TLC) NAND flash, which are key features to control data-lifetime flexibly at Si-chip level. The V_{TH}-distribution of 3D NAND flash decreases during the data-retention by the lateral charge migration and the vertical charge de-trap. Electrons trapped at oxide interface are more likely to be lost [4]. During the short-term data retention, electrons which are trapped at oxide interface are firstly lost by both lateral charge migration and vertical charge de-trap. Contrarily, in the long-term data retention, an increase/gradient of error rate becomes smaller because electrons trapped at oxide interface are reduced.

The lateral charge migration is caused by the electric field between WLs and not observed in the floating-gate 2D NAND flash. As shown in Fig. 4, when "P0"-states are adjacent to "P7"-state, V_{TH} of "P7"-state largely decreases and bit-error rate (BER) increases because lateral charge migration is enhanced. If adjacent WLs are "P0"-states, BER of "P7"-state is 26x higher than when those are "P0"-states. PDLCS utilizes the characteristics of both lateral charge migration and vertical charge de-trap to control the data-lifetime.

III. MODE 1 & 2 & 3: PROPOSED INVERSE HUFFMAN-CODING V_{TH} MODULATION (IHVM)

This paper proposes Inverse Huffman-Coding V_{TH} Modulation (IHVM) to schedule various data-lifetime by flexibly changing V_{TH}-distribution as shown in Fig. 5. When data-lifetime needs to be reduced, the V_{TH}-distribution is modulated to increase cells of the higher V_{TH}-states and BER increases by large charge migration. In contrast, when data-lifetime needs to be longer, lower V_{TH}-state cells are increased and BER is reduced by reduced charge loss.

IHVM can modulate V_{TH}-distribution flexibly and its process consists of four simple steps. First, estimate BER$_{Target}$ based on target data-lifetime as shown in Fig. 5. If ECC limit is 0.6 (a.u.) and privacy data should be collapsed after 1 day, BER$_{Target}$ is set to 0.6 (a.u.) at day 1 as privacy data are going to be collapsed in 1 day. BER$_{Target}$ is calculated for each data-retention time to optimize probabilities of each V_{TH}-state, p_{Pi}, which realizes target data-lifetime. Second, optimize p_{Pi} value, where overall BER of all states becomes BER$_{Target}$ at target data-lifetime (Fig. 6). If the privacy data are stored in WL(n), BER$_{Target}$ is calculated considering data of adjacent WL(n-1) and WL(n+1) (DLCs) because of the lateral charge migration. However, there are many combinations of p_{Pi} which achieve the same BER$_{Target}$. Therefore, optimized p_{Pi} pattern is calculated so that the entropy is made to be maximized as much as possible. Third, modulate data for optimized p_{Pi} and its results are shown in Fig. 7. For realizing optimized p_{Pi}, Huffman tree is created. Based on the Huffman tree, branches are assigned to input data, which is write request data before conversion, and leaf nodes are assigned to output data, which is after conversion, data are converted for approximately optimized V_{TH}-distribution. However, only 8 leaf nodes express the rough p_{Pi}, which means a few leaf nodes cannot realize detailed V_{TH} distribution.

To address the roughness of p_{Pi}, as described in Fig. 8, IHVM considers the combinations of 3 cells as output data because many leaf nodes express the detailed p_{Pi} thanks to deep depth of Huffman tree. Therefore, IHVM can realize expected p_{Pi}. Finally, measure actual BER (BER$_{Actual}$) by IHVM as shown in Fig. 9. Privacy data reach target BERs at target data-retention time because p_{Pi} is modulated as expected by IHVM. In SSD controller, only the relationship between input and output data is

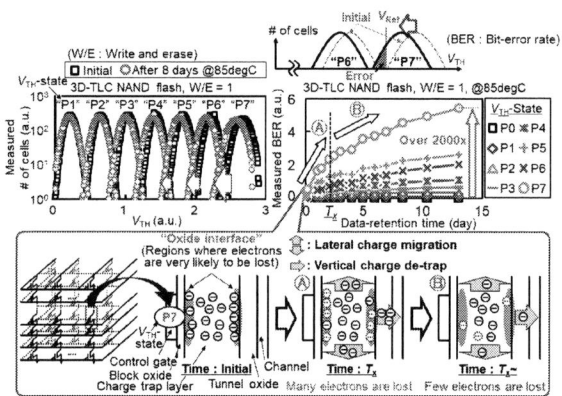

Fig. 3. Reliability problems of 3D-TLC NAND flash, lateral charge migration and vertical charge de-trap. By using adjacent WLs as DLCs, data-lifetime is controlled flexibly.

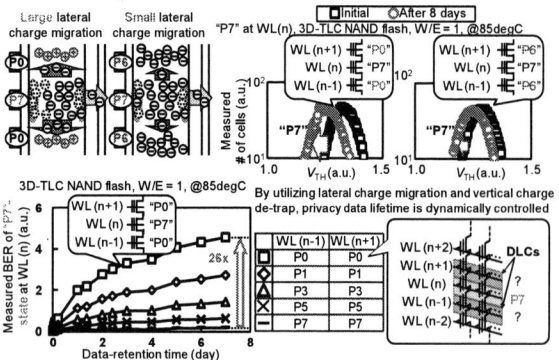

Fig. 4. Measured lateral charge migration. When "P0"-states are adjacent to "P7"-state, V_{TH} of "P7"-state largely decreases and bit-error rate (BER) increases because lateral charge migration is enhanced.

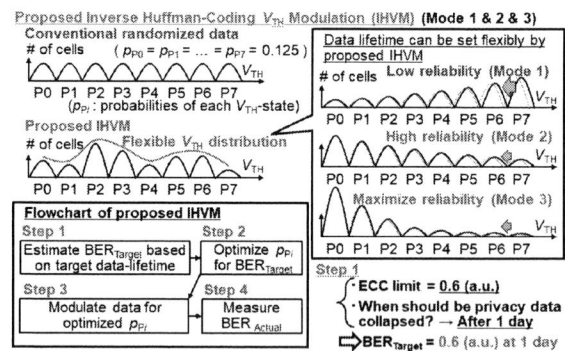

Fig. 5. Proposed Inverse Huffman-Coding V_{TH}-Modulation (IHVM), which modulates input data for flexible V_{TH}-distribution. Proposed IHVM consists of 4 steps. In step 1, IHVM estimates BER$_{Target}$ based on target data-lifetime (goal).

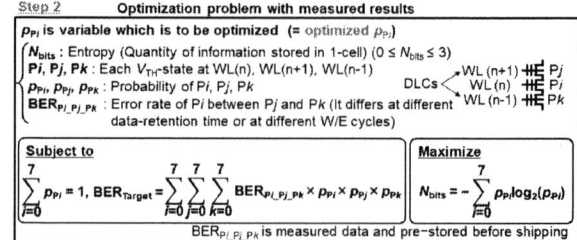

Fig. 6. In step 2, calculation of optimized p_{Pi} for BER$_{Target}$. Optimize V_{TH}-distribution, which can achieve the target BER and schedule data-lifetime, is calculated based on the procedure above.

needed for modulating data with IHVM. Therefore, encode and decode time of IHVM is negligibly short because of only referring to pre-stored tables.

IV. MODE 4: PROPOSED IMMEDIATE DELETE COMMAND (IDC)

Conventional Data Crush Technique [3] prevents privacy data from reading by outputting random data from the controller, therefore original data remain in NAND flash. Furthermore, data can be read such like text data and uncompressed images even when ECC fails. To solve this issue, this paper proposes Immediate Delete Command (IDC) as shown in Fig. 10. Proposed IDC overwrites the data as "P7"-state to prevent the illegal readout of the privacy data. Therefore, the privacy data are collapsed physically at the Si-chip level, and ensured from illegal readout by attacker.

V. MODE 5: PROPOSED ADAPTIVE LIFETIME EXTENSION (ALE)

This paper for the first time proposes data-lifetime control techniques while data are stored. When privacy data need to be

stored longer, proposed Adaptive Lifetime Extension (ALE) operates. ALE decreases BER and extends the data-lifetime as shown in Fig. 11. ALE writes data and injects electrons to DLCs, so that electrons are intentionally injected into target WL storing privacy data by the lateral charge migration and V_{TH} increases. Therefore, errors are recovered by proposed ALE. At the time T_1 in Fig. 11, if ALE writes "P1"-state or "P3"-state to WL(n+1), the data-lifetime is extended by 5.8 and 14-times, respectively.

VI. MODE 6: PROPOSED ADAPTIVE LIFETIME REDUCTION (ALR)

When privacy data need to be forgotten before scheduled time, proposed Adaptive Lifetime Reduction (ALR) operates. By ALR, an increase/gradient of error rate becomes harsher by rewriting the privacy data to another WL as shown in Fig. 12. In

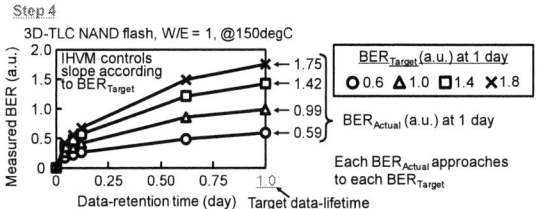

Fig. 9. In step 4, measure BER_{Actual}. IHVM flexibly controls data-lifetime so that privacy data archive various measured BER_{Actual}, depending on the target data-retention time.

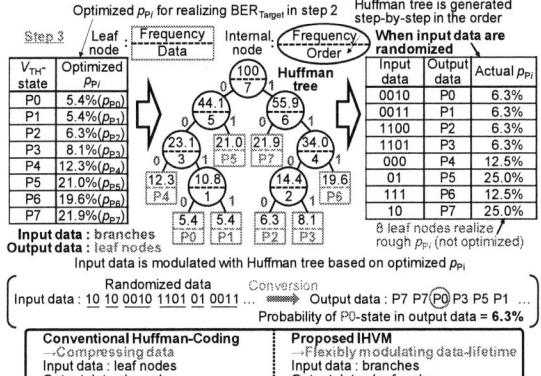

Fig. 7. In step 3, making Huffman tree by utilizing calculated optimized p_{Pi}. Each data is described as a leaf node and a Huffman tree is generated step-by-step from the bottom most infrequent node to the top most frequent node.

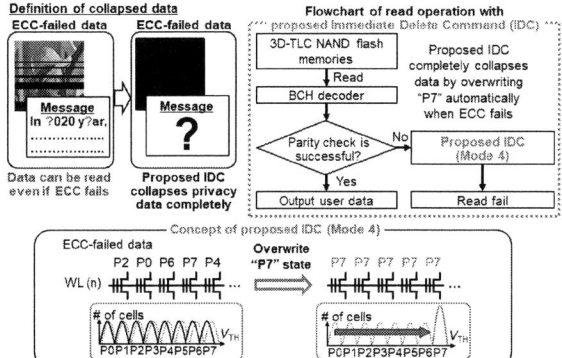

Fig. 10. Concept of proposed Immediate Delete Command (IDC). Proposed IDC overwrites the data as "P7"-state to collapse privacy data.

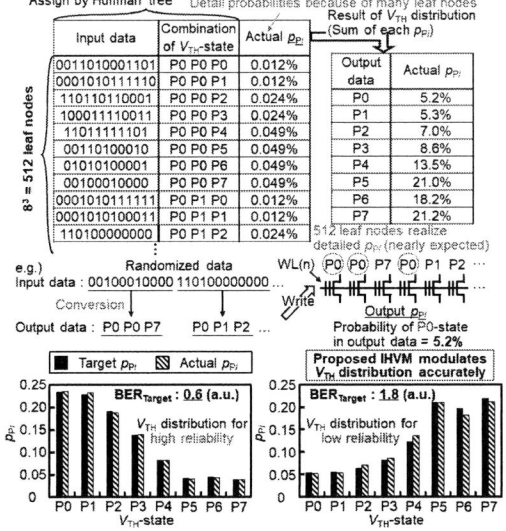

Fig. 8. Actual result of proposed IHVM in consideration of combinations of 3 cells. Proposed IHVM modulates V_{TH}-distribution (actual p_{Pi}) similar to target p_{Pi} because 512 leaf nodes realize more detailed p_{Pi}.

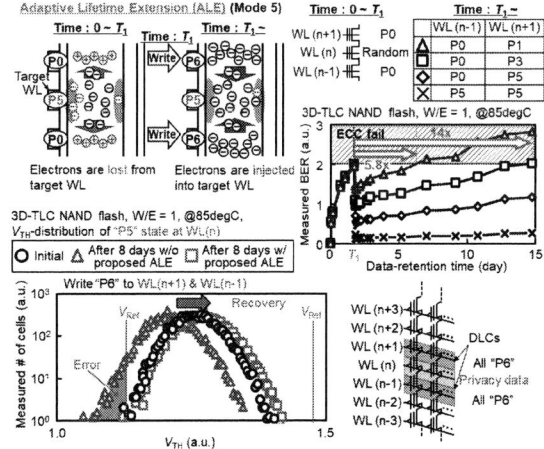

Fig. 11. Proposed Adaptive Lifetime Extension (ALE), which extends the data-lifetime by injecting electrons to memory cells. After applying ALE, V_{TH}-distribution is increased and memory cell errors are recovered.

IEEE Asian Solid-State Circuits Conference
November 4 – 6, 2019
The Parisian Macao, Macao SAR, China

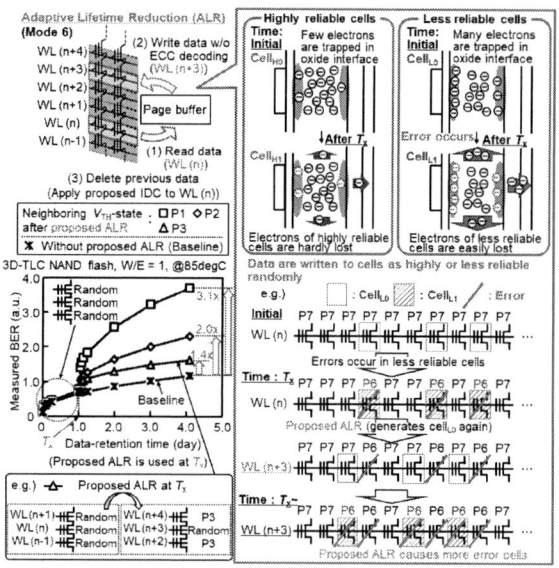

Fig. 12. Proposed Adaptive Lifetime Reduction (ALR), which accelerates data-lifetime by rewriting the data to other WL. By rewriting data, proposed ALR generates less reliable cells and data-lifetime becomes shorter.

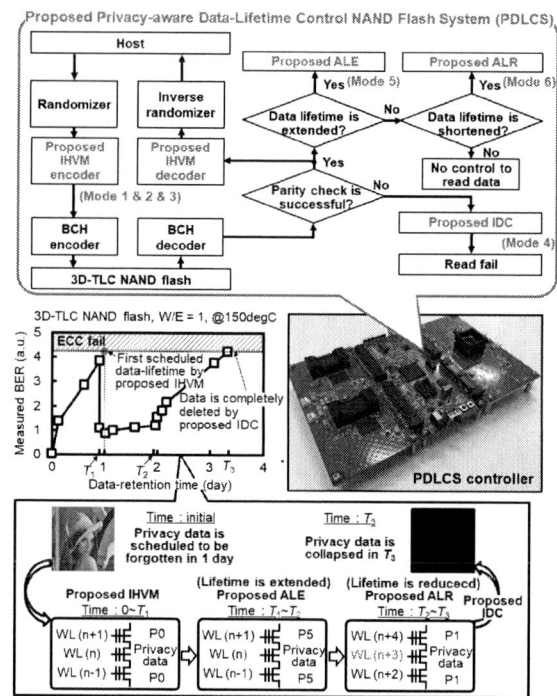

Fig. 13. Summary and photograph of the proposed SSD controller. Proposed PDLCS enables data-lifetime to be flexibly changed at any time.

TABLE I. SUMMARY OF THIS WORK

Functions	Conventional works [2][3]	This work
Mode 1 (Set short data-lifetime)	Yes (with only vertical charge de-trap)	Yes (with both vertical charge de-trap and lateral charge migration)
Mode 2 (Set long data-lifetime)	No	Yes (with both vertical charge de-trap and lateral charge migration)
Mode 3 (Maximize data-lifetime)	No	Yes
Mode 4 (Delete data immediately)	No	Yes
Mode 5 (Extend data-lifetime)	No	Yes (with lateral charge migration)
Mode 6 (Shorten data-lifetime)	No	Yes (with vertical charge de-trap)

case of highly reliable cells (cell$_{H0}$) where few electrons are trapped in oxide interface, electrons are hardly lost from charge-trap layer and error hardly occurs (cell$_{H1}$). On the other hand, in case of less reliable cell (cell$_{L0}$) where many electrons are trapped in oxide interface, electrons are likely to be lost from charge-trap layer and error occurs (cell$_{L1}$). When data are written to memory cells, cell$_{H0}$ and cell$_{L0}$ are generated randomly and become cell$_{H1}$ and cell$_{L1}$ as data-retention time increases. After the time T_x, an increase/gradient of error rate becomes smaller because both cell$_{H1}$ and cell$_{L1}$ trap few electrons in oxide interface. Then, ALR generates cell$_{L0}$ again and makes errors increase. The ALR process consists of three steps. First, read data where data-lifetime should be decreased from 3D NAND flash. Next, the read data are written to another WL, not DLCs, without BCH ECC decoding. By rewriting data without BCH ECC decoding, cell$_{L0}$ are generated again because cell$_{H1}$ and cell$_{L1}$ return to cell$_{H0}$ and cell$_{L0}$ by rewriting the data, and an increase/gradient of error rate becomes larger. Furthermore, an increase of error rate is also controlled by DLCs of rewritten WL. For example, as shown in Fig. 12, if DLCs of rewritten WL is written to "P1"-state, measured BER increases by 3.1 times. Contrarily, if "P3"-state is written to DLCs, measured BER increases by 1.4 times. In this way, ALR flexibly changes an increase/gradient of error rate by utilizing various V_{TH}-state written to DLCs. Finally, data-lifetime is shortened by ALR and privacy data need to be collapsed at the Si-chip level because of security problem. Thus, IDC operates and the previous data are erased/collapsed.

VII. CONCLUSION

Fig. 13 shows summaries of this work and a photograph [5] of proposed PDLCS controller. As shown in Table I, PDLCS consists of four techniques, IHVM, ALE, ALR and IDC, and it's possible to dynamically and flexibly control data-lifetime. This cannot be realized by conventional 2D-NAND flash because data-lifetime is controlled by the lateral charge migration in 3D-NAND flash. 3D-NAND flash secures privacy data and these proposals are universal solutions for the future security problems, which is not resolved by software techniques.

REFERENCES

[1] GDPR, REGULATION (EU) 2016/679 OF THE EUROPEAN PARLIAMENT AND OF THE COUNCIL of 27 April 2016, Article 17, pp 43-44, online available at: https://eur-lex.europa.eu/legal-content/EN/TXT/PDF/ ?uri=CELEX:32016R0679&from=EN

[2] S. Tanakamaru et al., "Privacy-Protection Solid-State Storage (PP-SSS) System: Automatic Lifetime Management of Internet-Data's Right to be Forgotten," VLSI Circuit, Jun. 2015, pp. C130 - C131.

[3] H. Yamazawa et al., "Privacy-Protection SSD with Precision ECC and Crush Techniques for 15.5× Improved Data-Lifetime Control," IMW, May 2016, pp. 137 - 140.

[4] Z. Lun et al., "Investigation of Retention Behavior for 3D Charge Trapping NAND Flash Memory by 2D Self-Consistent Simulation," SISPAD, Sep. 2014, pp. 141 - 144.

[5] T. Nakamura et al., "9.1x Error Acceptable Adaptive Artificial Neural Network Coupled LDPC ECC for Charge-trap and Floating-gate 3D-NAND Flash Memories," CICC, Apr. 2018, pp. 1 - 4.

978-1-7281-5107-6/19 $31.00 © 2019 IEEE

A 1.64mW Differential Super Source-Follower Buffer with 9.7GHz BW and 43dB PSRR for Time-Interleaved ADC Applications in 10nm

Yizhak Shifman, Yoel Krupnik, Udi Virobnik, Ahmad Khairi, Yosi Sanhedrai and Ariel Cohen

Mixed Signal IP Solutions Group, Intel Corporation

Jerusalem, Israel

yizhak.shifman@intel.com

Abstract— **A differential buffer for high bandwidth (BW), Time-Interleaved Analog to Digital Converters (TI ADCs) is described. The buffer, nested between Track and Hold (TH) circuits, drives the ADC input signal to the individual sub ADCs. A differential, super source-follower based architecture is utilized in the buffer. This architecture features high power supply rejection ratio (PSRR) together with high BW, low power consumption and low Inter-Symbol Interference (ISI). The buffer was fabricated in 10nm Intel process as a part of a 112Gb/s SerDes receiver. A unique method was developed to accurately measure the buffer's output waveform. Measurements show a BW of 9.7GHz, PSRR of 43dB and power consumption of 1.64mW, which is a 68% power reduction compared to the prior art.**

Keywords— Buffer, Source Follower, Time Interleaved ADC

I. INTRODUCTION

Many state-of-the-art wireline transceiver systems utilize an architecture based on high BW TI ADCs [1]-[5]. Such architectures deliver superior performance in terms of digital equalization capabilities and scalable design. In TI ADCs, multiple sub ADCs are instantiated in parallel, such that the input signal alternates between each sub ADC. A signal distribution tree of TH circuits is required to distribute the input signal to each of the interleaved sub ADCs in a timely manner. When the system's sampling frequency is high, a tree of depth n>1 is required, due to the limited system bandwidth and clock speed. Fig. 1 presents the distribution tree of [1], with n=2, for the PAM4 112Gb/s receiver described there. In this scheme, after the signal is sampled on one of the eight TH1, it is driven to one of eight alternating TH2, such that in total 64 Successive Approximation Register (SAR) sub ADCs are interleaved. Since the sampling element of the TH blocks is a capacitor, a nested analog buffer must be implemented between TH1 and TH2 to drive the signal to TH2 and charge its capacitor to the required voltage level during its limited track phase. This buffer is of a critical importance to the overall performance of the receiver, as any distortion it produces is translated to an error in the digital value converted by the ADC. Additionally, as the buffer is instantiated multiple times, its power consumption is an important contributor to the consumption the system. Typically, such nested buffer is based on the Source Follower (SF) architecture, as in [2][3]. SF amplifiers are frequently used as high-speed, low-power and close to unity gain voltage buffers, driving high speed signals into high capacitive loads [6]. The SF,

however, is limited in its ability to drive currents to its load. In the SF, the current through the g_m transistor depends on the load current, so its v_{gs} is not constant. This degrades the buffer's gain and linearity. Additionally, as discussed in [3], the input to output capacitance of the SF shifts the input voltage during the hold phase of TH1. This results in an impact from the current symbol to a future symbol, known as precursor ISI.

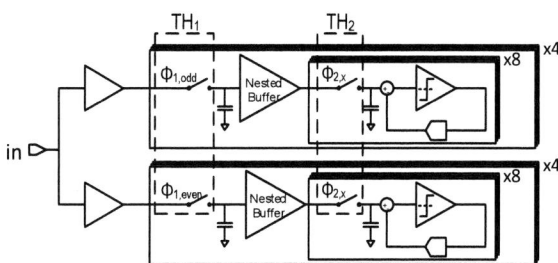

Fig. 1. Time-Interleave sampling structure of [1] with a nested buffer between TH1 and TH2

This paper presents the nested buffer implemented in the 112Gb/s ADC-based receiver published in [1]. The buffer utilizes a novel differential architecture, based on the class-AB Super Source Follower (SSF) [7]. While prior architectures have proven either too BW limited, power ineffective or noise / ISI sensitive, this architecture presents substantial improvements to all these aspects. Si measurements demonstrate large-signal BW of 9.7GHz, PSRR of 43dB and power consumption of 1.64mW, a 68% reduction compared to the prior art.

II. TI ADC NESTED BUFFER: CHALLENGES AND REQUIREMENTES

Apart from an apparent requirement to minimize its power consumption, a datapath buffer should minimize its impact on the signal it drives. Potential major contributors to such impact are supply noise, ISI due to BW limitation and ISI due to input to output capacitance. The impact of supply noise on the signal is defined in this context as $PSRR = \Delta Vcc/\Delta Vout$.

The BW of a nested buffer is required to be sufficiently high to enable charging TH2 sampling capacitor and prevent memory effects. As shown in Fig. 2, $\Phi_{1,x}$ is TH1 7Ghz clock with 25% duty cycle and $\Phi_{2,x}$ is TH2 0.875Ghz clock with 12.5% duty

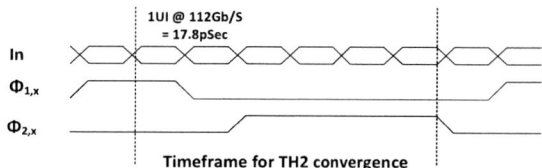

Fig. 2. timing diagram of TH1 and TH2. The timeframe for the buffer convergence ends at the sampling moment of TH2, at $\Phi_{2,x}$ falling edge.

cycle. The allowed timeframe for charging the capacitor of TH2 of [1] is approximately 6 Unit Intervals (UIs). If the buffer's BW is lower, a larger charge from the last conversion will add to the new symbol's charge. In this context, this effect is coined "symbol 64 ISI", as each TH2 samples a new symbol every 64 UIs.

Another type of ISI might be induced by the input to output capacitance of this buffer. This ISI, coined here as "symbol 8 ISI", is observed only on a buffer driven by a capacitor, as is the case for the discussed buffer during the hold phase of TH1. As shown in Fig. 3, if C_o hasn't completed its convergence during TH1 Track phase, i.e. when $\Phi_{1,x}$ is high, it will continue to charge also during TH1 hold phase, while the input is held on the capacitance C_i. The parasitic capacitance C_{PAR} between the input and the output terminals of the buffer couples these nodes together, such that buf_in node is pulled by buf_out node. Assume ΔV_{b1} is the voltage across the buffer at Track #1 falling edge (Fig. 3) and that its gain is approximately 1, then, by charge conservation, the voltage error on the buffer's output after convergence is given approximately by

$$\varepsilon_{symbol8} = \Delta V_{b1} \frac{C_{PAR}}{Ci} \tag{1}$$

This results in an ISI, as it depends on the signal: A large signal transition results in a large ΔV_{b1}, while if there is no transition, as in Track #2, $\Delta V_{b1} \approx 0$. In this context, this error is shown between a symbol and the symbol that appears 8 UIs later, since it is induced between two consecutive track phases of TH1.

III. CIRCUIT DESCRIPTION

A few modifications have previously been proposed to the conventional SF, all introducing an additional internal node, phase-inverted to the input node, which is used as a feedback reference voltage to additional devices. One of the most efficient of these architectures is the class-AB SSF, described in [7] (Fig. 4a), which sinks current through an added NMOS transistor. In this architecture, v_x is the phase-inverted internal node. M3 has its gate DC operation voltage set by v_{bp}, typically such that it mirrors a bias current from a bias circuit. This allows for an extended DC operation region of v_{out}, up to $V_{cc}-|V_{DSAT3}|$. In AC, however, this gate has the amplitude of v_x scaled by the capacitance ratio $\alpha = \frac{C_1}{C_1+C_g}$, where C_g is the parasitic capacitance of the gate node of M3. This allows g_{m3} to reduce the circuit's output resistance. Additionally, M4 allows for a class-AB operation by its current sinking capability and further reduces the circuit's output resistance, given by

$$R_{out,AB-SSF} \approx \frac{1}{g_{m1}r_{o1}(g_{m4} + \alpha g_{m3})} \tag{2}$$

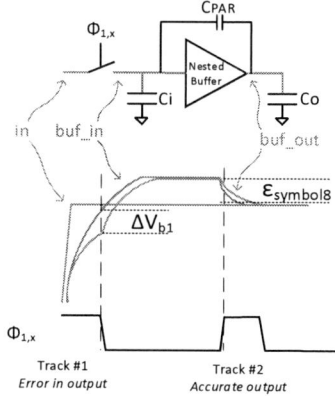

Fig. 3. Symbol 8 ISI on a traditional nested buffer, driven by TH. Large transition results in large Symbol 8 ISI, if the buffer's BW is not sufficiently high.

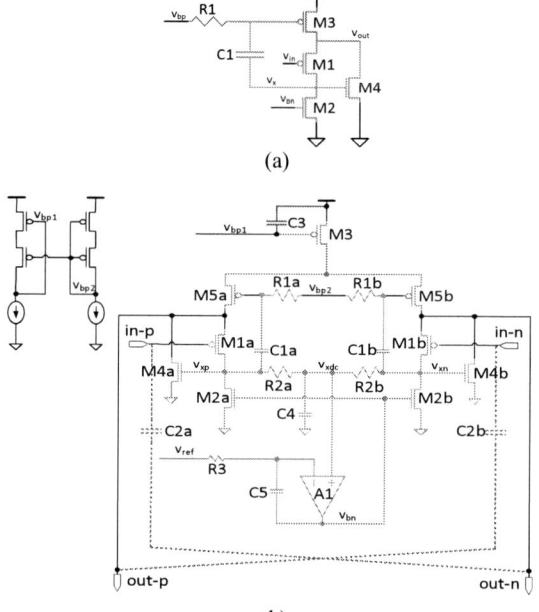

Fig. 4. (a) class-AB Super Source Follower (b) Differential class-AB Super Source Follower

with g_m the transconductance gain and r_o the small signal output resistance of the transistors. This is a substantially lower R_{out} relative to that of the SF, given approximately by $1/g_m$ [6]. The BW of the class-AB SSF, determined by the dominant pole at $f_p = 1/2\pi R_{out}C_{out}$, is thus higher relative to the original SF architecture, with only a small increment to the DC current dissipation. Alternatively, this improved R_{out} could be leveraged to a reduction of the power dissipation while maintaining the same BW as the SF.

Two main drawbacks, however, are remarkable in the class-AB SSF. First, its inferior PSRR. When analyzed for supply noise rejection, its M3 transistor could be viewed as a common-gate stage relative to the supply node, such that a positive gain might be observed from the supply node to the output node. This

978-1-7281-5107-6/19 $31.00 © 2019 IEEE   236

architecture, thus, is highly susceptible to supply noise. It should be mentioned that in the SF architecture, the gate and the source terminals of the current source transistor could be capacitively coupled, keeping its v_{gs} steady at high frequencies. Such capacitor is inadequate to the class-AB SSF as it will reduce α and consequently neutralize the desirable effect of M3 on R_{out}. The second drawback of the class-AB SSF is its susceptibility to mismatch between its two current sources, M2 and M3. v_x node in this architecture is a high impedance node, being driven by drain terminals only. As such, any mismatch between the currents sourced by M3 and that sinked by M2 may result in big variations in the DC voltage of this node, and either M3 or M2 may move out of saturation. In addition, the gate of M4 is driven by v_x, so variations of this node may either decrease its v_{gs} such that it will move to subthreshold / cutoff or increase it towards the linear region. The latter situation has a severe effect on the performance of the circuit, as it adds, effectively, a resistive load to the amplifier's output which significantly reduces the current driven to the actual load.

The circuit proposed in this work, a Differential class-AB SSF (DAB-SSF, Fig. 4b), maintains the BW/power advantages of the class-AB SSF and presents substantial improvements to that architecture which resolve the discussed drawbacks. Two separate amplifiers are integrated into one differential topology, such that three improvements are incorporated.

The first improvement is in the current source of the circuit. The P-channel current source, transistor M3 of Fig. 4a, is replaced with three transistors (colored red): M3, which fulfils its DC role, and M5a/b, which fulfill its AC role. M3 is the main current source, determining the overall current in the circuit. Transistors M5a and M5b are current steering transistors, determining whether this current flows to the positive branch or to the negative branch. For example, when in-n goes low, v_{xn} goes high, $|v_{gs}|$ of M5b drops, so less current flows to the load on out-n node, helping it to discharge. At the same time, out-p goes high, v_{xp} goes low, $|v_{gs}|$ of M5a increases and more current flows to the load on out-p, helping it to charge. The advantage of this separation is in PSRR. First, the circuit is now differential, and supply noise is mostly translated to a common mode noise on the differential signal out-p – out-n. Second, the gate of M3 is now capacitively coupled to the supply through C3. This results in a steady v_{gs} of M3 in high frequencies, such that even in the presence of supply noise, the current through M3 is steady. R_{out} of the circuit isn't impacted by this change, so the BW benefits of the class-AB SSF are still as effective. It could be observed that in small signal, the drain node of M3 is considered a virtual ground, and thus the small signal model of each of the branches of the DAB-SSF remains identical to this of the class-AB SSF.

A second improvement incorporated in the DAB-SSF is a DC regulation of v_{xn} and v_{xp} nodes (in orange in fig. 4b). v_{xn} and v_{xp} are filtered through R2a/b and C4, to generate their average DC voltage on v_{xdc} node. v_{bn} is driven by the amplifier A1, which together with R3 and C5 form an integrator. This integrator loop minimizes the difference between v_{xdc} and a voltage reference node, namely v_{ref}, generated by an external bias-generation circuit. This $v_{xn/p}$ regulation adapts the current through M2a/b to the current driven by M3, as well as keeps M4a/b in the saturation region.

The third improvement of the DAB-SSF is the rejection of Symbol 8 ISI. This improvement is intended to be taped out in the next Si revision of the circuit. Capacitors C2a and C2b couple between each input and its inverted output, and thus cancel the effect of C_{PAR}. At first order, Symbol 8 ISI is a linear effect, having the same impact on a rising and on a falling signal. Thus, cross capacitors (C2a/b) between out-p to in-n and out-n to in-p with the same size as C_{PAR} may completely cancel the Symbol 8 ISI, such that equation 1 is reduced to

$$\varepsilon_{symbol8} = \Delta V_{b1}\frac{C_{PAR}}{Ci} - \Delta V_{b1}\frac{C_2}{Ci} = 0$$

It should be noted that in the DAB-SSF architecture, C_{PAR} is approximately the parasitic capacitance between the gate and the source terminals of M1a/b transistors, and thus is a non-linear capacitor. Additionally, Symbol 8 ISI isn't fully linear at large signal. Thus, in an actual implementation, $\varepsilon_{symbol8}$ isn't fully cancelled. Simulations, however, show that with C2a/b implemented as metal-finger capacitors, even at the largest supported input signal, Symbol 8 ISI drops to less than 40% of its value without the capacitors. Also, note that TH1 has to drive the additional load added by these capacitors.

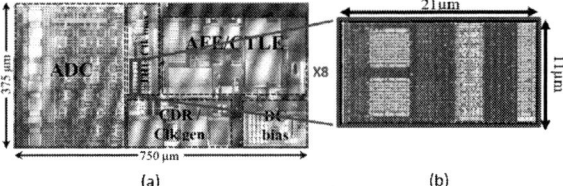

Fig. 5. DAB-SFF layout implementation in 10nm Intel process: die photo (a) and one buffer layout (b)

IV. RESULTS

This work was manufactured in Intel 10nm process together with the 112Gb/s receiver described in [1]. The total area of the circuit is 231 um^2 (Fig. 5) and its power consumption is 1.64mW. The lumped capacitance driven by the buffer is 200fF. C2a/b weren't implemented in this version and will be implemented in next Si version, without an increase to the overall area. The circuit was measured in an ambient temperature of 25°C and a supply voltage of 1.05V. The output waveform can't be directly measured in an oscilloscope, as the buffer is integrated inside the ADC's datapath and any attempt to probe would substantially downgrade its response. Thus, a special technique was developed to enable its measurements. At the falling edge of each TH2 Track signal, the voltage on its TH2 capacitor is captured and converted to a digital code by the sub-ADC. In a series of tests, the width of the Track pulse of $\Phi_{2,x}$ was swept across its range, from 35.7pS (2UI) to 92pS (5.2UI). For each test, an input pulse was injected to the ADC. After each conversion, the TH2 capacitors were discharged. Since the capacitor stops charging at the falling edge of $\Phi_{2,x}$, the capacitor voltage at that time represents a point on the output waveform of the buffer, when driving a non-switched capacitor. The sweep of the Track pulse width enabled the aggregation of multiple such points, such that the output waveform could be partially reconstructed. As discussed, the buffer has a dominant pole at its output node, so an exponential equation could be fitted to estimate the full output waveform. Fig. 6 presents the measured output waveform of the buffer, together with its exponential fitting. The blue dots represents the measured maximal code for each swept point. Based on the fitting, the large-signal BW of the buffer, F_c, is estimated as 9.7GHz. Fig. 6 also compares the reconstructed output waveform to the simulated one, in normalized voltage units. The similarity between the curves could be observed. Running the same fitting technique on the simulation waveform, a large-signal BW of 9.2GHz was calculated. It should be mentioned that the nature of the system doesn't allow actual voltage measurements, and thus DC gain wasn't measured. Simulations show a worst case DC gain of -1.5dB across all the valid operation range.

The measured PSRR is 43dB at DC. The measurement was performed with a large differential input DC signal, as a worst case scenario. This, since a low differential input would place both branches of the buffer at similar operation conditions, such that ΔV_{cc} would have a similar impact on both branches and a measured PSRR would be very high. In addition, a large ΔV_{cc} of 0.2V was imposed in order to accurately measure the high PSRR value. PSRR was measured only in DC conditions, since any high frequency supply noise is attenuated by the supply filtration network. Fig. 7 presents the measured DC PSRR together with a simulation of the PSRR across frequencies, under the same conditions. The similarity between the measured DC PSRR and the simulated PSRR at low frequency could be observed. PSRR reaches a peak of 56dB at a frequency range of 100MHz-1GHz. This range is where, empirically, most of the supply noise appears.

Table I presents a comparison to highest-BW previously published analog buffers and to a simple SF buffer. The simple SF buffer was optimized for the load and the BW requirements of the receiver, and was implemented on an early Si revision of the receiver. The DAB-SSF has the highest BW and PSRR, and the lowest area compared to the simple SF and to previously published buffers.

V. CONCLUSION

TI-ADC based receivers require high speed, low distortion and low power buffers in order to drive the received signal on time to each of the sub ADCs. In this work, a novel architecture for such buffer, the Differential class-AB SSF, was presented. The DAB-SSF is an evolution of the class-AB SSF [7], which exhibits a significantly reduced power consumption for a given BW relative to the SF, but suffers from low PSRR and vulnerability to mismatch. The DAB-SSF develops it to a differential architecture which delivers superior PSRR performance and mismatch immunity while maintaining its BW/power advantages. Additionally, the DAB-SSF rejects Symbol 8 ISI, a major issue in TI-ADC systems. The proposed DAB-SSF was implemented on Intel 10nm process and its measurements exhibit a superior large-signal BW of 9.7GHz, PSRR of 43dB and power consumption of 1.64mW, about 68% power consumption reduction compared to a SF amplifier with a smaller BW. The requirement to keep four stacked devices in saturation, however, is a disadvantage of the DAB-SSF. This property makes its PSRR advantage less prominent when the power supply is low or the required output common-mode voltage is high. In such cases, folding of the current source could be considered, similar to the concept of folded cascode, such that three stacked devices will have to be saturated instead of four.

TABLE I. COMPARISON WITH PRIOR ART ANALOG BUFFERS

	This work	Source Follower	[8]	[9]	[10]
Process (nm)	10	10	130	350	500
Supply (V)	1.05	1.05	1.3	1.5	3.3
Load	0.2pF	0.2pF	50Ω‖20 pF	10pF	30pF
Bandwidth (GHz)	**9.7**	8[b]	2	0.001	0.013
Power (mW)	**0.82[a,b]**	2.5[a,b]	7.34	0.15	0.2
PSRR at low frq. (dB)	**43**	42[b]	-	-	-
PSRR at 500MHz (dB)	**56[b]**	41[b]	-	-	-
Area (μm²)	**231**	-	6059	55K	17K

a. for one branch   b. simulated

REFERENCES

[1] Y. Krupnik *et al.*, "A 112 Gb/s PAM4 ADC-Based SERDES Receiver for Long-Reach Channels in 10nm FinFET Process " VLSI 2019.

[2] J. Hudner *et al.*, "A 112GB/S PAM4 wireline receiver using a 64-way time-interleaved SAR ADC in 16nm FinFET," VLSI 2018.

[3] K. Sun *et al.*, "A 56-GS/s 8-bit Time-Interleaved ADC With ENOB and BW Enhancement Techniques in 28nm CMOS," JSSC, March 2019.

[4] Y. Frans *et al.*, "A 56-Gb/s PAM4 Wireline Transceiver Using a 32-Way Time-Interleaved SAR ADC in 16-nm FinFET," JSSC, April 2017.

[5] D. Cui *et al.*, "3.2 A 320mW 32Gb/s 8b ADC-based PAM-4 analog front-end with programmable gain control and analog peaking in 28nm CMOS," ISSCC 2016.

[6] B. Razavi, "Design of analog CMOS integrated circuits." McGraw-Hill, 2001

[7] A. J. Lopez-Martin *et al.*, "Power-efficient analog design based on the class AB super source follower", Int. J. Circ. Theor. Appl. 2012

[8] K. Keikhosravy *et al.*, "A wideband unity-gain buffer in 0.13-μm CMOS," ICECS 2013

[9] J. M. Carrillo *et al.*, "Low-voltage wide-swing fully differential CMOS voltage buffer," ECCTD 2011

[10] A. J. Lopez-Martin *et al.*, "200 μW CMOS class AB unity-gain buffers with accurate quiescent current control," ESSCIRC 2010

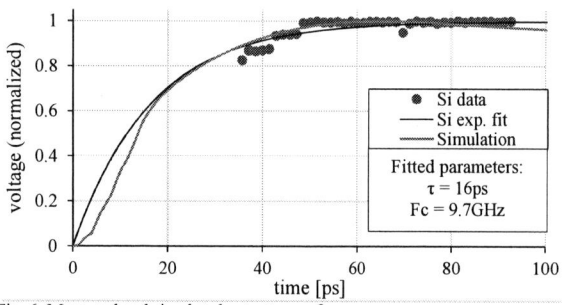

Fig. 6. Measured and simulated output waveform

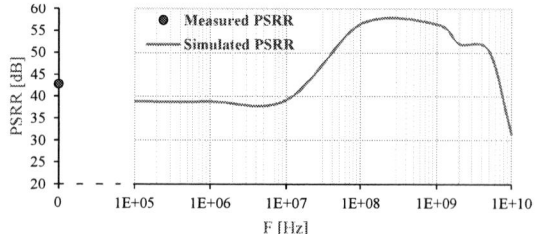

Fig. 7. Measured PSRR and simulated PSRR, under the same temperature and supply voltage

A 4.8pJ/b 56Gb/s ADC-Based PAM-4 Wireline Receiver Data-Path with Cyclic Prefix in 14nm FinFET

Gain Kim[1,2,3], Lukas Kull[2], Danny Luu[2], Matthias Braendli[2], Christian Menolfi[2], Pier-Andrea Francese[2], Hazar Yueksel[4], Cosimo Aprile[1,2], Thomas Morf[2], Marcel Kossel[2], Alessandro Cevrero[2], Ilter Ozkaya[1,2], Hyeon-Min Bae[3], Andreas Burg[1], Thomas Toifl[2], and Yusuf Leblebici[1]

[1]EPFL, Lausanne, Switzerland, [2]IBM Research Zurich Laboratory, Rueschlikon, Switzerland
[3]KAIST, Daejeon, South Korea, [4]IBM T.J. Watson Research Center, Yorktown Heights, NY

Abstract

This work presents an ADC-based receiver (RX) data-path for frame-based PAM-4 modulation with a cyclic prefix (CP). Similar to discrete multi-tone (DMT) modulation, a frame of PAM-4 symbols are protected from the channel delay spread by the CP taps. A PAM-4 frame window including CP taps is viewed as a DMT symbol and is equalized similarly to a DMT signal equalization, based on a discrete-time Fourier transform (DFT) and frequency-domain equalizer (FDE). The RX prototype implemented in 14nm FinFET achieves 56Gb/s data-rate at less than 3e-5 pre-FEC BER over a 19dB loss channel at 14GHz dissipating 270mW including the ADC and the DSP data-path excluding the inverse DFT and the BER checker.

Introduction

Rapidly increasing demands on wider bandwidth in wired networking infrastructure have been pushing the per-lane data-rate of wireline transceivers to 56Gb/s or further. ADC-based digital RXs have been demonstrating strong equalization capability using digital signal processing for mid-to-long-reach electrical link applications. Most of silicon-demonstrated >56Gb/s ADC-based links use PAM-4 symbol encoding [1,2] where their decision-feedback equalizer (DFE) necessitates tap speculation to meet the stringent timing constraints for closing the feedback loop in addition to a >10-tap feed-forward equalizer, resulting in area overhead and excessive power consumption. Discrete multi-tone (DMT) RX [3] does not require a feedback loop and features a simple equalization scheme with strong immunity to inter-symbol interference (ISI), but it suffers from a high output peak-to-average power ratio (PAPR). This work presents a digital RX forward-only-data-path using frequency-domain equalization for CP-inserted PAM-4 frames, mitigating the high PAPR issue that occurs in frequency-domain communication systems.

Frame of PAM-4 Sequence with Cyclic Prefix

The proposed scheme considers a sequence of time-domain PAM-4 with a CP. This is unlike in typical DMT systems where the time-domain waveform is obtained from the IDFT of frequency-domain symbols. The advantage of directly generating a time-domain waveform is that it avoids the large PAPR problem of DMT which increases dynamic-range requirements of TX DAC and RX ADC. While the waveform eye diagram of typical PAM-4 signal is illustrated in the bottom inset of Fig. 1, the upper-inset of Fig. 1 shows how the proposed PAM-4 frame is formed. The proposed scheme operates as a set of 32 PAM-4 symbol-basis. Having a set of 32 PAM-4 symbols, its four terminal symbols are copied and inserted in front of the beginning of the set as a CP. Hence, the constructed frame includes 32 data symbols and 4 CP taps, resulting in 1.78bits/symbol bit-efficiency. As long as the number of CP taps is enough to cover important ISI cursors, a frame of 32 PAM-4 data symbols undergo only a limited amount of residual ISI cursors, following the same principle as DMT (or OFDM) modulation. Since the ratio between the CP taps and the number of data symbols trades-off with the bandwidth efficiency, a limited number of CP taps should be chosen in a bandwidth-limited communication medium.

Fig. 2 visualizes signals on the TRX data-path for a set of 32 PAM-4 symbols. In time-domain (upper dotted box in blue color), the TX PAM-4 signal passes through the channel experiencing the frequency-dependent attenuation and the phase distortion. On the RX, the attenuated/distorted samples are captured by the ADC. The ADC should be locked to a proper sampling phase for a correct CP removal. On the TX side, a DMT symbol exists of which IDFT outputs the corresponding time-domain PAM-4 signal. Hence, the equalization of the PAM-4 on the RX side is equivalent to recovering the TX frequency-domain symbol from the received signal by a DFT and an FDE, then applying the IDFT for obtaining the time-domain PAM-4 samples. A purely real-numbered nature of the DFT input results in the complex-conjugate-symmetricity of the DFT output, hence 1+16 DFT output taps contain all the necessary frequency-domain information in case of 32-tap DFT. As can be seen in the plot at the bottom-right corner of Fig. 2, the RX frequency-domain symbols for $Tap_{0:16}$ are equalized to match with corresponding TX frequency-domain symbols. By transforming the equalized frequency-domain symbols to time-domain by a 32-tap IDFT, the PAM-4 samples are obtained. This architecture enables equalization without a feedback loop and/or timing-critical path, relaxing the digital RX design complexity and enabling pipelined data-path.

Receiver Architecture

Fig. 3 shows the overall block diagram of the implemented RX. A 10-bit ADC [4] receives a half-rate differential clock (C2) to internally generate 1/48-rate sub-ADC clocks (C48) for sub-ADCs data conversion. Differential input and external C2 clocks are on-die terminated with 100Ω parallel resistors, and the input bandwidth is extended by the T-coil. The ADC outputs raw 11b (including 1b of redundancy) time-interleaved data words which are synchronized to a single C48 clock domain before entering to the DSP. The DSP front-end consists of 11b-to-10b converter for the redundancy removal, an offset calibration circuitry, a FIFO for C48-to-C36 clock domain crossing, and a CP remover. The redundancy-removed 10b codes are then shortened to 7b by disregarding 3 lower-order bits for a compact DSP data-path. The cyclic-prefix-removed 7b RX PAM-4 symbols are then transformed into the frequency-domain by a 32-tap DFT processor that consists of two 16-tap Winograd DFT processors followed by a twiddling network for a compact area and low-power design. As the 32-tap DFT output has complex conjugate symmetric nature, only 1+16 output taps need to be equalized by one complex multiplier per tap. The per-tap-equalized frequency-domain constellations are captured by an on-chip equivalent-time digital oscilloscope and taken out of the chip through a bi-directional serial interface (BIDI). The IDFT is performed off-chip for the equalized PAM-4 reconstruction, with the IDFT input resolution of 12b and output resolution of 7b.

Measurement Results

The RX prototype is fabricated in 14nm FinFET process. To achieve a 56Gb/s effective data-rate taking the CP into account, 31.5GBaud/s PAM-4 symbols are transmitted by an 8-bit DAC-based PAM-4 TX [5] and received by the implemented RX chip. The reference C2 clocks of TX/RX are generated by external signal generators and synchronized by a 1GHz clock (external). On the PC, the TX PAM-4 symbols are constructed based on two independent PRBS13 patterns. Four CP taps are inserted for every 32 PAM-4 symbols and a 4-tap TX FIR filter is digitally-applied. The resulting pattern is loaded to the TX memory and repeatedly transmitted.

The RX performance is measured with a transmission line exhibiting 19dB loss at 14GHz (typical 56Gb/s PAM-4 Nyquist) and 23dB loss at 15.75GHz (Nyquist of the proposed scheme) where the channel characteristics is shown in Fig. 4(a). The insertion of 4 CP taps per 32 PAM-4 symbols brings a bandwidth overhead of 12.5%, resulting in a 4dB penalty in the tested environment. The TI-ADC samples the incoming data at Baud-rate consuming 181mW under 0.95V, 0.95V and 0.8V voltage supplies for the clock buffer, the interleaver, and the SAR array, respectively. The equalization coefficients are obtained off-chip and stored in the configuration memory of the RX chip. The DSP excluding IDFT runs at 875MHz dissipating 89mW under 0.85V supply. The energy efficiency of the implemented RX is 4.8pJ/b where 3.2pJ/b and 1.6pJ/b are consumed in the TI-ADC and the DSP, respectively. The non-equalized RX input distribution is shown in Fig. 4(b) while the histogram of the equalized RX PAM-4 samples is shown in Fig. 4(c), and the equalized 16K PAM-4 symbols in time-domain are shown in Fig. 4(d). The counted BER is less than 3e-5. The ADC area is about 0.12mm² and the DSP is 150um-wide and 400um-high (0.06mm²) which is 17% smaller than the DSP of [3]. The die photo and the layout are shown in Fig. 5, and Table I summarizes the RX performance and compares with other relevant works.

References

[1] P. Upadhyaya et al., "A Fully Adaptive 19-5o-56Gb/s PAM-4 Wireline Transceiver with a Configurable ADC in 16nm FinFET," *ISSCC*, Feb. 2018.

[2] L. Wang et al., "A 64Gb/s PAM-4 Transceiver Utilizing an Adaptive Threshold ADC in 16nm FinFET," *ISSCC*, Feb. 2018.

[3] G. Kim et al., "A 161mW 56Gb/s ADC-Based Discrete Multitone Wireline Receiver Data-Path in 14nm FinFET," *ISSCC*, Feb. 2019.

[4] L. Kull et al., "A 10-bit 20-40GS/s ADC with 37dB SNDR at 40GHz Input using First Order Sampling Bandwidth Calibration," *VLSIC*, Jun. 2018.

[5] C. Menolfi et al., "A 112Gb/s 2.6pJ/b 8-Tap FFE PAM-4 SST TX in 14nm CMOS," *ISSCC*, Feb. 2018.

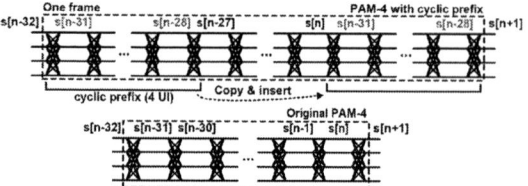

Fig. 1 Proposed PAM-4 frame formation with cyclic prefix (upper

inset) and the original PAM-4 stream (bottom inset).

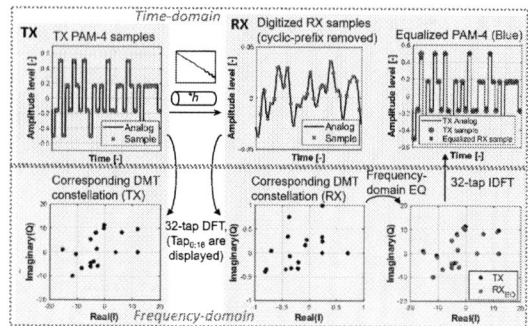

Fig. 2 Visualization of the signals in the TRX data-path with 4 CP taps and 32 PAM-4 data symbols per frame.

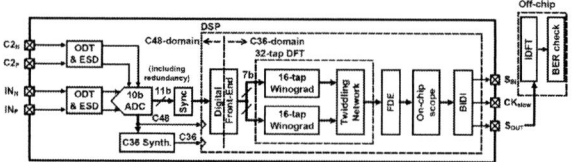

Fig. 3 Block diagram of the proposed RX.

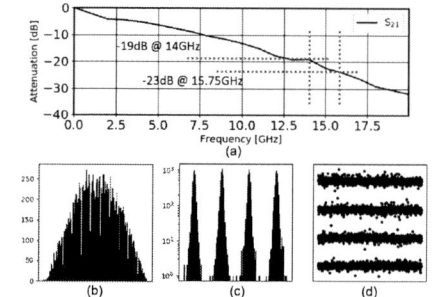

Fig. 4 Channel profile used for the measurement (a), non-equalized PAM-4 histogram at RX input (b), equalized PAM-4 histogram at proper ADC sampling point (c), equalized 16K PAM-4 symbols (d).

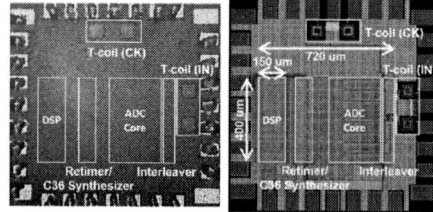

Fig. 5 Die photo and the layout of the implemented RX.

Table I. Comparison with relevant prior-arts.

	ISSCC18[1]	ISSCC18[2]	ISSCC19[3]	**This work**
Data Rate (Gb/s)	56	64.375	56	**56**
Modulation	PAM4	PAM4	DMT	**PAM4 w/ CP**
Analog EQ	CTLE/VGA	CTLE/VGA	–	**–**
ADC Res. (bits)	7	6	10/8*	**10/7****
Fs (GS/s)	28	32.1875	22.4	**31.5**
DSP Power (mW)	220	N.A.	68.22	**89**
RX Tot. Power (mW)	545	283.9***	161.46	**268.8**
Energy (pJ/b)	9.7	4.41***	2.88	**4.8**
Supplies (V)	0.85/0.9/1.2/1.8	0.9/1.2	0.8/0.7/0.7/0.75	**0.95/0.95/0.8/0.85**
DSP Area (mm2)	0.38****	–***	0.072	**0.06**
RX Area (mm2)	1.76****	0.1625***	0.3	**0.288**
Max. Ch. Loss (dB)	32 (@14GHz)	29.5 (@16GHz)	28 (@14GHz)	**19 (@14GHz)**
Max. Pre-FEC BER	<1e-12	<1e-4	<2e-4	**<3e-5**
Technology	16nm FinFET	16nm FinFET	14nm FinFET	**14nm FinFET**

*DSP receives 8 MSBs for equalization out of 10 bits

**DSP receives 7 MSBs for equalization out of 7 bits

***DSP is not implemented in silicon; hence DSP area and power are not included.

****Estimated from die photo reported in [1]

17-3 (7045)

A 32-Gb/s 0.46-pJ/bit PAM4 CDR Using a Quarter-Rate Linear Phase Detector and a Low-Power Multiphase Clock Generator

Zhao Zhang, Guang Zhu, Can Wang, Li Wang and C. Patrick Yue
HKUST-Qualcomm Lab, Department of Electronic and Computer Engineering
Hong Kong University of Science and Technology, Clear Water Bay, Hong Kong SAR, China

Abstract—This paper presents a low-power low-jitter PAM4 clock and data recovery circuit. A novel quarter-rate linear phase detector (QLPD) is proposed to lower the recovered clock jitter. A self-biased PLL based multiphase clock generator (MCG) is proposed to reduce power consumption. Fabricated in 40-nm CMOS process, the prototype achieves error-free operation at 32-Gb/s input data rate with 0.46-pJ/bit bit efficiency and 352.6-fs integrated jitter of the 4-GHz recovered clock. The measured jitter tolerance at BER of $<10^{-12}$ is higher than 0.35 UI$_{PP}$ with the corner frequency at about 10 MHz.

Keywords—PAM CDR, quarter-rate linear phase detector, multiphase clock generator, low-power, self-biased PLL

I. INTRODUCTION

Recently PAM4 signaling has become popular for serial links over 28 Gb/s [1-4]. Quarter-rate CDRs are attractive for PAM4 transceivers because their power consumption can be significantly reduced due to the lower recovered clock frequency [1-4]. However, such CDRs suffer from several design challenges. First, large dithering jitter stems from the nonlinearity of the widely used bang-bang phase detector (BBPD). This significantly degrades the jitter of the recovered clock. Second, the multiphase clock path, which consists of a LC-based voltage-controlled oscillator (VCO) with x2 frequency of the recovered clock and a multiphase clock generator (MCG) including a divide-by-2 divider (DIV2) and several phase interpolators (PI), is power-hungry. Lower power can be achieved by using ring VCO as the MCG [4] but comes with the penalty of degraded jitter of the recovered clock. A TDC is introduced to reduce dithering jitter in [3]; however, it consumes excessive power. This work aims to simultaneously reduce the recovered clock jitter and power consumption of a PAM4 CDR by utilizing a quarter linear phase detector (QLPD) and a self-biased PLL based MCG (SBPLL-MCG).

II. PAM4 CDR DESIGN

Figure 1 shows the proposed PAM4 CDR architecture. The four data paths are implemented to receive incoming data. Each data path consists of a sample/hold stage (S/H), an isolation buffer (ISOBUF), 3 slicers, a thermal-to-binary (T2B) converter and 3 offset calibrators for the slicers. To reduce the number of clock phases and thus save power consumption, only two edge paths are used to form the QLPD. Each edge path consists of a S/H, a voltage-to-current converter (V2I) and an edge selection logic circuit (ESL). Compared with the BBPD, the proposed QLPD is free of the dithering jitter, and consumes less power by removing the slicers in the BBPD. The V2I output voltage (V_C) is connected to the loop filter (LPF) and the LC-VCO to control the frequency and phase of the

This work was supported by the Key Areas Research & Development Program of Guangdong Science and Technology Department (HKUST contract no. GDST19EG05).

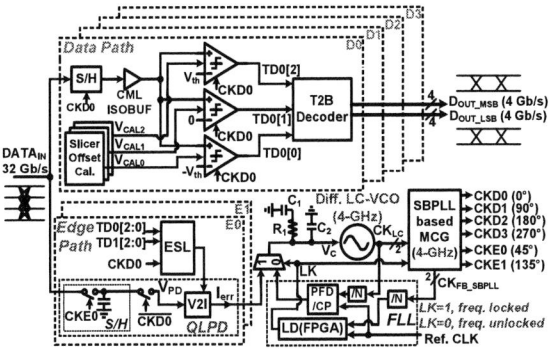

Fig. 1. Block diagram of the proposed PAM4 CDR.

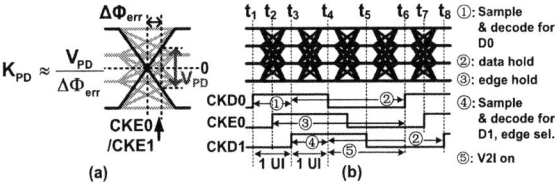

Fig. 2. (a) Operation principle and (b) timing diagram of the proposed QLPD.

recovered clock. The LC-VCO instead of ring VCO is adopted to generate low-jitter recovered clock. The SBPLL-MCG converts its 2-phase input clock from the LC-VCO (CK$_{LC}$) into a 6-phase clock for the data and edge paths. Since the input and output frequency of the SBPLL-MCG are the same, the LC-VCO only needs to run at 1/4 baud rate (4 GHz in this design), and the power-hungry DIV2 and PIs in prior MCG are omitted. So the power consumption of the LC-VCO and MCG can be reduced simultaneously. A frequency-locked loop (FLL) is adopted to achieve initial frequency acquisition. A counter-based lock detector (LD) implemented in FPGA is used to control the FLL operation mode by the output signal LK as shown in Fig. 1. The reference frequency for FLL is 125 MHz.

Figure 2(a) shows the operation principle of the QLPD. Since the PAM4 eye has 3 zero crossing points, we only select the major transition for phase detection purpose, as shown in Fig. 2(a). This can maximize the QLPD gain K$_{PD}$ and avoid jitter degradation. When the crossing point of the major transition and the edge sampling clock (CKE0 or CKE1) are misaligned, the S/H in the edge path holds a differential voltage V_{PD}, which is proportional to the phase difference $\Delta\Phi_{err}$ between the input data and the sampling clock. The V2I generates a current pulse I_{err} that is proportional to V_{PD} and $\Delta\Phi_{err}$. I_{err} charges or discharges the LPF to control the phase

978-1-7281-5107-6/19 $31.00 © 2019 IEEE

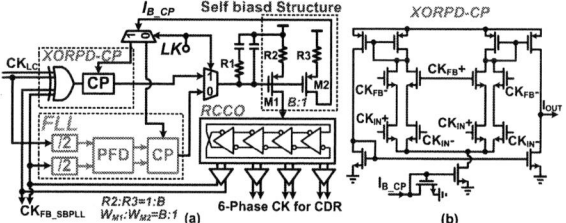

Fig. 3. Schematic of (a) proposed SBPLL-MCG and (b) XORPD-CP.

and frequency of the LC-VCO to make the edge sampling clock aligned with the crossing point of the major transition. Based on the slicer thermometer-code output of two neighbor data path (TD0[2:0] and TD1[2:0]), the major transition is selected by the ESL circuit.

The timing of the QLPD is illustrated in Fig. 2(b) by analyzing the edge path (E1) signals. One UI is allocated for the data path D0 and D1 to sample and decode the incoming PAM4 data. D0 and D1 hold the decoded output for three UIs. The edge sample V_{PD} is valid from t_2 to t_6. The edge selection is done also from t_3 to t_4, and the V2I is enabled from t_4 to t_6.

Figure 3 shows the schematic of the proposed SBPLL-MCG. The multiphase clock is generated by the current controlled ring oscillator (RCCO) in the SBPLL. In order to sufficiently suppress the RCCO phase noise and reduce the SBPLL-MCG jitter, a very wide loop bandwidth (LBW) of about 600 MHz is selected. This greatly relaxes the phase noise requirement of the RCCO and thus significantly reduces the SBPLL-MCG power. It is crucial to set the SBPLL LBW to be much higher than the overall CDR LBW to keep the CDR stable because the second-order SBPLL is inside the CDR loop, as shown in Fig. 1. Since the stability of a ring VCO based PLL with very wide LBW is sensitive to process, temperature and supply (PVT) variation, the self-biased architecture is adopted by setting the charge pump current proportional to the control current of the RCCO. This can keep the LBW almost constant and make the SBPLL stable over PVT variation [5].

The exclusive-OR (XOR) based PD (in Fig. 3(b)) with merged charge pump (XORPD-CP) is utilized to operate at high speed with low power consumption by removing all the high-speed logic gates in the traditional phase/frequency detector (PFD), as shown in Fig. 3(b). Due to the limited locking range of the XORPD based PLL, a FLL with a PFD and CP and two DIV2s is used for initial frequency locking. When FLL mode control signal LK (shown in Fig. 1) is high, the XORPD-CP is enabled and the FLL in the SBPLL is turned off so to save power since the SBPLL is in lock condition.

III. MEASUREMENT RESULTS

Figure 4(a) shows the 40-nm CMOS PAM4 CDR with a core active area of 0.281 mm². The power consumption excluding output buffers is 14.7 mW (also in Fig. 4(a)), of which only about 10% is consumed by the SBPLL-MCG. The measurement was performed under 1-V power supply with 32-Gb/s PAM4 PRBS-9 input signal (in Fig. 4(b)). The PAM4 CDR achieves error-free operation at 32-Gb/s data rate. The measured jitter tolerance (JTOL) is higher than 0.35 UI$_{PP}$ at

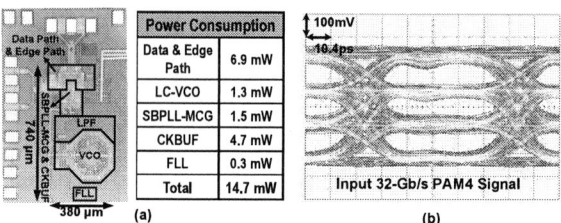

Fig. 4. (a) Die photo and power consumption breakdown, (b) input data eye.

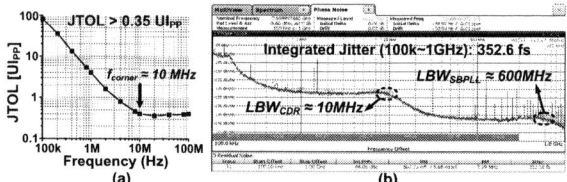

Fig. 5. Measured (a) JTOL and (b) phase noise of the recovered clock.

TABLE II. PERFORMANCE SUMMARY AND COMPARISON

	This work	ASSCC'17 [1]	JSSC'19 [2]	JSSC'18 [3]	TCAS-II' 19 [4]
Architecture	Quarter-Rate	Quarter-Rate	Quarter-Rate	Quarter-Rate	Quarter-Rate
Process	40-nm	65-nm	65-nm	65-nm	28-nm
Supply (V)	1	1.2	1.2	1.2	1.2
Data rate	32 Gb/s	51 Gb/s	56 Gb/s	28 Gb/s	32 Gb/s
Osc. Type	LC	LC	LC	Ring	Ring
PD Type	Linear PD	BBPD	BBPD	BBPD	BBPD
MCG type	SBPLL	Divider+PI	Divider+PI	Ring Osc.	Ring Osc.
Freq. of CK$_{OSC}$	4 GHz	12.55 GHz	14 GHz	3.5 GHz	4 GHz
Jitter$_{rms}$ of CK$_{REC}$	0.3526 ps	1.08 ps	NA	0.513 ps	3.4 ps
Jitter Tolerance	0.35UI$_{PP}$@20MHz @BER<10^{-12}	NA	0.12UI$_{PP}$@10MHz @BER<10^{-9}	0.2UI$_{PP}$@40MHz @BER<10^{-9}	0.05UI$_{PP}$@10MHz @BER<10^{-12}
P$_{DC}$ (mW)	14.7	321	49.2*	47*	32
Bit Effi.(pJ/bit)	0.46	6.29	0.88	1.68	1

*Only including the power consumption of the CDR part for fair comparison

BER of <10^{-12} with the corner frequency at about 10 MHz, as presented in Fig. 5(a). The integrated jitter of the 4-GHz recovered clock is 352.6 fs, as shown in Fig. 5(b). The measured SBPLL LBW is about 600 MHz, which is designed to be much wider than the CDR LBW (about 10 MHz), as highlighted in the measured phase noise profile. Thanks to the QLPD and the SBPLL-MCG, the proposed PAM4 CDR achieves the best bit efficiency and the lowest recovered clock jitter among the recently reported PAM4 CDRs, as summarized in Table I.

REFERENCES

[1] N. Qi, et al., "A 51Gb/s, 320mW, PAM4 CDR with baud-rate sampling for high-speed optical interconnects," IEEE Asian Solid- State Circuits Conference (A-SSCC), pp. 89-92, Nov. 2017

[2] A. R.-Zamir, et al, "A 56-Gb/s PAM4 Receiver With Low-Overhead Techniques for Threshold and Edge-Based DFE FIR- and IIR-Tap Adaptation in 65-nm CMOS," IEEE J. Solid-State Circuits, vol. 54, no. 3, pp. 672-684, Mar. 2019.

[3] Aurangozeb, et al, "Channel-Adaptive ADC and TDC for 28 Gb/s PAM-4 Digital Receiver," IEEE J. Solid-State Circuits, vol. 53, no. 3, pp. 772-788, Mar. 2018.

[4] D.-H. Kwon, et al, "A 32-Gb/s PAM-4 Quarter-Rate Clock and Data Recovery Circuit With an Input Slew-Rate Tolerant Selective Transition Detector," IEEE Trans. on Circuits and Syst. II: Express Breifs, vol. 66, no. 3, pp. 362-366, Mar. 2019.

[5] W. Y. Jung, et al, " "A 1.2mW 0.02mm² 2 GHz current-controlled PLL based on a self-biased voltage-to-current converter," in ISSCC Dig. Tech. Papers, Feb. 2007, pp. 310-311.

17-4 (7024)

IEEE Asian Solid-State Circuits Conference
November 4 – 6, 2019
The Parisian Macao, Macao SAR, China

A Maximum-Eye-Tracking CDR with Biased Data-Level and Eye Slope Detector for Optimal Timing Adaptation

Hye-Yoon Joo[1,2] and Deog-Kyoon Jeong[1]

[1] Department of Electrical and Computer Engineering, Seoul National University, Seoul, Korea
[2] Samsung Electronics, Hwaseong, Korea
henijoo@gmail.com, dkjeong@snu.ac.kr

Abstract—In this paper, a maximum-eye-tracking CDR (MET-CDR) for minimum bit error rate (BER) is presented. The proposed CDR does not require a BER counter or eye-opening monitor to find the optimal sampling phase. The biased data-level obtained from the weighted sum of UP and DN is proposed to extract the actual eye height information considering the pre-cursor ISI. Two error samples with small time spacing detect the current eye height and the slope of the eye height so that the CDR tracks the maximum eye height where the slope becomes zero. Measured results prove that the maximum eye height phase and the minimum BER phase matches well. A prototype receiver fabricated in 28-nm CMOS process operates at 26Gb/s with an eye-opening of 25% UI and consumes 87mW while equalizing 21dB of loss at 13GHz.

Keywords—*clock and data recovery (CDR), pre-cursor intersymbol interference (ISI), sampling point control, timing adaptation.*

I. INTRODUCTION

Clock and data recovery (CDR) circuits are essential in many high-speed serial link applications. Traditional CDR techniques such as bang-bang CDR (BB-CDR) [1] are widely used because of the simplicity. However, depending on the single bit response (SBR) of the channel, BB-CDR may not converge on the optimal phase. For optimal clock recovery, BER-based [2], [3] and eye-opening monitor (EOM) based approaches [4], [5] are proposed. However, a BER estimation and iterative eye-opening check by sweeping of sampling phase require a large amount of time, silicon area, and off-chip assistance. To further simplify the optimal clock recovery, we propose a maximum-eye-tracking CDR (MET-CDR). With similar hardware as the BB-CDR, it automatically searches for the optimal sampling phase where the eye height is maximized.

II. PROPOSED MET-CDR

A. Eye height vs. sampling phase

Fig. 1 (a) shows an SBR and the locking phase of the BB-CDR where $h_{+0.5}=h_{-0.5}$. Fig. 1 (b) shows the optimal sampling time determined by the maximum eye height. The eye height of the sampler's input can be obtained by subtracting the sum of the pre-cursor ISI from the main-cursor. For simplicity, we assume that the decision feedback equalizer (DFE) is ideal and

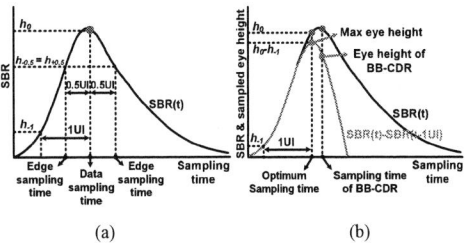

Fig. 1. (a) SBR with its data and edge sampling time with BB-CDR. (b) Optimum sampling time determined by eye height curve.

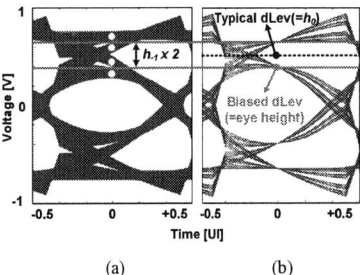

Fig. 2. Simulated eye diagram (a) during and (b) after the DFE adaptation with typical dLev and the proposed biased dLev.

the number of taps is large enough to cover all the post-cursors. The maximum eye height appears earlier than the locking phase of the BB-CDR because the pre-cursor ISI is reduced faster than the main cursor as the sampling position is pulled forward.

B. Eye height information from biased data-level

The typical data-level (dLev) obtained by adding the UP and DN signals from the error sampler at the same rate means h_0. To extract the actual eye height information, we define a 'biased dLev' using a weighted sum of UP and DN. Fig. 2 shows the simulated eye diagrams during and after the DFE adaptation with one pre-cursor. The two red lines indicate the levels generated by the pre-cursor ISI. Assuming that the data pattern is random, the residual ISI errors represented by the four yellow dots in Fig. 2 (a) contain the same number of hits. Therefore, the value of the lower level or biased dLev can be

978-1-7281-5107-6/19 $31.00 © 2019 IEEE   243

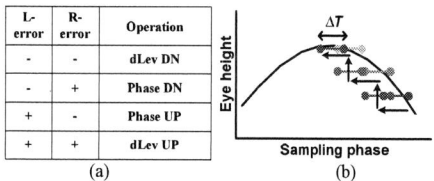

(a)　　　　　　　　(b)

Fig. 3. (a) Adaptation algorithm. (b) Procedure of convergence.

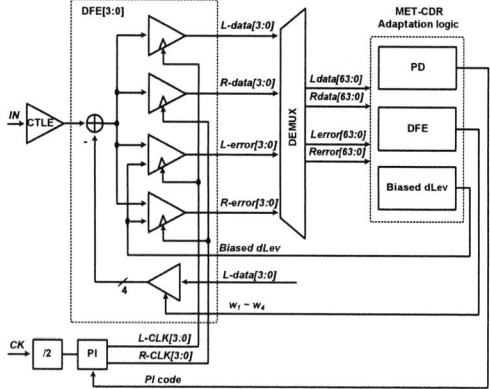

Fig. 4. Architecture of the proposed receiver with MET-CDR.

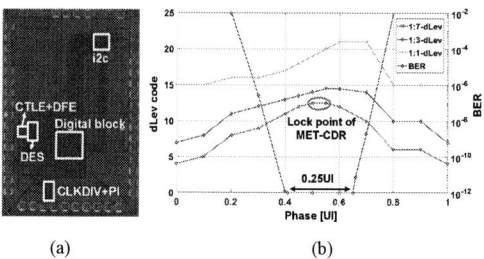

(a)　　　　　　　　(b)

Fig. 5. (a) Die photograph. (b) Measured three dLevs with different weighting factors (1:1, 1:3, 1:7) and the bathtub curve with 1:7.

consists of CTLE, four-tap quadrature DFE with four samplers, phase interpolator (PI) and digital logic including the proposed adaptation algorithm. It adds only one sampler compared to the BB-CDR per one clock phase.

III. MEASUREMENT RESULTS

The proposed MET-CDR was designed and fabricated in a 28-nm CMOS process as shown in Fig. 5 (a). The silicon area of the CDR core is $0.089mm^2$. The overall measurements are conducted at 26Gb/s with a PRBS7 pattern. Fig. 5 (b) shows three dLevs with different weighting factors (1:1, 1:3, 1:7), the lock point of the MET-CDR with 1:7, and the bathtub curve with 1:7. The peak of 1:7-dLev occurs before the peak of 1:3-dLev and 1:1-dLev because of the two pre-cursors. We can find that the peak of 1:7-dLev, or the lock point of the MET-CDR, matches well with the center of the region where the BER is less than 10^{-12}. It proves that the proposed MET-CDR effectively finds the optimal sampling phase to minimize BER. It is noteworthy that the lock point of the BB-CDR, which generally appears later than the peak of 1:1-dLev, deviates from the center of the bathtub curve. The measured power consumption of the CDR core is 87mW. Table I shows comparisons between various types of CDR architectures. The proposed MET-CDR locks at the optimal sampling phase with very simple hardware and short processing time.

obtained by adding UP and DN at a ratio of 1:3 in the sign-sign LMS (SS-LMS) algorithm. The biased dLev is equal to the eye height as shown in Fig. 2 (b). In a similar approach, for 2 pre-cursors, the weighting factor that achieves the eye height is 1:7.

C. Adaptation algorithm

Two error samples with ΔT spacing are used for phase detector (PD) and the DFE adaptation. Fig. 3 (a) and (b) show the adaptation algorithm and an example of the procedure of convergence. The proposed adaptation logic simultaneously detects the current eye height and the slope of the eye height, so that it converges on the maximum eye height phase at which the eye height slope becomes zero. A tradeoff exists when we determine ΔT. The smaller ΔT, the more accurate the result of DFE and peak dLev, but the smaller PD gain for noisy eye diagram, and vice versa.

D. Architecture

Fig. 4 shows the architecture of the proposed MET-CDR. It

REFERENCES

[1] J. Alexander, Electron. Lett., vol. 11, no. 22, pp. 541–542, Dec. 1975.

[2] E. H. Chen et al., JSSC, vol. 43, no. 9, pp. 2144–2156, Sep. 2008.

[3] S. Son et al., JSSC, vol. 48, no. 11, pp. 2693–2704, Nov. 2013.

[4] H. Noguchi et al., JSSC, vol. 43, no. 12, pp. 2929–2938, Dec. 2008.

[5] H. W. Won et al., TCAS-I, vol. 64, no.3, pp. 664–674, Mar. 2017.

TABLE I
COMPARISON OF OPTIMAL TIMING ADAPTATION CDRs

	[2]	[3]	[5]	This work
Method	BER	BER	EOM	**Max-eye tracking**
Added hardwares for timing adaptation compared to BB-CDR	Adjustable sampler, XOR gate, BER counter, Logic for iteration	Adjustable sampler, XOR gate, BER counter, Logic for iteration	Adjustable sampler, Probability counter, Logic for iteration	**One sampler**
Processing time	'700x slower than SS-LMS' * @ BER 10^{-4}	1h 25min ** @ BER 10^{-5}	364ms *	**<10us ** @ BER 10^{-12}**
One-time adaptation	No	No	No	**Yes**
On/off-chip clock recovery	Off-chip	Off-chip	N/A	**On-chip**
Process	90nm	65nm	40nm	**28nm**
Data-rate	6.25Gb/s	5Gb/s	28Gb/s	**26Gb/s**
Channel loss	14.28dB	15dB	25dB	**21dB**

* For equalizer adaptation time only. ** For both equalizer and CDR adaptation.

A 200-MHz Wide Input Range CMOS Passive Rectifier with Active Bias Tunning

Xiaofei Li[1], Fangyu Mao[1], Pyungwoo Yeon[2], Yan Lu[1]*, Maysam Ghovanloo[2], Rui P. Martins[1,3]

[1]Institute of Microelectronics / State Key Laboratory of AMS-VLSI and FST-ECE, University of Macau, Macao, China.
[2]School of ECE, Georgia Institute of Technology, Atlanta, GA, USA; [3]IST, Universidade de Lisboa, Lisbon, Portugal.
*E-mail: yanlu@umac.mo

Abstract—A high efficiency wireless power transfer system that operates in the very high frequency (VHF) band is desirable for miniaturized implantable medical devices (IMD), but also challenging. This work presents an integrated VHF passive rectifier with closed-loop active bias tuning (ABT). Power PMOS and NMOS transistors are separately cross-connected, such that the DC bias voltages of the NMOS gates can be dynamically adjusted. We employed a search scheme for V_{OUT}-peak to obtain the optimum bias voltage over a wide input range. A prototype fabricated in 65nm std-CMOS process measured a power convention efficiency of 64% at 200MHz with 3kΩ loading.

Keywords—Active bias tuning, wireless power transfer, rectifier, implantable medical devices, VHF.

I. Introduction

A series of mm-sized implantable medical devices (IMD) are under development with the promise of extended longevity for being less invasive [1]-[3]. Wireless power transfer (WPT) to these devices, however, needs new solutions to enable near-field WPT that operate at hundreds of MHz to improve the power transfer efficiency. In such high frequencies, active rectifiers can barely function, and the cross-connected (CC) CMOS passive rectifier, shown in Fig. 1a, is commonly used. However, the passive CC rectifier suffers from shoot-through and reverse currents, resulting in poor power conversion efficiency (PCE). This is highly affected by the MOSFETs' threshold voltage, V_{TH}, evident from the fact that the half-CC half-diode passive rectifier, presented in Fig. 1b, can effectively remove the shoot-through current [4]. Even though V_{TH}-compensation and floating-gate techniques [5] have been proposed to help reduce the diode drop, the static compensation optimizes only one operating point, and is very sensitive to PVT variations, as plotted in Fig. 1c. A dual-path architecture can extend the input range [6], but the challenge remains for the optimum path selection. In this paper, we propose a wide input range CMOS passive rectifier with active bias tuning.

II. Circuit And System Implementations

Fig. 2 shows the implementation of one stage of the proposed rectifier with active bias tuning (ABT). Those NMOS gate voltages, V_C and V_D, which are DC-biased by V_{Bi} through R_1 and R_2, are AC-coupled to V_B and V_A through C_2 and C_1, respectively. This allows the adaptive DC bias voltage, $V_{Bi}-V_i$, actively tunes the operating point of the i-th rectifier stage to its optimal point. C_{Bi} is dynamically charged or discharged by C_i in the ABT charge pump. The V_{DC} source used to refresh C_i can be obtained from any DC potentials of the 3-stage rectifier,

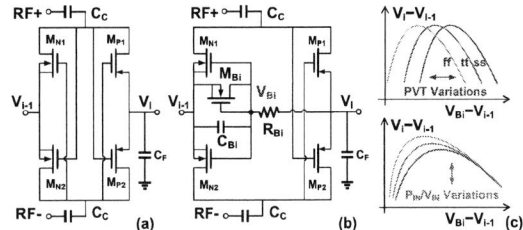

Fig. 1. (a) Conventional cross-connected passive rectifier; (b) V_{TH}-compensated rectifier and (c) its DC output versus gate bias voltage.

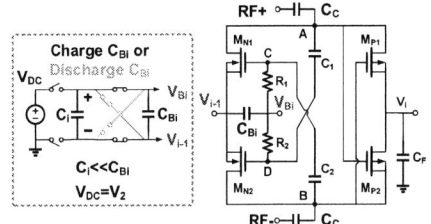

Fig. 2. One stage of the proposed rectifier with the adaptive gate bias tuning.

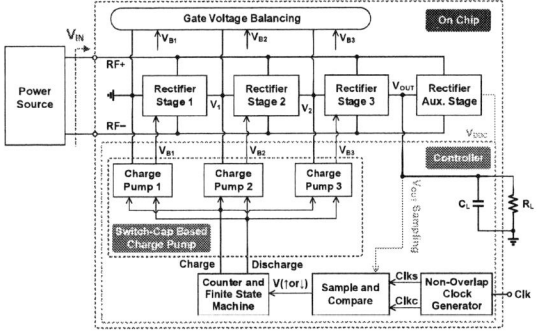

Fig. 3. System overview of the proposed three-stage rectifier with active bias tuning and V_{OUT} peak search controller.

while $V_{DC}=V_2$ in this design, which is the rectifier stage2's output in Fig. 3. The NMOS transistors are implemented in deep N-well to make their V_{TH} the same in each stage.

Fig. 3 presents the block diagram of a 3-stage rectifier with the proposed active bias tuning for the gate voltages of each stage. Aside from the three main stacked rectifier stages that deliver power to the load, there is an auxiliary rectifier stage that generates a higher supply for the active control block. Two non-overlapped signals Clk_C and Clk_S that are half the

978-1-7281-5107-6/19 $31.00 © 2019 IEEE

IEEE Asian Solid-State Circuits Conference
November 4 – 6, 2019
The Parisian Macao, Macao SAR, China

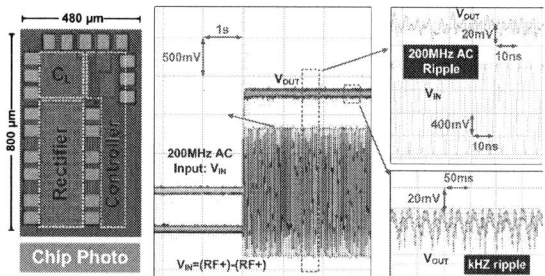

Fig. 4. The measured V_{OUT} waveforms during the startup and the steady state.

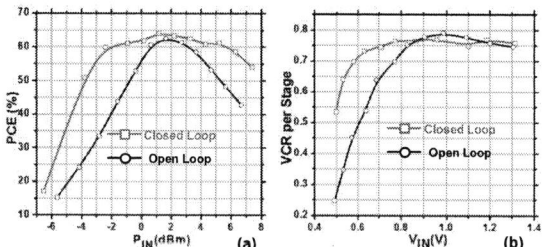

Fig. 5. Measured (a) PCE and (b) VCR of the rectifier with or without the proposed closed-loop active tuning @ R_L=3kΩ and f_c = 200MHz.

frequency of external clock frequency, Clk, are used to control the sample and compare phases, respectively. For the closed-loop gate bias voltage control, V_{OUT} is sensed and compared with its previous value. The comparison result directs a finite state machine to signal either "Charge" or "Discharge" to the charge pumps. Each charge pump generates a corresponding gate bias voltage V_{Bi} (i = 1, 2, or 3) to stage-i. Considering different DC potentials at each rectifier stage, the actual gate bias voltages applied to stages 1, 2, and 3, are V_{B1}, V_{B2}–V_1, and V_{B3}–V_2, respectively. To make sure these gate bias voltages are stable and equal (V_{B1}=V_{B2}–V_1=V_{B3}–V_2) across PVT corners and compensate for the mismatches between three stages, we use an additional block for dynamic gate bias voltage balancing.

III. MEASUREMENT RESUTLS

The prototype was fabricated in a 65nm CMOS process. 1.8-V devices are used for efficient AC-DC power conversion and low leakage current, while 3.3-V devices are used in the controller to withstand the higher supply voltage, V_{DDC}. Fig. 4 (left) exhibits the chip micrograph, which occupies 480μm×800μm silicon area.

A vector network analyzer (VNA) provides the 200MHz AC input signal of the rectifier, plus an external clock Clk=1kHz applied to the active controller. Fig. 4 also shows the V_{OUT} waveforms during the startup process as well as the steady state with R_L=3kΩ and P_{IN}=2.05dBm, while the obtained DC output is 1.78V. With an on-chip load capacitor C_L of 80pF, we measured the 200-MHz V_{OUT} AC ripple to be around ±10mV, and the V_{OUT} low-frequency ripple to be close to ±18mV.

To demonstrate the effect of the proposed ABT, we compare in Fig. 5 measured PCE and one-stage voltage

TABLE I. BENCHMARKING AGANST THE STATE-OF-THE-ARTS

	[1] ISSCC'18	[2] VLSIC'16	[6] TCAS-II'17	This Work
Freq. (MHz)	60	144	900	200
Process (nm)	350	180 SOI	65	65
Chip Area (mm²)	N. A.	0.078	0.084	0.384
Rectifier Type	Passive	Active	Passive	Passive
No. of Stage	2	1	5	3
Load Cap (nF)	10,000	0.25	100	0.08
Load (kΩ)	N. A.	N. A.	147	3
Input Power Range (dBm)	N. A.	N. A.	−16 to −5	−7 to 7
Max PCE	40%*	66.5%	36.5%	64%

*Estimated from figure.

conversion ratios (VCR) with or without the ABT, with V_{Bi} in the open loop rectifier setting to 0.5(V_i+V_{i-1}). Here, the PCE is defined as PCE=P_{OUT}/P_{IN}, where P_{OUT}=V_{OUT}^2/R_L, P_{IN}=P_S(1-$|S_{11}|^2$), P_S is the maximum available power from the VNA, and S_{11} is the reflecting factor. The measured results clearly show that the PCE with active bias tuning is higher over the entire input power range, and the VCR in the low input voltage range has been significantly improved. The maximum PCE is 64% with P_{IN}=1.16dBm, R_L=3kΩ, and V_{OUT}=1.58V. Table I benchmarks this work against the state-of-the-Art. We achieve high PCE in a wide input power range with a 3-stage passive rectifier.

IV. CONCLUSIONS

This paper presented a CMOS passive cross-connected rectifier, which the NMOS gate DC bias voltage is actively tuned by a sample and compare loop for enhancing the PCE. We demonstrated clear performance improvements on silicon, over a wide input range, with the maximum PCE of 64% with 3kΩ loading at 200MHz RF input.

ACKNOWLEDGMENT

This work is funded by the RC of the University of Macau (under MYRG2018-00102-AMSV), and The Science and Technology Development Fund, Macau SAR (SKL-fund).

REFERENCES

[1] Y. Jia et al., "A mm-sized free-floating wirelessly powered implantable optical stimulating system-on-a-chip," in *IEEE International Solid - State Circuits Conference - (ISSCC), San Francisco*, CA, Feb. 2018.

[2] C. Kim et al., "A fully integrated 144 MHz wireless-power-receiver-on-chip with an adaptive buck-boost regulating rectifier and low-loss H-Tree signal distribution," *2016 IEEE Symposium on VLSI Circuits (VLSI-Circuits)*, Honolulu, HI. Jun 2016.

[3] R. Muller et al., "A Minimally Invasive 64-Channel Wireless μECoG Implant," in *IEEE Journal of Solid-State Circuits*, vol. 50, no. 1, pp. 344-359, Jan. 2015.

[4] M. Kiani et al., "A Q-Modulation Technique for Efficient Inductive Power Transmission," in *IEEE Journal of Solid-State Circuits*, vol. 50, no. 12, pp. 2839-2848, Dec. 2015.

[5] S. Mandal and R. Sarpeshkar, "Low-Power CMOS Rectifier Design for RFID Applications," in *IEEE Transactions on Circuits and Systems I: Regular Papers*, vol. 54, no. 6, pp. w1177-1188, June 2007.

[6] Y. Lu et al., "A Wide Input Range Dual-Path CMOS Rectifier for RF Energy Harvesting," in *IEEE Transactions on Circuits and Systems II: Express Briefs*, vol. 64, no. 2, pp. 166-170, Feb. 2017.

978-1-7281-5107-6/19 $31.00 © 2019 IEEE

A 7.5 - 42V Input High-VCR Monolithic DC-DC Converter Using Stacked Isolated SC Cores

Elly De Pelecijn and Michiel Steyaert

KU Leuven ESAT-MICAS

Kasteelpark Arenberg 10, 3001 Heverlee, Belgium

Email: elly.depelecijn@esat.kuleuven.be

Abstract—A fully integrated switched-capacitor DC-DC power converter that converts input voltages up to 42 V into 3 V is presented. The converter achieves a wide input operating range with multiple VCRs based on stacking of several isolated switched-capacitor converters. Stacking simplifies the design to a low voltage unit converter with only a few VCRs, while the wide input range high voltage converter achieves the efficiency of the unit converter. The DC-DC converter is designed in a 0.35 μm CMOS technology and achieves an output power of 2.1 mW.

I. INTRODUCTION

Fully integrated switched-capacitor (SC) DC-DC converters have gained large interest in recent years. Since SC converters only use switches and capacitors, they are perfectly fit for integration in standard CMOS technologies. As a result, small, low-cost and efficient SC converters are available in different flavors. Gearbox and recursive converters overcome the efficiency limitation of the single topology converters by implementing multiple voltage conversion ratios (VCRs), as the efficiency of one topology is limited to $V_{OUT}/(V_{IN} \cdot VCR)$.

Furthermore, the applications in [1] and [2], as well as mW-range industrial applications demand for a high voltage (HV) capability of the DC-DC converter. In these applications, the high input voltage goes hand in hand with a high VCR and requires bootstrap circuits and high voltage switches and capacitors that decrease the efficiency and power density.

In general, combining numerous VCRs with a high input voltage and thus a large conversion ratio in a monolithic way, is challenging. Therefore, this work proposes the concept of 'stacked switched-capacitor power conversion' for HV mW-range applications. Here, it is possible to achieve a monolithic wide input range converter by only designing a single low voltage isolated converter with a few VCRs. These converters can be connected in series at their inputs to tolerate the high input voltage, while no degradation in efficiency occurs. The principle of converter stacking is introduced in section II, while the isolated SC converter needed for stacking is discussed in section III. Sections IV and V cover the implementation details and measurement results. Finally, a conclusion is drawn.

II. A STACKED POWER CONVERTER

The concept of stacked power conversion is illustrated in Fig. 1. N identical power converters (PCs) are connected in series at their input ($V_{IN,i}$), while the outputs are connected in parallel to the converter output V_{OUT}. Since the converters

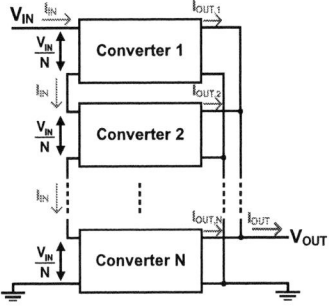

Fig. 1: High level concept of stacked power converters.

are identical, the input voltage V_{IN} divides equally across the different converters (1) while the input current I_{IN} is shared. At the output, the converters' currents $I_{OUT,i}$ are summed towards the load (2). Supposing a VCR of M:1 for the single PCs, the total stacked converter achieves a high VCR of $(M \cdot N) : 1$.

$$V_{IN,i} = V_{IN}/N \tag{1}$$

$$I_{OUT} = \sum_{i=1}^{N} I_{OUT,i} = N \cdot I_{OUT,i} \tag{2}$$

$$\eta_{STACK} = \frac{V_{OUT} \cdot I_{OUT}}{V_{IN} \cdot I_{IN}} = \frac{V_{OUT} \cdot (N \cdot I_{OUT,i})}{(N \cdot V_{IN,i}) \cdot I_{IN}} = \eta_i \tag{3}$$

Equation (3) shows that the efficiency η_{STACK} of the stacked converter equals the efficiency η_i of the single PCs. Since these PCs face a low input voltage (1) and thus need a low VCR, the result of (3) holds two advantages when considering SC converters for the unit cells. First, converters with a small conversion step are typically more efficient than those with a large step, so that the latter can be designed at the efficiency of the former. Second, the design simplifies to the implementation of one unit cell instead of a high VCR-converter.

The stacked converter achieves multiple VCRs in two ways. First, it is possible to implement multiple ratios in the converter units. Second, it is possible to reconfigure the connection of the inputs of the units, so that the number of stacked cells, N, changes. The second implementation is more efficient than the first as it only requires static switches. This means that they are on or off depending on the VCR, but that they do not switch continuously compared to the first case. Figure 2 illustrates the influence of stacking with multiple ratio converters on the ideally achievable efficiency of the converter.

978-1-7281-5107-6/19 $31.00 © 2019 IEEE

IEEE Asian Solid-State Circuits Conference
November 4 – 6, 2019
The Parisian Macao, Macao SAR, China

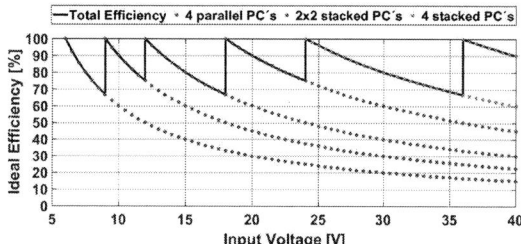

Fig. 2: Ideal efficiency vs. V_{IN} for stacked power conversion.

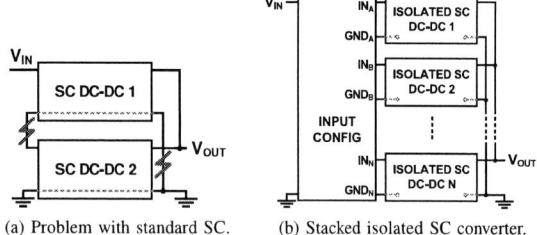

(a) Problem with standard SC.

(b) Stacked isolated SC converter.

Fig. 3: High level definition of the stacked power converter.

Four PCs with an ideal 2:1 and 3:1 ratio are considered. Connecting all PCs in parallel results in the red curves with a high efficiency drop for high input voltages. Stacking 2 times 2 converters or all 4 converters generates the blue and green curves, so that a high efficiency for high inputs appears. By correctly switching between the different converter modes ($\frac{1}{2}$, $\frac{1}{3}$) and changing the 'stack-setting', a stacked converter can handle a wide range of input voltages at the efficiency shown in black. Additionally, the designer only needs to optimize the unit cells, which have less VCRs, a low $V_{IN,i}$ and low ratio.

III. A STACKABLE SWITCHED CAPACITOR CONVERTER

A. The need for an isolated SC-converter

Despite the advantages of stacking, the known SC converters cannot directly be plugged into the converter cells of Fig. 1. Since the input and output of a SC converter share a common ground, the converter should be considered as a 3-port element. Consequently, stacking results in the shorts marked in Fig. 3a. A SC-based output combiner, which shifts the outputs of the different converters to the GND-V_{OUT} domain, solves this problem. However, this technique has a number of drawbacks. First, unequal loading of the PCs by the combiner results in an unequal input voltage division. Second, the combiner does not contribute to the voltage conversion, as it realizes a 1:1 step. Therefore, extra losses occur while the power density drops due to the capacitors needed in the combiner. Third, the capacitors in the combiner face a high bottom plate swing by the conversion between the stack-domains and the output.

As a result, this work introduces an isolated switched capacitor converter that isolates the ground terminals so that a 4 terminal converter is obtained. Here, a compromise is made between the losses the additional combiner would insert and the fact that the unit cells in the stacked converter of

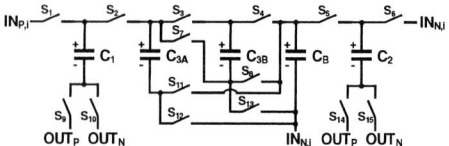

Fig. 4: Isolated switched capacitor converter core.

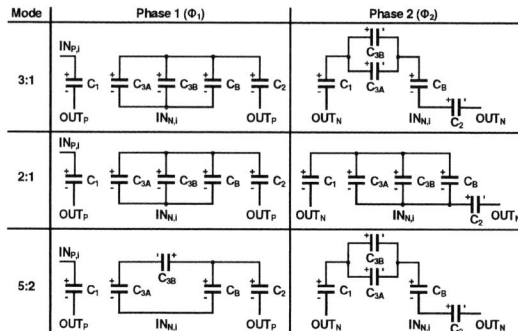

Fig. 5: Core configuration in different phases and modes.

Fig. 1 only needed low-voltage capacitors. In the isolated SC converter, some capacitors face a higher voltage when realizing the combiner-function. Figure 3b illustrates the use of isolated DC-DC converters in the stacked converter, while it also shows the input configuration block needed to reconfigure the converter inputs for realization of the different VCRs.

B. Isolated Switched Capacitor Converter Core

Fig. 4 shows the designed isolated SC core, which transfers power from the input ($IN_{P,i}-IN_{N,i}$) to the output (OUT_P-OUT_N). It realizes a 2:1, 5:2 and 3:1 voltage conversion, illustrated in Fig. 5. The number of stacked converters N is 4, so that the converters can be configured all in parallel, in a double stack of 2 cores or all in stack-mode. The ratios are chosen in conjunction with the number of converters, so that no overlap between the ratios in the different stack configurations appears. Meanwhile, the output voltage of 3 V is considered and all voltages are limited to the tolerable range of the used technology. As a result, the VCRs of table I are possible.

The 3:1 topology is based on the Dickson topology, which explains the configuration of capacitors C_1 and C_2. In all modes, these capacitors face a voltage ($IN_{P,i}-OUT_P$) and ($IN_{N,i}-OUT_N$) respectively. $IN_{P,i}$ and $IN_{N,i}$ depend on the core's position in the stack. C_3 ($C_{3A}//C_{3B}$) realizes the ground isolation while C_B keeps the charge balance. Both capacitor voltages are limited to V_{OUT}. Since C_B is a low voltage non-flying capacitor, it does not lower the converter performance. Reconfiguring or splitting C_3 realizes the other VCRs.

Compared to a 12:1 Dickson converter that implements the mentioned ratios, the design optimization relaxes strongly. Only one core with less capacitors, switches and VCRs needs to be considered, while the high conversion ratios are automatically achieved by stacking. In addition, the stacked converter in 12:1 mode uses 3 high voltage capacitors less than the standard Dickson, so that the power density increases.

978-1-7281-5107-6/19 $31.00 © 2019 IEEE   248

IEEE Asian Solid-State Circuits Conference
November 4 – 6, 2019
The Parisian Macao, Macao SAR, China

TABLE I
Operating Modes of Stacked Power Converter

Stack-setting	All //	2x2 stacked	4 stacked
VCR	2:1, 5:2, 3:1	4:1, 5:1, 6:1	8:1, 10:1, 12:1
V_{IN}-range	7.5 V - 15 V	15 V - 30 V	30 V - 42 V*

* limited by technology, theoretically V_{IN} up to 60 V possible

IV. IMPLEMENTATION

The proposed converter with isolated switched capacitor cores is implemented in a $0.35\,\mu m$ CMOS technology and is fully integrated on 1 die. The implementation details in this section focus on the challenges caused by the high input voltages, multiple VCRs and the stacked-topology.

A. Stackable Power Converter Unit Cell

The stacked power converter consists of 4 stackable unit cells. Since bottom plate (BP) losses are dominant in high voltage converter designs, this work makes sure that the maximum bottom plate swing of all capacitors is limited to V_{OUT}. To limit the BP-losses even further, charge recycling is implemented for C_1 and C_2 [3]. As a result, each unit cell consists of 8 phase shifted cores from Fig. 4. The capacitors facing a voltage higher than $2 \cdot V_{OUT}$ are realized as MOM-capacitors while the other capacitors are of the MIM- or MOS-type, depending on its bottom plate swing while optimizing the power density. All switches in Fig. 4 are realized as 2 serialized MOS-transistors to tolerate the applied voltages. The sizes of the capacitors and switches are optimized so that the efficiency and power density are as high as possible.

Correct biasing of the bulk of all switches demands attention during the design. Since a core is reconfigurable in different modes, the bulk connection is not always straightforward.
In order to drive the switches correctly, capacitive level shifters are used. These operate based on stable internal DC-voltages such as the voltage at the top-plate of C_B or voltages generated by connecting 2 anti-phase cores at the correct nodes.

B. Input Configuration Block

To reconfigure the stacked PC unit cells (PC_1-PC_4), an input configuration block (ICB) is added to the converter and implemented as shown in Fig. 6. The DMOS transistors S_1-S_9 configure the PCs in the correct stack-setting with control signals PS_1-PS_9 (top right Fig. 6). The position of the transistors between the unit cells is optimized so that the blocking voltages are as low as possible, while still tolerable by the transistors. The ICB has a low influence on the converter efficiency: first, the transistors in this block do not add a switching loss as reconfiguration takes place at a lower speed than the core converter operation; second, these transistors are made as large as possible to limit their conduction losses and voltage drop. More specifically, this voltage drop influences the input voltage of the different PCs. To equalize these input voltages, the transistors are not sized equally.

The extra circuit elements in Fig. 6 perform the following function. When the converter transitions from an all parallel setting (all PCs //) to the 2x2 stack-setting (stack connection of (PC1//PC2) and (PC3//PC4)), charge flows back from the PCs to the input. To lead this charge in the correct direction and avoid latch-up, multiple diodes are added. Transistors S7-S9 demand a bulk bias (BB) circuit as shown in Fig. 6. In order to avoid latch-up, the bulk of these PMOS devices always needs to be connected to a high potential that avoids the drain- and source-bulk diodes to become forward biased. Therefore, depending on the setting, the bias circuit connects the bulk to either the source or the drain while special care is taken of the setting-transitions by adding switching transistors in the bias circuit. Finally, small capacitors are added to ensure an equal voltage division of the input voltage across the different cores.

The switch-control signals are generated in a 0-3 V domain, which cannot directly be applied to the switches, as the gate-oxide would break. To this end, capacitive level-shifters (LS) shift the control signals between the correct voltage rails, which are generated in the cores. Diodes are added to the LSs to ensure correct operation during start-up, so that the ICB enters a valid state. Delay elements and buffers ensure synchronization between the switches of the ICB.

C. Control Strategy

The output V_{OUT} of the converter is regulated to 3 V by a controller based on [3] switching at f_{clk}. This controller generates 8 non-overlapping phases and 20 charge recycle phases that are fed into the 4 isolated power converters. These control signals are divided in the PC-units across the 8 cores. An extra controller consisting of slow-switching comparators ($\ll f_{clk}$), high impedant resistive dividers and digital logic, chooses the correct mode of table I based on the input voltage.

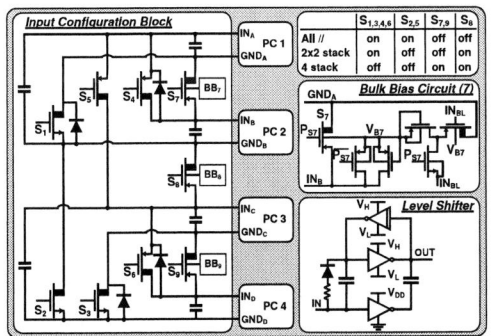

Fig. 6: Implementation of Input Configuration Block.

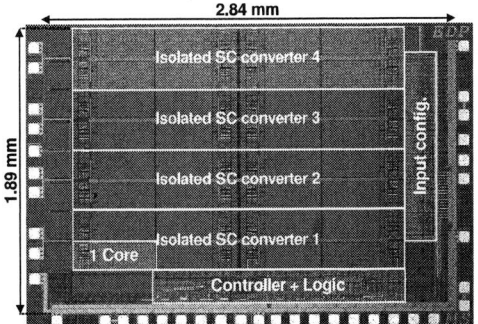

Fig. 7: Chip Micrograph.

978-1-7281-5107-6/19 $31.00 © 2019 IEEE

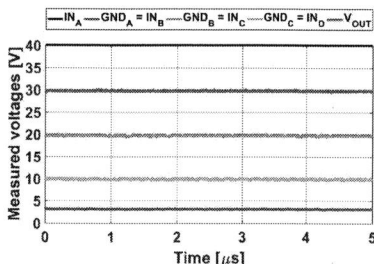

Fig. 8: Measured voltages for $V_{IN} = 40\,\text{V}$ at peak P_{OUT}.

TABLE II

Comparison to Monolithic HV SC DC-DC Converters

	[4]	[5]	[1]	This work
Technology	90 nm	0.35 μm	0.35 μm	0.35 μm
# VCRs, max	8, 5:1	17, 16:9	1, 4:1	9, 12:1
V_{IN} [V]	2.5 - 8	2 - 13	17	7.5 - 42
V_{OUT} [V]	1.2	5	3.3	3
P_{dens} [mW/mm²]	13.3	1.47	0.44*	0.4
I_{peak} [mA/mm²]	11	0.29	0.13*	0.13

* estimated

V. MEASUREMENT RESULTS

The presented converter implementing the new idea of stacking with isolated SC converters is fabricated in a $0.35\,\mu\text{m}$ CMOS technology and measures $5.4\,\text{mm}^2$. The stacked converters as well as the 32 separate SC-cores are visible on the chip micrograph in Fig. 7. For input voltages ranging from $7.5\,\text{V}$ to $42\,\text{V}$, the converter delivers a stable $3\,\text{V}$ DC-output at a peak output power of $2.1\,\text{mW}$, while operating at $20\,\text{MHz}$.

Figure 8 shows the measured voltages across the stacked cores for an input voltage of $40\,\text{V}$ at peak load. The input divides equally across the different stacked cores and therefore validates the correct operation of the input control block.

Figure 9 indicates the measured efficiency η of the converter depending on V_{IN}, while operating at peak load. The graph shows a repeating efficiency curve. More specifically, the efficiency of the single isolated converter, indicated in blue, is repeated when the stack-setting changes to a double stack of 2 converters and the full-stack mode. This confirms the predicted efficiency trend of (3). Therefore, stacking can be considered as a good design technique to implement high voltage wide input range power converters with good efficiency, since only the optimization of the low VCR isolated converter should be considered. The black dots in Fig. 9 indicate operation outside the technology's safe operating range. Fig. 10 shows a more detailed η-curve of the stacked converter. It plots the contours of constant η for a varying input voltage and output power.

Table II compares the presented converter to the state-of-the art monolithic SC converters for high input voltages (HV). The stacked design converts the highest reported DC-input at power and current densities comparable to [5] and [1]. [4] achieves the highest power density, but uses a much smaller technology node while V_{IN} is 5 times smaller than for the presented design, thereby showing the typical trade-off

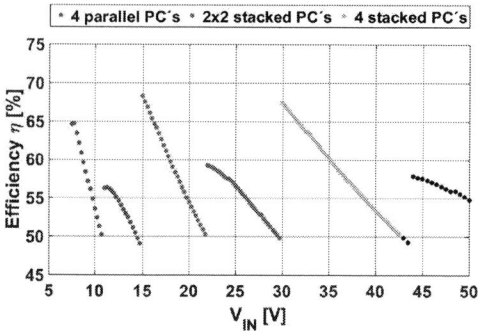

Fig. 9: Efficiency at peak P_{OUT} for wide input range.

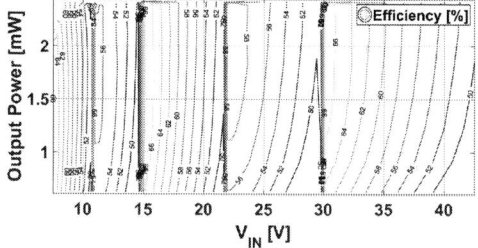

Fig. 10: Measured η-contours for variable V_{IN} and P_{OUT}.

between input voltage and performance [1]. Nevertheless, the stacked converter achieves the same performance as [1] while even converting a more than 2 times higher input voltage.

VI. CONCLUSION

This work showed a monolithic DC-DC converter able to handle input voltages from $7.5\,\text{V}$ up to $42\,\text{V}$ while delivering $2.1\,\text{mW}$ at $3\,\text{V}$. The presented 'stacking-technique' enables the high voltage compatibility and allows a multi-VCR design by reconfiguration of the stacked isolated SC converter cells. In addition, stacking simplifies the design since it only requires an optimized low voltage, low VCR isolated converter and thereby increases the efficiency across the whole V_{IN}-range.

ACKNOWLEDGMENT

The authors thank Research Foundation-Flanders (FWO) for the SB-fellowship (1S42517N) that supports this work.

REFERENCES

[1] D. Lutz *et al.*, "An Integrated 3-mW 120/230-V AC Mains Micropower Supply," *IEEE Journal of Emerging and Selected Topics in Power Electronics*, vol. 6, no. 2, pp. 581–591, June 2018.

[2] M. Steyaert *et al.*, "When hardware is free, power is expensive! Is integrated power management the solution?" in *ESSCIRC Conference*, Sep. 2015, pp. 26–34.

[3] N. Butzen and M. Steyaert, "Scalable Parasitic Charge Redistribution: Design of High-Efficiency Fully Integrated Switched-Capacitor DCDC Converters," *IEEE Journal of Solid-State Circuits*, vol. 51, no. 12, pp. 2843–2853, Dec 2016.

[4] A. Sarafianos *et al.*, "A folding dickson-based fully integrated wide input range capacitive DC-DC converter achieving Vout/2-resolution and 71% average efficiency," in *IEEE Asian Solid-State Circuits Conference (A-SSCC)*, Nov 2015, pp. 1–4.

[5] D. Lutz *et al.*, "12.4 A 10mW fully integrated 2-to-13V-input buck-boost SC converter with 81.5 % peak efficiency," in *IEEE International Solid-State Circuits Conference (ISSCC)*, Jan 2016, pp. 224–225.

A 918MHz Wide-Range CMOS Rectifier with Diode-Feeding and Switch-Capacitor-Based Load Modulation Technique

Chen-Yi Kuo, *Student Member, IEEE*, Chun-An Lu, *Student Member, IEEE*, and Yu-Te Liao, *Member, IEEE*

Abstract—This paper presents a wide-input-range rectifier design with a diode-feeding (DF) architecture and load modulation. Dynamically-biased diodes are connected to the gates of the rectifying transistors to create bias voltage differences for PMOS and NMOS devices and the isolation of the reverse currents at different input power levels. Also, the switch-capacitor-based load modulation scheme is proposed to achieve optimal efficiency over wide-range load current deviations automatically. The proposed four-stage rectifier design was fabricated using 65nm CMOS technology and occupies a silicon area of 0.92 x 0.27mm². The measured sensitivity of the rectifier is −15dBm at 1MΩ load, and the peak power conversion efficiency is 28% at a load of 22kΩ. With the load modulation scheme, the conversion efficiency is enhanced by >10% in a load current from 7μA to 35μA, when compared to a design without the load modulation scheme.

Keywords—*CMOS, rectifier, wide input range, load compensation*

I. Introduction

The increasing demands of data acquisition from the environment necessitate a large number of compact and mobile sensor nodes. Radio frequency (RF) energy-harvesting techniques have been widely used in powering wireless sensors for many batteryless Internet-of-Things applications, such as radio-frequency identification (RFID)-based sensors [1] and bio-implants [2]. The path loss and environment uncertainty make the received power at the rectifier fluctuate significantly. This varying power impedes the operation range of the sensors on mobile devices. Therefore, the rectifier needs to scavenge energy efficiently over a wide input range to stabilize the power supply of the sensor nodes in various environments.

Recent research [3][4][5] has been conducted to improve the sensitivity and efficiency of rectifiers to extend the range of wireless-powered sensors. Conventionally, cross-coupled (CC) rectifiers have been widely used to achieve high sensitivity and efficiency at a low power input by differentially compensating the threshold voltage of the rectifying metal–oxide–semiconductor (MOS) transistors; however, a lower threshold voltage causes considerable reverse leakage at a high power input, resulting in efficiency degradation, power waste, and operational range constraints.

A self-biased rectifier [6] adopts feedback bias to enforce the tight turnoff of the rectifying transistors when the output voltage is high. In addition, the conversion efficiency of the rectifier is affected not only by the input power but also by the load conditions [7]. Therefore, an optimal load condition at different input power levels is essential in the CMOS rectifier design. To overcome the issues of a constrained input power range in the CC rectifier [8], this paper proposes a rectifier with a diode-feeding (DF) technique and an adaptive load

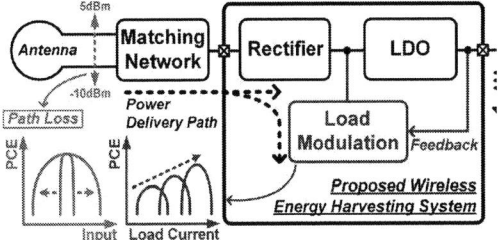

Fig. 1 Block diagram of the proposed RF-DC converter.

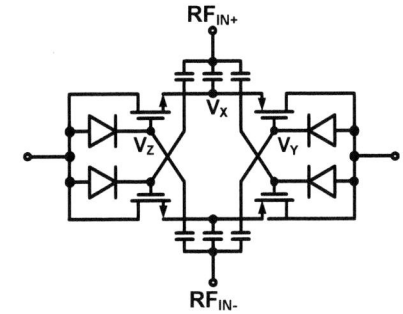

Fig. 2 Schematic of the proposed DF rectifier.

adjustment in response to the power and load current fluctuations.

II. The Proposed Rectifier with an Automatic Load Adjustment Scheme

Fig. 1 shows the conceptual block diagrams of the proposed reconfigurable rectifier. The design includes a rectifier with a voltage detector, a switch-capacitor load modulation circuit, a low-dropout (LDO) regulator, bias circuitry, clock generator, and a low-power diode for power isolation and delivery. To attain greater efficiency over a wide input power range, the rectifier adopts a DF architecture, which adjusts the bias voltage of the rectifying transistors dynamically, according to the input power. After voltage rectification, an LDO regulator follows the rectifier output to provide a stable supply voltage. Since the rectifier efficiency is affected by the load current, a switched-capacitor resistor parallel to the actual load from the LDO is added. A current sensing circuit detects the load current and converts it to a corresponding frequency that changes the equivalent switched-capacitor resistance. The effective load of the rectifier can be adjusted to maintain the RF-DC conversion efficiency by adopting this mechanism. When the input power becomes insufficient to supply the system, the energy passing through the path of the switched capacitor is stored in a

IEEE Asian Solid-State Circuits Conference
November 4 – 6, 2019
The Parisian Macao, Macao SAR, China

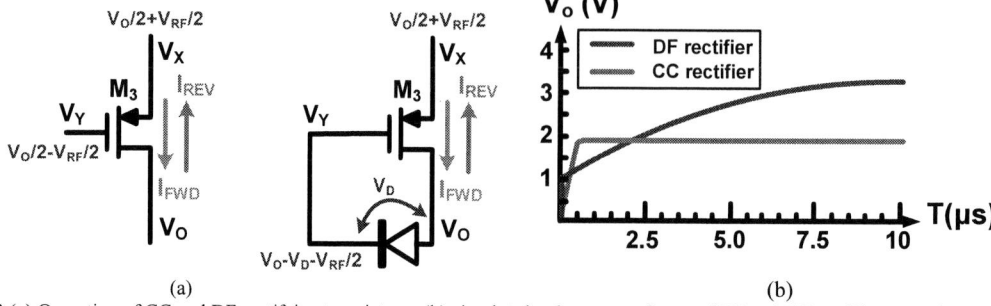

(a)   (b)

Fig. 3 (a) Operation of CC and DF rectifying transistors ; (b) simulated voltage waveforms of DF and CC rectifiers at an input power of -5 dBm (918MHz)

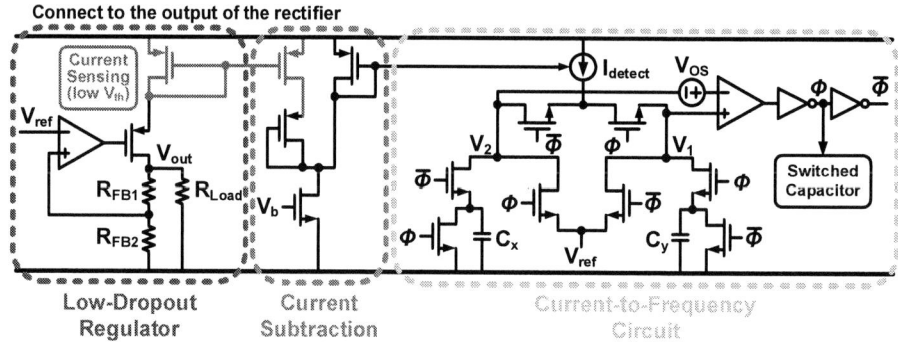

Fig. 4. Schematic of the load-tracking circuits.

capacitor and then delivered to the actual load through an ultra-low power (ULP) diode [9]. The design details of key circuit blocks are as follows.

A. Design of the DF rectifier

Fig. 2 shows the schematic of the proposed rectifier, employing the DF bias technique for reverse leakage reduction and sensitivity improvement. Diode-connected transistors are added at the gate of the rectifying transistors. To separate the bias voltage of the NMOS and PMOS rectifying transistors, the RF signal is capacitively coupled to V_x, V_y, and V_z nodes. The diodes at the NMOS and PMOS rectifying transistors are connected to the output of the previous stage and current stage, respectively, which increases the threshold voltage to reduce leakage and reversed current at high input power and to reduce the turn-on threshold voltage at low power input. The analysis of this design is shown in Fig. 3(a).

As observed in the simplified circuit model, the diode is inactive at a low power input ($V_o < 2V_D$). Then, V_X of the CC and DF rectifiers is $V_O/2 + V_{RF}/2$, whereas the signal at V_Y of the CC and DF rectifiers is $V_O/2 - V_{RF}/2$ and $V_O - V_D - V_{RF}/2$, respectively. The turn-on condition of the rectifying transistor is derived for CC and DF rectifiers, respectively.

$$V_{SG} = V_{RF} > |V_{thp}| \tag{1}$$

$$V_{SG} = V_{RF} + V_D - \frac{V_o}{2} > |V_{thp}| \tag{2}$$

where the threshold voltage depends on the input signal amplitude, V_{RF}, and the diode threshold voltage, V_D. From (1) and (2), it can be observed that the DF rectifier can achieve

lower equivalent threshold voltage at low input voltage ($V_o/2 < V_D$) and the low voltage across the transistor at high input voltage ($V_o/2 > V_D$) when compared with the CC rectifier. Therefore, the DF rectifier has a higher sensitivity and lower reverse leakage potential than does the CC rectifier, which can extend the operation range of the wireless energy harvester. Fig. 3(b) shows the simulated output waveforms of CC and DF rectifiers at -5 dBm 918MHz input power. Initially, the DF rectifier has a faster voltage increase at the output than the CC rectifier does. When the output voltage (Vo) rises until it reaches $2V_D$, the diode at the DF rectifier starts to turn on, and the output voltage increases slower than the CC rectifier, whereas the reverse leakage is reduced. Therefore, DF rectifier can achieve a higher output voltage compared to the CC rectifier at the same input power level.

B. Load compensation mechanism

In addition to rectifier efficiency, the load impedance affects the conversion efficiency [10]. In this design, a current redistribution architecture is inserted between the rectifier and the regulator to help maintain an optimal load by adjusting currents to the load and the storage capacitor. Fig. 4 shows the schematic of the proposed load adjustment architecture. A switch-capacitor load modulator is employed to control the current through the regulator according to the load current at the LDO regulator. In the switch-capacitor load modulator (M_{15}, M_{16}, C_1), the effective resistance is inversely proportional to the switching frequency and capacitance. Thus, for the automatic load adjustment, a feedback loop monitors the load current of the LDO regulator and controls the ratio of the current delivered to the load and

978-1-7281-5107-6/19 $31.00 © 2019 IEEE   252

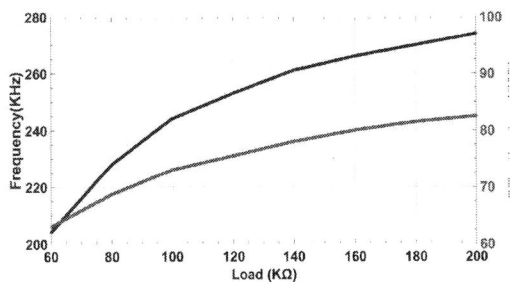

Fig. 5. The output frequency of I-to-F and corresponding effective load

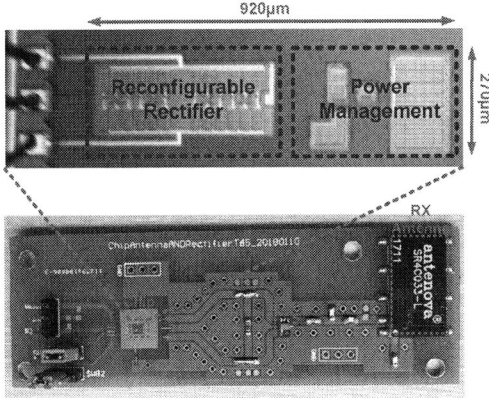

Fig. 6. Chip micrograph and PCB integration.

storage capacitor. A current-sensing transistor using a low-Vt device is stacked on top of the pass transistor of the LDO regulator, where the current is the sum of the load current and the constant DC bias current while the output voltage is regulated. Current subtraction compensates for the significant current variations, resulting in a suitable current range for the following current-to-frequency (I-to-F) converter. The current-to-frequency converter[10] is realized with a charge-based oscillator, which has a better linear range than does a ring oscillator. In this design, an offset cancellation technique is applied to reduce the process-variation-caused frequency drifts. The period of the I-to-F converter is expressed as

$$T_{OSC} = T_{\emptyset=1} + T_{\emptyset=0} = 2\left(\frac{CV_{ref}}{I_{detect}} + t_{delay}\right) \quad (3)$$

According to (3), since the period (the frequency of which is about 200kHz) is much larger than the inverter delay (~nano seconds), the delay is negligible. Therefore, the oscillator frequency is controlled by the subtraction current from the current detection circuit while C and V_{ref} are fixed. Fig. 5 shows the simulated results of the variable frequency and equivalent resistance at the rectifier output. From the simulation, with the feedback compensation loop, the effective load at the output of the rectifier varies by 15kΩ when the oscillator frequency is adjusted from 204kHz to 273kHz, while the actual load resistance changes dramatically from 60kΩ to 200kΩ. Thus, the rectifier is less sensitive to the load variations and can then be operated at the desired load resistance over an extensive load resistance range.

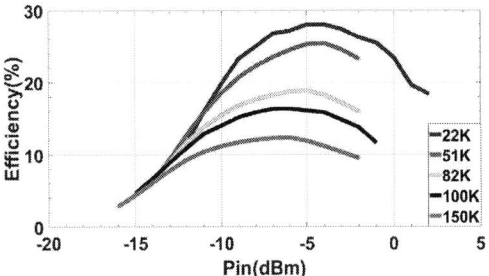

Fig. 7 Measured efficiency at different loads versus input power.

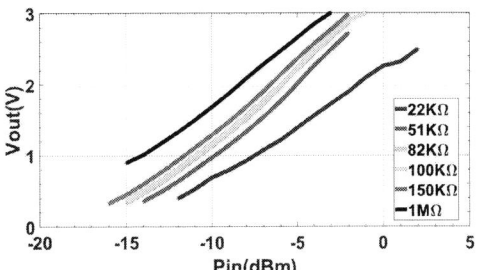

Fig. 8 Measured output voltage at different loads versus input power.

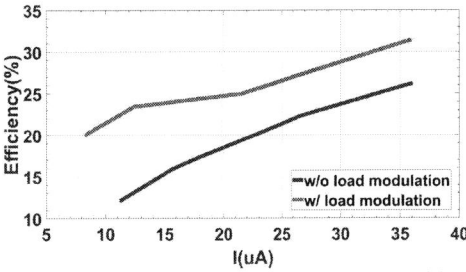

Fig. 9 Measured efficiency over various load currents of the design, with and without the load modulation control loop, at −5dBm input power.

III. IMPLEMENTATION AND MEASUREMENT RESULTS

Fig. 6 shows the chip micrograph of the proposed reconfigurable rectifier design. The chip was fabricated in a 65nm CMOS process. The core chip area is 0.92 x 0.27 mm². The chip is integrated on a printed circuit board (PCB) and tested with a PCB antenna for wireless powering characterization.

The rectifier was tested with a 918MHz RF signal, while an on-board matching network was utilized. Fig. 7 shows the measured conversion efficiency of the proposed rectifier at different loads over a wide input power range. The measured peak efficiency is 28% at a load of 22kΩ. Note that the conversion efficiency of the rectifier varies with output load significantly. The efficiency drops with the increase of the load at a fixed input power. Therefore, a load-tracking and modulation circuit is needed to optimize the conversion efficiency over wide-range load deviations. Fig. 8 shows the

978-1-7281-5107-6/19 $31.00 © 2019 IEEE

Table I Performance summary and comparison to other works.

	JSSC 17'[4]	JSSC 14'[5]	TCAS II 17'[8]	TMTT 14'[11]	This Work
Technology	180 nm	90 nm	65 nm	Discrete component	65 nm
Frequency	915MHz	868MHz	900MH	900MHz	918MHz
No. of Stages	1-2-4-8	5	5	1	4
Peak PCE	25%	40% @330kΩ	36.5% @147kΩ	40%	28.1% @22kΩ
High-PCE Range (>20%)	10dB	12dB	11dB	>15dB	11dB
Sensitivity	-14.8dBm @1.2V (1MΩ)	-18dBm @1V (200kΩ)	-16.5dBm @1V (147kΩ)	-10dBm @1V	-15dBm @1V (1MΩ)
Output Voltage	Non-Regulated	Non-Regulated	Non-Regulated	Non-Regulated	1V (Regulated)
Full Integration	Yes	No (Off-Chip Control Circuit)	Yes (only Rectifier)	No	Yes
Load Modulation	Yes	No	No	Yes	Yes
Regulation Type	N.A.	N.A.	N.A.	Boost converter	LDO
Chip Area(mm²)	5.29 (with matching network)	0.029	0.048	N.A	0.2484

measured results of the output voltage of the rectifier at various loads and input power. The measured sensitivity is −15dBm and −10dBm at 1MΩ and 51kΩ, respectively. Fig. 9 shows the measured conversion efficiency of the design over a range of load currents at an input power of −5dBm, with and without load modulation.. The proposed designachieves a conversion efficiency of greater than 20% over a load current of 7µA to 35µA. Compared to the design without load modulation, this design improves the power conversion efficiency by more than 10%, especially in a heavy load condition. The rectifier can attain a 1V output from a 4-meter distance, excited by an equivalent isotropically radiated power of 1W. Table I shows the performance summary and comparisons to other works. The techniques proposed in this work maintains high power efficiency over wide input and load variations as well as input RF sensitivity to achieve 1V output at −15dBm.

IV. CONCLUSION

This paper presents a 918MHz CMOS rectifier with a DF bias technique and a load modulation scheme to achieve operation range extension and robustness in various environments. The DF rectifier reduces the turn-on voltage of the rectifying transistors at a low power input and increases the threshold voltage at high power input, resulting in good sensitivity and low leakage performance. The load modulation using the switch-capacitor method effectively improves the efficiency over a load current range of 7µA to 35µA (5X) by around 10 %. The proposed CMOS rectifier can be used for mobile batteryless sensor devices for ubiquitous sensing and data collection.

ACKNOWLEDGEMENT

The authors would like to thank the Ministry of Science and Technology, Taiwan, for financial support under the grants 107-2221-E-009-127-MY3 and 108-2636-E-009-008. The chip fabrication was supported by Taiwan Semiconductor Company through the University Shuttle Program.

REFERENCE

[1] D. Yeager, F. Zhang, A. Zarrasvand, N. T. George, T. Daniel and B. P. Otis, "A 9µA, Addressable Gen2 Sensor Tag for Biosignal Acquisition," in IEEE Journal of Solid-State Circuits, vol. 45, no. 10, pp. 2198-2209, Oct. 2010.

[2] Y. Zhang et al., "A Batteryless 19uW MICS/ISM-Band Energy Harvesting Body Sensor Node SoC for ExG Applications," in IEEE Journal of Solid-State Circuits, vol. 48, no. 1, pp. 199-213, Jan. 2013.

[3] Atsushi Sasaki, Koji Kotani and Takashi Ito, "Differential-drive CMOS rectifier for UHF RFIDs with 66% PCE at −12 dBm Input," 2008 IEEE Asian Solid-State Circuits Conference, Fukuoka, 2008, pp. 105-108.

[4] M. A. Abouzied, K. Ravichandran and E. Sánchez-Sinencio, "A Fully Integrated Reconfigurable Self-Startup RF Energy-Harvesting System With Storage Capability," in IEEE Journal of Solid-State Circuits, vol. 52, no. 3, pp. 704-719, March 2017.

[5] M. Stoopman, S. Keyrouz, H. J. Visser, K. Philips, and W. A. Serdijn, "Co-Design of a CMOS Rectifier and Small Loop Antenna for Highly Sensitive RF Energy Harvesters," IEEE Journal of Solid-State Circuits, vol. 49, pp. 622-634, 2014.

[6] M. H. Ouda, W. Khalil and K. N. Salama, "Wide-Range Adaptive RF-to-DC Power Converter for UHF RFIDs," in IEEE Microwave and Wireless Components Letters, vol. 26, no. 8, pp. 634-636, Aug. 2016.

[7] K. Gharehbaghi, F. Koçer and H. Külah, "Optimization of Power Conversion Efficiency in Threshold Self-Compensated UHF Rectifiers With Charge Conservation Principle," in IEEE Transactions on Circuits and Systems I: Regular Papers, vol. 64, no. 9, pp. 2380-2387, Sept. 2017.

[8] Y. Lu, H. Dai, M. Huang, M. Law, S. Sin, U. S, et al., "A Wide Input Range Dual-Path CMOS Rectifier for RF Energy Harvesting," IEEE Transactions on Circuits and Systems II: Express Briefs, vol. 64, pp. 166-170, 2017.

[9] D. Levacq, V. Dessard and D. Flandre, "Low Leakage SOI CMOS Static Memory Cell With Ultra-Low Power Diode," in IEEE Journal of Solid-State Circuits, vol. 42, no. 3, pp. 689-702, March 2007.

[10] A. Paidimarri, D. Griffith, A. Wang, G. Burra and A. P. Chandrakasan, "An RC Oscillator With Comparator Offset Cancellation," in IEEE Journal of Solid-State Circuits, vol. 51, no. 8, pp. 1866-1877, Aug. 2016.

[11] D. Masotti, A. Costanzo, P. Francia, M. Filippi and A. Romani, "A Load-Modulated Rectifier for RF Micropower Harvesting With Start-Up Strategies," in IEEE Transactions on Microwave Theory and Techniques, vol. 62, no. 4, pp. 994-1004, April 2014.

18-4 (7140)

A CMOS Switched-Capacitor Boost Mode Envelope Tracking Regulator with 4% Efficiency Improvement at 7.7dB PAPR for 20MHz LTE Envelope Tracking RF Power Amplifiers

Neha Kumari, Shang-Hsien Yang, and Ke-Horng Chen
National Chiao Tung University
Institute of Electrical and Computer Engineering
Hsinchu, Taiwan

Ying-Hsi Lin, Shian-Ru Lin, and Tsung-Yen Tsai
Realtek Semiconductor Corp
Hsinchu, Taiwan

Abstract—**In this paper, the switched-capacitor boost (SCB) mode envelope tracking (ET) technique with a switched-capacitor circuit and a low dropout (LDO) regulator are used to replace conventional envelope tracking regulators. At a Peak-to-Average Power Ratio (PAPR) of 7.7dB common for LTE communication, the efficiency is 72%, which is 4% higher than the theoretical limit of conventional ET techniques. When PAPR increases from 6.1dB to 7.3dB, the conventional ET efficiency decreases from 82% to 69% whereas the SCB mode increases by 2%.**

Keyword: envelope tracking (ET), Peak-to-Average Power Ratio (PAPR), switched-capacitor boost (SCB).

I. INTRODUCTION

A radio frequency (RF) power amplifier (PA) is a sensitive circuit that requires a well-regulated voltage supply. A standalone buck regulator generates a fixed supply voltage Vreg to meet the maximum output voltage Vout required by the PA but high peak-to-average power ratio (PAPR) results in a Vreg significantly higher than the averaged Vout. Significant power loss occurs across the RF PA due to the difference between Vreg and Vout through heat dissipation, as shown in Fig. 1(a).

Thus, envelope tracking technique [1] - [8] increases the power efficiency of the RF PA as illustrated in Fig. 1(b) by modulating the supply voltage with a variable Vet. Conventional envelope tracking technique tracks the envelope signal of Vout and biases the PA with its output Vet to minimize power dissipation of the PA. A linear amplifier (LA) is added to track high frequency energy components, the energy delivered by IL is reduced, and the buck converter tracks the average power requirement of the PA. The buck converter has high efficiency but the linear amplifier dissipates energy directly proportional to the voltage difference between Vbat and Vout. Considering that modern communication protocols often compress the data sent through the uplinks, the majority of the envelope frequency

components are larger than tens of megahertz that exceed the bandwidth of the buck converter, significant energy is lost by the dissipative push-pull process of the linear amplifier. Furthermore, efficiency is influenced by Vbat. When Vbat is high, the linear amplifier dissipates more power, and when Vbat is low, the envelope tracking mechanism fails and results in malfunction of the PA when instantaneous Vet is larger than Vbat as shown in Fig. 1(b). In this paper, a switched-capacitor boost (SCB) mode envelope tracking technique as depicted in Fig. 1(c) is proposed, aiming to reduce push-pull power dissipation. A switched-capacitor (SC) circuit and a low dropout (LDO) regulator are used to replace conventional linear amplifiers used in ET PA.

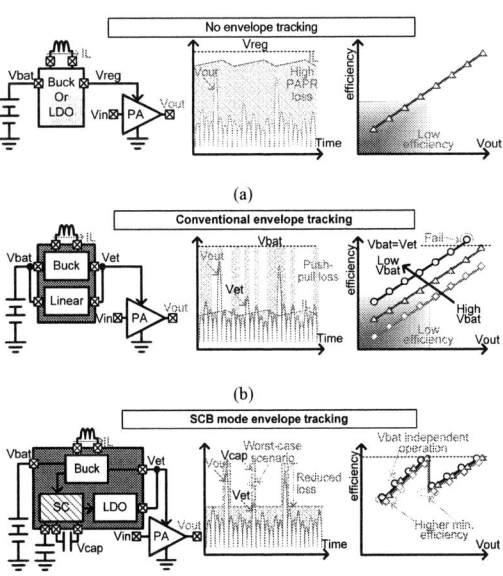

Fig. 1. (a) Conventional fixed supply for RF PA, (b) conventional hybrid buck and linear amplifier architecture for RF PA, and (c) the proposed SCB technique.

978-1-7281-5107-6/19 $31.00 © 2019 IEEE

When the Vout is low, the buck regulator supplies voltage Vcap to the LDO and delivers a low Vet. Contrarily, when Vout is high, the SC circuit doubles Vcap for the LDO to generate a high Vet while the buck regulator also directly feeds its IL to the PA. This proposed SCB technique ensures the operation of the envelope tracking regulator independent of the Vbat to avoid the failure of the envelope tracking mechanism as well as dissipating power with the linear amplifier. High efficiency is maintained throughout the entire Vout range. Compared to conventional ET techniques, the drawback of low efficiency at low Vout can be effectively improved at the expense of only one additional capacitor.

The paper is organized as follows. Section II describes the SCB technique. The circuit implementation is illustrated in Section III. Experimental results are presented in Section IV. Finally, conclusions are made in Section V.

II. PROPOSED SCB TECHNIQUE

Fig. 2(a) shows the buck regulator of the SCB mode envelope tracking regulator operating in the PWM mode and is compensated with a proportional-integral-derivative (PID) circuit. When the digital signal Vbst=0, Mm1, Mm2, and Msc2 are ON and Msc1 is OFF, C1 and C2 are connected in parallel and voltages Vc1+ and Vc2+ are regulated to the value. When Vbst=1, Mm1, Mm2, and Msc2 are OFF and Msc1 is ON, C1 and C2 are in series and Vc2+ is doubled. These power transistors has long transition time, Vbst has to be generated from an original envelope signal with a phase lead over the signal tracked by the LDO. The finite state machine (FSM) in Fig. 2(b) initialize at state a0. CNTa accumulates at every clock edge from a0 to a3 as long as it is not reset by threshold signal TH. When the envelope reaches TH, the FSM is reset to a0. A second counter CNTb count from b0 to b3 whenever CNTa ≠ a3. This mechanism ensures that the SC does not switch back and forth between two modes when high Vet peak values appear in adjacent sequentially. When CNTb reaches state b3, Vbst becomes logic 1 and activates Msc1 and Mm3 to boost Vc2+. When TH is not triggered for two clock cycles, CNTa reaches state a3 and resets CNTb to b0, leading Vbst to logic 0 and activating Mm1, Mm2, and Msc2 to return Vc2+ to its nominal voltage. Fig. 2(c) shown the waveform corresponding to Fig. 2(b) to obtain the delayed envelope.

III. CIRCUIT IMPLEMENTATION

The operation of the SCB mode is exemplified in Fig. 3. There are two modes used in the SCB technique shown in Fig. 3(a). In Mode I, Vbst=OFF, Vc2+ = Vc1+. Vc2+ is sensed by the PID compensator and is regulated by applying PWM IL flowing into both C1 and C2. The LDO drains Ilp from both C1 and C2 to generate Vet. To maintain the value of Vc2+, the PID compensator senses undershoot and overshoot conditions of Vc1+ and adjusts IL to reflect the averaged Ilp. This configuration resembles a single inductor multiple output (SIMO) converter with the cross-regulation characteristics [7].

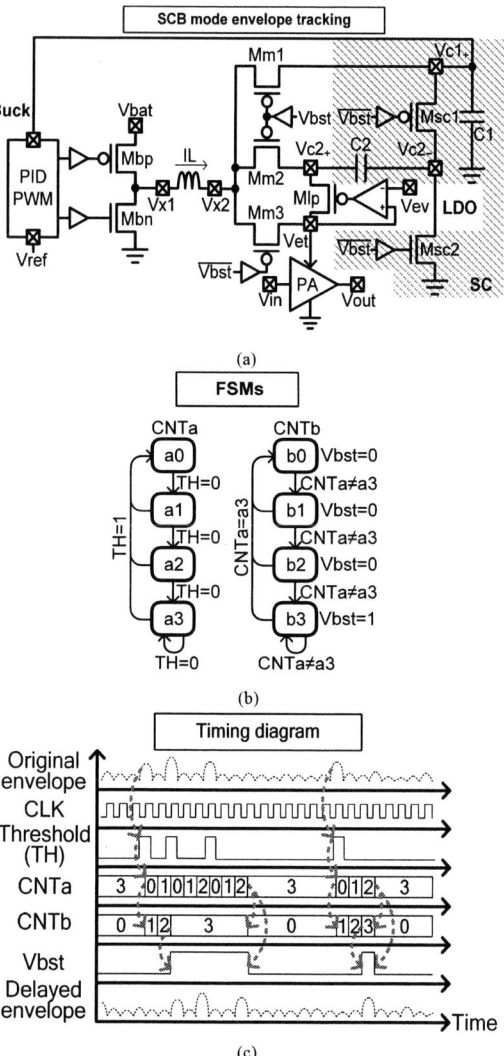

Fig. 2. (a) SCB mode envelope tracking regulator, (b) FSM operation, and (c) operation timing diagram.

The advantage is explained as follows. For most applications, cross-regulation is a negative side effect, but for this particular application, it is utilized as a method for IL to reflect Ilp and Vet in Mode I. Considering that a buck converter inherently exhibits higher efficiency than an LDO, the overall efficiency is enhanced when the proportion of IL is increased immediately in comparison with Ilp. Because the enhancement is caused by the inherent cross regulation of the SIMO converter, no additional loop is required to speed up the tracking efficiency. To maximize cross regulation, in Mode II when Vbst=ON as shown in Fig. 3(b), Vc2+ = 2×Vc1+, a sample and hold (S/H) circuit breaks the feedback loop of the

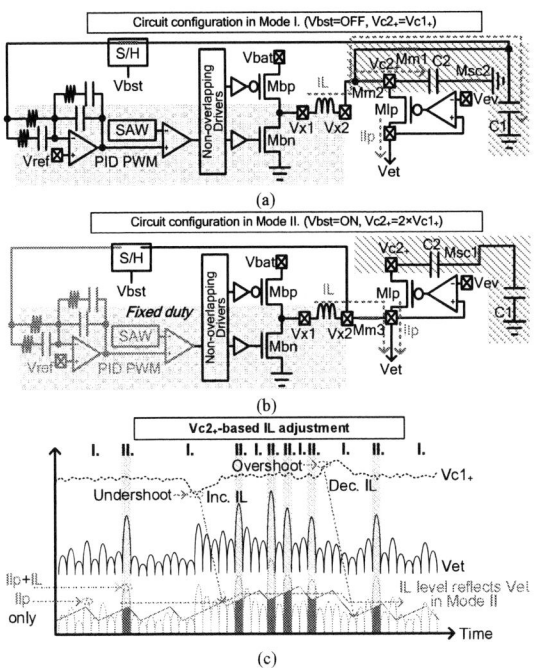

Fig. 3. (a) SCB mode envelope tracking regulator in Mode I. (b) SCB in Mode II. (c) The timing diagram in SCB technique.

buck regulator and holds the previous feedback voltage of $Vc1_+$. This will fixate the duty and IL of the buck regulator to reflect the averaged value of Ilp and Vet. Both IL and Ilp are delivered to the PA simultaneously. Thus, the driving capability of the proposed SCB technique is enhanced to avoid the failure of the envelope tracking mechanism. It solves the drawback of conventional ET techniques when an instant Vout occurs as illustrated in Fig. 3(c).

The LDO has to operate in both Mode I and II, and so a bias circuit dependent on both modes in Fig. 4(a) is included. A voltage-to-current generator converts the reference voltage Vref to a bias current based on the resistor R selected to compensate for process variation. The current is copied through Mb2 to Mb3 and Mb4. Mb4 is always on and Mb3 is activated in Mode II to allow higher bias current on Mb6. Hence, the LDO error amplifier consisting of Ml1 to Ml8 will be given higher current driving ability and slew rate to achieve higher Vet. Smooth transition between Mode I and II is achieved with core devices with smaller parasitic capacitance and improved efficiency. In Mode II when Msc1 is switched on, $Vc2_-$ becomes equivalent to $Vc1_+$, and the driver of Msc1 is biased between $Vc2_+$ and $Vc2_-$ to limit the voltage across the Vgs of Msc1 and the driver itself to the value of $Vc1_+$.

In Fig. 4(b), a level shifter (LS) implemented using I/O devices shifts the voltage level up. Non-overlapping switching of Msc1 and Msc2 prevents C2 from accidentally shorted. Mm3 is activated when Msc1 turns on, and IL is delivered to the PA. When Mode II transitions to Mode I, the LDO has to

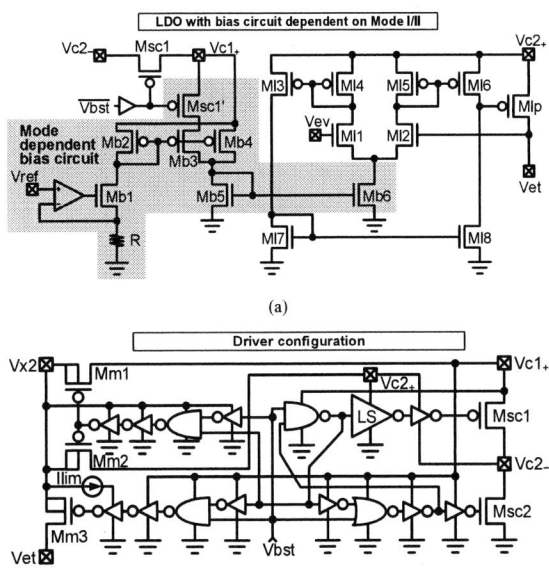

Fig. 4. (a) LDO with bias circuit dependent on Mode I/II. (b) Driver configurations.

handle all current demand, and a current limit Ilim is imposed on the driver of Mm3 to decrease the speed and switch off Mm3 to ensure a proper transition.

IV. EXPERIMENTAL RESULTS

The prototype in Fig. 5 includes three individual chips fabricated with a 0.18μm CMOS process. Experimental results are shown in Fig. 6. The proposed SCB mode envelope tracking regulator is capable of delivering up to 3.5V of envelope tracking voltage Vet (equivalent to the maximum voltage of HBT devices) with a slew rate exceeding 2V/ns, with a peak and valley inductor current of 400mA and 320mA, respectively. The voltage drop occurring on mode transitions is minimized to 0.3V. Mode I/II control and Vbst signal durations are computed in the digital domain clocked at 20MHz. The efficiency of envelope tracking regulators is heavily influenced by the envelope signal. To quantify actual efficiency improvements, rectified cosine waves, which are more stringent to the envelope tracking regulator compared to the slew-rate limited signal, are applied to allow for the accurate estimation of real world instantaneous efficiency rather than ideal peak efficiency. Fig. 7 compares the measured efficiency versus theoretical values with various amplitude distribution and PAPR by evaluating the minimum amount of energy which must be dissipated by the LDO. Assuming Vbat=3.6V (nominal Li-Ion battery voltage) and Vavg is equivalent to the DC component (1V minimum to maintain PA linearity) + 0.637 × peak AC component. η_{SBM} and η_{ET} are the theoretical efficiency limit (assuming Buck and SC converters have 100% efficiency) for SCB mode and conventional envelope tracking regulators; η_{MEAS} is the

978-1-7281-5107-6/19 $31.00 © 2019 IEEE

IEEE Asian Solid-State Circuits Conference
November 4 – 6, 2019
The Parisian Macao, Macao SAR, China

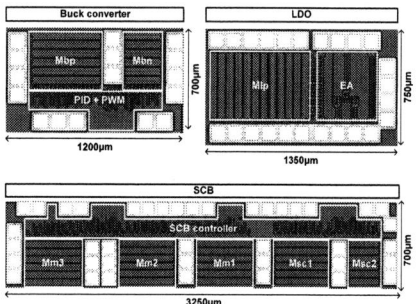

Fig. 5. Chip micrographs of the Buck converter, LDO, and the proposed SCB technique.

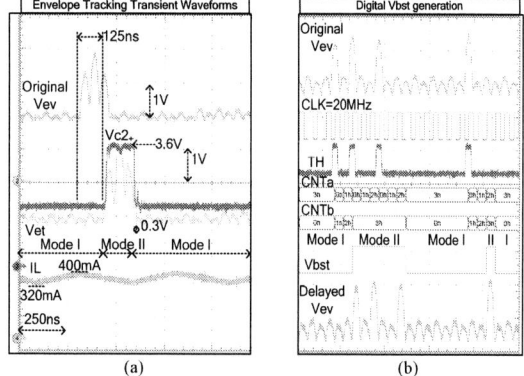

Fig. 6. Measurement results. (a) Envelope tracking transient waveforms. (b) The digital Vbst control generation and the timing diagram.

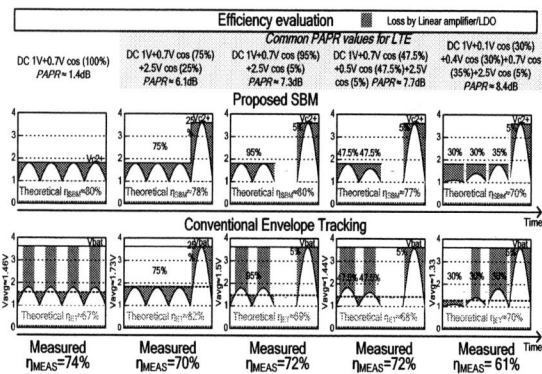

Fig. 7. Statistic data compares the measured efficiency versus theoretical values with various amplitude distribution and PAPR.

Table I: Comparison table.

Comparison with Prior Envelope Tracking Regulators									
ISSCC year	[1] 2013	[2] 2013	[3] 2014	[4] 2015	[5] 2016	[6] 2017	[7] 2017	[8] 2017	This work
CMOS process	0.15μm	0.18μm	0.13μm	0.13μm	0.13μm	28nm	0.5μm	65nm	0.18μm
Protocol/ bandwidth	HSUPA 5MHz	LTE 20MHz	LTE 70MHz	LTE 10MHz	LTE 40MHz	802.11n 40MHz	LTE 20MHz	LTE 20MHz	LTE 20MHz
Switching frequency	Fixed 1.6MHz	Max unknown	Up to 140MHz	Up to 250MHz	Up to 40MHz	Max unknown	Fixed 2MHz	Fixed 25MHz	Buck:2MHz SC:20MHz
Peak efficiency full power	80%	83%	86%	82%	83%	94%	82%	88.7%	Depends on PAPR
Avg. efficiency at half power (-3dB)	75%* estimated	<80%	72%	79%* estimated	79%	75%	68%	84%	72%
Max. power	0.66W	0.9W	1.8W	1.2W	0.54W	0.209W	2W	0.8W	1.42W

from 6.1dB to 7.3dB. However, conventional ET efficiency decreases from 82% to 69%.

measured efficiency. Notably, when PAPR increases from 6.1dB to 7.3dB, η_{ET} of the conventional method decreases from 82% to 69% whereas the SCB mode increases from 69% to 80. At a PAPR of 7.7dB (common for a modulated envelope of LTE communication), the efficiency is 72%, which is 4% higher than the theoretical limit of conventional envelope tracking techniques. For typical high PAPR envelope components, η_{SBM} is commonly equivalent to or higher than η_{ET}. Other specifications are compared with prior arts as well in Table I. The proposed SCB technique has improved efficiency over a wide variation of input PAPR and wide Vout range.

V. CONCLUSION

The proposed switched-capacitor boost (SCB) mode envelope tracking (ET) technique uses a switched-capacitor circuit and a low dropout (LDO) regulator to improve the efficiency. The conventional linear amplifier can be more efficient due to the SCB technique. The efficiency is 72% at a PAPR of 7.7dB common for LTE communication. It is 4% higher than the theoretical limit of conventional ET techniques. The SCB mode increases by 2% when the PAPR increases

REFERENCES

[1] P. Riehl, et.al, "An AC-coupled hybrid envelope modulator for HSUPA transmitters with 80% modulator efficiency," *ISSCC*, pp. 364-365, Feb. 2013.

[2] M. Hassan, et. al., "A CMOS dual-switching power-supply modulator with 8% efficiency improvement for 20MHz LTE Envelope Tracking RF power amplifiers," *ISSCC*, pp. 366-367, Feb. 2013.

[3] P. Amò, et. al., "Envelope modulator for multimode transmitters with AC-coupled multilevel regulators," *ISSCC*, pp. 296-297, Feb. 2014.

[4] S.-C. Lee, et. al., "A hybrid supply modulator with 10dB ET operation dynamic range achieving a PAE of 42.6% at 27.0dBm PA output power," *ISSCC*, pp. 42-43, Feb. 2015.

[5] J.-S. Paek, et. al., "An RF-PA supply modulator achieving 83% efficiency and -136dBm/Hz noise for LTE-40MHz and GSM 35dBm applications," *ISSCC*, pp. 354-355, Feb. 2016.

[6] D. Chowdhury, et. al., " A fully integrated reconfigurable wideband envelope-tracking SoC for high-bandwidth WLAN applications in a 28nm CMOS technology," *ISSCC*, pp. 34-35, Feb. 2017.

[7] S. Yang, et. al., "A single-inductor dual-output converter with linear-amplifier-driven cross regulation for prioritized energy-distribution control of envelope-tracking supply modulator" *ISSCC*, pp. 36-37, Feb. 2017.

[8] X. Liu, et. al., " A 2.4V 23.9dBm 35.7%-PAE -32.1dBc-ACLR LTE-20MHz envelope-shaping-and-tracking system with a multiloop-controlled AC-coupling supply modulator and a mode-switching PA," *ISSCC*, pp. 38-39, Feb. 2017.

978-1-7281-5107-6/19 $31.00 © 2019 IEEE

IEEE Asian Solid-State Circuits Conference
November 4 – 6, 2019
The Parisian Macao, Macao SAR, China

A Conversion-Ratio-Insensitive High Efficiency Soft-Charging-Based SC DC-DC Boost Converter for Energy Harvesting in Miniature Sensor Systems

Junyoung Park[12], Hyungmin Gi[1], Seungchul Jung[3], Sang Joon Kim[3], Yoonmyung Lee[1]

[1]Sungkyunkwan University, Suwon, Korea   [2]Samsung Electronics, Suwon, Korea
[3]Samsung Advanced Institute of Technology, Suwon, Korea
yoonmyung@skku.edu

Abstract – A switched-capacitor (SC) DC-DC boost converter suitable for energy harvesting in IoT sensor systems with varying harvesting source and battery voltages is presented in this paper. Unlike the conventional multi-staged SC DC-DC converters, where efficiency is optimized only for a few topology-dependent conversion ratios, soft-charging-based SC conversion is adopted to achieve conversion-ratio-insensitive high efficiency. With soft-charging flying capacitors with small voltage steps, charge re-distribution loss is minimized and evenly high conversion efficiency can be achieved for a wide range of input and output voltages. The proposed converter exhibited evenly high efficiency by achieving a peak efficiency of 85.3% and an average efficiency of 83.6% for generating regulated 1.8V output from a wide range of input voltages (0.9–1.8V). It also achieved a peak efficiency of 88.9% and an average efficiency of 87.6% for harvesting energy from a 1.2V source with a wide battery voltage range (3.0–4.2V).

Keywords – soft-charging, switched capacitor, boost converter, DC-DC converter, charge re-distribution

I. INTRODUCTION

Recently, mm-scale miniature sensor systems have been demonstrated for a variety of applications [1]. For such miniature systems, energy harvesting is often adopted to extend the battery lifetime or even achieve energy-autonomous operation. The harvested energy can be either stored in the battery for later use or directly consumed for sensor node operation. In both cases, efficient voltage conversion is necessary, and proper regulation is also required for direct use on the load circuit. However, achieving efficient voltage conversion is a non-trivial challenge, especially when the harvesting source voltage can vary with environmental conditions and the battery voltage can change over the battery's lifetime.

In miniature systems, switched capacitor(SC)-based voltage converters are often utilized since they allow the integration of their passive elements and compliance with the miniature sensor node's stringent volume restriction. In contrast, handling a few or hundreds of microwatts of power with an inductor-based voltage converter requires a bulky inductor or extremely high operation frequency, making it an unviable solution. Since many harvesting sources have low output voltages, as low as 0.64V to 1.4V, whereas many lithium-based high energy density batteries have voltages ranging from 3.0V to 4.2V, an SC-based boost converter is required for energy harvesting in such systems.

The conversion efficiency of a conventional SC-based boost converter is very sensitive to the voltage conversion ratio (VCR) as shown in Fig. 1. Ideally, achieving evenly high conversion efficiency for a wide range of VCRs, as shown by

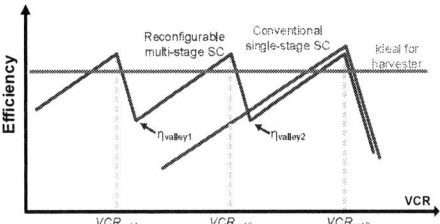

Fig. 1. VCR vs Efficiency graph of harvester topologies

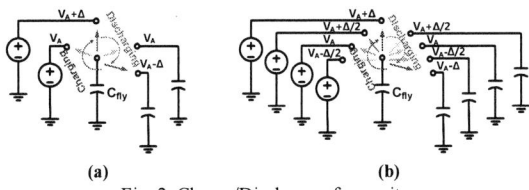

(a)   (b)

Fig. 2. Charge/Discharge of capacitor

the red line, is desired. However, a conventional single-stage SC structure has a single optimal VCR, marked as VCR_{opt3} in Fig. 1, which is determined by the SC network topology. If the SC converter operates with a VCR far from VCR_{opt3}, the flying capacitor in the SC network must be charged and discharged by a large voltage step, significantly increasing the charge re-distribution loss (CRL) and thereby reducing the efficiency. Many reconfigurable multi-stage SC-based boost converters have been proposed [3][4] that address this issue by providing additional optimal VCRs (VCR_{opt1} and VCR_{opt2} in Fig. 1) by dynamically adjusting the SC network topology. However, to provide evenly high conversion efficiency for a wide range of VCRs, many VCR_{opt} are required, which eventually degrades efficiency with more conduction loss due to the additional cascaded SC stages.

In this paper, a fully integrated soft-charging-based SC boost converter that enables VCR-insensitive evenly high conversion efficiency is presented. In the proposed SC boost converter, soft charging is adopted to minimize CRL during charge transfer between the integrated flying capacitors. By allowing flexible charging/discharging step voltage, CRL is suppressed for wide ranges of input and output voltage changes. A test chip is fabricated in 180nm technology to demonstrate the proposed converter's operation for charging the battery (3.0V-4.2V) and providing regulated voltage (1.8V) to directly supply power for circuit operation.

978-1-7281-5107-6/19 $31.00 © 2019 IEEE

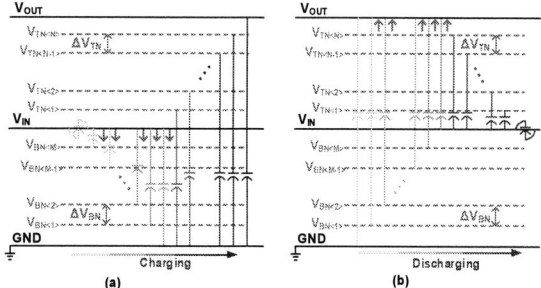

Fig. 3. Soft (a) charging and (b) dis-charging sequences of a flying capacitor for boost conversion

II. SOFT CHARGING FOR BOOST CONVERSION

A. Charge redistribution loss reduction with soft charging

Soft charging [5] is an effective SC operation scheme that can significantly reduce CRL. Fig. 2 conceptually shows the charge transfer process of a flying capacitor in an SC converter. In Fig. 2(a), the flying capacitor is charged and discharged by a voltage step of ΔV in every step: the initial voltage is V_A, which is then charged to $V_A + \Delta V$, and then discharged to V_A, and $V_A - \Delta V$. In each charging step, the amount of charge transferred to the flying capacitor is $C_{fly} \cdot \Delta V$ and the CRL is $\frac{1}{2} C_{fly} \cdot (\Delta V)^2$, where C_{fly} is the capacitance of the flying capacitor. In each discharging step, the amount of charge transferred from the flying capacitor to the load capacitor and CRL are identical with those in the charging steps. Therefore, after one cycle of charging/discharging operation, $2 \cdot C_{fly} \cdot \Delta V$ of charge is transferred from the voltage sources to the load capacitors with a CRL of $4 \cdot \frac{1}{2} \cdot C_{fly} \cdot (\Delta V)^2$. Fig. 2(b) shows the case where the voltage step is reduced to $\Delta V/2$ for each charging/discharging step, but the flying capacitor is still charged up to $V_A + \Delta V$ and discharged down to $V_A - \Delta V$. In this case, the amount of charge transferred and CRL in each step is now $C_{fly} \cdot \Delta V/2$ and $\frac{1}{2} C_{fly} \cdot (\Delta V/2)^2$, respectively. With one cycle of charging/discharging operation, $2 \cdot C_{fly} \cdot \Delta V$ of charge is still transferred to the load capacitors, but the CRL is reduced to $8 \cdot \frac{1}{2} \cdot C_{fly} \cdot (\Delta V/2)^2$. Therefore, if the charging/discharging step is set to $\Delta V/N$, the total amount of charge transferred to the load would remain unchanged, whereas the CRL can be reduced to $4N \cdot \frac{1}{2} \cdot C_{fly} \cdot (\Delta V/N)^2 = 2\,C_{fly} \cdot (\Delta V)^2/N$. Therefore, in general, to take advantage of CRL reduction with soft charging, N should be made larger. However, excessively increasing N would also incur significant control overhead and related switching loss; hence, a reasonable N value and corresponding SC network topology must be determined for the proposed soft-charging-based SC boost converter.

B. Soft-charging/discharging sequence for boost conversion

Fig. 3 shows how a flying capacitor can be charged or discharged with small voltage steps (ΔV_{BN}, ΔV_{TN}). V_{IN} is the input voltage from the harvesting source, and V_{OUT} is the output voltage, i.e. the battery voltage. Between V_{IN}, V_{OUT} and GND, the virtual voltage levels ($V_{BN<1...M>}$, $V_{TN<1...N>}$) are assumed to be available. The solid lines represent the real voltages supplied by the harvesting source or battery, and the dotted lines represent the virtual voltage levels that are assumed to be generated with soft-charging operation. The

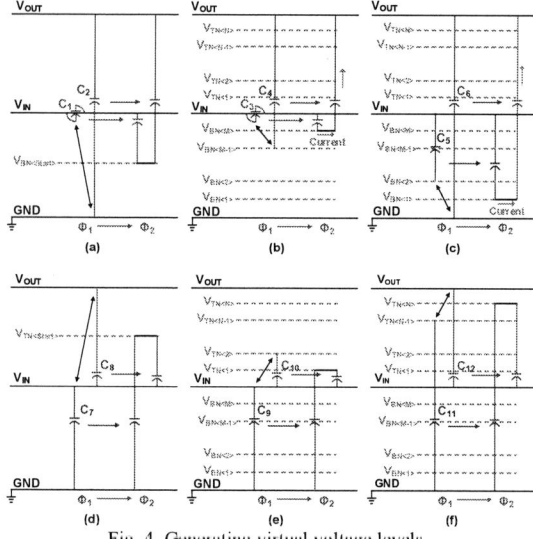

Fig. 4. Generating virtual voltage levels

vertical location represents the magnitude of the voltages: V_{OUT} is above V_{IN} to represent boost conversion.

Assuming the availability of virtual voltage levels, a flying capacitor can be charged gradually by using these virtual voltage levels. The capacitor's connection diagram shows how the top and bottom plates of the flying capacitor should be connected in each phase. By following the sequences shown in Fig. 3(a), a flying capacitor can be gradually charged from zero voltage to V_{OUT} using these virtual voltage levels. Note that many early phases draw charge from V_{IN}, which is desired for harvesting operation. As discussed earlier, CRL during charging operation can be minimized by keeping the voltage differences between the virtual voltage levels small.

Fig. 3(b) shows the discharging sequence of the capacitor. Similar to the charging sequence, with virtual voltage levels, the capacitor can be discharged with small voltage steps to reduce CRL. In addition, the flying capacitor's bottom plate parasitic capacitance is charged and discharged with small voltage steps, reducing parasitic loss as well [6]. During the discharging process, many early phases deliver charge to V_{OUT}, which is desired for harvesting operation as well. By repeating these charging and discharging sequences, the flying capacitors can collect charges from the harvesting source and efficiently deliver them to the battery.

C. Generating virtual voltage levels

To operate as described in Fig. 3, virtual voltage levels should be generated with soft-charging operation, and the voltage steps (ΔV_{TN}, ΔV_{BN}) should be kept small. In the beginning of SC operation, the capacitors can be only connected to real voltage rails, and all other virtual voltages should be generated from these voltages. To generate a voltage between V_{IN} and ground, two capacitors can be initially connected between (V_{OUT}, GND) and (V_{IN}, V_{IN}) as shown in Φ_1 of Fig. 4(a). Then, as the phase is switched to Φ_2, C_1, C_2 are connected together to charge-share and generate intermediate voltage $V_{BN<start>}$. A similar operation can be performed to generate voltage between V_{IN} and V_{OUT} as

978-1-7281-5107-6/19 $31.00 © 2019 IEEE

IEEE Asian Solid-State Circuits Conference
November 4 – 6, 2019
The Parisian Macao, Macao SAR, China

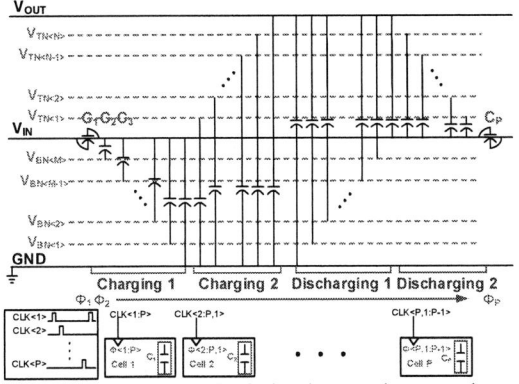

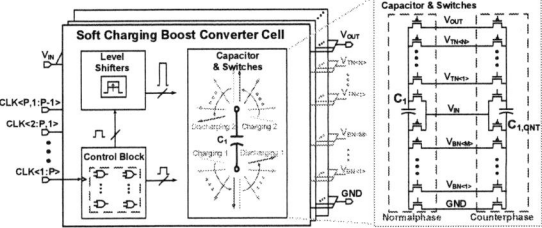

Fig. 5. Block diagram of soft-charging capacitor network

Fig. 6. Details of flying capacitor cell

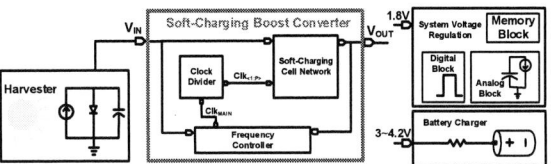

Fig. 7. Two implementation scenarios for propose boost converter make all intermediate voltage levels available for soft-charging operation. With these procedures, the flying capacitors can be charged/discharged with small voltage steps while transferring energy from input to output. To realize these operations, the corresponding SC network topology and switching sequence should be carefully designed.

III. SOFT-CHARGING-BASED SC BOOST CONVERTER DESIGN

Fig. 5 shows the flying capacitor phase allocation scheme for boost converter implementation. Each capacitor (C_1, C_2, … C_P) in the top corresponds to a flying capacitor cell at the bottom. Each cell is in a different phase, and it is guaranteed that two flying capacitors are paired to form intermediate voltage levels at any given time. Each capacitor cell is fed with a set of non-overlapping shifted clocks (CLK<1..P>) in a shifted manner. As one of the clocks is asserted from CLK<1> to CLK<P>, the capacitor's phase is changed from Φ_1 to Φ_2, Φ_3, and eventually to Φ_P, and this sequence is then repeated from Φ_1. Each capacitor goes through a set of 'Charging 1, 2' and 'Discharging 1, 2' phases and transfer charges from V_{IN} to V_{OUT}.

Fig. 6 shows the details of the flying capacitor cell, which consists of a flying capacitor, a set of connection switches, a control block and level shifters. The control block generates the required control signals for the switches. For efficiency, input voltage V_{IN} is used as a power supply for most of circuits, except level shifted portion. Therefore, when lower V_{IN} is used, V_{gs} of switches is decreased, resulting in higher resistive loss. This limits the minimum of V_{IN}. To control switches connected to voltage levels higher than V_{IN}, level shifters are utilized to generate high-voltage-swing control signals. The phase-antiphase control scheme in [6] is adopted to reduce the number of cells and the switching loss.

The proposed boost converter is implemented for the two scenarios as shown in Fig. 7: as a battery charging harvester and as a power management unit (PMU) for directly supplying regulated power to a sensor system. As a battery charger, a 104 phase SC network is implemented for 11 intermediate voltage levels between V_{IN} and GND, and 39 levels between V_{IN} and V_{OUT}, where V_{OUT} is expected to be 3.0V-4.2V depending on the battery condition. As a PMU, a 72 phase SC network is implemented for 19/15 intermediate voltage levels below/above V_{IN}, respectively. The output voltage is monitored to regulate the voltage supplied to the system by adjusting the switching frequency. The frequency controller adjusts the main clock frequency by monitoring feedback, and the multi-phase clock generator splits the clock into P phases so that they can be directly fed to the SC cells.

IV. MEASUREMENT RESULTS

Two types of soft-charging-based SC boost converters, a battery charger and a PMU, are fabricated in 180nm CMOS technology. The PMU is designed with standard transistors,

shown in Fig. 4(d). After Φ_2, if one of the capacitors in Φ_2 is connected to one of the capacitors in Φ_1, another voltage level between $V_{BN<start>}$ and V_{IN} or GND can be generated as shown in Fig. 4(b) and (c). In a similar fashion, other intermediate voltage levels between V_{IN} and V_{OUT} can be generated as shown in Fig. 4(e) and (f). Repeating these sequences allows the generation of many virtual voltage levels with small voltage steps.

Fig. 4(b), (c) also shows how the charge and energy can be transferred from V_{IN} to V_{OUT}. Initially, the bottom plate voltages of C_3 and C_4 in Φ_1 of Fig. 4(b) differ by 2 ΔV_{BN}. The top plates of the capacitors are connected to static power rails (V_{IN}, V_{OUT}). Therefore, when the bottom plates of C_3 and C_4 are connected as shown in Φ_2, charge sharing charges C_3 by ΔV_{BN} and discharges C_4 by ΔV_{BN}. Such operation generates a new voltage level $V_{BN<M>}$, which is the average of the bottom plate voltages. During this operation, charge is drawn from V_{IN}, and same amount of charge is delivered to V_{OUT}, achieving energy transfer from V_{IN} to V_{OUT}. As another example, C_5 and C_6 in Fig. 4(c), whose bottom plate voltages are at different level, but difference is also $2\Delta V_{BN}$, can charge-share just like C_3 and C_4. An average voltage level $V_{BN<1>}$ is generated, and charge transfer from V_{IN} to V_{OUT} is achieved. By adding more capacitors for these sequences, more virtual voltage levels can be generated, and efficient voltage conversion can be achieved by reducing CRL.

Fig. 4(e), (f) shows how virtual voltage levels between V_{OUT} and V_{IN} are maintained. Similar to that shown in Fig. 4(b) and (c), two capacitors with a top plate voltage difference of $2\times\Delta V_{TN}$ are charge shared to generate average top plate voltages. Although these cases do not directly transfer charges from V_{IN} to V_{OUT}, these cases are required to charge capacitor to full V_{OUT} range and discharge capacitor to zero voltage to

978-1-7281-5107-6/19 $31.00 © 2019 IEEE

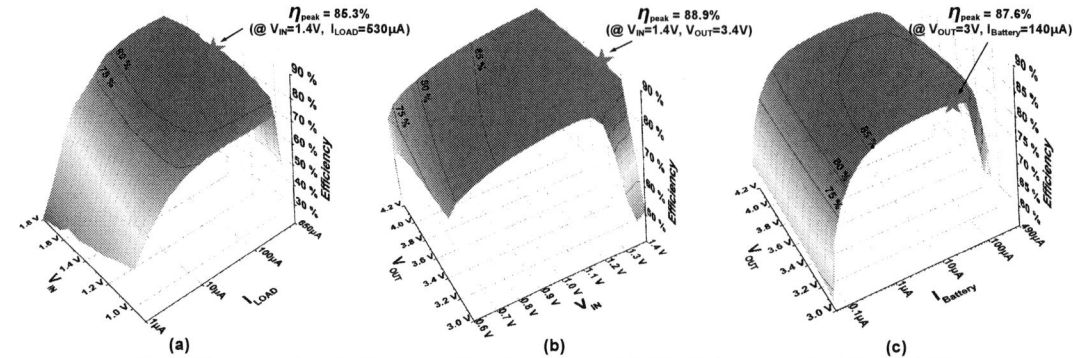

Fig. 8. Measurement results for proposed boost converter: (a) PMU efficiency with fixed V_{OUT} = 1.8V
(b) battery charger efficiency with fixed F_{clk}=100kHz (c) battery charger efficiency with fixed V_{IN} = 1V

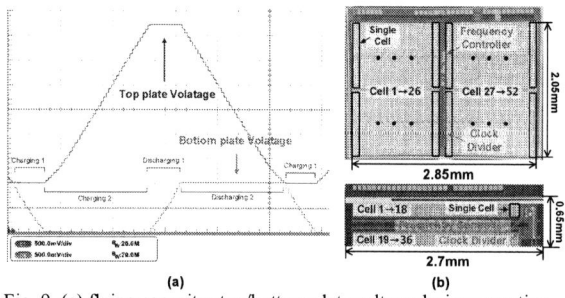

Fig. 9. (a) flying capacitor top/bottom plate voltage during operation
(b) chip micrographs : Battery charger(top), PMU(bottom)

whereas the battery charger is designed with I/O transistors to deal with high battery voltage. For the battery charger, 19.8nF integrated capacitor (8nF MIM, 11.8nF MOS) is used. For the PMU, 5.5nF capacitance is used.

Fig. 8 shows the efficiency measurement results for the proposed converter. In Fig. 8(a), the PMU efficiency is measured with 1.8V regulated output voltage. Evenly high efficiency is observed for wide V_{IN} and I_{LOAD} ranges. The average efficiency for the entire voltage range at the target load current (300µA) is 83.6%. Fig. 8(b), (c) shows the measurement results for the battery charger. With the clock frequency set to 100kHz, the efficiency is measured for a wide range of V_{IN} (0.64−1.4V) and V_{OUT} (3−4.2V) in Fig. 8(b). Efficiency higher than 75% is achieved for more than 93% of V_{IN}-V_{OUT} configurations for the measured voltage range, confirming the evenly high efficiency of the proposed soft-charging-based boost converter. For 1V input, an average efficiency of 85.8% is achieved across the entire battery voltage range. The efficiency plotted as a function of the battery charging current in Fig. 8(c) shows that >80% efficiency is maintained even when the charging current is as low as 1µA. Table 1 summarizes measurement results and compares with recently published SC-based boost converters. With reconfigurable SC topologies, there exists valley efficiencies, as shown in Fig. 1, and the difference between valley and peak efficiencies are as high as 13% in [3], [4]. Such valleys are undesirable for applications that require operation with continuous VCRs. Sensitivity of efficiency on VCR is estimated in Table 1, and the proposed topology significantly reduced the sensitivity, representing evenly high efficiencies across wide range of VCRs. Fig. 9 shows the

Metric	[2]		[3]	[4]	This Work	
Technology	0.18µ		28n FDSOI	0.35µ	0.18µ	
Topology	Charge pump		Reconfigurable	Reconfigurable	Soft-charging	
Application	PMU	Battery Charger	PMU	PMU	PMU	Battery Charger
V_{IN} (V)	0.35 – 0.60		0.7, 1	1.4 – 3	0.95 – 1.8	0.64 – 1.4
V_{OUT} (V)	0.86 – 1.8	2.5 – 5.2	0.75 – 1.5, 1.2 – 2.4	4.8	1.8	3.0 – 4.2
Area(mm²)	1.75		0.114	3.8 *	1.75	5.84
Power(W)	< 396 µ	< 114 µ	< 830 µ	< 48 m	< 1.17 m	< 2.1 m
VCR Range	3.0	9.0	1.1 – 2.14, 1.2 – 2.4	1.6 – 3.43	1.0 – 1.89	2.14 – 6.56
VCR Sweep Dimension	Fixed	Fixed	1-D (V_{OUT})	1-D (V_{IN})	1-D (V_{IN})	2-D (V_{IN}, V_{OUT})
Peak Efficiency	75.8% @0.59V V_{IN} 1.41V V_{OUT}	49.1% @0.62V V_{IN} 4.54V V_{OUT}	88% @1.0V V_{IN} 1.9V V_{OUT}	82% @2.6V V_{IN}	85.3% @1.4V V_{IN}	88.9% @1.4V V_{IN} 3.4V V_{OUT}
Average Efficiency	-**		77%*** @1.0V V_{IN} 100µA	70%*** @10mA	83.6% @300µA	87.6% @1.2V V_{IN} 23.8µA
Valley Efficiency	-**		< 75%@1.0V V_{IN}*** < 70%@0.7V V_{IN}***	63%*** @1.7V V_{IN}	N/A	N/A
VCR Sensitivity****	> 8.33 @ 1.0V V_{IN} > 16.8 @ 0.7V V_{IN}		10.6	3.70 @4.8V V_{OUT}	3.13 @ 0.8V V_{IN} @1.8V V_{OUT}	3.57 @ 3.2 V_{OUT}

*use external passive elements **fixed VCR ***estimate based on graphs
**** Voltage Conversion Ratio Sensitivity = $(\eta_{peak} - \eta_{min})/\Delta VCR$ [%/VCR]

Table. 1. Measurement result summary and comparison

measured flying capacitor top/bottom plate voltages during conversion operation and chip micrographs.

V. CONCLUSIONS

A soft-charging-based SC DC-DC boost converter for energy harvesting in miniature sensors is presented in this paper. VCR-insensitive high conversion efficiency is achieved by adopting a soft-charging scheme and minimizing CRL. Our measurement results show that the proposed converter achieves evenly high conversion efficiency for a wide range of input/ output voltages and currents, making it an ideal solution for use in energy harvesting in miniature sensor systems.

REFERENCES

[1] I. Lee, et al "A 10mm³ Light-Dose Sensing IoT² System with 35-to-339nW 10-to-300klx Light-Dose-to-Digital Converter," IEEE Symposium onVLSI Circuits, June 2019..

[2] T. Ozaki, et al "Fully-Integrated High-Conversion-Ratio Dual-Output Voltage Boost Converter With MPPT for Low-Voltage Energy Harvesting," IEEE Journal of Solid-State Circuits, Oct. 2016.

[3] Avishek Biswas, et al. "A 28 nm FDSOI Integrated Reconfigurable Switched-Capacitor Based Step-Up DC-DC Converter With 88% Peak Efficiency" IEEE Journal of Solid-State Circuits, July 2015

[4] H. Lee, et al "A Reconfigurable 2x/2.5x/3x/4x SC DC–DC Regulator With Fixed On-Time Control for Transcutaneous Power Transmission" IEEE Transactions on VLSI Systems, April 2015.

[5] Y. Lei, et al. "Split-Phase Control: Achieving Complete Soft-Charging Operation of a Dickson Switched-Capacitor Converter," IEEE Transactions on Power Electronics, Jan. 2016.

[6] N. Butzen, et al, "Design of Single-Topology Continuously Scalable-Conversion-Ratio Switched- Capacitor DC–DC Converters," IEEE Symposium onVLSI Circuits, April 2019.

33us, 94uJ Optimal Ate Pairing Engine on BN Curve over 254b Prime Field in 65nm CMOS FDSOI

Makoto Ikeda, Tadayuki Ichihashi[*1], Hiromitsu Awano[*2]

VLSI Design and Education Center, the University of Tokyo
2-11-16, Yayoi, Bunkyo-ku, Tokyo, JAPAN
ikeda@silicon.u-tokyo.ac.jp
[*1]Now with Konami Corp., Tokyo Japan, [*2]Now with Osaka Univ. Osaka Japan

Abstract— We have designed optimal ate paring engine on BN curve over 254bit prime field in 65nm CMOS FDSOI process. Algorithm is broken down into 2nd extension of the prime field (Fp2) and optimized for pipelined multiplier of Fp2, results in global optimum design. Measurement results demonstrate world fastest implementation of 33us paring time with as low as 94uJ per paring. Energy consumption can be minimized down to 13.7uJ per pairing for 490mV operation. The design consists of 2.8MG and 85% of functional units are active throughout the paring operation. The demonstrated results are promising to realize functional encryptions such like ID based encryption, attribute-based encryption and searchable encryption.

Keywords— *Optimal ate paring, BN curve, 254bit prime field, 65nm CMOS, crypto engine design, and functional encryption*

I. Introduction

In the ICT era, the importance of cryptography not only for preventing information leakage during data transfer and data storage but also for the authenticity of data, is increasing more and more. Shared-key cryptography such like AES and SM4 have been widely studied for high-throughput and smaller hardware realizations for data encryption[1]. While the shared-key cryptosystems are compact and can realize high-performance for enough security strength, key management in many-to-many communication such as IoT equipment and vehicle-to- vehicle communication has become a big problem. Public-key cryptography, on the other hand, such like RSA and ECC utilizes two keys, public key and secret key, and data encryption is carried out only by the public key, and decryption is possible with the secret key. Security of RSA relies on the discrete logarithmic problem (DLP), which comes from the difficulty of calculating a and k from $x=a^k$, in a finite field with a set of numbers by a reminder of a certain prime number p. RSA, however, requires longer key length for the same security level as AES, so that elliptic curve based encryption is becoming major stream.

In the era of IoT, where billions of network nodes access to the networks, not only securing the data transfer, but also functionality and convenience for cryptosystems are expected. To meet such requirements, functional encryption such like ID-based encryption, attribute-based encryption and searchable encryption, which is usually based on bi-linear mapping, are emerging. Bilinear mapping which calculates number in extended field from two points on a elliptic curve is called paring. Most of functional encryption requires paring operation which is usually heavily computational complexity, which is not practical for IoT edge devices, nor even for cloud data center applications. Especially, number of paring required for the searchable encryption is in proportion to the number of records. Latest results of paring engine design by software on the high-performance processor is ~200us, and by latest FPGA is 100us~60us, which is far below the required performance for such application like searchable encryption with more than 1 million records. To cope with this situation, scalable energy efficient as well as high-performance paring hardware is highly expected.

II. Paring Operation on BN Curve

Set of curves expressed as $y^2=x^3+ax+b$ is called elliptic curve. For the elliptic-curve based cryptography, all the parameters and variables, x,y,a,b are the member of a finite field Fp, and all the defined points over the elliptic curve and the infinity point O form set of points E(Fp). Barrento-Naehrig curve (BN curve) $y^2=x^3+b$ had been proposed as a paring friendly curve. Where key-length $p=36u^4+36u^3+24u^2+6u+1$, and number of points on the curve $r=36u^4+36u^3+18u^2+6u+1$. BN curve has a feature that the embedding degree $k=12$, which is defined as the minimum extension field where a group with order r independent from E exists. Assuming number of bits for key is l, pairing output results in $k \times l$ bits, which is 3,048bits for $k=254$bit. ECC friendly curve such like FourQ has large $k\sim3.7\times10^{73}$ bit, results in more than 1,063TB of results database, which is not practical for the paring. Note that there have been several other curves proposed claiming as pairing friendly curves, like MNT curve, $k=6$, KSS curve, $k=16$, and BLS curve, $k=24$.

Pairing has non-degeneracy, along with bilinearity. Optimal Ate paring[3] is an algorithm enables efficient calculation of paring operation, which consist of the following three part: 1) Miller loop, 2) Final Addition, and 3) Final Exponentiation, listed as Fig.1. Miller loop is to establish a rational function with poles at the r-th position. For the Optimal Ate paring, as Miller loop calculates l-th poles, instead of r-th, Final addition corrects so that results to have r-th poles. For the case of the embedding degree $k=12$, Fp_{12} calculation and $E(Fp_2)$ calculation dominate. Final Exponentiation will eliminate the optionality of Miller function, where Fp_{12} calculation and inverse calculation dominate. Nagashima[6] optimized Fp_{12} calculation by global optimization of pipelined Fp multipliers, results in 18,151 cycles for entire Optimal Ate paring operation. Awano[10] optimized Fp_{12} calculation by global optimization of pipelined Fp_2 multipliers, results in 9,270 cycles The target of this study is to

IEEE Asian Solid-State Circuits Conference
November 4 – 6, 2019
The Parisian Macao, Macao SAR, China

Input: $P \in \mathbb{G}_1, Q \in \mathbb{G}_2, l = |6u + 2| = \sum_{i=1}^{\log_2 l} l_i 2^i$

Output: $a_{opt}(Q, P)$

   Miller Loop:

1: $d \leftarrow g_{Q,Q}(P),\ T \leftarrow 2Q,\ e \leftarrow 1$

2: **if** $l_{\lfloor \log_2 l \rfloor - 1} - 1$ **then** $e \leftarrow g_{T,Q}(P),\ T \leftarrow T + Q$

3: $f \leftarrow d \cdot e$

4: **for** $i = \lfloor \log_2 l \rfloor - 2$ **downto** 0 **do**

5: $f \leftarrow f^2 \cdot g_{T,T}(P),\ T \leftarrow 2T$

6: **if** $l_i = 1$ **then** $f \leftarrow f \cdot g_{T,G}(P),\ T \leftarrow T + Q$

7: **end for**

Final Addition:

8: $Q_1 \leftarrow \pi_p(Q),\ Q_2 \leftarrow \phi_p^2(Q)$

9: **if** $u < 0$ **then** $T \leftarrow -T,\ f \leftarrow f^{p^6}$

10: $d \leftarrow g_{T,Q_1}(P),\ T \leftarrow T + Q_1$

11: $e \leftarrow g_{T,-Q_2}(P),\ T \leftarrow T - Q_2$

12: $f \leftarrow f \cdot (d \cdot e)$

Final Exponentiation:

13: $f \leftarrow f^{(p^6-1)(p^2+1)(p^4-p^2+1)/r}$

14: **return** f

Fig. 1. Optimal Ate Paring Algorithm.

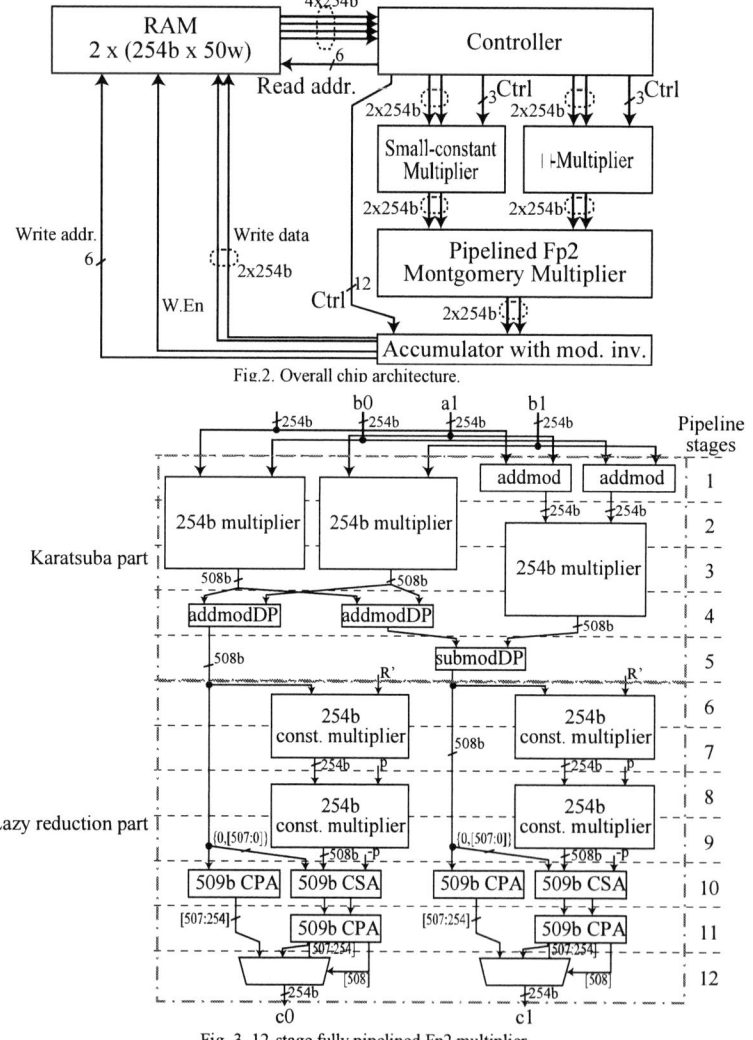

Fig.2. Overall chip architecture.

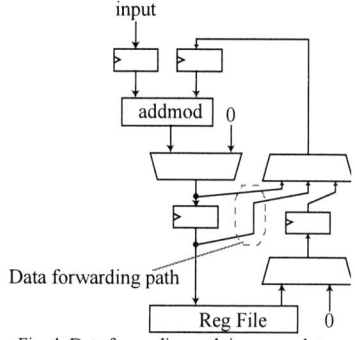

Fig. 4. Data forwarding path in accumulator.

Fig. 3. 12-stage fully pipelined Fp2 multiplier.

optimize number of cycles by fully pipelined Fp_2 multiplier and also minimize control steps to realize shorter control critical paths.

III. ARCHITECTURE, OPTIMIZATION AND MEASUREMENT RESULTS

Fig. 2 shows entire chip architecture. Here Fp_{12} calculation consists of 144 Fp operations, which have less data dependence to each other. According to data bandwidth and hardware efficiency, such highly parallel realization is nothing practical. On the other hand, Fp_{12} operations have strong data dependency, which is not practical for scheduling optimization. In this study, we have implemented fully pipelined 12-stage Fp_2 multiplier, with Karatsuba method and Lazy Reduction method, as shown in Fig. 3. Karatsuba method calculate multiplication of $\gamma = \alpha \times \beta$: $\alpha \in Fp_2 = a_0 + a_1 i, \beta \in Fp_2 = b_0 + b_1 i, \gamma \in Fp_2 = c_0 + c_1 i$ as $c_0 = a_0 b_0 - a_1 b_1, c_1 = [(a_0 + a_1)(b_0 + b_1) - a_0 b_0 - a_1 b_1]$, which reduces number of

978-1-7281-5107-6/19 $31.00 © 2019 IEEE   264

Table I. Schedule optimization results.

Name	Operation	#ops	Cycles/op	Operation efficiency
Init-PADD	Initialize-S12-PDBL-SM12-PADD-SM12	1	93	72.0%
PADD	S12-PDBL-SM12-PADD-SM12	4	107	90.7%
PDBL	S12-PDBL-SM12	58	56	100%
Final Addition	PADD Phase-PADD-SM12-PADD-SM12	1	162	78.8%
FE-EP	Final Exp. Easy Part incl. Inv.	1	431(254)	25.3%
SQR_{012345}	S12 by SQR	192	17	76.4%
M12	Mult. in Fp_{12}	20	39	92.3%
Frobenius	1st and 2nd Frobenius	7	12	58.3%
CONJ12	Conjugate in Fp_{12}	2	12	50.0%

Table II. Total Clock cycles for one Optimal Ate Paring.

Name	Cycles	Operation efficiency
Miller Loop	3,697	99.3%
Final Addition	162	78.8%
Final Exponentiation Easy Part	431	25.3%
Final Exponentiation Hard Part	3,885	88.4%
Total	8,175	84.6%

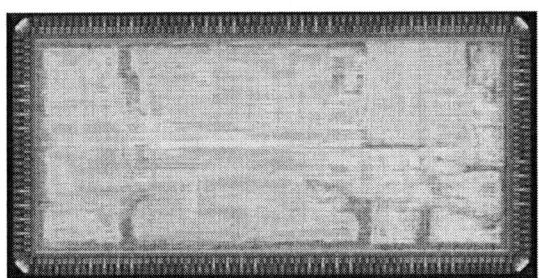

Fig. 5. Chip layout in 6mmx3mm die.

multiplications. We break-down entire paring algorithm into Fp_2 operations. We carried out global schedule optimization of Fp_{12} operation with Fp_2 operations, by using number of clock cycles for entire Fp_{12} operation as the optimization metrics, and number of pipeline stages, as well as number of 254b-multipliers of Fp_2 operations as the parameter. For relaxing data dependency restrictions, we designed accumulator with data forwarding mechanism as shown in Fig. 4. We employ radix-4 unified division[11] for the Montgomery inversion operation, which usually takes $3l$ cycles, results in at most l cycles. This results in maximum 500 cycles reduction. For the Final Exponentiation, we employ a kind of compressed squaring, SQR_{012345}[12], which results in over 2,000 cycles reduction. Table I summarizes required clock cycles for operations and Table II summarizes total number of clock cycles required for one Optimal Ate paring operation, which is reduce to 8,175 cycles, 12% decrease from the prior art, with keeping hardware efficiency of 85%.

The optimized design has been implemented in 65nm CMOS FDSOI, with 2,793kG occupying 6mm x 3mm die, as shown in Fig. 5. We carried out functional and at-speed measurements by logic tester T-2000 by Advantest in room temperature. Fig. 6 shows measurement results of Paring time, t_{Paring}, power consumption during paring operation, P_{Paring}, and energy consumption required for one paring operation, E_{Paring} according to power supply change. t_{Paring}=111us at nominal operating voltage of Vdd=0.75V with P_{Paring}=240mW and E_{Paring}=26.5uJ/Paring, with V_{BP}=0V and V_{BN}=Vdd body bias conditions. Measurement results show the fastest operation of t_{Paring}=33us at Vdd=1.33V with P_{Paring}=2.85W and E_{Paring}=94.0uJ/Paring, with V_{BP}=-0.3V and V_{BN}=0.4V body bias conditions, and the minimum energy of E_{Paring}=13.7uJ/Paring at Vdd=0.49V with t_{Paring}=792us and P_{Paring}=17.2mW, with V_{BP}=0V and V_{BN}=Vdd body bias conditions.

The paring chip operates down to Vdd=0.478V with t_{Paring}=908us, P_{Paring}=16.8mW and E_{Paring}=15.3uJ/Paring, with V_{BP}=-0.3V and V_{BN}=0.4V body bias conditions.

The measured results show 3 times faster than ever reported, in terms of measured results, order of magnitude smaller energy consumption ever reported. Comparison results to the prior art are listed in Table III. Technology scaling forecasts 6.5us paring operation with 2.3uJ per paring for 12nm FINFET process with Vdd=550mV.

One of the promising applications for this Optimal Ate Paring engine is searchable encryption. For searchable encryption, not only simple paring operation but further complicated computation, such like n-times Muller Loop, followed by one Final Exponentiation operation. The presented chip consists of Fp_2 based data path with controlling sequencer, can be applied to such applications without any redesign. For a case of n=7, instead of 8,175x7=57,225 cycles, only (3,697+162)x7+(431+3,885)=31,329 cycles are required, which results in further performance improvement for such advance encryption applications.

IV. CONCLUSIONS

In this study, we have designed and fabricated Optimal Ate Pairing Engine on BN Curve over 254b Prime Field in 65nm FDSOI, demonstrates the fastest ever reported performance of 33us and smallest energy consumption of 94uJ ever reported. Energy consumption is minimized down to 13.7uJ, power consumption of 16mW at Vdd=490mV, which can be applied to IoT edge devices as well. The demonstrated results are promising to realize functional encryptions such like ID based encryption, attribute-based encryption and searchable encryption for both performance hangry cloud servers as well as power hangry IoT devices.

ACKNOWLEDGMENT

This paper is based on results obtained from a project commissioned by the New Energy and Industrial Technology

978-1-7281-5107-6/19 $31.00 © 2019 IEEE

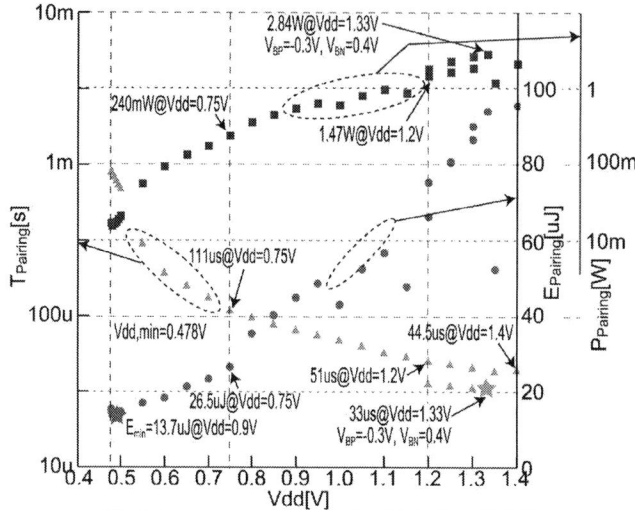

Fig. 6. Measurement results of paring chip in 65nm CMOS.

Table III. Comparison results of Optimal Ate Paring engine on BN curve over 254bit prime field.

Platform	#Gates[kG]	Area [mm²]	#Clk	Vdd [V]	Freq [MHz]	t_{Pair} [us]	P_{Pair} [mW]	E_{pair}[uJ]
Mobile Device[4] Apple A5 32nm	-	-	9,909,000	-	1,000	9,905	-	-
HighendPC[5] Corei7-6700K 14nm	-	-	840,000	-	4,000	210	91,000	19,110
HighendFPGA[6] KintexUltra 20nm	14,463slice+460DSP	-	18,151	-	170	107	-	-
ASIC*[7] 130nm CMOS	94	-	5,340,400	-	338	15,800	-	-
ASIC[8] 65nm CMOS	354	2.51	512,541	-	800	640	255	163
ASIC**[9] 65nm CMOS	323	-	330,053	-	633	521	-	-
ASIC*[10] 65nm CMOS FDSOI	3,205	12.8	9,270	-	147	202	-	-
This work 65nm CMOS FDSOI	2,793	12.8	**8,487**	1.33	250	**33**	2,850	94.0
				0.75	74.6	110	240	26.5
				0.49	9.2	792	**17.2**	**13.7**

* Synthesized results, ** Post-layout result

Development Organization (NEDO). The VLSI chip is designed and fabricated through VDEC, the University of Tokyo in collaboration with Cadence, Synopsys, Mentor Graphics, and Renesas Electronics Corp.

REFERENCES

[1] S. Satpathy, V. Suresh, S. Mathew, M. Anders, H. Kaul, A. Agarwal, S. Hsu, and R. Krishnamurthy, "220mV ‐ 900mV 794/584/754 Gbps/W Reconfigurable GF(24)2 AES/SMS4/Camellia Symmetric ‐ Key Cipher Accelerator in 14nm Tri ‐ gate CMOS," 2018 IEEE Symposium on VLSI Circuits, C16-4, pp. 175-176.

[2] E.Barker, "Recommendation for Key ManagementPart 1: General," NIST Special Publcation 800-57 Part 1, Revision 4, http://dx.doi.org/10.6028/NIST.SP.800-57pt1r4, 2016.

[3] D. F. Aranha, K. Karabina, P. Longa, C. H. Gebotys, and J. Lopez, "Faster Explicit Formulas forComputing Pairings over Ordinary Curves," 30th Annual International Conference on the Theory and Applications of Cryptographic Techniques, (Advances in Cryptology – EUROCRYPT 2011), pp. 46-68, 2011.

[4] P. Grabher, R. Azarderakhsh, P. Longa, S. Hu, and D. Jao, "Efficient Implementation of Bilinear Parings on ARM Processors," 19th International Conference on Selected Areas in Cryptography 2012, v. 7707, pp.149-165, 2012.

[5] J.L. Beuchat, J.E.G. Diaz, and S. Mitsunari, "High-speed software implementation of the optimal ate pairing over Barrreto-Naehrigcurves," International Conference on Pairing-Based Cryptography(PAIRING 2010), Lect. Note Comput. sci., vol. 6487, pp. 21-39,2010.

[6] Y. Nagahama, D. Fujimoto, and T. Matsumoto, 2018 Symposium on Cryptography and Information Security (SCIS 2018), 2D4-1, 2018.

[7] D. Kammler, D. Zhang, P. Schwabe, H. Scharwaechter, M. Langenberg, D. Auras, G. Ascheid, R. Mathar, "Designing an ASIP for Cryptographic Pairings over Barreto-Naehrig Curves," International Workshop on Cryptographic Hardware and Embedded Systems (CHES 2009), pp.254-271, 2009.

[8] Y. Li, J. Han, S. Wang, D. Fang, and X. Zeng, "An 800Mhz Cryptographic Pairing Processor in 65nm CMOS," 2012 IEEE Asian Solid State Circuits Conference (A-SSCC 2012), 8-6, pp. 217-220, 2012.

[9] J. Han, Y. Li, Z. Yu and X. Zeng, "A 65 nm Cryptographic Processor for High Speed Pairing Computation," IEEE Transactions on Very Large Scale Integration (VLSI) Systems, vol. 23, no. 4, pp. 692-701, April 2015.

[10] H.Awano, T. Ichihashi, and M. Ikeda, 2018 Symposium on Cryptography and Information Security (SCIS 2018), 2D4-3, 2018.

[11] Y. Chen, J. Lee, P. Liu, H. Chang and C. Lee, "A dual-field elliptic curve cryptographic processor with a radix-4 unified division unit," 2011 IEEE International Symposium of Circuits and Systems (ISCAS 2011), pp. 713-716, 2011.

[12] K. Karabina, "Squaring in cyclotomic subgroups," Cryptology ePrint Archive, Report 2010/542 (2010), http://eprint.iacr.org/.

An IoT Sensor Node SoC with Dynamic Power Scheduling for Sustainable Operation in Energy Harvesting Environment

Yuji Yano[1], Seiya Yoshida[1], Shintaro Izumi[1], Hiroshi Kawaguchi[1], Tetsuya Hirose[2], Masaya Miyahara[3],
Teruki Someya[4], Kenichi Okada[4], Ippei Akita[5], Yoshihiko Kurui[6], Hideyuki Tomizawa[6], and Masahiko Yoshimoto[1]

1 - Graduate School of System Informatics, Kobe University, 1-1 Rokkodai-cho, Nada-ku, Kobe, Japan
2 - Graduate School of Engineering, Osaka University, 2-1 Yamada-oka, Suita, Osaka, Japan
3 - High Energy Accelerator Research Organization (KEK), 1-1 Oho, Tsukuba, Ibaraki, Japan
4 - School of Engineering, Tokyo Institute of Technology, 2-12-1 Ookayama, Meguro-ku, Tokyo, Japan
5 - Advanced Industrial Science and Technology (AIST), 1-1-1 Umezono, Tsukuba, Ibaraki, Japan
6 - Corporate R&D Center, Toshiba Corporation, 1 Komukai Toshiba-Cho, Saiwai-ku, Kawasaki, Kanagawa, Japan
E-mail: yuji.yano@cs28.cs.kobe-u.ac.jp

Abstract— **This paper describes a low-power IoT sensor node SoC that can be used for factory automation and wearable healthcare applications. It features dynamic power scheduling method suitable for the environment that works with an energy harvester. Our proposed SoC architecture is composed of a low power (60 nA) RTC, a normally-off RF which has fast start-up time (0.5 μs), and other low-power components. A mixed-signal SoC has been fabricated using the TSMC 65 nm LP process. The sensor node system equipped with dynamic power scheduling consumes about 3.4 μW when the activity ratio is 0.1%. Evaluation results show that the proposed method can reduce about 51% of the power consumption compared with the case without dynamic power scheduling in sensor node SoC. Also, the parameters of the system in energy harvesting environment can be adjusted under the trade-off with power consumption and measurement accuracy according to the application requirements.**

Keywords— *IoT, energy harvesting devices, low-power devices*

I. INTRODUCTION

To realize Internet of Things (IoT), in which a large number of devices are connected to the network, it is necessary to develop a small and low-power edge sensor node. Furthermore, to implement a batteryless sustainable edge sensor node, the power consumption of the circuit must be controlled according to the harvested energy that is dynamically fluctuating.

We propose a dynamic accuracy control (DACC) that monitors the power supply voltage level and adjusts the bias current of analog front-end (AFE) and word length of analog-digital converters (ADC) automatically, as well as a dynamic sampling-rate control (DSRC) that adjusts the ADC sampling rate when the power supply voltage decreases. In addition, the power consumption for wireless communication occupies a significant part of the power consumption of the edge sensor node system[1]. If the data analysis performed in the conventional base station is performed by the edge sensor node, the amount of communication data between the station and the sensor node can be reduced. So we propose a dynamic data transmission-rate control (DDRC) that dynamically adjusts the

data reduction process according to requests from the base station and the supply voltage level fed from the power supply circuit.

To realize the above, we implemented system-on-a-chip (SoC) equipped with various low-power consumption technologies for sensor circuits, communication circuits, and arithmetic circuits.

II. DYNAMIC POWER SCHEDULING

In an environment where the amount of harvested energy changes with time, such as transfer from solar sunshine to shade, the amount of power supplied from the energy harvester is unstable, and in some cases, sufficient power may be not obtained for circuit operation. To achieve a sustainable edge system even in an environment where the supplied energy is insufficient, sophisticated power controls should be utilized according to the amount of supplied energy.

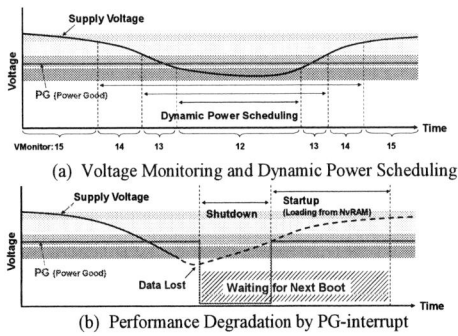

(a) Voltage Monitoring and Dynamic Power Scheduling

(b) Performance Degradation by PG-interrupt

Fig. 1. High Stability System by using Dynamic Power Scheduling

Fig. 1(a) illustrates a voltage fluctuation of a system that operates continuously by dynamic power scheduling. Here the supply voltage that changes with time in the energy harvester environment is measured using the monitor signal VMonitor which is an output of a voltage monitor circuit integrated in the power manager circuit and has 16 gradations by 4-bit length.

978-1-7281-5107-6/19 $31.00 © 2019 IEEE

The operation control is performed according to the VMonitor output. Fig. 1(b) shows an example of system operation when dynamic power scheduling is not applied. When the supply voltage falls below the system operable range, the PG (Power Good) indicator falls, thereby losing the acquired sensor data, which results in the performance deterioration due to the overhead period for system restart. Our dynamic power control consists of three control techniques as shown below.

A. Dynamic Accuracy Control (DACC)

The supply voltage from the energy harvester or the output voltage of the power supply circuit is monitored, and the bias current of the AFE circuit of the sensor device and the operation word length of the ADC circuit are adjusted according to the voltage level.

Although lowering the bias current of the AFE reduces the SN ratio, power consumption can be reduced instead, and the system operating period can be extended. When the SN ratio of the AFE circuit decreases, the accuracy of the digital data obtained by the ADC also can be decreased. Therefore, by adjusting the arithmetic word length according to the AFE bias current, it is possible to reduce the power consumption of the circuit.

Fig. 2 illustrates a conceptual block diagram a chip (A companion chip developed together with the SoC in this paper) on which AFE and ADC are mounted. It contains multiple set of control registers for AFE and ADC, so that multiple sensor devices can be supported. Device parameter control for dynamic accuracy control is updated by rewriting the register map from the microcontroller within the SoC via SPI.

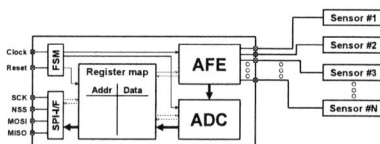

Fig. 2. Configurable AFE and ADC for Dynamic Power Scheduling

B. Dynamic Sampling-Rate Control (DSRC)

Similar to dynamic accuracy control, by reducing the sampling rate of the ADC via FSM control in Fig. 2 when the amount of power supplied from the energy harvester decreases, not only the power consumption of the control circuit but also that of the sensor device can be reduced. The voltage regulator is configured to be controlled by the Enable signal from the controller, such that the power supply voltage to each circuit block can be shut off as required. Figure 3 illustrates the power profile of the SoC in the sensing period. When the sampling rate is low, most of the sensing period remains in the standby period, thus reducing the average power consumption.

C. Dynamic Data Transmission-Rate Control (DDRC)

The sensing data collected by the sensor node are transmitted to the base station via wireless communication, and the power of wireless communication power is dominant in the system. So the packet size was extended as much as possible to suppress the power overhead during the on / off transition. In addition, instead of directly transmitting raw data, the amount

of communication data can be reduced through data compression techniques or on-node feature extraction. In this design, the amount of communication data can be reduced to less than 40 % of raw sensing data by using compressed sensing techniques or AR model-based compression algorithms. In practical applications, sensing data are buffered in memory, and signal processing and wireless transmission can be performed in a short time as shown in Figure 4.

The data amount is adjusted via signal processing at the edge according to both the power supply amount from the energy harvester and the amount of data requested by the base station, so that the wireless communication power is reduced. Although power overheads for digital circuits including microcontrollers exist for data compression and feature extraction, the power consumption of the entire system is reduced by employing a highly efficient DSP accelerator whose operating power consumption can be reduced to about 21.4 % of MCU.

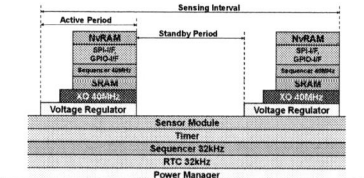

Fig. 3. Diagram of Power Profile for Sensor Data Acquisision

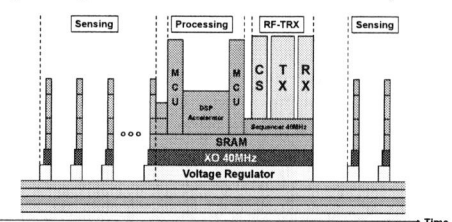

Fig. 4. Diagram of Power Profile for Data Processing

III. DESIGN OF THE INTEGRATED SOC

An ultra-low-power SoC optimized for the proposed dynamic power scheduling has been designed. In order to implement dynamic power scheduling on SoC efficiently, it has been designed so as to manage power domains and clock domains adaptively. According to the operation rate of each circuit component, clock supply control and power supply control that can be started and stopped at high speed have been introduced. By doing this control with the hardware sequencer, the power overhead of the control was minimized. In addition, in a system with the normally-off operation, the on / off switching time may affect the power overhead greatly when the activity ratio is low. Therefore, in order to make the effect of dynamic power scheduling maximum, high-speed boot IPs such as RF[2] and oscillator[3] were employed.

The mixed signal SoC integrates a microcontroller based on ARM Cortex M0, a hardware sequencer, a power supply circuit (power manager), two clock oscillators, an RF circuit, an SRAM memory, a DSP accelerator, and other peripheral circuits illustrated in Fig. 5. The power manager supports two types of energy harvester connections. One is connection with

978-1-7281-5107-6/19 $31.00 © 2019 IEEE

energy harvesters that require boosting such as solar power generation, rectenna power generation, and thermal power generation. The other is harvester connections that requires buck conversion such as vibration power generation[4]. It has a power supply voltage monitoring function and outputs 4-bit digital VMonitor signals for the operation control to the hardware sequencer. There are the supply voltage regulators composed by REG.X and four REG.D in the system. Where REG.D drive the supply voltage lines including VDDD_XO for the 40 MHz system clock oscillator (XO), VDDD_MCU for the microcontroller and its peripheral circuits, and VDDD_RF/VDDA_RF for RF circuit, respectively. Each REG.D has an enable port to cutoff the power supply to connecting circuits, the current consumption of the circuits can be effectively suppressed to zero during REG.D disabled. On the other hands, REG.X always works to supply stable power to a 32 kHz real-time clock (RTC) oscillator and the hardware sequencer. The RTC oscillator consumes only 60 nA in operating mode, and the oscillator XO features very fast settling time of up to 64 µs in intermittent operation. XO also has a clock supply cutoff function achieved by enable control. The system clock line is partitioned into many clock domains which have an AHB (Advanced High-speed Bus) bus clock, APB (Advanced Peripheral Bus) bus clocks, a proprietary CBUS (Custom Bus) bus clock, and the clocks for microcontroller's peripheral circuits. Since the clock supply to unused circuit area is cutoff by clock gating, the current consumption in that area can be suppressed to the standby leakage current level.

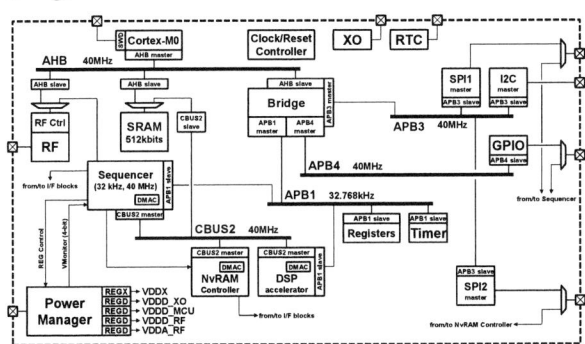

Fig. 5. IoT Sensor Node SoC Architecture

The RF circuit transmits and receives a 0.5-Mbps GMSK modulated signal with a 920-MHz carrier frequency, which is capable of about 0.5 µs high-speed start-up and normally-off operation to reduce power consumption[2]. The SRAM is a 64 KB memory in which four 16 KB SRAM blocks with a power switch are arranged, and the power supply can be shut off in 16 KB units. Consequently, the standby current can be suppressed by up to 1.03 µA (25%). The DSP accelerator is a hardware accelerator that primarily aims to handle bio-signals among digital signal processing, and it supports 8-bit product–sum operations and 16-bit product–sum operations. The standby leakage current of the DSP accelerator is up to 90 nA when the clock supply is stopped. The SoC also contains timers, GPIO, SPI, and I2C, which allows direct connection with commercial IoT sensor modules including photoplethysmography (PPG) sensors and tri-axial acceleration sensors.

Thus, the total power consumption of the SoC has been reduced to 2.9 µA including both the operating power of RTC and the hardware sequencer, and the standby power of SRAM and other digital components in order to realize the energy harvesting sensor node SoC. In addition, the start-up times of the normally-off RF and XO are reduced to up to 0.5 µs.

As shown in Fig. 6, the SoC is accompanied with an external sensor device. The low-power sensor device[5, 6] has an AFE capable of adjusting the bias current, and an ADC capable of changing the word scale and sampling rate, which are both controlled by the SoC through SPI.

Fig. 7 shows a core circuit diagram of the DSP accelerator. One general purpose 16-bit product-sum operation or two 8-bit product-sum operations can be executed as IoT edge processing, and most of the circuits are shared by 8-bit processing and 16-bit processing, which allows the circuit area minimization. By utilizing this computing unit, part of microcontroller operation can be replaced with hardware processing, and energy consumption of the system can be saved significantly[7]. The core circuit is applied to error correction processing, autocorrelation coefficient calculation, and frequency analysis such as AR model-based spectrum analysis[8, 9], for IoT sensing data.

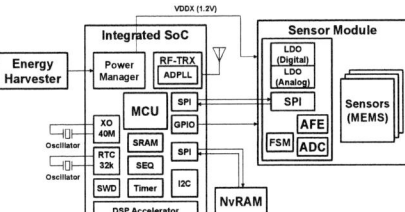

Fig. 6. Energy Harvesting IoT Sensor Node

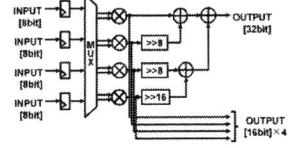

Fig. 7. Core Circuit Diagram of DSP Accelerator

IV. IMPLEMENTATION AND PERFORMANCE EVALUATION

The chip layout image of the implemented integrated SoC is shown in Fig.8. The power supply domain is separated by dividing the power supply circuit into several voltage regulators and arranging them in the vicinity of the circuit block. Two types of clock oscillator circuits are located at the corner of the chip to suppress the influence of the noise generated by other circuit blocks. The 32 kHz oscillator (RTC) and sequencer unit that operate in always-on mode and the regulator that drives them are placed on the left side of the chip. On the other hand, circuit blocks such as the 40 MHz oscillator (XO) and the RF block are mainly placed on the left side of the chip. There are a lot of decoupling capacitors for power supply voltage stabilization around the RF block. A photograph of the fabricated chip and the evaluation board is shown in Fig.9.

978-1-7281-5107-6/19 $31.00 © 2019 IEEE

IEEE Asian Solid-State Circuits Conference
November 4 – 6, 2019
The Parisian Macao, Macao SAR, China

Fig. 8. Layout Image of the Integrated SoC

TABLE 1. INTEGRATED SOC SPECIFICATIONS

Parameter	IoT Sensor Node SoC
Process	65nm LP
Chip size	4.0mm x 3.0mm
Logic gate	83,000 gates
SRAM density	64 KB (4 x 16 KB)
Supply Voltage	1.2V (Core), 3.3V (I/O)
MCU Frequency	40 MHz

Fig. 9. Fabricated Test Chip and its Evaluation Board

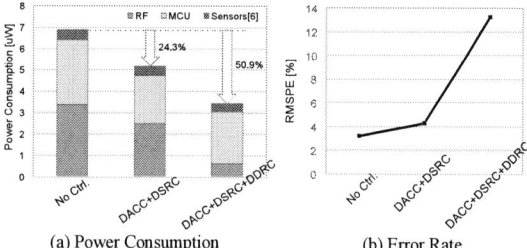

(a) Power Consumption   (b) Error Rate

Fig. 10. Performance of Integrated SoC with Dynamic Power Scheduling

Fig. 10 shows the evaluation results of the power reduction and the error rate by the dynamic power scheduling consisting of DACC, DSRC, and DDRC. In Fig. 10(a), compared to the case where dynamic power scheduling is not applied, power consumption of 24.3% can be reduced when DACC and DSRC on sensor devices is performed. In addition, it can be reduced by 50.9% when applying DDRC. Here, the active rate is set to 0.1%.

Also, the root mean square percentage error (RMSPE) in the frequency spectrum of the PPG sensor data acquired by the sensor node system with the dynamic power scheduling is shown in Fig. 10(b). In this figure, RMSPE for the system that samples data at 256 Hz without the dynamic power scheduling is about 3.2%, because the reference (golden) data has the ideal parameters which consist of a 1024 Hz sampling rate and a 14 bits word length. RMSPE in the system running DACC and DSRC is raised to 4.3%, and the difference between the above

and the reference is negligible. On the other hand, RMSPE for the system equipped with DACC, DSRC and DDRC can be about 13.2%, measurement accuracy of the sensor data is degraded. Therefore, the parameters and the sampling rate of the system in energy harvesting environment can be adjusted under the trade-off with power consumption and measurement accuracy according to the application requirements.

V. CONCLUSION

A dynamic power scheduling method for the low-power sensor node system in the energy harvesting environment has been proposed. Also very low-power mixed-signal SoC whose architecture and circuit design was optimized for the above method, has been fabricated using the TSMC 65nm LP process. It can realize IoT sensor node which can be used for factory automation or wearable healthcare applications. The fabricated test chip consumes approximately 3.4μW when the active rate is 0.1% and can extend the period of sustainable use under the trade-off with SNR and measurement accuracy.

ACKNOWLEDGMENT

This paper is based on results obtained from a project commissioned by the New Energy and Industrial Technology Development Organization (NEDO).

REFERENCES

[1] O. Landsiedel, "Accurate prediction of power consumption in sensor networks," in Proc. The 2nd IEEE Workshop on Embedded Network Sensors, Sydney, Australia, May. 2005.

[2] B. Liu, Y. Zhang, J. Qiu, W. Deng, Z. Xu, H. Zhang, J. Pang, Y. Wang, R. Wu, T. Someya, A. Shirane, and K. Okada, "An HDL-described Fully-synthesizable Sub-GHz IoT Transceiver with Ring Oscillator Based Frequency Synthesizer and Digital Background EVM Calibration," IEEE Custom Integrated Circuits Conference (CICC), Apr. 2019.

[3] M. Miyahara, Y. Endo, K. Okada, and A. Matsuzawa, "A 64μs Start-Up 26/40MHz Crystal Oscillator with Negative Resistance Boosting Technique Using Reconfigurable Multi-Stage Amplifier," 2018 Symposium on VLSI Circuits, pp. 115-116

[4] S. Kanzaki, T. Hirose, H. Asano, Y. Nakazawa, N. Kuroki, and M. Numa, "Switched-capacitor voltage buck converter with step-down-ratio and clock-frequency controllers for ultra-low-power IoT devices," Proc. of IEEE International Conference on Electronics Circuits and Systems (ICECS 2018), 2018, pp. 209-212.

[5] I. Akita, T. Okazawa, Y. Kurui, A. Fujimoto, and T. Asano, "A 181nW 970μg/√ Hz accelerometer analog front-end employing feedforward noise reduction technique," in Proc. Symp. VLSI Circuits, pp. 161-162, June 2018.

[6] H. Tomizawa, Y. Kurui, I. Akita, A. Fujimoto, T. Saito, A. Kojima, and H. Shibata, "High-sensitivity and low-power inertial MEMS-on-CMOS sensors using low-temperature-deposited poly-SiGe film for the IoT era," 2018 symposium on VLSI technology, pp.41-42

[7] K. Kajihara, S. Izumi, S. Yoshida, Y. Yano, H. Kawaguchi, and M. Yoshimoto, "Hardware Implementation of Autoregressive Model Estimation Using Burg's Method for Low-Power Spectral Analysis," 2018 IEEE International Workshop on Signal Processing Systems (SiPS), Oct. 2018.

[8] J.M. Spyers-Ashby, P.G. Bain, S.J. Roberts, "A comparison of fast Fourier transform (FFT) and autoregressive (AR) spectral estimation techniques for the analysis of tremor data," Journal of Neuroscience Methods, vol.83, no.3, pp. 35–43, October. 1997.

[9] P. Stoicssa, R. L. Moses, "Spectral analysis of signals," Upper Saddle River, New Jersey, USA: Prentice Hall, pp.90–91, March 2005.

978-1-7281-5107-6/19 $31.00 © 2019 IEEE

19-3 (7066)

A 3.01 mm² 65.38Gb/s Stochastic LDPC Decoder for IEEE 802.3an in 65 nm

Qichen Zhang, Yun Chen,
Xiaoyang Zeng
State Key Lab of ASIC & System
Fudan University
Shanghai, China

Keshab K. Parhi
Department of Electrical and Computer
Engineering
University of Minnesota
Minneapolis, MN, United States

Borivoje Nikolic
Electrical Engineering and Computer
Sciences
University of California, Berkeley
Berkeley, CA, United States

Abstract—**A fully-parallel high-throughput LDPC decoder architecture leads to high power consumption and large area. Using stochastic logic, this paper proposes three novel strategies to improve throughput and reduce power consumption; these include: variable node initialization, bit-flipping post-processing and posterior-information-based hard decision. Moreover, a random number based probability stochastic sequences generator is proposed to reduce hardware resources. Chip test results from an LDPC decoder for the 10GBASE-T standard (2048, 1723) code using 65 nm CMOS process demonstrate a 74.3% reduction in average decoding cycles at 4.4 dB with satisfactory decoding performance. The decoder supports 65.38 Gb/s throughput at 420 MHz and requires 1.1W power consumption. Compared with other works, the proposed decoder can achieve lower power and average decoding cycles with similar error performance.**

Keywords—Stochastic computation; High throughput decoder; Bit-Flipping algorithm

I. INTRODUCTION

Low-density parity-check (LDPC) codes [1] have been applied in many high throughput communication systems, such as the 10GBASE-T [2]. The fully-parallel architecture of LDPC decoder can satisfy high throughput requirement in these systems. However, the complex architectures of conventional decoders, for example, the sum product algorithm (SPA) and min sum algorithm (MSA), occupy large hardware resources and limit the chip logic utilization. Therefore, new decoding approaches need to be introduced to eliminate this bottleneck.

Stochastic computing was first be used in LDPC decoder in 2003 [3]. Various stochastic decoder architectures have been proposed, such as the Edge Memory (EM) [4], Majority-based Tracking-Forecast Memory (MTFM) [5] and Delayed stochastic (DS) [6]. EM and MTFM introduce different types of probability tracers which can enhance the convergence speed of the decoder. However, the area efficiency of EM is low and the latency of the MTFM is too long. DS simplifies the structure of variable node (VN) significantly and improves throughput with performance loss.

In this paper, we demonstrate a stochastic LDPC decoder architecture to accelerate the convergence speed by using three different methods: a VN initialization method, a bit-flipping (BF) algorithm as a post-processing strategy, and a posterior-information-based hard decision. A new Bernoulli sequence generator is used for resource reduction. For the (2048, 1723) LDPC code in the 10GBASE-T standard, hardware

This work is supported by the National Natural Science Foundation of China (Grant No.61774049), the Shanghai Pujiang Talent Science funding (Grant No.16JD008), Science and Technology Commission of Shanghai Municipality (Grant 2019-jmrh1-kj24), the Program of Shanghai Academic/Technology Research Leader under Grant 16XD1400300. (Corresponding author: Yun Chen, email: chenyun@fudan.edu.cn)

implementation shows that, compared with EM, this decoder can achieve about 74.3% reduction in the average number of decoding cycle (DC) at 4.5dB. Moreover, the decoder supports a throughput of 65.38 Gb/s at 420 MHz and consumes 1.1W power consumption, achieving lower power and average decoding cycles (ADC) at similar error performance.

II. COMPUTATION NODES IN STOCHASTIC LDPC DECODER

A. Check Node (CN) Update

A CN, c_j, receives sequences which have probability $P_{i \to j}$ from the corresponding VN, v_i, and outputs sequences which have probability $Q_{j \to i}$ as

$$Q_{j \to i} = P_{l \to j}(1 - P_{m \to j}) + P_{m \to j}(1 - P_{l \to j}). \tag{1}$$

The CN function can be implemented by using XOR gate.

B. VN Update

A VN, v_i, receives sequences which have probability $Q_{j \to i}$ from the corresponding CN, c_j, and outputs sequence which have the probability $P_{i \to j}$ given by

$$P_{i \to j} = \frac{Q_{e \to i}Q_{f \to i}}{Q_{e \to i}Q_{f \to i} + (1 - Q_{e \to i})(1 - Q_{f \to i})}. \tag{2}$$

The degree-2 VN structure, which contains a 64-bit EM [4], is shown in Fig. 1. EM is a type of special shift memory where bits in it can be chosen by the input address. The two inputs come from CN and the channel bit. If the two inputs are the same, i.e., u = 1, the EM is updated by r, and the input is sent to output directly. Otherwise, the EM doesn't change, and the output is equal to bit c, which is chosen randomly from the EM. VNs in the 10GBASE-T standard have 7 inputs (6 inputs from CNs and 1 input from channel information) and 6 outputs to corresponding CNs.

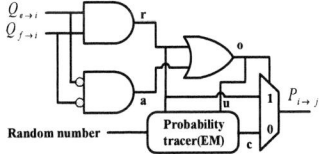

Fig.1. Degree-2 VN[4].

978-1-7281-5107-6/19 $31.00 © 2019 IEEE 271

III. PROPOSED DECODER ARCHITECTURE

A. VN Initialization

The proposed VN initialization is designed to reduce the initialization period that is required for the proposed stochastic decoder and provides more accurate channel values to accelerate the decoding convergence.

The VN initialization block in [7] contains 2048 LUTs to convert log-likelihood-ratio (LLR) channel message to initialization message. However, these 2048 LUTs occupy too much hardware resources, about 11.6% of the whole decoder [7]. New linear fitting block is implemented to convert the LLR number more efficiently which can cut 73.2% resource usage of LUTs-based VN initialization.

A counter-based probability tracer unit is used to replace EM-based probability tracer in Fig.1. The counter-based probability tracer replaces the EM with a 7-bit up/down saturation counter, which stops counting after reaching its saturated values. Initializing flag and initializing value are added to the inputs of the probability tracer. The up/down counter can be initialized in only one DC instead of dozens in EM. When the initializing flag turns to 1 in first DC, the counter is initialized with the initializing value from VN initialization block. A comparator will compare the counter value with random number and then generate signal c in Fig.1. The output value is set to 0 when the random number is larger than the counter value; otherwise, it is set to 1. Compared with EM, the counter-based probability tracer can achieve better area efficiency.

The corresponding initialization value of channel LLR message for a 7-bit counter can be calculated by following equation, $P = \left(e^{L}/\left(e^{L}+1\right)\right) \times 128$, which is shown by the solid line in Fig.2. This curved line can be expressed approximately by five red dashed linear lines. Table I lists these five linear equations. From Fig.2, it can be seen that, when $-2 < x < 2$, the maximum difference between the solid line and the dashed line is less than 3. For LLR with 4-bit decimal part, the quantization precision for the probability value is 8, so the dashed line can be used to replace the solid line. When $2 < |x| < 8$, the value of y is constant. Simulation shows that this approximation will not decrease the decoding performance.

Because all parameters in these equations are powers-of-2, instead of adders and multipliers, these can be realized by shifters and some simple logic gates. For example, if the first bit is 1 and other bits are 0 in the integer part of x, then the value of x will satisfy $1 \leq x < 2$. Then, left shift of y yields $y = 16x$, which satisfies $16 \leq y < 32$ in which the fifth bit is 1 and the sixth bit is 0. Finally, by making the fifth bit to 0, the sixth bit and the seventh bit to 1, we get $y = 16x + 16 + 64$.

B. BF-based Post-Processing

Generally, belief propagation iterative LDPC decoders suffer from error floor at high SNR region because of quantization error [8] and some special structures called trapping sets. In stochastic decoder, the bits propagating in decoder do not represent the exact value. So, compared with SPA or MSA, a pure stochastic LDPC decoder has nearly 0.5dB performance lo-

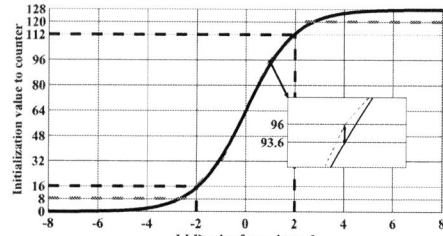

Fig.2. The range of initialization value to counter.

TABLE.I. Five Linear Equations in Fig.2

LLR range (x)	Linear equation (y)
$2 \leq x < 8$	$y = 120$
$1 \leq x < 2$	$y = 16x + 16 + 64$
$-1 \leq x < 1$	$y = 32x + 64$
$-2 \leq x < -1$	$y = 16x + 16 + 32$
$-8 \leq x < -2$	$y = 8$

ss and higher error floor [9].

Introducing post-processing is a good way to solve this problem. Simulation results show that BF algorithm can achieve better decoding performance at high SNR when the number of bit errors is less than a certain threshold. By switching stochastic computing to BF algorithm under the threshold, a better BER performance can be achieved.

As shown in Fig. 3, the proposed architecture consists of four parts. 384 parity check results from CN are used to determine if the decoding is successful or not. All 384 results are 0 means that all parity check equations are satisfied, and the current code word can be deemed to be correct. Otherwise, the sum of wrong parity check results can be treated as a flag to trigger the BF post-processing in the Post Enable block. If the total is less than the threshold of 30, which is chosen based on simulations, the BF post-processing operation will be activated. The Bit Flip block calculates the results of the six parity checks corresponding to each sub-node in one VN. When the sum is larger than 3, the hard decision is considered to be wrong and the corresponding bit will be flipped. Computer simulation results show that the proposed decoding procedure can, in most situations, be decoded successfully within 3 DCs after the post-processing stage. The decoding operation is terminated if the sum of all parity check results is 0. If the post-processing decoding doesn't terminate in 3 DCs, the stochastic decoding will re-decode this code once and may get the correct result due to non-deterministic nature of stochastic computation [9].

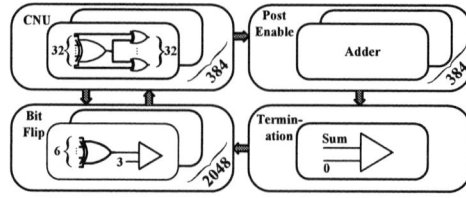

Fig.3. Post-processing block.

978-1-7281-5107-6/19 $31.00 © 2019 IEEE

C. Random Number Probability Bernoulli Sequences Generator

In order to maximize the throughput of a LDPC decoder, the most common architecture is the fully-parallel decoder, namely, the whole decoder is a hardware copy of the Tanner graph and every VN and CN is built in the decoder chip. Meanwhile, other modules also need to be fully-parallel. For example, in [7], the fully-parallel LLR to probability block contains 2048 LUTs. According to the area analysis of chip in [7], these 2048 LUTs occupy nearly 7.5% area of the whole chip.

In LLR to probability block, the inputs are LLR messages from channel and the output is corresponding probabilities. These probabilities will be used in the channel bit generation block and then compared with random number in probability domain. Actually, as long as the channel message and random number are in the same domain, the output stochastic sequences will be the same. So, the comparator can use channel messages in LLR domain and random number in LLR domain to generate the same sequence. Therefore, the probability to LLR converter for random number only contains 64 LUTs because the proposed decoder only has 64 LFSR random number generators. By using this method, 96.875% LUTs can be reduced.

D. Overall Decoder Architecture

Fig. 4 shows the proposed decoder architecture. Linear fitting VN initialization converts LLR messages from channel to VN for probability tracer. 64 LFSRs are responsible for generating random numbers to 2048 VNs. These random numbers are converted into LLR domain and compared with LLR messages in channel bit generator to output Bernoulli sequences for VN. Instead of using VN to CN messages, each VN contains a posterior-information-based hard decision block to gather all inputs from CN and channel bits to generate more precise hard decision bits. As shown in Fig. 4, which is similar to the VN sub-node, by adding another degree-2 sub-node, all bits from CN can be used for hard decision. Sub-nodes group in [10] is used to reduce VN area. The Hard Decision block generates the codewords and sends them to the Post Processing block to determine whether the post processing can begin or not. If the post-processing does not start, the VNs and the CNs continue to exchange message bits.

IV. PERFORMANCE EVALUATION

A. The Reduction of Average Decoding Cycles

Fig. 5 shows the ADC reduction ratio for the three proposed

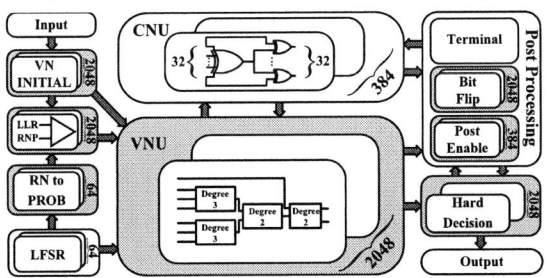

Fig.4. Proposed decoder architecture.

strategies. The proposed initialization scheme can initialize the number of counters in VNs in one DC, which can save a lot of decoding time and cut 39.9% ADC compared with EM. By using proposed post-processing, if the code can be decoded, the decoding can be completed in less than 3 clock cycles which can get 26.4% ADC reduction. The whole CN information can be used in the proposed hard decision scheme which makes more accurate calculation with 8.0% ADC reduction. All in all, the ADC can be reduced 74.3% by using the proposed design.

B. Measurement Results

Fig. 6 shows the BER performance of the (2048, 1723) LDPC code. The results from the EM, the DS, NMSA floating, [7], [11], normalized probabilistic MSA (NPMSA) [12], and the proposed decoder are chosen for comparison. It can be observed that the proposed decoder can achieve comparable performance among all these architectures.

The die graph of proposed decoder is shown in Fig. 7, the core area of the test chip is 3.01mm². Inside the chip core, 0.18 mm² clock generator and 0.21mm² input & output buffer are implemented for chip test. By using the new VN initialization block and stochastic sequences generator to eliminate 73.2% VN initialization block and the whole LLR to Probability block, 15.46% more chip area can be reduced. The maximum frequency of this decoder is 420MHz, and the chip can achieve a throughout of 65.38 Gb/s and 1.1W power consumption at room temperature under 1.2V supply.

In Table. II, a comparison between the different decoders is

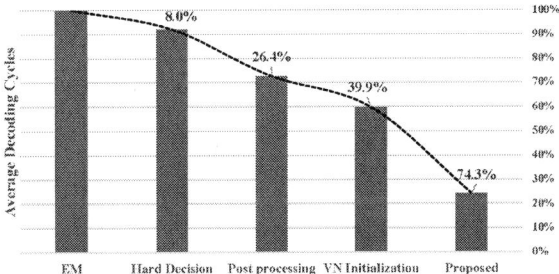

Fig.5. ADC reduction rate at 4.5 dB.

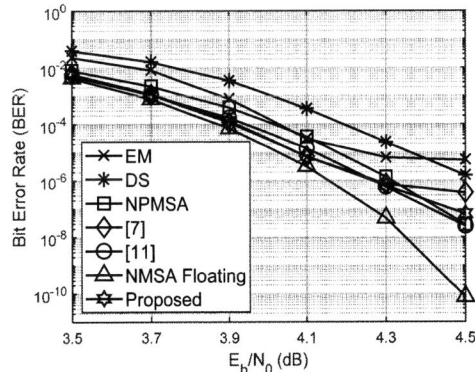

Fig. 6. BER performance using AWGN and BPSK modulation.

TABLE. II. Comparison with state of the art fully parallel (2048,1723) LDPC decoder

	Stage	Technology	Scaled Area in 65nm (mm²)	Decoder Area (mm²)	Frequency (MHz)	Power (W)	Throughput (Gb/s@dB)	Throughput/Area (Gb/s/mm²@dB)	ADC (DCs@dB)
This Work	Chip	65nm	3.01	2.62	420	1.10	25.56@4.4 65.38@5.5	9.76@4.4 24.95@5.5	34@4.4 13@5.5
This Work (new blocks)	Post-sim	65nm	2.60	2.21	-	-	25.56@4.4 65.38@5.5	11.57@4.4 29.58@5.5	34@4.4 13@5.5
MTFM [5]	Post-sim	90nm	3.33	3.33	500	-	21.30@4.4 61.30@5.5	6.39@4.4 18.41@5.5	48@4.4 17@5.5
DS [6]	Post-sim	90nm	2.05	2.05	750	-	32.70@4.4 172.40@5.5	15.95@4.4 84.09@5.5	47@4.4 9@5.5
[11]	Post-sim	90nm	2.15	2.15	749	1.34	39.30@4.4 69.70@5.5	18.28@4.4 32.42@5.5	39@4.4 22@5.5
NPMSA[12]	Chip	90nm	5.01	5.01	199.6	1.11	45.42@4.4	9.07@4.4	9@4.4

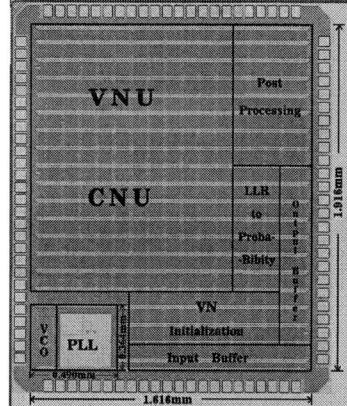

Fig. 7. Die graph of the proposed decoder in 65nm CMOS.

presented. MTFM, DS and [11] are all stochastic LDPC decoders. Among these decoders, DS has the smallest chip core area and best throughput/area efficiency. However, Fig.6 shows the BER performance of DS is not as good as others. The proposed decoder has the smallest ADC among stochastic decoders, but the NPMSA decoder still leads in decoding latency due to LLR message decoding. As a result, the NPMSA decoder has the highest absolute throughput under the 4.4dB. However, considering the much smaller chip area, the proposed decoder can achieve better throughput to area ratio than the NPMSA. In future work, higher frequency can be achieved with small ADC loss by adding pipeline in post-processing block.

V.   CONCLUSION

This paper presents a high throughput and small area stochastic LDPC decoder for 10GBASE-T. The proposed linear fitting VN initialization block can initialize probability tracers in one DC. Quicker converge speed can be achieved by BF-based post processing and posterior-information-based hard decision. Moreover, hardware resources can be reduced by linear fitting VN initialization block and new stochastic sequences generator. Chip test shows that the average decoding cycles are reduced by 74.3% at the SNR of 4.5 dB in comparison with the originally-described EM decoder.

REFERENCES

[1] R. Gallager, "Low-density parity-check codes," in *IRE Transactions on Information Theory*, vol. 8, no. 1, pp. 21-28, January 1962.

[2] IEEE Standard for Information technology--Telecommunications and information exchange between systems--Local and metropolitan area networks--Specific requirements Part 3: Carrier Sense Multiple Access with Collision Detection (CSMA/CD) Access Method and Physical Layer Specifications," in *IEEE Std 802.3-2008 (Revision of IEEE Std 802.3-2005)*, vol., no., pp.1-2977, 26 Dec. 2008

[3] V. C. Gaudet and A. C. Rapley, "Iterative decoding using stochastic computation," in *Electronics Letters*, vol. 39, no. 3, pp. 299-301, 6 Feb 2003.

[4] S. Sharifi Tehrani, S. Mannor and W. J. Gross, "Fully Parallel Stochastic LDPC Decoders," in *IEEE Transactions on Signal Processing*, vol. 56, no. 11, pp. 5692-5703, Nov. 2008.

[5] S. Sharifi Tehrani, A. Naderi, G. A. Kamendje, S. Hemati, S. Mannor and W. J. Gross, "Majority-Based Tracking Forecast Memories for Stochastic LDPC Decoding," in *IEEE Transactions on Signal Processing*, vol. 58, no. 9, pp. 4883-4896, Sept. 2010.

[6] A. Naderi, S. Mannor, M. Sawan and W. J. Gross, "Delayed Stochastic Decoding of LDPC Codes," in *IEEE Transactions on Signal Processing*, vol. 59, no. 11, pp. 5617-5626, Nov. 2011.

[7] D. Wu, Y. Chen, Q. Zhang, Y. Ueng and X. Zeng, "Strategies for Reducing Decoding Cycles in Stochastic LDPC Decoders," in *IEEE Transactions on Circuits and Systems II: Express Briefs*, vol. 63, no. 9, pp. 873-877, Sept. 2016.

[8] X. Zhang and P. H. Siegel, "Quantized Iterative Message Passing Decoders with Low Error Floor for LDPC Codes," in *IEEE Transactions on Communications*, vol. 62, no. 1, pp. 1-14, January 2014.

[9] K. Huang, V. C. Gaudet and M. Salehi, "Trapping sets in stochastic LDPC decoders," *2015 49th Asilomar Conference on Signals, Systems and Computers*, Pacific Grove, CA, 2015, pp. 1601-1605.

[10] Q. Zhang, Y. Chen, D. Wu, X. Zeng and Y. Ueng, "An area-efficient architecture for stochastic LDPC decoder," *2015 IEEE International Conference on Digital Signal Processing (DSP)*, Singapore, 2015, pp. 244-247.

[11] Y. Ueng, C. Wang and M. Li, "An Efficient Combined Bit-Flipping and Stochastic LDPC Decoder Using Improved Probability Tracers," in *IEEE Transactions on Signal Processing*, vol. 65, no. 20, pp. 5368-5380, 15 Oct.15, 2017.

[12] C. C. Cheng, J. D. Yang, H. C. Lee, C. H. Yang and Y. L. Ueng, "A Fully Parallel LDPC Decoder Architecture Using Probabilistic Min-Sum Algorithm for High-Throughput Applications," in *IEEE Transactions on Circuits and Systems I: Regular Papers*, vol. 61, no. 9, pp. 2738-2746, Sept. 2014.

A Millimeter Wave Digital CMOS Baseband Transceiver for Wireless LAN Applications

Kang-Lun Chiu, Hsun-Wei Chan, Wei-Che Lee, Chang-Ting Wu, Henry Lopez, Hung-Chih Liu, *Meng-Yuan Huang,
Chun-Yi Liu, Tsai-Hua Lee, Hsin-Ting Chang, Chih-Wei Jen, °Nien-Hsiang Chang, *Pei-Yun Tsai, +Yen-Cheng Kuan, and Shyh-Jye Jou

Department of Electronics Engineering
National Chiao Tung University
Hsinchu, Taiwan

+ *International College of Semiconductor Technology*
National Chiao Tung University
Hsinchu, Taiwan

* *Department of Electrical Engineering*
National Central University
Taoyuan, Taiwan

° *Taiwan Semiconductor Research Institute*
National Applied Research Laboratories
Hsinchu, Taiwan

Abstract—**In this paper, we introduce a 10 Gbps digital baseband transceiver with 1V supply voltage, 16-QAM, 3/4 code rate single carrier mode using 28 nm CMOS process to do the implementation. In millimeter wave communications, well-defined standard, IEEE 802.11ad is referenced for our system design and simulation. We target at modified chip rate 2.5 GHz with 4 times parallelism hardware at clock rate 625 MHz. The overall transceiver architecture design with hardware implementation considerations will be proposed. With the smart shared memory allocation, we can reduce 31% of the total amount. According to the referenced specifications, we achieve the required bit error rate 3×10^{-7} before SNR 18.5 dB. After the system algorithm design, we implement the hardware with the core area 2.92 mm² with measured power 1.8 Watt for baseband transceiver chip. The performance of energy per bit is 0.023 nJ and 0.211 nJ per bit for Tx and Rx, including ADC/DAC.**

Keywords—*mmWave, Single Carrier (SC), Digital Baseband, IEEE 802.11ad/ay*

I. INTRODUCTION

For the large amount of data transmission requirements in the wireless communications, developing high speed system toward high carrier frequency and wide bandwidth is necessary. The millimeter wave (mmWave) band, 28-86 GHz, have clear advantage of large bandwidth. In 60 GHz band, standards for indoor wireless communication as known as IEEE 802.11ad/ay [1][2] focus on mobile offloading, wireless docking, display connectivity and outdoor backhauling. Channel bonding and aggregation technology are used to further enlarge the equivalent channel bandwidth for higher transmission requirement.

This work is based on IEEE 802.11ad/ay single carrier (SC) transmission due to the specification has already been defined clearly and the channel information is modeled completely. We propose a new mmW_10Gbps specification based on 802.11ad/ay as shown in Table I. The chip rate is

modified from 1.76 GHz to 2.5 GHz through four times parallelism design. Furthermore, higher order quadrature amplitude modulation (QAM) is considered. With the progress of advanced CMOS technique, we can achieve the physical data throughput rate up to 10 Gbps to fulfill the requirement stated in IMT-2020 [3].

The paper is organized as follows. In section II, we give the overall transceiver architecture with the module allocations and signal flow, followed by the shared module designs and implementation result. Section III presents the fabrication and measurement results. Finally, the conclusions are given in section IV.

II. TRANSCEIVER ARCHITECTURE

The proposed digital baseband (BB) transceiver is illustrated in Fig. 1. The non-idealities from radio frequency (RF) circuits become worse to the performance of transmitter (Tx) and receiver (Rx) when the carrier frequency raise to mmWave, such as phase noise (PN) [4][5], IQ-imbalance effect. It is important to have models considering industrial feasibility for RF components when evaluating the performance of baseband designs. There are several key modules including the 4-in-1 synchronizer (SYNC), non-line-of-sight equalizer (EQ) [6], two-stages phase noise cancellation (PNC) and tracking, and self-healing IQ-imbalance compensation [7] in the inner receiver. A routing aware low-density parity check (LDPC) channel decoder is implemented in the outer receiver.

A. Module Allocations and Signal Flow

The bit stream data delivered from medium access control (MAC) layer will be aligned with each other by the FIFO and pass through the scrambler and LDPC encoder in outer Tx. In the inner Tx, modulation mapping and pilot word insertion are carried out. The IQ imbalance parameters evaluated off line will be used for IQ pre-distortion module. Next, a four times up-sampling square-root-raised-cosine filter (SRRCF) is used as the Tx pulse shaping filter (PSF).

For the Rx, signal will first pass through SRRCF as the Rx filter with two times down-sampling and the SYNC module does also two times down-sampling. Then, goes to the IQ Compensation module whose parameters are also evaluated off line. Before the equalization, the signal will be transferred to frequency domain by the FFT and transferred back to time domain after the channel equalization. After the PNC, we will remove the pilot words and demodulate for the original data. Finally, the signal will pass through the outer Rx including LDPC decoder and de-scrambler. Then deliver to the MAC layer by the first-in-first-out (FIFO) interface. The direct digital frequency synthesizer (DDFS) is used for analog to digital converter (ADC) and digital to analog

TABLE I. SUMMARY OF THE PARAMETERS IN MMW_10GBPS.

Mode	mmW_10Gbps
Transmission	SC
Constellation	$\pi/2$ BPSK, $\pi/2$ 16-QAM
Channel Coding	1/2 or 3/4 LDPC
Training sequence	Golay code
Beamforming	Yes
Chip rate	2.5 GHz
Bandwidth	2.5 GHz
Frequency tolerance	±20 ppm
Maximum frequency offset	40 ppm

IEEE Asian Solid-State Circuits Conference
November 4 – 6, 2019
The Parisian Macao, Macao SAR, China

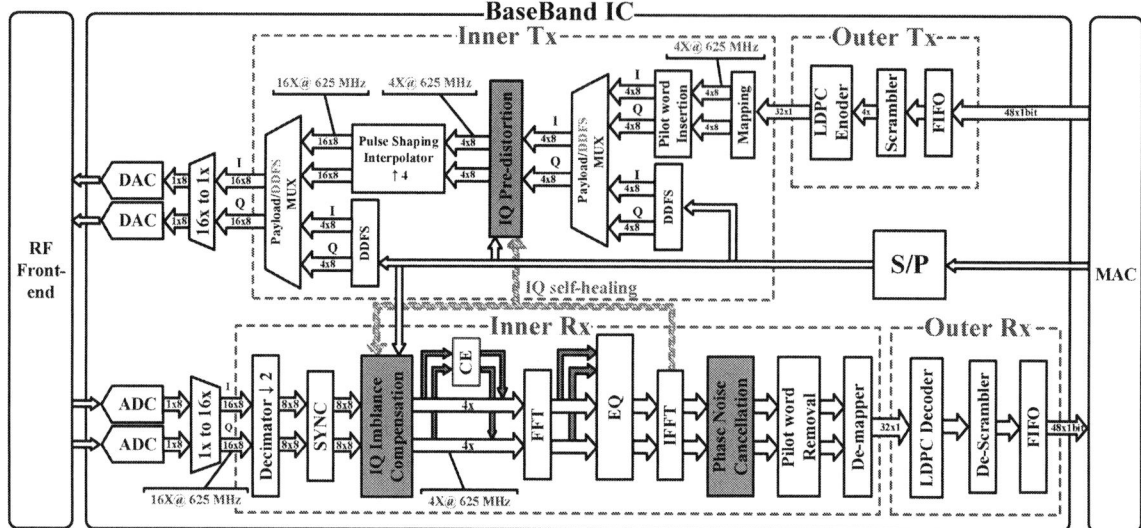

Fig. 1. Block diagram of proposed BBIC

converter (DAC) calibration and parameter training for IQ-imbalance com-pensation.

Both Tx and Rx are operated under 625 MHz clock rate with four times parallelism to achieve equivalent symbol rate 2.5 Gs/s which is equivalent to 10 Gbps with 16 QAM. All modules are developed through feedforward algorithm under pipelining and parallelization consideration. The enhanced real time IQ imbalance compensation is also designed in the ongoing version.

B. System Non-Ideal Effects Simulation Flow

The system non-ideal effects models are crucial to simulation performance. It is important to have models considering industrial feasibility for RF components when evaluating baseband designs performance. The whole simulation system includes PN, multipath channel model (line-of-sight/non-line-of-sight), Additive white Gaussian noise (AWGN), carrier frequency offset (CFO), initial symbol timing offset (STO), sampling clock offset (SCO), IQ-imbalance and DC voltage offset as shown in Fig. 2. All these effects are solved to meet the system requirement, for example, a reformed slicer for the two-stage PNC as shown in Fig. 3.

C. Shared Module Design

In the inner receiver, almost every key module need to access memory. Hence, shared memory arrangement is also taken into account. The size of the shared memories should be the largest required memory size for all modules. Consequently, we set the size of memory 2 (M2) as 512 samples. Because of 4 times parallelism for our system, we set the size of memory 1 (M1) as 932 samples.

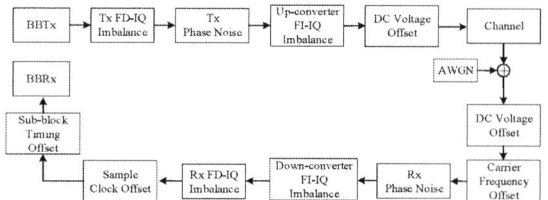

Fig. 2. Simulation system block diagram with non-ideal effects

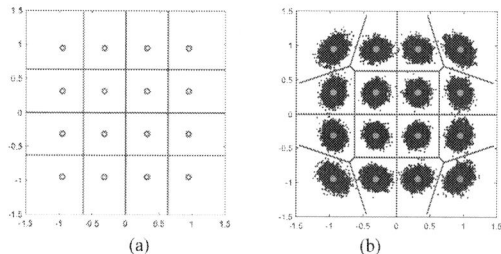

Fig. 3. (a)conventional 16-QAM slicer (b) reformed PN aware slicer

The overall memory unit that can be replaced by shared memory is shown in Table II. The reduction is 31% (1444/2083 samples). Note that the memory size in entries is the size in samples divided by the degree of parallelism which is 4.

D. Implementation Results

Taking IEEE 802.11ad standard [1] as reference, the performance requirement for packet error rate (PER) shall be less than 5% for an uncoded payload length of 256 bytes and less than 1% for 4096 bytes, which is equivalent to 3×10^{-7} BER for code rate of 3/4. The input levels consider around 10 dB noise figure and 5 dB implementation loss which includes non-ideal receiver effects such as channel estimation errors, tracking errors, quantization errors imperfect synchronization and PN effect. The input levels can be derived to the required baseband input SNR through:

$$SNR\ (dB) = S_i - 10\log_{10}(kT_0B) - 30 - NF , \quad (1)$$

where S_i (dB) is the receiver sensitivity, k is the Boltzmann's constant, T_0 is the absolute temperature of the receiver, B is the bandwidth, NF is the noise figure. The require SNR for receiver is 3.52-18.52 dB for SC mode. In this work, the system specification is $\pi/2$-16QAM with 3/4 coding rate, so our target is that pre-LDPC decoding BER should reach 3×10^{-3} with SNR requirement less than 18.5 dB since the LDPC decoder with hard decision results from inner receiver can reduce BER by 4 orders of magnitude.

978-1-7281-5107-6/19 $31.00 © 2019 IEEE

TABLE II. Memory Sharing Arrangement and Comparison

Module	Master of shared memory	Required memory size	Request for memory access (fields in a frame)			
			STF	CE	Header	Payload
SYNC	NGC	128	M1			
EQ	Shared OGC	256		M1		
	CIR memory	192		M2		
	WNC	64			M1	
	CFR memory	512			M2	M2
PNC	PNC	931				M1
Total		2083	Total		Entry	
Shared memories	M1	512	1444		128	
	M2	932			233	

The verification result after implemented is shown in Fig. 4.

We implement the digital BB transceiver with *28nm CMOS* process. In order to achieve the high chip rate of the target mmW_10Gbps specification, we must adopt parallelism for the hardware implementation of the BBRx. The clock rate for inner Rx part is 625 MHz and chip rate is 2.5 GHz so the degree of parallelism is 4 for the non-oversampled part of the inner Rx. The parallelism of all the circuits is realized through unfolding technique. The ADC performs 4 times oversampling and has 16 times parallelism outputs with 8 bits one another. The decimator is a SRRC PSF with 8 times parallelism outputs. After the SYNC module, the signal is further decimated without oversampling with 4 times parallelism outputs. The rest of the modules in the inner receiver is 4 times parallelized.

The synthesis gate count and power consumption of the proposed baseband receiver is listed in Table III. Due to the power consumption requirement, "double I/O ring" and "stacking bonding pad" techniques are adopted in order to increase the amount of the power I/O buffer and electrostatic discharge (ESD) modules. The whole BB integrated circuit (BBIC) core area is 2.92 mm^2 under the pad limit constraint, and it is able to achieve 2.20 mm^2 without pad limitation. The ADC/DAC is target at 8 bit resolution and 10 Gs/s sample rate with 88/150 mW power consumption. The detail gate count and power consumption are listed in Table III.

To achieve a flexible and reusable hardware design, intellectual property (IP) based implementation methodology is adopted. With these modularized and parameterized IPs, we are able to adjust the system architecture, such as the word length, QAM number, pipeline stage number, or parallelism number to meet higher transmission capacity requirement. We have proposed an inner Rx only design with SC/OFDM dual mode and implementation which is 8-parallelisms architecture [8]. According to the theoretical analysis and previous experience, 30 Gbps throughput rate is achievable with 64-QAM and 8-parallelisms approaching which needs roughly two-times aggregated bandwidth.

III. Experimental Results

The BBIC chip microphotograph and the package figure of this work are shown in Fig. 5. In order to verify the digital function correctness, we allocate a part of data bit stream, control signals and chip clock as the I/O pins for testing.

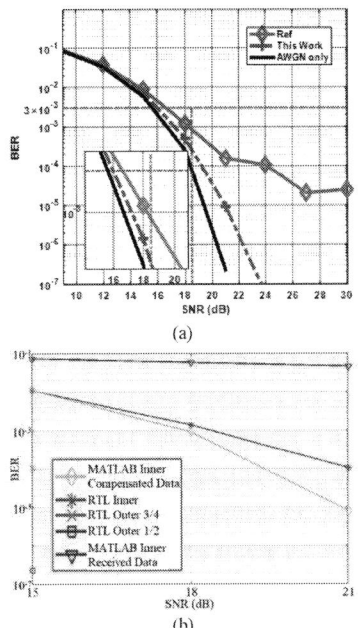

Fig. 4. BBIC BER performance for (a) inner transceiver (b) outer transceiver

TABLE III. The Measurement Results of the BBIC @ 625 MHz

Module			Gate Count (K Gates)	Power (mW)
Tx			350.5	50
Rx	Inner Rx	Decimator	58.6	47
		IQ Calibration	44.5	22
		FFT	364.0	375
		IFFT	552.0	700
		EQ	686.8	108
		SYNC PNC Memory Others	505.8	258
		Total	2211.4	1510
	Outer Rx	LDPC (main module)	629.5	284
		Total	643.0	286
Control, clk_gen			52.8	13
Total			3257.2	1810

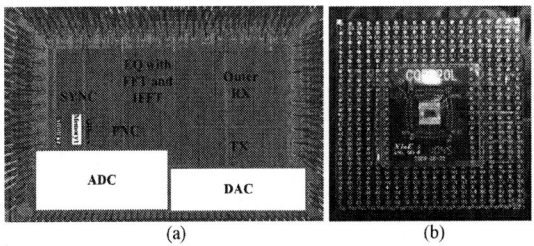

Fig. 5. (a) BBIC Chip microphotograph (b) Chip with package

Hence, an auto testing system platform, *ADVANTEST V93000 PS1600*, is used for the testing. Fig. 6 shows the testing environment. For the functional correctness, we pass all the test bench. Due to the current limited by the platform and the input clock loading, the maximum input clock rate of testing is 300 MHz. The performance of energy per bit (energy/bit) is outstanding as an overview with other works which shown in Table IV.

IV. CONCLUSIONS

A mmWave digital BB transceiver has been implemented in a 28nm HPC plus CMOS process with 1V supply voltage. Based on IEEE 802.11ad/ay standard, we proposed a specification whose chip rate is 2.5 GHz with physical throughput rate up to 10 Gbps under 16-QAM and 3/4 code rate SC mode. For the practical concern, 4 times parallelism is adopted with 625 MHz clock rate. Non-ideal effects including PN and IQ-imbalance due to the RF circuit are also considered and compensated in the digital BB. The target BER which is 3×10^{-3} before SNR 18.5 dB at pre-LDPC is achieved. After the auto placement and routing (APR), we implement a $3.26 \times 2.12 mm^2$ chip (digital BB core area is 2.20 mm²) with total power consumption 1.8 Watt. The performance of energy/bit, including ADC/DAC, achieves 0.023 nJ/bit and 0.211 nJ/bit for Tx and Rx, respectively. We also pass the functionality verification through the measurement in *ADVANTEST V93000 PS1600* platform. With modularized and parameterized IP design concept, our system parameter, such as QAM number, channel bonding/aggregation number, and word length are adjustable. The hardware architecture retain flexibility to further extend parallelisms or pipeline stages for higher data rate requirement. Moreover, an adaptive self-organized system approach is achievable for the diversified transmission scenario as shown in Fig. 7. This mmWave transceiver can be a capstone for the next generation wireless communications including the IQ-imbalance all-self-healing, multi-antenna/multi-stream techniques, and artificial intelligence assisting operation acceleration and optimization.

ACKNOWLEDGMENT

The authors would like to thank the University Shuttle Program of Taiwan Semiconductor Manufacturing Company, MediaTek, Quanta Computer, Ministry of Science and Technology, and Taiwan Semiconductor Research Institute, Taiwan R.O.C.

TABLE IV. PERFORMANCE OVERVIEW

Items		ISSCC'12[9]	JSSC'13[10]	This Work
Distance (meters)		1.7	1	**25****
Throughput Rate (Gbps)		3.1	1.8	**10**
Power Consumption* (W)	*TX*	0.196	0.441	0.226
	RX	0.398	0.710	2.11
Max. Consumption Energy/bit (nJ/bit)	*TX*	0.063	0.245	0.023
	RX	0.128	0.394	0.211

* The power consumption of this work include ADC/DAC data, but without MAC information.
** The distance of this work is interpretated from BER verification and link budget equation.

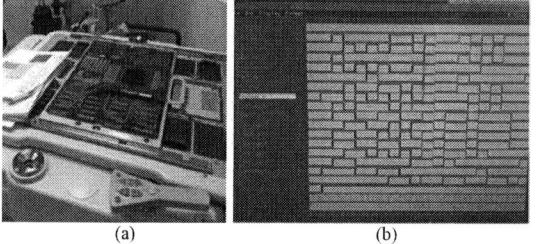

Fig. 6. (a) Testing system platform (b) Functional verification

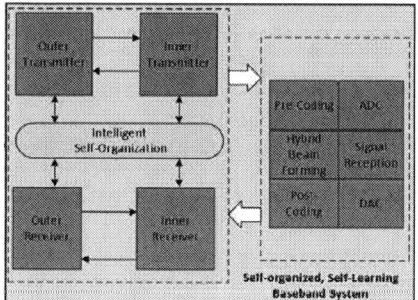

Fig. 7. Intelligent self-organized learning BB system structure

REFERENCES

[1] Wireless LAN Medium Access Control (MAC) and Physical Layer (PHY) Specifications Amendment 3: Enhancements for Very High Throughput in the 60 GHz Band, IEEE Std. 802.11ad, 2012.

[2] P802.11ay/D3.0, Feb 2019 - IEEE Draft Standard for Information Technology--Telecommunications and Information Exchange Between Systems Local and Metropolitan Area Networks--Specific Requirements Part 11: Wireless LAN Medium Access Control (MAC) and Physical Layer (PHY) Specifications--Amendment: Enhanced Throughput for Operation in License-Exempt Bands Above 45 GHz.

[3] "IMT-Vision - Framework and overall objectives of the future development of IMT for 2020 and beyond," Recommendation ITU-R M.2083-0, Sep.2015. [Online] Available: https://www.itu.int/dms_pubrec/itu-r/rec/m/R-REC-M.2083-0-201509-I!!PDF-E.pdf

[4] Nokia, Alcatel-Lucent Shanghai Bell, "On the evaluation of PN model," 3GPP TSG-RAN WG1 Meeting #85, Nanjing, P. R. China, Tech. Rep. R1-165005, May. 2016.

[5] P. Nicholas *et al.*, "Correlation based phase noise compensation in 60 GHz wireless systems," *IEEE 28-th Convention of Electrical and Electronics Engineers*, pp. 1-5, Dec. 2014.

[6] H. W. Chan, C. T. Wu, C. W. Jen, C. Y. Liu, W. C. Lee, and S. J. Jou, "A Pseudo MMSE Linear Equalizer for 60GHz Single Carrier Baseband Receiver," *in 2017 IEEE 12th International Conference on ASIC (ASICON)*, 2017, pp. 643-646.

[7] S. D'Souza, F. Hsiao, A. Tang, S. Tam, R. Berenguer, M. F. Chang, "A 10-Bit 2-GS/s DAC-DDFS-IQ-Controller Baseband Enabling a Self-Healing 60-GHz Radio-on-Chip," *IEEE Transactions on Circuits and Systems II*, vol 60, no.8, pp. 457-461, Aug. 2013.

[8] W. Liu, T. Wei, Y. Huang, C. Chan, and S. Jou, "All-Digital Synchronization for SC/OFDM Mode of IEEE 802.15.3c and IEEE 802.11ad," *IEEE Trans. Circuits Syst. I, Reg. Papers*, vol. 62, no. 2, pp. 545-553, 2015.

[9] K. Okada *et al.*, "A full 4-channel 6.3 Gb/s 60 GHz direct-conversion transceiver with low power analog and digital baseband circuitry," *in IEEE ISSCC Dig. Tech. Papers*, 2012, pp. 218–219.

[10] N. Saito *et al.*, "A Fully Integrated 60-GHz CMOS Transceiver Chipset Based on WiGig/IEEE 802.11ad With Built-In Self Calibration for Mobile Usage," *IEEE Journal of solid-state circuits*, vol. 48, no. 12, pp. 3146-3159, Dec. 2013.

A 23-mW 60-GHz Differential Sub-Sampling PLL with an NMOS-Only Differential-Inductively-Tuned VCO

Bingwei Jiang, Howard C. Luong

Hong Kong University of Science and Technology, Clear Water Bay, Hong Kong

Abstract — A 60-GHz sub-sampling PLL (SSPLL) employs a differential tuning loop to suppress the reference spur. The VCO frequency is differentially tuned only by NMOS transistors. An injection-locked frequency divider (ILFD) with a 4th-order resonant tank is employed between the VCO and the phase detector (PD) to further minimize the spur even without an isolation buffer. Fabricated in 65nm CMOS, the prototyped differential SSPLL measures RMS jitter of 236 fs and reference spur from -43 to -52-dBc over a frequency tuning range from 55.5 to 60 GHz while consuming 23mW from 1.2-V supply corresponding to FOM of -238.9dB.

Index Terms—Sub-sampling PLL, phase noise, differential tuned VCO, injection-locked divider

I. INTRODUCTION

Due to the reduced phase noise contribution of frequency dividers and the phase detector/charge pump (PD/CP), an SSPLL with a large loop bandwidth can suppress the VCO phase noise and achieve low RMS jitter and low power dissipation [1] [2]. However, the spur induced by the PD/CP cannot be sufficiently suppressed by the loop filter. The problem becomes more severe in an mm-wave SSPLL due to the poor isolation between the PD and the VCO.

In this paper, a differential SSPLL is proposed to reduce the spurious tones caused by the common-mode interferences. In addition, an ILFD with a 4th-order transformer coupled resonant tank is employed between the VCO and the PD to reduce the phase sensitivity to minimize the spur over the tuning range. Finally, a 60-GHz VCO is designed with the tank inductance being differentially and finely tuned by only NMOS-based variable resistors to avoid degradation of tuning range and quality factor Q by PMOS or P-type varactors.

II. PROPOSED DIFFERENTIAL SSPLL

Fig. 1 shows the block diagram of the proposed differential SSPLL. The system consists of a frequency-locked loop with a dead zone [1] and a sub-sampling PLL. For the sub-sampling loop, to relax the sampling bandwidth issue and to meet the trade-off between phase noise contribution from the PD/CP and the frequency divider, the reference input is sampled by the output of a divide-by-3 (/3) ILFD followed the VCO. The transformer in Fig.1 couples the output of the /3 ILFD and provides an optimized DC bias for the sampler while distributing the parasitic capacitive loading to the divider. The detected

phase error within the differentially sampled outputs are sent to the differential charge pump, which is implemented as a fully-differential transconductance gain (G_m) stage. The VCO's frequency is differentially tuned by the loop filter's outputs.

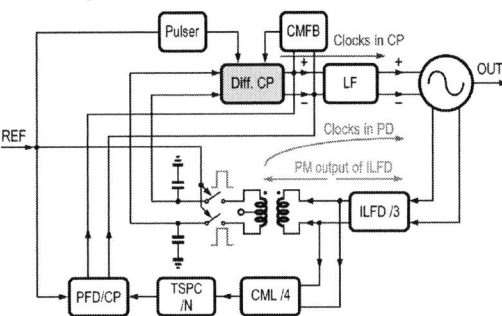

Fig. 1. Proposed differential SSPLL.

The potential main sources of the spur are shown in Fig. 1: the clocks for sampling PD, the pulsed clocks in the CP, and the phase modulation on the ILFD. The spur caused by clock feedthrough and charge sharing of the switches in PD and the differential CP appears as common-mode interferences (CMIs) and thus can be adequately suppressed by differential tuning. The differential CP is sampled by a pulsed reference to reduce the loop gain of SSPLL, as shown in Fig. 1. Compared to the differential tuned PFD/CP PLL in [3], the differential SSPLL is less sensitive to static phase error, including those caused by the UP/DN current mismatch. Moreover, the phase-modulated (PM) output of the ILFD could also be coupled to VCO or sampled by the PD and introduce spurs. To minimize the coupling at high frequencies, a power-hungry isolation buffer is adopted in [2]. In this work, the buffer is removed to save power and area, and the PD and the /3 ILFD with a 4th-order resonant tank are co-designed to maintain the good isolation.

III. CIRCUIT DESIGN

A. 60-GHz Differential Inductively Tuned VCO

The differentially-tuned VCO is one of the key building blocks for the proposed mm-Wave PLL. The high operation frequency limits the application of varactors for differential

978-1-7281-5107-6/19 $31.00 © 2019 IEEE

tuning [4]. By controlling the current flowing through the coupled inductors, the magnetic tuning in [5] introduces more noise sources, including both thermal and flicker noise, from those actively-biased transistors for tuning. PMOS devices together with NMOS devices in conventional inductive tuning methods [6] can be adopted to realize the differential tuning but inevitably degrade the tuning range due to larger parasitic capacitance.

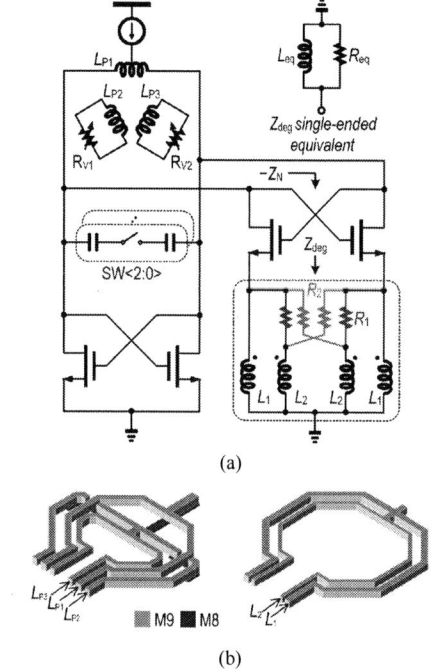

(a)

(b)

Fig. 2. (a) Proposed differentially-tuned VCO and (b) design of the coupled coils for coarse tuning (left) and fine tuning (right).

Fig. 2 shows the schematic of the proposed NMOS-only differential-inductively-tuned VCO. Two non-uniform variable resistors R_{v1} and R_{v2} are employed to achieve three frequency bands together with a swapping control scheme to avoid low-Q region [6], and 3-bits switched-capacitors array (SCA) is used for coarse frequency tuning. The impedance Z_{deg} looking into the port of the fine-tuning network essentially implements the differential inductive tuning through the two coupled inductors, L_1 and L_2, and two groups of variable resistors, R_1 and R_2. With a pair of cross-coupled transistors, the negative inductance included in the negative impedance $-Z_N$ is realized for two main purposes. Firstly, small on-chip inductors but with high Q-factors at the operating frequency can be realized by adding negative and positive inductors in parallel. Secondly, the

DC bias voltage of the NMOS transistors that implement R_1 and R_2 is set at zero to ensure that the differential inductive tuning operates with a practical tuning voltage range.

From Fig. 2, varying R_1 and R_2 in the same direction tends to oppositely change the direction of the currents that flows through L_2 and thus the sign of mutual inductance between L_1 and L_2, $M = k\sqrt{L_1 L_2}$. As a result, the single-ended parallel equivalent inductance looking into the fine-tuning network, L_{eq}, is differentially controlled by R_1 and R_2. Assuming that $L_1 \approx L_2 = L$ and $(1 - k^2)\omega L \gg R_1 \| R_2$, L_{eq} can be derived as:

$$L_{eq} = \frac{1}{2} \frac{(1 - k^2) L (R_1 + R_2)}{(1 - k) R_1 + (1 + k) R_2} \tag{1}$$

From (1), the variation of the equivalent inductance due to the common-mode variation of R_1 and R_2 is completely suppressed.

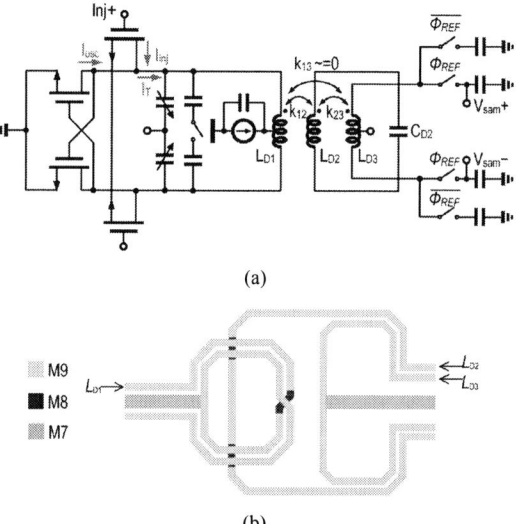

(a)

(b)

Fig. 3. (a) Proposed differential co-design of ILFD and PD and (b) design of the coupled coils for the co-design.

B. PD and /3 ILFD

The /3 ILFD and the PD are co-designed as shown in Fig. 3(a). The /3 ILFD incorporates a triple-coil 4th-order resonant tank. L_{D1} and L_{D2} are used to enhance the locking range, and 1-bit SCA and varactor are used to tune the frequency of the tank. L_{D3} scales down the output amplitude to guarantee the sampling speed over a large phase detection range. In this co-design, the isolation of the VCO

978-1-7281-5107-6/19 $31.00 © 2019 IEEE

from the modulated output of ILFD due to the on/off switching operation of the PD switches is embodied in two ways. First, the impedance variation is firstly scaled down by L_{D3}. Second, as compared to a conventional 2nd-order LC tank, the high-order resonant tank adopted by the /3 ILFD achieves a relatively flat phase response over a larger frequency range. The phase plateau is beneficial not only for locking range enhancement but also for suppression of the phase modulation over a wider frequency range.

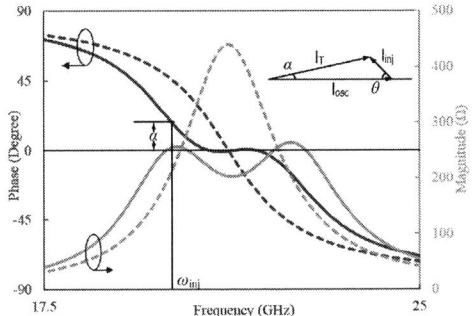

Fig. 4. Frequency responses of a 2nd-order LC tank (dashed line) and a 4th-order transformer resonant tank (solid line), and the relationship of the phasors in Fig. 3.

As shown in Fig. 3(a), I_{inj}, I_{osc} and I_T are the phasor representation of the injection current, the conduction current of the cross-coupled transistors, and the total current flows into the resonant tank, respectively. From the phasor diagram in Fig. 4, referring to the phase of I_{inj}, the phase θ of the /3 ILFD output voltage, which is out-of-phase from that of I_{osc}, can be derived using a simple triangle theorem. Consequently, the phase variation $\Delta\theta$ due to the periodically-varying capacitive loading is derived as:

$$\sin\theta = \sin\alpha \sqrt{1 + m^2 - 2m\cos\theta} \qquad (2)$$

$$|\Delta\theta| = \left|\frac{\partial\theta}{\partial\alpha}\right| \cdot \left|\frac{\partial\alpha}{\partial C}\right| \cdot \Delta C$$
$$\approx \left|\frac{\cos\theta - m}{(1 - m\cos\theta)^2}\right| \cdot \left|\frac{\partial\alpha}{\partial C}\right| \cdot \Delta C \qquad (3)$$

where $m = I_{inj}/I_T$ is the injection strength. Similar to the frequency domain analysis in [7], the assumption that m is small is used here to obtain the analytical results while providing sufficient design insights.

In the proposed co-design, the secondary coil is periodically loaded by the sampling capacitors. Compared to a 2nd-order resonant tank, the 4th-order tank has a small $|\partial\alpha/\partial C|$. Besides, it can also be proved that $|\partial\theta/\partial\alpha|$ is smaller with a smaller α. Thus, from (3), a 4th-order tank with zero phase plateau would achieve smaller phase

variation over a wider frequency range. For the same phase response of the tank impedance, a smaller secondary inductance would result in a smaller phase variation by the same capacitance variation. However, a larger capacitor would need to be added, and the trans-impedance gain would be reduced. Instead, a 3rd coil is used here to scale both the output voltage and also the tank impedance variation. Note that the effective injection current and the injection strength are dependent on the magnitude of the tank impedance. From (3) and Fig. 4, the spur level rises at the frequencies of zeros of the tank impedance transfer function due to the reduced magnitude of tank impedance and thus injection strength.

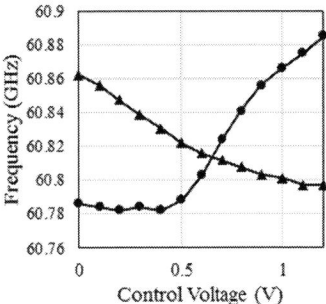

Fig. 5. Measured differential fine-tuning frequency vs control voltage on R_1 (bold dot) and R_2 (triangle).

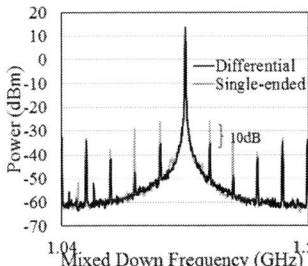

Fig. 6. Measured spur with differential tuning of R_1/R_2 and single-ended tuning (varying R_1 only) when the divider is turned off.

IV. MEASURED RESULTS

Fig. 5 shows the measured differential tuning curves of the proposed differential-inductively-tuned VCO with a common-mode voltage around 0.6-V. With the divider turned off, the spur from the CMI mentioned are compared for single-ended and differential tuning cases. From Fig. 6, the spur is suppressed by 10dB by differential tuning. Fig. 7 shows the closed-loop spur measurement of -43 to -52 dBc over the tuning range. Consuming 23-mW with a 1.2-V supply voltage, the PLL measures in-band phase noise of -86dBc/Hz, phase noise at 10MHz offset frequency of -113dBc/Hz and RMS jitter of 236-fs integrated from 1kHz to 100MHz, as shown in Fig. 8.

978-1-7281-5107-6/19 $31.00 © 2019 IEEE

The SSPLL is implemented in a 65-nm CMOS process with core area of 0.42 mm², and the chip photo is shown in Fig. 9. Performance is summarized and compared with state-of-the-art mm-wave PLLs in TABLE I. The proposed SSPLL demonstrates the best FoM$_J$ with the lowest power consumption while achieving comparable spur level even without an isolation buffer. In [8], the PD is operated at 5GHz and the isolation is relaxed to obtain a spur level of -72dBc. However, the in-band phase noise is significantly degraded due to the 20GHz divider which results in a higher jitter performance.

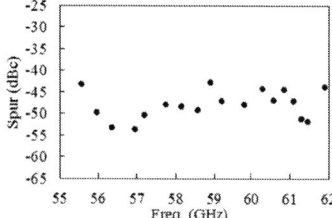

Fig. 7. Measured spur level of the proposed PLL.

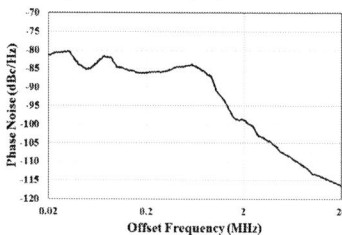

Fig. 8. Measured phase noise of the proposed PLL at 60GHz.

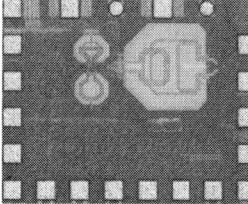

Fig. 9. Chip photo.

V. Conclusion

A differential SSPLL with an NMOS-only passive differential inductively tuned VCO is proposed to tackle the spur issue in mm-Wave frequency synthesis. In the proposed co-design of ILFD and PD, the high-order resonant tank of ILFD improves isolation between VCO and PD, and further suppress the spur level. By removing the buffer, the FoM in terms of jitter is improved.

Acknowledgement

This project was partially funded by the Hong Kong General Research Fund 16208617.

TABLE I. Performance Summary and Comparison

	[2]	[8]	[6]	This work
Tech. (nm)	40	65	65	65
Architecture	SSPLL	SSPLLx3 ILO	ADPLL	SSPLL
Freq. (GHz)	53.8-63.3 (16.2%)	58.3-65.4 (11.5%)	82-107.6 (27%)	55.5-62 (11.1%)
Ref. (MHz)	40	36	100	50
P$_{DC}$ (mW)	42	32	35.5	**23**
In-Band PN (dBc/Hz)	-92	-78.5	-87.1	-86
PN@10MHz	-108	-122	-108	-112
Jitter (fs)	202.9	290	276	236
Spur (dBc)	<-40	-72	-34~-52	-43~-52
FoM$_J$ (dB)	-237.7	-236	-235.7	**-238.9**
FoM (dB)	-167.8	-183.3	-171	-174
FoM$_T$ (dB)	-172	-184.5	-179	-174.9

$$\text{FoM}_J = 10\log[(\text{jitter}/1s)^2 \cdot (P_{DC}/1mW)]$$

$$\text{FoM} = \text{PN} + 20\log(\Delta f/f_0) + 10\log(P_{DC}/1mW)$$

$$\text{FoM}_T = \text{FoM} - 20\log(\Delta f/f_0)$$

References

[1] X. Gao, *et al.*, "A low noise sub-sampling PLL in which divider noise is eliminated and PD/CP noise is not multiplied by N²," *IEEE J. of Solid-State Circuits*, vol. 44, no. 12, pp. 3253-3263, Dec. 2009.

[2] V. Szortyka, et al., "A 42 mW 200 fs-jitter 60 GHz sub-sampling PLL in 40 nm CMOS," *IEEE J. Solid-State Circuits*, vol. 50, no. 9, pp. 2025-2036, Sep. 2015.

[3] L. Lin, L. Tee, and P. R. Gray, "A 1.4 GHz differential low-noise CMOS frequency synthesizer using a wideband PLL architecture," *IEEE Int. Solid-State Circuits Conf. (ISSCC)*, Feb. 2000, pp. 204-205.

[4] J. Kim, et al., "A 44GHz differentially tuned VCO with 4GHz tuning range in 0.12μm SOI CMOS," *IEEE Int. Solid-State Circuits Conf. (ISSCC)*, Feb. 2005, pp. 416-607.

[5] D. Pi, B. Chun, and P. Heydari, "A 2.5-3.2 GHz CMOS differentially-controlled continuously-tuned varactor-less LC-VCO," *IEEE Asia Solid- State Circuits. Conf.*, 2007, pp. 111-114.

[6] Z. Huang and H. C. Luong, "An 82-to-108GHz −181dB-FOMTADPLL employing a DCO with split-transformer and dual-path switched-capacitor ladder and a clock-skew-sampling delta-sigma TDC," *IEEE Int. Solid-State Circuits Conf. (ISSCC)*, Feb 2018, pp. 260-262.

[7] B. Razavi, "A study of injection locking and pulling in oscillators," *IEEE J. of Solid-State Circuits*, vol. 39, no. 9, pp. 1415-1424, Sep. 2004.

[8] T. Siriburanon, et al., "A low-power low-noise mm-wave subsampling PLL using dual-step-mixing ILFD and tail-coupling quadrature injection-locked oscillator for IEEE 802.11 ad," *IEEE J. of Solid-State Circuits*, vol. 51, no. 5, pp. 1246-1260, May 2016.

A 0.003-mm² 440fs$_{RMS}$-Jitter and -64dBc-Reference-Spur Ring-VCO-Based Type-I PLL Using a Current-Reuse Sampling Phase Detector in 28-nm CMOS

Zunsong Yang [1], Yong Chen [1], Pui-In Mak [1], and Rui P. Martins [1,2]

1 – State Key Laboratory of Analog and Mixed-Signal VLSI and IME/FST-ECE, University of Macau, Macau, China

2 – On leave from Instituto Superior Técnico, Universidade de Lisboa, Portugal {E-mail: ychen@um.edu.mo}

Abstract—**This paper reports a current-reuse sampling phase detector for a type-I phase-locked loop (PLL) to simultaneously achieve both wide loop bandwidth and low control voltage ripple, resulting in low RMS jitter and reference spur, while minimizing the chip area by avoiding an explicit loop filter. Fabricated in 28-nm CMOS, the PLL prototype measures a jitter of 440 fs$_{RMS}$, and a spur level of -64 dBc at 3.296 GHz. The die area is 0.003 mm².**

Keywords—**Ring voltage-controlled oscillator (VCO), phase locked loop (PLL), reference spur, RMS jitter, phase detector.**

I. INTRODUCTION

Low-jitter and low-spur RF phase-locked loops (PLLs) are the key cornerstone of most high-tier wireless transceivers. LC-oscillator-based Type-II PLL [1] offers an excellent phase noise (PN) performance, but the area efficiency is low due to the oscillator's passive inductor, and the capacitor of the loop filter (LF). Differently, the ring voltage-controlled oscillator (VCO)-based PLL is attractive for its core area, e.g. <0.05 mm² in [2]-[4]. Yet, they entail extra power-and-area-hungry circuits to deal with their jitter-spur performance. In [2], the use of a master-slave sampling filter (MSSF) replaces the passive LF to enlarge the conversion gain (K_{PD}) of the phase detector (PD), and relax the tradeoff between the reference spur and loop bandwidth (BW). Regrettably, two sampling capacitors (e.g., 16 and 1 pF) and a harmonic-trap circuit are necessary to handle the reference spur from -47 to -65 dBc, as well as a pre-selected switch to control the K_{PD} variation. Thus, the XOR + MSSF topology raises the design complexity, and penalizes the area-and-power efficiency. Although the injection-locked clock multipliers (ILCM) can achieve a very low RMS jitter, it calls for complicated circuitries to calibrate the reference spur.

II. PROPOSED TYPE-I PLL USING A CURRENT-REUSE SAMPLING PD

The proposed PLL consists of a current-reuse sampling PD, a frequency generator (e.g., ring VCO and divider chain) and a non-overlap clock generator. Our PD detects the phase error ($\Delta\Phi = \Phi_{REF} - \Phi_1$) and outputs a dc voltage to control the ring VCO. With the divider chain, two differential non-overlap clocks ($\Phi_{1,2}$) are generated to activate the PD, and fix the $\Delta\Phi$ when the PLL is locked, without utilizing an explicit LF.

Instead of exploiting the XOR (i.e., digital linear phase detection) + MSSF structure, our sampling PD cascades 2 linearized current-reuse dynamic latches operating in a master-slave manner (Fig. 1). Without any extra calibration, we design a higher value of the overall phase-to-voltage conversion gain (K_{PD}) to suppress the ring VCO's PN contribution for the entire PLL, and keep high the linearity over the wide voltage range of

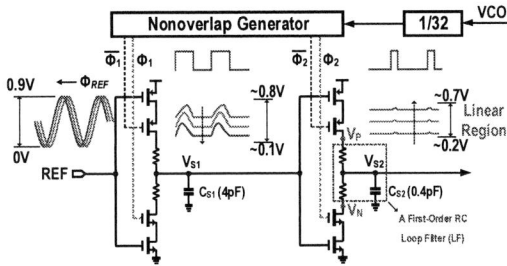

Fig. 1. Operating principle of the proposed current-reuse sampling PD.

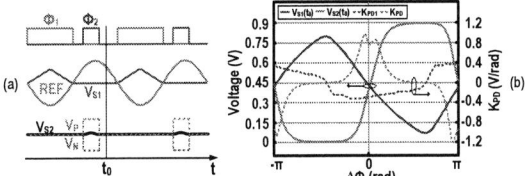

Fig. 2. (a) Timing diagram and (b) phase-detection characteristics.

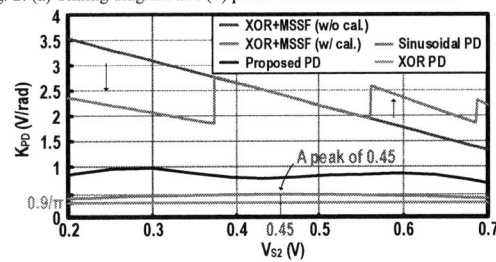

Fig. 3. Different K_{PD} characteristics over a wide range of V_{S2}.

V_{S2}, which corresponds to the ring VCO's tuning range. Φ_1 drives the 1st latch similar to a mixer PD. When the switch is turned on, it works as a current-reuse common-source amplifier with a 1st-order RC load to track the input reference (REF). When the switch is off, the charge/discharge path is cut-off, entering in the hold mode. The sampled voltage (V_{S1}) conveys the variation of Φ_{REF}. Thus, we implement quasi-linear phase-to-voltage conversion. The 2nd latch, switched by the narrow-pulse Φ_2, amplifies and transfers the dc voltage from V_{S1} to V_{S2}, indicating that K_{PD} raises the voltage-to-voltage conversion. The overall linearity is dominated by the 2nd latch, if without the series resistor, its voltage conversion characteristic has a very steep transition due to the push-pull configuration, leading to larger K_{PD} variation across the operating range of V_{S2}. To improve the linearity, we insert the series resistors to consume voltage headroom, rendering the PMOS or NMOS transistor to operate in the linear region. It incurs a narrow-pulse up and down jumps at V_P and V_N nodes [Fig. 2(a)], respectively. The

978-1-7281-5107-6/19 $31.00 © 2019 IEEE

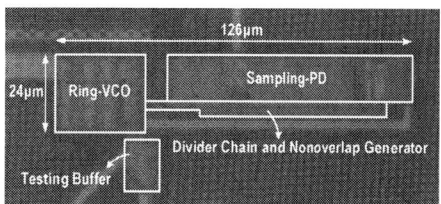

Fig. 4. Chip micrograph of the fabricated PLL in 28-nm CMOS.

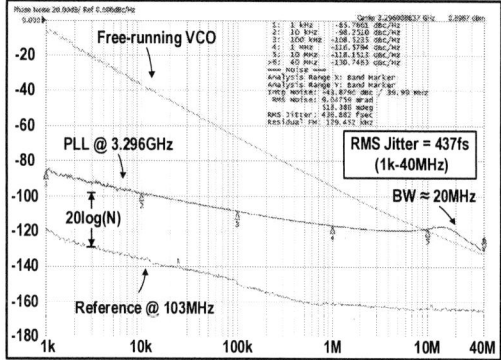

Fig. 5. Measured phase-noise profiles at 3.296 GHz.

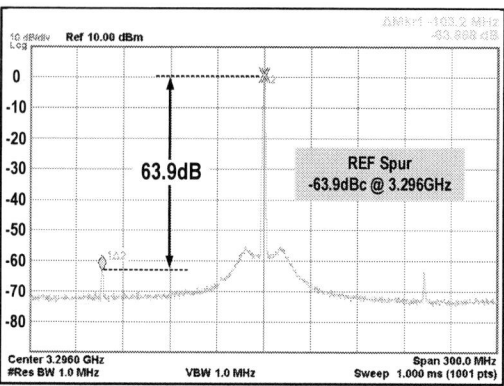

Fig. 6. Measured reference spur at 3.296 GHz.

TABLE I. PERFORMANCE COMPARISON.

	This Work	[2] ISSCC'15	[3] VLSI'17	[4] CICC'18
CMOS (nm)	28	45	65	65
Key Techniques	Ring Type-I Current-Reuse Sampling PD PLL	Ring Type-I Harmonic Traps MSPD-PLL	Ring Type-I FPEC SLF-PLL	Ring Type-II FFNC SSPD-PLL
Ref Freq. (MHz)	103	22.6	47	49.15
Ref. Spur (dBc)	-63.9	-65	-71	-55.2
Loop BW (Fref)	~0.2	~0.3	~0.5	~0.05
RMS Jitter Integrated Range	440 @3.296GHz (1k-40MHz)	970 @2.4GHz (1k-200MHz)	357 @3.008GHz (1k-80MHz)	630 @2.36GHz (1k-100MHz)
PN @ 1MHz (dBc/Hz)	-116.6	-113.8	-121.6	-119.1
FoM (dB)	-241.9	-234.2	-242.3	-236.3
Supply Voltage (V)	0.9	1	1.2/2.2	0.94
Total Power (mW)	3.3	4	4.6	5.86
Power Eff. (mW/GHz)	1	1.67	1.53	2.48
Freq. Range (GHz)	2.6 to 4.2 (47%)	2.0 to 3.0 (40%)	N/A*	2.0 to 2.8 (33%)
Active Area (mm²)	0.003	0.015	0.047	0.022

$$FoM = 10\log_{10}\left[\left(\frac{Jitter}{1\,sec}\right)^2 \cdot \frac{Power}{1\,mW}\right] \qquad \text{* No Capacitor Bank}$$

induced spur will be significantly reduced by the aid of the implicit LF (Fig. 1). Fig. 2(b) depicts the phase-detection characteristics over an interval $[-\pi, \pi]$ at the observed time (t_0). When compared with a ~0.3-V/rad K_{PD1} of the 1st latch, the overall K_{PD} of ~0.85 V/rad is enhanced, along with a linear mapping range (i.e., V_{S2} ranges from 0.2 to 0.7 V).

We briefly summarize different K_{PD} characteristics in Fig. 3. The conventional XOR-based PD has a linear and limited $K_{PD}=V_{DD}/\pi$ (e.g., $V_{DD} = 0.9$ V means a K_{PD} of $0.9/\pi$ V/rad). [2] inserts the MSSF after the XOR gate, resulting in a larger K_{PD} variation. It is also necessary a pre-selected switch to stabilize K_{PD} at 2.3 ± 0.2 V/rad. Compared to a sinusoidal K_{PD} with a peak of 0.45 V/rad at 0.45-V V_{S2}, our linearized PD offers an enhanced and stable value of 0.85 ± 0.15 V/rad. Meanwhile, it decouples the relationship between C_{S1} and C_{S2}. Here, a large C_{S1} can aid us suppressing the PD output noise and optimizing K_{PD} through the adjustment of the time constant at the 1st latch output. The 2nd latch output determines the loop BW and suppresses the voltage ripple on V_{S2}.

Interestingly, the series resistors in the 2nd stage of our PD, together with the 2nd sampling capacitor C_{S2} can serve as an implicit LF, which assists the PLL to maintain its stability and further suppress the impact of the clock feedthrough, e.g., V_P and V_N in Fig. 2(a), on the control side of the ring VCO. Also, we use the layout skills to reduce the impact of the parasitic coupling on the control side of the ring VCO. These, together, lead to a spur level of <-60 dBc without any extra calibration.

III. EXPERIMENTAL RESULTS

Our PLL (Fig. 4) occupies a core area of 0.003 mm² in 28-nm CMOS. It consumes 3.3 mW (3 mW in the VCO, 0.15 mW

in the PD and 0.18 mW in the other blocks) at 0.9 V and at an operating frequency of 3.296 GHz. The PN was measured with a 103-MHz reference input (Fig. 5). Due to the wide loop bandwidth of ~20 MHz, the ring VCO's PN can be significantly suppressed. The PLL shows a 440-fs$_{RMS}$ jitter integrated from 1 kHz to 40 MHz, and a -116-dBc/Hz PN at 1-MHz offset. Measured at 3.296 GHz, the level of the reference spur is -63.9 dBc (Fig. 6). Table I summarizes the performance of this work and compares it with the recent arts. Without any calibration, smaller core area (0.003 mm²) and better power efficiency (1 mW/GHz) are achieved concurrently in this work.

REFERENCES

[1] Z. Yang et al., "A 25.4-to-29.5GHz 10.2mW isolated sub-sampling PLL achieving -252.9dB jitter-power FoM and -63dBc reference spur," ISSCC, pp. 270-272, Feb. 2019.

[2] L. Kong and B. Razavi, "A 2.4GHz 4mW inductorless RF synthesizer," ISSCC, pp. 1-3, Feb. 2015.

[3] T. Seong et al., "A -242-dB FOM and -71-dBc reference spur ring-VCO-based ultra-low-jitter switched-loop-filter PLL using a fast phase-error correction technique," VLSI, pp. 186-187, Jun. 2017.

[4] S. S. Nagam and P. R. Kinget, "A -236.3dB FoM sub-sampling low-jitter supply-robust ring-oscillator PLL for clocking applications with feed-forward noise-cancellation," CICC, pp. 1-4, Apr. 2018.

A 360-456 MHz PLL frequency synthesizer with digitally controlled charge pump leakage calibration

Peilin Yang, Yanshu Guo, Hanjun Jiang and Zhihua Wang
Institute of Microelectronics, Tsinghua University, Beijing, China
Email: jianghanjun@tsinghua.edu.cn

Abstract—A charge pump leakage current calibration circuit has been proposed for the phase-locked loop frequency synthesizer. A digitally controlled feedback loop has been built for the adaptive calibration. A 360-456 MHz ring oscillator based fractional-N PLL has been designed and fabricated in 65 nm CMOS technology. The prototype chip occupies a die area of 0.064 mm², and consumes 1.11 mA current from a 1 V supply. Measurement results show that the reference spur is suppressed by 25 dB with the leakage calibration enabled, when outputting 432 MHz clock. The relative level of the reference spur is at least 10 dB lower that other designs in literature.

Keywords—PLL, charge pump, leakage current, calibration

I. INTRODUCTION

For the charge pump phase-locked loop (CP-PLL) based frequency synthesizer, the leakage current of the charge pump is one major reason for the output spur [1]. The leakage current is mainly caused by the current source mismatch. More severely, the charge pump leakage current is voltage dependent, which may further cause some high-order non-idealities. The direct charge pump leakage current compensation methods, such as the replicated leakage cancellation [2], the current pulse position randomization [3] and the pulse split method [4], cannot efficiently handle the leakage issue with the presence of process, voltage and temperature (PVT) variations. The analog feedback compensation method [5] adopts a feedback control to dynamically calibrate the leakage current. However, it suffers from the leakage current of the analog calibration circuit itself, and the static power consumption associated with the analog circuit. In this work, a CP leakage current calibration circuit with a digitally controlled feedback loop is proposed, with the advantages of small extra power consumption and high calibration accuracy. A simulation-based study of the CP leakage compensation method has been presented in [6], and this paper provides the circuit implementation results. The 360-456 MHz fractional PLL circuit has been designed, fabricated and measured to validate the proposed calibration circuit.

II. CIRCUIT IMPLEMENTATION

The proposed frequency synthesizer is composed of the PLL core and the calibration circuit is shown in Fig. 1. The PLL core consists of the two-stages ring oscillator (VCO), the phase frequency detector (PFD), the differential charge pump (CP), the loop filter, and the delta-sigma divider. The leakage

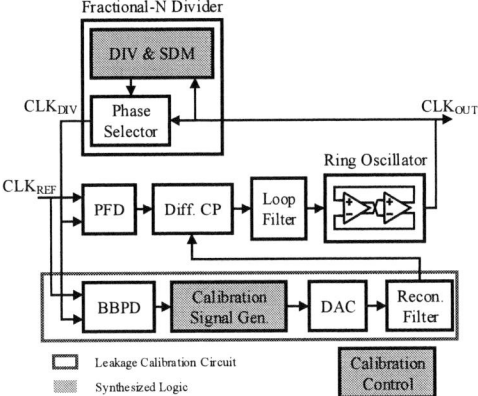

Fig. 1. Proposed frequency synthesizer

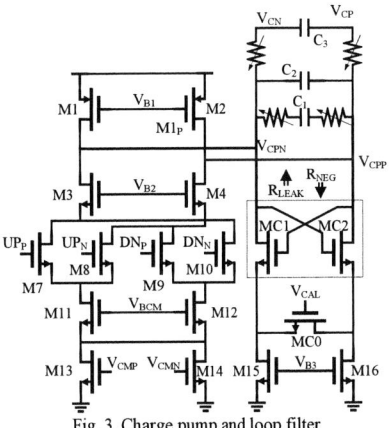

Fig. 3. Charge pump and loop filter

calibration circuit mainly consists of the bang-bang phase detector (BBPD) to detect the output phase error, the logic circuit to generate the calibration signal, and the digital-to-analog converter (DAC) and reconstruction filter to convert the signal into the analog calibration voltage.

A. Charge pump and loop filter

The differential CP and the PLL loop filter are shown in Fig. 3. The current mismatch between M1(M2) and M3(M4) leads to the leakage current, and this leakage is dependent on

978-1-7281-5107-6/19 $31.00 © 2019 IEEE

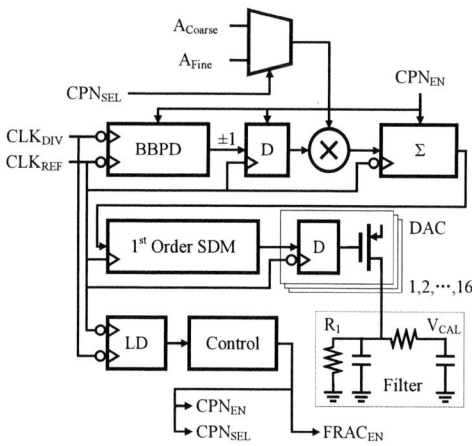

Fig. 3. Circuit to generate the calibration voltage

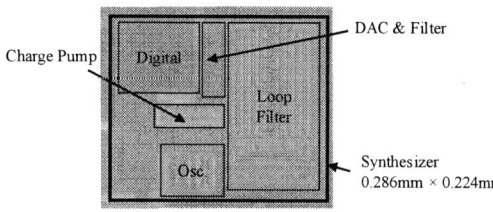

Fig. 4. Die micrograph

the CP output voltage V_{CPP}/V_{CPN}. A virtual resistor R_{LEAK} can be used to model the leakage effect. The cross-coupled transistors MC1 and MC2 in the loop filter are used to generate the negative resistance R_{NEG}, and R_{NEG} can be tuned by changing the calibration voltage V_{CAL}. The leakage calibration is realized by using the negative resistance R_{NEG} to cancel the positive resistance R_{LEAK}.

B. Calibration voltage generation with feedback loop

A bang-bang phase detector (BBPD) is used to detect the phase difference between the PLL output in the presence of charge pump leakage and the 24 MHz reference clock. The binary output of the BBPD is integrated and its 16-bit output code is sent to a first-order sigma-delta modulator to convert it into a 1-bit stream. The 1-bit data stream passes through a shifter register delay chain which serves as a rectangle finite impulse response (FIR) filter. Finally, a DAC with 16 equally weighted current source elements converts the digital signal into the analog calibration voltage V_{CAL}. A closed-loop calibration is consequently formed.

III. MEASUREMENT RESULTS

The proposed 360-456 MHz PLL has been designed and implemented in a 65 nm CMOS technology. The die micrograph is shown in Fig. 4. It occupies a die area of 0.064 mm². With a 1.0 V power supply, the fractional-N PLL consumes 1.1 mA current while outputting 360-456 MHz clocks. The calibration circuit only adds ~ 40μW extra power consumption. The effectiveness of the leakage calibration is

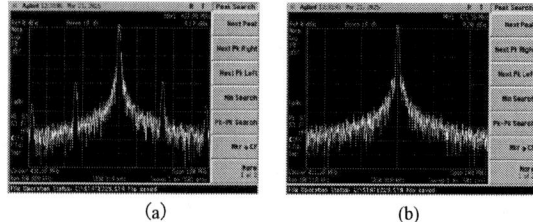

Fig. 5. PLL output spectrum: (a) no calibration, (b) calibration enabled

TABLE I.   COMPARSION TO OTHER RING OSCILLATOR BASED PLL'S

	This work	TCAS-II 2010[2]	JSSC 2012[5]	ICCE 2015[7]
Process	65nm	65nm	65nm	65nm
Supply (V)	1	1.2	1.2	1.2
Ref. Freq. (MHz)	24	50	40	3
Output Freq. (MHz)	360-456	1000	640	576
Power Cons. (mW)	1.11	8.4	1.2	4
Ref. Spur (dBc)	**-68.9**	-50.92	-54.82	N.A.
PN @1MHz (dBc/Hz)	-98.92	-97.12	-88.6	-72.57
FoM** (dB)	-218.7	-217.6	-219.56	-216.3

*: Measured for 432 MHz output
**: FoM = 20log10(Jitter$_{RMS}$[ps]) + 10log10(Power[mW])

shown in Fig. 5. As shown in Fig. 5(a), without the calibration, the highest spur is about -39 dBc compared to the 432 MHz clock. With the calibration circuit enabled, the major spurs are suppressed by at least 25 dB. In Table I, the presented design is compared to other ring oscillator based PLL's in literature. With the proposed charge pump leakage current calibration, this work shows the lowest reference spur, which is >10 dB lower than other designs.

ACKNOWLEDGMENT

This work was funded, in partial, by National Natural Science Foundation of China #61661166010, Suzhou-Tsinghua Innovation Leadership Program #2016SZ0214, National Key R&D Program of China #2016YFC0105603, and Beijing Engineering Research Center No. BG0149.

REFERENCES

[1] W. Rhee, "Design of High-performance CMOS Charge Pumps in Phase-locked Loops," Circuits and Systems, IEEE Int. Circuit and System Conference (ISCAS), vol.2, pp. 545-548, 1999.

[2] J. Chang, et al, "A Phase-Locked Loop With Background Leakage Current Compensation," IEEE Trans. on Circuits and Systems II: Express Briefs, vol. 57, no. 9, pp. 666-670, 2010.

[3] C. Thambidurai, et al, "Spur Reduction in Wideband PLLs by Random Positioning of Charge Pump Current Pulses," Proceedings of 2010 IEEE Int. Symposium on Circuits and Systems, pp. 3397-3400, 2010.

[4] Y. Fan, et al, "19.5 Digital Leakage Compensation for a Low-Power and Low-Jitter 0.5-to-5GHz PLL in 10nm FinFET CMOS Technology," IEEE Int. Solid- State Circuits Conference (ISSCC), pp. 320-322, 2019.

[5] I. Lee, et al, "A Leakage-Current-Recycling Phase-Locked Loop in 65 nm CMOS Technology," IEEE Journal of Solid-State Circuits, vol. 47, no. 11, pp. 2693-2700, 2012.

[6] P. Yang, et al, "A 400MHz Fractional-N Frequency Synthesizer with Charge Pump Output Leakage Balancing," in Proc. 14th IEEE ICSICT. pp. 1-3, 2018.

[7] S. Park, et al. "A Fully Integrated Phase-locked Loop with Leakage Current Compensation in 65-nm CMOS Technology," IEEE Int. Conf. on Consumer Electronics (ICCE), pp. 587-588, 2015.

20-4 (7100)

IEEE Asian Solid-State Circuits Conference
November 4 – 6, 2019
The Parisian Macao, Macao SAR, China

A 12-GHz All-Digital Calibration-Free FMCW Signal Generator Based on a Retiming Fractional Frequency Divider

Zhengkun Shen, Heyi Li, Haoyun Jiang, Zherui Zhang, Junhua Liu and Huailin Liao

Key Laboratory of Microelectronic Devices and Circuits (MOE)
Institute of Microelectronics, Peking University, Beijing 100871, China
Email: zkshen2013@pku.edu.cn; junhua.liu@pku.edu.cn; liaohl@pku.edu.cn

Abstract—A 12-GHz all-digital calibration-free frequency-modulated continuous-wave (FMCW) signal generator is presented in this paper based on a retiming fractional frequency divider (FFD). Instead of modulating a multi-modulus divider (MMD) by a $\Delta\Sigma$ modulator, a fractional divider is utilized to release the narrow loop bandwidth limitation and achieve better phase noise without active noise cancellation techniques. Thus, the FMCW signal generator can achieve low root-mean-square (RMS) frequency error under a fast chirp slop. A calibration-free retiming FFD architecture which is not sensitive to process, voltage and temperature (PVT) variations is proposed to avoid complex and slow calibration process. Implemented in 40-nm CMOS, the generator consumes 33.8 mW power and 0.32 mm² chip area. Measurement results show that the generator achieves 78 MHz/µs maximum chirp slope and 6.0 kHz RMS frequency error when 300 MHz frequency is swept in 2 ms. The phase noise from 12 GHz carrier is -113.6 dBc/Hz at 1 MHz offset.

Keywords—All-digital PLL; Calibration-free; Frequency-modulated continuous-wave (FMCW); Radar; Fractional frequency divider

I. INTRODUCTION

Millimeter-wave (MMW) radars play an important role in various applications such as autonomous-driving and micro unmanned aerial vehicle (UAV) imaging, and a frequency-modulated continuous-wave (FMCW) chirp generator is the key block of a radar transceiver. Charge pump based fractional-N phase-locked loops (PLLs) are typically employed to implement FMCW generator. But it suffers from the limitation on chirp slope and needs bulky off-chip loop filter [1]. Compared to the analog PLL, an all-digital PLL (ADPLL) has compact and scalable size. Besides, the flexibility of an ADPLL makes it possible to meet different application needs with one FMCW generator. However, conventional fractional-N ADPLL with a multi-modulus divider (MMD) modulated by a $\Delta\Sigma$ modulator needs narrow loop bandwidth to suppress the out-of-band quantization noise. As a consequence, the maximum chirp slope is limited. Though utilizing two-point modulation (TMP) technique may provide infinite bandwidth and release the chirp slope limitation theoretically, a wide loop bandwidth is still needed in reality to compensate the gain or timing mismatch between the two modulation paths. A counter-assisted digital PLL (CDPLL) does not need a MMD thus releases the loop bandwidth restriction, but it needs a high linearity, wide detection range time-to-digital convertor (TDC)

This work was supported by the National Natural Science Foundation of China under grant 61574008 and 61831006

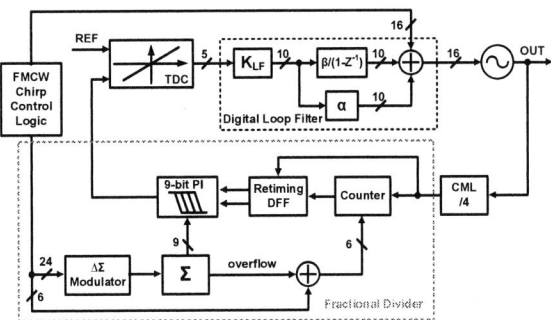

Fig. 1. Architecture of the proposed FMCW signal generator.

to estimate the phase error [1]. However, wide range TDC usually has low resolution and poor linearity which will introduce high level in-band phase noise and spurs. Moreover, glitches and the rising and falling time mismatch of the delay chain in TDC also deteriorate CDPLL's performance.

This paper presents a calibration-free all-digital FMCW generator. A fractional frequency divider (FFD) is utilized to obtain low phase noise and wide loop bandwidth simultaneously. To avoid complex calibration methods, a retiming FFD architecture which is not sensitive to process, voltage and temperature (PVT) variations is proposed. A high resolution Vernier TDC and a high frequency resolution digitally-controlled oscillator (DCO) with a differential tapped inductor is employed to suppress the quantization noise. TMP technique is also utilized to obtain fast chirp slope. The FMCW generator prototype is implemented in 40-nm CMOS.

II. CIRCUIT DESCRIPTIONS

A. Architecture Overview

The block diagram of the proposed digital FMCW signal generator is shown in Fig. 1. The 12 GHz output is 4× divided by a 2-stage current-mode logic (CML) frequency divider and sent to a retiming FFD. The FFD has a 6-bit integer part implemented by a programmable counter and 9-bit fractional part implemented by a phase interpolator (PI) in order to provide a fine fractional frequency division step of 4/512 = 1/128. Besides, a 3rd-order 1-1-1 MASH $\Delta\Sigma$ modulator is utilized to further expand the fractional frequency division

978-1-7281-5107-6/19 $31.00 © 2019 IEEE

287

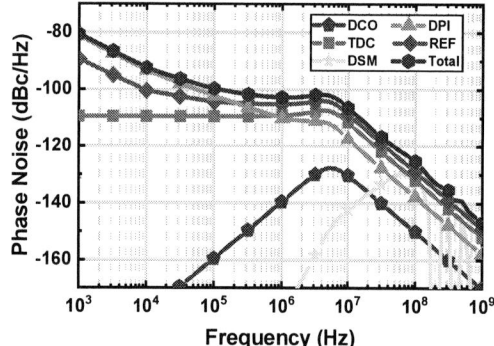

Fig. 2. Simulated phase noise contribution.

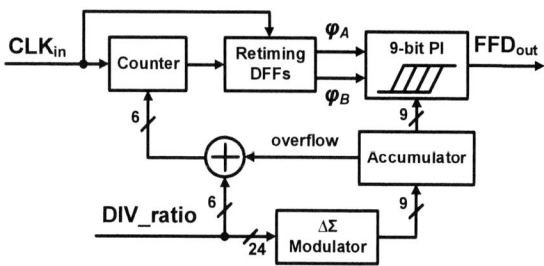

Fig. 3. Architecture of the proposed retiming FFD.

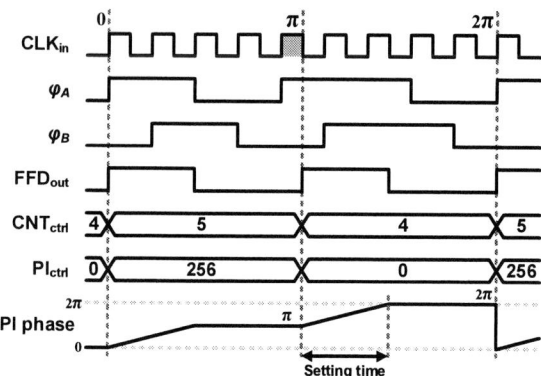

Fig. 4. The waveform and control logic when the division ratio is 4.5.

resolution to 24 bit. As a consequence, the quantization noise is $20\log(512) = 54\text{dB}$ lower than those PLLs in which a $\Delta\Sigma$ modulator modulates a MMD directly. Besides, the loop bandwidth is extended and the maximum chirp slope is improved. The detection range requirement of the TDC is also released by 512 times thanks to the FFD. So a narrow detection range, high resolution TDC is adopted to obtain better in-band phase noise performance. According to simulation results shown in Fig. 2, the FMCW generator achieves lower than -100dBc/Hz in-band phase noise at 1 MHz offset with 5 MHz loop bandwidth. Besides, a TPM scheme is utilized in order to further improve the chirp slope.

B. Retiming Fractional Frequency Divider

Delay line based digital-to-time convertor (DTC) and 360° PI are typically employed to implement a FFD. The gain of a DTC is sensitive to PVT variations and complex calibration algorithm like least-mean-square (LMS) regression is necessary in order to suppress spurs. Besides, LMS regression takes a long time to fit different output frequency. As a consequence, when the output frequency of a FMCW generator chirps quickly, the effect of LMS regression may degrade. PI based FFD is naturally adaptable to frequency changes because the scale range of a PI is determined by the input clock's period, and a PI is usually robust to PVT than a DTC if the current mirror is well designed. However, most of the PI needs 8-phase input while the phase mismatch between the 8 input signals will directly impact the linearity of the PI and introduce spurs to the output spectrum.

To overcome the drawbacks of the current FFDs, a retiming FFD architecture is proposed as shown in Fig. 3. The FFD consists of a programmable counter, two retiming digital flip-flops (DFFs) and a 9-bit PI controlled by an accumulator. The single phase input signal CLK_{in} is divided by the counter at first and this process achieves the integer part of frequency division. Then the output signal of the counter is retimed by CLK_{in} using two DFFs in order to obtain two signals φ_A and φ_B which have adjacent phases. The output of the FFD is finally generated by the 9-bit PI by interpolating phase between φ_A and φ_B. Thus, a FFD with 1/512 frequency division step is realized. The last 3 bits of PI's control word is modulated by a 3rd-order 1-1-1 MASH $\Delta\Sigma$ modulator and the fractional frequency division resolution is expanded to 24 bit.

To further understand the workflow of the FFD, the waveforms and control logic outputs are shown in Fig. 4. The input division ratio is 4.5 so the control word of the counter CNT_{ctrl} is 4 and the phase accumulation rate of the PI controller is π. The input range of the PI is 0 to 511 which means the PI cannot provide 2π phase interpolating. So the division ratio of the counter should plus one when the output of the accumulator PI_{ctrl} overflows to provide proper output phase.

The proposed FFD has the follow advantages compared to the FFDs mentioned above. The FDD only needs a single phase input which avoids the mismatch problem between 8-phases input thus enhanced the linearity of the FFD. By utilizing the retiming technique, the PI covers 2π phase interpolating range corresponding to the input signal (3 GHz in this case) with 9-bit resolution while the PI is working at reference frequency (100 MHz in this case) which is much lower than the input frequency. Working at a lower frequency makes the PI less sensitive to parasitic capacitance or resistance and provides enough setting time which will improve the PI's performance. With these advantages, the proposed FFD achieves good linearity and free from complex calibration.

C. Parastic Insensitive PI

A parasitic insensitive PI shown in Fig. 5 is proposed to improve the FFD's linearity. The PI consists of two complementary current allocation branches and the interpolation weight is adjusted by 9-bit control word D. Conventional charge-based PIs have poor linearity due to

IEEE Asian Solid-State Circuits Conference
November 4 – 6, 2019
The Parisian Macao, Macao SAR, China

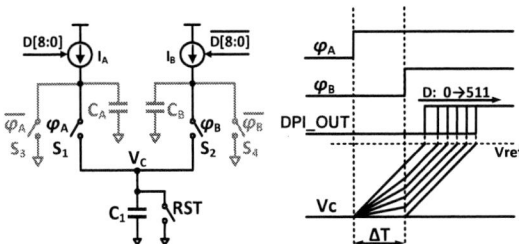

Fig. 5.   Circuit implementation of the parasitic insensitive PI.

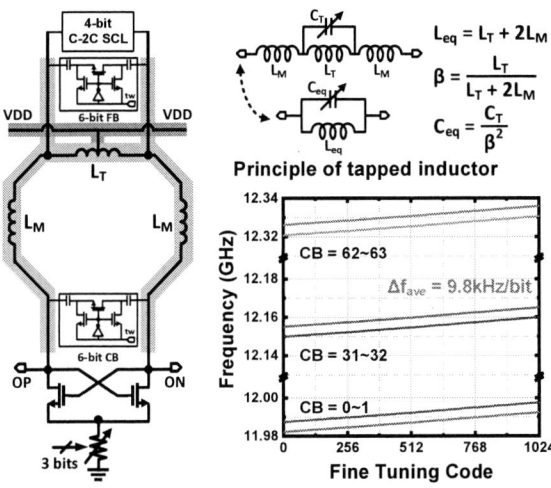

Fig. 6.   Implementation of the high frequency resolution DCO and measured DCO frequency tuning curves.

charge sharing effect caused by parasitic capacitances C_A and C_B. If there is any charge accumulated in C_A and C_B before S_1 and S_2 are closed, the rising edge of Vc will be influenced due to the charge sharing effect. So the accuracy of phase interpolation may be degraded. To eliminate this effect, two switches S_3 and S_4 are added in order to make sure that there is no charge accumulated in C_A and C_B before φ_A's and φ_B's rising edges coming. This method guarantees that the charging rate of C_1 is only determined by the control word D and reduces the PI's sensitivity to parasitic capacitance. Therefore, the PI's linearity is improved.

D. High Time Resolution TDC

A high time resolution Vernier TDC is utilized in the FMCW generator to detect the phase error. When the PLL is at locked state, the transient phase error introduced by the FFD is

$$(1 \times 8) / (3\text{GHz} \times 512) = 5.2\text{ps} \qquad (1)$$

after considering the $\Delta\Sigma$ dithering. So, a TDC with detection range larger than 5.2ps is suitable for the generator. A Vernier TDC can achieve relative high resolution in such narrow range. A 31-stage TDC with time resolution of 2.3 ps is adopted in the generator and the quantization noise introduced by the TDC is only -110 dBc/Hz.

Fig. 7.   Chip microphotograph of the proposed FMCW signal generator.

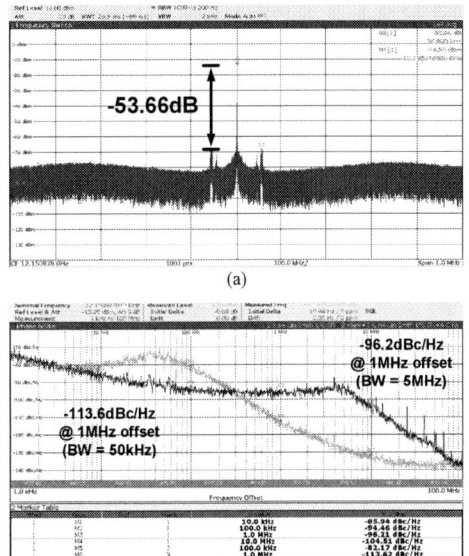

Fig. 7.   (a) PLL output spectrum and in-band fractional spurs. (b) Measured phase noise under different loop bandwidth.

E. High Frequency Resolution DCO

A high frequency resolution DCO (shown in Fig. 6) is adopted to minimize the quantization noise introduced by the DCO. The DCO tuning is segmented into coarse bank (CB), fine bank (FB) and a 4-bit C-2C switched-capacitor ladder (SCL) [2]. The 6-bit CB provides over 300 MHz frequency tuning range for FMCW chirp generation. In order to get high resolution and frequency tuning linearity simultaneously, a differential tapped inductor based on our previous work [3] which features good scaling linearity is utilized to scale the 6-bit FB by approximately 32× with LM = 178 pH and LT = 79 pH. Besides, a 4-bit SCL is adopted to scale down the equivalent switched capacitor by 16× so as to further improve the frequency resolution. The switched capacitors used in FB and SCL are identical to get a smooth frequency tuning curve. Measurement frequency tuning curves shown in Fig. 6 indicate high tuning linearity and 9.8-kHz/bit average frequency resolution. The overlaps between different FB tuning curves are pre-compensated for linearization. The NMOS cross-coupled differential pair is implemented by thick gate oxide transistors to obtain better phase noise performance. A 3-bit tail

978-1-7281-5107-6/19 $31.00 © 2019 IEEE   289

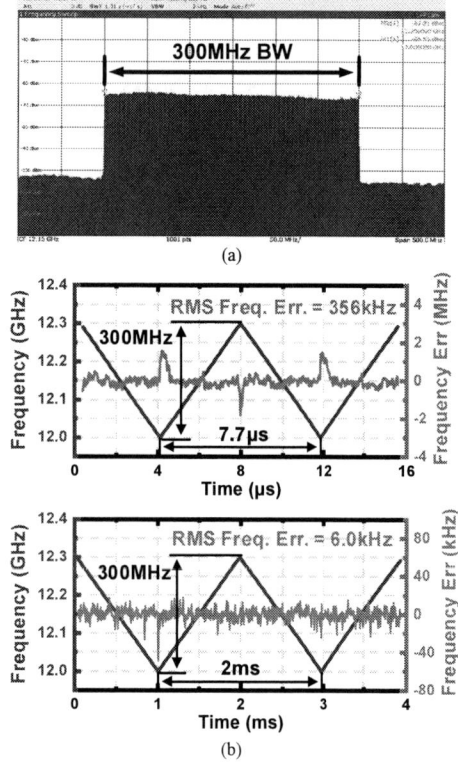

(a)

(b)

Fig. 8. (a) FMCW chirp spectrum. (b) Measured FMCW chirp performance.

TABLE I PERFORMANCE COMPARISON

	This Work	[4]	[5]	[6]
Architecture	ADPLL + TPM	DPLL + TPM	SSPLL + TPM	Frac-N
Technology (nm)	40	65	28	45
Freq. (GHz)	11.7~13.5	20.4~24.6	14.7~17.2	19~20.25
Ref. Freq. (MHz)	100	40	80	1000
PN @ 1MHz (dBc/Hz) *	-113.6	-106.4	-109.0	-110.2
BW/T_{mod} (GHz/µs)	Fast: 0.3/7.7 Slow: 0.3/2000	0.2/1.2	1.5/100	1/40
RMS Freq. Err. (kHz)	Fast: 356 Slow: 6	124**	46**	600***
Power (mW)	33.8	19.7	44	N.A.

* Normalized to 12 GHz carrier.
** Without the turn-around points.
*** Only maximal error reported.

frequency error is only 6.0 kHz including turn-around points (TAPs) when 300 MHz frequency is swept in 2 ms. Table I summarizes the performance of the proposed FMCW signal generator and makes a comparison with the previously published works.

IV. CONCLUSION

A 12 GHz all-digital FMCW signal generator is presented in this paper. A calibration-free retiming FFD is proposed to get better phase noise performance and expand the loop bandwidth. The FFD also releases the detection range of a TDC and a narrow-range 2.3-ps resolution TDC is utilized to suppress the quantization noise. A high resolution DCO is proposed to further improve the phase noise performance. Fabricated in 40-nm CMOS, the FMCW generator achieves only 6.0 kHz RMS frequency error and 78 MHz/µs chirp slope.

REFERENCES

[1] J. Wu et al., "A 77-GHz Mixed-Mode FMCW Generator Based on a Vernier TDC with Dual Rising-Edge Fractional-Phase Detector," 2018 IEEE Asian Solid-State Circuits Conference (A-SSCC), Tainan, 2018, pp. 79-82.

[2] Z. Huang and H. C. Luong, "A dithering-less 54.79-to-63.16GHz DCO with 4-Hz frequency resolution using an exponentially-scaling C-2C switched-capacitor ladder," 2015 Symposium on VLSI Circuits (VLSI Circuits), Kyoto, 2015, pp. C234-C235.

[3] F. Yang, H. Guo, R. Wang, Z. Zhang, J. Liu and H. Liao, "A low-power calibration-free fractional-N digital PLL with high linear phase interpolator," 2016 IEEE Asian Solid-State Circuits Conference (A-SSCC), Toyama, 2016, pp. 269-272.

[4] D. Cherniak, L. Grimaldi, L. Bertulessi, R. Nonis, C. Samori and S. Levantino, "A 23-GHz Low-Phase-Noise Digital Bang–Bang PLL for Fast Triangular and Sawtooth Chirp Modulation," in IEEE Journal of Solid-State Circuits, vol. 53, no. 12, pp. 3565-3575, Dec. 2018.

[5] Q. Shi, K. Bunsen, N. Markulic and J. Craninckx, "26.1 A Self-Calibrated 16GHz Subsampling-PLL-Based 30s Fast Chirp FMCW Modulator with 1.5GHz Bandwidth and 100kHz rms Error," 2019 IEEE International Solid- State Circuits Conference - (ISSCC), San Francisco, CA, USA, 2019, pp. 408-410.

[6] . P. Ginsburg et al., "A multimode 76-to-81GHz automotive radar transceiver with autonomous monitoring," 2018 IEEE International Solid - State Circuits Conference - (ISSCC), San Francisco, CA, 2018, pp. 158-160.

resistor array is utilized to adjust the power consumption of the DCO with negligible phase noise contribution compared to a tail current source.

III. MEASUREMENT RESULTS

The proposed FMCW signal generator is implemented in 40-nm CMOS technology as shown in Fig. 7 and the core area is 0.32 mm². The DCO draws 13 mA from a 1.8-V supply and the remain circuits works under a 1.3-V supply. The total power consumption is 33.8 mW excluding the test buffer.

The PLL's output spectrum and phase noise performances measured at the output of the DCO are shown in Fig. 7 and the reference frequency is 100 MHz. The in-band fractional spurs level are below -53.66 dB which shows that the proposed FFD has a great linearity without calibration. The out-of-band phase noise under a relative narrow loop bandwidth is -113.6 dBc/Hz at 1 MHz offset. With the help of the proposed FFD, the generator can achieve a wide loop bandwidth up to 5 MHz without ΔΣ noise hump and the in-band phase noise at 1 MHz offset is -96.2 dBc/Hz. The all-digital FMCW generator can generate FMCW signals between 11.7 GHz to 13.5 GHz, and the maximum chirp bandwidth is 300 MHz as shown in Fig. 8(a). Thanks to the TMP scheme and wide loop bandwidth, the generator can achieve a chirp slope up to 78 MHz/µs. To verify the quality of the chirp signal, different triangular FMCW chirp slope is generated as shown in Fig. 8(b), and the RMS

A 100Mb/s 3.5GHz Fully-Balanced BFOOK Modulator Based on Integer-N Hyrbrid PLL

Cong Ding, Haixin Song, Woogeun Rhee, and Zhihua Wang

Institute of Microelectronics, Tsinghua University, Beijing, China

Abstract—**This paper presents an energy-efficient high data rate constant-envelope modulator by employing a binary frequency-domain on-off keying (BFOOK) method. Thanks to the fully-balanced FSK feature, high data rate BFOOK modulation is performed based on an integer-N hybrid PLL with VCO modulation only. A carrier spreading technique based on low-frequency wideband triangular modulation can also be added as an optional fractional-N mode in case carrier power suppression is needed to have better spectrum compliance. A prototype 3.5GHz modulator is implemented in 65nm CMOS. The BFOOK modulator achieves the maximum data rate of 100Mb/s, maintaining averaged center frequency regardless of the data pattern. The modulator consumes 9.6mW from a 1V supply, achieving the energy efficiency of 96pJ/b.**

Keywords—phase-locked loop; frequency modulation; OOK; FSK; FOOK; two-point modulation; transimitter

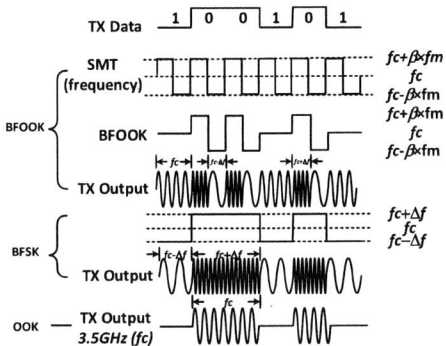

Fig. 1. FOOK modulation.

I. INTRODUCTION

Existing standards for wireless connectivity such as Bluetooth Low Energy (BLE) or WiFi encounter the problem of limited data rate or poor energy efficiency. Compared with the OOK modulation that features high energy efficiency, the FSK modulation is known to be resilient to interferers, while obtaining good energy efficiency [1]-[2]. The FSK modulation, however, has difficulty in achieving high data rate performance unless the modulation is done with an open-loop voltage-controlled oscillator (VCO) or digitally controlled oscillator (DCO) that is highly vulnerable to noise coupling or frequency pulling. The high data rate FSK modulation based on a phase-locked loop (PLL) suffers from the limited loop bandwidth. Even if the two-point modulation whose rate is independent of the PLL bandwidth is used, a delay time mismatch between a low-pass modulation path and a high-pass modulation path would be a bottleneck in case of the high data rate modulation [3]. Another way is to use a narrowband integer-N PLL with VCO modulation only. However, the narrowband PLL does not give strong benefit compared with the open-loop VCO modulation since the VCO with the narrowband PLL is still sensitive to the frequency pulling in addition to poor phase noise performance. On the other hand, the VCO modulation within a wideband PLL suffers from data-pattern-dependent performance when the FSK data contains a long sequence of data "1" or "0" since the PLL promptly reacts to the VCO modulation.

Recently, a frequency-domain OOK (FOOK) modulation that uses a pre-defined symmetric triangular modulation template (STMT) is proposed to achieve good bandwidth efficiency by occupying narrower bandwidth than the binary FSK (BFSK) modulation [4]. The FOOK is considered a fully balanced FSK since it makes the average carrier frequency insensitive to the data pattern. However, the use of the STMT makes it difficult for the FOOK transceiver to achieve a high data rate.

For high data rate FOOK transmission, we employ a binary FOOK (BFOOK) modulation method as illustrated in Fig. 1. Thanks to the fully-balanced feature, the BFOOK modulator does not suffer from data pattern dependency, meaning that average carrier frequency is fixed regardless of the data pattern. Hence, it is shown that the high data rate FSK modulation can be performed based on an integer-N PLL by modulating the VCO only, namely 1+-point modulation. As long as the PLL bandwidth is less than the data rate by 10 times, decent modulation performance can be achieved. In [5], simulation results show that the fully-balanced FOOK can achieve the data rate as high as 1Gb/s, while the BFSK modulation cannot due to a carrier drift problem when a long sequence of "1" or "0" is present. The non-OOK ultra-wideband (UWB) modulation, FM-UWB, is based on low-frequency triangular modulation, thus not being suitable for high data rate modulation.

II. FULLY-BALANCED FSK MODULATION

A. BFOOK vs. BFSK

For robust constant-envelope modulation, only a PLL-based modulation is considered. We first investigate modulation benefits of the BFOOK in comparison with the BFSK for high data rate. We assume the 1+-point modulation,

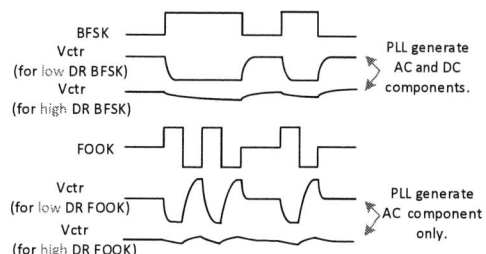

Fig. 2. PLL response to BFSK and BFOOK with low and high data rates.

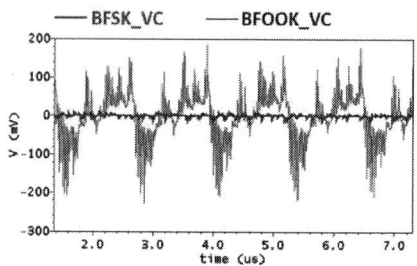

Fig. 3. PLL response to 100Mb/s BFSK and BFOOK baseband signals

that is, the VCO based modulation in the PLL whose bandwidth is much lower than the data rate [4]. As illustrated in Fig. 2, the BFOOK generates AC component only in a loop filter voltage of the PLL since the average center frequency of the carrier of BFOOK modulation is fixed regardless of data pattern, resulting in no carrier frequency drift. To the contrary, the BFSK modulation generates both AC and DC components, making the carrier frequency deviated depending on the data pattern of the baseband.

Behavior simulations are done to verify the data-patter-dependent performance. Fig. 3 shows a transient settling behavior of the PLL with 100Mb/s BFOOK and BFSK baseband signals whose modulation index equal 1. The baseband data is generated by 2^7-1 pseudo-random binary sequence (PRBS). Differential control voltages of the VCO with two modulations are plotted for comparison. Due to the limited sequence length of the 2^7-1 PRBS, periodic tone is observed in the control voltage. Even though both control voltages are settled at the same value, the control voltage with the BFSK modulation exhibits large fluctuation since averaged carrier frequency has data pattern dependency, while the BFOOK having negligible DC component. Fig. 4 shows modulated output spectra with the BFOOK and the BFSK at 100Mb/s data rate. As expected, the BFOOK spectrum shows carrier power clearly, while the BFSK spectrum exhibiting unbalanced side lobes. In Fig. 5, the PLL-based modulation spectra of the BFSK and the BFOOK are compared to spectra with direct modulation of the open-loop VCO. In the BFSK, severe suppression of two carriers are observed. The BFOOK maintains nearly the same spectrum, proving that the PLL-based modulation works well.

B. Carrier Spreading

One potential problem of the BFOOK modulation is that high carrier power could violate the spectrum mask. To overcome the carrier leakage issue, a carrier spreading (CS) technique is used. The CS voltage can be added in the integral path of the PLL to modulate the VCO with a very low rate triangular modulation to suppress the carrier power, which is analogous to the spread-spectrum clock generation (SSCG) in wireline communication. In this work, a fractional-N mode with a $\Delta\Sigma$ modulated frequency divider is employed simply to utilize an existing fractional-N PLL. The effect of such a slow varying triangular modulation has a negligible effect on the RF front-end performance when a noncoherent receiver based on a frequency derivative circuit is used. For complete

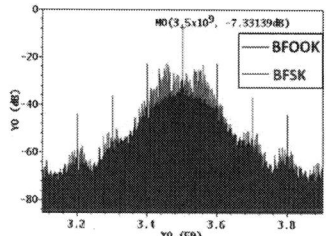

Fig. 4. Modulated spectra with 100Mb/s BFSK and BFOOK baseband signals.

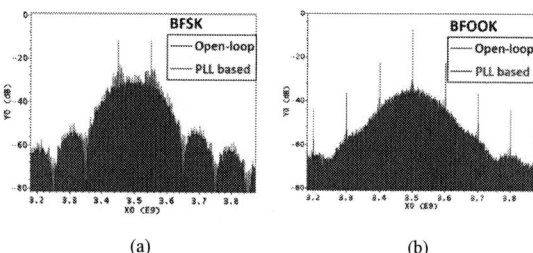

(a)   (b)

Fig. 5. Modulation spectra with PLL based and open-loop VCO based: (a) BFSK, (b) BFOOK.

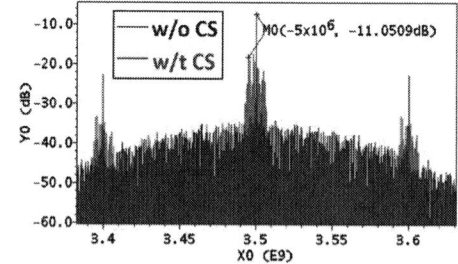

Fig. 6. BFOOK spectrum with carrier spreading.

rejection of the CS modulation, a carrier recovery loop can also be considered in the receiver at the cost of additional power. Fig. 6 shows simulated carrier suppression with the CS method. To reduce the simulation time, a 0.5Mb/s triangular wave with a large modulation index is used to spread the carrier power. The simulation result shows that the carrier power is reduced by about 13dB.

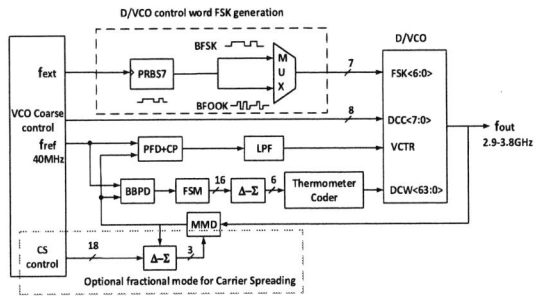

Fig. 7. Proposed modulator based on integer-N hybrid PLL with optional fractional mode for carrier spreading.

III. IMPLEMENTATION

A. Architecture

Fig. 7 shows the block diagram of an integer-N PLL based 1^+-point modulator in which the BFOOK modulation is performed by modulating the VCO only, assuming that the modulation rate is much higher than the PLL bandwidth. A hybrid type-2 PLL architecture consisting of an analog proportional gain path and a digital integral path is used to replace a large integration capacitor with a digital control block [6]. A differential charge pump with a differential loop filter without the integration capacitor is designed, which is followed by a digital/voltage-controlled oscillator (D/VCO). The digital control of the D/VCO consists of the fine control with a 64-bit thermometer code $DCW<63:0>$ and the coarse control with a 8-bit binary code $DCC<7:0>$. The BFOOK and BFSK modulations are performed by switching a single capacitor of the capacitor array, and the magnitude of frequency deviation is controlled by a 7-bit control word $FSK<6:0>$. An optional fractional-N mode is added to have the CS function, which can also be done by directly adding a modulation voltage to the loop filter of the integer-N PLL as an alternative way.

B. Buiilding Bloks

Fig. 8 shows the schematic of the charge pump. In the hybrid PLL, the analog proportional-gain path is the same as the loop filter of a type-1 PLL. Hence, it is good to have a fully differential topology with a differential loop filter to avoid a separate DC voltage for the loop filter and to minimize noise coupling. An NMOS transistor M_1 is in triode region for common-mode feedback to set common-mode voltage V_C.

Since it is difficult to provide high-frequency data from an off-chip source, an on-chip baseband signal generator is implemented. Fig. 9 shows a block diagram of the on-chip BFSK/BFOOK modulator based on 2^7-1 PRBS generator. The modulator achieves the maximum data rate of 400Mb/s. An external clock CLK_{ext} is used to provide an internal clock CLK_{PRBS} for the PRBS generator after a divide-by-two circuit, then directly modulating the D/VCO. The control word is given by serial peripheral interface (SPI).

Fig. 10 shows the schematic of the D/VCO. To achieve good power supply rejection, a current bias and a cross-

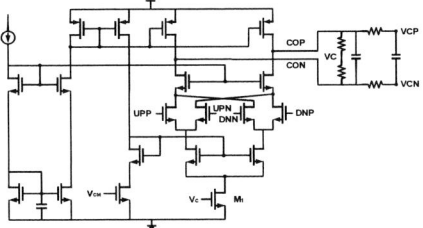

Fig. 8. Schematic of differential charge pump.

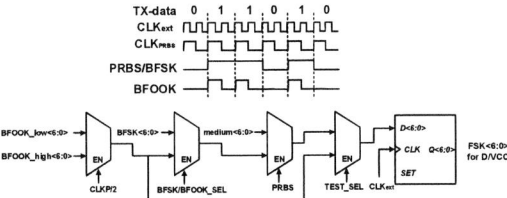

Fig. 9. On-chip BFSK/BFOOK modulation circuit.

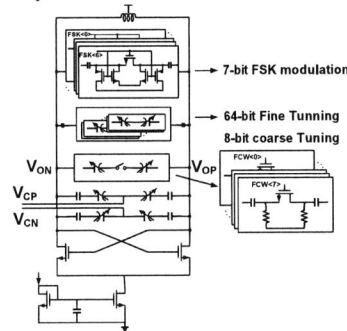

Fig. 10. Schematic of D/VCO.

coupled pair are designed with NMOS transistors. The hybrid tuning circuit consists of an 8-bit coarse-tuning capacitor array, a 64-bit fine tuning capacitor array, a 7-bit BFSK tuning capacitor array, and a pair of analog varactors. The coarse-tuning capacitor array is used to tune the center frequency over process variation. The 7-bit FSK tuning array is utilized for on-chip BFSK and BFOOK modulations. The FSK tuning array have a resolution of 1MHz/LSB, so that different modulation index can be set for the BFSK and the BFOOK modulations.

IV. MEASUREMENT RESULTS

A prototype of 3.5GHz BFOOK modulator is fabricated in 65nm CMOS, and a chip micrograph is shown in Fig. 11. The active area of the modulator is 1.1.mm x 0.45mm. The area of the analog loop filter is negligible with the hybrid PLL architecture. The measured tuning range of the D/VCO is 2.91-3.82GHz under 1V supply.

Fig. 12 shows the measured phase noise and the reference spur at the 3.2GHz output with the reference clock of 40MHz. The measured in-band phase noise is -84.48dBc/Hz at 10kHz and the out-of-band phase noise is -130dBc/Hz at 10MHz offset. Because of the high data rate modulation, the in-band phase noise performance is not critical for the overall BFOOK

978-1-7281-5107-6/19 $31.00 © 2019 IEEE

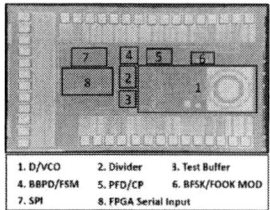

Fig. 11. Chip micrograph.

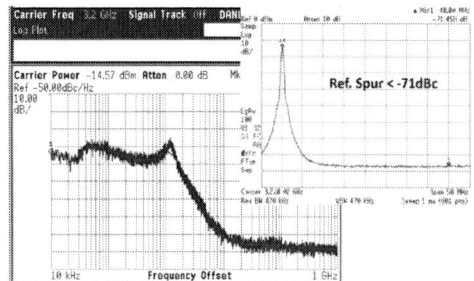

Fig. 12. Measured phase noise performance at 3.2GHz output.

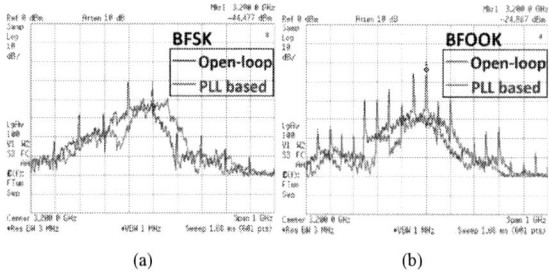

(a)   (b)

Fig. 13. Measured modulation output spectra with PLL based and open-loop VCO based: (a) BFSK, (b) BFOOK.

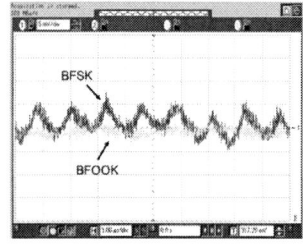

Fig. 14. Measured VCO control voltages (single-ended) with BFSK and BFOOK modulations at 100Mb/s.

modulation performance. The reference spur is un-estimable with a measured value less than -71dBc, showing that the hybrid type-2 control works well.

Fig. 13 shows the measured modulated spectra with BFSK and BFOOK modulations with the data rate of 100Mb/s. With a proper control word of the digitally-controlled FSK capacitor array within the D/VCO, the frequency deviation $\pm\Delta f$ of 50MHz is obtained, having an effective modulation index of 1. Spurs at 100MHz and the multiples of 100MHz are observed for the BFOOK modulation, showing that the BFOOK modulation behaves like a narrowband frequency

TABLE I. PERFORMANCE COMPARISON

	This work	[1]	[2]	[4]	[7]
Technology (nm)	65	65	28	65	130
Supply (V)	1	1	0.2	0.8	1.2
Core Area (mm²)	0.53	1.64@TRX	0.53	0.41	0.11
RF Freq. (GHz)	3.5	2.4	2.4	2.4	3-5
Modulation Scheme	BFOOK BFSK	GFSK	GFSK	FOOK	IR-UWB
Architecture	1⁺-point DFM	Single-point DFM	Single-point DFM	1⁺-point DFM	OOK
Data Rate (Mb/s)	10-100	1	1	0.75	100
Power (mW)	9.6	5.3@0 dBm	4@0 dBm	1.9	14.7
Energy Efficiency (nJ/b)	0.096	5.2	4	1.9	0.147

modulation even with the modulation index of 1. Since the 2^7-1 PRBS does not provide a long sequence of identical bits, it is difficult to show clear modulation spectrum of the BFOOK to show much less data-pattern-dependency over the BFSK. The FPGA-based external PRBS generation is ongoing to fully characterize the modulator performance for future works. Fig. 14 shows the measured control voltages of the VCO, showing that the BFOOK exhibits less fluctuation as expected from Fig. 3.

Table I shows the performance summary in comparison with other works. The proposed 3.5GHz modulator achieves the energy efficiency of 96pJ/b, which is comparable to that of the pulse-based UWB modulator [7]. To the best of authors' knowledge, a sub-6GHz PLL-based FSK modulator with the highest data rate is presented in this work.

V. CONCLUSION

A 3.5GHz 100Mb/s fully-balanced BFOOK modulator is implemented in 65nm CMOS. The fully-balanced FSK feature of the BFOOK makes it possible for an integer-N PLL to achieve good modulation linearity with VCO modulation only. The modulator consumes 9.6mW from a 1V supply, achieving the energy efficiency of 96pJ/b.

ACKNOWLEDGEMENT

This work was supported in part by NSFC under contract #61774092.

REFERENCES

[1] H. Liu *et al.*, "A DPLL-centric Bluetooth Low-Energy transceiver with a 2.3-mW interference-tolerant hybrid-loop receiver in 65-nm CMOS," *IEEE JSSC*, vol. 53, no. 12, pp. 3672-3687, Dec. 2018.

[2] S. Yang *et al.*, "A 0.2-V energy-harvesting BLE transmitter with a micro power manager achieving 25% system efficiency at 0-dBm output and 5.2-nW sleep power in 28-nm CMOS," *IEEE JSSC*, 2019.

[3] N. Markulic *et al.*, "A self-calibrated 10 Mb/s phase modulator with -37.4 dB EVM based on a 10.1-to-12.4 GHz, -246.6 dB-FOM, fractional-N subsampling PLL," in *ISSCC*, Feb. 2016, pp. 176–177.

[4] Y. Zhang, R. Zhou, W. Rhee and Z. Wang, "A 1.9-mW 750-kb/s 2.4-GHz F-OOK transmitter with symmetric FM template and high-point modulation PLL," *IEEE JSSC*, vol. 52, no. 10, pp. 2627-2635, Oct. 2017.

[5] C. Ding, W. Rhee, and Z. Wang, "A Gaussian-filtered fully-balanced FSK modulator with integer-N PLL based 1+-point modulation," in *Proc. IEEE ISCAS*, May 2019, pp. 1-4.

[6] N. Xu, W. Rhee Z. Wang "A hybrid loop two-point modulator without DCO nonlinearity calibration by utilizing 1 bit high-pass modulation" *IEEE JSSC* , vol. 49, no. 10, pp. 2172-2186, Oct. 2014.

[7] R. Vauche *et al.*, "A 100 MHz PRF IR-UWB CMOS transceiver with pulse shaping capabilities and peak voltage detector," *IEEE Trans. Circuits Syst. I*, vol. 64, pp. 1612-1625, June 2017.

A 15-Ch. 0.019 mm²/Ch. 0.43% Gain Mismatch Orthogonal Code Chopping Instrumentation Amplifier SoC for Bio-Signal Acquisition

Jeong Hoan Park[1], Tao Tang[1], Lian Zhang[1], Kian Ann Ng[2], and Jerald Yoo[1,2]

[1]Dept. of ECE, National University of Singapore, [2]N.1 Institute for Health, Singapore

Email: {jhpark,jyoo}@nus.edu.sg

Abstract— This paper proposes a 15-Ch. low power, small area, Orthogonal Code Chopping Instrumentation Amplifier (OCCIA). Orthogonal codes directly modulate each channel and merge into a single IA while performing dynamic offset compensation. Digitization-Before-Demodulation (DBD) transmits the combined data directly, which alleviates ripple noise, and completely removes the demodulation and TX encoding overhead from the SoC. The proposed OCCIA SoC in 0.18μm 1P6M CMOS achieves the lowest gain mismatch (0.43%) among recent multi-channel IAs with one of the smallest areas (0.019mm²/ch.) while consuming 1.97μW/ch. and low crosstalk (<-53.5dB) at 490Hz bandwidth.

Keywords— *Bio-signal acquisition, demodulation-before-digitization, instrumentation amplifier, orthogonal code chopping.*

I. INTRODUCTION

Multi-channel bio-signal acquisition SoC is an essential instrument for physiological studies [1]- [4]. With increased channel count, the SoC requires aggressive area and power reduction strategies, and especially for power-constrained implantable applications, low-power data encoding and transmission is a must [1],[2]. To meet these requirements, we propose an Orthogonal Code Chopping Instrumentation Amplifier (OCCIA) with Digitization-Before-Demodulation (DBD) scheme (Fig. 1); it exploits the orthogonal code set that directly modulates and merges the multi-channel input data simultaneously into a single IA, while reducing dynamic offset noise, power consumption, and inter-channel crosstalk at once. Sharing only an IA among all channels fundamentally reduces the gain mismatch while achieving power and area efficiency. DBD performs digitization prior to demodulation between chopping transitions, which leads to intrinsically ripple-free as compared to conventional chopping. As the merged signal is digitized and directly transmitted outside, DBD relieves the demodulation and TX encoding overhead from the power-constrained SoC.

II. CIRCUIT DETAILS AND IMPLEMENTATIONS

A. Orthogonal Code Chopping (OCC)(Fig. 2)

Capacitive-coupled IA is commonly used in many bio-signal applications, and its area mainly consists of an input capacitor (C_{IN}) and an operational transconductance amplifier (OTA) ($Area_{IA} \approx Area_{CIN+OTA}$). In N-Ch.

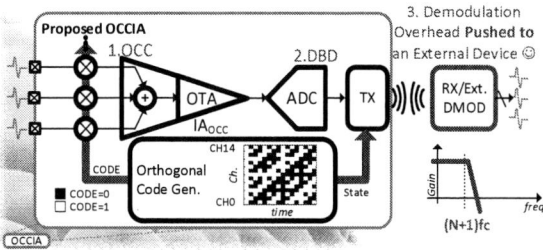

Fig. 1. Multi-channel Orthogonal Code-Chopping Instrumentation Amplifier (OCCIA) architecture.

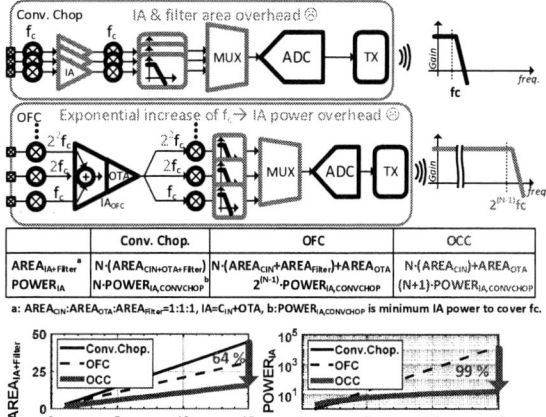

AREA$_{IA+Filter}$ POWER$_{IA}$	Conv. Chop.	OFC	OCC
	N·(AREA$_{CIN+OTA+Filter}$) N·POWER$_{IA,CONVCHOP}$[b]	N·(AREA$_{CIN}$+AREA$_{Filter}$)+AREA$_{OTA}$ 2$^{(N-1)}$·POWER$_{IA,CONVCHOP}$	N·(AREA$_{CIN}$)+AREA$_{OTA}$ (N+1)·POWER$_{IA,CONVCHOP}$

a: AREA$_{CIN}$:AREA$_{OTA}$:AREA$_{Filter}$=1:1:1, IA=C$_{IN}$+OTA, b:POWER$_{IA,CONVCHOP}$ is minimum IA power to cover fc.

Fig. 2 Analysis and comparison of IA & filter area and IA power among conventional chopping (Conv. Chop), orthogonal frequency chopping (OFC) and the proposed OCC.

conventional chopping IA, the IA and filter area increase proportionally to N, *i.e.*, AREA$_{IA,conv}$=N·(AREA$_{CIN+OTA+filter}$). Orthogonal Frequency Chopping (OFC) [5] shares an OTA (AREA$_{IA,OFC} \approx$ N·AREA$_{CIN}$+AREA$_{OTA}$) by using orthogonal frequency modulation, however, the IA power consumption (POWER$_{IA}$) increases exponentially with increased N, as the chopping frequency must be scaled ($2^{N-1} \cdot f_c$) to avoid harmonic distortion. In contrast, the proposed OCCIA uses orthogonal code (code length=N+1) set for chopping, which modulates and spreads the input spectrum while maintaining the orthogonality and dimming harmonic crosstalk. This leads OCC to take only the advantages of the OFC (superior area and gain mismatch) and the conventional IA (low power consumption compared to OFC at high N). Carefully considering the voltage margin, we chose 15-Ch. Fig. 2 table depicts the comparison, where the OCCIA shows 64% AREA$_{IA+Filter}$ reduction from conventional chopping, as well as 99% POWER$_{IA}$ saving from OFC (Fig. 2).

B. Digitization-Before-Demodulation (DBD)

In conventional signal acquisition chain, digitization of the output signal is done after the analog demodulator and multiplexer (Fig. 3). In contrast, the DBD directly digitizes the amplified outputs (prior to demodulation), therefore pushing the demodulator and filter overhead from the power-constrained SoC (such as in implantables) to an external device. Specifically, the ADC samples the outputs between chopping transitions, which significantly reduces the effect of chopping ripples (Fig. 3). 15×16b lookup table (LUT) structure enables to send orthogonal 4-bit code state, instead of the full 15-ch. orthogonal code directly, which further reduces transmission data amount for DBD operation.

III. MEASUREMENT RESULTS AND CONCLUSIONS

The 15-ch. OCCIA SoC in a 0.18μm CMOS 1P6M takes 0.019mm²/ch (Fig. 5 (a)). The clean demodulated output with single-channel input (CH [0]) can be seen in Fig. 4(a)&(b). With OCC on, measured output noise density with ADC and demodulation noise contribution is 0.047 LSB/√Hz(Fig. 4(c)). Fig. 4(d) shows the mean amplitude and signal distortion of all output with each same input stimuli. The GBP curve shows the bandwidth of 490 Hz and its low mismatch (0.43%), with the cross-channel distortion <-53.5dB, which fully meet the criteria of our signal and noise characteristics [3]. Multi-channel ECG is also successfully measured with the OCCIA (Fig. 5(b)). The chip performance (TABLE I) shows that OCCIA has the lowest gain mismatch, smallest area, and low power consumption, compared with the recent multi-channel IAs.

ACKNOWLEDGMENT

This work was funded by National University of Singapore (NUS) Hybrid-Integrated Flexible Electronic Systems (HiFES) Program.

REFERENCES

[1] R. Muller *et al.*, "A Minimally Invasive 64-Channel Wireless μECoG Implant," *IEEE Journal of Solid-State Circuits*, pp. 344–359, 2015.

[2] C. M. Lopez *et al.*, "An implantable 455-active-electrode 52-channel CMOS neural probe," *IEEE J. Solid-State Circuits*, pp. 248–261, 2014.

[3] V. Majidzadeh *et al.*, "Energy efficient low-noise neural recording amplifier with enhanced noise efficiency factor," *IEEE Trans. Biomed. Circ. Sys.*, pp. 262–271, 2011.

[4] M. Altaf *et al.*, "A 16-Ch. Patient-Specific Seizure Onset and Termination Detection SoC with Impedance-Adaptive Transcranial Electrical Stimulation," *IEEE J. Solid-State Circuits*, pp. 2728–2740, vol. 50, no. 11, Nov. 2015.

[5] Y. Tsai *et al.*, "A 2-Channel -83.2dB Crosstalk 0.061mm² CCIA with an Orthogonal Frequency Chopping Technique," *IEEE International Solid-State Circuits Conference (ISSCC) Dig. Tech. Papers*, pp. 92–94, 2015.

[6] F. Michel *et al.*, "On-chip gain reconfigurable 1.2V 24μW chopping instrumentation amplifier with automatic resistor matching in 0.13μm CMOS," *IEEE Int. Solid-State Circ. Conf. (ISSCC) Dig. Tech. Papers*, pp. 372–373, 2012.

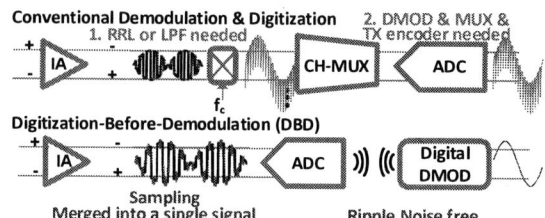

Fig. 3 Comparison between conventional chopped signal demodulation and digitization with Digitization-Before-Demodulation (DBD)

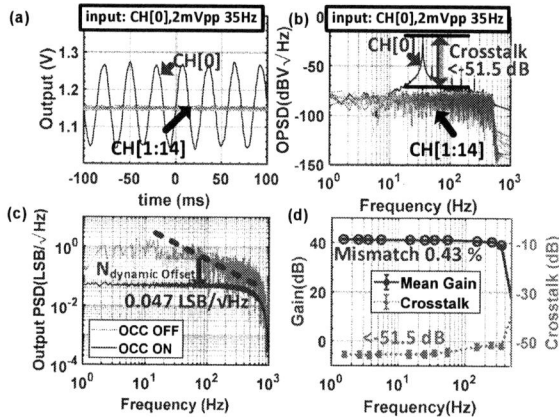

Fig. 4 (a) Transient response of output amplitude, (b) Output power spectrum density (crosstalk < -51.5 dB), (c) Output noise power spectral density including ADC, demodulation effect at OCC ON/OFF, (d) The mean gain and crosstalk variance of all channels.

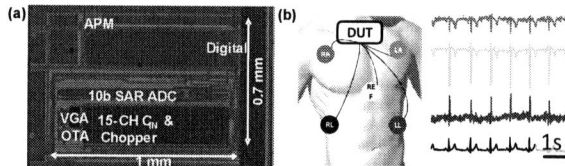

Fig. 5 (a) Chip photo, (b) 4-ch. ECG Measurements.

TABLE I Chip performance summary and its comparison.

	This Work	Y. Tsai [ISSCC'15,[5]]	F. Michel [ISSCC'12,[6]]	R. Muller [JSSC'15,[1]]	C. Lopez [JSSC'14,[2]]	V.Majidzadeh [TBCAS'11,[3]]
Methodology	OCC	OFC	R-trim feedback	ChopΔΣ rawData	ACT Electrode AC-couple	Ref. Shared
Area /ch. (mm²)	0.019[b]	0.031	0.465	0.025	0.022[a]	0.063[c]
Power/ch. (μW)	1.97[b]	40.5[c]	25.2	2.29[d]	7.02[c]	7.92[c]
Max. gain mismatch (%)	0.43(15-ch.)[e]	0.55(2-ch.)	0.70	15.0	-	-
N. of Ch	15	2	1	64	455	4
Crosstalk (dB)	-51.5	-83.2	-	-85.0	-44.8	-43.5
Supply (V)	1.2(analog) /1.5(digital)	3	1.2	0.5	1.8	1.8
Ch. gain (dB)	40-56	40	20-60	30	29.5-72	39.4
Bandwidth (BW)	490	1k	1k-100k	500	0.2,6k	7.2k
Noise density (nV/√Hz)	155[g]	26.0[c]	40.0	58.0	41.3[g]	39.4[g]
NEF	7.25[g]	3.74[c]	7.50	4.76	3.08[c]	3.35[c]
Process (nm)	180	350	130	65	180	180

a. Estimated by merging of IA and Electrode   b. IA+ADC+OCGen.   d. IA   e. IA+ADC
e. Mismatch includes ADC + Demodulation error   f. Noise of IA+ADC+Demodulation, current from IA+ADC+OCGen.
g. Estimated from IRN and BW with assumption of constant noise density.

A 10 μW -74.6 dB THD Arterial Pulse Waveform Sensing System with Automatic Bridge-Offset Calibration and Super Class-AB Output Stage

Yu-Pin Hsu, Zemin Liu, Mona Hella

ECSE, Rensselaer Polytechnic Institute, Troy, New York 12180

Email: hsuy3, hellam@rpi.edu

Abstract—This paper presents a power-efficient, highly-linear sensing system for arterial pulse waveform (APW) acquisition. The system consists of a front-end circuit connected to a packaged silicone-coated resistive-bridge pressure sensor for detecting APW through direct skin contact in a non-invasive manner. Compared to other pulse sensing approaches, the proposed front-end only requires an instrumentation amplifier (IA) and a low-pass filtering analog-to-digital converter (LPF-ADC), and consumes a total power of 10 μW. An automatic calibration circuit is proposed to compensate for the resistive-bridge offset and can reduce such offset down to 1 mV. High linearity in the IA is provided through the use of a super class-AB output stage. A LPF-ADC architecture is employed to extract and digitize the APW in a 100 Hz bandwidth, saving 90 % of the passives area. The front-end provides a measured input-referred noise of 7.2 μVrms with a gain of 32$-$40 dB, and a THD of -74.6 dB. Together with an on-chip low dropout regulator (LDO) and a buffer stage, the prototype fabricated in 0.18 μm CMOS technology has an active area of 0.52 mm^2.

I. Introduction

The blood pressure waveform is a fundamental medical signal that can provide critical information for the diagnosis of cardiovascular diseases clinically [1]. The majority of blood pressure detection systems provide either pulse wave velocity (PWV) and/or arterial pulse waveform (APW) information at the wrist or the finger. They are mostly based on light sensing, and have a dc power consumption ranging from 25 μW to 1280 μW depending on the details of the circuit architecture [2], [3]. Other approaches include ultrasound, catheter, electrocardiogram (ECG) with photoplethysmograpgy (PPG) [4], and impedance-based measurements [5]. The different approaches vary depending on cost, complexity, ease-of-use, and the level of comfort for long-term monitoring.

Recently, a discrete resistive-bridge sensing solution has been demonstrated in [6] to obtain PWV and APW information at the neck. Such technique can provide a more accurate clinical information as described in [1]. However, the inherent offset problem associated with bridge-based pressure sensors requires special compensation techniques ([7] as an example) and a customized front-end design. Besides, the resistive-bridge would draw a current that ranges from several μA to several hundreds μA (depending on the selected sensor resistance)

In this work, a power-efficient, highly-linear integrated solution for APW monitoring using resistive-bridge pressure

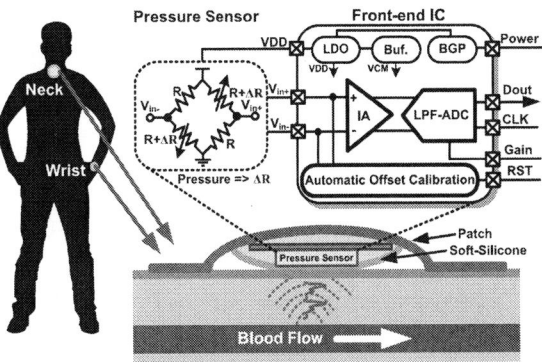

Fig. 1. System block diagram for skin-coupled arterial pulse waveforms sensing at the wrist or at the neck.

sensor with automatic sensor offset-cancellation for detection at both the wrist and the neck is presented.The The paper is organized as follows. Section II presents an overview of the system architecture. Section III provides details of the circuit implementation. Measurement results are shown in section IV, while conclusions are drawn in section V.

II. Proposed System Architecture

Fig. 1 shows the proposed APW detection system using resistive-bridge pressure sensors. The pressure sensor is coated with soft bio-compatible silicone for comfortable skin attachment, to facilitate long-term monitoring. When the sensor module is adhered to the skin-surface, for example at the wrist or at the neck, it can detect APW through the skin surface coupling.

The system details are shown in Fig. 2, which consists of an automatic bridge-offset calibration, an IA, and a low-pass filtering analog-to-digital converter (LPF-ADC). While bridge-type sensors are widely used in sensing temperature, humidity, pressure, etc, due to its low-cost, easy of use, and almost disposable nature, it suffers from dc offset from process variations, atmospheric pressure, and temperature [7]. Since this offset saturates the following front-end stages, and degrades the front-end linearity, an offset-calibration circuit is needed to ensure proper system operation.

IEEE Asian Solid-State Circuits Conference
November 4 – 6, 2019
The Parisian Macao, Macao SAR, China

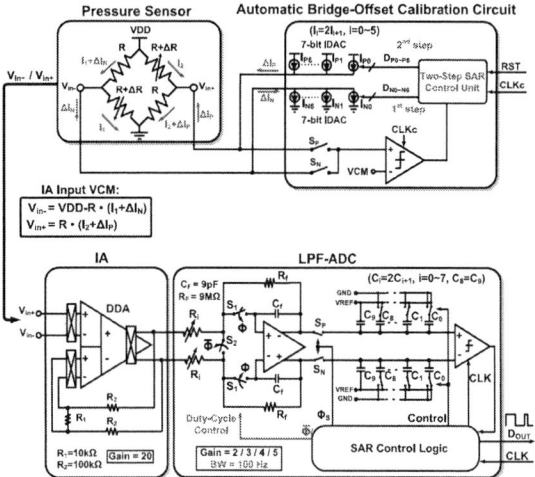

Fig. 2. Front-end system architecture.

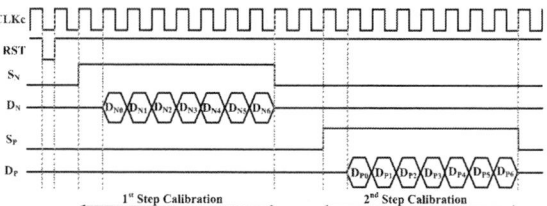

Fig. 3. Timing diagram of the automatic bridge-offset calibration.

Here, we propose a two-step successive-approximation-register SAR control for automatic calibration to reduce the bridge-offset voltage. Once the offset is removed, the calibration block is disabled to save the associated power consumption until a reset signal (RST) is triggered again. An LDO and a bandgap reference are integrated in the system, as show in Fig. 1, to provide a temperature independent supply voltage for both the sensor and the front-end circuit, thus eliminating any potential temperature drift. As shown in Fig. 2, the IA consists of a differential-difference amplifier (DDA) in a resistive-feedback configuration formed by R_1 and R_2. The resistive feedback can achieve higher linearity compared to capacitive feedback. However, it suffers from higher loading. A super class-AB output stage is employed in the IA to provide a high driving capability for both the resistive feedback in the IA, and the following LPF-ADC without the penalty of high power consumption. In addition, to achieve a low-noise performance, a chopping function is used to remove the 1/f noise and the DDA's offset. The LPF-ADC is designed with a 100 Hz bandwidth. A SAR-based control logic is shared between the ADC and a duty-cycle controlled filter. When the sampling frequency ω_s of the ADC is larger than $\frac{1}{R_f C_f}$, the filter can implement a bandwidth of $\frac{d}{R_f C_f}$ through a duty-cycle control method, where d ($0 < d \leq 1$) is the duty-cycle ratio of the control clock ϕ [8]. The LDO and the buffer stages provide the needed supply voltage (VDD) as well as the common-mode voltage (VCM) to each block. Details of the circuit blocks are given in the following section.

III. CIRCUIT IMPLEMENTATION

A. Automatic Bridge-Offset Calibration

As shown in Fig. 2, the calibration circuit consists of two switches S_P and S_N, two 7-bit current DACs (IDACs), I_{P0-P6} and I_{N0-N6}, a comparator, and a SAR control unit. Fig. 3

shows the timing diagram of the calibration circuit. A two-step calibration process starts after a falling pulse on the reset signal, RST, which triggers the calibration. In the first-step, S_N is on, and the clocked comparator compares the negative output of the sensor, V_{in-}, with the common-mode level, V_{CM}. The SAR logic generates corresponding codes D_{N0-N6} to adjust V_{in-} through the IDACs I_{N0-N6}, and $V_{in-} = R \times (I_1 + \Delta I_N)$ approaches V_{CM} after 8 clock cycles. After the first step is completed, in the second-step, S_N is OFF, S_P turns on, and IDAC I_{P0-P6} is adjusted until V_{in+} approaches V_{CM}. Finally, both S_P and S_N turns off, and IDACs hold the compensation current. The offset calibration is a one-time process, thus the overall power consumption is reduced compared to other techniques where the calibration is always running in the background [7].

B. Instrumentation Amplifier (IA)

As shown in Fig. 2, the IA employs a DDA and a resistive feedback of $R_1 = 10k\Omega$ and $R_2 = 100k\Omega$ to achieve a differential gain of 26 dB. The DDA has a high input impedance to avoid the loading effect at the front-end and to isolate the currents from the IDACs. Fig. 4 (a) shows the schematic of the DDA where the first stage, formed by $M_{0a,0b}$ to M_{10}, amplifies the differential difference input with a gain of 70 dB.

As shown in Fig. 4 (a), a super class-AB output stage is used to provide an output impedance of $R_{out} \approx 100$ kΩ. This is three orders of magnitude lower than conventional class-AB output stages ($R_{out} \approx 100$ MΩ), drawing the same amount of current from the power supply, thus providing a high driving capability. Fig. 4 (b) shows the simulated harmonics for the IA architectures with conventional class-AB output and with proposed super class-AB output. The IA with the proposed super class-AB output has less harmonic content, resulting in a lower THD of -82 dB. Despite the limited output swing of 0.4 V_{PP}, the signal can be further amplified by the following LPF-ADC.

C. LPF-ADC

Fig. 2 shows the block diagram of the LPF-ADC. The filter employs a duty-cycle controlled architecture with the digitization achieved through a monotonic-switching 10-bit SAR ADC [9] and a synchronous timing. The proposed architecture shares the same clocks from the SAR control logic. The timing diagram is shown in Fig. 5 (a). Using the duty-cycle control clocks ϕ and $\bar{\phi}$ in the filter, the system

978-1-7281-5107-6/19 $31.00 © 2019 IEEE   298

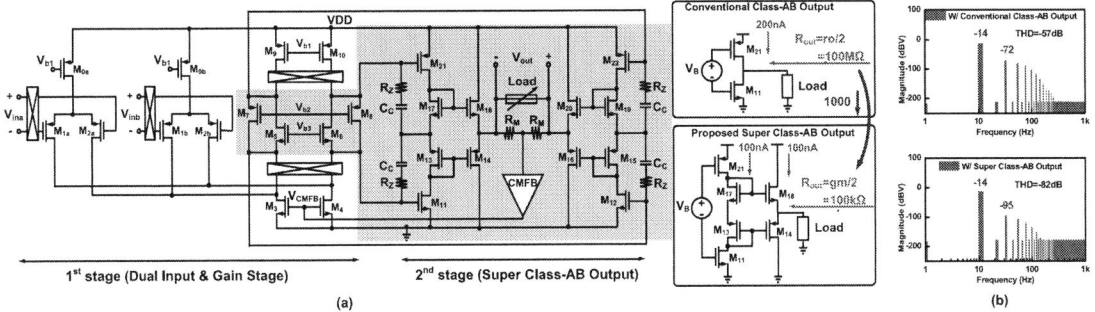

Fig. 4. (a) Circuit schematic of the differential-difference amplifier (DDA). (b) Simulated harmonics for IA architectures with conventional class-AB output and with the proposed super class-AB output at a differential output swing of 0.4 V_{PP}.

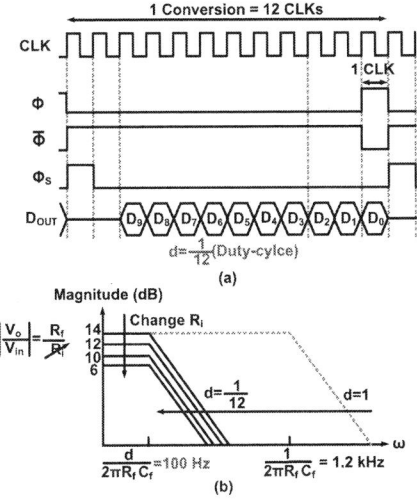

Fig. 5. (a) Timing diagram, and (b) the frequency response of the LPF-ADC showing a bandwidth reduction from $\frac{1}{R_f C_f}$ to $\frac{1/12}{R_f C_f}$.

can achieve a bandwidth reduction from $\frac{1}{R_i C_i}$ to $\frac{1/12}{R_f C_f}$ without using bulky resistors and capacitors. Moreover, the adjustable R_i provides a programmable gain ($\frac{R_f}{R_i}$) of 6 dB, 10 dB, 12 dB, 14 dB, as shown in Fig. 5 (b).

IV. MEASUREMENT RESULTS

Fig. 6 (a) shows the chip micrograph of the APW sensing system with an active area of 0.52 mm². The chip is designed and fabricated in 0.18 μm CMOS technology. Fig. 6 (b) and (C) show the experimental setup, where the resistive-bridge pressure sensor is connected to the chip via a cable. A signal analyzer and a logic analyzer are used to capture the output waveforms. The APWs are acquired by placing the pressure sensor at the wrist and at the neck, respectively. The on-chip LDO provides 1.8 V supply for the circuit blocks. Fig. 7 shows the measured output waveforms from the pressure sensor during offset calibration, where the two-step process can be clearly seen. The difference between V_{in+} and V_{in-} reduces

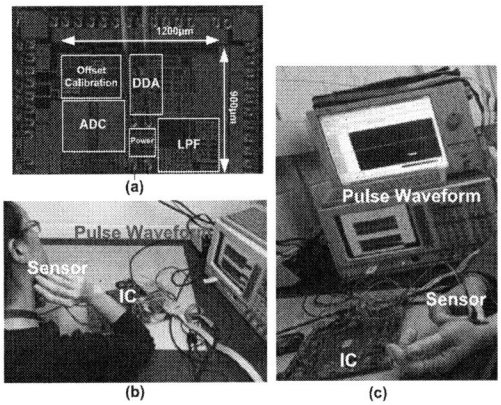

Fig. 6. (a) Die photo of the proposed APW monitoring system. Measurement setup for detecting APW at (b) neck and (c) wrist.

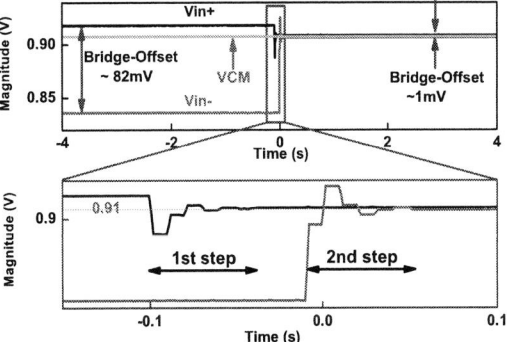

Fig. 7. Measured output waveforms from the pressure senor during calibration. The offset is reduced from 82 mV to 1 mV after calibration.

from 82mV to almost 1mV within 16 steps and both approach $V_{CM} = 0.91$ V after calibration. Fig. 8 (a) shows the measured transfer function of the front-end circuit. It can be seen that a 100 Hz filtering bandwidth is achieved with the duty-cycle control technique with a programmable gain from 32 dB to 40 dB. Fig. 8 (b) shows the measured linearity performance

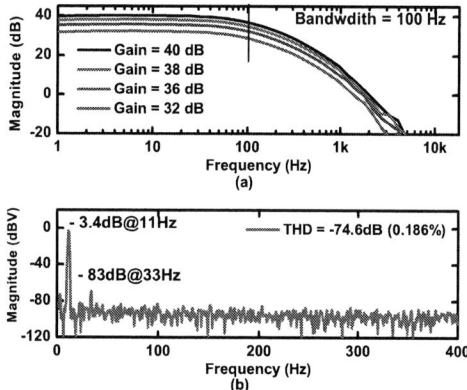

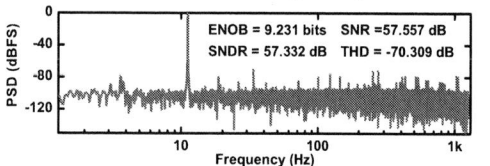

Fig. 8. Measured (a) transfer function, and (b) linearity of the front-end IC.

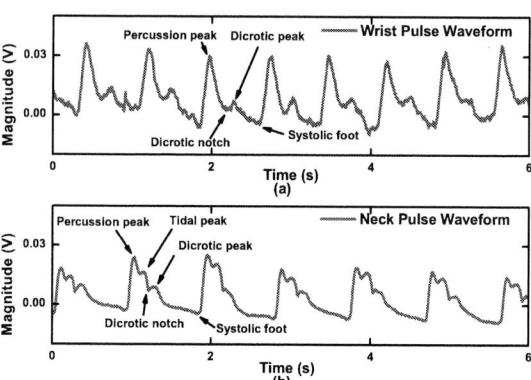

Fig. 10. Measured APWs at (a) wrist, and (b) neck.

TABLE I
PERFORMANCE SUMMARY OF THE FRONT-END IC

	[2]	[3]	[5]	This work
Technology	0.18 μm	0.18 μm	0.11 μm	0.18 μm
Method	PPG	PPG	Bio-impedance	Pressure sensor
Sensing position	Wrist/Finger	-	Wrist	Wrist/Neck
Pulse Type	Biracial	Biracial	Biracial	Central/Biracial
Power (μW)	172	25	1280	10
Bandwidth (Hz)	3.4	10	10k	100
Gain (dB)	66-93	-	50-80	32-40
THD (dB)	-	-50*	$-40^{\dagger}$	-74.6
Area (mm^2)	10	1.792	12.5	0.52

† Estimated from measured PSD [3]   * Estimated by [5]

Fig. 9. Measured ADC performance: 65,536−point FFT spectrum at 2.5 kS/s with a sinusoidal 11Hz input signal.

of the front-end IC. The system achieves a −74.6 dB THD, which ensures the high-fidelity of the APWs. Fig. 9 shows the measured power spectrum density of the ADC. With an input signal at 11 Hz, the ADC can achieve an ENOB of 9.231 bit at a 2.5 kS/s sampling rate. Fig. 10 shows the measured APWs at both the wrist and the neck. Due to the low-distortion performance of the APW front end, the large amplitudes of the percussion peak and systolic foot, as well as the small peaks, including the tidal and the dicrotic peaks, can be clearly identified. Table. I compares this work with recent publications. It can be seen that by using resistive-bridge pressure sensing, the fabricated chip can acquire APW signals including central and biracial pulse [1] at both wrist and neck at the lowest recorded power consumption of 10 μW, while consuming only a chip area of 0.52 mm^2.

V. CONCLUSION

In this paper, a power-efficient, highly-linear integrated solution for APW monitoring using resistive-bridge pressure sensor with automatic sensor offset-cancellation to provide both wrist and neck detection capability is presented. The sensor offset is reduced from 82mV to 1mV through the offset cancellation technique. A super class-AB output stage is proposed in the instrumentation amplifier (IA) to drive a load as low as 100 kΩ, and enable a front-end linearity of −74.6 dB at a power consumption of 10 μW. The co-design of the LPF-ADC utilizes a clock from the SAR ADC and triggers a duty-cycle controlled filter to achieve a 100 Hz filtering. The duty-cycle control approach saves almost 90 % of the passive elements area, which is evident by a competitive chip area of

0.52 mm×mm. The real-time APW at the wrist and at the neck verifies the functionality of the design.

REFERENCES

[1] Matthew R. Nelson et al., "Noninvasive measurement of central vascular pressures With arterial tonometry: clinical revival of the pulse pressure waveform ?," *Mayo Clinic Proceedings*, vol. 85, issue 5, no. 10, pp. 460-472, May. 2010.

[2] V. R. Pamula et al., "A 172 μW compressively sampled photoplethysmographic (PPG) readout ASIC with heart rate estimation directly from compressively sampled data," *IEEE TBioCAS*, vol. 11, no. 3, pp. 487-496, Jun. 2017.

[3] H. Kim and D. Jee, "A <25 μW CMOS monolithic photoplethysmographic sensor with distributed 1b delta-sigma light-to-digital converter," *IEEE ESSCIRC*, pp. 55-58, 2017.

[4] S. S. Thomas et al.,"BioWatch A wrist watch based signal acquisition system for physiological signals including blood pressure," *IEEE EMBS*, 2014, pp. 2286–2289.

[5] W. Lee and S. Cho,"Integrated all electrical pulse wave velocity and respiration sensors using bio-impedance," *IEEE JSSC*, vol. 50, no. 3, pp. 776–785, Mar. 2015.

[6] Y. Hsu and D. J. Young, "Skin-coupled personal wearable ambulatory pulse wave velocity monitoring system using Microelectromechanical sensors," *IEEE Sensors Journal*, vol. 14, no. 10, pp. 3490-3497, Oct. 2014.

[7] H. Jiang, S. Nihtianov and K. A. A. Makinwa,"An energy-efficient 3.7-nV/ $\sqrt{}$ Hz bridge readout IC with a stable bridge offset Compensation Scheme," *IEEE JSSC*, vol. 54, no. 3, pp. 856–864, Mar. 2019.

[8] P. Kurahashi, P. K. Hanumolu, G. C. Temes and U. K. Moon,"Design of low-voltage highly linear switched-R-MOSFET-C filters," *IEEE JSSC*, vol. 42, no. 8, pp. 1699–1709, Aug. 2007.

[9] C. C. Liu, S. J. Chang, G. Y. Huang and Y. Z. Lin, "A 10-bit 50-MS/s SAR ADC with a monotonic capacitor switching procedure," *IEEE JSSC*, vol. 45, no. 4, pp. 731–740, Apr. 2010.

21-3 (7063)

A 0.012 mm², 1.5 GΩ Z_{IN} Intrinsic Feedback Capacitor Instrumentation Amplifier for Bio-Potential Recording and Respiratory Monitoring

Lian Zhang[1,2], Tao Tang[2], Jeong Hoan Park[2], Jerald Yoo[1,2,3]

[1]NUS Graduate School for Integrated Sciences and Engineering, [2]Dept. ECE, National University of Singapore,
[3]N.1 Institute for Health
Singapore
lian.zhang@u.nus.edu, jyoo@nus.edu.sg

Abstract— **An Intrinsic Feedback Capacitor Instrumentation Amplifier (IFCIA) is presented for bio-potential recording and respiratory monitoring. Exploiting the intrinsic gate-to-drain capacitor to scale down input capacitor, combined with interdigitated MOM capacitors, the inverter-based instrumentation amplifier (IA) occupies smallest active area of 0.012 mm² with 1.5 GΩ input impedance (> 1 GΩ up to 100 Hz) and 0.65 μA current consumption, while maintaining die-to-die gain mismatch of 0.41%.**

Keywords— *Analog front-end, bio-potential recording, impedance boosting, intrinsic feedback capacitor instrumentation amplifier, respiratory monitoring.*

I.   INTRODUCTION AND MOTIVATION

To obtain high-fidelity physiological signals such as electrocardiogram (ECG) in wearable environment, large and stable input impedance, good gain-area efficiency, low flicker noise and offset are must. Chopper-stabilization achieves good noise and NEF performance in capacitively-coupled instrumentation amplifier, but the chopper switches at input pair reduce input impedance (Z_{IN}) to below MΩ range (near skin to dry electrode impedance), which degrades signal quality and is unsuitable for applications such as respiratory monitoring where the skin-electrode contact is dynamically changing. Input impedance boosting with a positive feedback loop [1] (Fig. 1) mitigates this issue, but it requires precise capacitor matching to obtain high impedance within a wide range and may suffer from stability issue. Designs such as [2] adopt capacitor bank to trim the capacitance of input impedance boosting loop, but the discrete capacitor values limit the trimming resolution and efficiency, which is strictly constrained by area. As in [3], an auxiliary chopping path is implemented to pre-charge the input capacitors, but the extra clock switching entails more ripples and the buffers on the auxiliary paths contribute to higher noise floor. Work in [4] utilizes T-shaped feedback capacitors to achieve a compact impedance boosting design. Unfortunately, the feedback capacitor cannot be scaled down indefinitely due to the fF-level device parasitic. Authors of [5] try to avoid these collateral issues by simply switching the position of chopper and input capacitor; however, CMRR degrades significantly due to off-

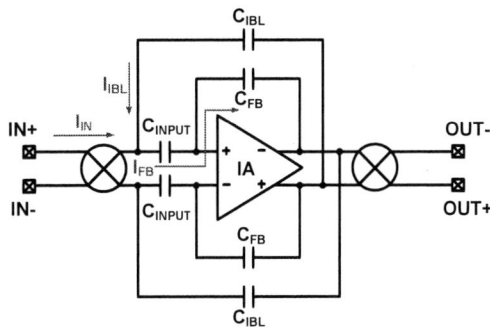

Fig.1. Conventional CS-CCIA with IBL

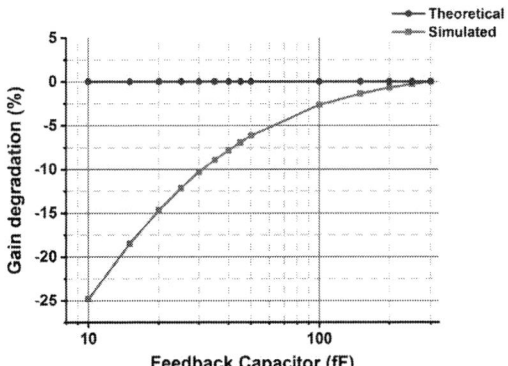

Fig.2. Gain A_{CL} degradation in a conventional IA structure, when scaling down C_{INPUT}, C_{FB}, C_{IBL}

chip capacitor mismatch.

In this paper, an Intrinsic Feedback Capacitor Instrumentation Amplifier (IFCIA) is proposed to resolve the above-mentioned trade-offs all at once. The feedback capacitor (C_{FB}) is replaced by the intrinsic gate-to-drain capacitance of the amplifier input pair so that input capacitor (C_{INPUT}) as well as impedance boosting loop capacitor (C_{IBL}) can be scaled down by 10x. This design simultaneously achieves high input impedance, small area and low input-referred noise without compromising closed-loop gain or power.

This work is funded by National University of Singapore (NUS) Hybrid-Integrated Flexible Electronic Systems (HiFES) Program.

978-1-7281-5107-6/19 $31.00 © 2019 IEEE

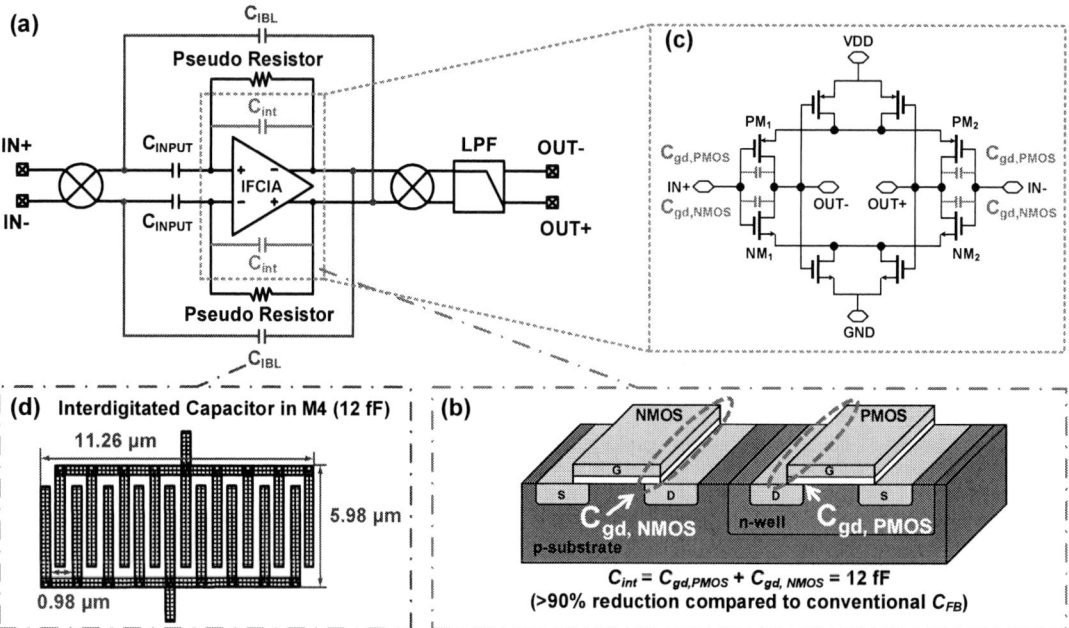

Fig.3. (a) IFCIA closed-loop architecture (b) Intrinsic feedback capacitor (C_{int}) from input pair (c) IFCIA structure (d) C_{IBL} implementation

II. CIRCUIT DESIGN AND IMPLEMENTATION

A. State-of-the-art Impedance Boosting Loop

As shown in Fig. 1, chopper-stabilized capacitive-coupled instrumentation amplifier (CS-CCIA) is commonly adopted for bio-potential recording, where an impedance boosting loop (IBL) is included to compensate Z_{IN} degradation. The principle is using positive current feedback to suppress the effective input impedance. The following equations derive the input impedance where I_{IN}, I_{FB}, I_{IBL} denote current from input, feedback and IBL, V_o denotes output voltage, A_{CL} denotes the closed-loop gain and t denotes an infinitesimal time interval.

$$I_{IN} = I_{FB} - I_{IBL} \tag{1}$$

$$I_{IN} = V_o[C_{FB} - C_{IBL}(1 - A_{CL}^{-1})]/t \tag{2}$$

$$Z_{IN} = t/\{C_{INPUT}[1 - C_{IBL}(1 - A_{CL}^{-1})/C_{FB}]\} \tag{3}$$

Assume A_{CL} is much larger than 1, from (3), the following strategies can be employed for IBL optimization:

a) C_{IBL} and C_{FB} should be carefully matched;

b) C_{INPUT} should be minimized in order to maximize Z_{IN}.

Most previous works utilizing IBL paid attention to strategy a); nevertheless, b) is usually neglected for 2 main reasons:

1. To maintain stable closed-loop gain, C_{INPUT}, C_{FB} as well as C_{IBL} should be scaled down simultaneously, which is constrained by the minimum MIM capacitance (C_{MIM}) value determined by standard CMOS process. (e.g. In 0.18 μm process, the minimum C_{MIM} is 35.6 fF)

2. As capacitance value decreases, parasitic capacitors on the input and feedback paths affect the total C_{INPUT} and C_{FB} more seriously, and thus impact gain accuracy.

To verify the above analyses, Fig. 2 shows the simulation result of closed-loop gain degradation in a conventional IA, when C_{INPUT}, C_{FB} and C_{IBL} scale down simultaneously. As C_{FB} is reduced to less than 100 fF, A_{CL} decreases exponentially due to the parasitic capacitors. When C_{FB} is 10fF, A_{CL} has been degraded by 25%. Therefore, simply shrinking C_{INPUT}, C_{FB} and C_{IBL} linearly would affect the overall gain, which will also be a detriment to the input referred noise.

B. Proposed IFCIA Structure

To scale down C_{INPUT} without sacrificing A_{CL} configuration, a novel structure is presented: intrinsic feedback capacitor instrumentation amplifier (IFCIA). Fig. 3(a) depicts the IFCIA closed-loop architecture, where C_{FB} is replaced by intrinsic feedback capacitor (C_{int}) formed by the dielectric layer between gate and drain of the amplifier input transistor pair ($C_{gd,NMOS}$ & $C_{gd,PMOS}$) as shown in Fig. 3(b) & (c). The effective value of C_{int} is 12fF, which is 1/3 of the minimum C_{MIM} (35.6fF), hence C_{INPUT} and C_{IBL} can be scaled down. Compared to traditional closed-loop designs [1][5][6], C_{INPUT} value (1.5pF), as well as active chip area is reduced by at least 10×.

Compared with the conventional C_{MIM} feedback structure, there are several benefits of utilizing C_{int} as C_{FB}:

1. Smaller capacitance value of C_{int} than C_{MIM}, and thus a much smaller C_{INPUT} and higher Z_{IN} is expected.

2. Stronger tolerance of process variation for C_{int} than C_{MIM}, and hence gain accuracy is enhanced.

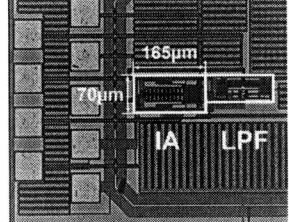

Fig. 4. Chip microphotograph

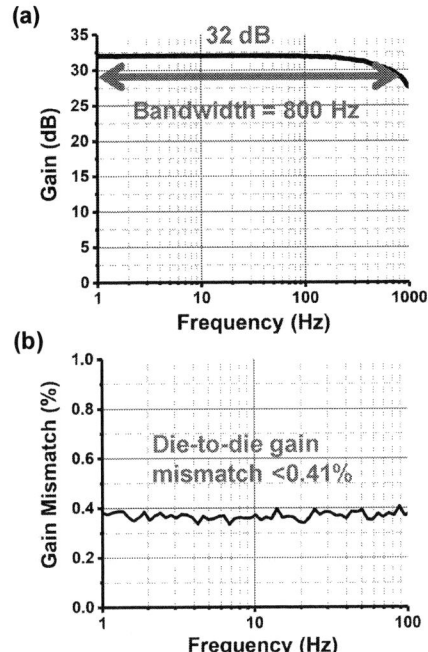

Fig. 5. (a) Gain-bandwidth plot (b) Die-to-die gain mismatch

3. Less complexity and area consumption of the closed-loop architecture for C_{int} than C_{MIM}, especially considering the routing of the feedback path.

Although serial-connected C_{MIM} can also achieve sufficiently small capacitance values theoretically, such configuration suffers more serious mismatch and occupies larger chip area. Therefore, the proposed intrinsic feedback capacitor is a preferable choice as the feedback capacitor.

To exploit the gate-to-drain capacitors as C_{FB}, a specified amplifier structure [7] should be adopted such that an intrinsic feedback path is between gate and drain of the input transistor. In this design, a single-stage inverter-based, self-biased amplifier is chosen as the core amplifier of IFCIA (Fig. 3(c)). The differential input pair is formed by the gates of NM_1/NM_2 and PM_1/PM_2 and the output nodes are the drain connection points shared by input PMOS and NMOS, hence the embedded $C_{gd,NMOS}$ and $C_{gd,PMOS}$ constitute the intrinsic feedback capacitor C_{int}. The single-stage, current-reuse topology also reduces thermal noise floor and maintains maximized

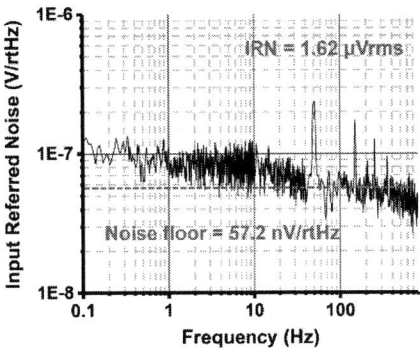

Fig. 6. Noise performance

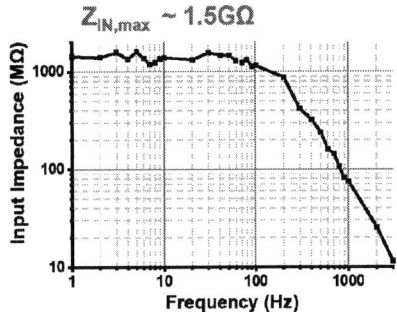

Fig. 7. Input impedance (Z_{IN}) over frequency

transconductance. The proposed IFCIA is followed by a switched-cap low-pass-filter (LPF) to suppress modulated flicker noise and intrinsic DC offset.

C. Impedance Boosting with Scaled C_{INPUT} and C_{IBL}

Since intrinsic feedback structure reduces C_{INPUT} to 1/10 of conventionally used size, Z_{IN} is automatically increased by 10×. To further boost Z_{IN}, an impedance boosting loop is added. Since the embedded C_{int} is only 12 fF, C_{IBL} should also be scaled down to match with C_{int}, therefore single C_{MIM} structure cannot be implemented. Additionally, shrinking capacitor size makes the circuit more susceptible to process variations, thus to maintain gain accuracy and impedance stability, an interdigitated structure using Metal 4 (Fig. 3(d)) is employed to mimic MOM capacitor for C_{IBL}. A fully symmetrical layout is implemented to minimize device mismatch. Also the output chopper up-modulates the mismatch and offset induced by C_{INPUT} and C_{int} to the chopping frequency of 2 kHz and the LPF with 100 Hz cut-off frequency filters out the unwanted harmonics. As a result, our design demonstrates low die-to-die gain mismatch, low noise and high Z_{IN} across low frequency range without extra capacitor trimming techniques.

III. MEASUREMENT RESULTS

The proposed IFCIA implemented in 0.18 μm 1P6M CMOS occupies an active area of 0.012 mm² (Fig. 4). Gain-BW curve shows the mid-band gain of 32 dB and bandwidth of 800 Hz (Fig. 5(a)). The maximum die-to-die gain variation across 5 test boards of 0.41% is obtained below 100 Hz (Fig.

IEEE Asian Solid-State Circuits Conference
November 4 – 6, 2019
The Parisian Macao, Macao SAR, China

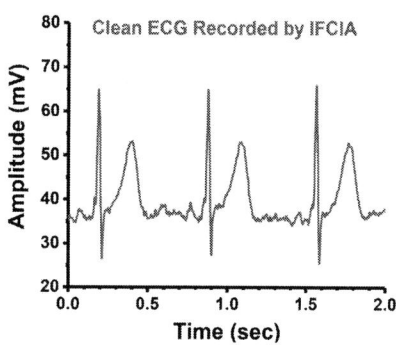

Fig. 8. *In-vivo* ECG readout

(a)

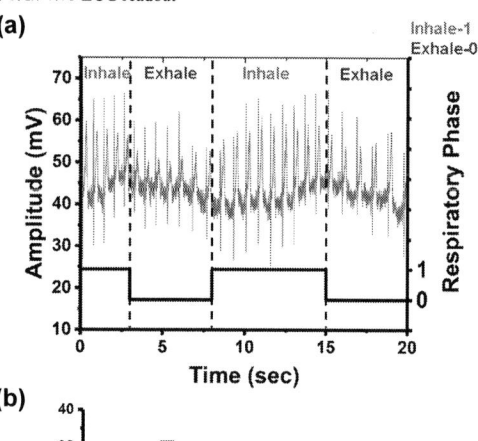

(b)

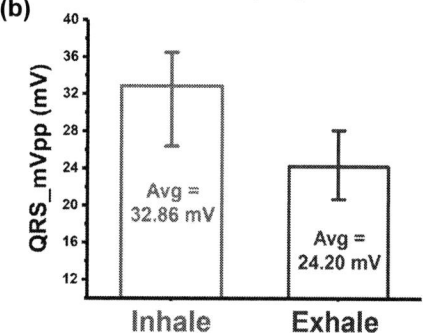

Fig. 9. (a) Real-time respiratory monitoring (b) Average peak-to-peak amplitude of inhalation and exhalation

TABLE I. PERFORMANCE COMPARISON.

	This work	[1]	[2]	[3]	[4]	[5]	[6]
Process (μm)	0.18	0.18	0.18	0.04	0.35	0.18	0.18
Area/Ch (mm²)	0.012	0.7	0.23	0.069	0.056	0.3	1
Supply (V)	1.2	1.8	1	1.2	3	1	0.2/0.8
Current/Ch (μA)	0.65	0.9	1.5	2.33	2	3.5	2.19
Gain (dB)	32	40	60	25.7	38.1	60	57.8
BW (Hz)	800	0.07-1k	0.37-750	0.12-5k	1.4-8.5k	0.5-100	670
Z_{in} @ 50Hz (Ω)	1.5G	>500M	560M	1.2G	N/A	>700M	N/A
IRN (μV$_{rms}$)	1.62 (0.5-800Hz)	0.9 (0.5-100Hz)	0.63 (0.5-100Hz)	1.8 (1-200Hz)	14.4 (1-8.5kHz)	1.3 (0.5-100Hz)	0.94 (0.5-670Hz)
NEF	1.77	3.29	N/A	7.4	8.5	9.33	2.1
Application	ECG	EEG	EEG	Neural recording	Neural recording	EEG	EEG/ ECG

IV. CONCLUSIONS

A low power, low noise, high input impedance IFCIA is demonstrated in this paper. The proposed structure utilizes intrinsic capacitance between MOS gate and drain as feedback path for a single-stage inverter-based amplifier. Both C_{INPUT} and C_{IBL} can be scaled down by 10x; hence the active area is shrunk to 0.012 mm² and Z_{IN} is boosted to 1.5 GΩ. The topology is robust against process variation as die-to-die gain mismatch is below 0.41%. The proposed IFCIA is suitable for bio-potential recording and respiratory monitoring applications.

ACKNOWLEDGMENT

The authors appreciate Prof. Changho Yoon, M.D., (Seoul National University Bundang Hospital) for his fruitful comments and suggestions.

REFERENCES

[1] M. A. Bin Altaf, C. Zhang and J. Yoo, "A 16-channel patient-specific seizure onset and termination detection SoC with impedance-adaptive transcranial electrical stimulator," in IEEE Journal of Solid-State Circuits, vol. 50, no. 11, pp. 2728-2740, Nov. 2015.

[2] T. Tang, W. L. Goh, L. Yao and Y. Gao, "A 16-channel TDM analog front-end with enhanced system CMRR for wearable dry EEG recording," 2017 IEEE Asian Solid-State Circuits Conference (A-SSCC), Seoul, 2017, pp. 33-36.

[3] H. Chandrakumar and D. Marković, "An 80-mVpp linear-input range, 1.6- GΩ input impedance, low-power chopper amplifier for closed-loop neural recording that is tolerant to 650-mVpp common-mode interference," in IEEE Journal of Solid-State Circuits, vol. 52, no. 11, pp. 2811-2828, Nov. 2017.

[4] K. A. Ng and Y. P. Xu, "A compact, low input capacitance neural recording amplifier with Cin/Gain of 20fF.V/V," 2012 IEEE Biomedical Circuits and Systems Conference (BioCAS), Hsinchu, 2012, pp. 328-331.

[5] N. Verma, et al., "A micro-power EEG acquisition SoC with integrated feature extraction processor for a chronic seizure detection system," in IEEE Journal of Solid-State Circuits, vol. 45, no. 4, pp. 804-816, April 2010.

[6] F. M. Yaul and A. P. Chandrakasan, "A noise-efficient 36 nV/√Hz chopper amplifier using an inverter-based 0.2-V supply input stage," in IEEE Journal of Solid-State Circuits, vol. 52, no. 11, pp. 3032-3042, Nov. 2017.

[7] M. Chae, J. Kim and W. Liu, "Fully-differential self-biased bio-potential amplifier," in Electronics Letters, vol. 44, no. 24, pp. 1390-1391, 20 November 2008.

5(b)). The input-referred noise floor is 57.2 nV/√Hz with an integrated noise of 1.62 μV$_{rms}$ (0.5- 800 Hz) (Fig. 6). With the proposed impedance boosting method, effective Z_{IN} is 1.5 GΩ from DC up to 50Hz and is maintained over 1 GΩ until 100 Hz (Fig. 7). Thanks to the IFCIA's wide bandwidth and dynamic range, *in-vivo* ECG can be recorded in real time as shown in Fig. 8. Inhaling and exhaling activities are clearly visible and distinguishable from Fig. 9, with average peak-to-peak QRS amplitude of 32.86 mV and 24.20 mV for inhalation and exhalation respectively. TABLE I summarizes the performance of the proposed IFCIA. Comparing with the state-of-the-art IAs, the IFCIA occupies smallest active area with high Z_{IN}.

21-4 (7010)

IEEE Asian Solid-State Circuits Conference
November 4 – 6, 2019
The Parisian Macao, Macao SAR, China

T/R-Switch Composed of 3 High-Voltage MOSFETs with 12.1 µW Consumption that can Perform Per-channel TX to RX Self-Loopback AC Tests for 3D Ultrasound Imaging with 3072-channel Transceiver

Shinya Kajiyama[1], Yutaka Igarashi[2], Toru Yazaki[2], Yusaku Katsube[2], Takuma Nishimoto[2], Tatsuo Nakagawa[1], Yohei Nakamura[1], Yoshihiro Hayashi[3] and Taizo Yamawaki[1]

[1]Research & Development Group, Hitachi, Ltd., Kokubunji, Tokyo, Japan
[2]Research & Development Group, Hitachi, Ltd., Yokohama, Kanagawa, Japan
[3]Healthcare Business Unit, Hitachi, Ltd., Kokubunji, Tokyo, Japan

Abstract—This paper presents an area- and power-efficient TX/RX-isolation switch implemented in a 3072-ch ultrasound transceiver IC. The proposed dynamic gate-source shunt topology, which utilizes a negative-HV-transmit-driven shunt switch, eliminates area- and power-hungry HV level shifters and ensures OFF during TX periods. In addition, source-driven HV-PMOS for ON/OFF control enables both gate charging and discharging using single HV-PMOS, thus contributing to area reduction. Moreover, by preparing an attenuation mode, a per-channel TX to RX self-loopback test can be performed. This function provides an on-wafer AC test without probing 3072 electrodes and can be applied to field diagnosis of assembled ultrasound probes.

Keywords—ultrasound, 3D imaging, 2D-array, matrix transducer, T/R-switch, HV-switch, loopback test

I. INTRODUCTION

Recently, real-time 3D ultrasound imaging for medical diagnosis has been highly demanded. To obtain a high-definition 3D image, a 2D-array matrix probe, which includes thousands of transducer (TD) elements and a per-element-channel (Ech)-transceiver integrated beamformer ASIC, is needed [1], [2]. The main challenges for the ASIC design are realizing a small Ech area and low power operation. We use a 3072-Ech 2D-array with a 0.3 mm pitch that is limited by constraint concerning grating lobes of the phased array [1]. Since an ultrasound probe is in direct contact with the human body surface, the ASIC power consumption is also regulated by self-heating. Moreover, ASIC testing is a challenging issue. Since directly probing 3072 electrodes is not practical in on-wafer testing nor post-assembly diagnosis, a per-Ech transmitter (TX) to receiver (RX) self-loopback conduction test is in great demand for per-Ech defect screening.

Our ultrasound 3D imaging system is composed of a 2D-array probe and a main unit. To interface the 3072-Ech 2D-array and 128 bundled cables, 24:1 channel reduction using in-probe delay-and-sum is conducted (Fig. 1a). An Ech includes a 3-level (HV+/0V/HV-) TX pulser, TX/RX isolation switch (T/R-SW), low-noise amplifier (LNA), and TX/RX-time-shared analog ring memory (ARAM) for delay control (Fig. 1b). The T/R-SW is absolutely necessary to protect the LV-LNA from HV bipolar pulses of up to 138 Vpp in TX, while transferring the small signal in RX with low Ron. To squeeze

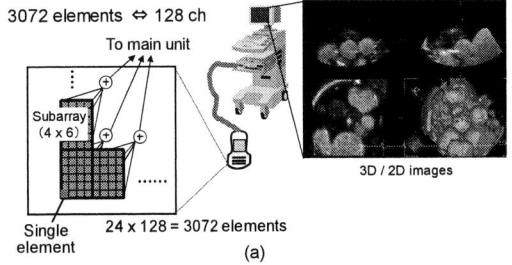

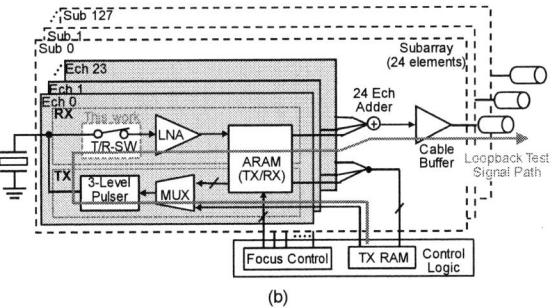

Fig. 1. (a) 3D ultrasound imaging system; (b) transceiver ASIC block diagram

the area-hungry HV-MOSs into the Ech circuit area of 0.3 mm x 0.3 mm, the number of them must be minimized in the pulser and the T/R-SW design.

II. DYNAMIC GATE-SOURCE-SHUNT T/R-SWITCH

Fig. 2 shows a comparison of the proposed T/R-SW with prior arts. The circled devices are HV-MOSs. To maintain the OFF state in both HV+ and HV- transmitting, a back-to-back connected HV-NMOS is generally used. Since the common source of M1 and M2 is driven down to near HV- via body diode of M1 in TX period, as shown in Fig. 2a, several topologies have been proposed in order to assure T/R-SW OFF state even in HV- transmission. Conventional T/R-SW topologies can be roughly categorized into three types. Zener diode bias is Vgs stable, whereas requires HV bias circuit and static power. The level shift type also requires the HV current

978-1-7281-5107-6/19 $31.00 © 2019 IEEE

305

IEEE Asian Solid-State Circuits Conference
November 4 – 6, 2019
The Parisian Macao, Macao SAR, China

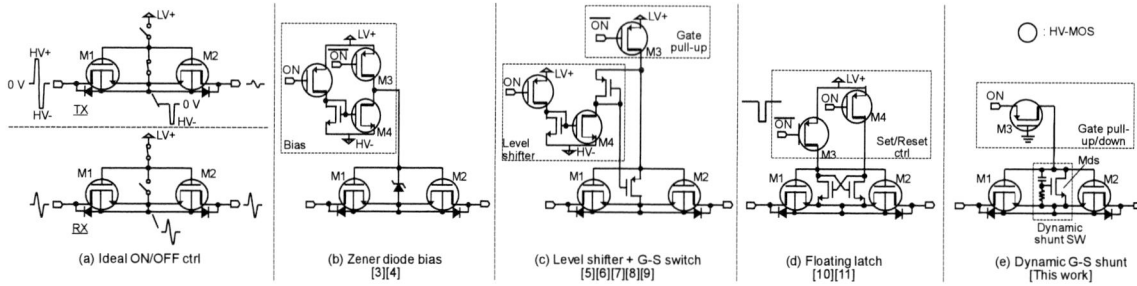

Fig. 2. Comparison of basic HV-switch topologies

path, though some state-of-the-art circuits realize zero power by using an HV-latch instead [7]. A floating latch can be implemented by LV-latch, however, the floating latch needs two or more HV-PMOSs for set/reset pulse control. To minimize the number of HV-MOSs, we propose dynamic gate-source shunt topology, as shown in simplified form in Fig. 2d and in detail in Fig. 4. The operation is shown in Fig. 3. The T/R-SW goes OFF in advance of the TX period. The LV-logic drives down the source of HV-PMOS, M3 (input port: "ON" in Fig. 2e) to pull the common gate of HV-NMOS, M1 and M2 down to near 0 V and change the gate and the source of M1 and M2 float to the OFF state. Due to this source-driven control, M3 is able to transfer the signal near the logic low level (|Vthp|) and rapidly discharge the gate of M1 and M2, as shown at 2 µs in Fig. 3, without extra HV-NMOSs for discharging.

Once the pulser drives the transducer and the T/R-SW input down to HV-, the body diode of M1 pulls the common source of M1 and M2 down, while the common gate of M1 and M2 tries to remain near 0 V. Accordingly, the Vgs of M1 and M2 almost increases. The LV-NMOS of the dynamic shunt switch, Mds, however, goes ON automatically and ensures the T/R-SW OFF state. As shown in Fig. 3, Vgs of Mds rises at every HV- transmission by responding to M1, M2 Vgs increase via the RC high-pass network in the dynamic shunt switch. Meanwhile in no pulse periods, Mds remains OFF due to a pull-down resistor. The RC cut-off frequency of the dynamic shunt switch should be chosen to be low to respond to the TX pulse slew rate. The bottom waveform in Fig. 3 is Vgs of M1 and M2 without the dynamic shunt switch. Since the gate of M1 and M2 is floating, Vgs of M1 and M2 increases at every HV- transmissions in this case. In the case with the dynamic shunt switch, this M1, M2 Vgs increase activates the dynamic shunt switch, then assures OFF state of the T/R-SW in a closed loop.

When transit to RX period, the common gate of M1 and M2 is charged to 5 V via M3 to turn the T/R-SW ON, thus Vgs of 5 V is applied to M1 and M2 during RX.

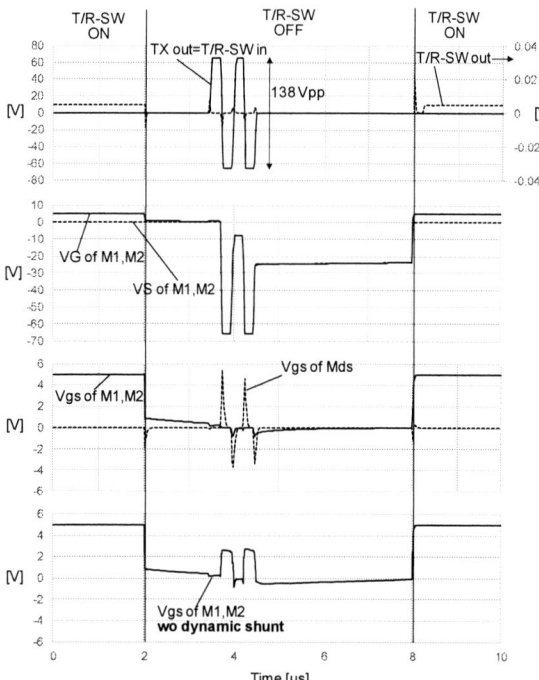

Fig. 3. Simulated waveforms of proposed dynamic shunt T/R-SW

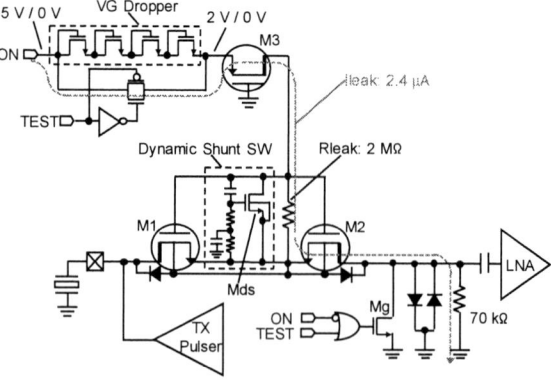

Fig. 4. Schematic of proposed T/R-SW capable of working as an attenuator for TX to RX self-loopback tests

978-1-7281-5107-6/19 $31.00 © 2019 IEEE

III. PER-CHANNEL TX TO RX SELF-LOOPBACK AC TEST

A per-Ech TX to RX self-loopback conduction test is in great demand, as stated already. Since the LNA and post-RX stages operate in the 1.8 V domain, the lowest 5 Vpp TX pulse needs to be attenuated through the T/R-SW.

The AC test signal path is shown in Fig. 1b. The test signal is routed from TX RAM. A low HV supply of ±5 V is chosen in the loopback test mode. The T/R-SW attenuates the continuous wave of the test signal by -16 dB down to 0.7 Vpp. When only a single Ech is activated, the loopback test signal is then amplified by LNA, passed through ARAM with 0 dB, and attenuated by 1/24 through the adder.

The Vgs of M1 and M2 in the T/R-SW is decreased from 5 V (RX) to 2 V to accept the 5 Vpp input by dropping the gate potential down, using a VG dropper including a series of diode-connected 5 V NMOSs (Fig. 4). Also, the 5 V NMOS at the T/R-SW output to GND, Mg, goes ON, and then the loopback signal is attenuated by dividing the increased M1, M2 Ron and Mg Ron. The 2 MΩ resistor, Rleak, gives the leak current to the VG dropper to provide a voltage drop of Vthn x 4. Rleak also contributes to preventing absolute floating of M1, M2 gate and giving finite impedance in the TX period. However, the static current Ileak of 5 V / (2 MΩ + 70 kΩ), flowing in the RX period, results in 12.1 µW power dissipation.

IV. MEASUREMENT RESULTS

Fig. 5 shows S_{21} measured using a TRSW + LNA test element. The OFF isolation of -64 dB at 10 MHz, that is sufficient for 138 Vpp pulse to decay to less than 1.8 Vpp, is obtained as the OFF/ON ratio.

A measured spatial distribution map and histogram of the 1.67 MHz continuous wave loopback output amplitude is shown in Fig. 6. No significant skew is found in the spatial distribution. As the histogram shows normal distribution, the randomness decreases as the Ech signals are averaged into the subarray and then the single beam.

The performance of the proposed T/R-SW is summarized in Table I and compared with the performance of prior works. Only this work achieves true 3-HV-MOS implementation without sharing any circuitry among Echs. That eliminates HV signal wiring over the array, thus allowing busy metal wiring of control signal buses over the large-scale array. The proposed dynamic shunt architecture eliminates HV supply for T/R-SW and only 5V should be supplied. The area of 9920 µm² is only 11 % of that of single Ech transceiver circuit. The proposed circuit topology does not consume any static power during TX. Only 12.1 µW is dissipated in RX, because of 2 MΩ Rleak shown in Fig. 4 that prevents absolute floating gate. The 2D-array transceiver ASIC including the proposed T/R-SWs is fabricated using 0.18 µm 140 V HV SOI CMOS.

The photograph of the 2D-array transceiver ASIC is shown in Fig. 7a. A subarray including 4 x 6 Echs are 16 x 8 arranged. T/R-SW ON/OFF control signal lines are buffered and distributed over the array from control logics located in peripheral blocks.

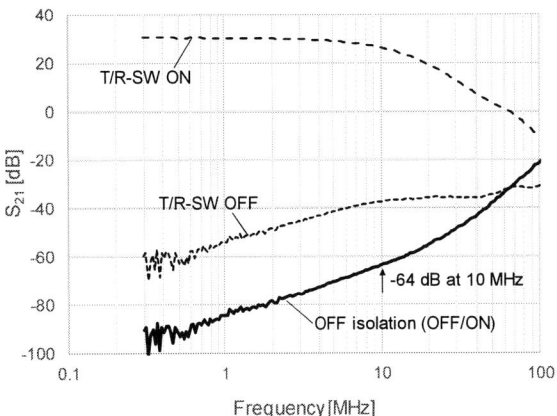

Fig. 5. Measured S_{21} of T/R-SW + LNA Test Element Group

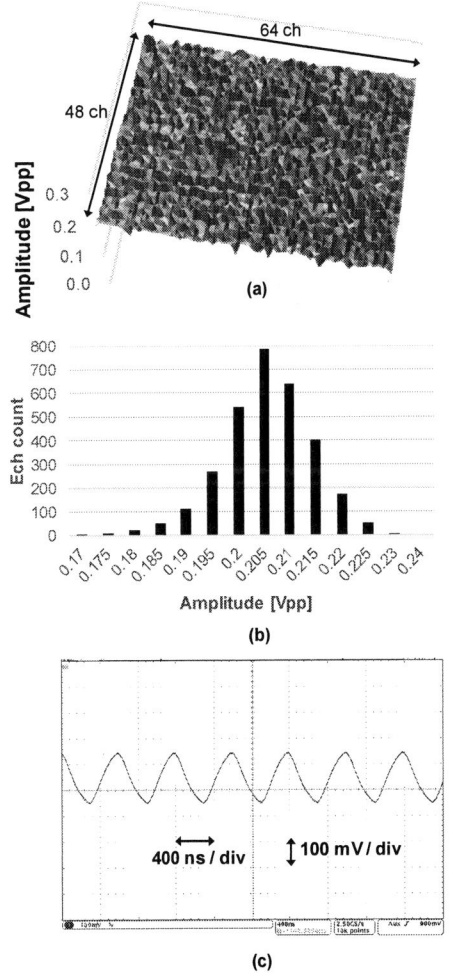

Fig. 6. (a) Measured spatial distribution map of loopback output amplitude; (b) amplitude histogram; (c) measured CW loopback output waveform

IEEE Asian Solid-State Circuits Conference
November 4 – 6, 2019
The Parisian Macao, Macao SAR, China

TABLE I.   HV-SWITCH COMPARISON WITH PRIOR WORKS

Parameter	This Work	[4] 2016	[7] 2014	[9] 2018	[11] 2018
Architecture	Dynamic shunt	Zener bias	Level shift	Level shift	Floating latch
Number of channels	3,072	16	1	64	512
HV transfer	No	Yes	No	Yes	Yes
Number of HV-MOSs/channel	3	12	17	≥ 5	5
SW supply	5 V	12 V	-25 V, 1.8 V	± 20 V, 3.3 V	N.A.
Area/channel	9,920 µm²	N.A.	5,525 µm²	150,000 µm²	70,312 µm²
Ron	290 Ω	18 Ω	N.A.	180 Ω	N.A.
OFF isolation	-64 dB@10 MHz	-54 dB@5 MHz	-53 dB	-35 dB@10 MHz	N.A.
Static power/channel	0 µW (TX) 12.1 µW (RX)	4.4 mW	0 mW	N.A.	N.A.
Technology	0.18 µm 140 V HV SOI CMOS	150 V HV SOI CMOS	0.18 µm 50 V HV CMOS	0.35 µm 120 V HV CMOS	LV +200 V HV CMOS

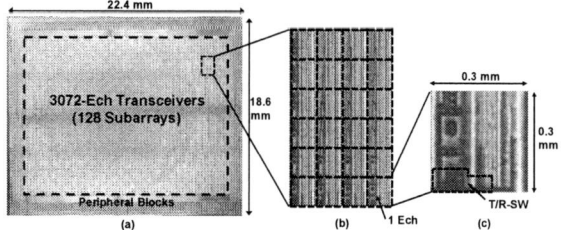

Fig. 7.   (a) Photograph of ASIC; (b) Subarray including 24-Echs

V.   CONCLUSION

We have proposed an area- and power-efficient T/R-SW, employing a LV dynamic shunt switch that assures OFF state during TX periods without any static power and extra HV-MOSs, for an ultrasound 3D imaging 3072-ch transceiver ASIC. The source-driven HV-PMOS for ON/OFF control also contributes to area reduction by making both charge and discharge possible using only single HV-PMOS. Thus the proposed T/R-SW requires only 3 HV-MOSs per Ech without sharing any circuits among Echs, and consume 0 µW and 12.1 µW in TX and RX, respectively. We have also demonstrated per-Ech TX to RX internal loopback AC conduction test function by preparing T/R-SW attenuation mode. A spatial distribution map of continuous wave loopback output amplitudes has been obtained without directly probing to 3072 electrodes.

REFERENCES

[1] Y. Igarashi, S. Kajiyama, Y. Katsube, T. Nishimoto, T. Nakagawa, Y. Okuma, Y. Nakamura, T. Terada, T. Yamawaki, T. Yazaki, Y. Hayashi, K. Amino, T. Kaneko, and H. Tanaka, "Single-Chip 3072-Element-Channel Transceiver/128-Subarray-Channel 2-D Array IC With Analog RX and All-Digital TX Beamformer for Echocardiography," IEEE JSSC, vol. 54, no. 9, pp. 2555-2567, Sep. 2019.

[2] E. Kang, Q. Ding, M. Shabanimotlagh, P. Kruizinga, Z. -Y. Chang, E. Noothout, H. J. Vos, J. G. Bosch, M. D. Verweij, N. de Jong, and M. A. P. Pertijs, "A Reconfigurable Ultrasound Transceiver ASIC With 24 × 40 Elements for 3-D Carotid Artery Imaging," IEEE JSSC, vol. 53, no. 7, pp. 2065-2075, Jul. 2018.

[3] B. Dufort, T. Letavic, and S. Mukherjee, "Digitally controlled high-voltage analog switch array for medical ultrasound applications in thin-layer silicon-on-insulator process," IEEE Int'l SOI Conference, pp. 78-79, Oct. 2002.

[4] F. Yamashita, J. Aizawa, and H. Honda, "A new compact, low on resistance and high off isolation high voltage analog switch IC without using high voltage power supplies for ultrasound imaging system," IEEE ISPSD, pp. 415-418, Jun. 2016.

[5] Y. -M. Li, R. Wodnicki, N. Chandra, and N, Rao, "An Integrated 90V Switch Array for Medical Ultrasound Applications," CICC, pp. 269-272, Sep. 2006.

[6] J. Borg and J. Johansson, "An Ultrasonic Transducer Interface IC With Integrated Push-Pull 40 Vpp, 400 mA Current Output, 8-bit DAC and Integrated HV Multiplexer," IEEE JSSC, vol. 46, no. 2, pp. 475-484, Feb. 2011.

[7] S. J. Jung, S. -K. Hong, and O. -K. Kwon, "Area-efficient high-voltage switch using floating control circuit for 3D ultrasound imaging systems," Electronics Letters, vol. 50, no. 25, pp. 1900–1902, Dec. 2014.

[8] S. Dai, R. W. Knepper, and M. N. Horenstein, "A 300-V LDMOS analog-multiplexed driver for MEMS devices," IEEE Trans. Circuits Syst., vol. 62, no. 12, pp. 2806–2816, Dec. 2015.

[9] H. Jung, R. Wodnicki, H. G. Lim, C. W. Yoon, B. J. Kang, C. Yoon, C. Lee, J. Y. Hwang, H. H. Kim, H. Choi, M. S. -W. Chen, Q. Zhou, and K. K. Shung, "CMOS High-Voltage Analog 1–64 Multiplexer/ Demultiplexer for Integrated Ultrasound Guided Breast Needle Biopsy," IEEE Trans. Ultrasonics, Ferroelectrics, and Frequency Control, vol. 65, no. 8, pp. 1334-1345, Aug. 2018.

[10] K. Hara, J. Sakano, M. Mori, S. Tamano, R. Sinomura, and K. Yamazaki, "A New 80V 32x32ch Low Loss Multiplexer LSI for a 3D Ultrasound Imaging System," IEEE ISPSD, pp. 359–362, May 2005.

[11] G. Ricotti and V. Bottarel, "HV Floating Switch Matrix with Parachute Safety Driving for 3D Echography Systems," ESSCIRC, pp. 271-273, Sep. 2018.

978-1-7281-5107-6/19 $31.00 © 2019 IEEE

A High DR High-Input-Impedance Programmable-Gain ECG Acquisition Interface with Non-inverting Continuous Time Sigma-Delta Modulator

Junhao Liang, Sai-Weng Sin[1], Seng-Pan U, Franco Maloberti[2], R. P. Martins[1,3], Hanjun Jiang[4]

1-State-Key Laboratory of Analog and Mixed-Signal VLSI, IME/ECE/FST, University of Macau, Macao, China
2 –University of Pavia, Pavia, Italy
3 – On leave from Instituto Superior Técnico/Universidade de Lisboa, Portugal
4-The Institute of Microelectronics, Tsinghua University, Beijing China
E-mail: terryssw@um.edu.mo

Abstract— **This paper presents a power-efficient, high dynamic range, high input impedance data acquisition circuit for an implantable electrocardiogram (ECG) detector in 65 nm CMOS, occupying an active area of 0.225 mm². To enhance the dynamic range, we utilize a 3ʳᵈ-order continuous-time sigma-delta modulator with a programmable input gain coefficient and an embedded antialiasing filter. The implementation of the non-inverting integrator allows a high input impedance for the ECG application. The measurement results show that it obtains 84.2dB SNDR in a 150Hz bandwidth with a 1V power supply, consuming 5.4µW at a nominal gain of 0dB. The maximum dynamic range is 99.3dB with the programmable gain activated from 0 to 18dB.**

Keywords—Programmable integrator, high impedance, non-inverting integrator, ECG, CT sigma delta AD converter.

I. INTRODUCTION

Each year a large number of emergency and syncope outpatients with unexplained dizziness, palpitations or fainting need medical treatment around the world [1]. However, doctors have a considerable proportion of patients with syncope to whom they don't have a clear diagnosis. The implantable ECG recorder, introduced in 1995, has a long-term monitoring function and can capture a sudden cardiac arrhythmia, which is an effective method to determine the cause of the above symptoms. As one of the key components of the implantable ECG detector, the acquisition circuit needs to tradeoff between bandwidth, sampling rate, dynamic range, and power consumption. Fig. 1(a) shows the conventional implementation [2], with each acquisition channel composed by an instrumentation amplifier (IA), a low-pass filter (LPF), and a high-resolution analog-to-digital converter (ADC). Therefore, the leakage caused by the low sampling rate and the accumulative noise of each component are the main challenges of the acquisition circuit in order to obtain a high DR. CT sigma-delta modulators can incorporate both the functions of a LPF and an ADC, simultaneously, then suppressing the accumulative noise and the leakage effect due to oversampling and the continuous-time operation of the RC integrator. Since the signal acquisition channel requires a high dynamic range for the weak signal detection, an instrumentation amplifier is necessary to amplify the signal in front of the ADC, as also shown in Fig. 1(a). Considering the characteristics of CT-ΣΔ ADCs, the programmable coefficient integrator [3] will be used in this work to achieve amplification and high dynamic range. Unfortunately, the input impedance is a significant issue of the implantable ECG application while the input of the CTSDM is

resistive. Therefore, we propose for the first time in an ECG application a non-inverting integrator, which has high input impedance.

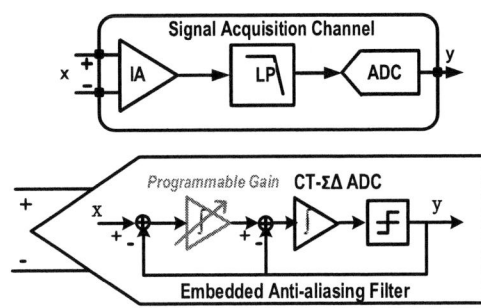

Fig. 1. (a) Signal acquisition circuit (b) Proposed acquisition circuit.

Fig. 1(b) exhibits the simplified system diagram of the proposed acquisition circuit, based on a 3ʳᵈ-order continuous-time sigma-delta modulator. Since we embedded the non-inverting integrator with a programmable coefficient in the sigma-delta loop, therefore, the programmable-gain-amplifier (PGA) function can be achieved with high input impedance and high dynamic range.

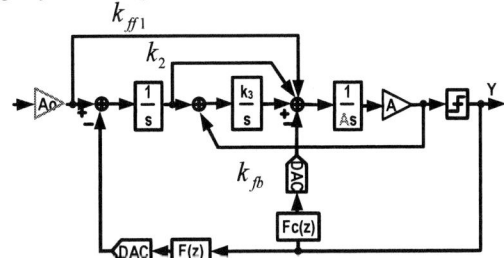

Fig. 2. The block diagram of the system.

II. THE PROPOSED ACQUISITION CIRCUIT

Fig. 2 presents the block diagram of the sigma-delta-based ECG signal acquisition system, where A_o is the coefficient of the programmable gain. The swing of the 3ʳᵈ integrator is relatively large because of the feed-forward path k_{ff1} and k_2. Therefore, the coefficient scaling 1/A and A are necessary in front of the 3ʳᵈ integrator and the quantizer. To further suppress the flicker noise, we use a chopper and a FIR DAC with F(z) and Fc(z) [4].

978-1-7281-5107-6/19 $31.00 © 2019 IEEE

A. Programmable Coefficient Integrator

Fig. 3 displays the schematic of the proposed acquisition circuit, where we apply a programmable coefficient integrator [3] due to the requirement of high dynamic range in the weak signal detection. To realize the adjustment of the coefficient A_0 which is related to $1/RC$, either the resistance or the capacitance value can be programmed. Moreover, the area of the capacitor array is larger than the resistor array due to the large time-constant needed in the ECG design. The programmable integrator with a series-connected resistor array can amplify the signal with an 8-fold maximum.

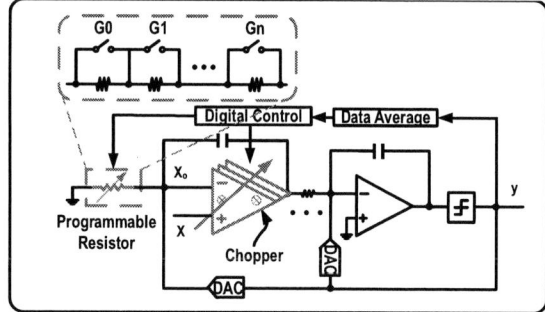

Fig. 3. The schematic diagram of the acquisition circuit.

However, when the gain A_0 increases, the feedback factor of the opamp decreases. To avoid the changing of the opamp bandwidth, its GBW needs to be programmed correspondingly. The overall programmable gain can be adjusted according to the average value of the digital output, and the resistor array and opamp scaling factor will be controlled by a 3-bit data.

B. Non-inverting Programmable Opamp

We propose the use of a non-inverting opamp due to the impedance issue of the ECG circuit. However, the use of a non-inverting integrator will cause a transfer function deviation from the ideal integrator, and the output of the non-inverting integrator will also contain signal components, in addition to the quantization noise. When compared with the inverting case, it has equivalently one more feed-forward path. Therefore, the output swing of the non-inverting integrator is larger than the inverting case, although, the swing limitation is not a severe issue for the ECG signal detection.

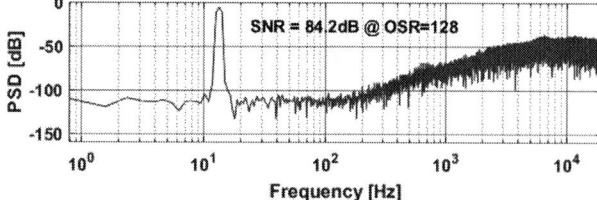

Fig. 4. Output Spectrum of 3rd order CTSDM

Another challenge of the non-inverting integrator is the feedback DAC. The feedback point of the inverting integrator is at the virtual ground while the feedback point of the non-inverting case is at X_0 in Fig.3, which tracks the input X. Therefore, we designed the feedback current steering DAC

with a cascode current source using the switches in the DAC implemented by a 2.5V thick-oxide transistor. They provide a switch function as well as another stage of the cascade, in such a way that it can provide a sufficiently high output impedance to isolate the signal swing at X_0.

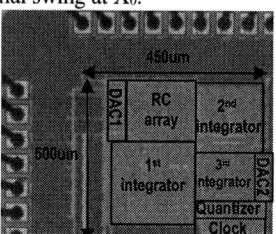

Fig. 5. Chip Micrograph and its area distribution summary.

TABLE I: PERFORMANCE SUMMARY

	Technology(nm)	65
Analog Front-End	V_{DD}(V)	1
	Max. Programmable Gain(dB)	18
	Current(uA)	5.4
	Input Impedance (Ω)	15G
	Bandwidth (Hz)	150
	ADC Architecture	CTSDM
	Input Referred Noise(uVrms)	2.73
	ADC Bits	14
	ADC sampling frequency(kHz)	38.4
	Power consumption G=1 (uW)	5.4
	Power consumption G=8 (uW)	20.8
	Max Dynamic Range(dB)	99.3
	Schreier FOM (DR) (dB)	167.88

III. MEASUREMENT RESULTS

Fig. 5 shows the chip micrograph of the ECG acquisition circuit with an active area of 0.225 mm² implemented in 65nm CMOS. The measurement results of Table I confirm that the peak SNDR is 84.2dB with 150Hz bandwidth and 1V power supply, consuming 5.4μW at a gain of 0dB. The maximum dynamic range achieves 99.3dB with the gain programmed from 0 to 18dB. According to the leakage current the input impedance is larger than 15GΩ.

ACKNOWLEDGMENT

This work was supported by the Research Committee of University of Macau and The Science and Technology Development Fund (FDCT) under Grant no 006/2016/AFJ.

REFERENCES

[1] R. Arzbaecher, D. Hampton, M.C. Burke, M.C. Garrett, "Subcutaneous ECG monitors and their field of view," in *Journal of Electrocardiology*, vol. 43, no. 6, pp. 601-605, Nov-Dec, 2010

[2] D. Jeon, et al. "24.3 An implantable 64nW ECG-monitoring mixed-signal SoC for arrhythmia diagnosis," in *IEEE ISSCC Dig. Tech. Papers*, pp.416-417, Feb. 2014.

[3] Yuan Ren, Sai-Weng Sin, Chi-Seng Lam, Man-Chung Wong, Seng-Pan U, & R.P. Martins, "A high DR multi-channel stage-shared hybrid front-end for integrated power electronics controller," in *Proceedings IEEE A-SSCC*, pp. 57-60, Nov. 2016.

[4] S. Billa, A.Sukumaran, & S. Pavan, "15.4 A 280uW 24kHz-BW 98.5 dB-SNDR chopped single-bit CTSDM achieving <10Hz 1/f noise corner without chopping artifacts," In *IEEE ISSCC Dig. Tech. Papers*, pp. 276-277, Jan./Feb. 2016.

IEEE Asian Solid-State Circuits Conference
November 4 – 6, 2019
The Parisian Macao, Macao SAR, China

Supplements for "A High DR High-Input-Impedance Programmable-Gain ECG Acquisition Interface with Non-inverting Continuous Time Sigma-Delta Modulator" by JunHao Liang *et al.*

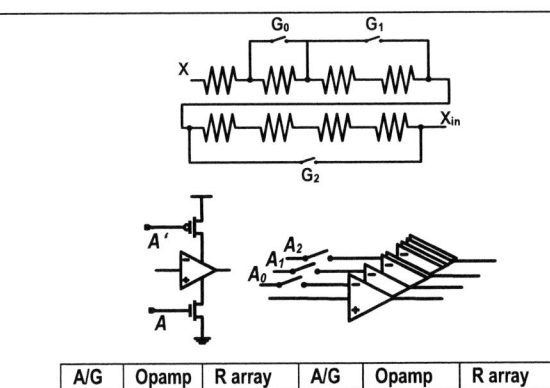

A/G	Opamp	R array	A/G	Opamp	R array
111	8 times	8 times	011	4 times	4 times
110	7 times	8/5 times	010	3 times	4/3 times
101	6 times	8/3 times	001	2 times	2 times
100	5 times	8/7 times	000	1 times	1 times

Fig S1. (a) Programmable resistor array (b)single unit of opamp (c) opamp array

TABLE S I: COMPARISON

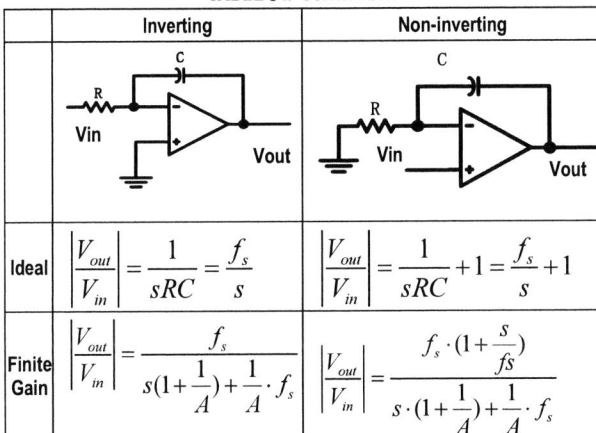

	Inverting	Non-inverting
Ideal	$\left|\dfrac{V_{out}}{V_{in}}\right| = \dfrac{1}{sRC} = \dfrac{f_s}{s}$	$\left|\dfrac{V_{out}}{V_{in}}\right| = \dfrac{1}{sRC} + 1 = \dfrac{f_s}{s} + 1$
Finite Gain	$\left|\dfrac{V_{out}}{V_{in}}\right| = \dfrac{f_s}{s\left(1+\frac{1}{A}\right) + \frac{1}{A}\cdot f_s}$	$\left|\dfrac{V_{out}}{V_{in}}\right| = \dfrac{f_s\cdot\left(1+\frac{s}{fs}\right)}{s\cdot\left(1+\frac{1}{A}\right) + \frac{1}{A}\cdot f_s}$

The difference in the transfer function between the inverting and non-inverting integrator. An equivalent feed-forward coefficient is generated in the non-inverting version.

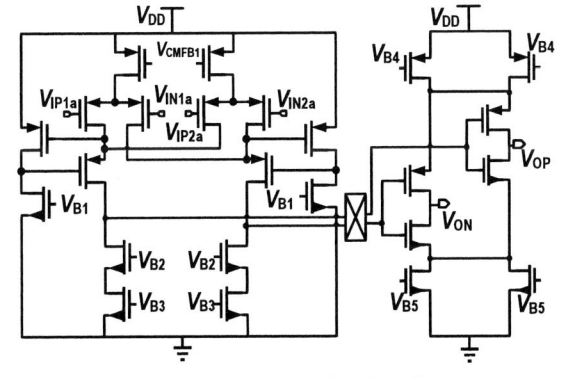

Fig. S2. Four Input Fully differential non-inverting Opamp

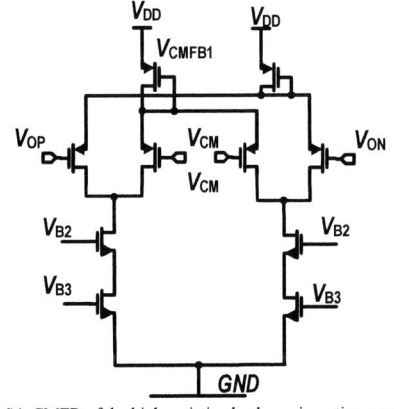

Fig S3. 1.5b Current Steering DAC with cascode switches.
P1 & P2 is the regular non-overlapping clock phase. A and B represents the code of "1" and "-1", and C represents the code for "0".

Fig S4. CMFB of the high resistive load non-inverting amplifier
Since the loading of the Opamp is resistive, the CMFB circuit should be with high input impedance, otherwise the coefficient of next stage will change.

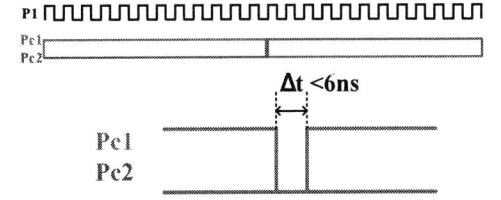

Fig. S5. The clock Pc1 and Pc2 of the chopper

The frequency of chopper is set 1/12 of the sampling clock [4] to suppress the aliasing and noise introduced by chopper in CTSDM. During the non-overlap time of the clock, the continuous-time RC integrator is floating. In such a low speed ECG application, Δt should be short enough to reduce the non-overlapping period. In this work, Δt<6ns.

978-1-7281-5107-6/19 $31.00 © 2019 IEEE

IEEE Asian Solid-State Circuits Conference
November 4 – 6, 2019
The Parisian Macao, Macao SAR, China

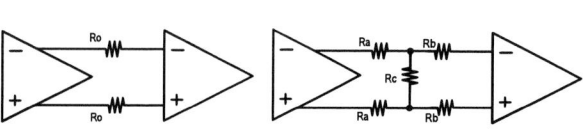

Fig S6. a) conventional connection b) Current shunt connection in second stage

$$\frac{Vx}{Ro} = \frac{Vx}{Ra + (\frac{1}{2}Rc \parallel Rb)} \cdot \frac{\frac{1}{2}Rc}{\frac{1}{2}Rc + Rb}$$

In low frequency application, RC value are quite large, that means the large chip area. Therefore, to squeeze the area, in non-sensitive path, the current shunt connection is utilized to separate a large resistor into two smaller resistors.[5]

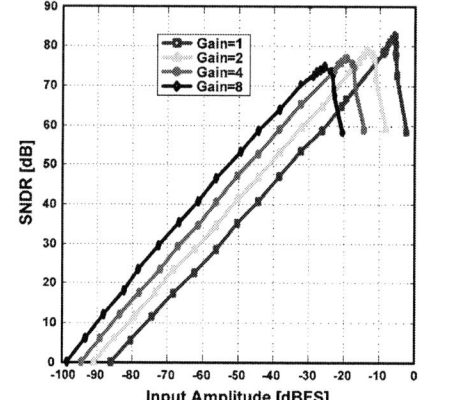

Fig S7. DR of the ECG acquisition circuit with programmable gain activated

TABLE S1I. PERFORMANCE COMPARISON WITH STATE-OF-THE-ART ECG ACQUISITION INTERFACES

		This work	ISSCC' 14[6]	ISSCC' 14[7]	ISSCC' 14[8]	ISSCC' 13[9]	ISSCC' 18[10]
Target Signals		ECG	ECG	ECG	ECG	ECG,bio-impedance	Neural
Technology(nm)		65	180	65	180	180	65
Analog Front-End	V_{DD}(V)	1	1.3-1.8	0.6	1.2	1.8	0.8
	Gain(dB)	18	31-43	51-96	-	40-64	-
	Current(uA)	5.4	0.68	0.031	-	-	1 x16
	Bandwidth (Hz)	150	130	250	250	500-1000	500
	ADC Architecture	CTSDM	SAR	SAR	DTSDM	SAR	Over sampling ADC
	ADC Bits	14	7-10	8	13.5	9.3	10.7
	ADC sampling frequency(kHz)	38.4	1.024	0.500	3200	-	32
	Power consumption(uW)	5.4/20.8	1.22	0.019	56	3.26	0.8
	Dynamic Range(dB)	99.3	-	-	76	82	92
	Schreier FOM (DR) (dB)	167.88	-	-	142.5	163.8-166.9	167.9

[5] Jang, M., Lee, S., & Chae, Y. "A 55μW 93.1 dB-DR 20kHz-BW single-bit CT ΔΣ modulator with negative R-assisted integrator achieving 178.7 dB FoM in 65nm CMOS," in *IEEE Proceedings of the Symposium on VLSI Circuits*, pp. C40-C41, Jun. 2017.

[6] Yan, L., Harpe, P., Osawa, M., Harada, Y., Tamiya, K., Van Hoof, C., & Yazicioglu, R. F. , "24.4 A 680nA fully integrated implantable ECG-acquisition IC with analog feature extraction," in *IEEE International Solid-State Circuits Conference Digest of Technical Papers (ISSCC)*, pp. 418-419, Feb. 2014.

[7] Jeon, D., Chen, Y. P., Lee, Y., Kim, Y., Foo, Z., Kruger, G., ... & Sylvester, D, "24.3 An implantable 64nW ECG-monitoring mixed-signal SoC for arrhythmia diagnosis." in *IEEE International Solid-State Circuits Conference Digest of Technical Papers (ISSCC)*, pp. 416-417, Feb. 2014.

[8] Van Helleputte, N., Konijnenburg, M., Kim, H., Pettine, J., Jee, D. W., Breeschoten, A., ... & Yazicioglu, R. F. , "18.3 A multi-parameter signal-acquisition SoC for connected personal health applications," in *IEEE International Solid-State Circuits Conference Digest of Technical Papers (ISSCC)*,pp. 314-315, Feb. 2014.

[9] Kim, S., Yan, L., Mitra, S., Osawa, M., Harada, Y., Tamiya, K., ... & Yazicioglu, R. F, "A 20μW intra-cardiac signal-processing IC with 82dB bio-impedance measurement dynamic range and analog feature extraction for ventricular fibrillation detection," In *2013 IEEE International Solid-State Circuits Conference Digest of Technical Papers (ISSCC)*, pp. 302-303, Feb. 2013.

[10] Kim, C., Joshi, S., Courellis, H., Wang, J., Miller, C., & Cauwenberghs, G, "A 92dB dynamic range sub-μV rms-noise 0.8 μW/ch neural-recording ADC array with predictive digital autoranging," in *IEEE International Solid-State Circuits Conference-(ISSCC)*, pp. 470-472, Feb. 2018.

978-1-7281-5107-6/19 $31.00 © 2019 IEEE

Author Index

A

Abou-El-Kheir, Nahla T.	6-2 (7054)	73
Abraham, Isaac	2-4 (7094)	21
Agarwal, Amit	11-4 (7023)	137
Agarwal, Meghna	2-1 (7025)	9
Akita, Ippei	19-2 (7080)	267
Alioto, Massimo	11-1 (7116)	125
An, Eun-Ji	14-1 (7133)	189
Anders, Mark	11-4 (7023)	137
Aprile, Cosimo	16-2 (7182)	219
	17-2 (7138)	239
Araga, Yuuki	3-1 (7037)	25
Arcot, Srinivas	14-4 (7069)	197
Awano, Hiromitsu	19-1 (7181)	263

B

Badami, Komail	2-2 (7152)	13
Bae, Chisung	11-6 (7205)	145
Bae, Hyeon-Min	17-2 (7138)	239
Bae, Seungyong	12-5 (7070)	165
Bhatia, Karan	2-1 (7025)	9
Billa, Sujith	10-6 (7086)	123
Bonetti, Andrea	16-2 (7182)	219
Braendli, Matthias	17-2 (7138)	239
Breen, Daniel	2-1 (7025)	9
Bryant, Carl	12-6 (7109)	169
Burg, Andreas	16-2 (7182)	219
Burg, Andreas	17-2 (7138)	239
Byun, Wooseok	15-1 (7206)	201
Byun, Younghoon	15-2 (7076)	205

C

Cai, Yan	10-5 (7134)	121
Carta, Corrado	12-4 (7031)	161
Castello, Rinaldo	12-6 (7109)	169
Cevrero, Alessandro	17-2 (7138)	239
Chan, Hsun-Wei	19-4 (7159)	275
Chandrakasan, Anantha P.	15-3 (7141)	209
Chang, Chih-Kai	5-1 (7062)	53

Chang, Dong-Jin	14-1 (7133)	189
Chang, Fu-Chun	3-3 (7172)	33
Chang, Hsin-Ting	19-4 (7159)	275
Chang, Kang-Yu	16-4 (7142)	227
Chang, Mau-Chung Frank	9-2	103
Chang, Meng-Fan	3-3 (7172)	33
	16-1 (7160)	217
Chang, Nien-Hsiang	19-4 (7159)	275
Chang, Yung-Chang	5-1 (7062)	53
Chao, Yen-Yu	7-4 (7022)	95
Chatterjee, Rohit	2-1 (7025)	9
Chen, Hong	15-4 (7122)	213
Chen, Hung-Ming	16-4 (7142)	227
Chen, Jia-Jing	16-1 (7160)	217
Chen, Ke-Horng	4-5 (7087)	49
	18-4 (7140)	255
Chen, Po-Hung	4-1 (7176)	37
	16-4 (7142)	227
Chen, Qian	3-4 (7090)	35
Chen, Shao-Qi	4-5 (7087)	49
Chen, Tsung-Shen	16-4 (7142)	227
Chen, Yen-Kai	3-3 (7172)	33
Chen, Yi-Ren	3-3 (7172)	33
Chen, Yong	20-2 (7093)	283
Chen, Yun	19-3 (7066)	271
Chen, Zhiming	12-2 (7020)	153
Chen, Zipeng	12-2 (7020)	153
Cheng, Lin	4-3 (7082)	43
Cherupally, Sai Kiran	11-6 (7205)	145
Chi, Baoyong	12-2 (7020)	153
Chi, Cathy	2-1 (7025)	9
Chiu, Kang-Lun	19-4 (7159)	275
Chiu, Yen-Cheng	16-1 (7160)	217
Cho, Youngsea	12-5 (7070)	165
Choi, Dooseok	12-5 (7070)	165
Choi, Seungkyu	5-2 (7033)	57
Choi, Soonwoo	12-5 (7070)	165
Choi, Yeongjae	5-2 (7033)	57
Chu, Li-Cheng	4-5 (7087)	49
Chun, Jung-Hoon	7-1 (7092)	83
Chung, Jae-Hyun	14-1 (7133)	189
Cohen, Ariel	17-1 (7125)	235

D

Das, Abhijit	14-4 (7069)	197
De, Vivek	11-4 (7023)	137
Deguchi, Yoshiaki	16-5 (7114)	231
Deng, Jian	2-2 (7152)	13
Ding, Cong	20-5 (7177)	291
Ding, Yingtao	12-2 (7020)	153
Dixit, Suhas	10-6 (7086)	123
Do, Anh Tuan	11-5 (7113)	141

E

Eberlein, Matthias	10-3 (7089)	113
Edayath, Vimal	2-1 (7025)	9
Ellinger, Frank	12-4 (7031)	161
ElShater, Ahmed	4-2 (7189)	39
Emery, Stephane	2-2 (7152)	13

F

Feng, Xiaoyu	5-3 (7074)	61
Francese, Pier-Andrea	17-2 (7138)	239
Frigerio, Nicolas	16-2 (7182)	219
Fuh, Jonathan	4-1 (7176)	37
Fukuoka, Kazuki	11-2 (7041)	129

G

Garvik, Harald	14-2 (7055)	193
Ge, Xinyuan	4-3 (7082)	43
Ghovanloo, Maysam	18-1 (7158)	245
Gi, Hyungmin	18-5 (7101)	259
Gielen, Georges	10-2 (7085)	109
Ginsburg, Brian	2-1 (7025)	9
Giterman, Robert	16-2 (7182)	219
Goto, Kenji	2-3 (7035)	17
Gu, Peng	6-3 (7131)	77
Guo, Yanshu	20-3 (7126)	285

H

Han, Byungki	7-2 (7104)	87
	12-5 (7070)	165
Hayashi, Yoshihiro	21-4 (7010)	305
He, Chen-Yu	16-4 (7142)	227
Hella, Mona	21-2 (7053)	297
Hirose, Tetsuya	19-2 (7080)	267
Hong, Je-Min	16-1 (7160)	217
Hong, Jiyoon	14-3 (7128)	195
Hsieh, Chih-Cheng	3-3 (7172)	33
	16-1 (7160)	217
Hsieh, Wayne	5-1 (7062)	53
Hsu, Steven	11-4 (7023)	137
Hsu, Tzu-Hsiang	3-3 (7172)	33
Hsu, Yu-Pin	21-2 (7053)	297
Huang, Heng	12-3 (7199)	157
Huang, Meng-Yuan	19-4 (7159)	275
Huang, Peng-Chang	13-4 (7061)	185
Huang, Wei-Hsing	16-1 (7160)	217
Huang, Yuxuan	13-1 (7166)	173
Hung, Yu-Ting	7-3 (7153)	91
Hwang, Seokha	15-2 (7076)	205

I

Ichihashi, Tadayuki	19-1 (7181)	263
Igarashi, Yutaka	21-4 (7010)	305
Ikeda, Makoto	19-1 (7181)	263
Ishii, Yuichiro	2-3 (7035)	17
Ishikawa, Jiro	2-3 (7035)	17
Itoh, Kiyoo	16-4 (7142)	227
Izumi, Shintaro	19-2 (7080)	267

J

Jain, Saurabh	11-1 (7116)	125
Jang, Jaeeun	13-2 (7193)	177
Jayaraj, Akshay	14-4 (7069)	197
Jen, Chih-Wei	19-4 (7159)	275
Jeong, Deog-Kyoon	17-4 (7024)	243

Jeong, Min-Gyu	4-4 (7012)	45
Jiang, Bingwei	20-1 (7171)	279
Jiang, Hanjun	20-3 (7126)	285
Jiang, Hanjun	21-5 (7118)	309
	20-4 (7100)	287
Jiang, Hui	10-1 (7064)	107
Jiang, Sijia	12-2 (7020)	153
Jin, Dong-Hwan	14-1 (7133)	189
Jin, KyoWon	1-2	5
Jin, Xuefan	7-1 (7092)	83
Joe, Junseo	15-2 (7076)	205
Joo, Hye-Yoon	17-4 (7024)	243
Jou, Shyh-Jye	19-4 (7159)	275
	16-4 (7142)	227
Ju, Chi-Cheng	5-1 (7062)	53
Jung, Jaehong	7-2 (7104)	87
	12-5 (7070)	165
Jung, Sangdon	7-2 (7104)	87
	12-5 (7070)	165
Jung, Seungchul	18-5 (7101)	259

K

Kadetotad, Deepak	11-6 (7205)	145
Kajiyama, Shinya	21-4 (7010)	305
Kang, Dong-Seok	7-1 (7092)	83
Kang, Jin-Gyu	4-4 (7012)	45
Kang, Myeonggu	5-2 (7033)	57
Kar, Monodeep	11-4 (7023)	137
Kargaran, Ehsan	12-6 (7109)	169
Karunarathne, Manupa	11-3 (7163)	133
Katsube, Yusaku	21-4 (7010)	305
Kaul, Himanshu	11-4 (7023)	137
Kawaguchi, Hiroshi	19-2 (7080)	267
Khairi, Ahmad	17-1 (7125)	235
Khan, Muhammad Bilawal	13-3 (7154)	181
Khan, Qadeer Ahmad	4-2 (7189)	39
Ki, Wing-Hung	4-3 (7082)	43
Kikuchi, Katsuya	3-1 (7037)	25
Kim, Bongjin	3-4 (7090)	35
Kim, Dokyu	15-1 (7206)	201

Kim, Eunhwan	16-3 (7016)	223
Kim, Gain	17-2 (7138)	239
Kim, Hyeonuk	5-2 (7033)	57
Kim, Hyunjoon	3-4 (7090)	35
Kim, Jae-Joon	16-3 (7016)	223
Kim, Ji-Hoon	15-1 (7206)	201
Kim, Jintae	14-3 (7128)	195
Kim, Jong-Pal	14-1 (7133)	189
Kim, Lee-Sup	5-2 (7033)	57
Kim, Sang Joon	11-6 (7205)	145
	18-5 (7101)	259
Kim, Seungjin	12-5 (7070)	165
Kim, Sung Yeon	15-1 (7206)	201
Kim, Taesu	16-3 (7016)	223
Kim, Ye-Dam	14-1 (7133)	189
Kitaji, Yuko	11-2 (7041)	129
Ko, Youngjun	7-1 (7092)	83
Koo, Jongeun	16-3 (7016)	223
Kossel, Marcel	17-2 (7138)	239
Krishnamurthy, Ram	11-4 (7023)	137
Krupnik, Yoel	17-1 (7125)	235
Kuan, Yen-Cheng	19-4 (7159)	275
Kulak, Ross	2-1 (7025)	9
Kulkarni, Aditi	11-3 (7163)	133
Kull, Lukas	17-2 (7138)	239
Kumar, Raghavan	11-4 (7023)	137
Kumar, Rushil K.	10-1 (7064)	107
Kumari, Neha	18-4 (7140)	255
Kuo, Chen-Yi	18-3 (7112)	251
Kuo, Tai-Haur	13-4 (7061)	185
Kurui, Yoshihiko	19-2 (7080)	267
Kwon, Kee-Won	7-1 (7092)	83

L

Leblebici, Yusuf	16-2 (7182)	219
	17-2 (7138)	239
Lee, Calvin Yoji	4-2 (7189)	39
Lee, Hyunhoon	15-2 (7076)	205
Lee, Jaehoon	12-5 (7070)	165
Lee, Jaehyuk	13-2 (7193)	177

Lee, Jihee	13-2 (7193)	177
Lee, Jongwoo	7-2 (7104)	87
	12-5 (7070)	165
Lee, Sunggu	15-2 (7076)	205
Lee, Tsai-Hua	19-4 (7159)	275
Lee, Wei-Che	19-4 (7159)	275
Lee, Yoonmyung	13-3 (7154)	181
	18-5 (7101)	259
Lee, Youngjoo	15-2 (7076)	205
Leuenberger, Spencer	4-2 (7189)	39
Li, An'an	12-2 (7020)	153
Li, Fei	11-5 (7113)	141
Li, Guolin	12-3 (7199)	157
Li, Heyi	20-4 (7100)	287
Li, Mingze	6-2 (7054)	73
Li, Qiang	10-5 (7134)	121
Li, Sih-Han	16-1 (7160)	217
Li, Xiaofei	18-1 (7158)	245
Li, Xueqing	5-3 (7074)	61
Liang, Junhao	21-5 (7118)	309
Liao, Huailin	20-4 (7100)	287
Liao, Yu-Te	18-3 (7112)	251
Lim, Won-Mook	14-1 (7133)	189
Lim, Yong	12-5 (7070)	165
Lin, Charlie	2-4 (7094)	21
Lin, Chien-Hung	16-4 (7142)	227
Lin, Chun-Yu	7-3 (7153)	91
Lin, Jia-Ying	5-1 (7062)	53
Lin, Longyang	11-1 (7116)	125
Lin, Shian-Ru	4-5 (7087)	49
	18-4 (7140)	255
Lin, Tsung-Hsien	7-3 (7153)	91
Lin, Ying-His	4-5 (7087)	49
	18-4 (7140)	255
Lin, Zixiao	10-5 (7134)	121
Liu, Chun-Yi	19-4 (7159)	275
Liu, Hung-Chih	19-4 (7159)	275
Liu, Junhua	20-4 (7100)	287
Liu, Ming	3-2 (7135)	29
Liu, Ren-Shuo	3-3 (7172)	33
	16-1 (7160)	217
Liu, Shen-Iuan	7-4 (7022)	95

Liu, Tsu-Ming	5-1 (7062)	53
Liu, Yongpan	5-3 (7074)	61
	13-1 (7166)	173
Liu, Zemin	21-2 (7053)	297
Lo, Chilun	12-5 (7070)	165
Lo, Chung-Chuan	3-3 (7172)	33
Lopez, Henry	19-4 (7159)	275
Lu, Cheng-Hsun	5-4 (7179)	65
Lu, Chun-An	18-3 (7112)	251
Lu, Yan	18-1 (7158)	245
Luong, Howard C.	20-1 (7171)	279
Luu, Danny	17-2 (7138)	239
Lv, Hangbing	3-2 (7135)	29

M

Mak, Pui-In	20-2 (7093)	283
Makinwa, Kofi A. A.	10-1 (7064)	107
Maloberti, Franco	10-4 (7005)	117
Maloberti, Franco	21-5 (7118)	309
Mani, Aarthy	11-5 (7113)	141
Manstretta, Danilo	12-6 (7109)	169
Mao, Fangyu	18-1 (7158)	245
Mao, Jingna	13-1 (7166)	173
Martins, R. P.	10-4 (7005)	117
	18-1 (7158)	245
	20-2 (7093)	283
	21-5 (7118)	309
Mason, Ralph D.	6-2 (7054)	73
Mathew, Sanu	11-4 (7023)	137
Matsuo, Yoshihiko	2-2 (7152)	13
Mavrogordatos, Themis	2-2 (7152)	13
Meng, Xiangyu	6-4 (7111)	79
Menolfi, Christian	17-2 (7138)	239
Miki, Takuji	3-1 (7037)	25
Mitra, Tulika	11-3 (7163)	133
Miura, Noriyuki	3-1 (7037)	25
Miura, Tomohiro	2-3 (7035)	17
Miyahara, Masaya	19-2 (7080)	267
Mizoguchi, Kyoji	16-5 (7114)	231
Mizushina, Keita	16-5 (7114)	231
Mochizuki, Yasunori	9-1	99

Moon, Seungsik	15-2 (7076)	205
Moon, Un-Ku	4-2 (7189)	39
Morf, Thomas	17-2 (7138)	239
Muljono, Harry	2-4 (7094)	21

N

Nagata, Shunya	2-3 (7035)	17
Nagata, Makoto	3-1 (7037)	25
Nagateja, T.	4-5 (7087)	49
Nakagawa, Tatsuo	21-4 (7010)	305
Nakamura, Daisuke	2-3 (7035)	17
Nakamura, Toshiki	16-5 (7114)	231
Nakamura, Yohei	21-4 (7010)	305
Narinx, Jonathan	16-2 (7182)	219
Nayak, Neeraj	2-1 (7025)	9
Ng, Kian Ann	21-1 (7071)	295
Nikolic, Borivoje	19-3 (7066)	271
Nishimoto, Takuma	21-4 (7010)	305

O

Oh, Seunghyun	7-2 (7104)	87
	12-5 (7070)	165
Okada, Kenichi	19-2 (7080)	267
Okidono, Takaaki	3-1 (7037)	25
Ouchi, Yukari	2-3 (7035)	17
Ozkaya, Ilter	17-2 (7138)	239

P

Paramasivam, Vishnu	11-5 (7113)	141
Parhi, Keshab K.	19-3 (7066)	271
Park, Chang-Un	14-1 (7133)	189
Park, Jeong Hoan	21-1 (7071)	295
	21-3 (7063)	301
Park, Jeongpyo	4-4 (7012)	45
Park, Jongmin	15-2 (7076)	205
Park, Junyoung	18-5 (7101)	259
Parthasarathy, Harikrishna	2-1 (7025)	9
Pavan, Shanthi	10-6 (7086)	123

Peh, Li-Shiuan	11-3 (7163)	133
Pelecijn, Elly De	18-2 (7052)	247
Peng, Kathy	2-4 (7094)	21
Polarouthu, Sudhir	2-1 (7025)	9
Prasad, Anjan	2-1 (7025)	9
Pretl, Harald	10-3 (7089)	113

Q

Qian, Yuan Cheng	7-4 (7022	95

R

Ram, Shankar	2-1 (7025)	9
Rentala, Vijay	2-1 (7025)	9
Reynaert, Patrick	12-1 (7119)	149
Rhee, Woogeun	20-5 (7177)	291
Rong, Bor-Doou	16-4 (7142)	227
Ruffieux, David	2-2 (7152)	13
Ryu, Seung-Tak	14-1 (7133)	189
Ryu, Sungju	16-3 (7016)	223

S

Sacco, Elisa	10-2 (7085)	109
Sachdev-Singh, Rittu	2-1 (7025)	9
Sahu, Debapriya	2-1 (7025)	9
Saif, Hassan	13-3 (7154)	181
Sanhedrai, Yosi	17-1 (7125)	235
Sanyal, Arindam	14-4 (7069)	197
Sato, Katsuyuki	16-4 (7142)	227
Seidel, Andres	12-4 (7031)	161
Seo, Jae-sun	11-6 (7205)	145
Seo, Min-Jae	14-1 (7133)	189
Sharma, Bhupendra	2-1 (7025)	9
Shen, Zhengkun	20-4 (7100)	287
Sheu, Shyh-Shyuan	16-1 (7160)	217
Shifman, Yizhak	17-1 (7125)	235
Shih, Shawn	5-1 (7062)	53
Shimamoto, Haruo	3-1 (7037)	25
Shin, Myeongcheol	12-5 (7070)	165

Si, Xin	16-1 (7160)	217
Sim, Jaehyeong	5-2 (7033)	57
Sin, Sai-Weng	10-4 (7005)	117
	21-5 (7118)	309
Someya, Teruki	19-2 (7080)	267
Song, Chunrong	2-4 (7094)	21
Song, Haixin	20-5 (7177)	291
Sonoda, Hiroki	3-1 (7037)	25
Standaert, Alexander	12-1 (7119)	149
Stärke, Paul	12-4 (7031)	161
Steyaert, Michiel	18-2 (7052)	247
Strange, Jon	12-6 (7109)	169
Su, Jian-Wei	16-1 (7160)	217
Subburaj, Karthik	2-1 (7025)	9
Sun, Linda	2-4 (7094)	21
Sun, Shiyan	12-2 (7020)	153
Sun, Wenyu	13-1 (7166)	173
Sung, Barosaim	12-5 (7070)	165
Suresh, Vikram	11-4 (7023)	137
Suzuki, Shun	16-5 (7114)	231

T

Takeuchi, Ken	16-5 (7114)	231
Tang, Kea-Tiong	3-3 (7172)	33
	16-1 (7160)	217
Tang, Kuei-Hua	16-4 (7142)	227
Tang, Tao	21-1 (7071)	295
	21-3 (7063)	301
Toifl, Thomas	17-2 (7138)	239
Tomizawa, Hideyuki	19-2 (7080)	267
Tsai, Chang-Hung	5-1 (7062)	53
Tsai, Pei-Yun	19-4 (7159)	275
Tsai, Tsung-Yen	4-5 (7087)	49
	18-4 (7140)	255
Tsujihashi, Yoshiki	2-3 (7035)	17
Tu, Yung-Ning	16-1 (7160)	217

U

U, Seng-Pan	10-4 (7005)	117
	21-5 (7118)	309
Uemura, Toshifumi	11-2 (7041)	129

V

Venkatachala, Praveen Kumar	4-2 (7189)	39
Vergauwen, Johan	10-2 (7085)	109
Virobnik, Udi	17-1 (7125)	235

W

Wang, Biao	10-4 (7005)	117
Wang, Bo	11-3 (7163)	133
Wang, Can	17-3 (7045)	241
Wang, Jing-Hong	16-1 (7160)	217
Wang, Jingyu	5-3 (7074)	61
Wang, Li	17-3 (7045)	241
Wang, Miaorong	15-3 (7141)	209
Wang, Mingyu	3-2 (7135)	29
Wang, Wei	12-2 (7020)	153
Wang, Yanzhi	5-3 (7074)	61
Wang, Zhibo	5-3 (7074)	61
Wang, Zhihua	12-3 (7199)	157
	20-3 (7126)	285
	20-5 (7177)	291
Watanabe, Hikaru	16-5 (7114)	231
Watanabe, Naoya	3-1 (7037)	25
Wei, Jinsong	15-4 (7122)	213
Wei, Shaojun	15-4 (7122)	213
Wei, Wei-Chen	3-3 (7172)	33
	16-1 (7160)	217
Wen, Tai-Hsing	3-3 (7172)	33
Wong, Ming Ming	11-5 (7113)	141
Wu, Chang-Ting	19-4 (7159)	275
Wu, Hui	15-4 (7122)	213
Wu, Yi-Chung	5-4 (7179)	65
Wulff, Carsten	14-2 (7055)	193

X

Xiao, Bohui	4-2 (7189)	39
Xu, Chenyu	6-1 (7178)	69
Xu, Yang	4-2 (7189)	39
Xue, Xiaoyong	3-2 (7135)	29

Y

Yagoub, M.C.E.	6-2 (7054)	73
Yamawaki, Taiz	21-4 (7010)	305
Yang, Chia-Hsiang	5-4 (7179)	65
Yang, Fu-Bin	4-1 (7176)	37
Yang, Huazhong	5-3 (7074)	61
	13-1 (7166)	173
Yang, Jianguo	3-2 (7135)	29
Yang, Peilin	20-3 (7126)	285
Yang, Shang-Hsien	18-4 (7140)	255
Yang, Yixiong	5-3 (7074)	61
	13-1 (7166)	173
Yang, Zunsong	20-2 (7093)	283
Yano, Yuji	19-2 (7080)	267
Yazaki, Toru	21-4 (7010)	305
Yeon, Pyungwoo	18-1 (7158)	245
Yin, Shihui	11-6 (7205)	145
Yokoyama, Yoshisato	2-3 (7035)	17
Yoo, Changsik	4-4 (7012)	45
Yoo, Hoi-Jun	13-2 (7193)	177
Yoo, Jerald	21-1 (7071)	295
	21-3 (7063)	301
Yoo, Seunghyun	16-3 (7016)	223
Yoon, Jae Sik	14-3 (7128)	195
Yoshida, Seiya	19-2 (7080)	267
Yoshimoto, Masahiko	19-2 (7080)	267
Ytterdal, Trond	14-2 (7055)	193
Yuan, Zhe	5-3 (7074)	61
Yue, C. Patrick	17-3 (7045)	241
Yue, Jinshan	5-3 (7074)	61
Yue, Patrick	6-4 (7111)	79
Yueksel, Hazar	17-2 (7138)	239

Z

Zahnd, Loic	2-2 (7152)	13
Zeng, Xiaoyang	3-2 (7135)	29
	19-3 (7066)	271
Zha, Yingyun	2-2 (7152)	13
Zhai, Pengfei	10-5 (7134)	121
Zhang, Fan	10-5 (7134)	121
Zhang, Jiajun	6-5 (7132)	81
Zhang, Jiaqi	6-4 (7111)	79
Zhang, Jilin	15-4 (7122)	213
Zhang, Lian	21-1 (7071)	295
	21-3 (7063)	301
Zhang, Milin	12-3 (7199)	157
Zhang, Qichen	19-3 (7066)	271
Zhang, Yuejun	3-2 (7135)	29
Zhang, Zhao	17-3 (7045)	241
Zhang, Zherui	20-4 (7100)	287
Zhang, Zhixiao	16-1 (7160)	217
Zhao, Dixian	6-1 (7178)	69
	6-3 (7131)	77
	6-5 (7132)	81
Zhao, Jian	13-1 (7166)	173
Zheng, Zhenpeng	6-4 (7111)	79
Zhou, Xiong	10-5 (7134)	121
Zhu, Guang	17-3 (7045)	241
Zhu, Yanjie	2-4 (7094)	21
Zhu, Yiming	1-1	1
Zhu, Zheng	10-5 (7134)	121

Panel Discussion

Are Analog and Mixed Mode Circuits the future solution of AI SoCs?

Date	November 5, 2019 (Tuesday)
Time	3:40 PM – 5:20 PM
Room	7402-7403 & 7502-7503
Organizer	Hoi-Jun Yoo, Korea Advanced Institute of Science & Technology, Korea Woogeun Rhee, Tsinghua University, China
Co-organizer	Jerald Yoo, National University of Singapore, Singapore
Moderator	Noriyuki Miura, Kobe University, Japan
Panelists	Mixed Mode AI Trends Overview: Jason Lee (UNIST) ADC and AI: Vanessa Chen (Carnegie Mellon University) Mixed Mode MAC: SeungTak Ryu (KAIST) Memory Based Analog AI: ShouYi Yin (Tsinghua University) Non-Volatile Memory and Analog AI: Yongpan Liu (Tsinghua University) Spike Neural Network: Samuel Tang (National Tsinghua University) Neuromorphic: Hiromitsu Awano (Osaka University)

Abstract:

DNN looks mature already, and Neuromorphic and Spike Neural Network are regarded as the next research area. The current AI SoCs are based on Digital DNN circuits with the limitations from the Von Neumann architecture. Analog/Mixed Mode circuits may implement Neuromophic SoCs and Spike Neural Network overcoming the Von Neumann bottleneck. Is it true? and if it is, how we can make it?

- Just ADC/DAC are enough to process AI operations
- Are Mixed Mode MACs better than Digital MACs?
- DRAM and SRAM can be the Processing Elements.
- Non-Volatile Memory is essential to implement Neuromorphic/SNN
- What kind of circuits do you need for Spike Neural Network?
- Neuromorphic AI and Mixed Mode Circuits.

978-1-7281-5107-6/19 $31.00 © 2019 IEEE

Biography

**Organizer
Hoi-Jun Yoo**

Hoi-Jun Yoo received the bachelor's degree from the Electronic Department, Seoul National University, Seoul, South Korea, in 1983, and the M.S. and Ph.D. degrees in electrical engineering from the Korea Advanced Institute of Science and Technology (KAIST), Daejeon, South Korea, in 1985 and 1988, respectively.

Since 1998, he has been the Faculty Member with the Department of Electrical Engineering, KAIST. From 2001 to 2005, he was the Director of the Korean System Integration and IP Authoring Research Center, Seoul. In 2007, he founded the System Design Innovation and Application Research Center, KAIST. Since 2010, he has been the General Chair of the Korean Institute of Next Generation Computing, Seoul. He is currently a Full Professor with KAIST. He has served as a member for the Executive Committee of ISSCC, the Symposium on VLSI, A-SSCC, the Far East Chair for the ISSCC from 2011 to 2012, the Technology Direction Sub-Committee Chair for the ISSCC in 2013, the TPC Vice Chair for the ISSCC in 2014, and the TPC Chair for the ISSCC in 2015. He has served as the TPC Chair for the A-SSCC 2008 and ISWC 2010. He has served as the IEEE Distinguished Lecturer from 2010 to 2011.

**Organizer
Woogeun Rhee**

Woogeun Rhee received the B.S. degree in electronics engineering from Seoul National University, Seoul, Korea, in 1991, the M.S. degree in electrical engineering from the University of California, Los Angeles, in 1993, and the Ph.D. degree in electrical and computer engineering from the University of Illinois, Urbana-Champaign, in 2001.

From 1997 to 2001, he was with Conexant Systems, Newport Beach, CA, where he was a Principal Engineer and developed low-power, low-cost fractional-N synthesizers. From 2001 to 2006, he was with IBM Thomas J. Watson Research Center, Yorktown Heights, NY and worked on clocking area for high-speed I/O serial links, including low-jitter phase-locked loops, clock-and-data recovery circuits, and on-chip testability circuits. In August 2006, he joined the faculty as an Associate Professor at the Institute of Microelectronics, Tsinghua University, Beijing, China, and became a Professor in December 2011. His current research interests include short-range low-power radios for next generation wireless systems and clock/frequency generation circuits for wireline and wireless communications. He holds 23 U.S. patents.

Dr. Rhee was an IEEE Distinguished Lecturer of the Solid-State Circuits Society (2016-2017) and currently serves as an Associate Editor for IEEE OPEN JOURNAL OF SOLID-STATE CIRCUITS. He has been an Associate Editor for IEEE JOURNAL OF SOLID-STATE CIRCUITS (2012-2018), IEEE TRANSACTIONS ON CIRCUITS AND SYSTEMS PART-II: EXPRESS BRIEFS (2008-2009) and a Guest Editor for IEEE JOURNAL OF SOLID-STATE CIRCUITS Special Issue in November 2012 and November 2013. He has served as a member of several IEEE conferences, including ISSCC (2012-2016), CICC, and A-SSCC.

**Co-organizer
& Moderator
Noriyuki Miura**

Noriyuki Miura received the B.S., M.S., and Ph.D. degrees in electrical engineering all from Keio University, Yokohama, Japan. He is currently an Associate Professor with Kobe University, Kobe, Japan, and concurrently a JST PRESTO researcher, working on hardware security and next-generation heterogeneous computing system. Dr. Miura is currently serving as a TPC Member for A-SSCC and Symposium on VLSI Circuits. He received the Top ISSCC Paper Contributors 2004-2013 and the IACR CHES Best Paper Award in 2014.

**Co-organizer
Jerald Yoo**

Jerald Yoo received the B.S., M.S., and Ph.D. degrees in Electrical Engineering from the Korea Advanced Institute of Science and Technology (KAIST), Daejeon, Korea, in 2002, 2007, and 2010, respectively.

From 2010 to 2016, he was with the Department of Electrical Engineering and Computer Science, Masdar Institute, Abu Dhabi, United Arab Emirates, as an Associate Professor. From 2010 to 2011, he was also with the Microsystems Technology Laboratories (MTL), Massachusetts Institute of Technology, MA, USA as a visiting scholar. Since 2017, he has been with the Department of Electrical and Computer Engineering, National University of Singapore, where he is currently an Associate Professor. He has pioneered researches on low-energy Body-Area Network (BAN) transceivers, wearable healthcare using embedded machine learning and the planar-fashionable circuit board. He was a recipient of several awards including the ISCAS2015 Best Paper Award (BioCAS Track), the Masdar Institute Best Research Award in 2015 and the A-SSCC Outstanding Design Award (2005). He is the Vice Chair of IEEE Solid-State Circuits Society (SSCS) Singapore Chapter. He served as the IEEE Solid-State Circuits Society' s Distinguished Lecturer (2017-2018), and is currently serving as the IEEE Circuits and Systems Society' s Distinguished Lecturer (2019-2020).

**Panelist
Kyuho Jason Lee**

Kyuho Jason Lee received the B.S. degree from the School of Electrical Engineering at Korea Advanced Institute of Science and Technology (KAIST), Daejeon, South Korea, in 2012, and the M.S. and Ph.D. degrees in the School of Electrical Engineering at KAIST in 2014 and 2017, respectively. His Ph.D. dissertation concerned Deep Learning based intelligent vision algorithm as well as it dedicated circuits and systems for Advanced Driver Assistance System.

He is an Assistant Professor of School of Electrical and Computer Engineering at Ulsan National Institute of Science and Technology (UNIST) since 2018. Before joining UNIST he was with Samsung Research America and uxfactory. His research interests include analog/digital mixed-mode neuromorphic SoC design, object matching processor and hardware-oriented algorithm, Machine Learning based network-on-chip (NoC) architecture, Deep Learning ASIC, and Artificial Intelligence System-on-Chips for mobile devices and automotive system.

Dr. Lee is currently a Technical Program Committee Member of the Asian Solid-State Circuits Conference (A-SSCC) and Design, Automation and Test in Europe (DATE).

**Panelist
Vanessa Chen**

Vanessa Chen is currently an Assistant Professor of Electrical and Computer Engineering at Carnegie Mellon University. She received her Ph.D. degree in Electrical and Computer Engineering from Carnegie Mellon University in 2013. Before joining Carnegie Mellon University, she was an Assistant Professor at The Ohio State University. Prior to that, she was with Qualcomm working on low-power data-acquisition systems for mobile devices. From 2010 to 2013, at Carnegie Mellon, she focused her research on self-healing systems and high-speed ADC designs, and held a research internship position at IBM T. J. Watson Research Center, Yorktown Heights, in 2012. Her work has appeared in top academic conferences and has been featured as highlights in the IEEE International Solid-State Circuits Conference (ISSCC). She was the recipient of the NSF CAREER Award in 2019, the Analog Devices Outstanding Student Designer Award in 2013 and the IBM Ph.D. Fellowship in 2012. Her research interests focuses on energy-efficient data conversion interfaces for machine learning and hardware security.

978-1-7281-5107-6/19 $31.00 © 2019 IEEE

**Panelist
Seung-Tak Ryu**

Seung-Tak Ryu (M'06–SM'13) received the M.S. and Ph.D. degrees from Korea Advanced Institute of Science and Technology (KAIST), Daejeon, Korea, in 1999 and 2004, respectively. He was with Samsung Electronics in Kiheung, Korea from 2004 to 2007. From 2007 to 2009, he was with the Information and Communications University, Daejeon, Korea, as an Assistant Professor. He has been with the Department of Electrical Engineering, KAIST, Daejeon, Korea, since 2009, where he is currently an Associate Professor. His research interests include analog and mixed-signal IC design with an emphasis on data converters.

Prof. Ryu served as a Technical Program Committee (TPC) member of the ISSCC, and he is now serving on the A-SSCC. He is now serving as an Associate Editor of the IEEE Solid-State Circuits Letters.

**Panelist
Shouyi Yin**

Prof. Shouyi Yin received the B.S., M.S., and Ph.D. degrees in electronic engineering from Tsinghua University, Beijing, China, in 2000, 2002, and 2005, respectively. He has worked with Imperial College, London, U.K., as a Research Associate. He is a full professor and vice director of Institute of Microelectronics in Tsinghua University. His research interests include reconfigurable computing, domain-specific reconfigurable architecture design and AI chips. He has published more than 100 journal papers and more than 80 conference papers. He has received Second Prize of China's State Technological Innovation Award (2015), China's Patent Golden Award (2015), First Prize of Technological Innovation Award of Ministry of Education, China (2014), Second Prize of Science and Technology Advancement Award of Jiangxi Province, China (2014) and Best Paper Award in China Communications IC Technology and Application Conference (2011). Prof. Shouyi Yin is the Secretary-General of EDA Chapter in Chinese Institute of Electronics. He is also the technical committee member of Asia Pacific Signal and Information Processing Association. He has been served as program committee member and organizer in the tops VLSI and EDA conferences such as A-SSCC, DAC, ICCAD and ASPDAC. He is the associate editor of ACM Transactions on Reconfigurable Technology and Systems, the associate editor of Integration, the VLSI journal and the editorial board member of Journal of Low Power Electronics.

**Panelist
Yongpan Liu**

Dr. Yongpan Liu received his B.S., M.S. and Ph.D. degrees from Electronic Engineering Department, Tsinghua University. He was a visiting scholar at Pennsylvania State University and City University of Hong Kong. He is now an associate professor in Dept. of Electronic Engineering Tsinghua University. His main research interests include energy efficient circuits and systems for artificial intelligent, emerging memory devices and IoT applications. He has published over 200 peer-reviewed conference and journal papers and developed several fast sleep/wakeup nonvolatile processors using emerging memory and artificial intelligent accelerators using algorithm-architecture co-optimization. His work has received Under 40 Young Innovators Award DAC 2017, Micro Top Pick 2016, Best Paper Award in ASPDAC2017, HPCA 2015, Design Contest Awards of ISLPED 2012, 2013 and 2019. He served as general chair for AWSSS 2016, IWCR 2018 and technical program chair NVMSA 2019 and a program committee member for DAC, DATE, ASP-DAC, ISLPED, ICCD, A-SSCC. He is an Associate Editor for IEEE Transactions on CAD, CAS-II and IET Cyber-Physical Systems. He is an IEEE Senior Member.

**Panelist
Kea-Tiong (Samuel) Tang**

Dr. Kea-Tiong (Samuel) Tang received the B.S. degree in electrical engineering from National Taiwan University, Taipei, Taiwan in 1996, and received the M.S. and Ph.D. degrees in electrical engineering from California Institute of Technology, Pasadena, CA, USA, in 1998 and 2001, respectively.

During 2001–2006, Dr. Tang was a Senior Electrical Engineer with Second Sight Medical Products, Inc., Sylmar, CA, USA. He designed mixed signal ASIC for the Argus® II Retinal Prosthesis System, which became the first FDA approved device for retinal prosthesis. Since 2006, he joined the Electrical Engineering Faculty at National Tsing Hua University, Hsinchu, Taiwan, and is currently a full Professor. His research interests include bio-inspired learning chip, miniature electronic system, and biomedical implantable prosthetic device. He has actively collaborated with researchers in Nanoengineering and Microsystems, Chemistry, Computer Science, Electrical Engineering, Life Science, and Medical doctors, and has published more than 140 peer-reviewed journal and conference papers in these research areas. He has led the largest electronic nose team in Taiwan to develop a system that can early detect and rapid diagnose ventilator-associated pneumonia. In 2017, Dr. Tang's team has joined the Dynamical Biomarkers Group to win for 2nd place of the $10 million Qualcomm Tricorder XPRIZE, the global competition to revolutionize digital healthcare. He is a recipient of Outstanding Young Scholar Award, Wu Ta-You Memorial Award, National Innovation Award, and Outstanding Electrical Engineering Professor Award. Dr. Tang is a senior member of IEEE. He is member of IEEE solid state circuit society (SSCS), circuits and systems society (CAS), electron device society (EDS), and Engineering in medicine and biology society (EMBS). He is also TC member of IEEE Biomedical and Life Science Circuits Systems Technical Committee (BioCAS), currently serves as TC Chair. He is an Associate Editor for IEEE Transactions on Biomedical Circuits and Systems (TBioCAS) and Guest Editor for IEEE Journal on Emerging and Selected Topics in Circuits and Systems (JETCAS). He is TPC member of ISCAS, BioCAS, and IEDM. He was IEEE CAS Chapter Chair of Taipei Section (2017-2018). He is now Vice President of IEEE Taipei Section.

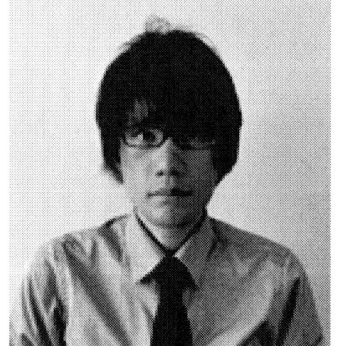

**Panelist
Hiromitsu Awano**

Hiromitsu Awano received his B.E. degree in Informatics and M.Sc. and Ph.D. degrees in Communications and Computer Engineering from Kyoto University in 2010, 2012, and 2016, respectively. He was with Hitachi, Ltd., Tokyo, Japan in 2016, and with the VLSI Design and Education Center, The University of Tokyo, Japan, from 2017 to 2018. In 2019, he joined the Graduate School of Information Science and Technology, Osaka University, Osaka, Japan, where he is currently an associate professor. His research interests include CAD for VLSI design and hardware accelerator for machine learning. He is a member of IEEE, IEICE, and IPSJ.

Committees

Steering Committee

Chair

Tadahiro Kuroda, Keio University, Japan

Members

Tzi-Dar Chiueh, National Taiwan University, Taiwan
Nicky Lu Etron, Technology, Inc., Taiwan
Makoto Ikeda, The University of Tokyo, Japan
Yoshio Masubuchi, Toshiba Memory Corp., Japan
Deog-Kyoon Jeong, Jeong Seoul National University, Korea
Stefan Rusu, TSMC North America, USA
Yong-Hyun Jun, Samsung Electronics, Korea
Toru Shimizu, Toyo University, Japan
Lawrence Loh, MediaTek, Taiwan
Jack Sun, TSMC, Taiwan
Yong Ping Xu, National University of Singapore, Singapore
Hoi-Jun Yoo, Korea Advanced Institute of Science and Technology, Korea
Zhihua Wang, Tsinghua University, China

Advisor

Takayasu Sakurai, The University of Tokyo, Japan
Chorng-kuang Wang, National Taiwan University, Taiwan

Liaison

Anantha Chandrakasan, Massachusetts Institute of Technology, USA
Bram Nauta, University of Twente, Netherlands

Organizing Committee

Conference Chair

Rui Martins, University of Macau, Macau, China

Organization Committee Chair

Seng-Pan (Ben) U, University of Macau & Synopsys Macau, Macau, China

Organization Committee Vice Co-Chairs

Leibo Liu, Tsinghua University, China
Howard Luong, Hong Kong University of Science and Technology, Hong Kong, China

Secretary

Chi-Hang Chan, University of Macau, Macau, China
Man-Kay Law, University of Macau, Macau, China

Treasurer

Sai-Weng Sin, University of Macau, Macau, China
Yan Lu, University of Macau, Macau, China

Publicity

Patrick Yue, Hong Kong University of Science and Technology, Hong Kong, China
Yan Zhu, University of Macau, Macau, China

Publication

Jun Yin, University of Macau, Macau, China
Sio-Hang Pun, University of Macau, Macau, China

978-1-7281-5107-6/19 $31.00 © 2019 IEEE

| Local Arrangement | Chi-Seng Lam, University of Macau, Macau, China |
| | Ka Meng Lei, University of Macau, Macau, China |

| Exhibition | Yong Chen, University of Macau, Macau, China |
| | Yanwei Jia, University of Macau, Macau, China |

Technical Support	Fan Ng, University of Macau, Macau, China
	Un-Pang Lei, University of Macau, Macau, China
	Chi-Wai Tang, University of Macau, Macau, China

| Professional Conference Organizer | Karen Neng, Moxlink Technology Group Ltd., Macau, China |

Technical Program Committee Charis

TPC Chair	Mototsugu Hamada, Keio University, Japan
TPC Co-Chair	Robert Chen-Hao Chang, National Chung Hsing University, Taiwan
TPC Vice-Chair	Jun Deguchi, Toshiba Memory, Japan
TPC Vice-Co-Chair	Po-Hung Chen, National Chiao Tung University, Taiwan

ACS Sub-Committee Chair	Po-Chiun Huang, National Tsing Hua University, Taiwan
DC Sub-Committee Chair	Kazuko Nishimura, Panasonic Corporation, Japan
DCS Sub-Committee Chair	Jun Zhou, University of Electronic Science and Technology of China, China
ETA Sub-Committee Chair	Woogeun Rhee, Tsinghua University, China
MEM Sub-Committee Chair	Junghwan Choi, Samsung Electronics, Korea
RF Sub-Committee Chair	Minoru Fujishima, Hiroshima University, Japan
SOC Sub-Committee Chair	Chi-Cheng Ju, MediaTek Inc, Taiwan
WLN Sub-Committee Chair	Chulwoo Kim, Korea University, Korea
FPGA	Shigeki Tomishima, Intel, USA

978-1-7281-5107-6/19 $31.00 © 2019 IEEE

Student Design Contest Co-Chair	Baoyong Chi, Tsinghua University, China
Student Design Contest Co-Chair	Jung-Hoon Chun, Sungkyunkwan University, Korea
Industry Program Chair	Stefan Rusu, TSMC North America, USA
Educational Program Co-Chair	Hoi-Jun Yoo, Korea Advanced Institute of Science and Technology, Korea
Educational Program Co-Chair	Woogeun Rhee, Tsinghua University, China
Invited Program Chair	Hoi-Jun Yoo, Korea Advanced Institute of Science and Technology, Korea

Analog Circuits and Systems (ACS)

Chair	Po-Chiun Huang, National Tsing Hua University
Members	Tetsuya Hirose, Osaka University Sai-Weng Sin, University of Macau Hao Yu, Southern University of Science and Technology Takeshi Ueno, Toshiba Corporation Po-Hung Chen, National Chiao Tung University Hyun-Sik Kim, KAIST Wanyuan Qu, Zhejiang University Vanessa H.-C. Chen, Ohio State University Milin Zhang, Tsinghua University Lin Cheng, University of Science and Technology of China (USTC)

Data Converters (DC)

Chair	Kazuko Nishimura, Panasonic Corporation
Members	Jong-woo Lee, Samsung Electronics Tsung-Heng Tsai, National Chung Cheng University Sanroku Tsukamoto, Fujitsu Laboratories Ltd. Jintae Kim, Konkuk University Seung-Tak Ryu, KAIST Yan Zhu (Julia), University of Macau Chih-Cheng Hsieh, National Tsing Hua University Zule Xu, University of Tokyo Qiang Li, University of Electronic Science and Technology of China (UESTC) Nan Sun, University of Texas at Austin

Digital Circuits and Systems (DCS)

Chair

Jun Zhou, University of Electronic Science and Technology of China

Members

Chia-Hsiang Yang, National Taiwan University
Massimo Alioto, National University of Singapore
Kenta Yasufuku, Toshiba Memory Corporation
Keiichi Kushida, Toshiba Memory Corporation
Yoonmyung Lee, Sungkyunkwan University
Leibo Liu, Tsinghua University
Koyo Nitta, NTT (Nippon Telegraph and Telephone Corporation)
Shouyi Yin, Tsinghua University
Xiaoyang Zeng, Fudan University

Emerging Technologies and Applications (ETA)

Chair

Woogeun Rhee, Tsinghua University

Members

Jerald Yoo, National University of Singapore
Noriyuki Miura, Kobe University
Minkyu Je, Korea Advanced Institute of Science and Technology (KAIST)
Shuenn-Yuh Lee, National Cheng-Kung University
Youngcheol Chae, Yonsei University
Shiro Dosho, Tokyo Institute of Technology
Zhichao Tan, Analog Devices Inc.
Ping-Hsuan Hsieh, National Tsing Hua University
Pui-In Mak, University of Macau
Masaki Sakakibara, Sony Semiconductor Solutions Corporation
Zheng Wang, University of Electronic Science and Technology of China (UESTC)

Memory (MEM)

Chair

Junghwan Choi, Samsung Electronics

Members

Atsushi Kawasumi, Toshiba Memory Corporation
Hung Jen Liao, TSMC
Kazutaka Miyano, Micron
Meng-Fan Chang, National Tsing Hua University
Chun Shiah, Etron
Ken Takeuchi, Chuo University
Tony T. Kim, Nanyang Technological University
Ik Joon Chang, Kyunghee University
Jun YANG, Southeast University
Shyh-Shyuan Sheu, Yuan Ze University
Song Junyoung, Incheon National University

978-1-7281-5107-6/19 $31.00 © 2019 IEEE

Radio Frequency (RF)

Chair

Minoru Fujishima, Hiroshima University

Members

Satoshi Tanaka, Murata
Huei Wang, National Taiwan University
Chien-Nan Kuo, National Chao Tung University
Baoyong Chi, Tsinghua University
Giovanni Mangraviti, IMEC
Tae Wook Kim, Yonsei University
Dixian Zhao, Southeast University
Taizo Yamawaki, Hitachi, Ltd.
Minjae Lee, Gwangju Institute of Science and Technology
Bo Zhao, Zhejiang University
Jun Yin, University of Macau

SoC and Signal Processing Systems (SOC)

Chair

Chi-Cheng Ju, MediaTek Inc.

Members

Kazutami Arimoto, Okayama Prefectural University
Chun Zhang, Tsinghua University
Yong Hei, Portland State University
Satoshi Shigematsu, NTT
Tsung-Te Liu, National Taiwan University
Pei-Yun Tsai, National Central University
Daisuke Mizoguchi, Renesas
JunYoung Park, UX Factory, Inc.
Kyuho Jason Lee, UNIST (Ulsan National Institute of Science and Technology)
Yun Chen, Fudan University

Wireline Communications (WLN)

Chair

Chulwoo Kim, Korea University

Members

Wei-Zen Chen, National Chiao -Tung University
Jun Terada, NTT
Jung-Hoon Chun, Sungkyunkwan University
Yasufumi Sakai, Fujitsu Laboratories Ltd.
Ching-Yuan Yang, National Chung-Hsing University
Ziqiang Wang, National Chung-Hsing University
Koichi Yamaguchi, Intel
Masum Hossain, University of Alberta
Gyungsu Byun, Inha University
Peng Liu, Zhejiang University
Qi Nan, Institute of Semiconductors, Chinese Academy of Sciences

Industry Program (IP)

Chair
Stefan Rusu, TSMC North America

Members
Toru Shimizu, Toyo University
Surhud Khare, Intel
Daisaburo Takashima, Toshiba
ShaoJun Wei, Tsinghua University
Ting Wu, Norel Systems
Ma Fan Yung, Infineon
Masaitsu Nakajima, Socionext
Yi Kang, University of Science and Technolog of China (USTC)

FPGA

Chair
Shigeki Tomishima, Intel / Intel Labs

Members
Tay-Jyi Lin, National Chung Cheng University
JI-HOON KIM, Ewha Womans University
Youngjoo Lee, POSTECH
Yong-Pan Liu, Tsinghua University
Hiromitsu Awano, Osaka University

IEEE
445 Hoes Lane
Piscataway, NJ 08854-4141

ISBN 978-1-7281-5107-6